Femtochemistry and Femtobiology

Ultrafast Dynamics in Molecular Science

Femtochemistry and Femtobiology

Ultrafast Dynamics in Molecular Science

University of Castilla-La Mancha, Toledo, Spain
September 2 – 6, 2001

Editors

Abderrazzak Douhal
University of Castilla-La Mancha, Toledo, Spain

Jesus Santamaria
University Complutense of Madrid, Spain

World Scientific
New Jersey • London • Singapore • Hong Kong

Published by

World Scientific Publishing Co. Pte. Ltd.
5 Toh Tuck Link, Singapore 596224
USA office: 27 Warren Street, Suite 401-402, Hackensack, NJ 07601
UK office: 57 Shelton Street, Covent Garden, London WC2H 9HE

British Library Cataloguing-in-Publication Data
A catalogue record for this book is available from the British Library.

Cover: The Figures on the cover illustrate the wide topics of femtochemistry and femtobiology covered by the book.

FEMTOCHEMISTRY AND FEMTOBIOLOGY
Ultrafast Dynamics in Molecular Science

ISBN-13 978-981-02-4866-6
ISBN-10 981-02-4866-0

FOREWORD

In the series of conferences on Femtochemistry, the first of the 21st century took place in Toledo, Spain on September 2–6, 2001. The conference was held one week before the tragic event of September 11, which was attributed by some writers as due to "clashes of civilizations". Toledo, "the City of Three Cultures" — the Jewish, Christian, and Muslim cultures, reminds us all that through dialogues of cultures we can live as civilized humans prospering with knowledge and progress. Indeed the new knowledge and the international makeup of the conference in Toledo is a clear example of this dialogue through Science.

The conference is unique in many dimensions — the scientific presentations of the latest in Femtochemistry and Femtobiology, the in-depth discussions that follow and the collegial and friendly interactions among the participants. We owe a great deal of appreciation to Professor Abderrazzak Douhal for his untiring effort to make the conference a most exciting venue for scientific and social exchanges. We also cannot forget the tremendous effort made by his colleagues and staffs. I would like to take this opportunity to thank the University Administration, the Rector and Vice-Rectors, for their warm welcome of the conference and of me.

At the conference almost all of us attended the sessions, for five days of "hard work", without being bored. The fundamental nature of the femtosecond time scale and the diversity in applications made us eager to learn about the nature of elementary dynamics, from isolated systems to complex biological assemblies. Equally important was the new frontiers and opportunities: new concepts in quantum control, advanced theoretical methods, development of new techniques for structural dynamics, and the next generation for light and electron sources to probe atoms and molecules over a whole range of wavelengths, at shorter times and in more intense fields.

The future promises to be as exciting as the past. In both Femtochemistry and Femtobiology we will continue to explore the landscape of molecules and make the fundamental bridging between structure and dynamics in the hope of understanding the function of complex systems. The next conference which is organized by Monique Martin and Casey Hynes will be held in Paris in July of 2003. We look forward to another exciting meeting.

Ahmed Zewail
Pasadena
January 2002

PREFACE

About fifteen years ago, Femtochemistry emerged as a new, important field for the study, in real time, of chemical and biological reactions. Since then, remarkable progress has been made. Femtosecond spectroscopy devoted to Chemistry and Biology has became a well established field of research, giving rise to Femtochemistry and Femtobiology. In 1999, the Scientific Community was extremely delighted by the remarkable achievement in this field of research with the award of the Nobel Prize in Chemistry to Ahmed H. Zewail.

The field has had impact all over the world with many conferences and publications. One particular series of conferences is devoted entirely to Femtochemsitry and is now in its tenth year. The first femtochemistry conference was organized by Jörn Manz and Ludger Wöste and it took place in Berlin in March 1993. The success of the Berlin Conference and the rapid growth of Femtochemistry and Femtobiology gave rise to this series of Femtochemistry Conferences. In the summer of the years 1995, 1997 and 1999, the second, third and fourth Femtochemistry Conferences were organized, respectively, by Majed Chergui in Lausanne, Villy Sundström in Lund, and Franz de Schryver in Leuven. The contributions made in these Conferences have crystallized in a complete series of books and Special Issues of the *Journal of Physical Chemistry.*

The first Femtochemistry Conference of the 21st century took place in Toledo, Spain during the period of September 2 – 6, 2001. The conference had some 250 participants from more than 30 countries, with a total of 45 lectures and 130 poster presentations in wide-ranging topics: gas phase, clusters, liquids, coherent control, new techniques, biological systems, polymers, nanostructures, surfaces and interfaces.

This book contains full reports of about 85 contributions presented at the Toledo Conference. Many of the contributions are article-reviews. The book highlights the most important achievements and the *state-of-the-art* in Femtochemistry and Femtobiology at the beginning of the new millenium. The scope of topics stresses several important features. First of all, it contains the latest developments for understanding ultrafast processes in gas, condensed and complex systems, including surfaces and nanostructures. Particular attention is paid to the latest developments in quantum control, both in theory and experiments, and to the experimental pulse shaping techniques. Finally, specific attention is paid to biological processes, an area where femtodynamics is becoming very useful to resolve the structural dynamics by using different techniques such as Electron Diffraction and X-ray Diffraction, and IR spectroscopy. The book consequently contains the following five sections: Gas Phase Clusters; Liquids; Control and Techniques; Nanostructures, Polymers, Surfaces and Interfaces; Biological Systems.

We are grateful to all the authors who devoted enough time to provide full accounts of their contributions to this Conference. We would like to thank in particular Professor Ahmed H. Zewail for his continual encouragement and support which greatly helped to shape and focus the Toledo Conference and also this book. We also wish to express our sincere thanks to Majed Chergui, Jörn Manz and Villy Sundström for their help, advice and support in the preparation of the Conference. Our thanks also to our colleagues, Rosa Fandos Paris, Maria José Ruíz Garcia, Leonor Chico Gómez, M. Moreno, J. A. Organero, L. Santos, A. Vargas, L. Tormo and P. Isorna, who were of great help in organizing and smoothing the conference.

Finally, we wish to express our gratitude to the University of Castilla-La Mancha for its generous financial and logistic support, and to the Ministry of Science and Technology of Spain, the European Science Foundation (Ultra Program, Femtochemistry and Femtobiology), the "Dirección General de Enseñanza Universitaria e Investigación" and the "Consejeria de Ciencia y Tecnología" in Castilla-La Mancha", "Cortes de Castilla-La Mancha" and to "Caja de Castilla-La Mancha" for financial support, and to the companies PicoQuant GmbH (Berlin), Innova Instrumentación (Madrid), Lasing s.a. (Madrid) and Spectra Physics for sponsorship.

Abderrazzak Douhal and Jesus Santamaria
TOLEDO, JANUARY 2002

CONTENTS

1. Gas Phase and Clusters

2. Liquids

3. Coherent Control and New Techniques

5. Biological Systems

1. Gas Phase and Clusters

ORIENTATION DYNAMICS AND MOLECULAR STRUCTURES FROM GAS PHASE TO CONDENSED MEDIA

J. SPENCER BASKIN AND AHMED H. ZEWAIL*

Arthur Amos Noyes Laboratory of Chemical Physics, California Institute of Technology, Pasadena CA 91125, U. S. A.

Abstract: Basic concepts of molecular orientation are discussed, and a unified picture is presented for rotational dynamics in different phases: coherent inertial in isolated molecules, partially coherent in dense fluids, and reaching the diffusive limit in liquids. Theory is compared to experimental studies from this laboratory on reactive and non-reactive systems, including biological systems. Important to molecular orientation is the time scale, and we compare quantum-state orientation (continuous wave experiment) and classical-like orientation achieved with femtosecond resolution.

I. Introduction

Chemical and many physical and biological properties of molecules are inherently dependent on spatial orientation, and therefore the degree of orientation and the evolution of orientation significantly affect studies of molecular systems in all physical phases, from dilute gases to solids. The type of orientational motion encountered varies from generally negligible in solids, to diffusive in liquids, to hindered in dense gases, to purely inertial and coherent in isolated molecules.

This article, which is based on an earlier more detailed review,[1] is concerned with some basic concepts of molecular orientational dynamics, with an emphasis on its role in femtosecond studies and its relationship to molecular structures. The femtosecond time scale is unique in that the orientation of molecular samples, even when undergoing free rotation, can be followed as it evolves, providing direct information on the nature of the motion and the forces that govern it.

The following topics will be addressed: free rotational motion of isolated molecules; orientation in the presence of collisions and the effect of solvent density in bridging the gap from inertial to diffusive motion, spanning gases and liquids; orientation in reactions; and some effects of restricted geometry in biological systems. The classical viewpoint, which provides an almost entirely adequate basis for understanding and discussing femtosecond experiments, will be relied upon heavily in this work. A brief discussion of the quantum origins of orientational dynamics and the differences to be expected between experiments using femtosecond pulses and those using nanosecond or longer pulses (continuous wave excitation) will be given in Section VII.

II. Orientation of Free Molecules

When an isotropic sample is excited by a linearly polarized pump pulse (polarization vector $\vec{\varepsilon}_1$) via an electric-dipole allowed vibronic transition, there is created at the instant of excitation ($t = 0$) a spatial alignment of the excited state population (Figure 1), because the probability for absorption depends upon the relative orientation of $\vec{\varepsilon}_1$ and the molecular transition dipole $\vec{\mu}_1$. The excited ensemble is created with its transition dipoles exhibiting a cosine-squared distribution around $\vec{\varepsilon}_1$ (Fig. 1, bottom). Subsequent rotational motion causes an evolution of that distribution, as represented in the figure. (Recurrences in the orientation are understandable classically only for molecules at a fixed angular momentum, but the recurrences shown in the figure at the indicated times occur in thermal samples as a quantum effect, as discussed later.)

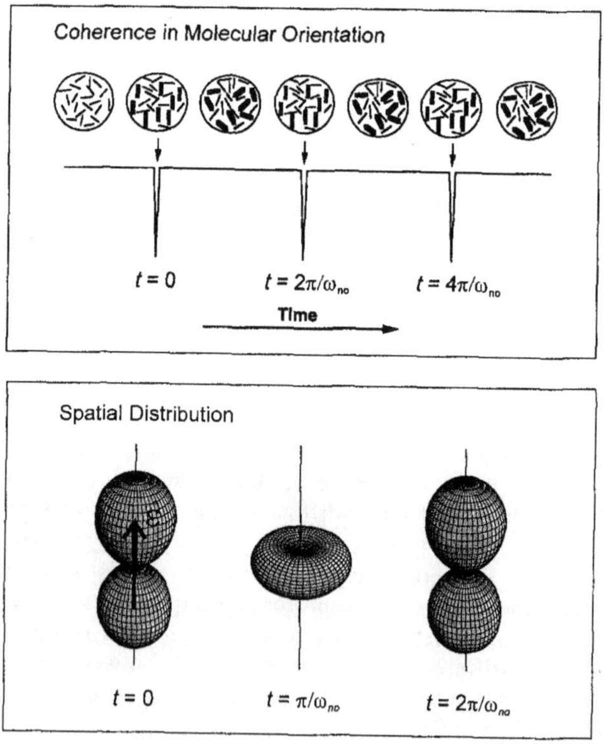

Fig. 1. (Top) General concept of time-resolved orientation experiments on isolated molecules. An isotropic distribution of dipoles is excited at $t = 0$ by a vertically polarized light pulse. The thicker lines represent transition dipoles of excited molecules, which follow individual rotational trajectories. (Bottom) Spatial distributions of transition dipole orientations for the excited state population at selected times for a simple rotor with a parallel dipole. The pump polarization vector is $\vec{\varepsilon}_1$.

To investigate the nature of molecular orientation, we proceed by measurement of the macroscopic transient polarization anisotropy, $r(t)$, given by

$$r(t) = \left(S_{||}(t) - S_{\perp}(t)\right)/\left(S_{||}(t) + 2S_{\perp}(t)\right) \tag{1}$$

where $S_{||}(t)$ and $S_{\perp}(t)$ are the transient pump-probe signal intensities measured for probe polarization vector parallel and perpendicular, respectively, to that of the pump. This quantity is equal to the time correlation function of the transition dipole unit vectors (see e.g. ref. 2):

$$r(t) = 0.4 \langle P_2[\hat{\mu}_1(0) \cdot \hat{\mu}_2(t)] \rangle, \tag{2}$$

where $\bar{\mu}_2$ is the probe-transition dipole vector; $\hat{\mu}_i = \bar{\mu}_i / |\bar{\mu}_i|$; $P_2(x)$ is the second-order Legendre polynomial; and the angle brackets denote a weighted average over the rotational state distribution of molecules in the sample.

An idealized case which nonetheless provides a useful approximation to the molecular orientation dynamics of many real molecules is that of a rigid symmetric top molecule undergoing classical free rotation. This rotation is describable[3] as a combination of two independent motions (Figure 2, inset): nutation of the rotor's symmetry or figure axis about the angular momentum vector $\vec{j}$ at angular frequency ω_n, and rotation about the figure axis at angular frequency ω_r. The angular frequencies are given classically[3] by $\omega_n = 4\pi B_{\perp} j = \omega_{no} j$ and $\omega_r = 4\pi(B_{||} - B_{\perp})\, j \cos\theta = \omega_{ro} k$, where θ is the angle between $\vec{j}$ and the figure axis. Here, $B_{||}$ and $B_{\perp}$ are the excited state rotational constants of the molecule about its figure axis and about any axis perpendicular to it, respectively, and $k = j \cos\theta$. Introduction of $\hbar$ in the definition of B requires that the values j and k which appear in the classical frequency expressions be the magnitudes divided by $\hbar$ of, respectively, the classical total angular momentum $\vec{j}$ and of its axial component $\vec{k}$.

Two common cases are $\hat{\mu}_1$ and $\hat{\mu}_2$ both either parallel or perpendicular to the figure axis of the molecule, as designated by the notations $(||,||)$ and $(\perp,\perp)$, respectively. In the first case, $r_{||,||}(j,\theta,t)$ (*i.e.* for a single rotational energy level) is independent of ω_r and strictly periodic at the fundamental nutation period, $2\pi/\omega_{n0}$, regardless of the value of j. A situation of particular interest and simplicity is that of $\vec{j}$ perpendicular to the figure axis, which holds to good approximation for all linear molecules. With $\theta = \pi/2$,

$$r_{||,||}(j,\pi/2,t) = 0.1 + 0.3 \cos[(2j+1)\,\omega_{no}\, t)] \tag{3}$$

which oscillates at $(2j+1)$ times the fundamental frequency and has a mean value of 0.1. For the $(\perp,\perp)$ dipole case, both nutation and rotation about the figure axis are important. The dependence of $r_{\perp,\perp}(j,\theta,t)$ on two independent fundamental frequencies means that this anisotropy generally displays no strict periodicity.

At the top of Figure 2 is shown a calculation for $j = 6$ and $k = 3$ when $2\pi/\omega_{n0}$ = 1948 ps and the dipoles are (||,||). (The rotational constants of S_1 *trans*-stilbene were used.)[4] The initial anisotropy is equal to 0.4, as is always the case when the pump and probe dipoles have the same orientation in the molecule-fixed frame.

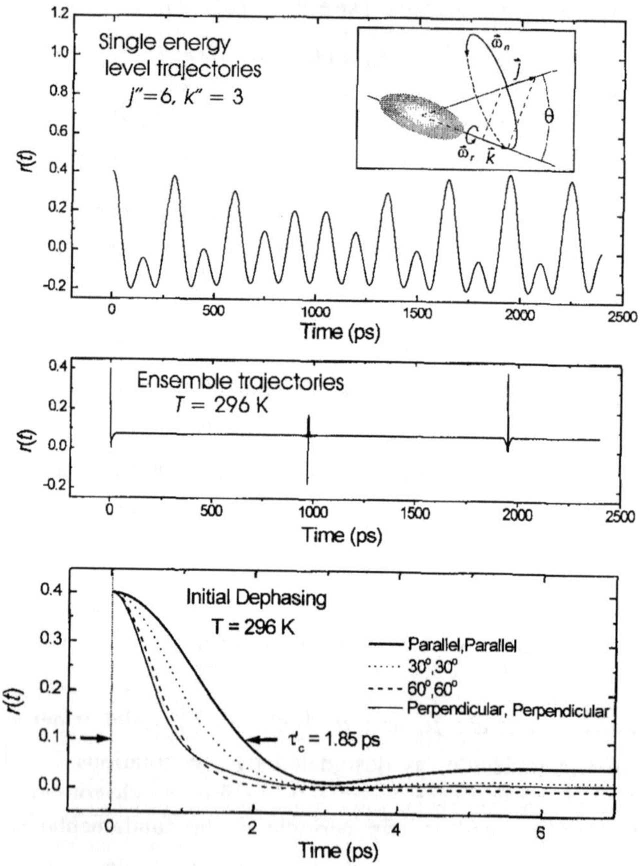

Fig. 2. (Top) Theoretical anisotropy for a single energy level of a symmetric top with $B_{||}$ = 2.603 GHz and $B_{\perp}$ = 0.2566 GHz and (||,||) transition dipoles. Inset: Classical motion of a rigid symmetric top molecule. (Middle) All as above, but for a thermal sample at T = 296 K. (Bottom) Expanded view of the early time anisotropy decays of the same molecule, again at T = 296 K, but for various dipole directions.

Given $r(j,\theta,t)$ for each rotational energy level, the macroscopic anisotropy will depend on the weight accorded to each in the ensemble average of eq. 2. When a thermal population is probed, a large number of different single-energy-level signals modulated by different frequencies are observed together (incoherently). An example of an anisotropy from such a thermal distribution is shown in the middle of Fig. 2 for the same molecule as used for the energy-resolved calculation at the top of the figure. At early time, the superposition of numerous cosinusoidal signals leads to a dephasing of the initial alignment, resulting in a decay of the anisotropy. The dephasing rate depends on the rotational constants, the dipole direction, and the rotational temperature. At the bottom of Fig. 2 are shown examples of the initial anisotropy decay as calculated for a room temperature rotational population distribution, with rotational constants as above.

Precise expressions can be derived[5,2] for the characteristic dephasing time of an ensemble of linear molecules in the $(||,||)$ dipole case (eq. 3); the coherence time, τ'_c, defined as the time for the anisotropy to fall to 0.1, is given by:

$$\tau'_c = 2.942\,(T B_{\perp})^{-1/2}\ \mathrm{ps} \tag{4}$$

where T is the temperature in degrees Kelvin and $B_{\perp}$ is expressed in cm^{-1}. This expression is found to hold to high accuracy for the $(||,||)$ dipole case of prolate nonlinear symmetric tops also, an example being the case shown in Fig. 2.

The dephasing described above requires that inertial rotation be uninterrupted for, typically, a few picoseconds. For isolated molecules, as in molecular beam studies, the possibility also exists for complete or partial *rephasing*, or recurrence, of the macroscopic alignment at much longer times. Rephasing will occur when the nutation and rotation periods of a significant fraction of the molecules are commensurable, or nearly so. Such is clearly the case for rigid symmetric tops, for which the nutation and rotation frequencies are integral (based on the quantization of j and $j\cos\theta \equiv k$) multiples of the fundamental frequencies, ω_{no} and ω_{ro}.

In the $(||,||)$ dipole case, complete rephasing occurs at times $m \cdot (2\pi/\omega_{n0})$ for integer m, with a partial negative recurrence at times $(m + \frac{1}{2}) \cdot (2\pi/\omega_{n0})$ (see Fig. 2, middle). In the $(\perp,\perp)$ dipole case, there are partial recurrences at times $m \cdot (2\pi/\omega)$ for the angular frequencies ω_{n0}, $2\omega_{r0}$, $\omega_{n0}+2\omega_{r0}$, and $2(\omega_{n0}+\omega_{r0})$. These times are directly related to the structural parameters $B_{\perp}$ and $B_{||}$, and recurrence measurement therefore provides information about the molecule's structure.

Experimentally, rotational coherence spectroscopy has proven to be a powerful, Doppler-free, technique for the determination of molecular structure; more than one hundred structures have been studied by rotational coherence spectroscopy (RCS) in laboratories around the world. As an example, a fluorescence anisotropy measurement of jet-cooled *t*-stilbene is shown in Figure 3, plotted with a quantum-theoretical asymmetric-top simulation.[6] The molecular structure deduced with the aid of such measurements[4] is also shown.

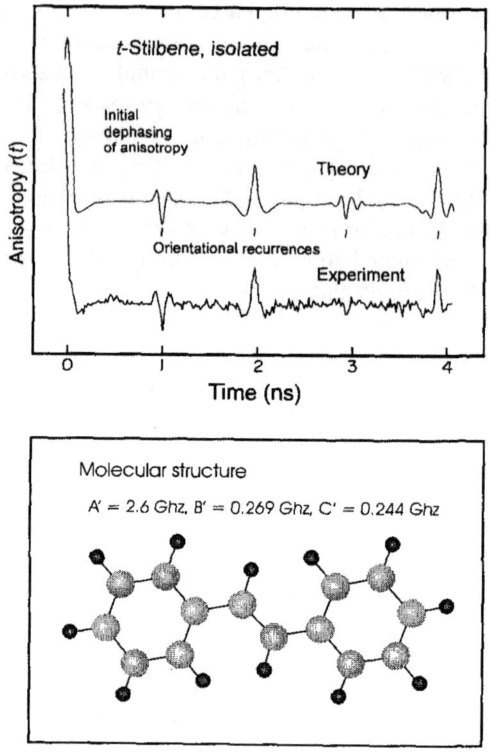

Fig. 3. Comparison of experimental and theoretical anisotropies of jet-cooled samples of *t*-stilbene [from ref. 6] and the structure deduced from rotational coherence measurements.[4]

III. Inertial to Diffusive Motion: Effect of Solvent Density

With an increase in the frequency of $\vec{j}$-changing collisions suffered by the molecules under study, the coherent and inertial rotational motion described in the preceding section survives for shorter and shorter times. At the molecular level, the rotation of a solute molecule in a bath of solvent particles is controlled by the rate of collisions and by the effectiveness of those collisions in transferring angular momentum to or from the solute.[7] With femtosecond time resolution, the rotational evolution of the solute can be precisely observed, and its dependence on solvent properties and densities can be characterized.[7] Such measurements probe directly the nature of microscopic friction, the interaction of solvent and solute at the molecular level, which plays a fundamental role in the dynamic evolution of solution phase chemical and physical processes.

(A) Langevin-hydrodynamic treatment

Theoretical models of the role of friction in controlling molecular rotation have a long history (see e.g. refs. 8-13). The reorientational motion of a solute molecule in high-density fluids may be treated as a form of rotational Brownian motion. Focusing on the motion of a linear rotor with two rotational degrees of freedom, the total torque acting on the molecule is separated into the frictional torque, $-\xi I\vec{\omega}$, and a random torque $\vec{T}$. Here, ξ is the reduced rotational friction coefficient, I is the unique moment of inertia of the rotor, and $\vec{\omega}$ is the angular velocity. A Langevin equation treatment,[14] $I\frac{\partial}{\partial t}\vec{\omega}(t) = -\xi I\vec{\omega}(t) + \vec{T}$, leads to an exponential decay of the angular velocity autocorrelation function ($\Omega(t) = \langle \vec{\omega}(t) \cdot \vec{\omega}(0) \rangle$) with a time constant of $\tau = 1/\xi$, and, for $t >> \tau$, an exponential decay of the anisotropy

$$r(t) \propto exp(-6D_r t), \tag{5}$$

using the Langevin-Einstein rotational diffusion constant, $D_r = kT/(\xi I)$, a well-known result for the diffusive limit. This treatment is invalid for large-angle free (inertial) rotation, however, and therefore is not useful under low friction conditions.

The above expressions require a value of ξ. Hydrodynamic theory allows one to calculate ξ for an ellipsoidal solute in a fluid continuum as a function of the ellipsoid dimensions, shear viscosity η of the solvent, and the boundary condition assumed at the solute-solvent interface.[10,15] This treatment is valid in the same limit as the Langevin treatment; that is, for small solvent/solute mass ratio and collision time short on the time scale of free rotation, and the choice of boundary condition appropriate to a particular system remains a source of uncertainty in the result.

(B) J- *and* m-*diffusion models*[13]

To overcome the restriction to small angular steps for free rotation between collisions, which clearly limits the preceding treatment to relatively high densities, Gordon developed the *J*- and *m*-diffusion models.[13] In these models, interaction of solute and solvent results solely from discrete, Poisson-distributed events ("collisions") with collision interval τ_{coll}. Between collisions, free rotation of the rotor pertains, and arbitrarily large angular excursions under free inertial motion are possible at large τ_{coll}.

In the Gordon models, the key to obtaining a closed form solution for $r(t)$ is the assumption that the direction of $\vec{j}$ is randomized at each collision. This assumption means that the angular velocities before and after a collision are uncorrelated ($\langle \vec{\omega}(t^-_{coll}) \cdot \vec{\omega}(t^+_{coll}) \rangle = 0$). Thus, the only contribution to the angular velocity autocorrelation function comes from the population which has suffered no collisions. According to the Poisson distribution, this population decays exponentially with time constant τ_{coll}; that is $\tau = \tau_{coll}$. The gradual transition of the polarization anisotropy from its distinctive and non-monotonic free-rotor form to a diffusion-controlled exponential decay at long time and non-zero pressure is approximated in the Gordon models, although not quantitatively and without explicit solvent dependence.

(C) J*-coherence model* [7]

The *J*-coherence model establishes a link between the physical parameters of the solute-solvent system in question, including the masses, moment of inertia, temperature, and the intermolecular potential, and the friction, or decay of $\Omega(t)$, by replacing the Gordon assumption of randomization of $\vec{j}$ by the specific distribution of final $\vec{j}$ which is produced by classical hard-sphere collisions of solute and

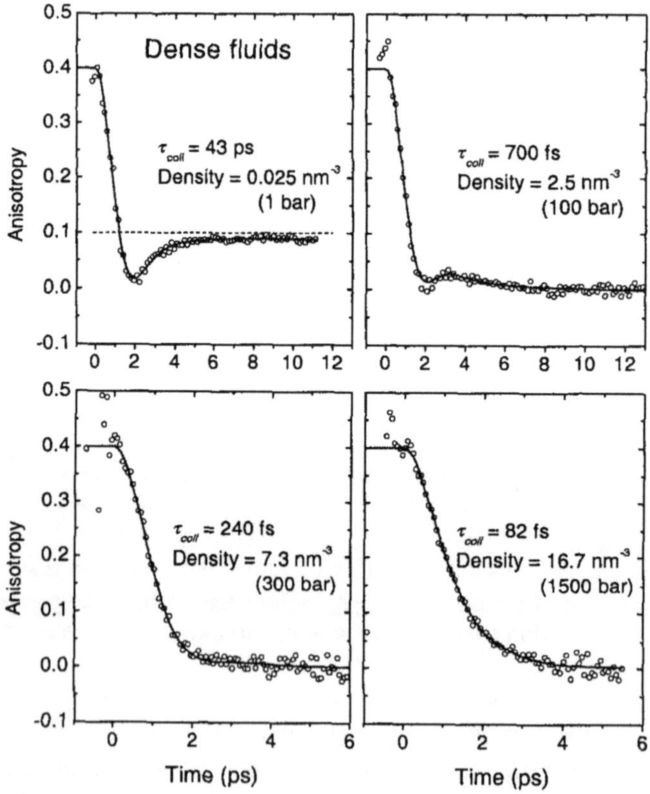

Fig 4. *J*-coherence fits of experimental polarization anisotropies of iodine in argon at a range of densities. λ_{pump} is 613 nm for the 300 bar data, and 623 nm for the other three.

solvent. The anisotropy decay of the sample is obtained by numerically tracking the rotational trajectories of a large number of molecules through binary collisions with a statistical selection of solvent atoms. In this way, the relationship between the true collision interval, as derived from kinetic gas treatments, and τ, the effective angular momentum scrambling lifetime, is determined.

In Figure 4, measured anisotropy decays of iodine in argon are shown with *J*-coherence model fits.[7] One sees that the general features of the experimental anisotropies across the full range of densities are well-reproduced by the theory, from free-rotor-like behavior at 1 bar argon to an exponentially decaying $r(t)$ at 1500 bar. The data in ref. 7 illustrate clearly, however, that the transition from low to high pressure limits has a very different density dependence in the solvent gases argon and helium.

From the *J*-coherence calculations, both the anisotropy and $\Omega(t)$ were obtained. $\Omega(t)$ was single exponential in form and its lifetime τ could be compared with τ_{coll} to quantify the $\vec{j}$ scrambling potency per collision of each solvent. It was found that, $\tau = 5.5\tau_{coll}$ for argon and $\tau = 43\tau_{coll}$ for helium. That is, the angular velocity of an average iodine molecule persists through ~5.5 collisions with argon and 43 collisions with helium.

In addition to accurately reproducing the temporal evolution of the anisotropy at arbitrary density, a primary goal of the *J*-coherence model is to be able to predict

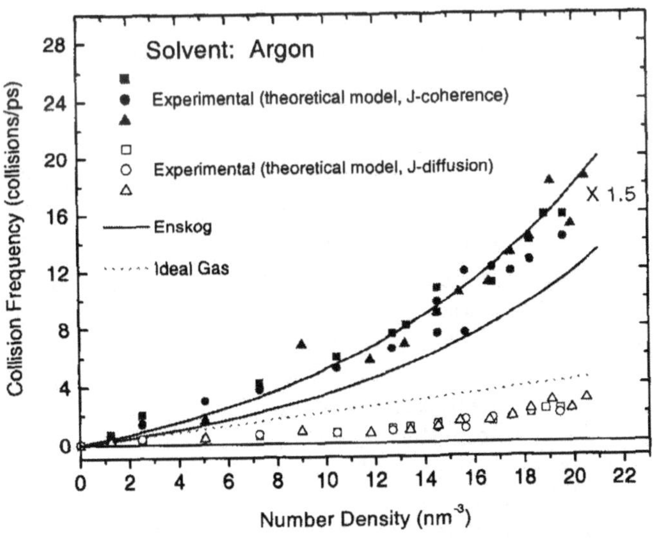

Fig. 5. Solvent density dependence of the collision rate of iodine in argon derived from experimental anisotropies by the *J*-coherence and *J*-diffusion models. Different symbols correspond to independent series of experiments. The hard-sphere ideal-gas and Enskog collision rates are shown for comparison. The Enskog curve is plotted a second time scaled by a factor of 1.5.

its dependence on the properties of the solvent. In Figure 5, the Enskog hard-sphere collision rate vs. density is shown as derived independently of the experiments. The experimental *J*-coherence collision rates in argon are only slightly higher than the Enskog rates, showing good agreement with a scaling of the rates by a factor of 1.5. For helium, in which collisions with iodine are well-represented by a hard sphere potential, the rates are in very good agreement with the Enskog rates without adjustment.[7]

In summary, the *J*-coherence bimolecular collision model provides a clear prescription for relating the hard-sphere properties of the solvent and solute to the friction and to the decay of solute anisotropy over the entire experimental density range, from free rotational motion to rotational diffusion.

IV. Orientation in Reactions

The concept of collision-induced angular momentum scrambling of Section III is relevant also to the dynamics of reactive systems, where now the recoil from fragmentation parallels the effect of impact with a solvent molecule. Thus, polarization anisotropy measurements are of value in providing a view of the vectorial dynamics in time-resolved studies of dissociation reactions.[16,17]

(A) Evolution of Coherence

Consider systems which dissociate into atom + molecule, *i.e.*,

$$M_1 \rightarrow M_1^* \rightarrow M_2 + A \qquad (6)$$

with M_2 ranging from a diatomic to a large polyatomic. Such a reaction is illustrated at the bottom of Figure 6. Here, the polarized pump pulse creates a spatial alignment of the *reactant* (M_1^*) population from an initially isotropic M_1 population, and the time evolution of alignment of the dipole moment, $\hat{\mu}_2$, of the *product* molecules determines the anisotropy of the signal. In the M_1^* center of mass frame, the centers of mass of M_2 and A separate with equal and opposite linear momenta along trajectories displaced from each other by the impact parameter *b*. The energy difference E' between the internal energy of M_1^* and that of the products appears as kinetic energy of rotation and translation.

To calculate the anisotropy for the reaction described by eq. 6, the rotational motion of each probe dipole must be traced continuously from $t = 0$, the moment when the aligned M_1^* population is created by the excitation pulse. In a "prompt, impulsive dissociation" approximation, the motion of M_1^* prior to dissociation and the duration of the "transition state" between reactant and free product are both taken to be of negligible duration when compared with the rotational periods entering significantly into the calculation. In this case, only the initial molecular structure, the direction of the impulse in the molecular frame, and the available energy E' need be specified.

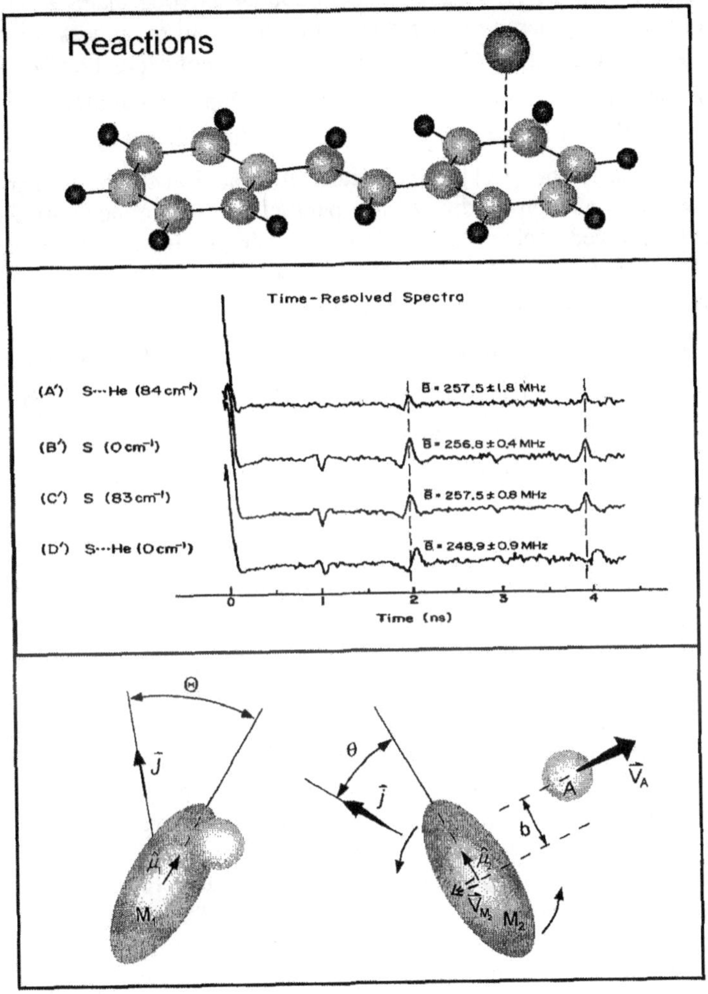

Fig. 6. Orientation in reactions. (Top) Approximate molecular structure of the *t*-stilbene-He_1 van der Waals complex. (Middle) Time-resolved anisotropies pumping different absorption bands of *t*-stilbene (S) and the *t*-stilbene–He_1 complex (S···He). Excitation energies relative to the respective electronic origins are given. $\overline{B}$ is ½(B' + C') of the complex or parent stilbene [see ref. 16]. (Bottom) Geometry of the dissociation problem (see text).

The orientational evolution which is pertinent to the pump/probe process is completely expressed in the function $\hat{\mu}_1(\{i\},0)\cdot\hat{\mu}_2(\{f\},t)$, as in eq. 2, but where now, $\{i\}$ and $\{f\}$ represent complete sets of rotational coordinates for M_1 and M_2, respectively, which generally differ due to the dissociation process. In the prompt, impulsive approximation, $\hat{\mu}_1(\{i\},0)\cdot\hat{\mu}_2(\{f\},t)$ does not depend on $\{i\}$, so the state resolved anisotropy does not either. Only the ensemble average of eq. 2 may depend on $\{i\}$ through the influence it exerts on $P_j(\vec{j})$, the distribution of final rotational state populations. In the $(||,||)$ dipole case illustrated at the bottom of Fig. 6, in which both pump and probe dipoles are parallel to the symmetry axis of M_2, the *energy-level*-resolved anisotropy following prompt, impulsive dissociation is identical to that for direct excitation of free M_2. However, the macroscopic anisotropies in the two cases may still be different, due to differing $P_j(\vec{j})$.

In the general case, the function $P_j(\vec{j})$ can be related directly to three physical sources of M_2 rotation: (i) initial rotation of M_1, (ii) the torque applied by the dissociation impulse, and (iii) initial vibration of M_1.[2] From a knowledge of $P_j(\vec{j})$, the macroscopic anisotropy of M_2 can be calculated. Such calculations may be compared with specific experimental results to infer a product rotational distribution and interpret the distribution in terms of the dynamics of dissociation.

(B) Theory vs. *Experiment*

The S_1 dissociation dynamics of the jet-cooled *t*-stilbene–He_1 complex (S–He) have been studied in our laboratory by picosecond time-resolved fluorescence detection.[18,19] To investigate the effect of dissociation on the dynamics of product rotation, stilbene product emission was polarization-analyzed.[16] The anisotropy for 84 cm^{-1} excitation is shown in Fig. 6. Also shown for comparison are anisotropies of parent stilbene for two excitation energies and of S–He excited to its S_1 origin.

In addition to the initial decay of the anisotropy, the data of Fig. 6 display prominent recurrences, though they are distinctly weaker in the anisotropy of the dissociation product than in the three other cases. The reduction in amplitude appears particularly severe for the negative or out-of-phase recurrences. The recurrence period is seen to match the nutation period of stilbene rather than that of the initially excited complex, thereby confirming the dissociation.

Calculations were performed[2] for product stilbene undergoing symmetric top motion following prompt, impulsive dissociation from the equilibrium geometry of the complex (Fig. 6, top). The theory indicates a reduction of the in-phase recurrence of about 30% due to dissociation, while that observed experimentally was about 40%.

Although the experimental reduction of the in-phase recurrence is similar to that predicted by the model calculation, the out-of-phase recurrences appear more severely damped in the measured anisotropies. A simple extension of the model to include a distribution of initial geometries was made to investigate the role of vibrational averaging and see if this might account for the above difference. Calculation of the anisotropy for a transverse displacement of the helium atom shows that contributions from off-axis geometries do produce noticeably smaller out-of-phase recurrences.

This example suggests the manner in which information about vectorial reaction dynamics may be obtained from polarization anisotropy measurements. Such measurements reflect the rotational distribution of the fragment, from which may be deduced information about: (i) the applied torque, and (ii) the geometry of the initial complex. The method should be particularly appropriate for use on large fragments for which the *shape* of rotational recurrences reflects the asymptotic rotational distribution, in the same way that their *separation* identifies the fragment molecular structure.

V. Single Bimolecular Collision—One-Atom Caging

When I_2 vapor at room temperature is excited to an energy above the *B*-state dissociation threshold (see wave packet in Figure 7 inset, top) in the presence of argon, a fraction of the initially excited I_2 population is caged by a single collision with an argon atom within ~ 200 fs of the excitation pulse.[20] Since both the *B* state and the $^1\Pi_{1u}$ state are reached in the pump step, the question of mechanism arises: Does caging take place entirely on the *B* state or does it involve a nonadiabatic crossing from the $^1\Pi_{1u}$ state? This question has resulted in many publications, both theoretical and experimental.

To address the question, $r(t)$ of I_2 was measured when wave packets were prepared *below* and *above* the *B*-state dissociation threshold.[20] In each case, the same energy range of the bound *B* state was probed, with a parallel transition dipole. For excitation within the bound *B* state, $r(t)$ is entirely consistent with theory for a $(\|,\|)$ dipole case, as expected (see Fig. 7, bottom). When the packet collisionally cages into the *B* state (Fig. 7, top), the anisotropy is similar and therefore also consistent with the $(\|,\|)$ case.

Since the transition from the ground state to the $^1\Pi_{1u}$ state has a transition dipole perpendicular to the internuclear axis, if I_2 were first excited onto the $^1\Pi_{1u}$ state, the anisotropy should resemble the theoretical calculation in Fig. 8 with $r(0) = -0.2$, which contrasts starkly with the observation. Therefore, the femtosecond anisotropy measurement shows conclusively that the caging dynamics occur adiabatically on the *B* state, without involvement of the $^1\Pi_{1u}$ state.

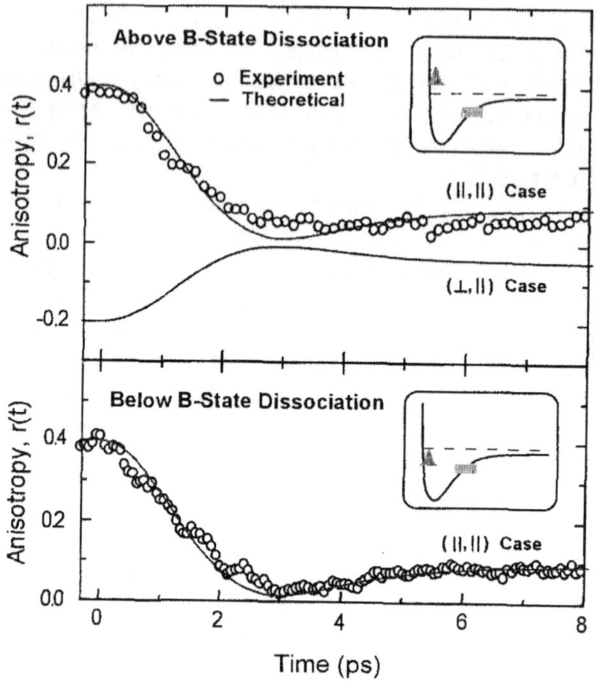

Fig. 7. Measured anisotropy of I_2 in 20 bar argon with excitation below (bottom; $\lambda_{pump} = 513$ nm) and above (top; $\lambda_{pump} = 495$ nm) the *B* state dissociation threshold.[20] In both cases, $\lambda_{probe} = 395$ nm. Solid lines represent theoretical anisotropies for two different transition dipole sequences.

VI. Biological Systems: DNA and Proteins

For biopolymers, anisotropy measurement is a powerful method for probing rigidity and mechanism of site recognition. In DNA, the rigidity was studied in this laboratory, and a rigidity constant was obtained with high accuracy—it depends on sequence and on helicity. The anisotropy decay in this case is not exponential in form, and theory and experiment confirm the involvement of all twisting (torsional) frequencies of the double helix, giving rise to $\sqrt{t}$ -type behavior,[21] as seen at the top of Figure 8. We have also studied, using the anisotropy decay, the effect of restricted geometry on the rigidity of a probe molecule in the hydrophobic pocket of proteins, and compared with results in chemical systems (micelles and sugars) and in bulk solvent[22] (see Fig. 8, bottom).

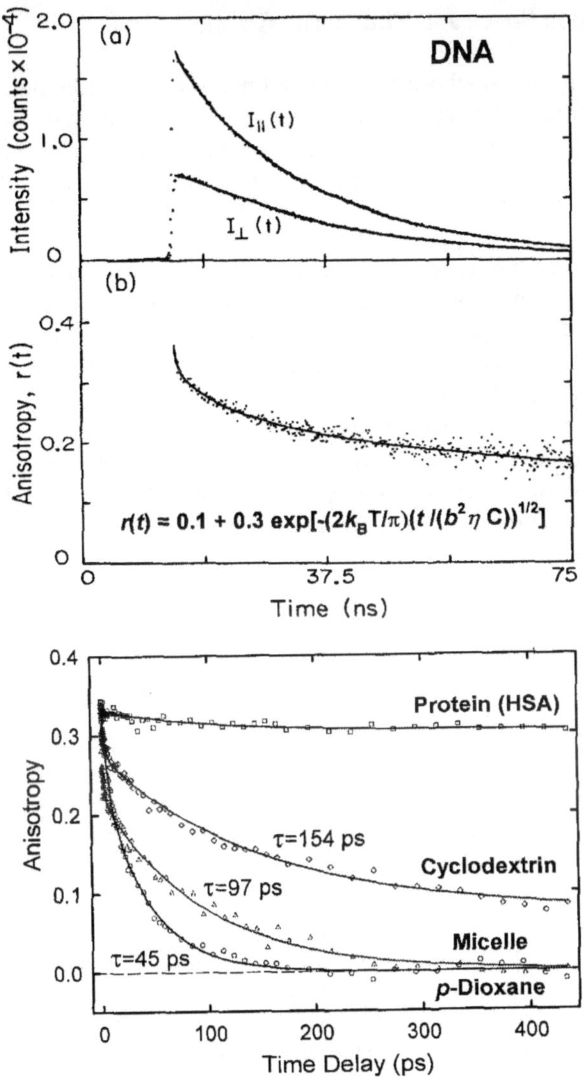

Fig. 8. The anisotropy decay of a probe molecule in DNA, protein, sugar, and micelle—the probe molecule is ethidium bromide for DNA (top) and HPMO for all others. The polarization-resolved transients are also shown for DNA. Note the nonexponential behavior in DNA and the drastic change of the anisotropy as the medium changes. In the equation given for the anisotropy in DNA, k_B is the Boltzmann constant, T is the temperature, b is the hydrodynamic radius of the helix, η is the solvent viscosity, and C is the rigidity constant. For more details, see references 21 and 22.

VII. Orientation in Space and Time : The Classical Limit

Consider a linear molecule whose orientational properties are completely described, from the quantum mechanical perspective, by rotational wavefunctions $\Psi_{j\,m}(\theta,\varphi)$, where θ and φ are polar coordinates for the orientation of the nuclear axis in the lab frame, and m is the quantum number for the projection of the angular momentum on the laboratory Z axis.

The transition probability for electronic excitation of a molecule depends on the transition dipole direction in the molecular frame, the laser polarization, and the initial and final rotational states.[23] As a specific example, we assume a transition dipole vector parallel to the nuclear axis and laser polarization parallel to the laboratory Z axis. In this case, only transitions $j' = j''-1$, $m' = m''$ (P branch) and $j' = j''+1$, $m' = m''$ (R branch) are possible.

An appropriately tuned narrow bandwidth (cw) laser will access a branch-specific transition of all m'' sublevels of a given j'', thereby creating an excited state ensemble with contributions of each excited state m' sublevel. Because each molecule occupies a single rotational eigenstate in the excited state, all distributions are *time-independent.*

On the other hand, a sufficiently short pulsed excitation with its broad bandwidth may coherently excite both P and R transitions in each molecule, creating an excited superposition state (rotational wave packet, $\Psi^{w.p.}_{rot}(t)$). Unequal temporal phase factors in the two terms of this superposition state lead to a time-dependent cross-term in the excited state probability distribution, $|\Psi^{w.p.}_{rot}(t)|^2$, that is modulated at the frequency $2B'(2j''+1)$ (*i.e.* $\omega = (2j''+1)\,\omega_{no}$). The initial total distribution is *exactly* equal to the classical cosine-squared, and it recurs periodically. At odd-multiples of half the recurrence period, the ensemble distribution approaches the classical ½ $\sin^2\theta$, even at low j, and equals it exactly in the limit of high j.

Note, however, that, when the excited state is short-lived, the absorption strength of rotational eigenstates is spectrally broadened, allowing rotational wave packets to be created *even with a cw laser.* Since these wave packets evolve as in the femtosecond excitation experiments, high levels of anisotropy can be measured, reflecting the classical orientation averaged over an excited state lifetime shorter than its rotation time. Based on this fact, steady-state anisotropy values and product angular distributions are commonly used to deduce how excited state lifetimes (or fragmentation lifetimes) compare with rotation times.

VIII. Conclusion

This contribution has summarized some basic concepts which have been important to the efforts of the Caltech research group in the area of time-resolved molecular anisotropy. Study of orientation in real time is powerful in elucidating the vectorial dynamics in reactive and non-reactive systems and in obtaining direct information on molecular structures. Studies of molecular orientation dynamics are also valuable in solution and for large biological systems. For achieving real classical orientation of molecules by creation of classical-like wave packets, the femtosecond time resolution is ideal, giving a direct view of dynamical phenomena such as coherent inertial motion and recurrences, collision dynamics, and microscopic friction/viscosity.

Acknowledgment

This research was supported by the National Science Foundation and Air Force Office of Scientific Research.

References

1. Baskin J. S. and Zewail A. H., *J. Phys. Chem. A* **105** (2001) pp. 3680–3692.
2. Baskin J. S. and Zewail A. H., *J. Phys. Chem.* **98** (1994) pp. 3337–3351.
3. Herzberg G., *Molecular Spectra and Molecular Structure* (Van Nostrand Reinhold, New York, 1945) Vol. II, p. 23.
4. Baskin J. S. and Zewail A. H., *J. Phys. Chem.* **93** (1989) pp. 5701–5717.
5. Zewail A. H., *J. Chem. Soc. Faraday Trans. 2* **85** (1989) pp. 1221–1242.
6. Baskin J. S., Felker P. M. and Zewail A. H., *J. Chem. Phys.* **86** (1987) pp. 2483–2499.
7. Baskin J. S., Chachisvilis M., Gupta M. and Zewail A. H., *J. Phys. Chem. A* **102** (1998) pp. 4158–4171; Baskin J. S., Gupta M., Chachisvilis M. and Zewail A. H., *Chem. Phys. Lett.* **275** (1997) pp. 437–444.
8. Edwardes E., *Quart. J. Pure Appl. Math* **26** (1893) pp. 70–78.
9. Debye P., *Polar Molecules* (Chemical Catalog Company, New York, 1929).
10. Perrin F., *J. Phys. Radium* **5** (1934) 497-511; **7** (1936) pp. 1–11.
11. Chuang T. J. and Eisenthal K. B., *J. Chem. Phys.* **57** (1972) pp. 5094–5097; Eisenthal K. B., *Acc. Chem. Res.* **8** (1975) pp. 118–124, and references therein.
12. Steele W. A., *J. Chem. Phys.* **38** (1962) pp. 2404–2410 ; 2411–2418.
13. Gordon R. G., *J. Chem. Phys.* **44** (1966) pp. 1830–1836.
14. Hansen J. P. and McDonald I. R., *Theory of simple liquids*; 2nd edition (Academic Press, London, 1986); Coffey W. T., Kalmykov Y. P. and Waldron J. T., *The Langevin Equation* (World Scientific, Singapore, 1996).
15. Hu C.-M. and Zwanzig R., *J. Chem. Phys.* **60** (1974) pp. 4354–4357.
16. Baskin J. S., Semmes D. H. and Zewail A. H., *J. Chem. Phys.* **85** (1986) pp. 7488–7490.
17. Dantus M., Bowman R. M., Baskin J. S. and Zewail A. H., *Chem. Phys. Lett.* **159** (1989) pp. 406–412.

18. Semmes D. H., Baskin J. S. and Zewail A. H., *J. Am. Chem. Soc.* **109** (1987) pp. 4104–4106.
19. Semmes D. H., Baskin J. S. and Zewail A. H., *J. Chem. Phys.* **92** (1990) pp. 3359–3376.
20. Wan C., Gupta M., Baskin J. S., Kim Z. H. and Zewail A. H., *J. Chem. Phys.* **106** (1997) pp. 4353–4356.
21. Millar D. P., Robbins R. J. and Zewail A. H., *Proc. Natl. Acad. Sci. USA*, **77** (1980) pp. 5593–5597; *J. Chem. Phys.* **74** (1981) pp. 4200–4201; *J. Chem. Phys.* **76** (1982) pp. 2080–2094.
22. Zhong D., Douhal A. and Zewail A. H. *Proc. Natl. Acad. Sci. USA*, **97** (2000) pp. 14056–14061.
23. See, *e. g.*, Cross P. C., Hainer R. M. and King G. W., *J. Chem. Phys.* **12** (1944) pp. 210–243.

DYNAMICS OF GAS PHASE CLUSTERS: INSIGHTS INTO SIZE AND SOLVATION EFFECTS ON ELECTRONIC RELAXATION AND HYDROGEN TRANSFER

E.S. Wisniewski, B.D. Leskiw, S.M. Hurley, T.E. Dermota, D.P. Hydutsky, K.L. Knappenberger, Jr., M.A. Hershberger‡ and A.W. Castleman, Jr.*

The Pennsylvania State University, Departments of Chemistry and Physics
University Park, PA 16803, USA

E-mail: awc@psu.edu

Presented herein are studies of several cluster systems undertaken to explore the effects of size and solvation on the dynamics of electronic relaxation and competitive reaction processes. The species selected for investigation using femtosecond pump-probe techniques include the unique metal-carbon cluster discovered in our group and termed Metallocarbohedrenes, and two classes of hydrogen containing molecules, namely methanol and acetone, solvated with water. The role of the degree of aggregation of a system and its evolving dynamics is well demonstrated by the findings from the present work.

1 Introduction

Clusters represent intermediate states of matter that often display behavior unlike that of either the gas or condensed phase, and their study gives insight into phenomena operative in systems of restricted size. In some cases the study of clusters of increasing size enables a bridging of behavior between the gas and condensed state, while in other situations new phenomenon such as that due to quantum confinement arises. In cases where the clusters are comprised of aggregates of stable compounds, interest also derives from the prospect that they may be useful as building blocks of new nanoscale materials. For example, the Metallocarbohedrenes (Met-Cars, for short) are a unique class of clusters that show promise as new materials. Interest in van der Waals and hydrogen bonded clusters arises in the context of elucidating the effects of degree of aggregation on photoexcited systems that can undergo atom transfer reactions, photodissociation reactions, and relaxation phenomena.

2 Experimental Findings

The pump-probe technique, pioneered by Zewail and coworkers [1], was implemented in our laboratory to study the reaction dynamics of various molecular clusters, was employed to investigate a series of separate reactions initiated by femtosecond irradiation. In the following three sections, we report recent findings of the relaxation of electronically excited states of Met-Cars, the relaxation and reaction dynamics of the hydrogen bonded methanol system and results of investigations of acetone-water clusters where the dynamics is found to have a dependence on the degree of aggregation.

2.1 The Electronic Relaxation of Photoexcited Met-Cars

Since the discovery in 1992 of a unique class of carbon-metal clusters, termed metallocarbohedrenes [2] (Met-Cars, for short), many studies have focused on elucidating the underlining properties that contribute to the enhanced stability observed in these species [3-7]. Several of the early investigations have utilized time-of-flight mass spectrometry showing that in addition to being very stable, Met-Car clusters also display interesting ionization behavior. Due to the fact that Met-Cars are very strongly bound [8], have a relatively low ionization potential [9-11], and possess the ability to effectively store excess energy, they undergo appreciable ionization under many experimental conditions [12-14]. In light of the continued interest in developing new materials, attention has focused on whether material of nanoscale dimension can exhibit the same properties in the bulk state.

In order for a cluster of limited size to display delayed ionization in a manner corresponding to bulk phase thermionic emission, i.e., the emission of electrons upon heating, two criteria must be fulfilled [15]. The first requirement is that the ionization potential of the cluster is less than its dissociation energy and secondly, all of the phase space is accessible and thus sampled by the system. The first condition ensures that ionization, as a mechanism to dissipate energy, is more favorable than dissociation, while the second excludes the possibility that the delayed response is due to one particular long-lived state. The process of delayed ionization has been previously observed to a lesser extent among other families of clusters as well, including the fullerenes [16], transition metal oxides [17,18], and some other transition metal carbides [18].

During early experiments focused on determining whether the delayed electron emission observed in Met-Car clusters is in fact attributable to a thermionic event, numerous studies have addressed the aspect whether a particular state was being accessed. One study in particular, for example, investigated a series of wavelengths and determined that since the propensity to delay ionize did not change, one particular state was not being repetitively populated and subsequently responsible for the process of delayed ionization [15].

Recent studies have also addressed whether changes to the pulse width of the ionizing laser would alter the process of delayed ionization. Much of the early work had utilized a relatively long pulse width (10 ns) where the incident energy had sufficient time to randomize among all of the applicable vibrational modes, while a more recent study, using femtosecond laser pulses, showed very different results [19]. By employing an ultrashort pulse width (50 fs), the energy applied to the species of interest is delivered too quickly, thus dictating that direct ionization will occur rather than the provided energy rapidly coupling among the various modes and causing delayed ionization. Further experiments utilizing ultrafast lasers have been performed on Met-Car clusters, and are beginning to shed more light on the applicable electron dynamics [20,21] and the origin of the delay ionization phenomenon.

2.1.1 Experimental

The instrument used in the delayed ionization experiments is a reflectron time-of-flight mass spectrometer with a singly modified double acceleration region. The entire apparatus has been previously described in detail [14], and only a short description follows. Briefly, a rotating and translating quarter inch metal rod is ablated with the second harmonic of a Nd:YAG (532nm) laser. The formed species are quenched via a buffer/reactant gas, and further cooled by subsequent supersonic expansion through a conical-shaped nozzle. Following expansion into vacuum, the supersonic jet, which is on axis with the flight tube of the mass spectrometer, is skimmed as it enters the ionization chamber housing the high voltage time-of-flight (TOF) grids. Pure methane is typically used as the reactant gas and in addition to using just pure metal rods (i.e.: Ti, V), ones containing a mixed metal can readily produce binary Met-Car clusters as well.

To ensure that the delayed ionization signal was only obtained from the photoionization of neutral species, those ions produced directly in the source were immediately deflected off axis by applying a voltage to a wire mesh mounted in front of the skimmer. Based on the design of Recknagel [22], a modified TOF acceleration lens assembly was employed to investigate the process of delayed ionization. Essentially, a set of deflectors is mounted in between the TOF repeller and extractor girds, and by using a series of high voltage pulses from Behlke switches, all prompt ions (those ions created during the photoionization laser pulse) are deflected off the axis and subsequently not detected. At time zero of the experiment, the neutrals from the laser vaporization source enter the TOF extraction region and are excited/ionized with a laser pulse. After a predetermined blocking time, referring to the time interval in which the prompt ions are knocked off axis through a voltage gradient on the deflectors, a "field-free" region is employed. During this time interval, both TOF grids and both deflector grids are set at the same voltage and any delayed ions that are newly formed continue to propagate along the axis of the mass spectrometer with their initial velocity from the cluster source. After a set drift time between the TOF grids, the delayed ions are extracted by the simultaneous pulsing of the first TOF grid to a higher voltage and the second grid to one of a lower potential. Any ions born between the time in which the blocking field was turned off and the extraction field was turned on will be extracted and subsequently detected by a set of microchannel plates three meters downstream.

The ion signal is recorded on a digital oscilloscope and analyzed with a personal computer.

The femtosecond laser pulses are generated in a mode-locked Ti:sapphire oscillator (Spectra Physics (SP) Tsunami), which is pumped by a 5 Watt Nd:YVO_4 continuous-wave laser (SP Millennia). The output of the Tsunami oscillator, typically centered at 800 nm, has a pulse energy of a few nJ/pulse and a pulse width of ~20 fs. The output of the Tsunami is subsequently amplified in a regenerative amplifier (SP Spitfire), using the second harmonic of a 10 Watt Nd:YLF laser (SP Merlin). The output of the regenerative amplifier, typically 1.2 mJ/pulse at 40 to 50 fs, is split with a 50/50 beamsplitter and is either directed onto a high precision delay stage (Aerotech) or to an optical parametric amplifier (SP OPA-800). The light that passes over the delay stage can be either frequency doubled, tripled, or quadrupled in two commercial harmonic generation boxes.

The optical parametric amplifier has a collinear double pass geometry and is capable of producing ~50 fs pulses of light between 240 nm and 3.1 μm. The pump pulse from the OPA and probe pulse from the harmonic box(es) are recombined and loosely focused, collinearly, intersecting the cluster beam perpendicularly between the extraction grids of our home built reflectron time-of-flight mass spectrometer. The zero delay time between the pump and probe pulses is identified by the known nearly prompt response of methyl iodide [23]. During pump-probe experiments, the fluences of the pump and probe beams are carefully adjusted to minimize or eliminate ion production when either the pump or probe beam is individually present.

2.1.2 Results

In the metal-carbon systems investigated thus far, the Met-Car stoichiometry (M_8C_{12} where M is either Ti, Zr, Nb) has been found to be the most dominant species to delay ionize, but for the Zr and Nb system, however, other metal carbides were also observed using the wavelengths employed [14]; see Figure 1. Interestingly, in addition to the fact that the intense delayed ion signal of the titanium-carbon series originates only from Ti_8C_{12}, there is also considerable evidence for delayed atomic ion emission from the Met-Car species alone [12]. Although the signal intensity of the niobium Met-Car was not as dominant among other Nb-carbides, the percentage of delayed $Nb_8C_{12}^+$ ions compared to prompt $Nb_8C_{12}^+$ ions is significantly greater than for any of the other niobium carbide clusters in the distribution.

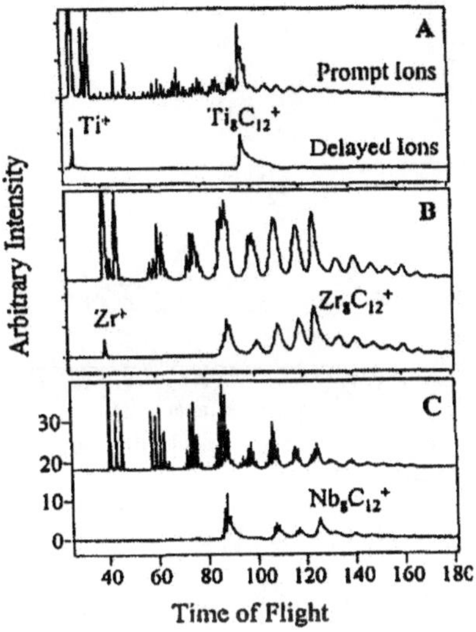

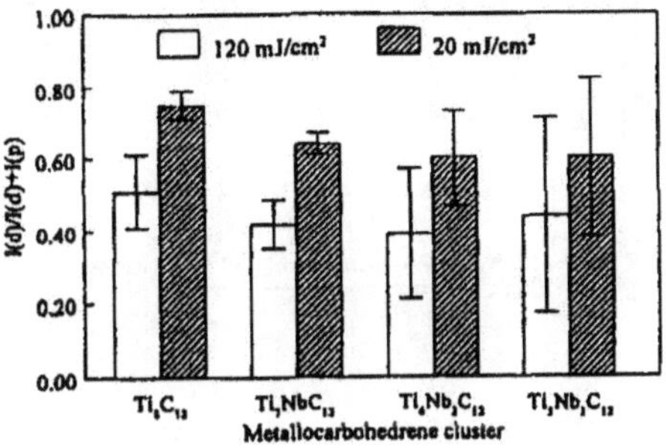

Figure 1 (Left). Prompt ion and delayed ion time-of-flight spectra in the (a) titanium-carbon, (b) zirconium-carbon, and (c) niobium-carbon systems produced by 355 nm photoionization of neutral cluster distributions generated by laser vaporization of an elemental rod in the presence of pure methane.

Figure 2 (Right). Plot of ratios of delayed ion versus total ion signal intensities under two different fluence conditions.

As shown in Figure 2, the ratio of Ti_8C_{12} species that preferentially delay ionize over prompt ionization approaches 70%, and other fluence effects are very evident as well [13]. Figure 2 also illustrates the ratio of delayed ions to prompt ions when substituting an Nb atom for a Ti atom of the binary Met-Car. To gain more insight on the bands of states that are present in the Met-Car series, studies have focused on elucidating the electron dynamics of various species [22].

A mass spectrum showing a typical distribution of vanadium-carbon clusters formed from photoionization (400 nm, 50 fs, 24 μJ) of neutral species produced in the laser vaporization source is shown in Figure 3. The ultrafast dynamics of

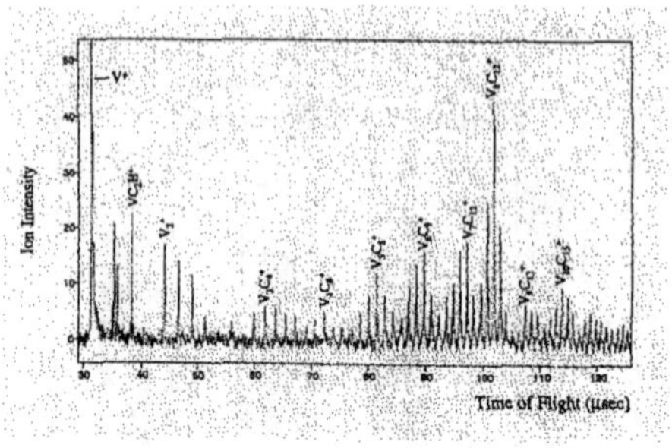

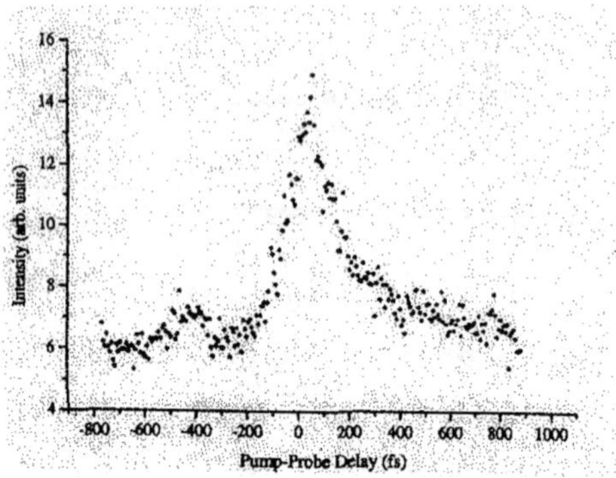

Figure 3 (Left). Mass spectrum of vanadium carbon clusters produced under 400 nm, 50 fs, photoionization of neutral clusters generated in the laser vaporization source.
Figure 4 (Right). Pump-probe transient (pump: 400 nm, 50 fs, 24 mJ; probe 620 nm, 50 fs, 20 mJ) observed for the vanadium Met-Car cluster between delay times of –1 ps and +1 ps.

vanadium-carbon clusters were investigated under 620 nm pump, 400 nm probe conditions and the corresponding transient of the vanadium Met-Car is illustrated in Figure 4. The cross-correlation of the two laser beams was measured to be ≤ 86 fs while the observed response of the V_8C_{12} cluster is 225 fs (FWHM). This distinct difference indicates that a state (or band of states) with an appreciable lifetime is being accessed and thus revealing the temporal dependence of the electronic relaxation. It should also be noted that the pump-probe transients for all of the observed vanadium-carbon clusters with more than four metal atoms show a similar pump-probe response, as seen in Figure 5, with exponential decays on the order of a few hundred femtoseconds [20]. This result suggests that a common chromophore [24], present in clusters as small as the V_4C_x series up through the Met-Car, V_8C_{12}, is being accessed.

Additionally, upon closer inspection and curve fitting, individual trends in the observed transients for V_4C_6, V_5C_8, V_6C_8, V_7C_{12}, and V_8C_{12} are evident. The trends observed in the fitted and normalized transients, as plotted in Figure 5, begin with the narrowing of the transient with increasing cluster size. This result suggests that the larger nanoscale complexes have more free-electron character and therefore the rate of energy relaxation is higher in the larger clusters. Secondly, the maxima of

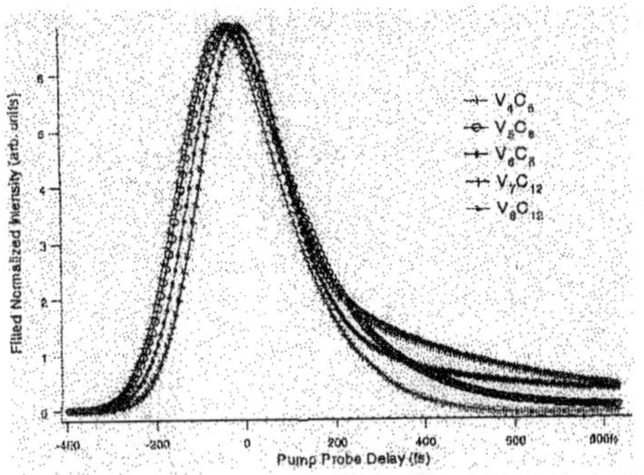

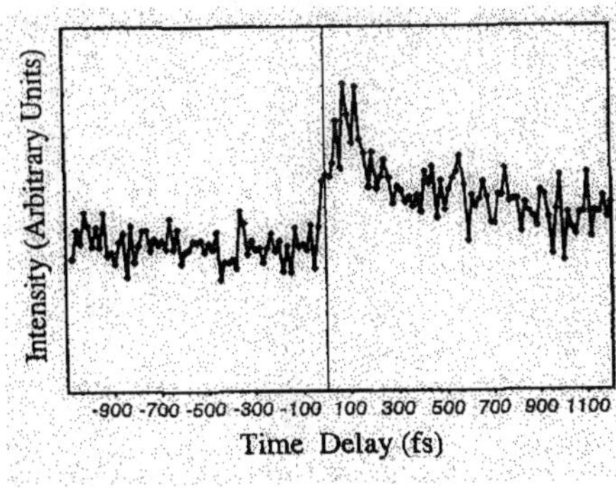

Figure 5 (Left). Summary of the fits of experimental pump-probe data for various vanadium-carbide clusters, including V_8C_{12}

Figure 6 (Right). 1260/800nm pump-probe transient observed for vanadium Met-Car.

the fitted transients shift to longer delay times with increasing cluster size. This shift suggests that the possible chromophore response is not initiated until the relaxation dynamics of the proximate states are affectedly relaxed [20]. The band of states that are influencing the delayed response could perhaps illustrate phonon relaxation modes and conceivably contribute during the process of delayed ionization.

The long-time tail response in the Met-Car pump-probe transient, again suggesting the relaxation into a band of states, is observable under other wavelengths, as well. For example, a typical pump-probe transient of the vanadium Met-Car at moderate laser fluence with photons of 800 nm and 1260 nm, is shown in Figure 6. These findings begin to uncover the complexity of the density of excited electronic states involved, and the concomitant relaxation dynamics. Observations indicate that small changes in the 1260 nm wavelength dramatically affect the behavior of the system pumped by the redder wavelength and probed at 800 nm [21]. Further evidence of efficient coupling among the electronic energy and lattice modes has been acquired with additional results supporting that the latter contributes to the thermionic emission we have observed. Importantly, these new findings show that the delayed ionization process, typically seen only under excitation via nanosecond pulses, might be revealed in experiments conducted with laser fluences that enable relaxation into electronic bands during a femtosecond excitation process.

Other recent investigations exploring the electronic excitation and relaxation properties of the vanadium Met-Car using short unchirped pulses also have enabled us to observe and characterize vibrational frequencies in the ionization dynamics;

see Figure 7. Periodic recurrences in the ionization intensities are readily seen, from

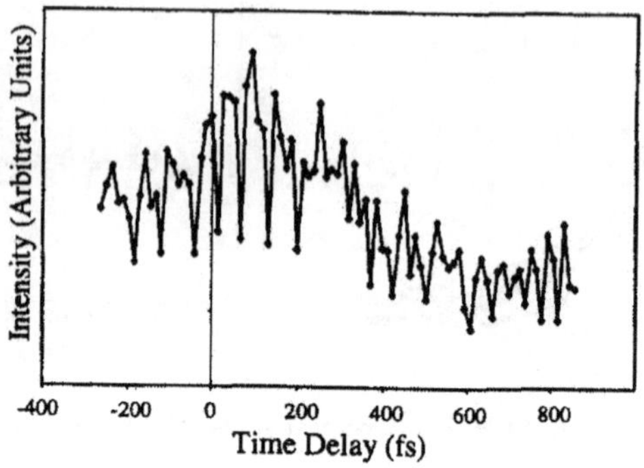

Figure 7. 1300/800nm pump-probe transient observed for vanadium Met-Car. Note the oscillatory frequency: 505 cm^{-1}.

which a frequency of 505 cm^{-1} is obtained [21]. Interestingly, this compares to a metal-carbon stretch frequency of 575 cm^{-1} reported for VC on vanadium surfaces [25]. Meijer and coworkers have extracted a vibration for Ti_8C_{12} with a frequency of 525 cm^{-1} [26].

2.2 *Investigations of Hydrogen Bonded Methanol-water Clusters*

The pump-probe scheme has been utilized in our laboratories for several years in studies of model cluster systems such as acetone, ammonia and 7-azaindole [27-30]. One common reaction pathway to the various systems is that related to hydrogen atom and/or proton transfer. In this section we present recent findings pertaining to the reaction mechanisms observed in methanol-water complexes following their excitation.

2.2.1 *Experimental*

Studies were conducted employing time-resolved femtosecond spectroscopy followed by ion detection with a reflectron time-of-flight mass spectrometer, arranged similarly to the Wiley-McLaren design [31], Figure 8.

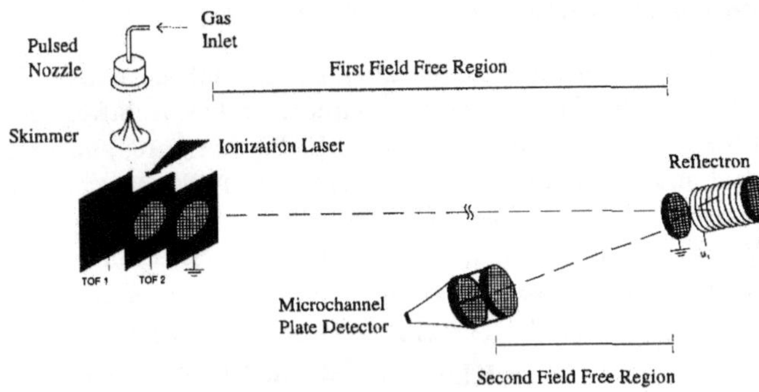

Figure 8. A typical reflectron time-of-flight mass spectrometer.

Ionization was achieved with an amplified colliding pulse mode-locked ring dye laser (CPM). In this arrangement, a gain jet containing rhodamine 590 tetrafluoroborate is pumped by a continuous wave argon ion laser (Coherent Innova 305). Continuous wave lasing from the gain jet is interrupted by a saturable absorber jet containing DODCI. Laser pulses on the order of 100 fs are generated at 90 MHz with pulse energies of only ~200 pJ. Amplification of the laser pulses is achieved in four stages using a six-pass bowtie amplifier and three successive Bethune cells where the beam is progressively expanded from a 2 mm initial diameter to a final beam diameter of 12 mm. All amplification is achieved by transverse pumping of sulphorhodamine 640 by the second harmonic of a 10 Hz Nd:YAG laser (Spectra Physics GCR-4/5). Recompression is performed using a prism pair to compensate for group velocity dispersion. The amplified laser pulse is 200 fs in duration determined by autocorrelation, 1.0 -1.5 mJ of energy and has a wavelength centered at 620 nm.

In order to perform two color pump-probe experiments, the fundamental laser beam is impinged upon a BBO crystal to generate the second harmonic (310 nm) for use as the pump pulse. The remaining light that is not converted to ultraviolet (UV) is harvested and used as the probe pulse. The UV light is sent through a delay stage so that the optical path lengths of the pump and probe lasers can be adjusted for temporal overlap. The laser pulses are loosely focused between the acceleration grids of the mass spectrometer with a 40 cm optical lens placed between the focal points of the pump and probe laser pulses.

Mixed methanol/water clusters were generated via supersonic expansion of a room-temperature mixture vapor seeded in helium at a pressure between 30 and 40 psig. The molecular beam produced in this fashion was collimated with a skimmer prior to ionization with femtosecond laser pulses, which were directed to intercept clusters located between the time-of-flight grids.

2.2.2 *Results and Discussion*

Methanol is one of the most versatile chemicals, used as industrial solvents and studied for potential use as an energy source and has rightfully garnered much interest from the scientific community [32-35]. Furthermore, since methanol is the simplest alcohol and has relatively few electrons, many theoretical simulations have been performed in an effort to elucidate some of its physical properties [36]. Such physical properties as the vacuum ultraviolet (VUV) spectrum of methanol have been obtained which shows several features in the region of 142-198 nm [37,38]. The lower energy region of 165-198 nm has been attributed to the $3s \leftarrow n$ Rydberg transition [38]. The high energy region has been attributed to two separate $3p \leftarrow n$ Rydberg transitions, at 161 nm and 149 nm, which have been referred to as the A and B states, respectively [37].

Experimental investigation of the 3*s* Rydberg state by Bersohn and coworkers determined that photodissociation of methanol mainly results in the loss of atomic hydrogen, with a quantum yield of 0.86 ± 0.1 [39].

$$CH_3OH + h\nu \rightarrow CH_3O + H \qquad (1)$$

Also, theoretical calculations have suggested that the 3*s* Rydberg state is exclusively dissociative along the O-H coordinate if photodissociation is initiated from the vibronic ground state [36]. These calculations predicted that no O-CH_3 stretch occurs following excitation from the ground state, and hence no dissociation into the CH_3 + OH channel:

$$CH_3OH + h\nu \rightarrow CH_3 + OH \qquad (2)$$

Wittig et al. questioned these theoretical findings based on studies made employing experimental photofragment translational spectroscopy of supersonically expanded methanol at 193.3 nm [40]. Results of this experiment yielded evidence of CH_3O fragments with "significant vibrational excitation." Wittig suggested this vibrational excitation partially accounted for other reaction products not identified by Eq. 1, though they found the dissociation of hydroxyl hydrogen dominates.

The 3*p* Rydberg manifold of the methanol molecule has not been as rigorously studied as the 3*s* Rydberg state. Yang and coworkers reported competing pathways from the photodissociation of methanol following excitation with 157 nm radiation [41]. The resulting photochemistry following the excitation to the 3*p* state is more complex than the photodissociation from the 3*s* Rydberg state. Both atomic and molecular hydrogen loss were reported from photofragment translational spectroscopy with the atomic loss being the dominant pathway (atomic:molecular loss ratio 1:0.15). Atomic hydrogen loss arose from methyl or hydroxyl elimination. Yang and coworkers also cited evidence from the VUV spectrum of methanol that hints that the 3*p* Rydberg state is much longer lived than the 3*s* state. They suggested that excitation to the 3*p* state could lead to internal conversion to the 3*s* state and that the hydroxyl loss pathway may proceed along this potential energy surface.

A typical mass spectrum of the mixed methanol water system is seen in Figure 9

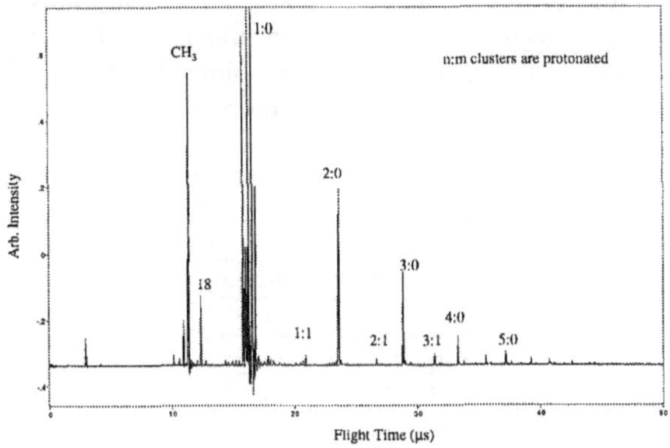

Figure 9. A typical mass spectral distribution of methanol and water clusters. The system is dominated by neat methanol clusters. x:y refers to $H^+(CH_3OH)_x(H_2O)_y$.

at zero time delay between the pump (310 nm) and probe (620 nm) laser pulses. As can be seen, the neat methanol clusters dominate the distribution while neat water clusters are relatively nonexistent. The mixed clusters of $H^+(CH_3OH)_n(H_2O)_m$ are of much smaller intensities and are generally dominated by the (n,1) series.

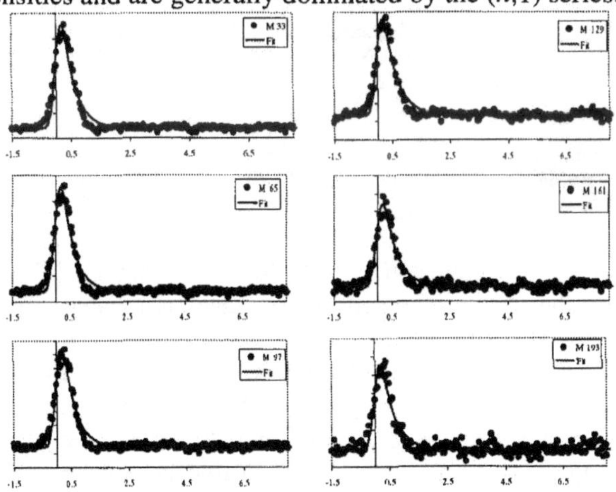

Figure 10. Pump-probe transients for protonated methanol clusters.

Figure 10 shows the pump-probe optical transients for neat methanol clusters. The transients have been mathematically fit [42] to determine the fast time decay but do not show any long lifetime behavior. The characteristic times for each of the neat methanol clusters are very similar to one another and are attributed to the dissociative lifetime of the hydroxyl proton loss following excitation to the 3p Rydberg state followed likely by repulsion along the 3s potential energy surface which has been previously determined to ≤ 1 ps [40]. Table 1 lists the time calculated for the neat methanol clusters. Ionized methanol clusters are known to rearrange to yield a protonated cluster, however; this process occurs much later than the dynamics imaged in these experiments and while this process is occurring, the dynamics are clearly due to femtosecond relaxation.

Table 1. The time constants for the neat protonated methanol clusters.

Cluster	t (fs)
$H^+(CH_3OH)_1$	404 ± 13
$H^+(CH_3OH)_2$	404 ± 13
$H^+(CH_3OH)_3$	394 ± 14
$H^+(CH_3OH)_4$	415 ± 15
$H^+(CH_3OH)_5$	419 ± 16
$H^+(CH_3OH)_6$	421 ± 22

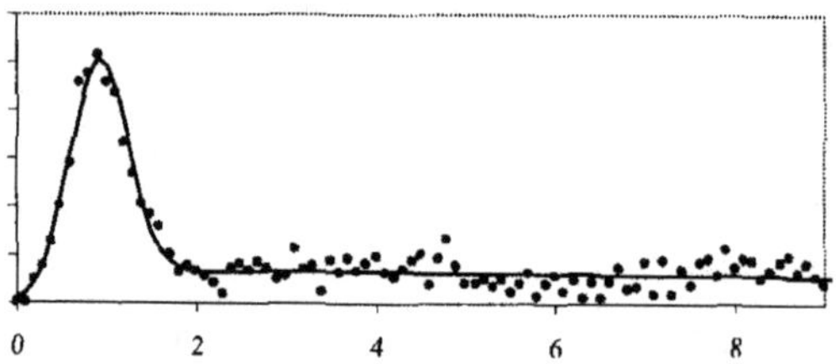

Figure 11. The pump-probe for the 1:1 mixed methanol/water cluster. The fast time determined from the mathematical fitting correlates to the hydroxyl hydrogen loss. The long time decay may be due to the formation of a neutral H_3O from hydrogen transfer.

The mixed methanol clusters exhibit different behavior. The optical transient in Figure 11 for the 1:1 protonated mixed methanol-water cluster shows a two-component decay after excitation by the pump laser pulse. It is likely that excitation to the 3*p* Rydberg state is achieved with 8 eV of energy acquired with two photons of the pump laser pulse. Generally, red shifts occur in the excited electronic states of solvated molecular systems, especially in hydrogen bonded clusters [43,44].

Considering these shifts, it is quite possible that the B state (149 nm for the methanol monomer) can easily be accessed in a solvated methanol/water cluster or a methanol solvated with other methanols. Theoretical calculations are necessary to determine the excited states shifts accurately. For the mixed cluster, the hydrogen source is donated from either an undetected water or methanol molecule. However, since it is likely that the 3*p* Rydberg state of the methanol is being excited, the undetected molecule must be a methanol that is acting as the chromophore. If excitation to the 3*p* state was followed by an internal conversion to the 3*s* Rydberg state as suggested by Yang and coworkers [41], the dissociation would proceed along the repulsive O-H coordinate and the hydroxyl proton would be separated from the CH_3O radical. This process accounts for the protonated mixed cluster detection in the mass spectrum.

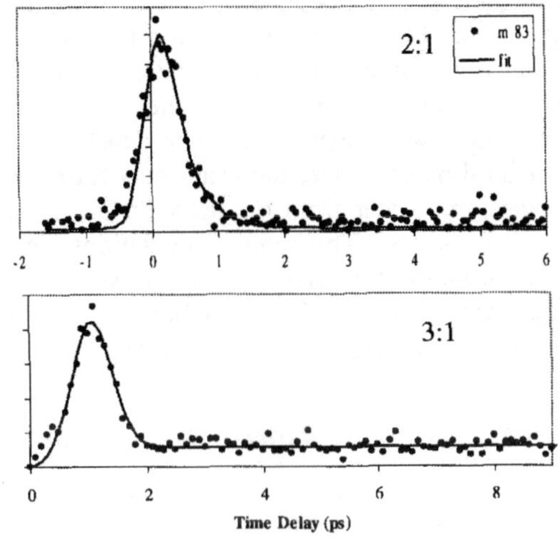

Figure 12. The pump-probe transients for the 2:1 and 3:1 mixed methanol/water clusters. The 2:1 cluster only exhibits a fast time decay for the hydroxyl hydrogen dissociation while the 3:1 cluster exhibits a two-component decay.

The optical transient in Figure 12 was fit to a series of kinetic rate equations for [B] in a consecutive reaction, A→B→C [45]. The first time component was determined to be 665 fs. This time is attributed to the dissociative pathway after internal conversion to the 3*s* Rydberg state. The long time decay of 34 ps is not so easily explained. One possible suggestion for the long time decay of the mixed (1:1) cluster is that the hydrogen readily associates with the water to form a neutral H_3O.

Porter *et al.* reported that the lifetime of the metastable H_3O neutral has a lifetime of substantially less than 10^{-7} s and could be on the order of a vibration [46]. While this range encompasses several orders of magnitude, it is plausible that this is is occurring at long time delays as seen in the mixed (1:1) cluster. The neutral methanol-hydronium is then ionized with relative ease considering the ionization potential of H_3O is only 5.1 eV.

Table 2. The time constants for the monohydrated mixed methanol/water clusters.

Cluster	t_1 (fs)	t_2 (ps)
1:1	665	34
1:2	320	n/a
1:3	450	95

The mixed (2:1) and mixed (3:1) cluster pump-probe transients can be seen in Figure 12. The mixed (3:1) cluster also shows both a fast and slow component corresponding to 450 fs and 95 ps. Surprisingly, the mixed (2:1) cluster does not show the long time decay. This may be due to geometrical effects that prevent the facile formation of the H_3O intermediate. Table 2 shows the times determined for the mixed cluster series. There is no progression to shorter lifetimes for the mixed cluster series as the solvation increases as has been observed in our laboratories for other systems [47]. Within error, the long time decays for the 1:1 and the 3:1 mixed clusters are in agreement with each other. The detected 1:1 cluster arises from a $(CH_3OH)_2(H_2O)$ neutral parent where the water may form a bridge between the two methanols. Likewise, the water from the detected 3:1 cluster may form a bridge between two methanol dimers. Theoretical calculations are necessary to determine the lowest energy structure for water clustered with more than one methanol. Continuing studies are underway in our laboratory to elucidate the dynamics associated with water-rich mixed clusters and will appear elsewhere [45].

2.3 *Dynamics of Acetone-Water Clusters: Relaxation and Photodissociation*

The acetone system has been one of long standing interest and controversy, where it has been found that excitation to some states leads to fast reactive events including step-wise Norrish Type I processes, while at other energies, reaction dynamics attributable to fast dephasing and slow dissociation events have been inferred from the findings. This section presents new findings on the behavior of acetone-water clusters excited to the π-π^* transition.

2.3.1 *Experimental*

The laser system that is used to monitor reaction dynamics consists of a mode-locked Ti:Sapphire laser (Spectra Physics, Tsunami 3955) that is pumped by an Ar-ion laser (Spectra Physics, Beam Lock 2060.) The output of the Tsunami oscillator is amplified in a regenerative Ti:Sapphire amplifier equipped with a pulse stretcher and compressor for chirped pulse amplification that is pumped by a Nd:YAG laser (Quanta Ray, GCR 150-10). The resulting laser light has a wavelength centered around 800nm (tunable between 720nm and 850nm) with a bandwidth of 16 nm at a repetition rate of 10 Hz, a pulse duration of about 75 fs, and a pulse energy of 3.0 mJ. Frequency doubling and tripling the output in BBO crystals creates the second and third harmonics centered around 400nm and 266nm, respectively, and generates the second wavelength necessary for the two-color pump-probe spectroscopy.

2.3.2 *Results and Discussion*

Studies of the dynamics of acetone have shown that the dissociation of the acetone monomer proceeds in a stepwise manner, [27,48-50], and other studies have investigated the interaction of acetone and water clusters [28,51,52]. However, the influence of solvation on the dynamics of acetone is still of interest. In the present study, mixed clusters of acetone and water were investigated to determine the influence of solvation on the excited state dynamics of acetone. Absorption of three photons of 407 nm light excites the $\pi^* \leftarrow \pi$ transition in acetone [53]. Photodissociation of the acetone molecule proceeds via a single pathway and occurs in a stepwise process to break the methyl carbonyl bonds. However, when clusters of acetone are present, several other reaction channels are available. Shown here are the possible products of the dissociation [54]:

$$(A)_x \rightarrow (A)_x^+ \qquad (3)$$
$$\rightarrow (A)_{x-1} \cdot H^+ + CH_3COCH_2 \qquad (4)$$
$$\rightarrow (A)_{x-1} \cdot (CH_3CO)^+ + CH_3 \qquad (5)$$
$$\rightarrow (A)_{x-1} \cdot (CH_3)^+ + CH_3CO \qquad (6)$$

where A represents an acetone molecule and only the products are depicted (not the pump-probe mechanism). From the mass spectra, it was found that channels 3-5 dominate while channel (6) is not detected under the conditions utilized in these experiments.

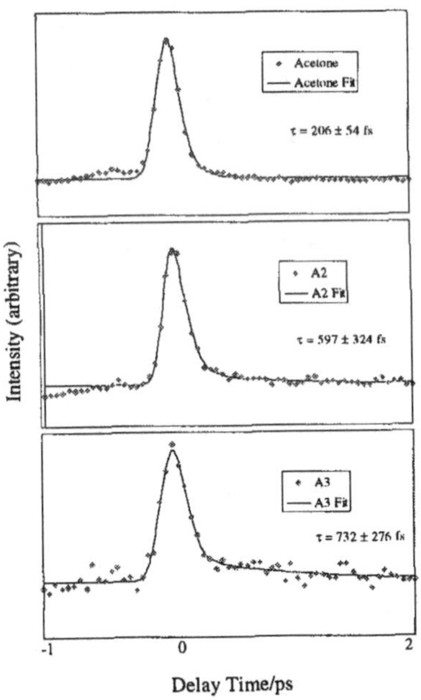

Figure 13. Pump-probe transients of protonated acetone clusters (A_n): A. monomer, B. dimer, and C. trimer.

Pump-probe traces of clusters of acetone are shown in Figure 13. The lifetime of acetone in this excited state represents the decay of the parent molecule into the various dissociation channels available (3-6). It can be observed in Figure 13 that as the degree of clustering increases, the acetone lifetime also increases. The protonated acetone monomer decays in 206 ± 54 fs. An increase in the decay time to 732 ± 276 fs is observed for the protonated acetone trimer. The data is fit to yield the lifetime using:

$$S(t) = \Sigma_i\, c_i \sigma \exp\left[\,(\sigma/2\tau_i)^2 - t/\tau_i\right]\left(1 - \mathrm{erf}\,(\sigma/2\tau_i - t/\tau_i)\right) \qquad (7)$$

where t is the delay time, τ_i is the lifetime, and σ is the FWHM pulsewidth/1.6651 [42].

The increase in the lifetime of the cluster could be due to solvent molecules altering the dissociation process of the acetone chromophore by two possible mechanisms, namely intramolecular rovibrational redistribution (IVR) or caging. IVR can occur causing the available energy to partition into the modes of the cluster, thus slowing down or preventing the dissociation process. If IVR (dephasing) is responsible for the observed dynamics, an increase in cluster size should decrease the lifetime due to the availability of additional modes that the energy may partition into, thus implying that since the lifetime of the excited state lengthened with size, dephasing is not very active. Another possibility is that caging by the acetone solvent molecules could be responsible for the increased lifetime.

Due to the femtosecond timescale of the dissociation, redistribution of the available energy follows the impulsive model [55], and the complete redistribution of energy is not possible. The second step in the dissociation of acetone is the cleavage of the C-C bond of the acetyl radical formed during the first dissociation step (channel 5). The lifetime of this species depends on the available excess energy. Absorption of three photons of 407 nm light corresponds to an energy of 211 kcal/mol. Cleaving he first methyl group from the acetone molecule requires 80 kcal/mol, [48] which leaves 131 Kcal/mol of available energy. Figure 14 shows the pump-probe traces of the acetyl photodissociation product. The impulsive model z

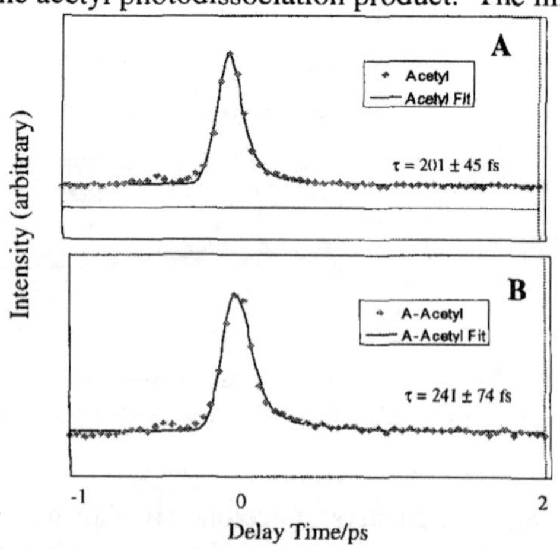

Figure 14. Pump-probe transients of A. acetyl radical (CH_3CO) and B. acetyl•acetone, $CH_3CO{\bullet}(CH_3COCH_3)$.

[55] predicts that 36% of the available energy goes into internal excitations of the acetyl fragment. The translational energy of the products accounts for 54% of the available energy and the methyl fragment contains the remaining 10% of the available energy as internal excitations. Figure 15 shows the acetyl lifetime versus excess energy for several studies. The observed trend is for the acetyl lifetime to decrease with increasing available excess energy. The influence of clustering can be seen in the lifetime exhibited by the acetone-acetyl cluster species, which is longer than the lifetime of the isolated acetyl radical (Figures 14 and 15).

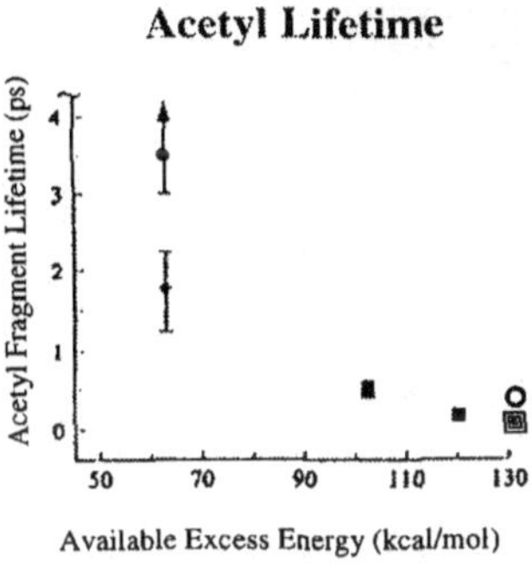

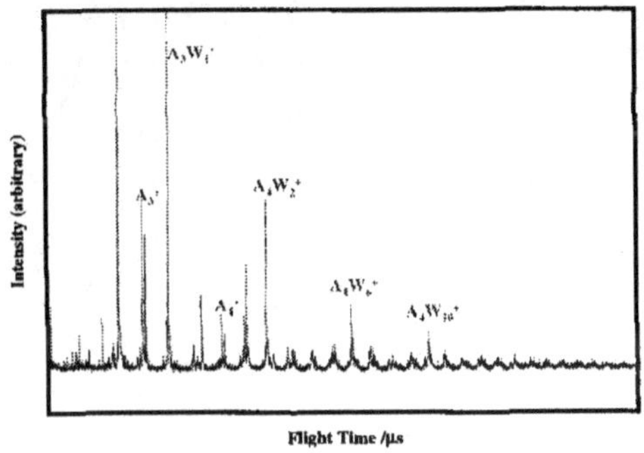

Figure 15 (Left). Section of time-of-flight mass spectrum of protonated acetone-water clusters A_nW_m.
Figure 16 (Right). Lifetime of acetyl and acetyl cluster intermediates: ■ Acetyl$^+$ (*J. Chem. Phys.* **103**, 477 (1995), ◆ Acetyl$^+$ ● A-Acetyl$^+$ (*J. Chem. Phys.* **104**, 5010 (1996), and current experiments ▪ Acetyl$^+$, o A-Acetyl$^+$.

To further study the effects of solvation on acetone dynamics, mixed clusters of acetone and water were also investigated. Figure 16 shows a section of the mass spectrum of the acetone-water system. The acetone trimer was the smallest acetone cluster observed that had any water molecules attached to it. Water molecules were not observed attached to either the acetone monomer or dimer. Based upon filled hydrogen bonding positions within the clusters, the mixed acetone-water clusters, $(CH_3COCH_3)_n{\bullet}(H_2O)_pH^+$, followed the trend $p = n - 2$ [28,52] for the attachment of the first water (p=1). In these mixed clusters, it was further observed that peaks corresponding to the addition of four water molecules displayed an enhanced intensity relative to all other water containing peaks, which can be seen in Figure 16. This indicates that these clusters could have a more stable structure. Figure 17 shows the transient dynamics of three of these enhanced species, and the observed lifetime is longer than the lifetime of the non-water containing cluster. This change in dynamics could be due to a more stable structure resulting in a longer lifetime, or

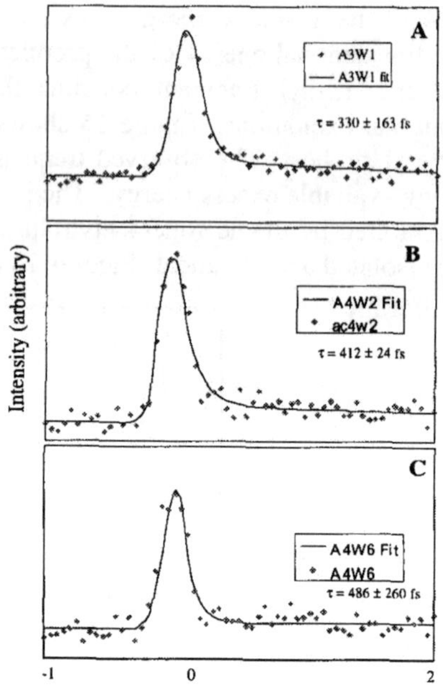

Figure 17. Transient dynamics of mixed clusters of acetone (A) and water (W).

the change in dynamics could be due to the influence of solvation by water molecules. A more stable structure most probably manifest itself in a change in the dynamics of neighboring species, for example the A_4W_1 and A_4W_2. The unenhanced water containing species were too small to measure the dynamics, so this determination was not possible. However, comparisons of peaks corresponding to enhanced water-containing species could be made. Panels B and C in Figure 17 show that the observed lifetime of the enhanced water containing species increases with degree of solvation. This is similar to what was observed in pure acetone clusters.

Studies of the influence of solvation on this system have shown that the degree of solvation has a pronounced influence on the dynamics. When the acetone is solvated by other acetone molecules, the lifetime of the parent is increased. Because of the timescale of the dissociation, the lifetime increase is probably due to caging by the acetone molecules. In the case of solvation by water molecules, the lifetime also increased with the degree of solvation. The acetone water clusters, $(CH_3COCH_3)_n(H_2O)_m$, exhibited favored stoichiometry in which clusters with n = 4, m = 2, 6, and 10 exhibited enhanced peak intensity. The enhanced stability of these magic clusters is probably responsible for their increased lifetime.

3 Assessment and Perspectives for the Future

Studies of the dynamics of clusters are providing information of unprecedented detail on the relaxation behavior and reactions of complex systems. Understanding various phenomena which arise following the photophysical excitation of systems of restricted size can provide new insights into behavior in the condensed state. Knowledge such as that obtained through utilization of the femtosecond pump-probe technique to interrogate clusters can be expected to have far reaching consequences including a more in depth understanding of processes of importance in the areas of materials science, atmospheric science, catalysis, and biology.

4 Acknowledgements

We gratefully acknowledge the U. S. Air Force Office of Scientific Research, Grant # F49620-01-1-0122 for support of the studies of the electronic excitation and relaxation dynamics of Met-Cars, and the U.S. Department of Energy Grant # DE-FG02-97ER14258 for support of the methanol-water studies. Financial support by the U.S. National Science Foundation, Atmospheric Sciences Division, Grant No. ATM-9711970 is gratefully acknowledged for support of the acetone photophysical studies.

References

1. Zewail, A.H., Femtochemistry – Ultrafast Dynamics of the Chemical Bond, vols. I and II, World Scientific, Singapore (1994).
2. Guo, B.C., Kerns, K.P., Castleman, Jr., A.W., *Science* **255** (1992) 1411.
3. Guo, B.C., Wei, S., Purnell, J., Buzza, S., Castleman, Jr., A.W., *Science* **256** (1992) 515.
4. Wei, S., Guo, B.C., Purnell, J., Buzza, S., Castleman, Jr., A.W., *J. Phys. Chem.* **96** (1992) 4166.
5. Wei, S., Guo, B.C., Purnell, J., Buzza, S., Castleman, Jr., A.W., *Science* **256** (1992) 818.
6. Guo, B.C., Wei, S., Chen, Z., Purnell J., Buzza, S., Kerns, K.P., Castleman, Jr., A.W., *J. Chem. Phys.* **97** (1992) 5243.
7. Chen, Z.Y., Guo, B.C., May, B.D., Cartier, S., Castleman, Jr., A.W., *Chem. Phys. Lett.* **198** (1992) 118.
8. Kerns, K.P., Guo, B.C., Deng, H.T., Castleman, Jr., A.W., *J. Chem. Phys.* **101** (1994) 8529.
9. Reddy, B.V., Khanna, S.N., Jena, P., *Science* **258** (1992) 1640.
10. Grimes, R.W., Gale, J.D., *J. Phys. Chem.* **97** (1993) 4616.
11. Sakurai, H., Castleman, Jr., A.W., *J. Phys. Chem. A* **102** (1998) 10486.

12. May, B.D., Cartier, S.F., Castleman, Jr., A.W., *Chem. Phys. Lett.* **242** (1995) 265.
13. Cartier, S.F., May, B.D., Castleman, Jr., A.W., *J. Chem. Phys.* **104** (1996) 3423.
14. Kooi, S.E., Castleman, Jr., A.W., *J. Chem. Phys.* **108** (1998) 8864.
15. Klots, C.E., *Chem.Phys. Lett.* **186** (1991) 73.
16. Campbell, E.E.B., Ulmer, G., Hertel, I.V., *Phys. Rev. Lett.* **67** (1991) 1986.
17. Nieman, B.C., Parks, E.K., Richtsmeier, S.C., Liu, K., Pobo, L.G., Riley, S.L., *High Temp. Sci.* **22** (1986) 115.
18. Amrein, A., Simpson, R., Hackett, P., *J. Chem. Phys.* **95** (1991) 1781.
19. Kooi, S.E., Leskiw, B.D., Castleman, Jr., A.W., in : Jena, P., Khanna, S.N., Rao, B.K. (Eds.), Cluster and Nanostructure Interfaces, World Scientific Publishers, Singapore (2000) 511.
20. Kooi, S.E., Leskiw, B.D., Castleman, Jr., A.W., *Nano Lett.* **1** (2001) 113.
21. Leskiw, B.D., Knappenberger, Jr., K.L., Castleman, Jr., A.W., *in preparation.*
22. Leisner, T., Athanassenas, K., Echt, O., Kreisle, D., Rechnagel, E., *J. Chem. Phys.* **99** (1993) 9670.
23. Janssen, M.H.M., Dantus, M., Guo, H., Zewail, A.H., *Chem. Phys. Lett.* **214** (1993) 319.
24. Bescós, B., Lang, B., Weiner, J., Weiss, V., Wiedenmann, E., Gerber, G., *Eur. Phys. J. D* **9** (1999) 399.
25. Chen, J.G. De Vries, B.D. Frühberger, B., Kim, C.M., Liu. Z.-M., *J. Vac. Sci. Technol. A* **13** (1995) 1600.
26. Heijnsbergen, D. Helden, G., Duncan, M.A., Roij, A.J.A., Meijer, G., *Phys. Rev. Lett.* **83** (1999) 4983.
27. Buzza, S.A., Snyder, E.M. and Castleman, Jr., A.W., *J. Chem. Phys.* **104** (1996) 5040.
28. Zhong, Q., Poth, L. and Castleman, Jr., A.W., *J. Chem. Phys.* **110** (1999) 192.
29. Snyder, E.M., Purnell, J., Wei, S., Buzza, S.A. and Castleman, Jr., A.W., *Chem. Phys.* **207** (1996) 355.
30. Folmer, D.E., Poth, L., Wisniewski, E.S. and Castleman, Jr., A.W., *Chem. Phys. Lett.* **287** (1998) 1.
31. Wiley, W.C. and McLaren, I.H., *Rev. Sci. Instrum.* **26** (1956) 1150.
32. Morgan, S. and Castleman, Jr., A.W., *J. Am. Chem. Soc.* **109** (1987) 2867.
33. Morgan, S., Keesee, R.G. and Castleman, Jr., A.W., *J. Am. Chem. Soc.* **111** (1989) 3841.
34. J. Feichtinger, R. Galm, M. Walker, K.-M. Baumgärtner, A. Schulz, E. Räuchle and U. Schumacher, *Surf. Coat. Technol.* **142-144** (2001) 181.
35. Raina, G. and Kulkarni, G.U., *Chem. Phys. Lett.* **337** (2001) 269.
36. Marston, C.C., Weide, K., Schinke, R. and Suter, H.U., *J. Chem. Phys.* **98** (1993) 4718.
37. Sominska, E. and Gedanken, A., *J. Mol. Spec.* **175** (1996) 234.

38. Nee, J.B., Suto, M., and Lee, L.C., *Chem. Phys.* **98** (1985) 147.
39. Satyapal, S., Park, J. and Bersohn, R., *J. Chem. Phys.* **91** (1989) 6873.
40. Wen, Y., Segall, J., Dulligan, M. and Wittig, C., *J. Chem. Phys.* **101** (1994) 5665.
41. Harich, S., Lin, J.J., Lee, Y.T. and Yang, X., *J. Chem. Phys.* **111** (1999) 5.
42. Fiebig, T., Chachisvilis, M., Manger, M., Zewail, A.H., Douhal, A., Garcia-Ochoa, I. And de La Hoz, A., *J. Phys. Chem.* **109** (1999) 7419.
43. Castleman, Jr., A.W. and Bowen, K.H., *J. Phys. Chem.* **100** (1996) 12911.
44. Nakajima, A., Hirano, M., Hasumi, R., Kaya, K., Watanabe, H., Carter, C.C., Williamson, J.M. and Miller, T.A., *J. Phys. Chem.* **101** (1997) 392.
45. Wisniewski, E.S., Hershberger, M.A. and Castleman, Jr., A.W., *in preparation.*
46. Williams, B.W. and Porter, R.F., *J. Chem. Phys.* **73** (1980) 5598.
47. Folmer, D.E., Wisniewski, E.S., Hurley, S.M. and Castleman, Jr., A.W., *Proc. Natl. Acad. Sci.* **96** (1999) 12980.
48. Merchán M., Roos B. O., McDiarmid R., and X. Xing, *J. Chem. Phys.* **104** (11996) 1791.
49. Tzeng W. B., Wei S., and Castleman Jr. A. W., *J. Am. Chem. Soc.* **111** (1989) 6035.
50. Kim S. K., Pedersen S., and Zewail A. H., *J. Chem. Phys.* **103** (1995) 477.
51. Owrutsky J. C. and Baronavski A. P., *J. Chem. Phys.* **110** (1999) 1206.
52. Stace A. J. and Moore C., *J. Phys. Chem.* **86** (1982) 3681.
53. Aviyente V., Iraqi M., Peres T., and Lifshitz C., *J. Am. Soc. Mass. Spectrom.* **2** (1991) 113.
54. Wei S., Tzeng W. B., Keesee R. G., and Castleman Jr. A. W., *J. Am. Chem. Soc.* **113** (1991) 1960.
55. Trentelman, K.A., Kable S.H., Moss, D.B., and Houston, P.L., *J. Chem. Phys.* **91** (1989) 7498.

Predissociation Dynamics of Acetylene Rydberg States

S. ZAMITH, V. BLANCHET, B. GIRARD
Laboratoire de Collisions, Agrégats et Réactivité (CNRS UMR 5589), IRSAMC
Université Paul Sabatier,
31062 Toulouse, France
sebas@irsamc.ups-tlse.fr and val@irsamc.ups-tlse.fr

J. NORIN, J. MAURITSSON, A. L'HUILLIER
Dept of Physics, Lund Institute of Technology, Sweden

J. ANDERSSON, S. SORENSEN
Dept of Synchroton Radiation Research, Lund University, Sweden

I. HJELTE, O. BJORNEHOLM
Dept of Physics, Uppsala University, Sweden

D. GAUYACQ
Lab. Photophysique Moléculaire (CNRS UPR 3361), Orsay, France

Acetylene dissociation and ionization is central to a large part of physical and chemical processes in the interstellar and circumstellar medium. The free radicals produced in these processes are the basis of a rich chemistry producing a number of carbon chains, which have been observed in these media over the past twenty years. Besides direct dissociation, a large contribution to these free radicals arises from the predissociation of Rydberg states, which carry most of the absorption strength in the VUV range (8-13 eV). We study ultrashort dynamics of acetylene Rydberg states by combining ultrashort VUV laser pulses, a molecular gas cell and an electron- and ion-spectrometer. One VUV photon transition populates Rydberg states in the supercomplex 4s-3d. The evolution of the system is probed by a second laser pulse via a photoionization transition close to the ionization threshold of the parent molecule. The photoelectron spectrum provides an insight on the internal conversions that happen before the dissociation. The experiments have been repeated with deuterated acetylene where the variations in coupling strengths result in different time constants which should be of great help in assigning the dynamics.

1 INTRODUCTION

In the last decade, the number of different molecules detected in the interstellar medium has exceeded the hundred, most of them are composed of carbon[1]. The acetylene molecule is the third one most abundant, after respectively H_2 and CO. The strong predissociation of acetylene in the vacuum ultraviolet (VUV) range is particularly important as it is a photochemical process yielding efficiently precursors of carbon chains. This dissociation dynamics may occur on time scales between tens of femtoseconds to picoseconds in a very small excitation energy range[2]. The first dissociation channels are :

$$
\begin{aligned}
C_2H_2 + h\nu &\rightarrow H + C_2H(\tilde{X}^2\Sigma^+)\ D_0 = 46074\ \mathrm{cm}^{-1} \\
&\rightarrow H + C_2H(\tilde{A}^2\Pi)\ D_0 = 49700\ \mathrm{cm}^{-1} \\
&\rightarrow H_2 + C_2(X^2\Sigma^+)\ D_0 = 50000\ \mathrm{cm}^{-1} \\
&\rightarrow 2CH(X^2\Pi)\ D_0 = 79800\ \mathrm{cm}^{-1}
\end{aligned}
$$

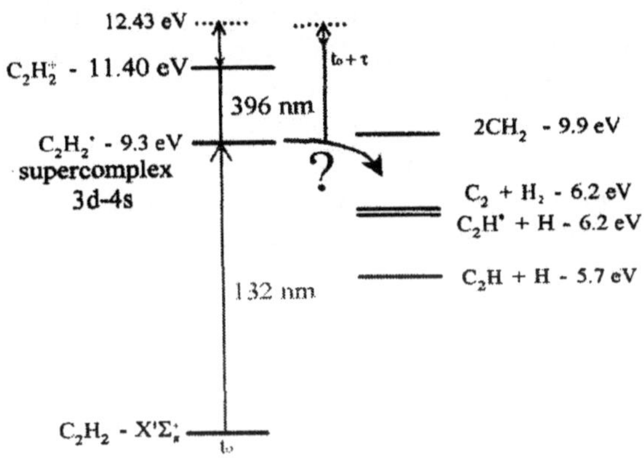

Figure 1: Level scheme in C_2H_2. The VUV pump pulse prepares the optically bright $F4_0^2$ state. The decay of this "zeroth" order state is probed by photoionization. Only two of the dissociation channels are opened- see the text.

In order to help the validation or not of the models developed for the interstellar photochemistry, we examine via time-resolved experiments, a few states of the Rydberg 3d-4s supercomplex. This supercomplex lies above 9 eV with the $F^1\Sigma_u^+ 3d\pi_g$ state as the strongest electronic component appearing through the absorption spectrum[3]. Apart from the origin bands, the absorption spectrum is mainly characterized by the excitation of the symmetric stretching mode ν_2, as expected from the slight increase of the CC equilibrium distance from the electronic ground state into the Rydberg states. The key point of this spectrum is the intriguing progression of the *trans* bending mode ν_4 of the Rydberg states. The presence of the E valence state with a non-linear equilibrium geometry, strongly coupled to the F state is believed to be responsible of this *trans* bending signature. The E valence state as well as the valence-Rydberg coupling have been studied via rotational analysis of double-resonance spectra[4], REMPI-photoelectron spectroscopy[5,6] as well as photofragment spectroscopy[7].

Its origin band lies close to the F ones (approximatively 130 cm^{-1} lower). Here, we investigate the dynamics from the $F4_0^2$ state (corresponding to the excitation of 2 quanta of the ν_4 mode, at 75578 cm^{-1}) as the "zeroth" order state prepared by a VUV pump pulse. At this energy only the two dissociation channels leading to the C_2H radicals are involved. The 2CH and C_2+H_2 channels are not open because of an energetic threshold and the presence of high dissociation barrier[9] respectively.
The energetic scheme as well as the pump-probe excitation are displayed in figure 1 : the Rydberg molecular dynamics is projected onto the ion continuum in order to probe it. The probe photon energy allows only the photoionization of the parent molecules. Thus, through the time-resolved photoelectron spectrum, as well as the parent photoion decay, we will examine the dynamics of the "zeroth" order state $F4_0^2$. Another goal of this experiment is the use of a femtosecond pump pulse lying in the VUV range. The requirement of a high energy fundamental pulse to produce a VUV pulse from harmonic generation through a gas jet[10,11] has limited during a long time the access to this excitation energy. In the present paper, it is the third harmonic of a 400 nm pulse that is used as a pump.

2 EXPERIMENTAL SET-UP

The general experimental setup is shown in Figure 2. We use the 10 Hz terawatt laser at the Lund high-power laser facility to generate infrared 100 mJ pulses with a duration of 110 fs. The 400 nm-pulse used to generate the harmonics and the time-delayed 400 nm-probe pulse are prepared in a Michelson interferometer by frequency doubling through two 1.5 mm-thick KDP cristals (FWHM of 400 nm is 160 cm^{-1}). A 2 mm-thick entrance window allows the two beams to propagate in vacuum before they become too narrow and the energy density too high. The first pulse is focused just under the nozzle of a Xenon gas jet, generating harmonics, whereas the probe pulse is focused about 1.3 m further, using a 3 mm-thick negative lens. The output energies are 7 mJ for the pulse generating the pump and 1.5 mJ for the probe. In the present experiment, the fifth harmonic (80 nm) is used to ionize acetylene or rare gases through one-photon absorption. This allows us to calibrate the electron spectrometer in energy, as well as to optimize and align the setup. All dynamics presented here are obtained, with the third harmonic (132.2 nm) as the pump pulse and the 400 nm as the probe pulse. When using the third harmonic, we eliminate all higher-order ones by inserting a 3 mm-thick Lithium-Fluoride (LiF) window after the gas jet. LiF absorbs radiation below 105 nm (11.8 eV).

The 400 nm-pulse generating the pump pulse is not eliminated, but since it is unfocused, its intensity in the interaction region is too low to generate ions through multiphoton processes. In addition, dispersion in the LiF window introduces a time delay between the 400 nm pulse and the generated harmonic pulse (which arrives later). The two pulses are not overlapped in time and the 400 nm-pulse can not ionize molecules excited by the VUV pulse. With the LiF window, we ionize the sample only when pump and probe pulses are properly synchronized.

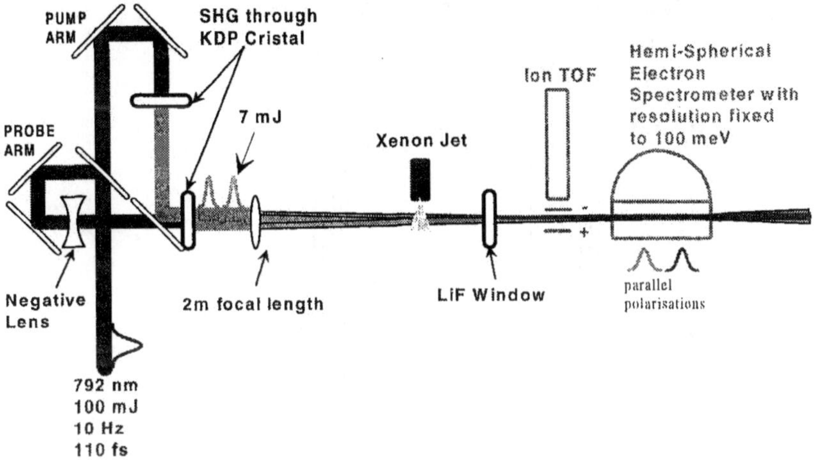

Figure 2: Experimental setup : the VUV pump pulse is the third harmonic from a 400 nm pulse. The harmonic is generated through a Xenon gas jet and is selected by a LiF window. The carrier wavelength of the probe pulse is 400 nm.

In our setup, we detect at the same time both ions and electrons produced in similar conditions. The probe pulse is focused just halfway between the two spectrometers, so that the interaction regions in both spectrometers are similar. Acetylene gas is introduced at the same static pressure in both chambers through leak valves.

Ions are separated in a 70 cm-time-of-flight tube and detected by an electron multiplier. There is an inevitable flow of Xenon from the harmonic generator into the ion spectrometer. When pump and probe pulses are synchronised, xenon atoms are ionised through a non-resonant two photon-ionization signal. This provides a cross-correlation curve, which is used to estimate the pulse durations of the pump and probe pulses. In the experiment, we thus record at each shot both acetylene and Xenon ion signals.

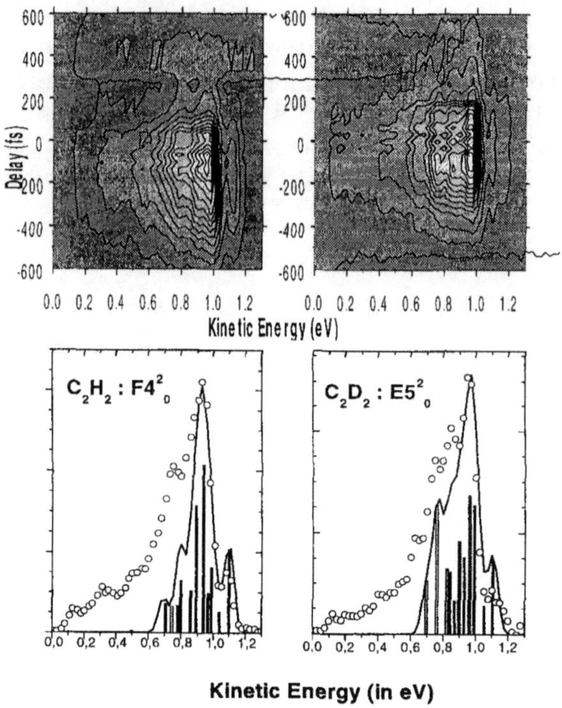

Figure 3: top : The photoelectron spectrum as function of the pump-probe delay in C_2H_2 (left) and C_2D_2 (right). bottom : empty circle data is one cut of the contour plot. this can be compared with the straight line that is the (3+1) ns-REMPI photoelectron spectrum convoluted by the 100 meV resolution. For information, the sticks point out the Renner-Teller ion component corresponding to the ns-REMPI photoelectron spectrum.

Since we use a considerable input energy of 100 mJ over a 2 cm diameter beam, the propagation through different materials (beam splitters, lens) leads to a significant temporal streching and frequency modulation of both pump and probe pulses, because of self phase modulation and group velocity dispersion. The chirp and duration of the pulses are estimated through a multifit operation described later. The electron spectrometer is a truncated hemispherical electrostatic analyzer employing a retarding lens and a two-dimensional multichannel detection system. The spectrometer resolution employed in this study is approximately 100 meV FWHM. The source is a continuously pumped gas cell with a pressure of approximately 0.01 mbar. The background pressure is 3.0 x10^{-8} mbar.

3 RESULTS

Figure 3 (top) shows the time-dependency of the photoelectron spectrum for both C_2H_2 and C_2D_2. Actually at this 132.2 nm wavelength, it is the $F4_2^0$ initial state that is prepared in C_2H_2, whereas it is the $E5_0^2$ in C_2D_2 (2 quanta of the *cis* bending mode are excited)[3]. Whatever the pump-probe delay, it is only the amount of signal and not the shape of the photoelectron spectra that changes for both isotopes. One typical cut of the photoelectron contour plot is displayed on the figure 3- bottom by empty circle symbols. We recognize the Rydberg character of the photoexcited state with an adiabatic photoelectron energy around 1.1 eV[15]. This fs-photoelectron spectrum can be compared to the photoelectron spectra recorded in (3+1) nanosecond Resonance Enhanced MultiPhoton Ionization (ns-REMPI) plotted by the straight lines, as well as the relevant Renner-Teller components of the ion ground state represented by stick[14,6]. These ns-REMPI spectra have been recorded with a 30 meV resolution and a laser on resonance with the $F4_0^2$ state (or $E5_0^2$ of C_2D_2). Since the total energies of (3+1) photons at 396 nm and of (VUV + 396 nm) are the same, the comparison between the ns and fs photoelectron spectra is possible after convolution of the ns-REMPI spectrum by the 100 meV resolution of our photoelectron spectrometer. Indeed, since the pump pulse prepared only one initial state coupled to a dissociative continuum, and the available internal energy of ion is identical for the two excitation schemes, the similitude between the ns and fs spectra is the signature of the Franck-Condon excitation window. Their differences at higher internal energy of the parent ion (i.e. lower kinetic energy of the emitted photoelectron) is mainly the imprint of the exit from the Franck-Condon window linked to the initial state decay. No photoelectron dynamics is visible, several reasons can be given: the available internal energy of the ion is not high enough (less than 1.3 eV), the resolution is too weak compared to the complex Renner-Teller structure of the ion ground state and especially the time-resolution is not good enough. In order to take into account the expected chirps of the pump and probe pulses due mainly to SPM as well as the LiF window, besides the C_2H_2 ions and photoelectrons signals, we collected also the Xe ion signal for a cross-correlation measurement and also the C_2D_2 signals since this molecule is also resonant with the VUV pulse. Actually, the five different time-dependencies (the photoion and photoelectron pump-probe signals of the two isotopes as well as the cross-correlation signals on Xenon) observed allow to determine the characteristics of the pump and probe pulses as well as the time constants of the initial state.

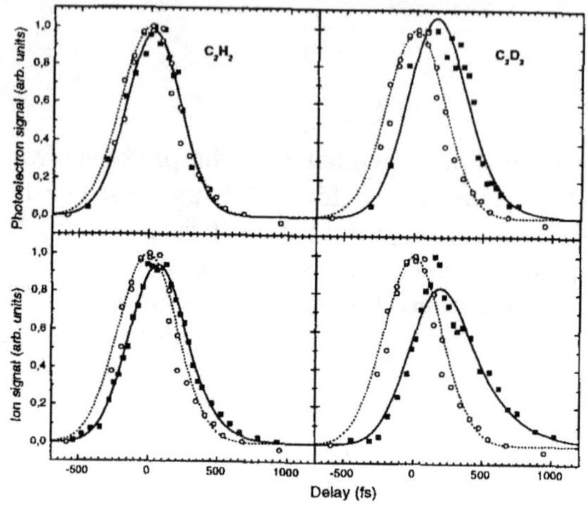

Figure 4: Results of multifit operation on the pump-probe signals on C_2H_2 and C_2D_2. There are two pump-probe signals for each isotope : photoelectron signal on the top and the ion signal on the bottom. The filled square symbols are the experimental pump-probe signals and the empty circle symbols are the experimental cross-correlation signals taken on Xenon. The fits are the straight line for the pump-probe signals and the dashed line for the cross-correlation signals.

The analysis of the time dependency is done through a "multifit" operation. Each pump-probe signal is treated as a decaying level signal with a formula which takes directly into account for the chirp of the pulses[16]:

$$S_{theo}(\tau) \propto e^{-\tau/\tau_e} e^{\frac{\tau_{0p}^2+\tau_{0s}^2}{8\tau_e^2}} e^{\frac{\delta_s\alpha_s\tau_{0s}^2\tau_s^2-\delta_p\alpha_p\tau_{0p}^2\tau_p^2}{2\tau_e}} \left|1-\phi(\frac{\gamma\sqrt{\beta_s+\beta_p}}{2\sqrt{\beta_s\beta_p}})\right|^2 \qquad (1)$$

with $\gamma = \frac{i\beta_p\delta_s-i\beta_s\delta_p-2\beta_p\beta_s\tau}{\beta_s+\beta_p} + 1/2\tau_e$ in which τ_e is the lifetime of the excited state, τ_{0p} and τ_{0s} are the FWHM of the pump and probe pulses, the δ_p and δ_s are the resonance detuning and the other coefficients characterize the chirps as follows :

$$\beta_p = \frac{1}{\tau_p^2} + i\alpha_p \quad \text{and} \quad \beta_s = \frac{1}{\tau_s^2} + i\alpha_s \qquad (2)$$

where τ_c and τ_s are the durations of pump and probe respectively. The terms α_p and α_p correspond to the linear chirp of the pump and probe pulse defined from the quadratic dispersion $\phi"$ as follows :

$$\alpha_p = \frac{2\phi"}{\tau_{0p}^4 + 4\phi"^2} \tag{3}$$

The multifit operation involves a simplex optimization of *Err* function :

$$Err = \sum_n \frac{Err_n}{N_n} \tag{4}$$

with the index n linked to the nth pump-probe signal, N_n the number of data points of this pump-probe signal and Err_n the error to minimize

$$Err_n = \sum_k^{N_n} \left(S_{exp}^n(\tau_k) - a_n S_{theo}^n(\tau_k) - b_n\right)^2 \tag{5}$$

with S_{exp}^n the experimental value of the nth pump-probe signal, recorded at a delay τ_k and compared to S_{theo}^n. The only parameters fixed as inputs are the Fourier limited duration of the VUV pump pulse (at 60 fs) and of the probe pulse (at 100 fs) as well as the pump wavelength at 132.2 nm and the position in energy of initial state ($F4_0^2$ state for C_2H_2 or $E5_0^2$ for C_2D_2). The figure 4 displays in filled squared symbols, the photoelectron time-dependency (on the top) already plotted in the figure 3, as well as the photoion pump-probe signal (on the bottom). The result from the multifit operation are the straight line curves, in good agreement with the experimental data for all the signals. The cross-correlation signal derived from the multifit (the dashed line curve) is also in good agreement with the experimental cross-correlation recorded through the Xenon ion signal. The time constants obtained through this multifit operation are given in **Table 1**. The cross-correlation time is then

Isotope	Signal	Lifetime (fs)	deduced from²
C_2H_2	ions	130	-
	photoelectrons	55	
C_2D_2	ions	230	140
	photoelectrons	140	
Pump duration (fs)		260	
Probe duration (fs)		420	

490 fs. The time-resolution is not sufficient to resolve the time-dependency of the C_2H_2 photoelectron spectrum, as observed from the similarity between the

cross-correlation and this latter signal. However it is interesting in the case of C_2D_2 to compare our measurement with the one deduced from the width of the H^+ photofragment band assuming it to be homogeneous[2]. The lifetime of the photoelectron signal that reflects the exit from the Franck-Condon windows is of the same order than the appearing time of the H^+ fragment. The dissociation is thus very rapid. Note that the ion signal is for both isotopes longer than the photoelectron signal. Indeed, due to a 1° acceptance angle, the photoelectron spectrometer detects only the photoelectrons emitted perpendicularly to the CC bound, whilst the ion time-of-flight detects all the ions. In this particular case, the two different time-scales of the two signals might reveal two different dissociation processes : one slower than the other. Indeed by recording the angular distribution of the photofragment action spectra of the radical C_2H, the group of Welge has already identified two dissociation processes in C_2H_2 : one involving an internal conversion from the initial state to a dissociative continuum and a second involving an internal conversion to the A or/and B state of C_2H_2 before decaying into a dissociative continuum. The correlation between these last observations and ours could be investigated through a time-resolved photoelectron angular distribution.

4 CONCLUSION

The access to the ultrashort VUV pulses allows the investigation of the photoinduced dynamics in small polyatomic molecules with highly lying electronic states. We have used here as a VUV pump pulse the third harmonic of an intense 400 nm pulse, generated in a gas jet. Another technic uses the cascaded nonlinear wave mixing (up to the third order[12]) in a hollow fiber filled of rare gas, or the two-photon resonant difference-frequency mixing[13], allow one to produce now a tunable VUV femtosecond pulses with only a fundamental pulse at mJ level. However up to now, it is relatively difficult to have enough energy in a recompressed VUV pulse. Therefore we developed a multifit approach to deal with the chirp of the pulses. This kind of analysis casts light on the time constants involved, even if the cross-correlation time is limiting the time-resolution. We illustrate this fitting method on the predissociation of acetylene.
A VUV probe pulse would also be of importance to photoionize efficiently the atomic photofragment (compared to the efficiency of the multiphoton ionization) and at the end to observe their kinetic energy distribution as well as the angular one. Furthermore it has been demonstrated that the analysis of non-adiabatic coupling via photoionization pump-probe experiment is elucidated by photoionization dynamics[17]. This is easily obtained with a VUV

probe pulse that gives access not only to the ground electronic state of ion, but also to the excited ion electronic states[18]. This could also be achieved by multiphoton ionization. However recent photoionization experiments with an intermediate intensity femtosecond pulse have reported, for some polyatomic molecules, a non-adiabatic multi-electron ionization mechanism that erases any previous dynamics initiated in a perturbative regime[19]. This last study leads to a reservation about multiphoton ionization as a probe step in time resolved experiments. The first importance of the ultrashort VUV pulse in photochemistry would be also to investigate the photoion dynamics as autoionization.

References

1. J.R. Cronin and S. Pizzarello, *Science* **275**, 951 (1997).
2. P. Löffler, D. Lacombe, A. Ross, E. Wrede, L. Schnieder and K. H. Welge, *Chem. Phys. Lett.* **252**, 304 (1996)
3. M. Herman and R. Colin, *J. Mol. Spectr.* **85**, 449 (1981).
4. J. K. Lundberg, D. M. Jonas, B. Rajaram Y. Chen and R.W. Field, *J. Chem. Phys.* **97**, 7180 (1992)
5. T. M. Orlando, S. L. Anderson, J. R. Appling and M. G. White, *J. Chem. Phys.* **87**, 852 (1987)
6. S. Boyé, S. Zamith, A. Campos, V. Blanchet, D. Gauyack, submitted to *J. Chem. Phys.*
7. P. Löffler, E. Wrede, L. Schnieder, J. B. Halpern, W. M. Jackson and K. H. Welge,*J. Chem. Phys.* **109**, 5231 (1998)
8. G. P. Agrawal, in "Nonlinear Fiber Optics", Academic Press, 1989.
9. S. Boyé and al, *accepted to J. Chem. Phys.*(2001).
10. F. Seifert, J. Ringling, F. Noack, V. Petrov and O. Kittelmann, *Opt. Lett.* **19**, 1538 (1994).
11. A. L'Huillier and P. Balcou, *Phys. Rev. Lett.* **70**, 774 (1993)
12. L. Misoguti, S. Backus, C. G. Durfee, R. Bartels, M. M. Murnane and H. C. Kapteyn, *Phys. Rev. Lett.* **87**, 013601 (2001).
13. A. Nazarkin, G. Korn, O. Kittelmann, J. Ringling and I. V. Hertel, *Phys. Rev. A* **56**, 671 (1997).
14. S. T. Pratt, P. M. Dehmer and J. L. Dehmer, *J. Chem. Phys.* **99**, 6233 (1993).

15. N. Shafizadeh, J.-H. Fillion, D. Gauyacq and S. Couris, *Phil. Trans. R. Soc. Lond. A* **355**, 1637 (1997).
16. S. Zamith, Thèse d'université, Université Paul Sabatier, Toulouse (2001).
17. V. Blanchet, M. Zgierski, T. Seidemann and A. Stolow, *Nature* **401**, 52 (1999).
18. L. Nugent-Glandorf, M. Scheer, D. A. Samuels, A. M. Mulhisen, E. R. Grant, X. Yang, V. M. Bierbaum and S. R. Leone, *Phys. Rev. Lett.* **87**, 193002 (2001).
19. M. Lezius, V. Blanchet, D.M. Rayner, D.M. Villeneuve, A. Stolow and M.Y. Ivanov, *Phys. Rev. Lett.* **86**, 51 (2001).

IR MULTIPHOTON DISSOCIATION OF GAS-PHASE METAL CARBONYLS BY FEMTOSECOND PULSES AT 5μm

L. WINDHORN, T. WITTE, D. PROCH, M. MOTZKUS, K. L. KOMPA AND W. FUSS

Max-Planck-Institut für Quantenoptik, D-85748 Garching, Germany

By focusing an MIR femtosecond laser in a small glass cell containing gas-phase metal carbonyls, the resonant infrared multiphoton dissociation of molecules was observed. $Cr(CO)_6$, $Mo(CO)_6$, $W(CO)_6$, and $Fe(CO)_5$ could easily be dissociated, which requires an excitation to at least v = 7 or 8 of the CO stretch vibration. After irradiation with ~150 fs pulses at 5 μm the metal carbonyl practically disappears in favor of free CO, as detected by the IR spectrum. The dissociation probability is independent of repetition rate and a steep function of the radiation energy density Φ. No dissociation is observed with wavelengths out of resonance. By comparing the power dependence of the total conversion with a model, we can infer that only few vibrational degrees of freedom are involved in the excitation process.

1 Introduction

To absorb an average number q of quanta in IR multiphoton excitation, a molecule requires a radiation energy density Φ(J cm^{-2}) depending on the (multiphoton) absorption cross section σ. The necessary Φ is minimum in the case of a linear absorber such as a harmonic oscillator:

$$q = \sigma \Phi / h\, \nu. \tag{1}$$

In the group of Heilweil [1], the CO stretch vibration (~2000 cm^{-1}) of $W(CO)_6$ in solution was excited with 2 ps pulses of 0.7 – 1.8 mJ cm^{-2} (~8 cm^{-1} bandwidth) and population of the v = 3 levels was detected. Maas et al. [2] excited the stretch vibration of gas-phase NO with 150 fs pulses at λ = 5.3 μm of a free electron laser (75 J cm^{-2}, ~100 cm^{-1} bandwidth) and detected population up to v = 5.

Transition metal carbonyls are ideal substrates for dissociation experiments due to their high absorption cross section and relatively small dissociation energies. We found that such molecules can easily be dissociated in the gas phase. This requires an excitation to at least v = 7 or 8 as inferred from bond energies for the dissociation reaction of a single M-CO bond [3].

The mononuclear metal carbonyls were previously dissociated by nanosecond pulses (frequency-doubled CO_2 laser) with Φ = 2 – 10 mJ cm^{-2}, and the absorption was measured [4]. With higher fluences up to 4 J cm^{-2} bare Mo and Cr atoms were detected after dissociation of their hexacarbonyl compound [5,6]. The same Φ, applied with the shorter fs pulses, implies a much higher intensity; the energy density actually required will, however, be probably higher since the bandwidth of the fs laser is substantially larger than that of the molecular absorption. Hence, only recent developments in laser technology to generate high pulse energies in excess of

10 μJ in the MIR allow for sufficient intensities to excite molecules with femtosecond pulses up to their dissociation threshold.

2 Experimental

The output of a standard Ti:Sapphire regenerative amplifier (Spitfire, Spectra Physics, 1 mJ at 800 nm, ~80 fs pulse duration, 1 kHz repetition rate) was used to pump a traveling-wave OPA (Topas, Light Conversion). Signal and idler pulses of the OPA were mixed noncollinearly in a $AgGaS_2$ crystal to generate mid-infrared pulses at 5 μm with energies up to 11 μJ (bandwidth >100 cm^{-1}). The pulse duration can be estimated to be in the range of ~150 fs. A CaF_2 lens of f = 100 mm was used to focus to a Φ up to ~70 mJ cm^{-2}. The incident power could be attenuated continuously with a gradient metallic filter.

Commercially available $Cr(CO)_6$, $Mo(CO)_6$, $W(CO)_6$, and $Fe(CO)_5$ were used without any further purification and filled into a small, closed glass cell of ~5 cm^3 volume and 5 cm path length with CaF_2 windows. Pressures ranged from 5 to 80 Pa, which is far from saturation pressures in all cases. Nevertheless, all metal carbonyls showed significant adsorption to the cell walls. Cells made of Teflon showed the same behavior. In a procedure intended to accelerate attaining the adsorption equilibrium, cells were filled at higher pressures and, after a few minutes, pumped to the desired pressures.

The concentration of metal carbonyls in the cell was determined by their IR absorption spectrum. For quantitative measurements a conventional FTIR spectrometer was used to monitor the concentration on a timescale of minutes. We thus accumulated microscopic dissociation in the focal volume of the laser to detect a macroscopic depletion of concentration in the cell, which could be observed after a few minutes of irradiation.

3 Results

Fig. 1A shows the change in absorbance of the fundamental CO stretch vibration at the absorption maximum (2000 cm^{-1}) of $Cr(CO)_6$ after 10 minutes of resonant irradiation at λ = 5000 nm (indicated by shaded area). The absorbance and thus the total concentration of free $Cr(CO)_6$ decreases by more than 80%. Note the change of absorbance due to adsorption before, and to desorption after dissociation. Part B shows the change in the absorption spectrum of $Fe(CO)_5$ after irradiation at λ = 4950 nm. The absorbance of the CO stretch vibration decreases by more than 90%. After the dissociation the rovibrational lines of free CO are observed.

Curves showing the successful dissociation of $Mo(CO)_6$ and $W(CO)_6$ are shown in Fig. 2A. All four metal carbonyls investigated in this work show very similar dissociation yields (within a factor of two). This is consistent with the bond energies for the dissociation reaction of a single M-CO bond, which all fall between 148 and

186 kJ mol[-1] (1.53 and 1.93 eV) [3] and with the comparable multiphoton absorption cross sections [4]. Thus, an energy corresponding to at least 7 or 8 absorbed photons at ~2000 cm^{-1} is required for the dissociation of any of the four metal carbonyl complexes.

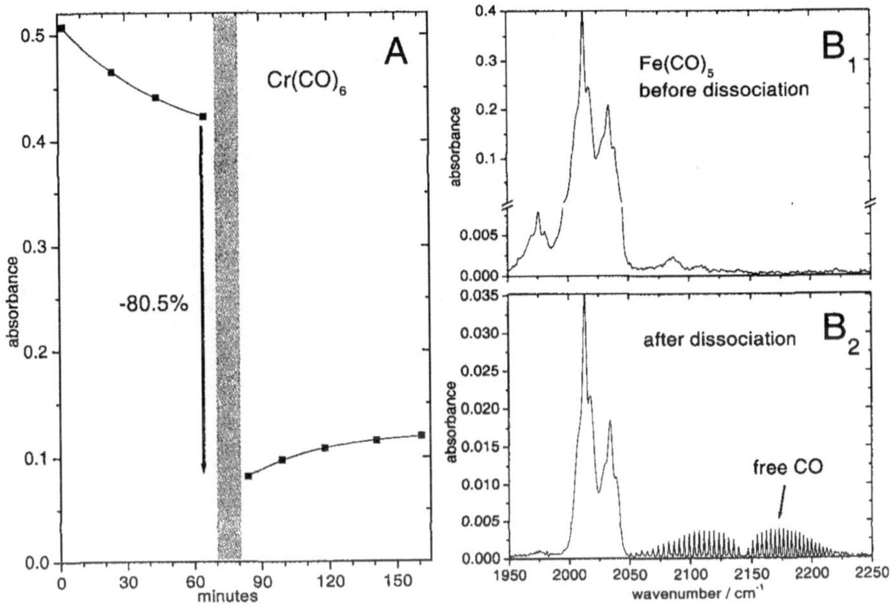

Fig. 1 Part A shows the change in absorbance of $Cr(CO)_6$ after 10 min (shaded area) irradiation with 11 μJ at λ = 5000 nm. Part B shows two absorption spectra of $Fe(CO)_5$ before and after 30 min irradiation with 11 μJ at λ = 4950 nm. Note the different scales in part B_1 and B_2. After the dissociation free CO is observed.

Dissociation occurs only as long as the excitation wavelength is in resonance with the fundamental CO stretch vibration (compare λ = 5 and 4.5 μm in Fig. 2B). The dissociation probability is independent of the laser's repetition rate (see Fig. 2C). Hence, the observed dissociation is a pure single shot process and not due to a long-term build up of energy.

The total conversion U was measured for $Fe(CO)_5$ for various incident energy densities. U is defined as the number of dissociated molecules per pulse divided by the gas density. Thus, it can be considered an equivalent dissociation volume per shot. It is a convenient unit since it is pressure independent.

The energy dependence of U for the dissociation of $Fe(CO)_5$ is shown in Fig. 3A where U was normalized with respect to the focal volume v_{focal}, which was calculated (assuming a Gaussian beam profile) to be ~0.13 mm^3 according to the formula [7]

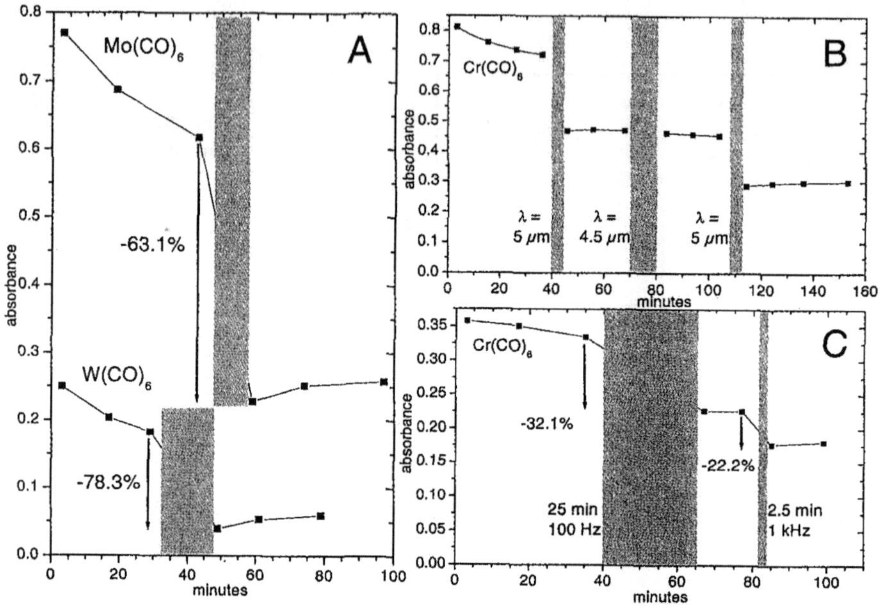

Fig. 2 Part A shows the change in absorbance of Mo(CO)$_6$ and W(CO)$_6$ after 10 min and 15 min irradiation respectively (shaded area) with 11 μJ at $\lambda = 5$ μm. Part B shows the change in absorbance of Cr(CO)$_6$ after three sequential irradiations with 8 μJ: 4 min at $\lambda = 5$ μm, 10 min at $\lambda = 4.5$ μm, and 4 min at $\lambda = 5$ μm. Part C shows the change in absorbance of Cr(CO)$_6$ after 25 min irradiation at 100 Hz repetition rate of the laser and 2.5 min irradiation at 1 kHz.

$$v_{focal} = 2\,(\pi w_0^2)^2 / 3\lambda \tag{2}$$

with w_0 the measured beam waist. Φ_{max} is the maximum energy density in the center of the focal volume and given by

$$\Phi_{max} = 2\,E / \pi w_0^2 \tag{3}$$

with E the pulse energy. U can be calculated from the dissociation rate which is determined by the measured concentration before and after the dissociation. The fraction of undissociated molecules after S pulses is

$$n_S / n_0 = exp(-S\,U / V_{cell}) \tag{4}$$

with V_{cell} the volume of the cell. The conversion U and thus the dissociation probability P_D is a steep function of Φ. Since $U / v_{focal} << 1$, we conclude that dissociation occurs only in the focal volume.

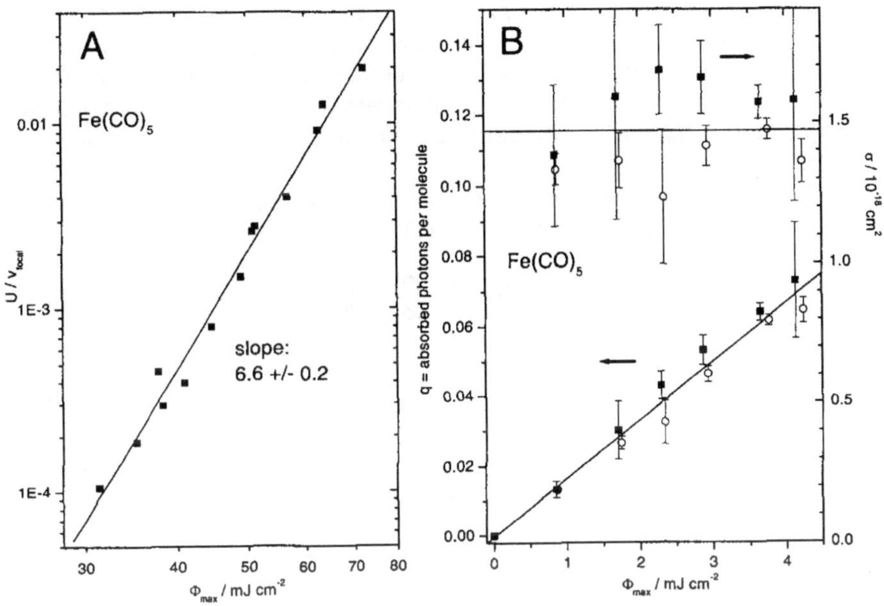

Fig. 3 Part A shows the total conversion U normalized by the focal volume v_{focal} for various incident energy densities Φ in a log-log chart. Dissociation occurs only in the focal volume. Part B shows q, the average number of absorbed photons per $Fe(CO)_5$ molecule, and the corresponding absorption cross sections σ measured in a parallel beam at $\lambda = 4950$ nm for different energy densities Φ. Black squares and open circles correspond to measurements taken at pressures of $p = 43$ Pa and 77 Pa, respectively.

Absorption (transmission through the cell) was measured with a parallel beam for different values of Φ up to 4 mJ cm^{-2} as shown in Fig. 3B. For $Fe(CO)_5$ an effective absorption cross section (eq. (1)) of $\sigma \sim 1.5 * 10^{-18}$ cm^2 was found. The same cross section was measured for two different pressures at $p = 43$ Pa and 77 Pa. Our laser has an approximately 10 times larger bandwidth than the molecule's multiphoton absorption band measured by a (narrow-band) frequency-doubled CO_2 laser. This spectrum had a maximum of $\sigma_{ns} \sim 2 * 10^{-17}$ cm^2 [4]. Therefore, the ratio of the absorption cross sections measured with fs and ns pulses, $\sigma_{fs} / \sigma_{ns}$, is just the ratio of bandwidth of the molecule to that of the fs laser, and the molecule seems to utilize just the resonant part of the energy density in the bandwidth offered. In ref. [4] the absorption cross section was found to be constant over a wide range of q up to $q \sim 20$ quanta/molecule. Hence, over a wide range of q, the molecules behave as linear absorbers.

4 Comparison with Model

A central question in the photodissociation is how many degrees of freedom are involved in the IR multiphoton absorption process. Specifically, is absorption confined to the CO stretch manifold(s), or does it include excitation of additional overtones and combination bands? To investigate this question of mode selectivity, we take a similar approach to the analysis in previous dissociation experiments with ns lasers [7,9].

If the dissociation probability $P_{diss}(\Phi)$ is known ($\Phi = \Phi(V)$ = local energy density), the total conversion U can be calculated from

$$U = \int_{V_{cell}} P_{diss}(\Phi)dV = \int_{\Phi_{min}}^{\Phi_{max}} P_{diss}(\Phi)g(\Phi)d\Phi \tag{5}$$

with $g(\Phi)d\Phi$ the differential volume in which the energy density is between Φ and $\Phi + d\Phi$. Φ_{max} and Φ_{min} are the maximum and minimum energy densities in the cell. Since dissociation occurs only in the focal volume, which is far from the cell walls, we can simplify the integration by the approximation $\Phi_{min} = 0$. If we assume our focused IR beam to be of Gaussian shape, $g(\Phi)$ can be calculated analytically [8]

$$g(\Phi) = v_{focal}\frac{1}{\Phi}(\frac{\Phi_{max}}{\Phi} - 1)^{1/2}(\frac{\Phi_{max}}{2\Phi} + 1)\,. \tag{6}$$

In the case of multiphoton infrared excitation, we choose a model function for $P_{diss}(\Phi)$ as in refs. [7,9]. This function assumes a stepwise absorption process with each vibrational energy level v exhibiting its own absorption spectrum with a cross section σ_v. The resulting rate equations can be solved analytically.

$$P_{diss}(\Phi) = e_{n-1}(\Phi_0/\Phi)\exp(-\Phi_0/\Phi) \tag{7}$$

$$e_n(x) = 1 + x + x^2/2 + ... + x^n/n! \tag{8}$$

Φ_0 is essentially a threshold energy density, for which the dissociation probability reaches some high value. The other parameter n is given by $n = s * m$ with s the mode degeneracy and m an exponent so that the absorbed energy is proportional to Φ^m [9]. For a linear absorber, the excited-state population is predicted to be exponential in the vibrational level v, with a preexponential factor v^{s-1}, i.e. depending on the number of participating modes s.

With Φ_0 and n, we have two independent parameters to fit the experimental data. Fig. 4 shows the theoretical curves for $U(\Phi)$ according to equations (5-8) for different parameters n. In the plot, both axes have been normalized: the conversion

U by the focal volume v_{focal} and Φ_{max} by $\Phi_0^* := \Phi_0 / (n + 0.2)$. Φ_0^* is the energy density at which $P_{diss} \sim 40 - 50$ % [7]. Also shown in the plot are the experimental data for the dissociation of $Fe(CO)_5$ as in Fig. 3A. Comparison with the theoretical curves yields n ~1 to 3 and Φ_0 ~300 mJ cm^{-2}.

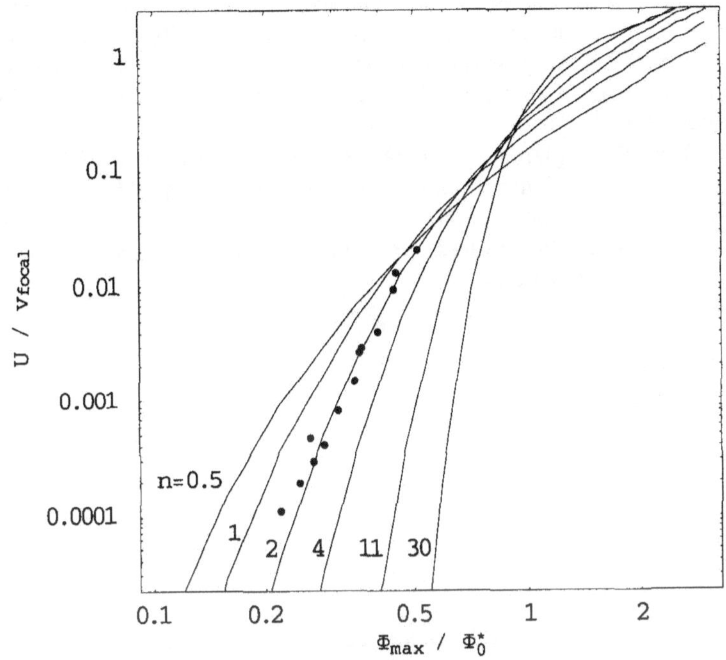

Fig. 4 Theoretical curves of the conversion $U(\Phi)$ calculated according to the model of equations (5-8) are shown for different parameters n. See text for details. The experimental data points shown are the same as in Fig. 3A for the dissociation of $Fe(CO)_5$. Good agreement with the model is obtained for the parameters n ~1 to 3 and Φ_0 ~300 mJ cm^{-2}.

As discussed in the previous section, we may consider the molecules as linear absorbers in the intensity range investigated. Hence, the number of absorbed photons q is proportional to Φ, i. e. $m \sim 1$. With $n = m * s$, we conclude that the number of vibrational degrees of freedom s involved in the excitation process ranges from ~1 to 3.

5 Conclusion

In this work, the resonant infrared multiphoton dissociation of molecules with femtosecond laser pulses has been demonstrated for the first time. MIR pulses of moderate energies E ~ 10 μJ focused to a Φ up to ~70 mJ cm^{-2} were sufficient to dissociate macroscopic quantities of $Cr(CO)_6$, $Mo(CO)_6$, $W(CO)_6$, and $Fe(CO)_5$ within minutes.

From a comparison of our experimental data with a model, we conclude that only few degrees of freedom s are involved in the excitation process. Previous studies with narrow-bandwidth ns lasers could not unambiguously answer this question, due to their limited spectral coverage of all rovibrational levels in the excitation process. By using broadband (>100 cm^{-1}) femtosecond pulses we avoid this problem and obtain s ~1 to 3. If all modes were participating in the excitation of $Fe(CO)_5$, s would be 27.

We are currently implementing a shot-to-shot probe for improved dissociation kinetics measurements. Furthermore, since the groups of Heilweil [10] and Noordam [11] report enhanced population transfer to higher vibrational levels with negatively chirped excitation pulses, we plan to introduce chirp in the excitation process to promote dissociation.

6 Acknowledgements

We thank H. Schröder and S. A. Trushin for helpful discussions.

References

1. S. M. Arrivo, T. P. Dougherty, W. T. Grupps and E. J. Heilweil, *Chem. Phys. Lett.* **235** (1995), 247.
2. D. J. Maas, D. I. Duncan, A. F. G. van der Meer, W. J. van der Zande and L. D. Noordam, *Chem. Phys. Lett.* **270** (1997), 45.
3. K. E. Lewis, D. M. Golden and G. P. Smith, *J. Am. Chem. Soc.* **106** (1984), 3905.
4. M. K. Au, P. A. Hackett, M. Humphries and P. John, *Appl. Phys. B* **33** (1984), 43.
5. S. A. Trushin, K. Sugawara and H. Takeo, *Chem. Phys. Lett.* **236** (1995), 402.
6. S. A. Trushin, K. Sugawara and H. Takeo, *Chem. Phys.* **203** (1996), 267.
7. C. D'Ambrosio, W. Fuss, K. L. Kompa and W. E. Schmid, *Opt. Soc. Am. B* **5** (1988), 1540.
8. W. Fuss and T. P. Cotter, *Appl. Phys.* **12** (1977), 265.
9. W. Fuss, *Chem. Phys.* **36** (1979), 135.
10. V. D. Kleiman, S. M. Arrivo, J. S. Melinger and E. J. Heilweil, *Chem. Phys.* **233** (1998), 203.
11. D. J. Maas, D. I. Duncan, R. B. Vrijen, W. J. van der Zande and L. D. Noordam, *Chem. Phys. Lett.* **290** (1998), 75.

OBSERVATION OF SCATTERING RESONANCES IN THE $H+D_2$ REACTION: DIRECT PROBE OF THE HD_2 TRANSITION-STATE GEOMETRY

F. J. AOIZ, L. BAÑARES AND J. F. CASTILLO

Departamento de Química Física, Facultad de Química, Universidad Complutense, 28040 Madrid, Spain

E-mail: banares@legendre.quim.ucm.es

D. SOKOLOVSKI

Theoretical and Computational Physics Research Division, Department of Applied Mathematics and Theoretical Physics. Queen's University of Belfast, BT7 1NN, UK

F. FERNÁNDEZ-ALONSO, B. D. BEAN, J. D. AYERS , A. E. POMERANTZ AND R. N. ZARE

Department of Chemistry, Stanford University, Stanford CA 94305-5080, USA

Measurement of differential cross sections over the collision energy range 1.39-1.85 eV for the elementary reaction $H+D_2 \rightarrow HD(v' = 3, j')+D$ shows pronounced forward scattering for $HD(v' = 3, j' = 0)$ which can be explained as arising from broad scattering resonances, whereas the $HD(v' = 3, \text{high } j')$ shows backward and sideways scattering attributable to direct dynamics. The behavior is quantitatively reproduced by quantum mechanical scattering calculations on the Boothroyd-Keogh-Martin-Peterson (BKMP2) potential energy surface. A Regge pole analysis of the scattering S-matrix allows a determination of the H-D and D-D bond lengths of the linear HD_2 transition-state complex.

One of the major goals of elementary reaction dynamics is to visualize how reactants come together and products fly apart in a bimolecular collision. Laser techniques have made it possible to study reactive scattering in unprecedented detail [1,2,3]. Of special interest is the detection of scattering resonances [4]. Resonance structures are related with the existence of nearly bound complexes formed by the collision partners. The width of the resonance is associated with the lifetime of the complex and its energy location with the energy of the quasibound state. Only very recently a sharp feature in the energy dependence of the integral cross section attributed to a relatively long-lived resonance has been observed for the $F+HD \rightarrow FH+D$ reaction [5]. For the hydrogen exchange reaction, Fernández-Alonso et al. [6] have reported a preliminary study of the $H+D_2 \rightarrow HD(v' = 3, j')+D$ reaction at a collision energy of 1.64 eV. This study showed the existence of a strong forward-scattering feature in the center-of-mass (CM) differential cross section (DCS) of the HD($v' = 3, j' = 0$) product, which is in marked contrast with the backward and sideways scattering found in HD($v' = 3$, high j') [6] and HD($v' = 0-2, j'$) [7,8]. This unexpectedly large forward-scattering feature could not be fully reproduced by quasiclassical (QCT) trajectory calculations although the QCT calculations did suggest the possibility of metastability of the HD_2 complex [6]. Shortly thereafter, Truhlar and coworkers [9] used a one-dimension vibrationally adiabatic model to interpret the experimental data as a barrier resonance corresponding to $HD_2(v_1 = 3, v_2 = 0, J = 20)$ where v_1 corresponds to the symmetric stretch motion, v_2 to the bend, and J is the total angular momentum of the three-atom complex. Since then interest in the H+D_2 reaction has fostered fully converged quantum mechanical (QM) scattering calculations by several groups [10,11,12,13] on the available *ab initio* potential energy surfaces.

By combining experimental measurements with QM scattering calculations we report the geometry of the transition state that leads to forward scattering in H+D_2 → HD($v' = 3, j' = 0$)+D, one of the channels in the most elementary of chemical reactions. From the theoretical analysis it is possible to extract the nature of the resonances that contribute to the scattering process and the structure of the transition-state region for this reaction channel.

The upper panel in Fig. 1 shows a schematic of the experimental setup. A 1:9 mixture of HBr and D_2 is coexpanded in a free jet using a pulsed nozzle valve. The H+D_2 reaction is initiated by photolysis of the HBr precursor with a 5-ns laser pulse operating in the wavelength range 203-225 nm. The HD($v' = 3, j'$) product is ionized 15 ns later by 2+1 resonance enhanced multiphoton ionization (REMPI) through the Q-branch members of the $EF^1\Sigma_g^+$-$X^1\Sigma_g^+$(0,3) band [14]. HD^+ ions travel inside a velocity-sensitive time-of-flight (TOF) spectrometer. A conical mask (core extractor) placed along the TOF axis in front of the detector selects only those ions with maximal speed projections along this axis. Figure 1 also presents TOF profiles for HD($v' = 3, j' = 0$) at various CM collision energies corresponding to different photolysis wavelengths. These profiles are symmetric about the center because there are as many reagent H atoms with speed projections toward the detector as those away from the detector. Conservation of energy and linear momentum in atom-diatom reactions guarantee a one-to-one mapping between laboratory speed and CM scattering angle provided that the diatomic product is detected state-selectively. In particular, the central portion of the TOF profiles shown corresponds to backward scattering whereas the signal on the periphery corresponds to forward scattering, all measured with respect to the direction of the incoming H atom. Examination of Fig. 1 clearly shows the apperance of forward-scattered wings, which are most prominent for a collision energy of 1.64 eV (see arrows). For HD($v' = 3$, high j') products only backward and sideways scattering is observed [6].

The QM scattering calculations have been performed on the *ab initio* Boothroyd, Keogh, Martin, Peterson (BKMP2) potential energy surface (PES) [15]. The QM calculations have been carried out using a time-independent coupled-channel hyperspherical method [16] at collision energies within the range 1.30-2.20 eV in 0.05-eV intervals [17]. The present QM calculations do not incorporate any effects from other PESs (geometric phase). As has been previously shown for the H+D_2 reaction, geometric-phase effects seem to cancel out when contributions from even and odd total angular momenta are added together[18]. Figure 2a shows the evolution of the QM DCS for the H+$D_2(v = 0, j = 0)$ →HD($v' = 3, j' = 0$)+D reaction as a function of collision energy. Scattering into this state is mainly confined into the extreme backward ($\theta > 150°$) and extreme forward ($\theta < 30°$) directions. The most striking feature, however, is the magnitude of the forward peak, which dominates the overall shape of the DCS. This forward peak grows from 1.5 eV collision energy, reaches a maximum at 1.9 eV and then declines rapidly. Similar behavior has been found for the HD($v' = 0-2, j' = 0$) product states in the form of successive maxima as a function of collision energy. Forward scattering seems to be transferred from one vibrational level to the following as energy increases.

A simulation of the experimental results can be readily obtained by convolution of the theoretical DCS with the instrumental resolution [19]. The solid curves in Fig. 1 show the results of the simulation along with the corresponding experimental profiles. The agreement is strikingly good with some small discrepancies [20]. Particularly, the peaks corresponding to forward scattering in the TOF profiles are well reproduced by the QM calculations. So far, no experiment with v', j' state resolution had been able to probe forward scattering for the title reaction. In the experiments by Welge, Schnieder and Wrede [7] using the D-atom Rydberg tagging technique, state resolution was achieved for the measurement of DCS for many individual HD(v', j') states. However, instrumental constrains prevented the measurement of scattering in the forward region. It is very reassuring that the present experimental results can be so well accounted for by the exact QM scattering calculations, specially in the forward region. QCT results also predict some forward scattering [6], but its magnitude is at least a factor of 3 less than that obtained in the QM calculations. The dashed curve in Fig. 1 shows the simulation obtained using the QCT DCS calculated at 1.64 eV collision energy [6]. The discrepancies be-

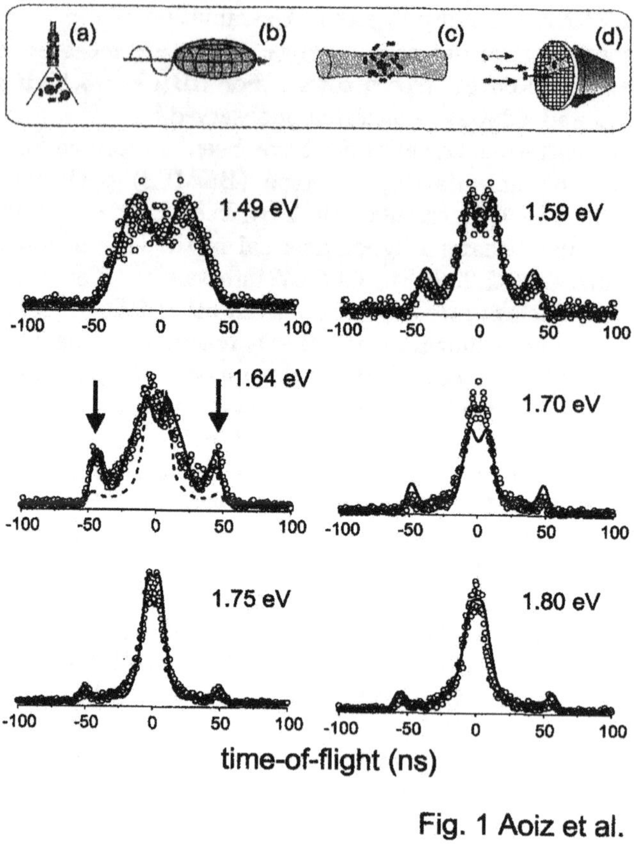

Figure 1. Upper panel. Sketch of the experimental setup: (a) a gas jet expansion produces translationally cold precursors in a collision-free environment; (b) a polarized *photolysis* laser generates "fast" H atoms from the HBr precursor; (c) a polarized *probe* laser ionizes $HD(v' = 3, j')$ product molecules via 2+1 REMPI after a fixed time delay; and (d) HD^+ ions are extracted, accelerated and detected in a Wiley-McLaren TOF spectrometer with a core extractor. Lower panel. Experimental TOF profiles (open circles) and theoretical simulations (*solid curves*) *based on the state-resolved* QM DCSs shown *in* Fig. 2a at the indicated collision energies. The data at 1.64 eV collision energy include the simulation based on the QCT DCS [6] (dashed curve). The arrows indicate the scattering into the forward region.

tween experimental and QCT results are significant, particularly with regard to forward scattering.

Since the very beginning of reaction dynamics forward scattering has been

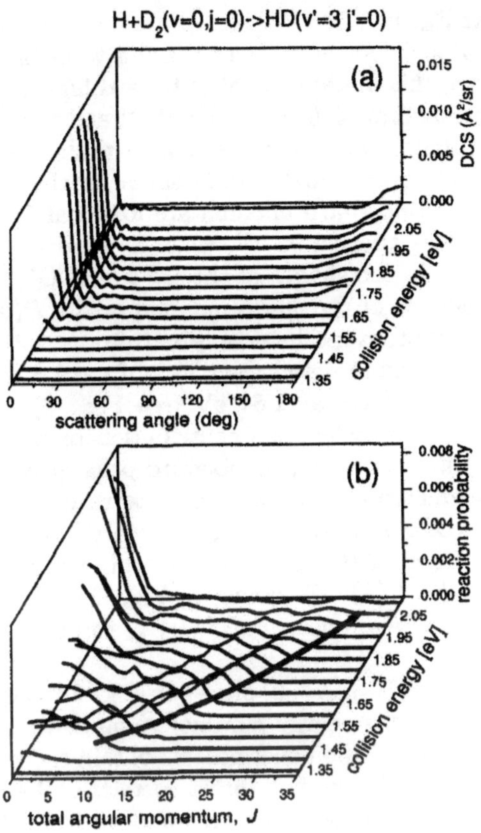

Figure 2. QM calculations for the $H+D_2(v=0, j=0) \rightarrow HD(v'=3, j'=0)+D$ reaction as a function of collision energy within the range 1.35-2.20 eV using the BKMP2 potential energy surface. (a) QM DCSs. Scattering angles $\theta < 90°$ correspond to the forward hemisphere; and (b) reaction probabilities as a function of total angular momentum J. The arrow indicates the outermost $E_{col} - J$ *ridge* (see text for details).

associated with large angular momenta or, in classical mechanics, with large impact parameters. To understand the origin of the observed forward scattering, Fig. 2b displays the evolution of the reaction probability $P(J)$ for $HD(v'=3, j'=0)$ as a function of the total angular momentum J and collision energy. At $E_{col} \approx 1.50$ eV the maximum in the reaction probability does not appear at $J = 0$, but is shifted toward a value of $J \approx 9$. As the collision energy increases, this maximum shifts toward higher J values. Additionally, new subsidiary maxima appear in the $P(J)$ at lower J values giving rise to a nested ridge structure. Comparison with the corresponding QCT $P(J)$ [6] (also see Fig. 3e) reveals further differences: in the QCT case all the $P(J)$s peak at $J = 0$ and no sharp maxima are found at higher values of J. Only at collision energies above 1.95 eV, when forward scattering in the DCS has already declined, do the shapes of the QCT and QM $P(J)$s become similar. Clearly the existence of the sharp maxima in the $P(J)$ shown in Fig. 2b is predominantly quantum mechanical in nature.

We examine in more detail the effect of partial-wave summation on the HD($v' = 3, j' = 0$) DCS at 1.64 eV (see Fig. 3). Forward scattering in the DCS only appears when the contribution of partial waves $J \geq 15$ are included (Fig. 3a). Moreover, the forward peak grows monotonically with J up to J values which no longer contribute to reactivity. This behavior is also found at other collision energies where forward scattering exists. Therefore, the dynamical origin of the forward scattering features can be traced to the high-J peak in the $P(J)$. Referring to Fig. 2b, the energy position of the $P(J)$ peak maxima as a function of $J(J+1)$ can be fitted to a second-degree polynomial [21].

If we take the $P(E_{col}; J = 0)$ shown in Fig. 3b and use the J-shifting approximation [22] with the energy parameters quoted in Ref. [21] to reconstruct the $P(J)$ at selected collision energies, we obtain the results shown as dashed curves in Figs. 3c-3f. There is a direct correlation between the peak labeled "1" in the $P(E_{col}; J = 0)$ and the high-J peak in $P(J)$ at all collision energies investigated. From this analysis we conclude that the observed forward scattering has its origin in the first peak of the $P(E_{col}; J = 0)$ (see Fig. 3b). The oscillations observed in the energy evolution of the reaction probability for $J = 0$ shown in Fig. 3b have been identified with scattering resonances [23].

It is possible to assign the structure in Fig. 3b as a bending progression of the HD_2 complex, characterized by regular spacings equal to $2v_2 = 1220$ cm^{-1}, as expected for a triatomic molecule with total angular momentum $J = 0$. From this assignment, it follows that forward scattering results from the $v_2 = 0$ quantum level (peak "1" in Fig. 3b), in agreement with the assignment given by Allison et al. [9] and consistent with the notion that the forward-scattered, rotationless product strongly correlates with no bending motion of the complex.

Scattering resonances are naturally associated with poles in the S matrix [24]. We have analyzed our S-matrix elements using a semiclassical approach

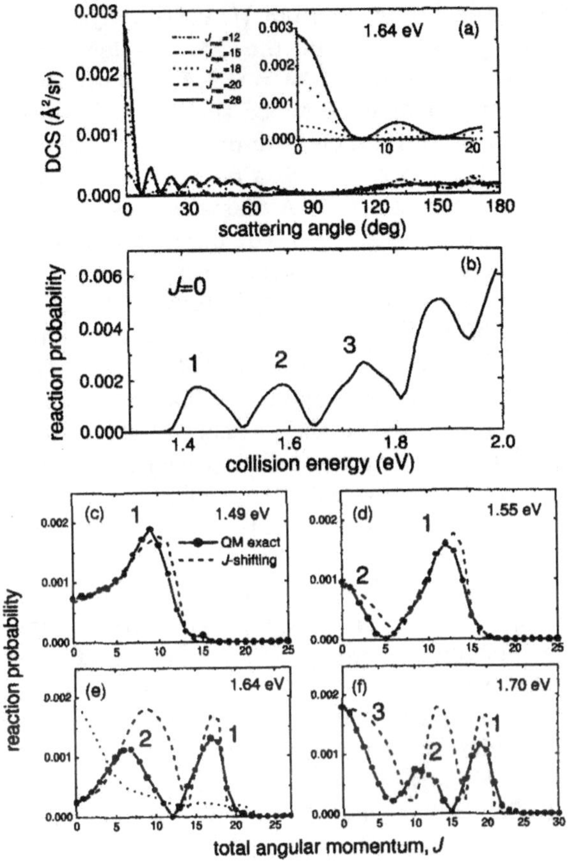

Figure 3. Partial wave analysis for $H+D_2(v=0, j=0) \rightarrow HD(v'=3, j'=0)+D$: (a) QM DCSs for different values of J_{max}, the number of partial waves included in the calculation; (b) reaction probability as a function of collision energy at total angular momentum $J=0$. The first three peaks in the reaction probability are labeled 1, 2 and 3; (c) through (f) reaction probability as a function of total angular momentum J at the indicated collision energies. Solid line with solid circles: exact QM calculation. Dashed line: results obtained by applying the J-shifting approximation to the data shown in panel (b) (see text for details). Dotted line in panel (e): QCT calculation at 1.64 eV collision energy.

based on complex angular momentum (CAM) analysis [25]. Briefly, the S matrix is analytically continued into the CAM λ-plane (λ is a semiclassical angular momentum related to J by $\lambda = J + 1/2$) with the help of the Padé approximation, that is, a rational function that depends on complex poles and zeroes of the S matrix. Regge poles are defined as poles of the S matrix in the CAM plane. A direct relation exists between the shape of the DCS and the position of these Regge poles [25]. A scattering resonance can be viewed as the formation of an intermediate complex of the atom-diatom system. This complex rotates a certain angle $\Delta\Theta$ before decaying into products, $\Delta\Theta$ being related to the imaginary part of λ by $1/(2\ \mathrm{Im}\lambda)$ [25]. This rotational motion affects the structure of the DCS even if the angle of rotation is small, *i.e.*, less than a full rotational period [25]. Our analysis for the $H+D_2$ reaction reveals that the forward peak in the DCS and its oscillatory structure in the forward hemisphere may be attributed to the presence of three Regge poles up to a collision energy of 1.90 eV. As the collision energy increases further, more poles come into play, coinciding with the decline of the forward peak in the DCS. Table 1 lists the positions of the three poles (Re λ, Im λ) at $E_{col} = 1.64$ eV. Notice that the three Regge poles lie in the region where the high-J peak in the $P(J)$ appears (compare Fig. 3e with the Re λ values shown in Table 1). Interestingly, Regge pole "2" contributes most to the forward peak, although it is necessary to include all three poles in order to reproduce precisely the scattering into the forward hemisphere (magnitude of the forward peak and oscillatory behavior in the DCS). By using the Re λ values found for each of the three poles at the different total energies calculated, we have obtained linear fits of the form $E = E_R + B[(\mathrm{Re}\ \lambda)^2 - 1/4]$ [25]. The parameters E_R and B are listed in Table 1. Pole "2" is characterized by a resonance energy $E_R = 1.61$ eV and a rotational constant $B = 5.7\ \mathrm{cm}^{-1}$. From this rotational constant, we obtain bond lengths corresponding to $r_{HD} = 0.78$ Å and $r_{DD} = 1.19$ Å[26]. Thus, the Regge pole analysis has yielded the geometrical structure of the HD_2 complex responsible for the observed resonance features.

Recalling that a complex on the collinear saddle point of the BKMP2 surface ($r^{\ddagger}_{HD} = r^{\ddagger}_{DD} = 0.93$ Å) is characterized by a rotational constant of 6.9 cm^{-1}, the quasibound complexes associated with the three Regge poles quoted above are consistent with elongated geometries in the D-DH exit channel. The assumption of vibrational adiabaticity then allows us to provide the quantum label $v_1 = 3$ because it correlates with formation of the $HD(v' = 3, j' = 0)$ product. These conclusions are in good agreement with those obtained by Allison *et al.* [9] from a one-dimensional analysis in terms of a local, vibrationally adiabatic model for the $H+D_2(v = 0, j = 0) \rightarrow HD(v' = 3, j' = 0)+D$ reaction on the DMBE surface [27]. They found a resonance energy of 1.616

eV and assigned the resonance to a complex in the D-DH exit valley with $(v_1 = 3, v_2 = 0)$ stretching and bending vibrational quantum numbers and a total angular momentum $J = 20$.

Table 1. Positions (Re λ, Im λ) of the Regge poles found from the analysis of the S matrix at 1.64 eV collision energy. The corresponding resonance energies (E_R), rotational constants (B) and geometries (r_{HD}, r_{DD}) of the collinear complexes are also listed. The geometry of the collinear saddle point of the BKMP2 surface is $r^{\ddagger}_{HD} = r^{\ddagger}_{DD} = 0.93$ Å $(B = 6.9\ cm^{-1})$, and the equilibrium distance of the $H_2(D_2)$ molecule is $r_e = 0.74$ Å.

	$E_{col} = 1.64$ eV					
Regge Pole	Re λ	Im λ	E_R/eV	B/cm^{-1}	r_{HD}/Å	r_{DD}/Å
1	14.6	2.2	1.50	5.3	0.77	1.26
2	18.8	1.3	1.61	5.7	0.78	1.19
3	23.3	0.9	1.68	6.7	0.86	1.00

The Padé approximation can also be used to locate poles of the S matrix in the complex energy plane [25]. Such analysis has been carried out for the S matrix corresponding to HD$(v' = 3, j' = 0)$ at $J = 0$ and $J = 18$ as a function of the total energy E. For $J = 0$, a pole with Re E = 1.6 eV and Im E = -0.027 eV has been found which corresponds to a resonance width $\Gamma = 0.054$ eV and a lifetime $\tau \approx \hbar/\Gamma = 12$ fs. For $J = 18$ the resonance parameters are Re E = 1.8 eV, Im E = -0.026, which yield $\Gamma = 0.052$ eV and $\tau \approx 12.6$ fs. These results are again in good agreement with the resonance parameters found by Allison *et al.* [9].

Another piece of evidence for the existence of scattering resonances is given by the time delay as defined by Smith [28], $\Delta\tau^J_{ij} = Re\left\{-i\hbar[S^J_{ij}]^{-1}(dS^J_{ij}/dE)\right\}$, which measures the time that an outgoing wavepacket in channel j is delayed ($\Delta\tau^J_{ij} > 0$) or advanced ($\Delta\tau^J_{ij} < 0$) relative to a freely evolving wavepacket at the same average energy (i denotes the input channel). Assuming that resonance dominates over direct scattering, the time delay at the resonance energy will be $\Delta\tau^J_{ij} \approx \hbar/\Gamma$, whereas it will be negligible at off resonance energies. Thus, the time delay is a direct measurement of the resonance lifetime. Time delays have been calculated for $J = 0$ and $J = 18$ as a function of total energy. For $J = 0$, negative time delays of about -20 fs are obtained in the energy range corresponding to the first maximum in the $P(E; J = 0)$. However, $J = 0$ corresponds to zero impact parameter, where it is expected that a considerable amount of direct scattering scrambles the resonance. At larger Js the resonance should be better observable, and this is actually the case, since a positive time-delay maximum has been found at $J = 18$ at a total energy of 1.82 eV with a time delay of 12.8 fs.

Our work shows that the measured forward scattering in the $H+D_2 \rightarrow HD(v' = 3, j' = 0)+D$ bimolecular reaction is caused by a small number of broad (short-lived) scattering resonances. The variation of the resonance features with energy provides the following information: (a) scattering in the forward-scattering direction is mainly quantum mechanical in nature yet it may be ascribed to classical collisions at large impact parameters temporarily trapped in the product valley; and (b) the geometry and lifetime of the HD_2 quasibound complex has been determined. This work demonstrates that scattering resonances are a exquisitely sensitive probe of the structure and dynamics of the transition-state complex in the most elementary of chemical reactions.

Acknowledgments

The experimental work has been supported at Stanford by the US National Science Foundation under grant number CHE-99-00305. The quantum scattering calculations were performed on the Centro de Supercomputación Complutense with the support of the Spanish Dirección General de Enseñanza Superior grant PB98-0762-C03-01. European Union support to the Spanish group through the Research Training Network HPRN-CT-1999-00007 is also gratefully acknowledged. Support through the Queen's University Research Fund is also acknowledged.

References

1. P. Casavecchia, N. Balucani, G. G. Volpi, *Annu. Rev. Phys. Chem.* **50**, 347 (1999).
2. P. Casavecchia, *Rep. Prog. Phys.* **63**, 355 (2000).
3. K. Liu, *Annu. Rev. Phys. Chem.* **52**, 139 (2001).
4. G. C. Schatz, *Science* **288**, 1599 (2000).
5. R. T. Skodje, D. Skouteris, D. E. Manolopoulos, S.-H. Lee, F. Dong, K. Liu, *J. Chem. Phys.* **112**, 4536 (2000); *ibid.*, *Phys. Rev. Lett.* **85**, 1206 (2000).
6. F. Fernández-Alonso, B. D. Bean, J. D. Ayers, A. E. Pomerantz, R. N. Zare, L. Bañares, F. J. Aoiz, *Angew. Chem. Int. Ed.* **39**, 2748 (2000).

7. L. Schnieder *et al.* *Science* **269**, 207 (1995); L. Schnieder, K. Seekamp-Rahn, E. Wrede, K. H. Welge, *J. Chem. Phys.* **107**, 6175 (1997); E. Wrede *et al.*, *ibid.* **106**, 7862 (1997); L. Bañares *et al.*, *ibid.* **108**, 6160 (1998); E. Wrede *et al.*, *ibid.* **110**, 9971 (1999).
8. F. Fernández-Alonso, B. D. Bean, R. N. Zare, *J. Chem. Phys.* **111**, 1022 (1999); *ibid.*, 1035 (1999); *ibid.*, 2490 (1999).
9. T. C. Allison, R. S. Friedman, D. J. Kaufman, D. G. Truhlar, *Chem. Phys. Lett.* **327**, 439 (2000).
10. S. C. Althorpe, *J. Chem. Phys.* **114**, 1601 (2001).
11. S. Der Chao and R. T. Skodje, *Chem. Phys. Lett.* **336**, 364 (2001).
12. F. J. Aoiz, L. Bañares, J. F. Castillo, *J. Chem. Phys.* **114**, 8237 (2001).
13. B. K. Kendrick, *J. Chem. Phys.* **114**, 8796 (2001).
14. K.-D Rinnen, D. A. V. Kliner, R. N. Zare, W. M. Huo, *Isr. J. Chem.* **29**, 369 (1989); W. M. Huo, K.-D. Rinnen, R. N. Zare,*J. Chem. Phys.* **95**, 205 (1991); K.-D. Rinnen, M. A. Buntine, D. A. V. Kliner, R. N. Zare, W. M. Huo, *J. Chem. Phys.*, **95** 214 (1991).
15. A. I. Boothroyd, W. J. Keogh, P. G. Martin, M. R. Peterson, *J. Chem. Phys.* **104**, 7139 (1996).
16. D. Skouteris, J. F. Castillo, D. E. Manolopoulos, *Comput. Phys. Commun.* **133**, 128 (2000).
17. The main key parameters that define the coupled-channel basis set are E_{max}, j_{max} and k_{max} (see Ref. [16]). At the collision energies of the present work, well converged integral and differential cross sections have been obtained using the parameters $j_{max} = 20$, $E_{max} = 2.85$ eV and $k_{max} = 10$, and including all partial waves up to total angular momentum $J = 36$. With these parameters the number of channels to be propagated ranges from 198 at $J = 0$ up to 1532 at $J \geq 10$. A maximum value of the hyperradius $\rho_{max} = 25a_0$ and a number of 250 sectors were selected to perform the calculations. Additional calculations were performed using the parameters indicated above for $J = 0$ at 360 total energies in the range 1.5217-1.8807 eV and for $J = 18$ at 43 total energies in the range 1.55-2.15 eV.
18. B. K. Kendrick, *J. Chem. Phys.* **112**, 5679 (2000). Erratum: *ibid.* **114**, 4335 (2001).
19. F. Fernández-Alonso, B. D. Bean, R. N. Zare, F. J. Aoiz, L. Bañares, J. F. Castillo, *J. Chem. Phys.*, in press.
20. Not shown are the data at 1.54 eV where the agreement is poorer but judged to be satisfactory. The agreement found at 1.39 eV and 1.85 eV is of the same quality as what is shown in Fig. 1.
21. The second degree polynomial is $E = 1.681 + 5.3 \times 10^{-4} J(J+1) + 3.9 \times$

$10^{-7}J^2(J+1)^2$. This fit yields a rotational constant for the HD_2 complex of 4.3 cm^{-1}. We note that this fit is subject to inaccuracies because the peaks in the reaction probability $P(J)$ are not sharp. This treatment, however, is sufficient to implement with success the J-shifting approximation on the $P(E_{col}; J = 0)$ reaction probability (see next paragraph in the main text). A Regge pole analysis gives a more accurate value of this rotational constant (see text).

22. J. M. Bowman, *J. Phys. Chem.* **95**, 4960 (1991).
23. See, for example, Fig. 3 of W. H. Miller and J. Z. H. Zhang, *J. Phys. Chem.* **95**, 12 (1991).
24. See, for instance, J. R. Taylor, *Scattering Theory* (John Wiley & Sons, New York, 1972).
25. D. Sokolovski, J. F. Castillo, C. Tully, *Chem. Phys. Lett.* **313**, 225 (1999); D. Sokolovski and J. F. Castillo, *Phys. Chem. Chem. Phys.* **2**, 507 (2000); D. Sokolovski, *Phys. Rev. A* **62**, 024702 (2000).
26. The geometries listed in Table 1 have been obtained from the collinear minimum energy path calculated for the $H+D_2$ reaction on the BKMP2 surface.
27. A. J. C. Varandas, F. B. Brown, C. A. Mead, D. G. Truhlar, N. C. Blais, *J. Chem. Phys.* **86**, 6258 (1987).
28. F. T. Smith, *Phys. Rev.* **118**, 349 (1960).

FEMTOSECOND INTERFEROMETRY WITH MOLECULAR WAVEFUNCTIONS - EXPERIMENTS ON THE QUANTUM DYNAMICS OF ROVIBRONIC MOTION

F. MILOTA[A], A. TORTSCHANOFF[A], CH. WARMUTH[A], W. JAKUBETZ[B], AND H. F. KAUFFMANN[A]

[A] ... *Inst. f. Physikal. Chemie der Universität Wien, Währingerstraße 42, 1090 Wien, Austria*

[B] ... *Inst. f. Theoret. Chemie der Universität Wien, Währingerstraße 17, 1090 Wien, Austria*

E-mail: kauffm@ap.univie.ac.at

A molecular wavepacket (WP) prepared in its Franck-Condon (FC) states is confined in space, but due to the anharmonicity of the potential, its positional coordinates spread as it moves from the FC-region and propagates between the two classical turning points. As free evolution proceeds, the (discrete) WP, instead of continuing to spread indefinitely into future, starts to interfere with itself, and after certain distinct times rephasing of the initial WP happens. These revival times [1] are of pure quantum-mechanical origin and do not have any classical analog. In this contribution we demonstrate first systematic interferometric studies on quantum-mechanical revivals of vibrational WP-motion in the electronic molecular B-state of thermal, gaseous iodine (I_2). Fluorescence noise interferograms ranging from 200 fs up to 40 ps are presented for various excitation tuning wavelengths which allow energy specific preparation of wavepackets in the B-potential state. The experiments using double pulses with random, relative phase ϕ probe the damping of classical WP oscillations and the rebirth of quantum-mechanical vibrational coherences due to vibrational-rotational coupling under less favorable thermal averaging conditions.

1 Introduction

Femtosecond pulses, shorter than 100 fs, provide an ideal tool to create coherent excitations for a large number of vibrational transitions between different electronic states, in simple diatomic molecules (i.e. iodine), thus launching initially well localized WPs in the Franck - Condon state. Since this WP is no eigenfunction, but a superposition of wavefunctions it starts to oscillate temporally and spatially in the excited state potential [2,3,4]. In the short-time domain the WP displays "classical"-like behavior, its motion is periodical and due to the anharmonicity of the potential, it is subject to dephasing and spreading. Going to longer time scales (>10 ps) theoretical analyses predict that the WP starts to interfere with itself, so the initial front of the WP becomes split into a number of localized fractions distributed over the motional coordinate [1]. This generic feature results in a dramatic deviation

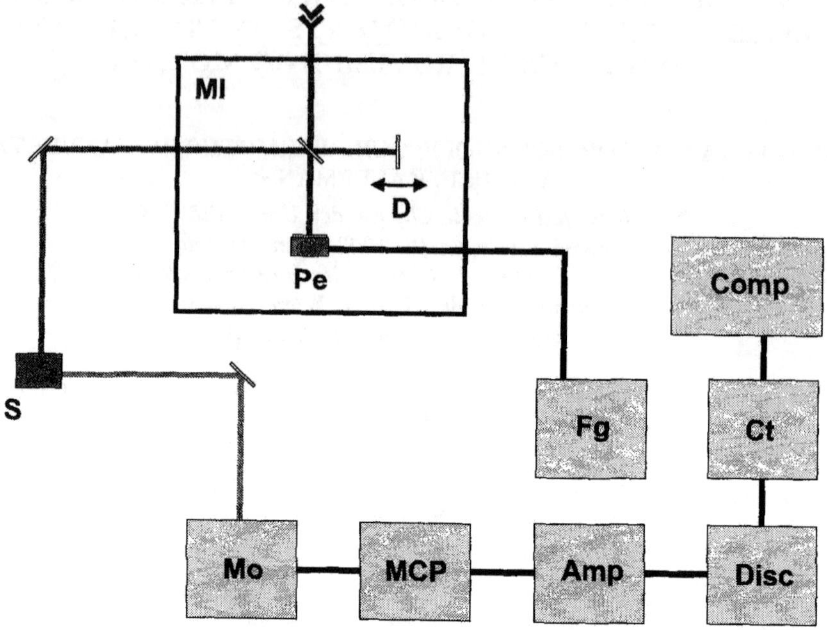

Figure 1. Schematic COIN-setup: MI... Michelson Interferometer, D... delay, Pe... piezo-element, S... sample, Mo... monochromator, MCP... multi channel plate, Amp... amplifier, Disc... discriminator, Ct... counter, Comp... computer, Fg... function generator.

from the behavior of the corresponding classical ensemble. As time goes on, these back-biting interferences trigger a tendency to restore the WP's initial shape after distinct periods, the revival times [1]. This phenomenon is of pure quantum-mechanical origin and does not have any classical analog.

A very promising way to monitor this behavior is fluorescence wavepacket-interferometry (WP-I) [5] which uses a sequence of two identical fs-laserpulses (1,2). The observable is the correlated mean intensity of the fluorescence F_{12} that arises from interfering populations of propagating WPs in the excited state. In the limit of one-photon absorption the wavefunction of the excited electron is a superposition of contributions that correspond to excitations by both pulse 1 and pulse 2. So the cross term, $F_{12}(\tau, \phi) \sim 2P_{12} =$

$Re\langle\Psi_{e,1}(\tau)|\Psi_{e,2}(\tau)\rangle$, determines the magnitude of the correlation that depends on the delay time τ and the relative phase ϕ of the exciting/probing pulses.

As the method of choice for this investigation we have used the concept of *Coherence Observation by Interference Noise* (COIN) [6]. COIN-interferometry relies on excitation through pulses with delay time τ and random relative phase ϕ that creates fluctuating interferences in the excited state population. So the correlated noise of fluorescence intensity - the variance $varF(\tau)$, $varF(\tau) = 2\sum_n |c_n|^2 \cos^2(\Delta_n\tau)varF(\tau) = 2\sum_n |c_n|^2 \cos^2(\Delta_n\tau)$ - directly mimics the dynamics of the propagating WP. The technique is experimentally simpler than methods with phase locked [5] or freely propagating pulses [5,7]. As the optical carrier frequency is annihilated, due to phase randomization of the excitation pulses, COIN is interferometrically stable at longer times, and therefore, a useful tool for the measurement of WP motion at the quantum-classical border.

2 Experiments and Results

Measurements were performed on gaseous iodine at a vapor pressure p$\approx$ 0.3 torr at room temperature. The motivation to use the gaseous iodine molecule as a test system is twofold: (i) spectroscopic information about I_2 is already available and the nature of the fluorescent B-state is well known, and (ii) the I_2 B-level has been closely scrutinized, both in numerous pump-probe (PP) [2,3,4] WP experiments and in phase-locked fluorescent interferometry (PL-FI) [5] studies, so the WP data of the fluorescent B-state detectable by COIN can be compared directly with those patterns obtained by PL-FI and PP, respectively.

The experimental setup (Fig. 1) consists of a Ti:S-lasersystem with an oscillator, regenerative amplifier, and an optical parametric amplifier yielding 100 fs pulses, which are tunable over the whole range of the electronic $X \rightarrow B$ transition of I_2. The pulses are sent through a Michelson interferometer, where the mirror in the static arm is mounted on a vibrating piezo, thus creating a collinear two-pulse train with well defined time delay τ and a randomized relative phase ϕ. These phase-randomized pulse pairs are used to excite iodine in a gas cell at room temperature and the fluorescence is collected in a 90° geometry, spectrally resolved with a monochromator and detected by means of a photomultiplier and photon counting electronics. For each time delay, typically 500 measuring values were accumulated, analyzed in terms of their variance $varF$, and plotted versus the delay time τ. Figure 1 shows the schematic setup of the COIN experiment, which basically consists

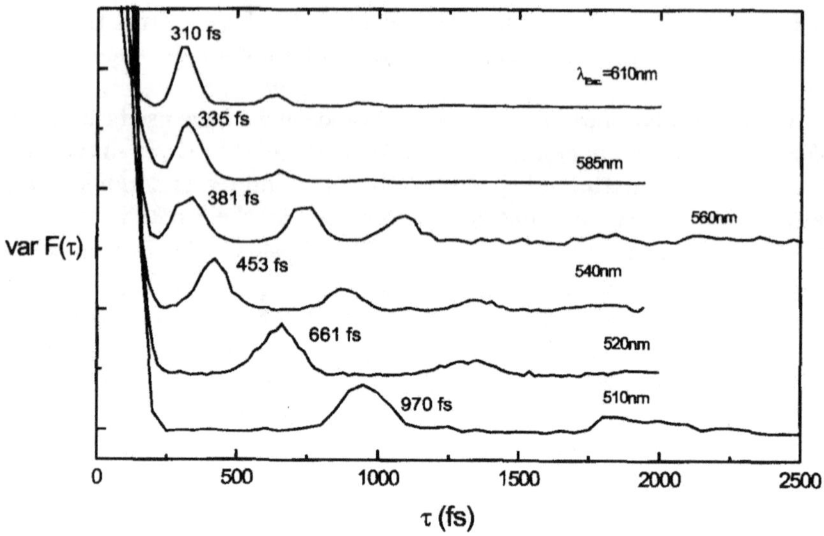

Figure 2. COIN-fluorescence interferograms for a variety of excitation wavelengths λ_{exc}. Patterns show the increase of the recurrence periods τ_R when tuning λ_{exc} from the bottom-bound states to the dissociative limit. The leading edge part has been cut off for better visualization.

of nothing more than an appropriate lightsource, a Michelson-interferometer and the recording equipment.

In previous work of our group [8] we investigated the short time behavior of thermal I_2. Figure 2 displays the results of similar experiments. All the typical signatures of WP motion become evident: initial propagation, periodic oscillations, rotational damping [2,9], and spreading due to anharmonicity of the potential. The excitation was precisely tuned to cover the bound part of the B-state surface up to the dissociative limit (i.e. from λ_{exc}=610 nm to λ_{exc}=510 nm). Preparing the WPs at increasingly higher vibrational levels, a systematic variation of the recurrence period with entirely different damping and spreading properties can be observed. These effects are demonstrated in Fig. 2 where the 200 fs leading edge part has been cut off for improved visualization. The recurrence times reflecting the shape of the B-state potential show the trend to be expected when tuning the excitation energy to the unbound limit. From this patterns it is obvious that the WP prepared in the COIN experiment does not last longer than 3 ps. These features, the theoretical analysis, and simulations concerning the WPs short-time behavior are given in full detail in Ref. [8]

In a further attempt to elucidate the quantum-dynamics of WP-motion we extended the interferograms to much longer time scales [10]. Here we report the first systematic COIN-observables of quantum-mechanical vibrational WP-revivals in a thermal ensemble of molecular iodine. In Figure 3 a typical COIN interferogram is presented with the variance $varF(\tau)$ plotted versus the delay time τ up to 20 ps. The excitation wavelength was 520 nm and the detection was performed at 560 nm. Again the high-intensity autocorrelation peak at τ=0 has been omitted for better visualization. The short-time behavior is consistent with previous investigations under similar conditions. The early-time oscillations correspond to the regime of "classical" WP motion in the potential of the bound B-state, with a period of approximately 660 fs that is related to the classical frequency ω_0 of the iodine molecular vibration excited by this wavelength (520 nm). A strong dispersion of frequencies constitutes the background oscillations, but no higher harmonics ($2\omega_0$, $3\omega_0$) can be observed by Fourier transform (FT) when analyzing the short-time domain with a cutoff at 4 ps (not shown). The main oscillation at about 660 fs persists for ~4 ps and decays, due to the thermally populated rotational levels in the I_2 vapor.

Inspection of Fig. 3 shows a different situation for delay times τ >5 ps. For times τ ~7 ps periodic behavior of moderate intensity with a more complicated structure reappears. Closer inspection through FT (lower display in Fig. 3) reveals that the bundle of spikes centered at ~9 ps shows a prominent period of about 350 fs which corresponds to twice the frequency of the fundamental recurrences, $2\omega_0$. In the terminology of Averbukh and Perel'man [1] this is the typical signature of a vibrational quarter-revival as previously observed by pump-probe measurements in energy-selected, cold diatomic molecules as Na_2 and Br_2 [11,12]. A quarter-revival at $\tau = T_{rev}/4$ exhibits itself in the splitting of a WP into two equal parts, which are moving out of phase with each other, and are shifted in time by a half of the vibrational period. Because both of the sub-WPs are oscillating with the classical vibration period $\tau_R = 1/\omega_0$, the detected signal carries the doubled frequency. At twice the time of the quarter-revival, $\tau = T_{rev}/2$, a half-revival is expected. In fact, a collection of well-defined spikes can be found in between 15 ps and 19 ps, with a period of 660 fs corresponding to the fundamental vibrational frequency. Based upon this preliminary assignment, the full-revival should be located at $\tau = T_{rev} \sim 35$ ps which was beyond the experimental time scale in these measurements. In all other cases, displayed in Figure 4, where the

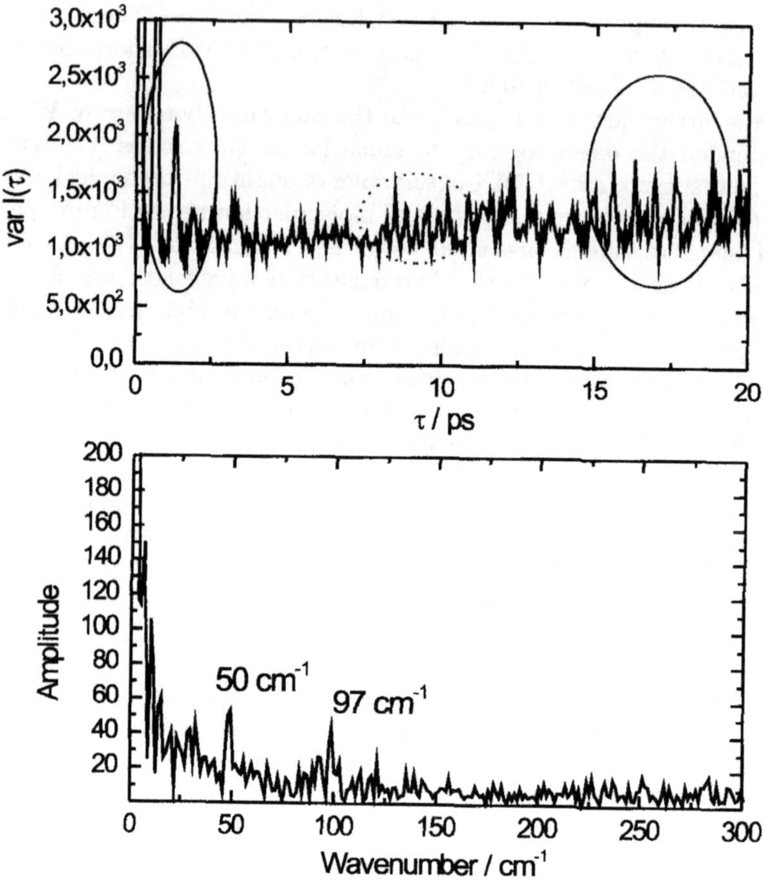

Figure 3. COIN-transient for a excitation wavelength $\lambda_{exc} = 520$ nm (upper panel) and the corresponding FT (lower panel). The full ellipse denotes the temporal region which corresponds to the 1/2-revival, the dotted ellipse the region of the 1/4-revival.

excitation wavelength was varied, a reappearance of the signal was observed. These measurements and the theoretical modelling of the experimental results have been given in much detail in Ref. 10.

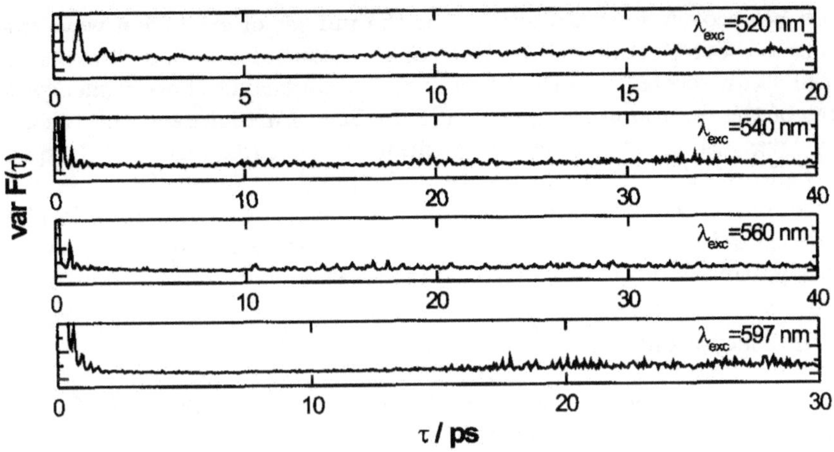

Figure 4. COIN-transients for various excitation wavelengths, showing the 1/2-revivals for all wavelengths and the 1/4-revival for $\lambda_{exc} = 520$nm.

3 Summary and Conclusions

In this work we have been concerned with the interferometric probing of the motional phenomena of molecular vibronic WPs in the thermal iodine vapor. Special emphasis was placed on the measurements of the long-time quantum-mechanical revival regime.

The COIN transients have revealed the quantum signature of fractional vibrational revivals. These revivals appear after a classical-like stage of strong damping and spreading of vibrational coherences due to rotational, inhomogeneous broadening and the anharmonicity of the potential. Owing to the discreteness of the vibrational levels, the almost overdamped vibrational signals are subject to a rephasing process that constitutes a renaissance of the initial pattern. Measurements were performed at various excitation wavelengths which all showed half-revival patterns on top of a multifrequency oscillatory background. For the special wavelength λ_{exc}=520 nm presented in this work the WP shows a quarter-revival, $T_{rev}/4$ around 8 ps followed by a half-revival near $T_{rev}/2 \sim 15$ ps. Simulations of the COIN signal have been performed for realistic RKR potential curves for the $X \rightarrow B$ transition in iodine under the assumption that the molecules are initially in a statistical mixture of vibrational and rotational energy states. In general the experimental patterns could be recovered satisfactorily. For the full set of excitation wavelengths, a simple theoretical model of vibrational-rotational coupling [10] explains the initial decay, reconstructs the spreading, and mimics the time-frequency structures of the quantum-mechanical WP regime. Furthermore, the theoretical model was also able to recover the positions and the frequencies of the quarter- and half-revival at λ_{exc}=520 nm. In closing, we have shown that probing rovibronic coherences by measuring correlated intensity fluctuations of incoherent fluorescence is a useful approach to tracking the time- and frequency evolution in a vibrating and rotating diatomic molecule.

Acknowledgments

This work was supported by the Austrian Science Fund within the framework of the Special Research Program F016.

References

1. I.Sh. Averbukh *et al*, Phys. Rev. Lett. **139**, 449 (1989).
2. A. Zewail, Science **242**, 1645 (1988).
3. M. Gruebele *et al*, J. Chem. Phys. **98**, 886 (1993).
4. T. Baumert *et al*, Phys. Rev. Lett. **67**, 3753 (1991).
5. N.F. Scherer *et al*, J. Chem. Phys. **45**, 4088 (1991).
6. O. Kinrot *et al*, Phys. Rev. Lett. **75**, 3822 (1995).
7. A. Tortschanoff *et al*, J. Chem. Phys. **110**, 4493 (1999)
8. Ch. Warmuth *et al*, J. Chem. Phys. **112**, 5060 (2000).
9. N.F. Scherer *et al*, J. Chem. Phys. **96**, 4180 (1992)
10. Ch. Warmuth *et al*, J. Chem. Phys. **114**, 9901 (2001).
11. T. Baumert *et al*, Chem. Phys. Lett. **191**, 639 (1992).
12. M.J.J. Vrakking *et al*, Phys. Rev. A **139**, R37 (1996).

EXCITONIC COUPLING IN BICHROMOPHORIC MOLECULES: CONFORMATIONAL INFORMATION FROM 2D OPTICAL PHOTON ECHO - A THEORETICAL STUDY

V. SZÖCS

Institute of Chemistry, Faculty of Natural Sciences, Comenius University, Mlynská dolina CH2, 842 15, Bratislava, Slovakia

E-mail: szocs@fns.uniba.sk

T. PÁLSZEGI

Institute of Physical Chemistry, Vienna University, Währingerstrasse 42, A-1090 Wien, Austria

E-mail: palszegi@ap.univie.ac.at

V.LUKEŠ

Department of Chemical Physics, Faculty of Chemical Technology, Slovak Technical University, Radlinského 9, 812 37 Bratislava, Slovakia

E-mail: lukes@chtf.stuba.sk

A.TORTSCHANOFF AND H.F.KAUFFMANN

Institute of Physical Chemistry, Vienna University, Währingerstrasse 42, A-1090 Wien, Austria

E-mail: kauffm@ap.univie.ac.at

In this paper the conformational information content of the 2D frequency domain electronic 2 pulse photon-echo (2-PPE) spectra of special systems of excitonically coupled symmetric dimers is compared to their 1D absorption counterparts. Different site orientations, determined by the molecular conformation, hidden in the 1D signal, can be decoded by comparing the relative heights of the diagonal and off-diagonal (cross) peaks in the frequency-frequency frame 2-PPE signal. The feasibility of this analysis rests upon the fact that two-excitonic effects are imprinted into the relative intensities of cross-peaks. The presented method can be used for conformational and excitonic coupling analysis of geometrically confusing systems of bichromophoric molecules.

1 Introduction

Symmetric two-site molecular systems (STS) built from two identical molecules or containing two identical molecular parts capable to be excited by an external electric field, represent the simplest class of multichromophoric ensembles. The electronic coupling in a chromophoric pair or bichromophoric molecule is usually described by the dipole-dipole interaction based on excitonic (resonance) coupling [1]. Such a type of coupling serves as a driving force for electronic excitation transport between chromophores and is related to the energetic splitting (gap) between the delocalised stationary electronic excited states of the multichromophoric molecular structures.

The stable conformation of a STS in a local energetic minimum determines the relative orientation and distance of transition dipoles of the chromophoric sites (i.e. of molecules or molecular parts). In condensed media the number of conformations is also affected by the multitude of solvent (matrix) local configurations. With the number of conformations the energetical and orientational disorder in the STS increases. As a consequence, in conventional spectroscopies (e.g. the linear absorption), which provide direct information on energy levels and their oscillator strengths, the conformationally disordered systems often have broad and complicated line-shapes.

Non-linear, multidimensional spectroscopic techniques are spreading the spectral information of congested spectra. Electronic 2D photon-echo (PE) can resolve and visualize the inhomogeneous line-shapes of chromophores [2] in the frequency-frequency, (ω,ω'), regime. For coupled chromophoric systems (aggregates) the intensities and positions of off-diagonal peaks in the 2D PE signal reflect the intermolecular coupling constants [3]. In ref.[4] 2D two-pulse PE (2-PPE) signals of symmetric dimers have been analyzed to find relationships between excitonic coupling, orientation, inhomogeneous and homogeneous line-broadenings, and to visualize the effects of these factors in the (ω,ω') plane.

In the past, the problem area of vibrational 2D (and more dimensional) spectroscopies have been the major focus of research in comparison with their electronic counterparts. A lot of investigations were performed on the linear and quadratic vibronic coupling strengths [5], the vibrational anharmonicity [6] and non-linearity of the coordinate dependencies of the dipole moment, polarizability, and hyperpolarizability [7], which are not available by 1D techniques. In analogy to the 2D excitonic coupling studies it was possible to analyze the coupling of localized vibrations of polyatomic molecules and aggregates [8]. The vibrational-exciton model helped to extract the vibrational-exciton population transfer and coherence dynamics from the time-dependent off-diagonal peak variations of the 3-pulse PE (ω,ω') spectrum [9].

In the present article the structural information content in the 1D absorption and 2D 2-PPE spectra of excitonically coupled symmetric dimers will be characterized analytically and calculated numerically. Our central concern will be the effects of the site transition dipole orientations and of the site-to-site excitonic coupling strength on the signal (peak) intensity. A special situation, where the conformational information of a simple system of STS, still hidden in the absorption spectrum, can be deciphered by the 2-PPE (ω,ω') spectrum, will be discussed in detail.

2 The absorption and frequency domain 2-PPE spectra of symmetric dimers

We model the symmetric dimer or STS in terms of two identical, energy degenerate, excitonically interacting pairs of singlet (S_0 and S_1) electronic states in the site-representation (4 level system). The two-site electronic state space is formed by the ground state $|G\rangle = |g_I\rangle|g_{II}\rangle$, the one-excitonic subspace E_1 of two states $\{|E_1(i)\rangle\}_{i=1}^{2} = \{|g_I\rangle|e_{II}\rangle, |g_{II}\rangle|e_I\rangle\}$ and by the two-excitonic subspace E_2, represented by the state $|E\rangle = |e_I\rangle|e_{II}\rangle$. Here, the ground and first excited singlet states of the sites are denoted by $|g\rangle$ and $|e\rangle$, respectively. The excitonic coupling J is defined in the standard way, (see explicitly e.g. in [1]) and depends on the mutual orientation θ of the transition dipoles, $\vec{d}_I$ and $\vec{d}_{II}$, and on the distance of the sites.

The 1D electronic absorption spectrum $\alpha(\omega_L)$ of our STS, is given by a standard linear response formula (see e.g. in [10])

$$\alpha(\omega_L) = K_2 \hbar\omega_L \left\{(1-\cos\theta)\delta(\hbar\omega_L - \omega_{eg} + J) + (1+\cos\theta)\delta(\hbar\omega_L - \omega_{eg} - J)\right\}.$$

Here, ω_L is the carrier frequency of the exciting electric field and we have averaged rotating the interacting sites and preserving the angle θ between the site transition dipoles. Further, K_2 is a constant, dependent among others on the absolute value of the site transition dipole moments and ω_{eg} is the transition energy of the site (given by the energy difference of the S_1 and S_0 states). The absorption peaks of our STS are split by $2J$. It follows, that the ratio of the heights of the higher and lower energy peaks is given by the formula $\alpha(\omega_2)/\alpha(\omega_1) = \{(\hbar\omega_L - J)(1+\cos\theta)\}/\{(\hbar\omega_L + J)(1-\cos\theta)\}$ where usually $J \ll \hbar\omega_L$. In this case we get

$$\alpha(\omega_2)/\alpha(\omega_1) = (1+\cos\theta)/(1-\cos\theta).$$

The heterodyne 2-PPE time-domain signal of STS was derived in [4] for δ-pulses by perturbative standard methods [11]. According to the result obtained in [4], the polarization $P_\delta(t,\tau_{12})$, which determines the 2-PPE signals, is given by the formula as follows:

$$P_\delta(t,\tau_{12}) = iC\exp(i\omega t')\exp(-t/T_2)\exp(-\sigma^2(t-\tau_{12})^2/2)$$
$$\left\{2F_{E_1}(\tau_{12},\tau_{12}+t,\tau_{12},0) - F_{E_2}(\tau_{12},\tau_{12}+t,\tau_{12},0)\right\}.$$

Here, C is a parameter related with the incoming electric fields (see in [4]); T_2 and σ are the standard parameters determining the homogeneous dephasing and the inhomogeneous line broadening, respectively. Further, τ_{12} is the time delay between the first and second pulse, t is the delay after the second pulse and $t' = \tau_{12} + t$. To get $P_\delta(t, \tau_{12})$ we have realized an orientational averaging in the same way as above in the case of the 1D absorption spectrum. The explicit forms of the four-point dipole correlation functions F_{E_1} and $F_{E_{21}}$, not involving and involving two-excitonic effects, respectively, are as follows:

$$F_{E_1}(\tau_{12}, \tau_{12}+t, \tau_{12}, 0)/d^4 = (8/5)\{\sin^4\theta \exp(-iJt_-) + \cos^4\theta \exp(iJt_-)\} + (2/15)\sin^2\theta\{\exp(-iJt_+) + \exp(iJt_+)\}$$

and

$$F_{E_2}(\tau_{12}, \tau_{12}+t, \tau_{12}, 0) = (2/15)\{(2+3\cos\theta+\cos 2\theta)\exp(-iJt_+) + (2-3\cos\theta+\cos 2\theta)\exp(iJt_+)\}.$$

Here, $t_+ = t + \tau_{12}$ and $t_- = t - \tau_{12}$; d is the absolute value of the transition dipole-moment of a site.

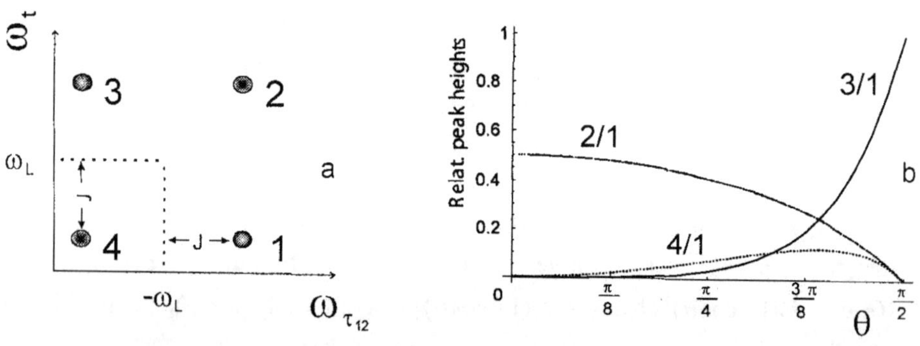

Figure 1: (a) The position of the 2-PPE peaks in the frequency plane. (b) The dependence of relative peak intensities (to peak 1) as a function of mutual site transition dipole orientation, θ.

The double, analytical Fourier inversion of the polarization $P_\delta(t, \tau_{12})$ in the frequency-frequency space, $S(\omega_t, \omega_{12})$, results in a long and complicated expression, giving us 2D heterodyne 2-PPE signals of STS in the region of the carrier frequency ω_L. It is formed by four peaks arranged at the edges of a square with distances determined by the excitonic inter-site coupling, J (see Fig. 1a). The two diagonal (d) peaks are of one-exciton origin, while the two off-diagonal (od, i.e. cross) peaks involve two-exciton effects. The heights $\{h_i\}_{i=1}^4$ of the four peaks are in distinct relation with the mutual orientation θ of the site-localized transition dipoles, as follows:

$$h_1 = \cos^4(\theta/2), \quad h_2 = (1/2)\cos^2(\theta/2)\cos\theta,$$
$$h_3 = \sin^4(\theta/2), \quad h_4 = (1/2)\sin^2(\theta/2)\cos\theta.$$

These formulas are serving as a structural probe for orientationally averaged systems of pairs of resonantly interacting groups (molecules). In Fig. 1b these orientational dependences of the relative heights of the 2D 2-PPE signal-peaks are given. The shapes of the peaks reflect the ratio between the homogeneous and inhomogeneous dephasing parameters.

3 Conformational analysis versus excitonic coupling of special bichromophoric systems by 2D 2-PPE

Our first study is concerned with the analysis and comparison of the information content of 1D absorption and the 2D-2PPE spectra of a simple model system formed by a mixture of two STS, denoted as A and B, where the excitonic coupling J is the same in both STS, i.e. $J_A = J_B$. Further, let the coupled chromophores in different STS have different mutual orientations of their site transition dipole moments ($\theta_A \neq \theta_B$). Then, from the constant excitonic coupling it follows, that the sites in the different STS must have unequal distances. We suppose also, that the two conformationally different STS have the same probability, i.e. $p_A = p_B$.

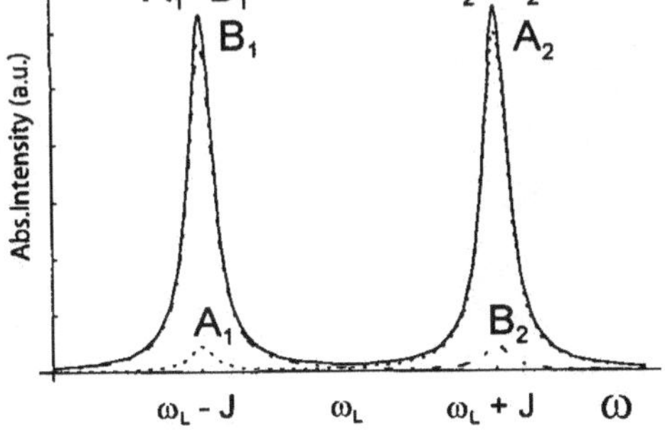

Figure 2: 1D absorption spectrum (solid line) of a system formed by two STS (A, B), having the same probability of conformations, i.e. $p_A = p_B$. Further, $J_A = J_B$, $\theta_A = \pi/6$ and $\theta_B = \pi - \theta_A$. Conformation A: dashed line; conformation B: dash-dotted line.

Let us compare now the $\alpha(\omega_L)$ and $S(\omega_l,\omega_{12})$ spectra of our model STS system in special case, if $\theta_A=\theta$ and $\theta_B=\pi-\theta$. The 1D absorption spectrum is a pair of equal height peaks (see Fig. 2). This follows from the fact, that $\cos(\pi-\theta)=-\cos\theta$, which is resulting in the absorption peak ratios $\alpha_A(\omega_2)/\alpha_A(\omega_1)=(1+\cos\theta)/(1-\cos\theta)$ and $\alpha_B(\omega_2)/\alpha_B(\omega_1)=(1-\cos\theta)/(1+\cos\theta)$. Then the sums $\alpha_A(\omega_i)+\alpha_B(\omega_i)$, giving the (normalized) absorption peak intensities of our system, are independent on θ, i.e. equal to 2, and on 1D basis no information on the conformational geometry can be obtained.

The situation is different in the case of the 2D 2-PPE spectrum (see Fig. 3). For our model system of two equimolar STS , with the same excitonic coupling, but different mutual orientation of site transition dipoles (i.e., $\theta_A=\theta$; $\theta_B=\pi-\theta$), the peak intensities are given by $h_1^A=\cos^4(\theta/2)$, $h_2^A=(1/2)\cos^2(\theta/2)\cos\theta$, $h_3^A=\sin^4(\theta/2)$, $h_4^A=(1/2)\sin^2(\theta/2)\cos\theta$ and by $h_1^B=\sin^4(\theta/2)$, $h_2^B=-(1/2)\sin^2(\theta/2)\cos\theta$, $h_3^B=\cos^4(\theta/2)$, $h_4^B=-(1/2)\cos^2(\theta/2)\cos\theta$, respectively. In Fig. 3a the absolute values of the sums of these peak intensities are shown, given by the values: $\sin^4(\theta/2)+\cos^4(\theta/2)$ (peak 1 and 3) and $(1/2)\cos\theta$ (peak 2 and 4). It follows, the 2D 2-PPE spectrum is sensitive on θ. This fact is illustrated by the Figs. 3a and 3b, showing the spectra for two different values ($\theta_1=\pi/6$ and $\theta_2=3\pi/8$) of the mutual site transition dipole orientations.

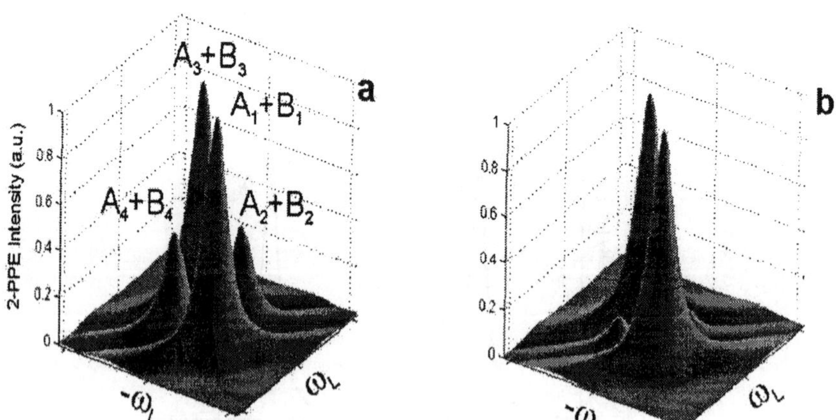

Figure 3: 2D 2-PPE spectra of a systems formed by two STS (A , B) with the same conformational probabilities and $J_A=J_B$. (a) $\theta_A=\pi/6$ and $\theta_B=\pi-\theta_A$, (b) $\theta_A=3\pi/8$ and $\theta_B=\pi-\theta_A$. $\{A_i\}_{i=1}^4$ and $\{B_i\}_{i=1}^4$ are the sets of peaks for conformation A and B, respectively.

Fig. 4 shows the θ dependence of the sums of diagonal and cross-peak intensities of the 2D 2-PPE spectrum of our special system (see above), composed of two STS with different conformations. Such dependencies can help in the conformational analysis of STS probes, fitting the structural parameters to simulate the experimental signal (in our special system it was θ, assuming $J_A = J_B$ and $p_A = p_B$, resulting in the condition $\theta_A = \theta$ and $\theta_B = \pi - \theta$). Changing the condition of equal conformational probabilities (i.e. $p_A \neq p_B$) would give non-equal 1D peak heights, which could be misinterpreted by assuming only one conformation (in this case the mutual angle of site transition dipoles could be deduced from the 1D spectrum it would follow). In such a case by fitting simultaneously the 1D absorption and 2D 2-PPE spectra to different sets of $\{\theta_i\}, \{p_i\}$ and distance parameters, it is possible to decode the conformational statistics of STS systems (see [12]).

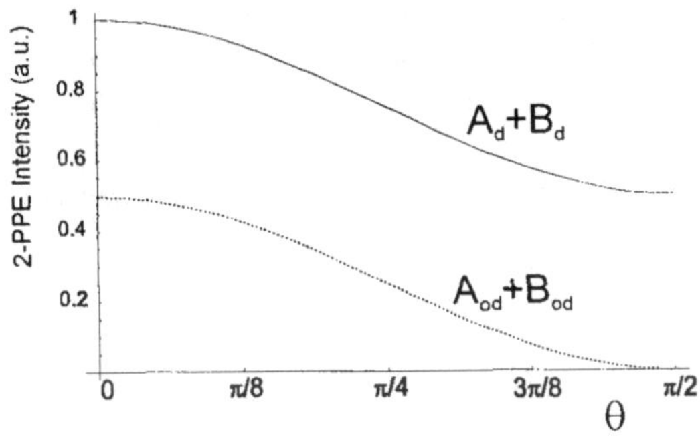

Figure 4: Sum of diagonal (d) and off-diagonal (od ; cross) peak intensities for mutual site transition dipole orientation, θ, dependence of the 2D 2-PPE signal. For the definition of the STS system see Figure 2.

4 Conclusions

Based on our very recent analysis [4], we presented and analyzed closed-form expressions of the heterodyne 2D 2-PPE signal of symmetric dimers and of their simplest (two-state) orientational, statistical averages. The diagonal and antidiagonal parts of the 2D spectrum of these systems, containing the diagonal- and cross-peaks, respectively, reconstruct the conformational information, which is masked in the 1D absorption spectra (Figs. 2-4).

The studied system of STS represents the simplest case of off-diagonally disordered systems of chromophoric aggregates. For similar, but larger, more complicated systems of chromophoric dimers (i.e. more conformations with different probabilities and different values of excitonic coupling) the presented technique could also be used with the help of an algorithm that starts with an hypothesis about the possible structures hidden in the one- and two-dimensional spectra [12].

In general, 2D electronic spectroscopy has a large potential for probing electronic inter-site coupling and energy transfer between molecules or molecular parts. Higher correlations of molecular polarizations from time-ordered pulses in nonlinear experiments provide a powerful vehicle to extract electronic coupling from homogeneous and inhomogeneous broadening and/or vibrations congested molecular spectroscopic data.

5 Acknowledgement

This work has been supported, in part, by the Austrian Science Foundation within the framework of the Special research Program ADLIS (F016). In addition, V.Sz, T.P. and V.L. thank for the financial support from the Austrian Ministry for Education, Science and Culture, bm:bwk (Ost-West-Kooperation-H, GZ.45.452). The support of the Slovak Ministry of Education (Vega 1/7355/20) is also acknowledged.

References

1. Stone A.J., *The theory of intermolecular interaction* (Clarendon Press, Oxford, 1996)
2. Gallagher Faeder S.M. and Jonas D.M., Two-dimensional electronic correlation and relaxation spectra: theory and model calculations, *J. Phys. Chem. A* **103** (1999) pp. 10489-10505
3. Mukamel S., Multidimensional femtosecond correlation spectroscopies of electronic and vibrational excitations, *Annu. Rev. Phys. Chem.* **51** (2000) pp. 691-729
4. Szöcs V., Pálszegi T., Tortschanoff A. and Kauffmann H.F., Electronic Coupling and Coherences in Disordered Polymers: Femtosecond 2D-Photon Echo Correlation Spectroscopy, Signatures of an Excitonic Two-Segmental Site System - A Theoretical Study, *J. Chem. Phys.*, (submitted)
5. Park K. and Cho M., Two-dimensional vibrational spectroscopy. VI. Higher-order contributions to the two-dimensional vibrational response functions, *J. Chem. Phys.* **112** (2000) pp. 10496-10509

6. Hamm P., Lim M. and Hochstrasser R.M., Structure of the amide I band of peptides measured by femtosecond nonlinear-infrared spectroscopy, *J. Phys. Chem. B* **102** (1998) pp. 6123-6138
7. Okumura K., Tokmakoff A. and Tanimura Y., *J. Chem. Phys.* **111** (1999) pp. 492-503
8. (a) Ren Y., Murakami T., Nishioka T., Nakashima K., Noda I. and Ozaki Y., Two-dimensional Fourier transform Raman correlation spectroscopy studies of polymer blends: conformational changes and specific interactions in blends of atactic polystyrene and poly(2,6-dimethyl-1,4-phenylene ether), *Macromolecules* **32** (1999) pp. 6307-6318; (b) Piryatinski A., Tretiak V., Chernyak V. and Mukamel S., Simulations of two-dimensional femtosecond infrared photon echoes of glycine dipeptide, *J. Raman Spectr.* **31** (2000) pp. 125-135
9. (a) Piryatinski A., Chernyak V. and Mukamel S., Vibrational-exciton relaxation probed by three-pulse echoes in polypeptides, *Chem.Phys.* **266** (2001) pp. 285-294; (b) Zanni M.T., Gnanakaran S., Stenger J. and Hochstrasser R.M., Heterodyned two-dimensional infrared spectroscopy of solvent-dependent conformations of acetylproline-NH2, *J. Phys. Chem. B* **105** (2001) pp. 6520-6535
10. May V. and Kühn O., *Charge and energy transfer dynamics in molecular systems*, (Wiley-VCH, Berlin, 2000)
11. Mukamel S., *Principles of nonlinear optical spectroscopy* (New York: Oxford Univ. Press, 1999)
12. Szöcs V., Pálszegi T., Lukeš V., Tortschanoff A. and Kauffmann H.F., Theoretical conformational analysis of bichromophoric molecules by 2D photon-echo spectroscopy (in preparation)

FEMTOSECOND COINCIDENCE IMAGING OF DISSOCIATIVE IONIZATION DYNAMICS IN CF_3I

ANOUK M. RIJS

Laser Center and Chemistry Department, Vrije Universiteit, de Boelelaan 1083, 1081 HV Amsterdam, the Netherlands

CARL C. HAYDEN

Combustion Research Facility, Sandia National Laboratories, P.O. Box 969, Livermore, California 94551-0696, USA

MAURICE H.M. JANSSEN

Laser Center and Chemistry Department, Vrije Universiteit, de Boelelaan 1083, 1081 HV Amsterdam, the Netherlands

The application of femtosecond coincidence imaging to study the dissociative photoionization of CF_3I is reported. Coincidence imaging measures the correlated energy and angular distributions of the reaction products. With coincidence imaging molecular frame photoelectron-photofragment angular correlations can be extracted. In this contribution some of the prominent features elucidating the multi-photon femtosecond photo-dynamics of CF_3I will be presented, illustrating the strength of the femtosecond photoelectron-photoion coincidence imaging technique to study complex multi-photon dissociation dynamics in polyatomic molecules.

1 Introduction

During the last decade the application of femtosecond lasers to chemical dynamics has enabled the real-time study of a wide variety of photon-induced dynamics [1]. Many of the femtosecond gas phase experiments use sensitive ion [2,3] or photoelectron detection [4-6] to study time-resolved dynamics. Because both the photoion and the photoelectron provide important information on the dynamics, it is advantageous to detect the ion and correlated electron in coincidence. This paper presents multi-photon induced dynamics in CF_3I using a new technique introduced recently [7] combining photoelectron-photoion coincidence detection with three-dimensional energy- and angle-resolved imaging [8-10].

Performing femtosecond coincidence imaging experiments, two types of correlations can be extracted from the coincidence data, namely energy and angular distributions. The energy distributions correlate the photoelectron kinetic energy with the photoion kinetic energy for a particular ion channel. These energy distributions provide information on the energetics of the excitation process. The angular correlations between the photoelectron and photoions provide more detailed information on the ionization and dissociation dynamics like the symmetry of potential surfaces and molecular orbitals involved in the dynamics. An important aspect that can be measured with these coincidence experiments is the molecular frame photoelectron angular distribution (PAD) [11], which correlates the velocity vector of the electron with a dissociation axis of the parent molecule. The photoelectron-photoion coincidence imaging technique provides the most complete information available in femtosecond time-resolved photodissociation experiments. Recently, velocity map ion imaging experiments on CF_3I using femtosecond pump-probe techniques were reported [12]. In this paper we present our first coincidence experiments on CF_3I, illustrating some of the capabilities of the coincidence imaging technique to disentangle complex multi-photon dynamics in a polyatomic molecule.

2 Experimental set-up

In the ultrafast energy- and angle-resolved photoelectron-photoion coincidence imaging experiments, performed at Sandia, Livermore, [7,8], laser pulses produced by a regeneratively amplified Titanium sapphire laser, operating at 700 Hz were used. The regen output at 792 nm, with a typical pulse width of about 120 fs, is frequency doubled and tripled. The pump pulse is centered at 264 nm and has a pulse energy of about 4 μJ. The probe pulse, centered at 396 nm, has a pulse energy of about 8 μJ. The pump and probe beams are recombined and focused (f = 1000 mm) into a ultrahigh vacuum chamber, where they intersect the continuous molecular beam. For all the data presented here, the polarization of both the pump and probe pulse was along the direction of the molecular beam velocity and in plane of the two detectors. The molecular beam was a mixture of 4% CF_3I in Argon.

The electrons and ions are accelerated and extracted in opposite directions towards time- and position-sensitive detectors. The acceptance solid angle of the detectors is 4π, such that full three-dimensional angular distributions are measured. In order to avoid false coincidences, a low ionization rate of approximately 1-3 events per 100 laser shots is used. Furthermore, an ionization event is recorded only if an single electron and a single ion are detected after one laser shot. After detecting 10^5 pairs of coincident ions and electrons the data was stored. Coincidence data was taken at various pump-probe delay times.

3 Kinetic Energy Distributions of CF_3I^+

Figure 1a shows the electron images measured in coincidence with the parent CF_3I^+ at 200 and 400 fs time delay. These two electron images are representative of the images observed at short delay times between 0 and 200 fs and at longer time delays between 300 and 1000 fs. Between 0 and 200 fs pump-probe delay, the distribution is isotropic. At 300 fs and longer delay times anisotropic images with polar caps along the direction of the polarization of the laser are observed. From these electron images we can extract the total electron kinetic energy distributions, shown in Figure 1b. The calibration of the absolute electron energies was done on both Xe and NO multi-photon ionization using the same femtosecond pulses at 264 and 396 nm by simply switching the gas line.

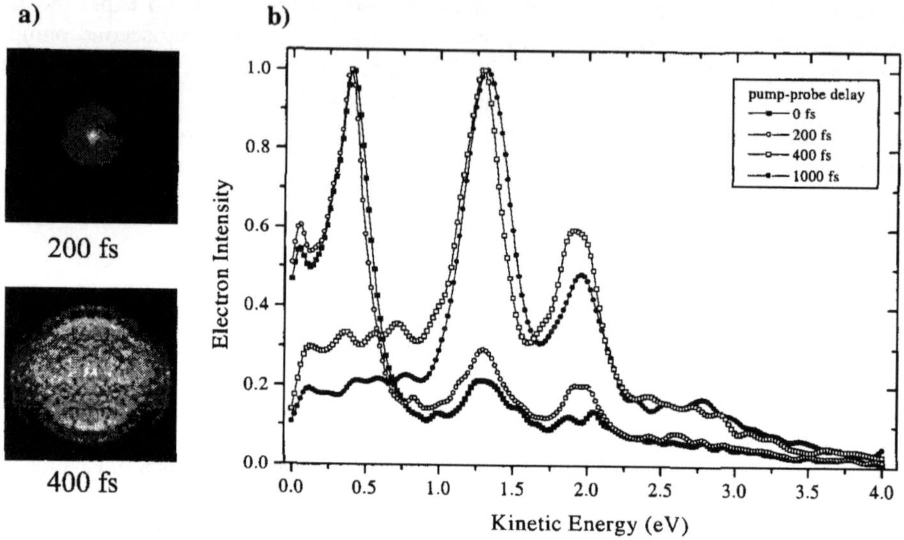

Fig 1. (a): Electron images of CF_3I^+ at 200 and 400 fs pump-probe delay;

(b): Electron energy distributions of CF_3I^+ at 0, 200, 400 and 1000 fs delay.

The photoelectron spectra change dramatically with the delay time. Up to 200 fs, the distribution is dominated by one single peak at 0.4 eV. However, at longer time delays two dominant peaks appear in the photoelectron spectrum, a strong peak at 1.3 eV and a smaller peak at 2.0 eV. We can assign the peaks using the known ionization energy of CF_3I and the photon energies. The electron peak at 0.4 eV, which is only observed at short time, results from an excitation process using one pump photon of 264 nm and two probe photons of 396 nm, a (1+2') ionization process. One-photon excitation to the dissociative A-band is followed by ionization with two probe photons to produce $CF_3I^+(X^2E_{3/2})$. At longer time delays this peak in the photoelectron spectrum can not be produced in a (1+2') ionization process anymore, because the CF_3I molecule dissociates rapidly and the wavepacket has moved away from the favorable geometry for ionization from the A-band to the ground state of ion.

The peaks at 1.3 eV and 2.0 eV, observed at longer delay times, originate from a (2+1') ionization scheme resulting in a total photon energy of 12.5 eV. The ionization energy of the ground state $CF_3I^+(X^2E_{3/2})$, is 10.40 eV and the energy of the spin orbit excited state, $X^2E_{1/2}$, is 11.13 eV [13, 14]. Ionization into these spin-orbit split ground state levels results in electrons ejected with 2.0 eV and 1.3 eV, respectively. At the two-photon level, the (2+1') excitation is resonant state with the 5pπ7sσ ($^2E_{1/2}$) Rydberg state [12]. The electron peak at 1.3 eV, corresponding to the $CF_3I^+(X^2E_{1/2})$ ionic state, is stronger because this ionic state has the same core as the intermediate 5pπ7sσ ($^2E_{1/2}$) Rydberg state.

4 Kinetic Energy Distributions of I^+

We will now focus on the energy distributions of the I^+ photofragment. Figure 2a shows the ion images of I^+ at 0, 400 and 1000 fs time delay. For all the images in Fig. 2 the polarization of both the pump and probe pulse were parallel to the recoil direction. At zero delay time, the ion image looks very anisotropic, the I^+ fragments recoil along the direction of the laser polarization. Between 200 and 400 fs delay the I^+ ion image changes and is concentrated at smaller distances from the center of the image. At long time delays, the I^+ fragments are recoiling with large velocities again. This dramatic change in I^+ kinetic energy was also observed in the velocity map ion imaging experiments by Roeterdink *et al.* [12]. The correlated electron images of I^+ at 0, 400 and 1000 fs delay are presented in Figure 2b. At zero and 1000 fs time delay, the distributions look similar and quite isotropic. However, at 400 fs a central spot of slow electrons is observed and anisotropic polar caps of fast electrons.

a)

b)

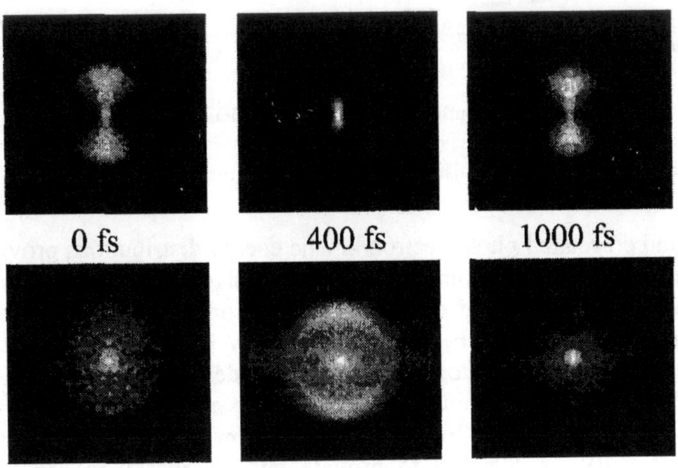

Fig. 2. (a) : Ion images of I^+ at 0, 400 and 1000 fs pump-probe delay;

(b) : Electron images of I^+ at 0, 400 and 1000 fs pump-probe delay.

From the ion and correlated electron data we can extract the total kinetic energy distributions at different delay times. Figure 3a shows the ion kinetic energy distribution of I^+ at 0, 400 and 1000 fs delay time. At 400 fs the kinetic energy release is very small and peaked at 0.05 eV. The recoil energy of the I^+ ion at zero and long time delays is quite similar and extends to energies up to 0.75 eV.

Figure 3b shows the photoelectron energy distribution of events correlating with all I^+ photofragments at 0, 400 and 1000 fs time delay. At 0 fs delay time, the photoelectron spectrum shows three peaks and a long tail up to 3 eV. At a pump-probe delay of 400 fs a dominant peak at 0.8 eV and a smaller peak at low electron energy is observed. The peak at 0.8 eV corresponds to the polar caps as observed in the electron images at 400 fs, while the small peak of slow electrons correspond to the central spot. At long delay times (~1000 fs) the energy distribution shows two peaks, a strong peak at low kinetic energies and a smaller peak at 0.8 eV.

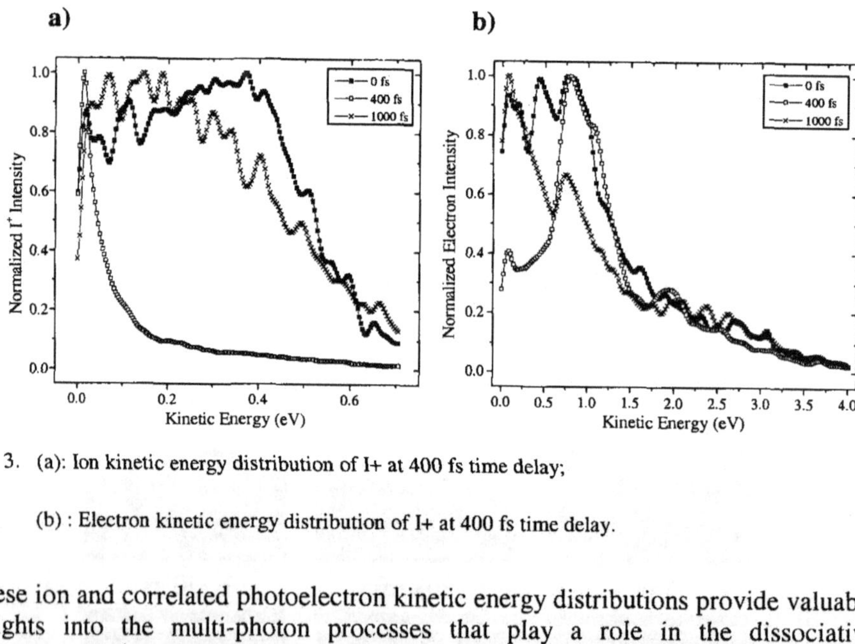

Fig. 3. (a): Ion kinetic energy distribution of I+ at 400 fs time delay;

(b) : Electron kinetic energy distribution of I+ at 400 fs time delay.

These ion and correlated photoelectron kinetic energy distributions provide valuable insights into the multi-photon proccsses that play a role in the dissociative ionization of I^+. The threshold energy for production of the lowest exit channel of the iodine ion, the $I^+(^3P_2) + CF_3$ channel, is 12.7 eV [14]. A (2+1') excitation, i.e. 2 photons of 264 nm and 1 photon of 396 nm, provides only 12.5 eV which is below the $I^+(^3P_2) + CF_3$ energy limit. This means that at least a (2+2') multi-photon excitation, i.e. 2 photons of 264 nm and 2 photon of 396 nm with total photon energy of 15.6 eV, is required to provide enough energy to access all the I^+ dissociation channels. Excitation via a (2+2') excitation scheme at zero and long time delays leads to neutral fragments with low internal energy. However, at a pump-probe delay between 200 and 400 fs, a much smaller kinetic energy release of

0.05 eV of the I^+ fragment is observed, see Figure 3a. The electron is ejected with an energy of 0.8 eV, similar as at zero and longer pump-probe delays. So, the neutral CF_3 fragment has a lot of internal energy, about 2 eV for the lower $I^+(^3P_2)$ + CF_3 channel and 1 eV for the higher $I^+(^3P_{0,1})$ + CF_3 channel. Thus, at 200 to 400 fs pump-probe delay the formation of the I^+ fragment appears to result from a different ionization process, than at longer or shorter delays. One pathway to produce low kinetic energy I^+ fragments may be a different multi-photon process, for instance a (1+3') excitation through the dissociative A-band. In a (1+3') ionization scheme, the parent molecule CF_3I is excited with one photon of 264 nm to the dissociative A-state and the CF_3I molecule starts to dissociate. One or two probe photons of 396 nm excite the dissociating CF_3I molecule to an intermediate ion-pair state. A third probe photon detaches the electron from a CF_3^- I^+ ion-pair state, resulting in two fragments CF_3 and I^+ with relatively low kinetic energy. However, the neutral CF_3 fragments will have a lot of internal energy. Another possibility is ionization after internal conversion of the initially excited Rydberg state. CF_3I is excited by two pump photons to the intermediate 5pπ7sσ Rydberg state. This state may internally converts to another state, possibly an ion-pair state. Two-photon ionization of these vibrationally excited CF_3I molecules can result in slow I^+, vibrationally hot CF_3 and a 0.8 eV electron. Currently, more extensive analysis and further experiments are underway to elucidate the mechanism responsible for the slow I^+ fragments at intermediate time delay [15].

5 Angular Distributions

5.1 Lab frame angular distribution of I^+

From the coincidence data we can also extract the angular distributions of the recoiling photoelectrons and ions. The laboratory angular distribution of I^+ at 400 fs delay time is shown in Figure 4a as a function of the angle, θ, between the recoil velocity of the I^+ fragment and the direction of the polarization of the lasers. The I^+ events were selected in two velocity groups, ions with kinetic energy between 0.0-0.1 and 0.45-0.75 eV, in both cases irrespective of the kinetic energy of the ejected electron. The energy of 0.75 eV corresponds to the energy cut-off value of the $I^+(^3P_2)$ + CF_3 exit channel via a (2+2') pathway and an electron kinetic energy release of 0.8 eV. The 0.0-0.1 eV energy region corresponds to the small peak in the ion energy distribution at 400 fs. The angular distributions are fitted to the following expression:

$$I(\theta)=\frac{\sigma}{4\pi}\left[1+\beta P_2(\cos\theta)+\gamma P_4(\cos\theta)+\ldots\right], \quad (1)$$

where $P_2(\cos\theta)$ and $P_4(\cos\theta)$ are Legendre polynomials and β, γ fit parameters representing the anisotropy. The angular distribution of I^+ at $\tau = 400$ fs is found to be more anisotropic for increasing ion kinetic energies. For the energy region between 0.45-0.75 eV, $\beta = 2.2$ and $\gamma = 0.6$ were obtained. These high values indicate that the fast I^+ fragments are ejected with a very anisotropic angular distribution and I^+ is distributed more anisotropically for increasing ion kinetic energies.

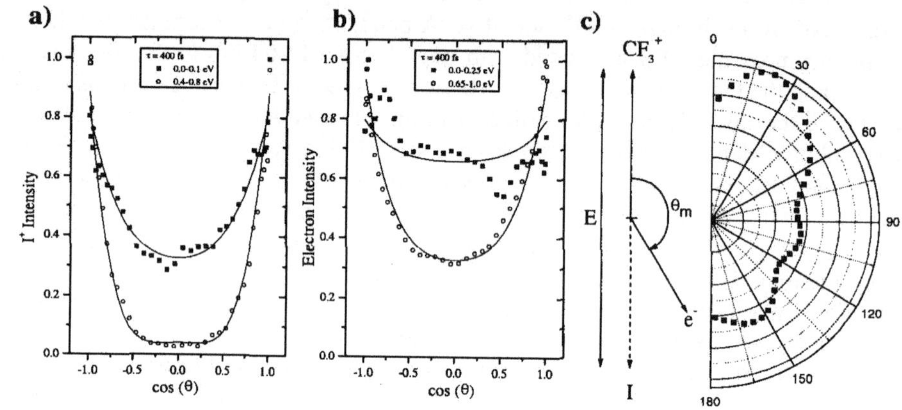

Fig. 4. (a) : Lab frame ion angular distribution of I^+ at 400 fs;
(b) : Lab frame electron angular distribution of I^+ at 400 fs;
(c) : Molecular frame angular correlation of CF_3^+ at 400 fs.

The photoelectron angular distributions measured in coincidence with the I^+ ion angular distributions are shown in Figure 4b for a pump-probe delay of 400 fs. Two curves are presented, one angular distribution for electrons ejected at high energies, 0.65-1.0 eV, and one at low electron energy, 0.0-0.25 eV. The lab frame electron angular distribution were fitted with a similar expression as equation (1), with θ_e the lab frame polar angle of the ejected electron resulting in anisotropy parameters β_e and γ_e. As is observed in Figure 4b the electrons at high energy, corresponding to the polar caps from the electron image, are ejected with a very anisotropic distribution. For the energy region between 0.65-1.0 eV we found for the anisotropy parameters $\beta_e = 0.8$ and $\gamma_e = 0.2$. However, at low kinetic energy region, 0.0-0.25 eV, the electron angular distribution is much more isotropic with $\beta_e = 0.1$ and $\gamma_e = 0.03$. It seems that these two electron channels originate from quite different dynamical processes.

5.2 *Molecular Frame Angular Correlations of CF_3*

Besides the lab frame angular distributions of selected ion and electron channels, with femtosecond coincidence imaging we can also obtain molecular frame angular correlations. This means that we can correlate the angular direction of the electron with respect to the recoil direction of the ionic fragment. The molecular frame angular correlation for CF_3^+ at a time delay of 400 fs for the high electron kinetic energy channel (0.3-1.0 eV) is presented in Figure 4c. In Figure 4c, the molecular frame angular correlations are displayed as a polar plot of the angle between the CF_3^+ ion and the electron. Only events with ions recoiling within 0-15 degrees of the direction of the polarization of the laser were selected. The angular correlation is forward-backward asymmetric and shows a lob at approximately 30 degrees. It appears that the probability is higher for photoelectrons to be ejected in the same direction as the CF_3^+ photofragments, than in the direction of the neutral I fragment. In contrast, photoelectrons with low kinetic energy recoil much more isotropically. The molecular frame PAD will be explored further in combination with theoretical calculations [15].

6 Acknowledgements

AMR and MHMJ acknowledge the visitors program at the Combustion Research Facility, Sandia Livermore, for support. The authors like to thank Mike Gutzler for excellent technical support. The authors like to thank W.G. Roeterdink for helpful discussions. The research of CCH is supported by Division of Chemical Sciences, Geosciences and Biosciences, the Office of Basic Energy Sciences, the U.S. Department of Energy. AMR acknowledges the Holland Research School of Molecular Chemistry for a PhD fellowship and for travel support.

References

1. A.H. Zewail, *J. Phys. Chem. A.* **104** (2000), 5660.
2. M.HM. Janssen, M. Dantus, H. Guo, and A.H. Zewail, *Chem. Phys. Lett.* **214** (1993), 281.
3. D.P. Zhong and A.H. Zewail, *J. Phys. Chem. A* **102** (1998), 4031.
4. V. Blanchet, M.Z. Zgierski, T. Scheideman, and A. Stolow, *Nature* **401** (1999), 52.
5. L. Wang, H. Kohguchi, and T. Suzuki, *Faraday Discussions* **113** (1999), 37.
6. A. Assion, M. Geisler, J. Helbing, V. Seyfried, and T. Baumert, *Phys. Rev. A* **54** (1996), R4605.
7. J.A. Davies, J.E. Leclaire, R.E. Continetti, and C.C. Hayden, *J. Chem. Phys.* **111** (1999), 1.
8. J.A. Davies, R.E. Continetti, D.W. Chandler, and C.C. Hayden, *Phys. Rev. Lett.* **84** (2000), 5983.
9. J. Ullrich, R. Moshammer, R. Dörner, O. Jagutzki, V. Mergel, H. Schmidt-Böcking, and L. Spielberger, *J. Phys. B* **30**(1997), 2917, and references therein.

10. K. A. Hanold, M. C. Garner, and R. E. Continetti, *Phys. Rev. Lett.* **77** (1996), 3335.
11. Y. Arasaki, K. Takatsuka, K. Wang, and V. McKoy, *J. Chem. Phys.* **112** (2000), 8871
12. W. G. Roeterdink and M.H.M. Janssen, *Chem. Phys. Lett.* **345** (2001), 72.
13. L.D. Waits, R.J. Horwitz, R.G. Daniel, J.A. Guest, and J.R. Appling, *J. Chem. Phys.* **97** (1992), 7263
14. P. Downie and I. Powis, *Faraday Discussions.* **115** (2000), 103.
15. A.M. Rijs, C.C. Hayden, and M.H.M. Janssen, work in progress.

High Resolution Femtosecond CARS Spectroscopy

Hans M. Frey, Paul Beaud*, Tobias Lang** and Marcus Motzkus**
University of Bern, CH-3012 Bern, Switzerland
*Paul Scherrer Institute, CH-5232 Villigen, Switzerland
**Max-Planck Institut für Quantenoptik, D-85748 Garching, Germany
(hans-martin.frey@iac.unibe.ch)

ABSTRACT

Time resolved femtosecond high resolution Raman spectroscopy (CARS, CSRS) was performed in the gas phase. The change of the polarizability is a direct probe of the evolution of the structure of the molecule in time. Results are presented for the diatomics H_2 and O_2, the linear molecules CO_2 and C_2H_2, the symmetric top $CHCl_3$ and the asymmetric top SO_2. With four-wave mixing (FWM), we prepare rotational coherences in the ground state within the sample. When delaying the probe pulse, we observe a periodic recovery of the signal. The analysis of the temporal behaviour of the transients yields fundamental molecular constants such as rotational and vibrational anharmonicity. A main advantage of the technique is the spatial resolution combined with a high temporal and spectral resolution. We present the possibility to measure temperatures in flames with a femtosecond CARS setup.

1. Introduction

Nonlinear optical detection schemes such as FWM and CARS have gained much of attention in the past years due to the availability of powerful and, most importantly for a nonlinear technique, stable laser systems. In the frequency domain the method has been successfully applied in gas phase, plasma diagnostics and condensed phase studies and theory has been extensively treated by Mukamel [1]. Recently, nonlinear methods have been used in the time domain to study chemical dynamics in liquids or in the gas phase with femtosecond time resolution. Motzkus et al. [2] have observed wave packets in molecular systems with degenerate four-wave mixing and two-colour four-wave mixing. Degenerate four-wave mixing can also be applied for rotational coherence spectroscopy. Linear schemes like fluorescence depletion have been used for years by Felker et al. to determine rotational constants, e.g., the structure of molecules in the gas phase [3]. In an earlier paper we have shown the application of degenerate four-wave mixing to determine the rotational constants of molecules in the gas phase [4]. The resolution of the experiment is strongly dependent on the length of the recorded transient. The duration of the signal is however limited due to collisional dephasing. Collisions are negligible for small molecules like H_2 but become strong for larger molecules with higher collisional cross section like CO_2 and $CHCl_3$. Collisions can be excluded when performing the experiment in a molecular beam with the constraint that the signal becomes weaker due to the square dependence of signal on the density in the beam [5]. In frequency domain CARS experiments the temperature is usually measured by monitoring the Q branch of tracer molecules like nitrogen or hydrogen. Recently single shot measurements were introduced by monitoring the spectral evolution of the signal by Lang et al. [6].

In this letter, we present high resolution studies of molecules in the gas phase by femtosecond four-wave mixing, that yield not only the rotational and vibrational constants but also directly determine the anharmonicity of rotation and vibration. We will also show that fs-FWM can be applied to determine temperatures.

2. Method

The setup for the experiments described in section 3 will be briefly outlined. The pulse from a Ti: Sapp laser system (100fs, 1mJ, 800nm) is divided into two beams. One beam pumps an optic parametric amplifier (OPA) which is frequency doubled and split into two beams to deliver pulses in a spectral range of 508-730nm at pulse energies up to 5 μJ/beam/pulse. In a two-colour experiment one beam from the OPA pumps the sample into a virtual level. The 800nm from the Ti:Sapp serves a Stokes beam and dumps the system back into the electronic ground state. The sample is then probed with the second pulse form the OPA. Phase matching condition is satisfied by using the folded BOXCARS configuration. A sketch of the experimental setup is given in Fig 1. The generated fourth beam is detected after spectral filtering through a monochromator by a near IR sensitive photomultiplier tube or an intensified CCD.

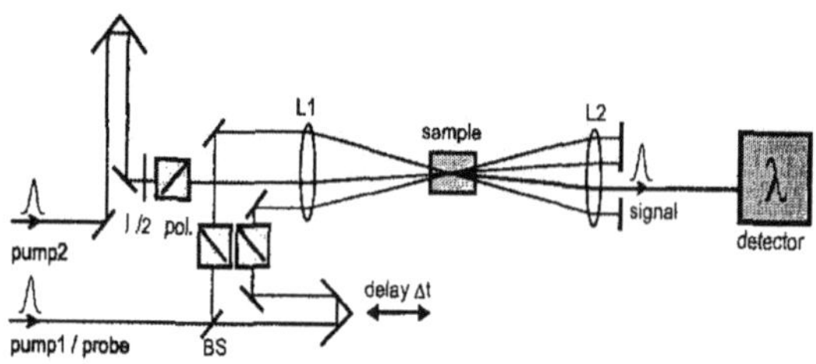

Figure 1. Setup of the FWM/CARS Experiment. Pump1 and Probe have the same wavelength, Pump2 is the Stokes beam. The sample is either a cell or a molecular beam.

The observed FWM process results from a third order polarization, primarily induced Raman transitions within the bandwidth of the laser. At short time delays an additional nonresonant contribution to the FWM signal due to the optical Kerr effect can be observed. The decay of the Kerr induced signal is fast (~fs) and will not interfere with signal originating from Raman transitions. At later times the signal shows recurrences that are due to the structural changes of the sample due to vibrations and/or due to realignment of the sample. Femtosecond CARS probes the square of the response function of the medium. This function is the Fourier sum over all contributing levels. The energy difference between two coherently excited states is added over all possible transitions, and as a result the second order and higher terms of the expansions of the solutions to the wave function can be directly measured. The signal is dependent on the delay of the probe pulse, and oscillates with a difference frequency of the states involved. The evolution of the signal is given in the equation below:

$$I_{CARS}(t) \approx \left| \sum_{J} \sum_{\Delta J=0,\pm 2} b_J^{\Delta J} \exp[(iE_J^{\Delta J}/\hbar - \Gamma)t] \right|^2$$

where:

$$\begin{aligned} E_J^{\Delta J} &= \omega_e - 2\omega_e x_e - \omega_e x_e v + \\ &B_e[(J+\Delta J)(J+\Delta J+1) - J(J+1)] - D_J[(J+\Delta J)^2(J+\Delta J+1)^2 - J^2(J+1)^2] - \\ &\alpha_e[3/2(J+\Delta J)(J+\Delta J+1) - 1/2J(J+1)] - \beta_e[3/2(J+\Delta J)^2(J+\Delta J+1)^2 - 1/2J^2(J+1)^2] \end{aligned}$$

and b are coefficients determined by the population difference, by the multiplicity of the molecular levels, by symmetry considerations (e.g. spin statistics) and by the J dependence of the Raman transition. B_e is the rotational constant, D_j the centrifugal distortion, α_e and β_e are is first and second order rotational anharmonicity, ω_e the vibrational frequency and $\omega_e x_e$ the vibrational anharmonicity.

Depending on the setup, the bandwidth and center wavelength of the laser, different molecular constants can be probed . In Table I we summarize the constants obtained from the different configurations. The rotational constants B_e and D_j can be extracted from the recurrences of a degenerate four-wave mixing experiment. All beating frequencies are a multiple of B_e. The transient shows recurrences that depend on the selection rules multiples of $1/nB_e$. Vibrational CARS yields besides the B_e and D_j the first and second order term of the rotational anharmonicity α_e and β_e. The transient is determined by the difference of the rotational spacings in the vibrational ground and excited state. It directly represents the rotational anharmonicity. The rotational contributions can be excluded by setting the polarization of the probe beam at the magic angle of 54.7 °. When more than one vibrational level of the molecule is populated we also gain information about the vibrational anharmonicity $\omega_e x_e$. Besides the rotational anharmonicity, the transient from vibrational CARS at high temperatures will also exhibit a structure that

DFWM	B, D_j, D_{jk}
CARS (T=295K)	α_e, β_e
CARS (T=1100K)	α_e, β_e, α_e, β_e

Table I. Accessible molecular constants for high resolution spectroscopy by femtosecond four-wave mixing.

represents the difference between vibrational levels that interfere and is a direct probe to the vibrational anharmonicity $\omega_e x_e$. In the following section we will present examples for the different excitation schemes and their interpretation.

3. Results and Discussion

Figure 2 shows the transients of molecules with different symmetry recorded with a degenerate setup. The center wavelength was set at 800 nm and the intensity was 80 μJ/pulse. The top transient represents oxygen recorded in a cell at 1 bar. The recurrences are separated by 1/8B.

O_2 has a $^3\Sigma_g^-$ ground state and a nuclear spin of 0. Therefore all rotational levels with even J are not populated. The analysis yields a rotational constant of 1.435 cm^{-1} and a centrifugal distortion of 4.839×10^{-6} cm^{-1}. The decay due to collisional dephasing is 76.9±4.3 ps. The second graph in Figure 1 shows the transient of CO_2 recorded at a reduced pressure of 50 mbar, because the collisional dephasing is stronger than for O_2 and only 4 recurrences could be recorded at 1 bar. As for O_2 every second level is missing (odd levels are not populated due the ground state symmetry of $^1\Sigma_g^+$ with positive parity). The second lowest graph shows the transient for the symmetric top chloroform. The larger moment of inertia, hence the smaller rotational constant results in a larger spacing of the recurrences. The last recurrence is clearly split due to centrifugal distortion as shown in the inset. The constants obtained are: B: $0.11012 \pm 5 \times 10^{-5}$, D_j: 5.07×10^{-8} and D_{jk}: -5.65×10^{-8}. The last graph in Figure 2 shows the transient of the asymmetric top SO_2. Transients of different nature can be observed; the strong transients with a spacing of 26 ps are, according to the labelling of Joireman et al. [8], J-type transients that represent the average of the

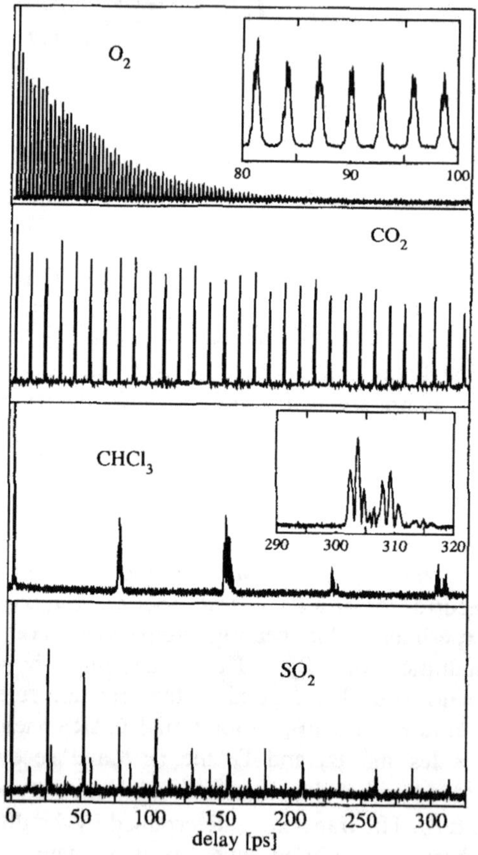

Figure 2. Transients recorded from femtosecond degenerate four-wave mixing. The transients show an increased complexity according to the structure of the molecule.

rotational constants 1/2(B+C). A second set of recurrences can be observed with a spacing of 28.6 ps, which is the C-type transient, which arises from beating of transitions in the low J limit. When increasing the pressure even A-type transients can be observed. A detailed analysis of the SO_2 transient is given in [5]. Rotational coherence spectroscopy by degenerate four-wave mixing with its, compared to fluorescence depletion, superb signal to noise ratio can be a valuable tool for determining structures of molecules.

With a two colour setup we can address vibrational modes in a molecule. In Figure 3 the transient of acetylene in a molecular beam is shown when the difference of the Pump1 and the Stokes beam is set to the symmetric C-H stretch at 3372 cm^{-1}. The lower graph is recorded with parallel polarization of all beams. Therefore we can observe rotational recurrences similar to the degenerate case. The structure to longer time delays is due to the rotational anharmonicity α_e. An analysis of the transient which extends up to 1ns yields a rotational anharmonicity for ν_1 of $\alpha_1=6.823\pm0.002\times10^{-3}$ cm^{-1} and a rotational constant of 1.178 $\pm$ 0.004 cm^{-1}. The population of the rotational levels is determined by the temperature, which is 99K. This is in agreement for an unseeded molecular beam considering the interaction of the laser was 1.5 mm downstream the nozzle. In Figure 4 a similar transient of the same molecule is shown but the wavelength is set to the C-C stretch at 1973 cm^{-1}. The probe beam was at magic angle and no rotational recurrences can be observed. The upper graph is a least square fit using equation 1 neglecting higher anharmonicity terms. The signal at long delays was too low to obtain accurate values for β_e as it is in the

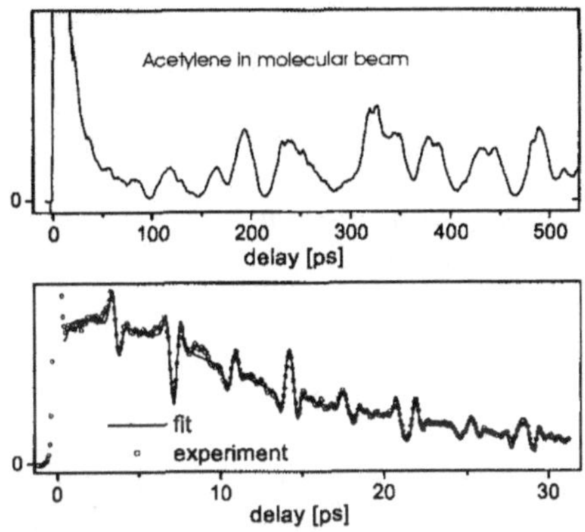

Figure 3. CARS transient of acetylene inducing the Raman transition on the symmetric C-H stretch. In the lower transient the peaks at multiples of ~4 ps are due to rotational

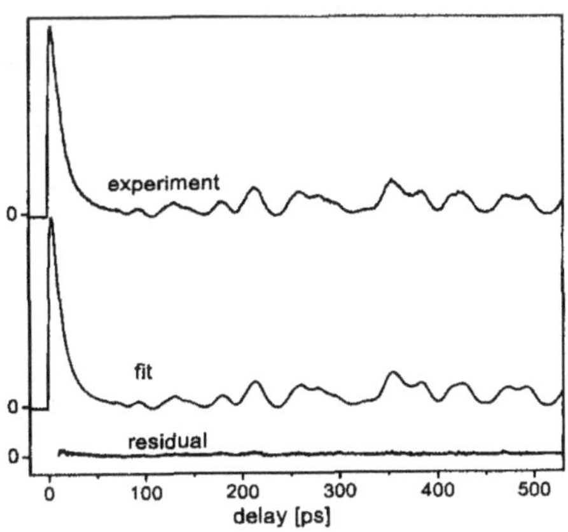

Figure 4. Vibrational CARS transient of acetylene on the C-C stretch. The polarization was at magic angle

case of H_2 [9]. The fit yields values of $\alpha_2 = 0.00616\ cm^{-1}$ and a temperature of 125 K. The difference in temperature is mainly because of different experimental conditions, but also due to different collisional cross sections of the vibrational levels.

Femtosecond CARS as a background free technique gives excellent results for high resolution spectroscopy on molecules in the gas phase. The drawback of collisional dephasing can be avoided when performing the experiments in a molecular beam. However for larger molecules the low density and volatility limit the application up to now to small polyatomics. Results for larger molecules have been reported [10] and are in progress in our laboratories.

4. Acknowledgement

We would like to thank the Swiss National Foundation and the Max Planck Society for financial support.

5. References

[1] S. Mukamel: *Nonlinear Optical Spectroscopy* (Oxford University Press, 1995)
[2] M. Motzkus, S. Pedersen and A. Zewail, *J. Phys. Chem.*, **100** (1996) 5620.
[3] P.M. Felker and A.H. Zewail, *J. Chem. Phys.*, **86** (1987) 2460.
[4] H.M. Frey, P.Beaud, T. Gerber, B. Mischler, P.P. Radi and A.P. Tzannis, *Appl. Phys. B.: Lasers Opt.*,**86** (1999) 735.
[5] H.M. Frey, P. Beaud, T. Gerber, B. Mischler, P.P. Radi, A.P. Tzannis, *J. Raman. Spectrosc.*, **31** (200) 71.
[6] T. Lang and M. Motzkus, *J. Opt. Soc. Am. B.*, accepted for publication.
[7] T. Lang, M. Motzkus, H. M. Frey and P.Beaud, *J. Chem. Phys.* **115** (2001) 5418.
[8] P.W Joireman, L.L. Connell, S.M. Ohline, P.M. Felker, *J. Chem. Phys.*, **96** (1992) 4118.
[9] T. Lang, K.-L. Kompa and M. Motzkus, *Chem. Phys.Lett.*, **310** (1999) 65.
[10] Ch. Riehn to be published

TUNNELING IN 5-METHYLTROPOLONE: COUPLING BETWEEN INTERNAL ROTATION OF METHYL GROUP AND PROTON TRANSFER

O.VENDRELL, M.MORENO, J.M.LLUCH
Departament de Química, Universitat Autònoma de Barcelona,
Bellaterra 08193, Spain
e-mail:mmf@klingon.uab.es

In this work we have theoretically studied the splittings experimentally observed in 5-methyltropolone in the first excited singlet state, S_1.Tropolone shows a vibrational level splitting due to delocalization of the wave function over the symmetric potential energy double well in the proton transfer coordinate. This splitting decreases dramatically when asymmetry is introduced. However, splitting is observed in 5-methyltropolone indicating a coupling between methyl internal rotational and vibrational degrees of freedom. We have developed a reduced-dimensionality model for the potential energy surface, *PES*, which maintains the main characteristics of the molecule and that can be fitted to electronic structure calculations. For this *PES* we have calculated stationary wave functions using a full quantum method. We have performed test calculations using different barrier heights in order to observe the splitting behaviour for this kind of molecules.

1 Introduction

It is well known that hydrogen is the lightest common element and that quantum effects will be very important in describing its behaviour. Quantum effects lead to experimentally measurable observations that cannot be explained by classical mechanics. These include *zero point energy, tunnel and diffraction effects, ...* In symmetric double wells vibrational wave functions are delocalized in both minima. The probability of finding an atom associated with such wave functions is equal at the two sides of the barrier. As lighter is the atom, as much the wave functions of the two sides interact, producing the splitting of energy levels in the same way that electronic orbitals produce a bonding and an antibonding orbital when they combine each other. This phenomenon may be thought as a particular kind of tunneling wich has been subject of much study showing that quantum effects of hydrogen atom play an important role in many mechanisms of chemical and biochemical reactions. [1]

Hydrogen tunneling is well observed in tropolone and related molecules, so they have been extensively studied theoretically and experimentally and for this reason there are many available data and information. Tropolone is a medium sized, pseudoaromatic molecule, with an intramolecular hydrogen

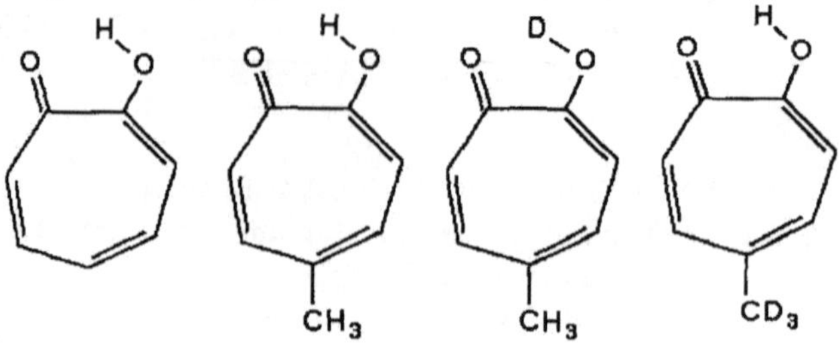

Figure 1: From left to right: Tropolone, 5-methyltropolone(OH), 5-methyltropolone(OD) and 5-methyltropolone(CD_3)

bond between two oxygen atoms in which proton transfer occurs. This proton transfer can be described by a symmetric double well. The ground electronic state, S_0, has been studied at the *ab initio* level by Rios *et al.*[2] and Redington *et al.*[3] An estimation of the splitting is obtained using a monodimensional polinominal interpolation for the reaction coordinate.[3] Paz *et al.* have performed *ab initio* calculations for both S_0 and the first excited singlet electronic state, S_1, using a bidimensional model in their splitting calculations.[4] Paz *et al.* have also studied 5-hydroxytropolone[5] and 5-aminotropolone[6] in both S_0 and S_1 states, showing the behaviour of the splitting when asymmetries are introduced in the potential double well. Their studies include also the effect of isotopic substitutions, which may play a very important role as will be seen later. Many experimental studies are also available mainly based on the obtention of electronic spectra of *jet-cooled* molecules.[7] Those studies include 5-methyltropolone.[8] In the spectra the observed splitting leading to a doublet corresponds in fact to the difference in splitting between S_1 and S_0 states. However as the splitting in the S_0 state is very small, that difference can be usually identified with the splitting in S_1. Experiments have shown that the observed splitting is smaller in 5-methyltropolone than in the parent tropolone, so that some kind of coupling exists between methyl rotation and proton transfer. It is important to note that in 5-methyltropolone experiments, the doublets obtained are centered on the methyl rotational band frequencies. Coupling between internal methyl rotation and proton transfer has been already proposed as a possible mechanism of the *Intramolecular Vibrational energy Redistribution* observed in femtochemistry proton transfer experiments.[9] In the next section we will expose a model which is able to reproduce those facts, giving insight in how the atom transfer and methyl rotation may be coupled. In order to calculate vibrational energy levels of our molecules, we have used the *Discrete Variable Representation*, *DVR*, to generate the Hamiltonian matrix of our system.

The structure of the next sections is as follows: In section 2.1 we will explain the model used. In section 2.2 we will give some details on the *DVR* method. In section 3 we will expose our results and discuse them. Section 4 will be dedicated to conclusions.

2 Theoretical Frame

2.1 Potential Energy Surface Model

In order to calculate nuclear properties of molecules, it is not enough to obtain geometries and energies of stationary points. More information is needed about the *Potential Energy Surface*, *PES*. *PES* is in general $3N-6$ dimensional where N is the number of atoms of the system. This situation plus the fact that high *ab initio* levels of calculation need to be used for treating excited states makes the full dimensional *PES* impossible to obtain. On the other hand, it would be virtually impossible to fit a full dimensional *PES* with an analitical function. For these reasons it is necessary to develop a low dimensional model for the *PES* which retains the main features of the full dimensional one.

All the process(i.e. proton transfer and methyl rotation) has been modeled using the sum of two potential functions. The total low dimensional *PES* is then:

$$V_{tot} = V(x,y) + V(x;\Theta) \tag{1}$$

Proton-transfer process has been modeled by using a function of two mass-weighted Cartesian coordinates (x,y)[11] that describes in a simple but realistic way this process between two heavy atoms. The $V(x,y)$ potential has been constructed as a symmetric double well potential. This implies that no methyl rotation has been taken into account in the calculation of geometrical parameters that fit such potential and that the minimum energy in this potential is the arithmetic mean of the molecule energies when methyl group is in its minimum and maximum rotational energy geometries. Methyl rotation is introduced, as will be seen later, via the $V(x;\Theta)$ term in order to reproduce in our $3D$ model what is happening in the full dimensional *PES*.

$$V(x,y) = V_0(x) + \frac{1}{2}m\omega^2\left(y - \frac{g(x)}{m\omega^2}\right)^2 \tag{2}$$

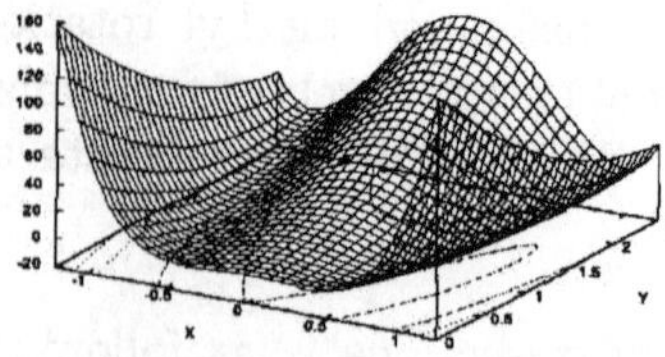

Figure 2: Function $V(x, y)$, Eq.2, representing a symmetric double well wich describes the proton transfer

where

$$V_0(x) = -ax^2 + bx^4 \tag{3}$$

$$g(x) = cx^2 \tag{4}$$

The polynomial used for $V_0(x)$ is a simple model for a symmetric double well potential. The $g(x)$ formula is taken to represent a quadratic coupling. As mass-weighted Cartesian coordinates are used, m in Eq.2 is taken to be the unity. Mass is implicit in x and y coordinates. Coordinates x and y cannot be associated directly to two concrete molecule geometrical parameters. Cordinate x represents all the geometrical changes that take place going from minimum to minimum by the shortest (i.e. linear) path, which in most of cases imply a high energy barrier. Coordinate y takes into account the geometrical rearrangements that lead the molecule along the *Minimum Energy Path, IRC.* It mainly involves the approximation of the H-atom donor and H-atom acceptor in the transition state. When the molecule goes from one minimum to the other the geometrical parameters described by the y variable recover their initial value. The function is constructed in order to preserve the energetic as well as the geometric characteristics of the full dimensional proton transfer proces. At this point the *Linear Reaction Path, LRP*, concept has to be introduced. The *LRP* is the path between two geometric structures when linear interpolation is made using mass-weighted Cartesian coordinates. [12] Its lenght is simply given by:

$$\Delta s = |\vec{x}_f - \vec{x}_i| \tag{5}$$

After calculating the lenght of the *LRP* between the two structures when proton goes from donor to acceptor oxygen atom(methyl rotation has not yet been introduced), the low dimensional $V(x, y)$ is forced to have the same distance between the two structures than the hypothetical full dimensional *PES*. The distance between the highest energy point in *LRP* and the highest energy point in *IRC* is also calculated and maintained in the model. From the energetic point of view the $V(x, y)$ potential is constructed to have the same transition state energy(taking into account the methyl rotational energy as has been already pointed) and *LRP* highest energy than the ones in the full dimensional case. With this four conditions we set the four parameters in Eq.2, 3 and 4

Methyl internal rotation $V(\Theta; x)$ has been modelled taking into account the following points: when moving in the x direction from reactants to products with the Θ coordinate frozen, the potential changes from being a minimum to being a maximum. That means that methyl rotation potential is dephased 60^0 in one side of the proton transfer barrier with respect to the other. When proton is located at reactants or products, methyl rotation has a 120^0 period. When proton is at the transition state the methyl rotation period is 60^0. This behaviour is explained on the basis that proton transfer occurs at the same time that the reorganization of the tropolone double bonds. This reorganization changes the environment of methyl group, passing from being eclipsed with a single C-C bond to being eclipsed with a double one. Such observations were made by Sekiya *et. al.* [8] and are based on experimental data and structural *ab initio* calculations of Paz *et. al.* [4] In this sense the potential has been constructed as a function of the methyl rotation angle Θ and the x coordinate. It can be thought as a parametrical dependence of Θ on x:

$$\begin{aligned} V(\Theta; x) = &\Big[\frac{V_3}{2}(1 - \cos 3\Theta) + \frac{V_6}{2}(1 - \cos 6\Theta)\Big]\Big(\frac{x + x_0}{2x_0}\Big) \\ &-\Big[\frac{V_3}{2}(1 + \cos 3\Theta) + \frac{V_6}{2}(1 - \cos 6\Theta)\Big]\Big(\frac{x - x_0}{2x_0}\Big) \\ &+\Big[\frac{V_3}{2} - \frac{V_{ts} - V_6}{2}(1 - \cos 6\Theta)\Big]\Big(\frac{x^2 - x_0^2}{x_0^2}\Big) \end{aligned} \tag{6}$$

In this expression x_0 represents the value of the x coordinate in reactants. If we take a look on this equation, we see three blocks between brackets. They are periodic functions that represent the methyl rotation. V_3, V_6 and V_{ts} are directly related to the methyl rotational barriers in minima and the transition state. The second and third blocks vanish at products, first and third blocks vanish at reactants and first and second blocks vanish at transition state. With these two parts of the low dimensional *PES* constructed, we can evaluate the behaviour of V_{tot} (Eq.1). If we keep frozen Θ coordinate and move along the x coordinate we go from a minimum to a rotational transition state. This is

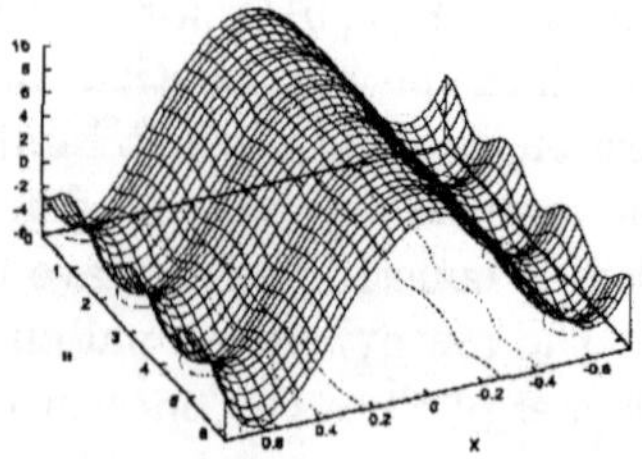

Figure 3: Function V_{tot} with y=const. Coordinate x represents the proton transfer. It can be seen that when going from reactants to products in the x coordinate, Θ passes from being a minimum to become a maximum

due to the dephasing of methyl rotation in one side of proton transfer barrier with respect to the other side. On the other hand, the symmetric path that connects two minima in V_{tot} implies a x coordinate displacement as well as a 60^0 change in Θ coordinate(see Fig. 3 for better understanding).

Note that we have not included an y dependence of Θ in our model. This may lead to think that this model misses something. However, y coordinate only implies some molecular rearrangements that make the proton transfer easier and these motions are not affecting double bonds on the ring. Double bonds will change only with the x coordinate and they are the main responsibles of the methyl dephasing during the proton transfer.

2.2 Stationary vibrational levels calculation. DVR *method*

Once we have the *PES*, we need to obtain the stationary vibrational levels for it. One possible approach to this problem is to choose an orthogonal-functions basis set $\{\phi_N\}$. For this basis set we then calculate the Hamiltonian matrix elements. We are obtaining the projection of the Hamiltonian operator in the choosen basis set:

$$H_{ij} = \langle \phi_i | \widehat{H} | \phi_j \rangle \tag{7}$$

The Hamiltonian matrix diagonalization leads to the obtention of the energy levels, eigenvalues, and the coefficients for the basis functions, eigenvectors, for each vibrational state. This approach is called *Finite Basis Representation, FBR.* When using this methodology, a large amount of computer time is consumed in the evaluation of the Hamiltonian matrix elements. Note that a numerical integral is customarily needed in order to obtain the potential part of the Hamiltonian as the integrals involved cannot be solved analitically.

An alternative to this methodology, the *Discrete Variable Representation* (*DVR*), was first pioneered by Light *et. al.*[13] Miller *et. al.*[14] developed some years later a general formulation which we are going to use. The *DVR* method is based on the obtention of the matrix elements by numerical *quadrature* over a defined grid of points $\{\vec{x_N}\}$. In the general treatment, the expressions are derived for an equally spaced grid on the x, y and Θ coordinates. Hamiltonian operator, $\widehat{H}$, is a sum of the kinetic and the potential operators:

$$\widehat{H} = \widehat{T} + \widehat{V} \tag{8}$$

so, the Hamiltonian matrix elements will be a sum of a kinetic and a potential part. In the *DVR* representation, the potential energy matrix $V_{ii'jj'kk'}$ is simply the potential energy V_{ijk} at the grid point that we are evaluating:

$$V_{ii'jj'kk'} = \delta_{ii'}\delta_{jj'}\delta_{kk'}V_{ijk} \tag{9}$$

note that there are no cross terms between different points on the grid, so that the resulting potential energy part of the Hamiltonian matrix is diagonal. The kinetic elements of the Hamiltonian matrix are evaluated as follows:

$$T_{ii'jj'kk'} = T_{ii'}\delta_{jj'}\delta_{kk'} + T_{jj'}\delta_{ii'}\delta_{kk'} + T_{kk'}\delta_{ii'}\delta_{jj'} \tag{10}$$

in the kinetic part of the Hamiltonian matrix, cross terms between different points are only possible when these two points have the same value for two of the coordinates and are different on the other. This makes the generated Hamiltonian matrix very sparse. Kinetic terms are different if they are of Cartesian or angular type:

Cartesian coordinates elements:

$$T_{ii'} = \frac{\hbar^2}{2m\Delta x^2}(-1)^{i-i'}\begin{cases} \pi^2/3 & i = i' \\ \frac{2}{(i-i')^2} & i \neq i' \end{cases} \tag{11}$$

Angular coordinate elements:

$$T_{kk'} = \frac{\hbar^2}{2m}(-1)^{k-k'}\begin{cases} \frac{N(N+1)}{3} & k = k' \\ \frac{\cos[\pi(k-k')/(2N+1)]}{2\sin^2[\pi(k-k')/(2N+1)]} & k \neq k' \end{cases} \tag{12}$$

Within this method evaluation of the matrix elements only takes a few seconds of computer time, even when we are working with a Hamiltonian matrix of thousands of rows and columns. The computer-time demanding step is the matrix diagonalization. However, as it has been previously pointed, *DVR* produces very sparse Hamiltonian matrices which are easier to be diagonalized with respect to some other kinds of matrices. On the other hand, we do not need all the eigenvalues of the matrix. For this calculations we have linked our software to the *dspevx* subroutine of the *LAPACK*(*Linear Algebra Pack*) library[15] which iteratively finds the eigenvalues starting from the lowest. This avoids the calculation of high energy eigenvalues that do not have any physical significance in our case.

3 Results and Discusion

Our calculations on this system have been oriented to stablish the degree of coupling between the proton transfer and the methyl internal rotation and how this coupling behaves in front of energetic as well as isotopic modifications. The analysis of the wave functions have shown that the first energy levels in the S_1 state correspond essentially to methyl internal rotational levels. In Table 1 these rotational levels are denoted as $0A, 1E, 2E, 3A$ following its energetic order.[8] A and E denote the rotational symmetry of the methyl group quantum state. A levels are singlets while E levels are doubly degenerate (without considering the posterior level splitting due to proton transfer). $0A$ and $1E$ levels are almost of the same energy and $2E$ and $3A$ are located 80 cm^{-1} above. We have performed three sets of vibrational level calculations using the 3D model, each set being composed of three calculations with three different energy barriers for H-atom transfer. The parameters a, b, c and ω appearing in equations 2 to 4 have been adjusted to reproduce the choosen energy barriers using also for this purpose the *LRP* distance between both minima corresponding to *CIS* geometry optimizations. In the first set of calculations the transfering atom is a proton, in the second set the transfering atom is a deuterium and in the third one the CH_3 group has been substituted by a CD_3 group. The parameters that control the $V(\Theta; x)$ (Eq.6) have been kept constant in all the calculations. With that parameters the methyl rotational barrier is 0.6 $kcal/mol$, the value obtained in the analysis of experimental spectra of 5-methyltropolone.[8]

As already noted in previous works,[5,6] the splitting is a magnitude very sensible to the barrier height and double well asymmetry. It is important to note that although splittings are smaller than in the parent tropolone molecule, they have still an appreciable magnitude. In our calculations the splitting decreases dramatically as atom-transfer barrier height increases. When the transferring atom is a deuterium, the splitting also decreases dramatically with respect to the transferring proton case. When a deuterium atom is transferred, the

Table 1: Splittings of the different rotational levels of 5-methyltropolone and related isotopomers in the S_1 state (cm^{-1})

	Transfer Barrier (kcal/mol)								
	Proton transfer			*Deuterium transfer*			*Proton transfer (CD_3)*		
Rot.Lev.	2.00	3.00	4.00	2.00	3.00	4.00	2.00	3.00	4.00
$0A$	10.31	3.48	1.26	2.61	0.52	0.12	4.20	1.46	0.69
$1E$	5.22	1.77	0.63	1.33	0.27	0.07	2.07	0.66	0.24
$2E$	10.19	3.78	1.44	2.64	0.61	0.26	8.57	2.99	1.34
$3A$	26.26	9.26	3.40	6.62	1.39	0.34	18.93	7.27	3.98

distance between both sides of the barrier in mass-weighted Cartesian coordinates becomes larger, so that tunneling decreases. A couple of $2D$ calculations has also been done to help us to understand how the tunneling process takes place. In the first one we have calculated the vibrational energy levels in a double well with an energy difference between minima equal to the methyl rotational barrier. In this case the asymmetry introduced leads to the complete disappearence of the splitting. The second one has been the calculation of vibrational energy levels on a symmetric double well corresponding to the linear reaction path between two minima of the $3D$ *PES*. This symmetric path involves a 60^0 methyl rotation, so a large distance in mass-weighted Cartesian coordinates separes both minima. In this case the splitting decreases to 17% of the value observed in the corresponding $3D$-model calculations. These results show that at least a $3D$ model is needed to deal with this system.

The results already seen show that splittings of the rotational levels calculated with the $3D$ model are greater than the splitting obtained from a $2D$ calculation involving the double symmetric well between two minima. Such observation leads to the idea that methyl delocalization over its rotational energy barrier makes the interaction of the rotational wave functions at both sides of the proton transfer barrier easier, so that delocalizaton of methyl group is facilitating the proton transfer process. To test such idea we have calculated the splitting in a 5-methyltropolone with a fully deuterated methyl group. In this case the distance between both sides of the H-transfer barrier is the same than in the parent 5-methyltropolone. The splitting, however, clearly decreases. Such observations can be explained without thinking in the methyl rotation in the classical sense but considering the methyl group as a delocalized entity over its rotational barrier. Thus, tunneling is always possible in a symmetric way with a certain *probability*, involving only proton transfer. In Table 1 it can be seen that high methyl rotational energy levels experiment a greater splitting. This is due to the larger wave function delocalization in these levels. The reason that makes E levels split less than the corresponding A levels may be found in the smaller delocalization of the degenerated E states over the methyl rotational potential. From our nuclear calculations and comparing with experimentally obtained splittings,[8] it is inferred that the proton transfer barrier for 5-methyltropolone lays in a range between 3 and 4 $kcal/mol$. The splittings for the 5-methyltropolone*(OD)* molecule cannot be resolved experimentally because they are below the resolution limit of the techniques used, about 3 cm^{-1}.[8] We have also obtained very small splittings for such molecule so that our results are in well agreement with experimental facts.

4 Conclusions

In this work we have calculated the splittings of the methyl internal rotational levels of 5-methyltropolone and related isotopomers in the first excited singlet state. We have developed a $3D$ model *PES* that keeps the main features of the multidimensional one. Those rotational levels have been calculated using the *DVR* method to obtain the Hamiltonian matrix of our system which has been diagonalized. Our results for 5-methyltropolone*(OH)* clearly agree with the experimental ones.[8] According to our calculations the splitting of the rotational levels is due to methyl group wave function delocalization over its rotational potential, being such splitting directly related to the proton transfer process. We have also shown that explicit introduction of the methyl rotation degree of freedom is essential in order to understand and calculate how methyl rotation is coupled with proton transfer. This study give insight in a phenomenon which seems to be related with the internal redistribution energy processes[9,16] in proton transfer reactions. Further work devoted to clarify the different behaviour of A and E levels as well as *ab initio* electronic structure calculations in order to fit a *PES* from first principles is now in progress in our laboratory.

Acknowledgment

Financial support from the DGESIC through Poject PB98-0915 and the use of the computational facilities of the CESCA and CEPBA coordinated by the C^4 are gratefully acknowledged.

References

1. *Proton Transfer Reactions*, ed. E. F. Caldin and V. Gold (Chapman & Hall, London, 1975); M. T. Page, *The Chemistry of Enzyme Action* (Elsevier, Amsterdam, 1974); R.P. Bell, *The Tunnel Effect in Chemistry*, (Chapman & Hall, London, 1980).
2. M. Rios, J. Rodriguez, *Can. J. Chem.* **69**, 201 (1991)
3. R.L. Redington, C.W. Bock, *J. Phys. Chem.* **95**, 10284 (1991)
4. J.J. Paz, M. Moreno, J.M. Lluch, *J. Chem. Phys.* **103**, 353 (1995)
5. J.J. Paz, M. Moreno, J.M. Lluch, *J. Chem. Phys.* **107**, 6275 (1997)
6. J.J. Paz, M. Moreno, J.M. Lluch, *J. Chem. Phys.* **108**, 8114 (1998)
7. H. Sekiya, H. Hamabe, H. Ujita, N. Nakano, Y. Nishimura *J. Chem. Phys.* **103**, 3895 (1995); R. Rossetti and L.E. Brus, *J. Chem. Phys.* **73**, 1546 (1980); T. Tsuji, H. Sekiya, Y. Nishimura, A. Mori, H. Takeshita *J. Chem. Phys.* **95**, 4802 (1991)
8. K. Nishi, H. Sekiya, *J. Chem. Phys.* **111**, 3961 (1999)
9. J.L. Herek, S. Pedersen, L.Baares, A.H. Zewail, *J. Chem. Phys.* **97**, 9046 (1992)
10. K. Fukui, *Accounts of Chemical Research* **14**, 363 (1981)
11. E. Bosch, M. Moreno, J.M. Lluch, *J. Am. Chem. Soc.* **114**, 2072 (1992)
12. W.H. Miller, B.A. Ruf, Y. Chang, *J. Chem. Phys.* **89**, 6298 (1988)
13. Z. Bacic and J.C. Light, *J. Chem. Phys.* **85**, 4594 (1986)
14. D.T. Colbert, W.H. Miller, *J. Chem. Phys.* **96**, 1982 (1992)
15. *LAPACK* Users' Guide, Second Edition, SIAM, Philadelphia, PA (1995)
16. M. Ito, *J. Phys. Chem.* **91**, 517 (1987)

Femtosecond Photodissociation of the Ethyl Radical C_2H_5

Gregor Knopp, Paul Beaud, Peter Radi, Marek Tulej and Thomas Gerber
Paul Scherrer Institute, 5232 Villigen, Switzerland
(gregor.knopp@psi.ch)

ABSTRACT

The first observation of the ultrafast H atom release of the ethyl radical is presented. Once electronically excited, C_2H_5 undergoes a rather complicated dissociation process, which is still not complete understood. With the pump probe multiphoton ionization technique we were able to show, that at least one product channel is present in the ultrafast regime (fs). Further, a second 'slower' channel in the ps time regime was observed.

1. Introduction

Alkyl radicals constitute a class of unstable hydrocarbon species with central importance in combustion processes. Especially the dynamics of the H atom loss and the building of a double bond is of major interest. As one of the simplest alkyl radicals, ethyl (C_2H_5) represents an important prototype for even more dynamically complicated radicals. The lowest accessible electronic excited states ($A^2A_1$3s-Rydberg and $A^2A_1$3p-Rydberg) of C_2H_5 tend to dissociate and lie in the ultraviolet. The first unstructured absorption band of ethyl refers to the 3s-Rydberg state, which is expected to decay very rapidly by internal conversion to the ground electronic state upon absorption of a UV photon [1]. After the excitation with a 250 nm photon, the molecule has an excess energy above the dissociation threshold of approximately 100 kcal/mol. Calculations predicted reaction rates for the H atom loss in the order of 10^{11} s^{-1} to 10^{12} s^{-1} [2]. Gilbert *et al.* [3] have measured reaction rates of $1.8*10^7$ s^{-1} by nanosecond photo excitation and ion detection of the released hydrogen atom. This discrepancy of 5 orders of magnitude between calculation and experiment is still not understood. Recently Amaral *et al.* [4] investigated the photodissociation of ethyl measuring the H atom energy-dependent angular distribution after UV excitation of the molecule and concluded that at least two channels for the dissociation exist. A 'slow' channel, which is isotropic in character for H atom release and therefore is expected to be comparable (or longer) in lifetime with a rotational period (~ ps) and a fast anisotropic channel with high kinetic energy release, suggesting a prompt and direct dissociation from the excited $^2A'(3s)$ state via its C_{2v} configuration $^2A_1(3s)$.

2. Method

We report multiphoton induced ion signals of ethyl radicals after photo excitation from the X^2A' ground electronic state into the 3s Rydberg state on a femtosecond time scale, yielding a first direct measurement of the fast dissociation process of C_2H_5. For the deliverance of the radicals, ethyliodide (C_2H_5I) is used as precursor. The direct photodissociation of $C_2H_5I \rightarrow C_2H_5 + I$ and further excitation and dissociation of $C_2H_5 \rightarrow C_2H_4 + H$ upon absorption of a 250 nm (265 nm) photon is used to extract information about the ethyl radical dynamics as depicted in Figure 1. The 250 nm (240 fs) pulse simultaneously serves as 'photolytic pump' for the mother molecule and as excitation for the radical. It is expected, that C_2H_5I dissociates rapidly (τ_1<60 fs) after photo excitation to its electronic-A state ($^1Q_1$, $^3Q_0$, $^3Q_1$ states) into C_2H_5 in its ground electronic state and an I or I* atom [5]. The branching ratio because of the potential crossing between the $^1Q_1$ and $^3Q_0$ state is ~3:1 towards the I* channel. Therefore, it has to be considered that the released ethyl radicals are energetically not cold. The 'initial' excess energy varies between 0.41 eV (I* channel/ 265 nm) and 0.78 eV (I channel / 250 nm), respectively 0.78 eV to 1.48 eV for the kinetic energy release [6]. The final excitation energy of the ethyl radical upon absorption of an additional UV photon within the same laser pulse is between 5.07 eV and 5.74 eV. The excitation energy is not high enough to excite the 3p-states of the radical.

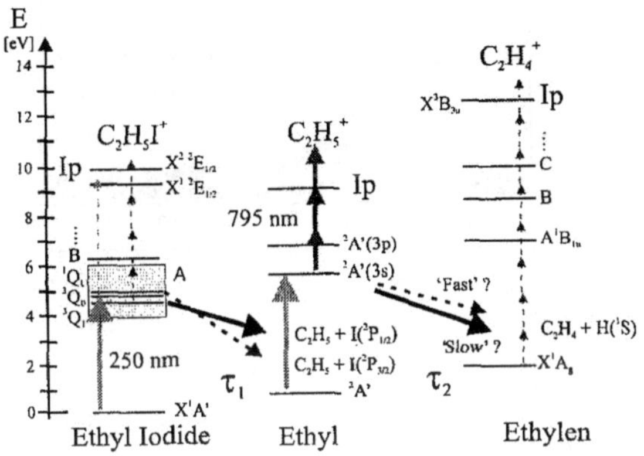

Figure 1. Energy level scheme of the C_2H_5 pump probe multiphoton ionization.

3. Results and Discussion

Figure 2 shows two typical ion signals for C_2H_5 and C_2H_5I, as function of the probe delay. The data are presented together with their least square optimised fitted curves. The fit-model interactively fits the mother and the ethyl signal, including the two coupled time constants (τ_1,τ_2). These time constants reflect the dissociation of the mother molecule into ethyl and iodine (τ_1) and the dissociation from the $^2A'(3s)$ – Rydberg state of the radical (τ_2). For a 250 nm pump pulse these decay times could be determined to $\tau_1 = 40$ fs and $\tau_2 = 370$ fs.

The strong contributions at coincidence most probably originate from excitations to the C_2H_5I ion states, followed by fragmentation to ethyl ions and neutral iodine. Considering the pump pulse intensity, a double absorption process should give a power dependence of 2, while the probe laser dependence is of the order of ~ 2 to 3. This was confirmed for delays > 0 fs. At coincidence the signal shows a linear pump and ~5 to 6 photon probe power dependency, and therefore could be well distinguished. The ethyl iodine signal itself shows a linear pump power dependency independent from the delay.

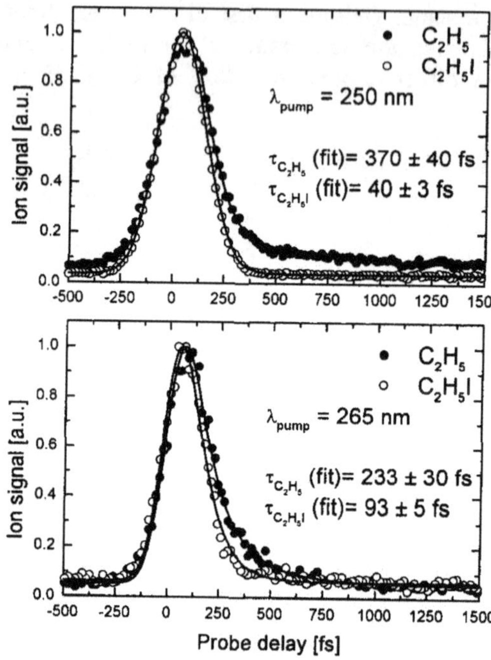

Figure 2. C_2H_5 ion signal as function of the probe delay and their best fitted curves. The time constants reflect the dissociation C_2H_5I into ethyl and iodine (τ_1) and the dissociation from the 3s–Rydberg state of the radical (τ_2).

Upon the absorption of the UV photon, the 3s state has the same classical C_s structure as the ground state ethyl and the excited molecule appears in A^2A' geometry. This state has no local minima in the vicinity of the ground state equilibrium structure. After photo excitation one proton from the CH_3 part slides to the C_{2v} bridging position over the centre of the CC-bond. The resulting A^2A_1 state is unstable with respect to H detachment and dissociates without any barrier to the C_2H_4 (X^1A_g) and H (^2S) products [7]. We believe that the τ_2 dissociation time observed in our experiments is the first direct observation of this fast dissociation channel. It can be related to the anisotropic signal contributions of the experiments from Amaral *et. al.*. [4] The 370 fs decay is much faster than the typical rotational periods of the radical (which are in the order of several picoseconds (B,C~0.73cm^{-1}, A~3.4cm^{-1})). Changing the excitation wavelength from 250 to 265 nm reduced the fitted time to τ_2 = 233 fs (τ_1 = 93 fs). In this case the final excitation of the radical is ~3000 cm^{-1} lower in energy. The molecules seem to dissociate faster with decreasing excitation energy.

Even though a good correspondence between fit and experiment has been found by using a single (τ_2) dissociation process for ethyl in the model, it might not be sufficient enough for longer delay times. It is expected that the molecule undergoes

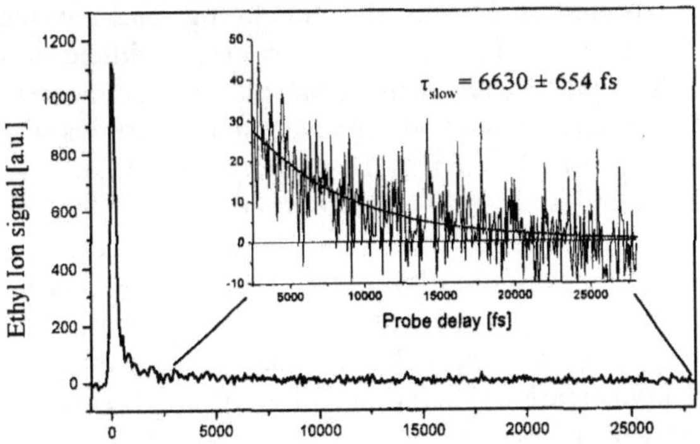

Figure 3. Ethyl ion signal from 2.5 to 28 ps probe delay. The signal shows evidence of a slow dissociation process.

an avoided conical intersection on its way to the C_{2v} configuration (between 70° and 85° degrees of CCH angle) [7]. This could fairly disturb the direct dissociation and the dissociation after internal conversion to the ground state. For example it could reflect the H atom back to its classical position after hopping onto the ground state PES. Therefore an additional second ('slower') dissociation rate has to be considered. The conversion probability is a function of the coupling strength, the energy gap between the diabatic PES's, the velocity in the tunnel region and in general is difficult in treatment. Considering the concept of transition rates in the form of the 'Golden Rule' expression [8], the overlap integral of the initial (A^2A'state) and final (X^2A'state) vibrational wave function mainly determines the rate.

Amaral *et. al.* have measured a branching ratio between their 'anisotropic' and 'isotropic' dissociation channel of 0.2 and defined them as 'fast' and 'slow' channel. This would lead to a tunnel probability of ~80%. Increasing the delay times in our experiments to several pico-seconds gives some clear evidence for an additional decay process as shown in Figure 3. Considering the high 'tunnel'- rate, this 'slow' decaying contribution appears unexpected weak in intensity. But, the dependence of the photo ionization matrix element on the geometry of the molecule is not negligible and particularly when the wave packet moves through a region of an avoided crossing, a change in the ionization probability is expected [9]. This second time constant could be fitted to τ_{slow} ~ 6-7 ps, which is comparable to the rotational periods of the molecules. The internal energy might not be completely randomised after the UV excitation and a lifetime of a few picoseconds is expected

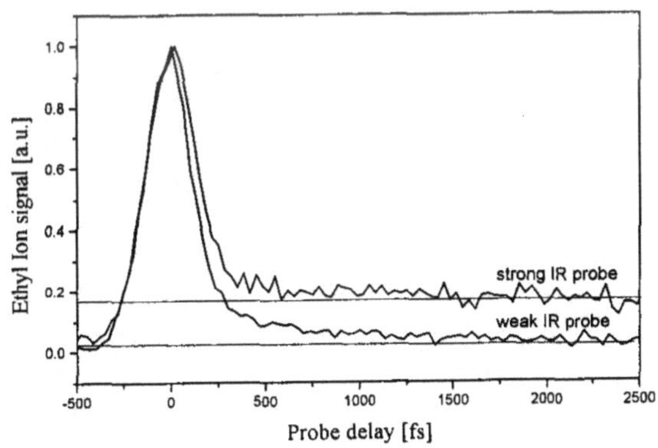

Figure 4. Ethyl ion signal for a 'weak' and a 'strong' probe pulse. A clear offset for delays > 0 fs can be observed.

[4,10]. This 'slow' time can be interpreted as time for the 'tunnelling' and moving out of the 'ionization window'. So far in our investigations, no clear evidence for life times, much longer than the rotational periods of the molecule or even close to ns in duration [3], has been found. The only indication for such 'long' lifetimes was given in experiments, we performed with high intensity IR probe pulses (Figure 4). Here a clear additional offset of the ethyl ion signal was observed. To our believe, this contribution mainly comes from ground state ethyl molecules, directly produced by C_2H_5I dissociation and probed via n-multiphoton ionization (n~5). But, it is also plausible that radicals with completely randomised internal energy after the UV excitation contribute to the offset. For a more stable statement concerning the 'slow' decay processes additional investigations including different precursors are required. However, our results confirm the presence of a fast dissociation channel with a dissociation rate in the order of $\sim 3*10^{12}$ s^{-1} and a 'slower' process of $\sim 2*10^{11}$ s^{-1} as proposed in Ref. [2,4,7].

4. Acknowledgement

We would like to thank Ingo Fischer for helpful discussions and the Federal office of Energy for financial support.

5. References

[1] M. Blomberg, B. Liu, *J. Chem. Phys.*, **83** (1985) 3995.
[2] W. Hase, B. Schlegel, V. Balbyshev, M. Page, *J. Chem. Phys.*, **100** (1996) 5354.
[3] T. Gilbert, T. Grebner, I. Fischer and P. Chen, *J. Chem. Phys.*,**110** (1999) 5485.
[4] G. Amaral, K. Xu, and J. Zhang, *J. Chem. Phys.*, **114** (2001) 5164.
[5] N. Knoblauch, A. Strobel, I. Fischer and V. Bondybey, *J. Chem. Phys.*, **103** (1995) 5417.
[6] W. Kang, K. Jung, D. Kim, K. Jung and H. Im, *Chem. Phys.* **196** (1995) 363.
[7] A. Zyubin, A. Mebel and S.Lin, *Chem. Phys. Lett.*, **323** (2000) 441.
[8] B.Bixon, J. Jortner, *J. Chem. Phys.*, **176** (1993) 467.
[9] Y. Arasaki, K. Takatsuka, *J. Chem. Phys.*, **112** (2000) 8871.
[10] W. Hase, D. Buckovski, K. Swamy, J. Phys. Chem., **87** (1983) 2754.

FEMTOSECOND TIME-RESOLVED ANALYSIS OF THE LASER-INITIATED INTRACLUSTER REACTION IN BA...FCH_3

V. STERT, H.-H. RITZE, W. RADLOFF*
Max-Born-Institut für Nichtlineare Optik und Kurzzeitspektroskopie
Max-Born-Str. 2A, D-12489 Berlin, Germany

K. GASMI AND A. GONZALEZ-UREÑA
Instituto Pluridisciplinar,
Universidad Complutense de Madrid,
Juan XXIII-1^0, E-28040 Madrid, Spain

The model of the intracluster reaction in Ba...FCH_3 initiated by a femtosecond laser pulse at 618 nm was confirmed applying two different probe pulse wavelengths at 400 nm and 267 nm for detection.

1 Introduction

The photoinduced intracluster reaction Ba...FCH_3 $\rightarrow$ BaF + CH_3 is a typical bimolecular reaction of femtochemistry where the reactants are combined in a weakly bound Van der Waals complex. In this way the start time of the photochemical reaction is exactly defined by the femtosecond start pulse [1] in contrast to experiments with crossed molecular beams. The intracluster reaction in Ba...FCH_3 is initiated by resonant electronic excitation of the cluster at 618nm. This resonant wavelength as well as the ionization potential IP= 4.5 eV of the complex have been determined by depletion spectroscopy with nanosecond laser pulses [2]. After initiation the excited valence electron of Ba acts as the harpoon to catch the F atom of the fluoromethane component. A necessary condition for this harpooning process is a stretch of the C-F bond in the complex by vibrational excitation. Only in this way a positive electron affinity of the F atom can be realized. The formation of the stable BaF product can be followed on the femtosecond time scale by ionization with a time-delayed probe pulse in a pump-probe experiment with femtosecond laser pulses.

In our experiments the intracluster reaction was initiated by resonant excitation of the Ba...FCH_3 complex to its electronic $\tilde{A}$ state with the pump

*AUTHOR TO WHOM CORRESPONDENCE SHOULD BE ADDRESSED.

pulse at λ_1 = 618 nm (2eV). For the detection of the BaF reaction product and the Ba fragment time-delayed probe pulses of two different wavelengths $\lambda_2^{(1)}$ = 400 nm (3.1eV) and $\lambda_2^{(2)}$ = 267 nm (4.65eV) were applied. Preliminary results for each of the pump-probe combinations have been already published in previous papers [3, 4]. In the present paper the experimental results of both pump-probe experiments will be directly compared and discussed on the basis of the reaction scheme recently developped by our theoretical and experimental studies [5].

2 Experimental

The experimental setup has already been described in detail [3, 5] and will be characterized here only briefly. The weakly bound Ba...FCH_3 complexes were formed in a molecular beam by adiabatic expansion of a gas mixture containing laser desorbed Ba vapor and CH_3F molecular gas (10%) in He (1 bar) carrier gas. To enlarge the cluster concentration we have cooled the interaction channel of the vaporization source down to about -130^{0}C. The energy of the focused radiation (diameter: 0.2...0.3mm) of the vaporization laser (SHG of a Nd:YAG laser) is restricted to low values (0.8mJ) in order to reduce chemical reactions in the gas mixture near the laser heated Ba rod. The molecular beam is crossed by two weakly focused copropagating laser beams in the interaction region of a time-of-flight (TOF) mass spectrometer. The laser system used is a commercial Ti:sapphire laser and amplifier system (Clark MXR) tuned to 800nm which is combined with a commercial OPA system (Clark I-GOR-SHG). The second harmonic of the OPA signal wave is tuned across the resonance of the Ba...FCH_3 complex at 618nm and is used to pump the clusters. The second or third harmonic of the Ti:sapphire laser at $\lambda_2^{(1)}$ =400nm or $\lambda_2^{(2)}$ =267nm is applied to probe the excited clusters and the products by ionization. The width (FWHM) of the laser pulses is about 130fs.

The ion signals are detected by a microchannel plate detector in the TOF mass spectrometer, digitized by a fast digital oscilloscope (Tektronix TDS 520A) and stored and processed by a PC.

A standard delay line is used to scan the delay time between the pump and the probe pulses. At a repetition frequency of 33 Hz the mass spectra are accumulated typically for 50 laser pulses at each delay time τ and averaged over 50 up- and down scans of the delay line.

3 Results and discussion

In Figs. 1a,b the time-dependent ion signals of Ba...FCH_3 and BaF are shown as a function of the delay time τ between the pump pulse at 618 nm and the probe pulse at 400 nm. At this probe wavelength, i.e. for a photon energy of 3.1 eV, no transient Ba^+ signals were observed. The parent ion signal can be fit very well by a single-exponential decay with the time constant of 270 fs. The solid line in Fig. 1a as well as those for all ion signals in Fig. 1 represents the theoretical fit curve obtained by describing the resonant coherent excitation of the parent complex by optical Bloch equations. For $\tau < 0$ the parent ion signal is zero because the probe photon energy is below the ionization potential of Ba...FCH_3. The BaF^+ ion signal (Fig. 1b) is superimposed by two contributions: the contribution (2) due to fragmentation of the parent ion which resembles the parent ion signal but is shifted somewhat to larger delay

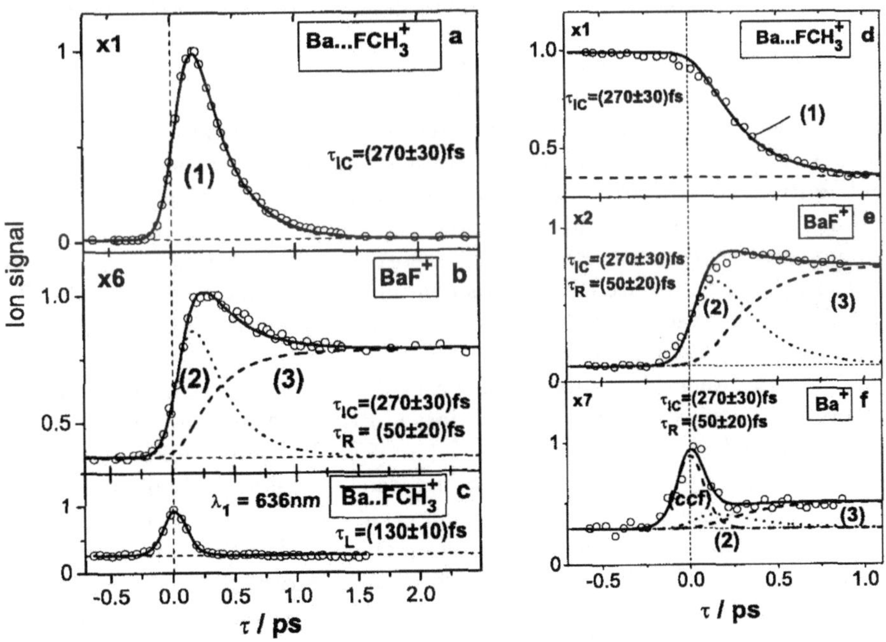

Figure 1. Ion signals versus delay time τ between the pump pulse at 618 nm and the probe pulse at 400 nm (a, b, c) or 267 nm (d, e, f).

times, and the contribution (3) which describes the formation of the stable product. The detailed analysis [5] shows that the BaF product is formed in two successive steps with the time constants of 270fs and 50fs.

The accuracy of the time analysis is ensured by the exact knowledge of the delay time zero. Here the time zero is determined by the weak nonresonant pump-probe signal of the parent complex (Fig. 1c) measured at the detuned pump wavelength of 636nm which represents the cross correlation function (ccf) of the laser pulses used.

In Figs. 1d,e,f the time-dependent ion signals of Ba...FCH_3, BaF and Ba are given as obtained for the identical pump wavelength of 618 nm but the shorter probe wavelength of 267 nm. The higher probe photon energy of 4.65 eV leads to the ionization of the parent complex by single-probe-photon absorption ($h\nu_2^{(2)}$ >IP) and, thus, to the decay of the ion signal after irradiation with the pump pulse at $\tau = 0$. The decay time obtained by the fit procedure agrees with the result of Fig. 1a within the limits of the experimental error.

The BaF^+ signal (Fig. 1e) is superimposed by the two contributions (2) and (3) which are interpreted analogously as for Fig. 1b. The time constants derived from the fit procedure do not differ from those in Fig. 1b.

At the probe photon energy of 4.65 eV a time-dependent Ba^+ signal is observed which is superimposed by three contributions (Fig. 1f): the contributions (2) and (3) show the identical time dependence as the corresponding contributions to the BaF^+ signal (Fig. 1e), thus, demonstrating their identical origin. The third contribution denoted ccf which is symmetrical to the delay time zero reflects the nonresonant absorption of one pump and one probe photon by free nonexcited Ba atoms contained in the molecular beam. This contribution is absent for the lower probe photon energy of 3.1 eV because the ionization potential of Ba is 5.21 eV.

The experimental results displayed in Fig. 1 are in full agreement with our present knowledge of the laser initiated intracluster reaction in Ba...FCH_3 as schematically represented in Fig. 2 (cf. [5]). The initial excitation of the parent complex by the pump photon $h\nu_1$ leads to the vibrationless electronically excited Ã state which decays by internal conversion with the time constant τ_{IC} = 270 fs to the next lower electronic Ã' state [6]. The latter state is vibrationally excited which (a) leads to the dissociative ionization of the parent complex and (b) is sufficient to overcome the reaction barrier of the harpooning reaction which afterwards proceeds in competition to the non-reactive fragmentation channel with the resultant time constant τ_R = 50 fs [5]. The vibrationally excited parent ion resulting from ionization of the (Ã', v≠ 0) state represents the common origin for the contributions (2) of the

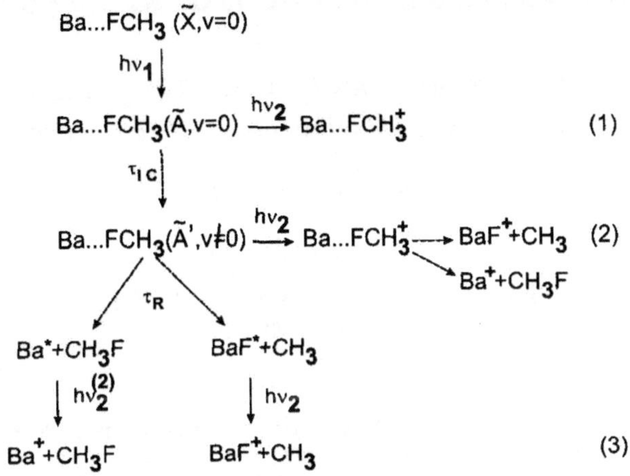

Figure 2. Scheme of reaction initiated by $h\nu_1$ and probed by $h\nu_2$.

BaF^+ signal (Fig. 1e) and of the Ba^+ signal (Fig. 1f). The stable Ba^+ ion signal representing the contribution (3) in Fig. 1f can only be observed for the higher probe photon energy $h\nu_2^{(2)} = 4.65$ eV.

Because the time constant τ_R is much smaller than the conversion time τ_{IC} we obtained an effective BaF formation time only slightly larger than the time constant τ_{IC}. Hence, in conclusion, the internal conversion represents the rate-determining process of the intracluster reaction in Ba...FCH_3.

Acknowledgments

Financial support by the Deutsche Forschungsgemeinschaft through Sonderforschungsbereich 450 is gratefully acknowledged.

References

1. A.H. Zewail, J. Phys. Chem. **100**, 12701 (1996).
2. S. Skowronek et al., J. Phys. Chem. **A101**, 7468 (1997).
3. V. Stert et al., Phys. Rev. **A59**, R1727 (1999).
4. P. Farmanara et al., Chem. Phys. Lett. **304**, 127 (1999).
5. V. Stert et al., Phys. Chem. Chem. Phys. **3**, 3939 (2001).
6. V. Stert et al., Chem. Phys. Lett. **337**, 299 (2001).

FIRST EXPERIMENTAL OBSERVATIONS OF RYDBERG SERIES IN C_{60}

M.BOYLE, K.HOFFMANN, C.P.SCHULZ, I.V. HERTEL
Max Born Institut, Max Born Str. 2A, D-12489 Berlin, Germany
E-mail: boyle@mbi-berlin.de

E.E.B.CAMPBELL
Göteborg University and Chalmers University of Technology, SE-41296 Göteborg Sweden

Photoelectron spectra of C_{60} taken with femtosecond laser pulses have illuminated a highly reproducible structure, which has been attributed to Rydberg series. Variations of experimental parameters have yielded new information about the population and the emission from these states. Theoretical modeling has been made using a jellium-like potential and compared to the experimental results.

1 Introduction

Fullerenes are extremely interesting model systems for studying the dynamics and energy coupling in relatively complex systems with a high degree of symmetry. Time-of-flight (TOF) photoelectron spectroscopy of C_{60} in the gas phase has provided many important clues about the ionization process occurring on ultrafast laser irradiation. The pulse duration dependence of the photoelectron spectra (PES) has shown three distinct regimes of ionization: a prompt, direct regime and two thermal regimes [1]. In previous experiments, the pulse duration was varied between 25 fs and a few ps by temporally chirping the pulse. Below 70 fs, the ionization is predominantly multi-photon in character, and is characterized by above threshold ionization (ATI) peaks in the photoelectron spectra. The first thermal regime occurs with intermediate pulse durations between ca. 200-700 fs, and the photoelectron spectra (PES) are characterized by a predominant statistical emission from a hot electron bath. In the final regime with still longer pulses, the electronic excitation couples to the vibrational excitation and delayed ionization [2] is observed. The electron distribution is still thermal in this regime, however it possesses a much lower temperature [3].

It has proved very difficult to obtain high-resolution absorption spectra of C_{60}, and there are practically no available resolved spectra of highly excited states in molecular beams[4-7]. This is a result of the high vibrational excitation of C_{60} from oven evaporation, which has proven to be difficult to efficiently cool [15,16].

In the following, the experimental section will provide information about the population and electron emission mechanisms under variable experimental parameters. The modeling will then be described and compared to the experimental results.

2 Experimental

The apparatus has been described in detail elsewhere [1,8]. Briefly, the main excitation source was a Ti:Sapphire regenerative amplifier laser system providing a bandwidth-limited pulse duration of 180 fs (energy bandwidth ~ 10meV). We used both the fundamental (800 nm) and the second harmonic (400 nm) for the excitation of C_{60}. The pulse duration was controlled by detuning the compressor in the amplifier. An effusive C_{60} beam was provided by resistively heating a ceramic oven to 500 °C and was perpendicular to both the laser beam axis and the electron and ion TOF detection axes. The electron spectrometer and laser intensity at the focal spot were calibrated using Xenon.

At relatively low intensities (10^{11}-10^{12} W/cm^2), an electronic structure emerged from the thermal background, as can be seen in the kinetic energy plots of Figure 1.

Column (a) of figure 1 shows the resultant kinetic energy electron spectra for irradiation with a 1.5 ps laser pulses at 800 nm. The intensities are displayed on the far right hand side. As the intensity is decreased, the temperature and thus the thermal background is reduced, which is seen clearly by the emergence of regularly spaced ATI peaks separated by the energy of one photon in the spectra at lowest intensity. The column (b), on the right, are enlargements of the energy region between 0 and 2 eV (the first ATI peak), and here the distinctive electronic structure of the Rydberg series is clearly seen. The four most prominent peaks, which lie in the region between the vertical lines, will be called the 'main Rydberg peaks' in this article.

Two questions that immediately arise about these features are: 'How are the states populated?' and 'How is the electron emitted?'.

To answer these questions, four different systematic experiments have been performed to date, where the polarization, chirp, laser bandwidth, and wavelength were varied. Each result yields a significant piece of information towards solving the puzzle. All experiments were performed at 800 nm unless otherwise noted.

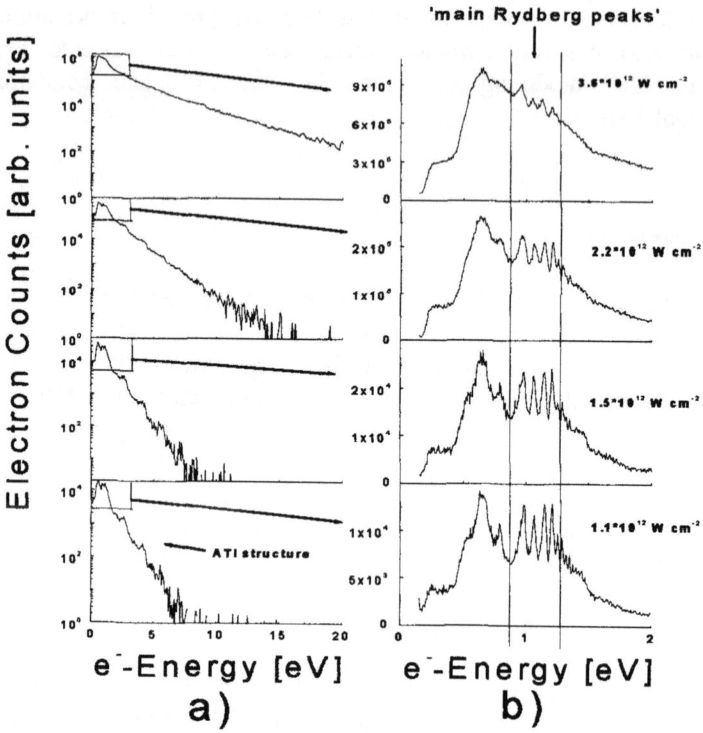

Figure 1. Electron Kinetic energy spectra for 1.5 ps pulse duration at 800 nm. a.) Shows a large range of electron kinetic energies. b.) a closer look at the region between 0 and 2 eV clearly showing the Rydberg structure.

Direct electron emission occurs most probably along the laser electric field direction, while thermal electronic emission is isotropic. Thus, it is expected when the plane of polarization is perpendicular to the detector, the resultant spectra will be primarily of thermal nature, however, when the polarization is parallel to the detector, both thermal and direct emission will be observed. The result from this experiment, Figure 2, indicates that the Rydberg structure develops from a direct emission of the electrons, which further implies that the electrons are emitted during the laser pulse.

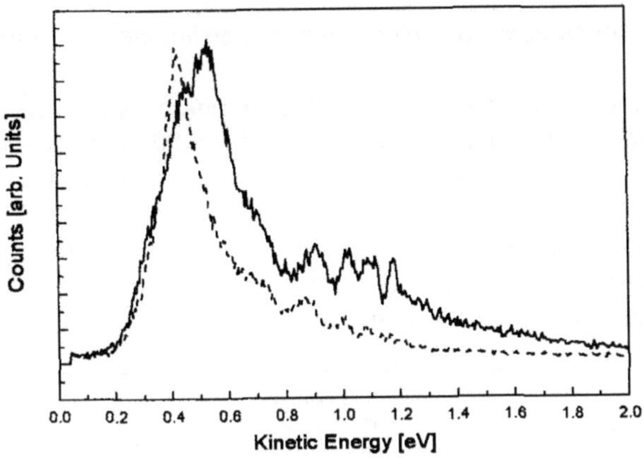

Figure 2 Polarization dependence on the Rydberg Structure. The solid and dashed lines denote polarization oriented parallel to detector axis and perpendicular axis, respectively.

The second confirmation of direct emission within the pulse duration was found by comparing the PES resulting from a positively and negatively chirped pulse. The energy of a blue photon is greater than the energy of a red photon, and thus results in a higher electron kinetic energy of the emitted electron. The positively chirped pulse, in which the red frequencies lead the blue frequencies, leads to higher kinetic energies by an amount equal to the energy bandwidth of the laser. This is consistent with a direct emission of electrons caused by single-photon ionization at the end of the laser pulse.

In the third experiment, three different lasers with varying bandwidth (10meV, 20meV, and 85meV) were used to understand the effect of bandwidth on the photoelectron spectra. As the bandwidth was increased, the Rydberg peaks became less resolved. This is, of course, expected since the measured energy width is only a convolution of the energy width of the decay time and the energy width of the laser, $\Delta E_{measured} = \Delta E_{laser} \otimes \Delta E_{lifetime}$, however, two notable pieces of information resulted from this experiment. The first is that the width of the measured peaks from the highest energy levels was as wide at the energy width of the laser for those states lying closest to the ionization potential (i.e. BE < 0.5 eV), thus the energy width of the state is narrow in comparison to the laser, i.e $\Delta E_{measured} \approx \Delta E_{laser} \Rightarrow \Delta E_{laser} \geq \Delta E_{lifetime}$.

This indicates relatively long-lived states of 400 fs or longer. The other piece of information acquired, in combination with the chirped pulse and polarization experiments, is that the complete process of population and electronic emission occurs on a very fast time scale (within the scale of the laser pulse). The 'main Rydberg peaks' are clearly resolvable down to pulse durations as short as 100 fs, and residual structure can be determined to the shortest pulse durations used of 30 fs [9].

In the final experiment, the second harmonic was used to excite the C_{60}. To properly compare the two different wavelengths of 800 nm and 400 nm, the kinetic energy converted to a binding energy assuming a 1-photon ionization of the electronically excited state at the end of the pulse (i.e. BE = hν–KE). This conversion does not include correction for emission resulting from a higher number of absorbed photons. The structure for binding energies less than 1 eV is very similar for the two wavelengths.

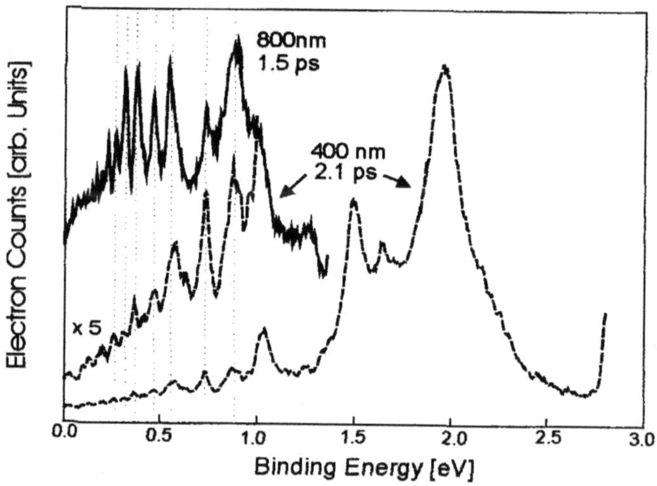

Figure 3. Comparison of wavelengths 800 nm (solid line) and 400 nm (dashed lines). The BE was calculated assuming a 1 photon transition.

The spectral peaks were fit to Lorentzian curves and the peak positions were then compared to the theoretical model as discussed in the following section.

3 Modeling of C_{60} Rydberg States

To aid in the understanding of these spectra, we have solved the Schrödinger equation [10] for the bound energy levels of C_{60} using a simple one electron model with a jellium-like potential [11,12]. The binding energy of the HOMO (n=2, l=5) was found to be 7.58 eV, in good agreement with the ionization potential of C_{60}. In Fig.4, the three best fitting angular momentum series are shown together with the energetic position of the Lorentzian fits of the experimental data.

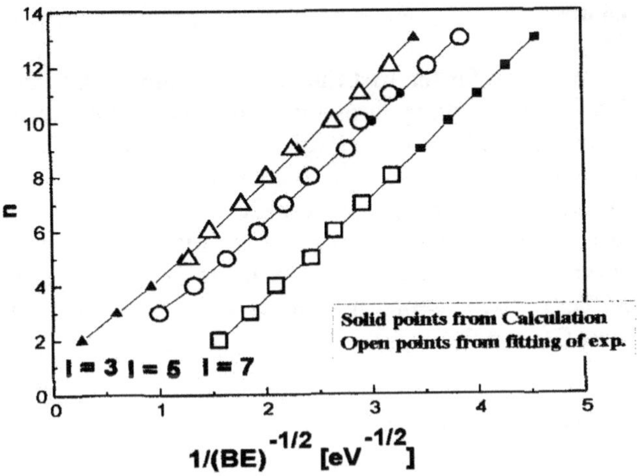

Figure 4. Comparison of calculated binding energies and experimental peak positions extracted from the Lorentzian fits.

The three series are associated with the angular momentum quantum numbers of l = 3, 5, and 7 (f, h, j states). The HOMO electron is in the l = 5 (h shell) orbital. Following angular momentum conservation rules, the observed population of the Rydberg series indicate the absorption of an even number of photons (e.g. 4 photons for 800 nm or 2 photons for 400 nm). However, it is not as simple as a pure 4- (or 2-) photon absorption process because it is impossible to populate the diversity of states seen in the photoelectron spectra resonantly. The energy range covered by the states is 1-2 eV, which is far too broad to be covered by the energy bandwidth of the laser (10meV) or the AC - Stark shift (60 meV for these experiments).

The states we are exciting are in the intermediate energy range where neither the Born Oppenheimer nor the Inverse Born Oppenheimer approximation holds, but rather where states of the same total energy are mixed [13]. Thus, C_{60} molecules containing a vibrational content between 0 and the difference between the IP and the absorption of 4 photons (7.6 eV-6.2 eV)= 1.4 eV can then access all Rydberg states with binding energies less than 1.4 eV. These states can then be one photon ionized within one pulse duration. Molecules with a higher vibrational energy than 1.4 eV will be excited above the ionization level and contribute to the broad background underneath the Rydberg structure.

This mechanism is clearly different from that suggested by Schick, et al. [14] for similar observations with phenol. Their experiments show a similar structure in the PES, however the levels are assumed to be populated by non-radiative decay from resonantly excited super excited states.

4 Conclusions

We have observed for the first time rich, structured electronic spectra for C_{60} using ultrafast photoelectron spectroscopy, which is attributable to Rydberg energy levels. Experimental and theoretical calculations, which are in good agreement, indicate the population mechanism is predominantly via a 4-photon process (fro 800 nm). Variation of experimental parameters shows the electron emission to be direct and fast, happening within the pulse duration. Future studies will focus on two-color pump-probe experiments to gain information on the lifetime and dynamics of the Rydberg Series in C_{60}, as well as comparison to other model systems such as C_{70} or NC_{59}.

5 Acknowledgements

We thank R.D. Levine and R. Stoian for most helpful discussion and F.Schöberl for supplying his Mathematica program to solve the Schrödinger equation. Financial support was received from the Sonderforschungsbereich 450, the Access to Large Scale Facilities Programme of the European Community, the EUIHP 'Cluster Cooling' Network, and Verenskåpsredet are gratefully acknowledged.

References

1. Campbell, E.E.B, et al., PRL **84**, 2128 (2000).
2. Campbell, E.E.B and R.D. Levine, Annu. Rev. Phys. Chem. **51**, 65 (2000).
3. Campbell, E.E.B., Hoffmann, K., and Hertel, I.V., Euro. Phys. J. D, in press
4. Brady, B.B. and Beiting, E.J., J. Chem. Phys. **97**, 3855 (1992).
5. Dai, S., et al., J. Chem. Phys. **101**, 4470 (1994).
6. Yasumatsu, H., et al., J. Chem. Phys. **104**, 899 (1994)..
7. Coheur, P.F., et al., J. Phys. B **29**, 4987 (1996).
8. Tchaplyguine, T., et al., J. Chem. Phys. **112**, 2781 (2000).
9. Boyle, M. et al., PRL submitted.
10. Lucha, W. and Schöberl, F., Int. J. Mod. Phys. C **10**, 607 (1999).
11. Puska, M.J. and Niemenen, R.M., Phys. Rev. A **47**, 1181 (1993).
12. Frank, O. and Rost, J.-M., Chem. Phys. Lett. **271**, 367 (1997).
13. Remacle, F. and Levine, R.D., Int. J. Quant. Chem. **67**, 85 (1998).
14. Schick, C.P. and Weber, P.M., J. Phys. Chem. A **105**, 3725 (2001)
15. Haufler, R.E., et al., J. Chem. Phys. **95**, 2197 (1991).
16. Hansen, K., et al. ZPD, **42**, 153 (1997)

TESTING THE APPLICABILITY OF CLASSICAL MECHANICS TO DESCRIBE THE ULTRAVIOLET PHOTOLYSIS OF HYDROGEN-BONDED CLUSTERS

J.C. JUANES-MARCOS, R. PROSMITI, and A. GARCÍA-VELA

Instituto de Matemáticas y Física Fundamental, C.S.I.C., Serrano 123, 28006 Madrid, Spain

Quantum interference effects are found to play an important role in the ultraviolet photolysis of $Ar-HCl$. The applicability of classical dynamics simulations to this process is tested against an exact wave packet treatment. The photolysis product fragment distributions are investigated for different excitation energies of $Ar-HCl$ in the range of the absorption spectrum. The validity of the classical description is discussed in the light of the results.

1 Introduction

Ultraviolet (UV) photolysis of small hydrogen-bonded clusters of the type Rg_n-HX, $(HX)_n$, and Rg_n-H_2Y (X=halogen, Rg=rare gas, $Y=O,S$) has been the subject of an increasing interest in the last years. On the one hand, experimental evidence has been found of formation of radical complexes like $Ar-I$,[1] $I-HI$,[2] $Cl-HCl$,[3] and $Ar-SH$ and Ar_2-SH,[4] as products of the UV photolysis of $Ar-HI$, $(HI)_2$, $(HCl)_2$, and Ar_n-H_2S ($n \leq 2$) clusters, respectively. On the other hand, these systems allow one to investigate the possible manifestations in small clusters of effects typical of condensed matter, like the cage effect[5–7].

From the theoretical point of view, the $Ar-HCl$ cluster has been the most extensively studied system[7–11]. It was found[10,11] that, upon UV photodissociation, this system can follow two fragmentation pathways,

$$Ar-HCl+h\nu \rightarrow H+Ar-Cl, \tag{1}$$

$$Ar-HCl+h\nu \rightarrow H+Ar+Cl. \tag{2}$$

The partial fragmentation (PF) pathway of Eq.1 occurs through direct dissociation of the H fragment. The total fragmentation (TF) pathway of Eq.2 involves (at least) one collision of the recoiling hydrogen with Ar, and subsequent energy transfer to the $Ar-Cl$ bond. Energy-resolved studies[11] have shown that the TF pathway is dominant at most of the excitation energies of the $Ar-HCl$ absorption spectrum. It has been recently found that quantum interference effects between fragment states populated by the two fragmentation pathways play a central role in the photolysis of $Ar - HCl$[12].

Quantum-mechanical calculations for the UV photolysis of $Rg - HX$ triatomic clusters are very demanding, due to the double-continuum character of the TF pathway of Eq.2. Such calculations become extremely costly, or even out of reach for larger systems. A less demanding classical description could be applied to clusters larger than triatomic ones. However, the presence of interference effects poses some doubts about the validity of such a description. In this sense, it is worth investigating the validity of classical simulations of the photolysis process, and this is the aim of the present work. To this purpose, energy-resolved classical distributions of the fragments produced after UV photodissociation of $Ar - HCl$ are tested against quantum-mechanical ones.

2 Methodology

Photolysis of $Ar - HCl$ occurs upon excitation of the HCl chromophore from its ground electronic state $X^1\Sigma^+$ to the repulsive excited state $A^1\Pi$. Such an excitation is simulated by means of a vertical Franck-Condon transition of the calculated vibrational ground state of the cluster in the $X^1\Sigma^+$ potential surface to the upper $A^1\Pi$ surface. In the excited surface the H atom dissociates in a time scale of a few tens of fs. The simulations have been carried out for a time $t_f = 80fs$. Quantum mechanically the photolysis process is simulated by means of a 3D wave packet treatment (assuming $J = 0$). Product distributions are obtained by projecting the asymptotic wave packet out onto the fragment states corresponding to different excitation energies of the parent cluster in the region of the absorption spectrum. Detais on the quantum calculations have been given elsewhere[10,11].

Initial conditions for the classical trajectory simulations were sampled using the quantum probability distribution associated with the initial wave packet. A sampling method proposed recently[13] was applied where an initial quantum phase-space distribution is defined as $|\Phi\tilde{\Phi}|^2 f$. In the above definition Φ and $\tilde{\Phi}$ are the initial state in coordinate and in momentum space, respectively, and f is the coordinate dependence of the volume element (in the Jacobi coordinates used $f = \sin\theta$, being θ the Jacobian angular coordinate). The initial conditions sampled are required to correspond with specific excitation energies E of the parent cluster in the excited potential surface, within an energy dispersion $\Delta E = \pm 10cm^{-1}$. This ensures that the classical fragment distributions are energy-resolved and comparable to the quantum-mechanical ones. The demanding criterion of such a small energy dispersion is needed to obtaind meaningful state distributions of the $Ar - Cl$ fragment, since their energy domain is $\simeq 200cm^{-1}$.

3 Results and Discussion

Classical product fragment distributions were calculated for different excitation energies of the parent cluster in the region of the absorption spectrum (which covers the range of energies $E = 1 - 6eV$, where the limit $E = 0$ corresponds to three separated atoms). Specifically, the excitation energies $E = 1.60eV$, $1.87eV$, $2.87eV$, and $4.97eV$ were investigated, for which 1105, 11960, 42867, and 6013 trajectories were integrated, respectively. The number of initial conditions corresponding to a specific energy is related to the probability given by the quantum phase-space distribution for that energy. In the following the classical distributions associated with PF and TF products are discussed and compared to quantum-mechanical ones.

3.1 Partial Fragmentation Distributions

Quantum and classical rovibrational state distributions of the $Ar - Cl$ radical fragment are shown in Fig.1 for three different excitation energies of $Ar - HCl$. Population at energies $E_{v,j} < 0$ is associated with $Ar - Cl$ bound states, while population at $E_{v,j} > 0$ corresponds to quasibound states of $Ar - Cl$ (supported by rotational barriers). Good qualitative agreement is found between the classical and quantum distributions for low energies $E < 1.9eV$. Such an agreement deteriorates gradually with increasing E, as shown by the distributions corresponding to $E = 2.87eV$.

For low energies ($E < 2eV$), as E increases the $Ar - Cl$ population shifts gradually towards the highest quasibound states[11]. This is expected since the energy available for the $Ar - Cl$ fragment increases with E, and the highest quasibound states can accommodate the largest amount of the available energy. Such a behavior is reproduced by both the quantum and the classical descriptions, giving rise to qualitatively similar distributions.

As E increases ($E > 2eV$) the above trend is inverted in the quantum-mechanical distributions, and the population in quasibound states (particularly in the highest ones) gradually decreases. It has been found[12] that this vanishing of population is due to interference between PF states (both bound and quasibound ones) and TF states of the fragments, caused by the first collision of H with Ar. Upon this collision, population initially (before the collision) in PF states is promoted to TF states. Population in the highest quasibound states of $Ar - Cl$ is most likely to be excited to three-fragment states, which explains the vanishing of this population found in the quantum distributions.

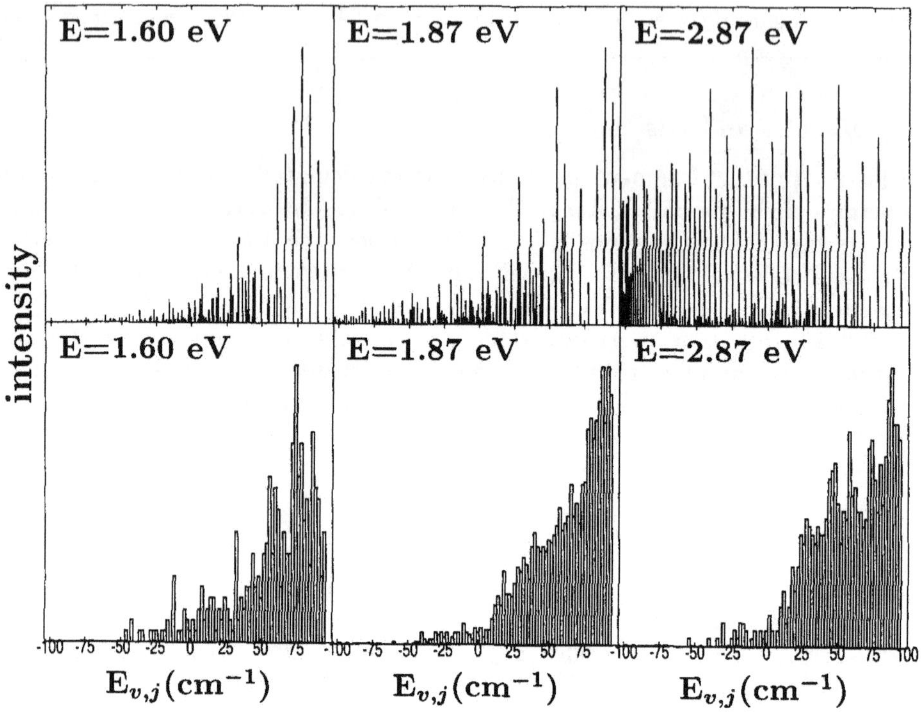

Figure 1: Rovibrational energy distributions of the $Ar - Cl$ fragment for three different excitation energies calculated quantum mechanically (upper panels) and classically (lower panels). Normalization of the quantum and classical distributions is different.

The interference effects compete with and even become dominant over the effect of the energy available for $Ar - Cl$ as E increases, due to the increase of the probability of the TF pathway with E.

Classical mechanics is unable to take into account interference effects, and it can only describe the effect of the energy available for $Ar - Cl$ (which is a rather classical effect). For this reason, as the energy available for $Ar - Cl$ increases (with increasing E), the classical rovibrational distributions concentrate in the region of the highest quasibound states. Thus, as the interference effects become gradually intense, the quantum and classical results diverge and they predict the opposite behavior.

Angular distributions of the H fragment are displayed in Fig.2. The angle θ is that formed by the vectors associated with the $Ar - Cl$ distance and the

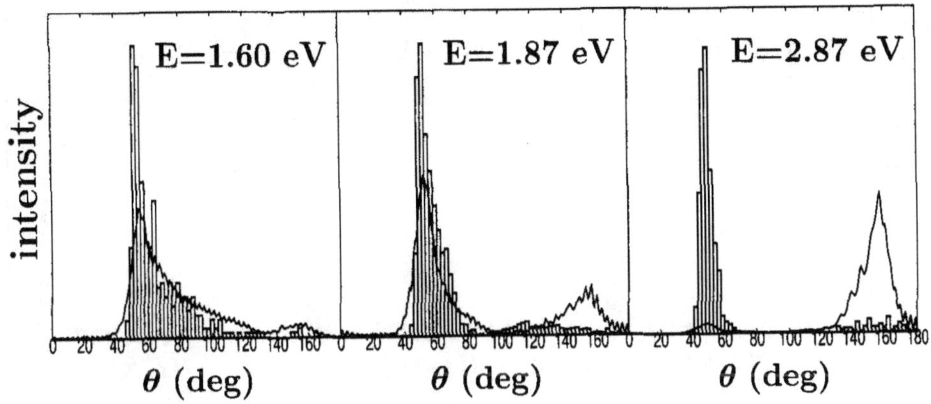

Figure 2: Classical (histograms) and quantum-mechanical (solid line) angular distributions of the H fragment produced by the PF pathway for three excitation.

distance between H and the $Ar-Cl$ center-of-mass ($\theta = 0$ corresponds with the collinear configuration $Ar-H-Cl$, with the H position vector pointing towards the Ar atom). Similarly as in Fig.2, the quantum and classical distributions agree reasonably well for $E < 2eV$, and they predict that most of the angular intensity appears at relatively low angles (around $\theta = 50°$).

As E increases the quantum and classical distributions become more and more different. The intensity of the peak at lower angles in the quantum distribution decreases gradually and even vanishes. Such a peak is mainly associated with the vanishing population in quasibound states. With increasing E only a peak at large angles survives in the quantum distributions. That peak is associated with population in bound and low quasibound states of $Ar - Cl$. By contrast, as E increases the classical angular intensity concentrates in the peak at lower angles, and only a small portion of the distribution remains at large angles, consistently with the classical distributions of Fig.1.

3.2 Total Fragmentation Distributions

Classical kinetic-energy distributions (KED) of the H fragment produced by the TF pathway are shown in Fig.3. In the case of $E = 2.87eV$ the quantum-mechanical KED is compared to the classical one. A similar level of agreement between the classical and the quantum KED as that displayed for $E = 2.87eV$ was also found for other excitation energies in the range of the $Ar - HCl$ absorption spectrum[12].

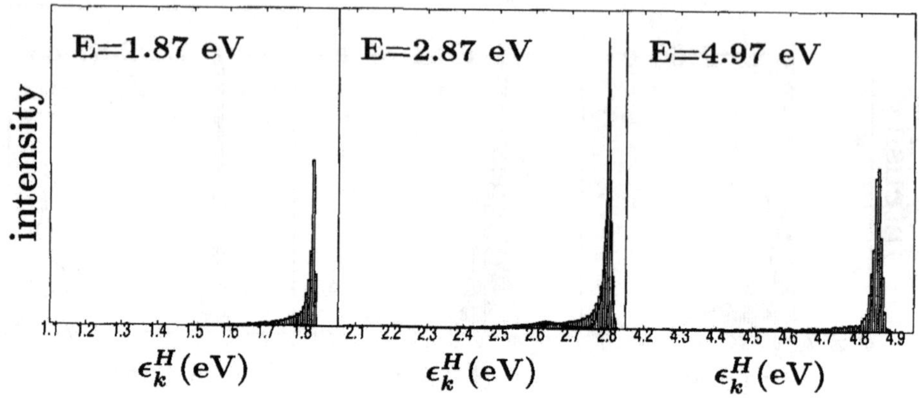

Figure 3: Classical (histograms) and quantum-mechanical (solid line) kinetic-energy distributions of the H fragment produced by the TF pathway for three excitation energies of the cluster. All the distributions have been normalized to unity.

The distributions consist of a main peak at high H kinetic energies ϵ_k^H, and a tail extending towards lower kinetic energies. The main peak is associated with nearly direct dissociation of the hydrogen, involving a weak, side collision of H with Ar. The tail at lower kinetic energies is a manifestation of cooling of the H fragment by a first hard H/Ar collision, and further collisions with Cl and again with Ar. The main difference between the quantum and classical distributions is that the quantum tail shows structure, while the classical tail is unstructured. Up to three broad peaks were found[12] in the tail of the quantum KED for different excitation energies (at the scale of Fig.3 only the first of these three peaks, at $\epsilon_k^H = 2.65 eV$ is appreciated). These three peaks are associated with successive collisions of hydrogen with Ar, Cl, and Ar again, respectively. The structure in the quantum tail is a manifestation of interference between TF states upon the successive hydrogen collisions. Again, the absence of structure in the classical tail is due to the absence of interference effects in the classical description.

Except for the tail structure, the classical KED reproduces well the quantum result (the main peak position and the extension and intensity of the tail) in the whole range of excitation energies of the absorption spectrum. This behavior is in contrast with that of the classical PF distributions. The reason is that the main interference effects in the KED occur mainly only between TF states[12], while in the PF distributions both the PF and the TF pathways interfere. Such a result is very interesting, since a quantum-mechanical calculation of the H KED is very costly even for a triatomic system (much more than that of the PF distributions), particularly as E increases. Thus, a classical description of these distributions could be an alternative.

4 Conclusions

A classical description of the UV photolysis of $Ar - HCl$, subject to intense interference effects, is tested against an exact wave packet treatment. The classical distributions of the partial fragmentation products are found reasonably good as long as interference effects are not very strong. In the case of $Ar-HCl$ this is the situation for low excitation energies. As the excitation energy increases, the increasing intensity of interference effects between states populated by both the partial and the total fragmentation pathways, causes the quantum and classical predictions for the partial fragmentation distributions to diverge. By contrast, the classical kinetic-energy distributions of the H fragment produced by total fragmentation of the cluster presents a nearly semiquantitative level of accuracy for all the energies accessible by UV excitation. In this case the main interference effects occur only between total fragmentation states. A classical description of the total fragmentation pathway upon UV photolysis becomes an alternative for triatomic and larger systems.

Acknowledgments

This work was supported by D.G.I.C.Y.T., Spain, Grant No. PB95-0071. R.P acknowledges a postdoctaral fellowship by MEC, Spain.

References

1. C. Jaques, L. Valachovic, S. Ionov, Y. Wen, E. Böhmer, J. Segall, and C. Wittig, *J. Chem. Soc. Faraday Trans.* **89**, 1419 (1993).
2. J. Zhang, M. Dulligan, J. Segall, Y. Wen, and C. Wittig, *J. Phys. Chem.* **99**, 13680 (1995).
3. K. Liu, A. Kolessov, J.W. Partin, I. Bezel, and C. Wittig, *Chem. Phys. Lett.* **299**, 374 (1999).
4. S.R. Mackenzie, O. Votava, J.R. Fair, and D.J. Nesbitt, *J. Chem. Phys.* **105**, 11360 (1996); *J. Chem. Phys.* **110**, 5149 (1999).
5. J. Segall, Y. Wen, R. Singer, C. Wittig, A. García-Vela, and R.B. Gerber, *Chem. Phys. Lett.* **207**, 504 (1993).
6. R. Baumfalk, U. Buck, C. Frischkorn, N.H. Nahler, and L. Hüwel, *J. Chem. Phys.* **111**, 2595 (1999).
7. A. García-Vela and R.B. Gerber, *J. Chem. Phys.* **98**, 427 (1993).

8. A. García-Vela, R.B. Gerber, D.G. Imre, and J.J. Valentini, *Phys. Rev. Lett.* **71**, 931 (1993).
9. T. Schröder, R. Schinke, M. Mandziuk, and Z. Bačić, *J. Chem. Phys.* **100**, 7239 (1994); T. Schröder, R. Schinke, and Z. Bačić, *Chem. Phys. Lett.* **235**, 316 (1995).
10. A. García-Vela, *J. Chem. Phys.* **108**, 5755 (1998).
11. J.C. Juanes-Marcos and A. García-Vela, *J. Chem. Phys.* **112**, 4983 (2000); *J. Chem. Phys.* **115**, xxxx (2001) (E).
12. J.C. Juanes-Marcos and A. García-Vela, to be submitted.
13. A. García-Vela, *J. Chem. Phys.* **112**, 8302 (2000).

THE DYNAMICS OF THE ENERGIZED SC_3H RADICAL: A THEORETICAL STUDY OF THE S + l-C_3H REACTION

E. MARTÍNEZ-NÚÑEZ, J.R. FLORES* AND F.J. GÓMEZ

Departamento de Química Física, Facultad de Ciencias, Universidad de Vigo, Lagoas-Marcosende, 36200-Vigo (Pontevedra), Spain

* *E-mail: flores@uvigo.es*

The dynamics of the S + l-C_3H reaction, which may play an important role in the generation of sulfur-bearing carbon chains in interstellar clouds and circumstellar envelopes, is studied through classical trajectory methods. The reaction is divided into two parts: the capture step and the unimolecular evolution of the SC_3H complex. For the capture step we use two-dimensional trajectory computations following basically the centrifugal sudden approximation. The results of this treatment are the capture rate as well as energy and angular momentum distribution functions. The evolution of the complex formed in the collision is treated by full-dimensional direct-dynamics method that relies on a PM3–SRP potential improved with some analytical functions. The results suggest that dissociation of the energized complex into the reactants or even S+c-C_3H is not an important process. There are two other exothermic reaction channels, which lead to SC_3+H and SC+C_2H. The SC_3+H/SC+C_2H branching ratio is very sensitive to the sampling procedure, but there is little doubt that both channels have comparable importance. The total reaction rate increases rather slowly with decreasing temperature, while the branching ratio is essentially unaltered. Our conclusion is that the S+l-C_3H reaction could be a source of SC_3 in media of astrophysical importance although its impact is very much limited by the existence of an important alternative channel, i.e. the generation of SC+C_2H.

1 Introduction

The sulfur-carbon chains SC_n (n = 1–3, 5) are of astrophysical importance and have been the subject of a number of theoretical and experimental studies.[1-7] There are several Fourier-transform microwave studies by the groups of Lovas[2], Oshima[3], Hirahara[4], Kasai[5] and Saito.[6] Fourier-transform infrared absorption spectra have also been recorded by Vala and coworkers[7] as well as by Maier and coworkers in the case of SC_3[8]. There is a wealth of theoretical work. Lee[9] has studied the linear SC_n systems n = 2–9 through a (density functional) BLYP method, while Vala and coworkers[7] have used both the B3LYP and MP2 (second-order Mpller-Plesset) methods. Maclagan and Sudkeaw[10] have computed the ionization potential and proton affinities of SC_2 and SC_3 while Peeso *et al.*[11] and two of us[12] have studied the excited states of SC_3. Murakami[13] has computed spectroscopic properties of the ground states of SC_n, n = 1–3. Botschwina and Maier and coworkers have made a detailed theoretical study of SC_3.[14]

The critical role played by neutral-neutral reactions in the generation of sulfur-carbon chains has been strongly advocated[15,16]. It has been argued that SC_3 could be generated by the following reaction [15,16]: $S+C_3H \rightarrow C_3S+H$. However, its rate coefficient or even the products are unknown. For this reason we have made a theoretical study of the dynamics of the energized SC_3H system generated in S+l-C_3H collisions.

2 Features of the *ab initio* potential energy surface

Two of us have performed a theoretical study of the potential energy surface (PES) of the SC_3H system[17] at relatively high levels of computation. The most relevant reaction paths and the relative energy of the stationary points are shown in Figure 1. It is readily seen that the reaction of S with l-C_3H or c-C_3H is exothermic for the generation of $SC(^1E^+)+C_2H(^2E^+)$ and $SC_3(^1E^+)+H(^2S)$. The results of a study of the excited states of SC_3[12] indicate that none of them can be formed in these reactions for the reaction would be endothermic, although at least the lowest-lying excited state of C_2H (^2A in linear geometries) could be formed[18]. We have found two types of reaction path. There first type involves only the ground state of the SC_3H system (M1) as the intermediate and the other involves several other intermediates:

1. S + l-$C_3H \rightarrow$ M1 – TS1H$\rightarrow$ $SC_3(^1E^+)$ + H
2. S + l-$C_3H \rightarrow$ M1 $\rightarrow$ $SC(^1E^+)$ + $C_2H(^2E^+)$
3. S + l-$C_3H \rightarrow$ M1 –TS12$\rightarrow$ M2 –TS23$\rightarrow$ M3 –TS3H$\rightarrow$ $SC_3(^1E^+)$ + H
4. S + l-$C_3H \rightarrow$ M1 –TS12$\rightarrow$ M2 –TS23$\rightarrow$ M3 –TS3H$\rightarrow$ M4 –TS4H$\rightarrow$ $SC_3(^1E^+)$ + H

(the reactants are in their electronic ground states). The reactive collisions between S and l-C_3H should give initially M1 (a SCCCH adduct), which could dissociate directly into $SC_3(^1E^+)$+H or $SC(^1E^+)+C_2H(^2E^+)$; the former process has a transition state (TS1H). Alternatively, M1 may isomerize into M2 and generate $SC_3(^1E^+)$+H through the intermediates M3 or M4 by surmounting TS3H or TS4H respectively. M1 and M2 may dissociate into S + C_3H, the former intermediate would give l-C_3H while the latter would produce c-C_3H

5. S + l-$C_3H \rightarrow$ M1 $\rightarrow$ S + l-C_3H
6. S + l-$C_3H \rightarrow$ M1 –TS12$\rightarrow$ M2 $\rightarrow$ S + c-C_3H

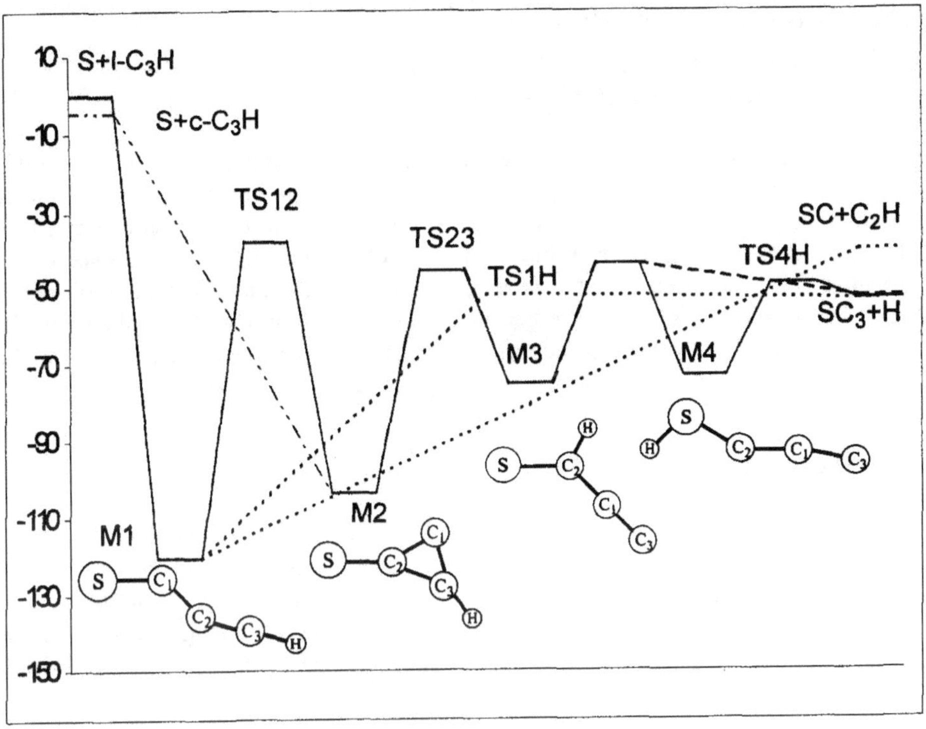

Figure 1. Schematic representation of the potential surface of the S+C_3H reaction. The relative energies (in kcal/mol) have been computed at the QCISD(T)/6-311+G(3df,2p)//QCISD/6-311G** + ZPVE(B3LYP/6-311G**) with the corrections described in reference 17: S+c-C_3H (−4.5), M1 (−120.5), M2 (−103.4), M3 (−75.0), M4 (−73.0), TS12 (−37.8), TS23 (−45.3), TS34 (−43.6), TS1H (−51.8), TS3H (−43.5), TS4H (−48.9), SC_3 +H (−52.9), SC+C_2H (−40.0).

Of course, all other reverse processes are possible in the reaction schemes 1–6; taking this fact into account would obviously lead to many other possibilities. Although processes 5 and 6 are likely to be unimportant in comparison to production of $SC_3(^1E^+)$+H, it is hard to say in advance whether the production of $SC(^1E^+)+C_2H(^2E^+)$ will be relevant for the latter products are higher in energy but, on the other hand, their generation from M1 does not imply a saddle point, (unlike that of $SC_3(^1E^+)$ + H from M1, M3 and M4).

We will reduce even further the number of reaction paths in order to cast the results of our trajectory computations into a very simple scheme:

a. M1 –TS1H→ $SC_3(^1E^+)$ + H

b. M3, M4 $\rightarrow SC_3(^1E^+) + H$
c. M1 $\rightarrow SC(^1E^+) + C_2H(^2E^+)$
d. M1 $\rightarrow$ S + l-C_3H; M2 $\rightarrow$ S + c-C_3H

We have split the computation of the rate coefficients into two steps, which we will call the capture step and the unimolecular evolution step. Given that our computed upper limit to the dissociation rate of the SC_3H system into the reactants happens to be quite small, the capture step basically accounts for the overall rate constant. The treatment of the capture step also provides energy and angular momentum distribution functions, which are needed in one of the approaches employed in the unimolecular step. The most important outcome of the present treatment of the unimolecular evolution step is the bracketing of the branching ratios.

3 The capture step

Our treatment of the capture step is based on a simple classical trajectory computation in which the dimensionality is reduced drastically and the emphasis is put on the treatment of angular momentum couplings. We will distinguish two regions of the potential surface according to whether long-range or short-range (chemical) forces dominate the potential. We have employed the following Hamiltonian for the long-range region:

$$H_{lr} = \frac{p_r^2}{2\mu} + V_{min}(r) + \frac{|\vec{J}-\vec{j}|^2}{2\mu r^2} + \frac{p_\theta^2}{2I} + \frac{(p_\phi - p_\psi \cos\theta)^2}{2I\sin^2\theta} + \frac{p_\psi^2}{2I_3} + V_o(r,\theta) \qquad (1)$$

The vectors $\vec{J}$ and $\vec{j}$ represent the total collision and rotation angular momentum, respectively, $V_{min}(r)$ is the minimum long-range potential for a particular value of r, the distance between the centers of mass, and $V_o(r,\theta)$ represents the orientation potential. The total potential would be, of course, $V(r,\theta) = V_{min}(r) + V_o(r,\theta)$; θ represents the angle of the dipole moment with respect to the line connecting the centers of mass, I is the moment of inertia, μ is the reduced mass of the collision and p_r is the momentum associated with the coordinate r. The other momenta correspond to the rotations of C_3H group in the present case and have well known expressions.[19] The momentum p_ψ would classically represent the rotation of a molecule, in this case the C_2H group, with respect to its symmetry axis; in our case it will be invariant and made equal to $\Omega\hbar$. Given that the ground electronic of C_3H is (^2A) this momentum is either $3/2\hbar$ or $1/2\hbar$; we finally used $1/2\hbar$, but this choice has virtually no effect in the results. Besides, in the long-range region, the projection of the rotational angular momentum on the line connecting the centres of masses is given by the quantum number M; consequently the fifth term on the right-hand side of Eq. (1) reduces to $(M - \Omega)^2 \hbar^2 / 2I\sin^2\theta$, which can be viewed as a θ-dependent effective potential. This Hamiltonian is singular for $\theta = 0$ or $\theta = \pi$; but these values are never reached for $M - \Omega \neq 0$. In the long-range region we have chosen to consider both J and j good quantum numbers; the corresponding term in the Hamiltonian is then written as follows:

$$\frac{|\vec{J} - \vec{j}|^2}{2\mu r^2} = \frac{\hbar^2}{2\mu r^2}\left[J(J+1) - 2M^2 + j(j+1)\right] \tag{2}$$

This choice is consistent with the centrifugal sudden approximation.[20] A more complete description and the details of the sampling procedure are given in reference 22.An adequate representation of the long-range potential is crucial for a meaningful computation of the capture rates. The potential is split into long and short-range parts such that the total potential has the following expression

$$V_{cap} = (V_m + V_d)w + V_{lr}(1 - w) \tag{3}$$

where w is a switching function, V_{lr} is the long range potential and $V_m + V_d$ is the short-range potential. V_m is a Morse-type function and V_d accounts from the potential anisotropy at short range:

$$V_d(r,\theta) = A\, e^{-\eta(r-r_0)}(1 + \cos\theta) \tag{4}$$

The long-range potential has been derived from the expressions of the electrostatic, induction and dispersion interactions obtained by Buckingham[21] through perturbation theory. Specifically, we used the expressions corresponding to the interaction of linear molecules, adequately simplified for our case. We will distinguish three parts, namely induction, dispersion and electrostatic:

$$\begin{aligned}
V_{ind} &= -\frac{1}{2}\alpha_2\mu_1^2(3\cos^2\theta + 1)/r^6 - 6\alpha_2\mu_1\Theta_1\cos^3\theta/r^7 \\
V_{dis} &= -[3U_1U_2\alpha_2/2(U_1 + U_2)r^6]\{\alpha_1 + \frac{1}{6}(\alpha_{1,\parallel} - \alpha_{1,\perp})(3\cos^2\theta - 1) + \\
&\quad 2A_{1\parallel}\cos^3\theta/r + \frac{4}{3}A_{1\perp}(3\cos\theta - 2\cos^3\theta)/r\} \\
V_{el} &= \frac{3}{2}\mu_1\Theta_2(1 + \cos\theta)/r^4
\end{aligned} \tag{5}$$

The indexes 1 and 2 refer to C_3H and S respectively; μ, Θ and α, represent the dipole and quadrupole moments and the polarizabilities, $\alpha_{\|}$ and $\alpha_{\perp}$ represent the polarizabilities corresponding to the symmetry axis and the direction perpendicular to it, respectively; $A_{1\|}$, ($A_{1\perp} = -A_{1\|}/2$), $U_{12} = U_1U_2/(U_1 + U_2)$, A, η, r_0, D_e, β and R_e were determined by a fitting procedure similar to the one described in reference [22] to a RHF-UCCSD(T)/6-311G**/QCISD/6-311G**[23] energy profile corrected for basis set superposition error, which was computed through the counterpoise procedure.[24] The rest of the parameters were obtained at the B3LYP/6-311++G**//B3LYP/6-311++G** level, except α_2 which was given the experimental value (see Table 1). Having determined the parameters we have computed the long-range part of the potential as $V_{lr} = V_{ind} + V_{dis} + V_{el}$. The success of the density functional method in the computation of the electric properties has already been established[25].
The results of the evolution of the capture rate with temperature are given in Table 2. The computed values have been scaled by a factor of 0.44, which represents the fraction states that are attractive according to the behaviour of the PES, as suggested by Clary[26]. The behaviour of the PES has been established through CASSCF computations in a previous work[17].

Table 1. Parameters of the potential employed in the capture step[e].

Parameter	Value
$\mu(C_3H)^a$	3.7414 D
$\Theta_1 = \Theta_{zz}(C_3H)^a$	0.5480 DÅ
$\alpha(C_3H)^a$	4.7900 Å^3
$\alpha_{\|} - \alpha_{\perp}(C_3H)^a$	5.3988 Å^3
$\Theta_2 = \Theta_{zz}(S)^a$	3.0771 DÅ
$\alpha(S)^b$	2.90 Å^3
A^c	23.824 kj/mol
η^c	1.748 Å^{-1}
r_0	4.00 Å
D_e^c	482.512 kj/mol
β^c	2.085 Å^{-1}
R_e^c	1.560 Å
U_{12}^c	602803 cm^{-1}
$A_{\perp}^{c,d}(C_3H)$	-0.128×10^{-5} F m^4

[a]B3LYP/6-311++G** computation. [b]Experimental value taken from Ref. [27]. [c]Obtained by fitting to the RHF-UCCSD(T)/6-311G**//QCISD/6-311G** energy profile. [d]$A_{\|} = -2A_{\perp}$. [e]The switching function is $w = 1/\{1 + \exp[-\gamma/(R - R_e)^2]\exp[b(r - r_x)]\}$; $\gamma = 5$, $b = 1.9$, $r_x = 3.0$

Table 2. Capture rate coefficients in units of 10^{-10} cm^3s^{-1} for the S+l-C_3H reaction.

T/K	60	100	140	180	220	260	300
k	2.40	2.32	2.31	2.21	2.13	2.08	2.06

4 The unimolecular step

The model potential energy surface (PES) employed in the unimolecular dynamics study is the result of fitting the PM3 Hamiltonian with specific reaction parameters (PM3–SRP)[28,29,30,31] and some analytical functions to the *ab initio* PES* shown in Fig. 1

$$V_{uni} = V_{PM3\text{-}SRP} S_{C1C2} S_{C3H} + V_{SC1+C2C3H}(1 - S_{C1C2}) + V_{SC1C2C3+H}(1 - S_{C3H}) - D_{SC1} S_{SC1} S_{SC2} \quad (6)$$

$V_{PM3\text{-}SRP}$ is the PM3–SRP potential, $V_{SC1+C2C3H}$ and $V_{SC1C2C3+H}$ are potential energy functions describing the $SC(^1\Sigma^+)+C_2H(^2\Sigma^+)$ and $SC_3(^1\Sigma^+)+H(^2S)$ dissociation channels, respectively. S_{C1C2}, S_{C3H}, S_{SC1}, S_{SC2} are switching functions that depend on the C1–C2, C3–H, S–C1, and S–C2 bond distances, respectively (for the atomic labelling see Fig. 1). Finally, D_{SC1} is a parameter that corrects the S–C1 dissociation energy.

Switching functions S_{C1C2} and S_{C3H} were used to switch on the PM3–SRP potential with the asymptotic SC+C_2H and SC_3+H potentials

$$S_{C1C2} = exp[-a(r_{C1C2} - r_{C1C2}^{eq})^b] \quad (7)$$

$$S_{C3H} = exp[-c(r_{C3H} - r_{CH}^{eq})^d] \quad (8)$$

with r_{C1C2}^{eq} and r_{CH}^{eq} being the C1–C2 and C3–H equilibrium distances of M1. Switching functions S_{SCX} (with x = 1, 2) have the following form

$$S_{SCX} = 0.5\{1 + tanh[e(r_{SCX} - r_{SC}^{s})]\}\{1 - \exp[-f(r_{SCX} - r_{SC}^{s})]\}^2 \quad (9)$$

where r_{SCX} is the *SCX* distance and r_{SC}^{s} is a reference distance.

The asymptotic exit channel potentials have the following form

---

* We also used some B3LYP/6-311G** points of the M1 → SC_3 + H and M1 → SC + C_2H energy profiles and re-scaled them with QCISD/6-311G** energies.

$$V_x = D_x - \chi_x \frac{C_6}{(r_x - r_x^{eq})^6} + V_{intra} \tag{10}$$

where the subscript x refers to one of the exit channels, D_x is the reaction endothermicity, χ_x are damping functions

$$\chi_{SC1+C2C3H} = (1 - S_{C1C2})^2 \tag{11}$$

$$\chi_{SC1C2C3+H} = (1 - S_{C3H})^3 \tag{12}$$

and r_x and r_x^{eq} are the C1–C2 (or the C3–H) bond distance and its equilibrium value at the M1, respectively. For the SC_3 + H channel the following exponential term was included to reproduce saddle point TS1H in Fig.1

$$V = g\,exp[-h(r_{C3H} - r_{CH}^{eq})] \tag{13}$$

The intramolecular potentials [V_{intra} in Eq. (10)] comprise the following harmonic functions fitted to the force field of the corresponding products

$$\begin{aligned}
V_{SC} &= 0.5k_{SC}(r_{SC} - r_{SC}^{eq,prod})^2 \\
V_{C_2H} &= 0.5k_{CC}(r_{CC} - r_{CC}^{eq,prod})^2 + 0.5k_{CH}(r_{CH} - r_{CH}^{eq})^2 + 0.5k_{CCH}(\theta_{CCH} - \theta_{CCH}^{eq,prod}) \\
V_{SC_3} &= 0.5k_{SC1}(r_{SC1} - r_{SC}^{eq,prod})^2 + 0.5k_{C1C2}(r_{C1C2} - r_{C1C2}^{eq,prod})^2 \\
&\quad + 0.5k_{C2C3}(r_{C2C3} - r_{C2C3}^{eq,prod})^2 + 0.5k_{SC1C2}(\theta_{SC1C2} - \theta_{SC1C2}^{eq,prod})^2 \\
&\quad + 0.5k_{C1C2C3}(\theta_{C1C2C3} - \theta_{C1C2C3}^{eq,prod})^2
\end{aligned} \tag{14}$$

where the bond distances, bond angles and their equilibrium values refer to the corresponding product molecules, except for the CH distance, which, for simplicity, was given the same value as in M1. The optimized parameters involved in the global PES are gathered in Table 3.

Our model PES predicts, for the entrance channel (S+l-C_3H), a reaction endothermicity of 114.1 kcal mol^{-1}, a value in exact accord with the *ab initio* one.[17] The SC_3+H and SC+C_2H exit channels are also accurately described by our model PES from the M1 intermediate. The endothermicities of both channels are in exact accord with the *ab initio* values (72.5 kcal mol^{-1} and 82.9 kcal mol^{-1} for the SC_3+H and SC+C_2H channels, respectively). Also, a saddle point of index one was predicted in the SC_3+H exit channel 2 kcal mol^{-1} higher in energy than the products as in the *ab initio* PES (TS1H in Fig. 1). Other saddle point of index one was found in the minimum energy path between the M1 and M2 intermediates with an energy of 81.9 kcal mol^{-1} above M1 (in good agreement with the *ab initio* value: 83.8 kcal mol^{-1}). Additionally, M2 was predicted to be 15.1 kcal mol^{-1} higher in energy than M1 vs 15.9 kcal mol^{-1} in the *ab initio* PES. The saddle point of index one, which is identified in Figure 1 as TS23, was found to be 83 kcal mol^{-1} above M1. The simplification made in the dynamics study was that when the system surmounts TS23 no barrier recrossings are possible, and therefore the SC_3+H channel is attained (mechanism b of section 2). Furthermore, the values of the *ab initio* geometries and frequencies of the relevant stationary points (not included for simplicity) are reasonably well reproduced by our model PES.

The trajectories were initialized from the M1 structure in three different ways. In the most simplest approach, the average vibrational energy evaluated from the capture simulation corresponding to the temperature of interest is distributed among the normal modes of vibration with the microcanonical normal mode sampling[32] (model A). In the second approach (model B), the vibrational energy was deposited nonrandomly in particular normal modes of vibration, according to the results of the capture computations. Particularly, different amounts of energy were deposited in the S–C1 stretching mode,[33] and in the lowest frequency S–CCCH bending modes. Additionally, in both models (A and B) the rotational energy assigned to each particular axis corresponds to the average values obtained in the capture step. Finally, only for the highest temperature (300 K), we also employed an efficient microcanonical sampling (EMS)[34] procedure to distribute the vibrational energy among the vibrational normal modes, with the rotational energy distributed as in models A and B. For the EMS, 500000 incubation steps for the warm-up walk and 100000 steps between trajectories were employed, with all the atoms moved at each step up to a maximum of 0.125 Å, and all the bond distances being lower than 2 Å during the Markov walk, where the acceptance ratio was 0.45.

For each temperature (60, 100, 140, 180, 220, 260 and 300 K) and type of initial conditions, one hundred trajectories were run on the above described full-dimensional PES. The trajectories were integrated by using the fourth-order Runge-Kutta-Gill routine, as implemented in a program that interfaces the classical trajectory GenDyn[35] program and the molecular orbital package MOPAC[36], with a step size of 0.1 fs. The first derivatives of the PM3–SRP part of the PES with respect to the Cartesian coordinates were evaluated analytically using the Dewar-Liotard algorithm in MOPAC. The trajectories were propagated until one of the following distances: S–C1, C3–H or C2–C3 reached 5 Å and the results were recorded.

Table 3. Parameters of the full-dimensional PES· Only those parameters that differ from its PM3 original value have been collected in this table.

Parameter	**value**	**Parameter**	**Value**
U_{ss} (H)	−13.000000	G_{pp} (C)	10.506292
β_s (H)	−5.500000	G_{p2} (C)	8.942566
Z_s (H)	1.067807	H_{sp} (C)	2.190980
α (H)	3.000000	U_{ss} (S)	−55.995371
G_{ss} (H)	14.894208	U_{pp} (S)	−43.592583
U_{ss} (C)	−49.800320	β_s (S)	−8.027465
U_{pp} (C)	−36.300000	β_p (S)	−7.291415
β_s (C)	−11.900015	G_{ss} (S)	9.164667
β_p (C)	−9.902755	G_{sp} (S)	6.585936
Z_s (C)	1.600000	G_{pp} (S)	9.868164
G_{ss} (C)	10.860708	H_{sp} (S)	3.941836
G_{sp} (C)	10.035027		
Parameter	**value**	**Parameter**	**value**
a	10 $Å^{-6}$	$r_{C2C3}^{eq,prod}$	1.287 Å
b	6	$\theta_{CCH}^{eq,prod}$	180.0 degrees
c	30 $Å^{-6}$	$\theta_{SC1C2}^{eq,prod}$	180.0 degrees
d	6	$\theta_{C1C2C3}^{eq,prod}$	180.0 degrees
e	12 $Å^{-1}$	k_{SC}	8.8 mdyn $Å^{-1}$
f	5 $Å^{-1}$	k_{CC}	16.4 mdyn $Å^{-1}$
g	50 kcal mol^{-1}	k_{CH}	6.43 mdyn $Å^{-1}$
h	2.5 $Å^{-1}$	k_{CCH}	0.05 mdyn
r_{C1C2}^{eq}	1.413 Å	k_{SC1}	7.5 mdyn $Å^{-1}$
r_{CH}^{eq}	1.060 Å	k_{C1C2}	2.0 mdyn $Å^{-1}$
r_{SC}^{S}	2.700 Å	k_{C2C3}	15.76 mdyn $Å^{-1}$
$r_{SC}^{eq,prod}$	1.546 Å	k_{SC1C2}	0.20 mdyn
$r_{CC}^{eq,prod}$	1.216 Å	k_{C1C2C3}	0.27 mdyn
$r_{C1C2}^{eq,prod}$	1.298 Å	D_{SC1}	26 kcal mol^{-1}

5 Discussion

The results of Table 2 suggest that the present reaction is relatively fast, considering it is of neutral-neutral type. The temperature dependence is weak for the results indicate a rather small increase with decreasing temperature. The question of whether re-dissociation of the energized SC_3H complex formed in the collision is an important process is answered in table 4, for the two models employed in the unimolecular step agree in a very small branching ratio, of about 10%. It must also be noted that since the zero-point energy is not adequately conserved in the trajectories of the unimolecular step, the latter value is likely to be an upper limit.

Table 4. Statistics obtained in the direct dynamics simulations[a,c]

T[b]	Model A				Model B			
	S–C1	C3–H		C1–C2	S–C1	C3–H		C1–C2
60	7 (3)	63 (6)	26	30 (5)	10 (4)	27 (5)	50	63 (6)
100	5 (3)	66 (6)	21	29 (6)	5 (3)	27 (6)	47	68 (6)
140	6 (3)	60 (6)	22	34 (6)	9 (4)	25 (5)	35	66 (6)
180	13 (4)	56 (6)	11	31 (6)	10 (4)	26 (5)	44	64 (6)
220	15 (5)	47 (6)	7	39 (6)	9 (4)	21 (5)	27	70 (5)
260	14 (4)	50 (7)	18	36 (7)	7 (3)	31 (6)	52	62 (6)
300	12 (4)	52 (6)	12	35 (6)	6 (3)	24 (5)	19	71 (6)

[a]Meaning of the acronyms : S–C1: percentage of trajectories that redissociate to S + l-C_3H (mechanism d). C3–H is the percentage of trajectories leading to SC_3 + H (mechanisms a and b); the numbers in parentheses are 68% uncertainties. The first number of the second column below C3–H indicates the fraction of SC_3 + H dissociations going through the cyclic intermediate M2 (mechanism b). C1–C2 is the percentage of trajectories leading to SC + C_2H (mechanism c).
[b]Temperature in K.
[c]The lifetime of M1 ranges between 600 fs and 800 fs depending on the model and the temperature; the temperature dependence is very weak.

Two main features of Table 4 must be pointed out: the weak temperature dependence of the branching ratios and the fact that the likely smooth variations are mostly hidden by the errors of the sampling procedure. Employing substantially higher numbers of trajectories would have been desirable but was not computationally feasible. It must be noted that the EMS procedure, (which was employed only for T=300K), gives practically the same results as model A; since the latter is computationally faster we used it for the whole set of temperatures.
The two models differ significantly in the branching ratios of the channels a, b and c (see section 2). At T=300K model A predicts that about 52% of the trajectories lead to SC_3+H and 35% lead to SC+C_2H, while model B gives percentages of 24% and 71% respectively. The reason of this discrepancy is easy to understand. In model B, the fact that one is placing large amounts of the available energy in the S-C stretching and S-CCCH bending normal modes deprives the C-H and C_1-C_2 vibrations of most of the energy they get in model A. Model B is obviously too extreme; we have used it to have an idea of the effect that the lack of ergodicity

would have in the branching ratios. One would foresee that the right amount of energy to be given to the C-H or C_1-C_2 stretching normal modes is probably average between that provided by both models, so the true branching ratios would be somewhere in between.

However, both models agree in the fact that production of SC_3+H takes place mostly in a direct way (mechanism a), the reaction path involving M3 or M4 (mechanism b) is clearly less important, especially in the case of model A . It must even be noted that, because of the approximations introduced in the dynamics, the percentages of the trajectories corresponding to mechanism b should be viewed as upper bounds (see above, section 4). Finally, it must be noted that we could not give a percentage to channel 6 (see section 2) for too few trajectories correspond to this process. An accurate computation of the S+l-C_3H/S+c-C_3H ratio would require treating adequately the ZPE problem.

One may conclude that the S+l-C_3H reaction is moderately fast $k.k_c=2.1\times10^{-10}$ cm^3s^{-1} but the branching ratio of the channels giving SC_3+H is not very high ranging between 52% and 24% at T=300 K. The S+l-C_3H reaction is almost surely a source of SC_3, but these numbers may not guarantee a very strong impact in its production. Experimental studies and astrochemical simulations are necessary for a definitive assessment of its importance.

6 Acknowledgements

We are glad to acknowledge financial support from government of the autonomous community of Galicia (projects PGIDT99PXI30102B and PGIDT00PXI30104PN) and the Spanish ministry of education (project PB98-1085). E.M.-N also thanks Ministerio de Ciencia y Tecnología, Spain, for financial support (BQU2000-0462).

References.

[1] Smith, D. *Chem.Rev.* **1992**, *92*, 1473
[2] Lovas, F.J.; Suenram, R.D.; Ogata, T.; Yamamoto, S. *Astrophys.J.* **1992**, 399, 325
[3] Ohshima, Y.; Endo, Y. *J.Mol.Spectrosc.* **1992**, *153*, 627
[4] Hirahara, Y.; Ohshima, Y.; Endo, Y. *Astrophys.J.* **1993**, *408*, L113
[5] Kasai, Y.; Obi, K.; Ohshima, Y.;Hiahara, Y.;Endo,Y.;Kawaguchi, K.;Murakami, A. *Astrophys. J.* **1993**, 410, L45
[6] Tang, J.; Saito, S.; *J.Mol.Spectry.* **1995**, *169*, 92
[7] Szczepanski, J.; Hodyss, R.; Fuller, J.; Vala, M. *J.Phys.Chem.* **1993**, *103*, 1975
[8] Maier, G.; Reisenauer, H.P.; Schrot, J.; Janoschek, R. *Angew.Chem.Int.Ed.Engl.* **1990**, *29*, 1464
[9] Lee, S. *Chem.Phys.Lett.* **1997**, *268*, 69

[10] Maclagan, R.G.A.R.; Sudkeaw, P. *Chem.Phys.Lett.* **1992**, *194*, 147
[11] Peeso, D.J.; Ewing, D.W.; Curtis, T.T. *Chem.Phys.Lett.* **1990**, *166*, 307
[12] Flores, J.R.; Pérez Juste, I.; Carballeira, L.; Estévez, C.; Gómez, F. *Chem.Phys.Lett.* **2001**, *343*, 105
[13] Murakami, A. *Astrophys.J.* **1990**, *357*, 288
[14] Seeger, S.; Botschwina, P.; Flügge J.; Reisenauer, H.P.; Maier, G. *J.Mol.Struct.(Theochem)* **1994**, *303*,213
[15] Millar, T.J.; Flores, J.R.; Markwick, A.J. *Mont.Not.R.Astron.Soc.* in press
[16] Petrie, S. *Mont.Not.R.Astron.Soc.* **1996**, *281*, 666
[17] Flores, J.R.; Gómez, F.J. *J.Phys.Chem.A.* submitted
[18] Jacox, M.E. "Vibrational and Electronic Energy Levels of Polyatomic Transient Molecules" in NIST Chemistry WebBook, NIST Standard Reference Database Number 69, Eds. W.G. Mallard and P.J. Linstrom, February 2000, National Institute of Standards and Technology, Gaithersburg MD, 20899 (http://webbook.nist.gov).
[19] Goldstein, H. *Classical Mechanics*, Adisson Wesley Inc . Reading, 1959
[20] Pack, R.T. *J. Chem. Phys.* **1974**, *60*, 633
[21] Buckingham, A.D. in *'Intermolecular Forces'*, Hirschfelder, J.O. ed. , John Wiley & Sons, 1967
[22] Flores, J.R.; Estevez, C.M.; Pérez I.; Carballeira, L. *J.Phys.Chem.A.* **2001**, *105* 4716
[23] Knowles, P.J., Hampel, C. and Werner, H.-J. *J.Chem.Phys.* **1993**, *99*, 5219
[24] Boys, S.F.; Bernardi, F. *Mol.Phys.* **1970**, *19*, 553
[25] Dickson, R.M.; Becke, A.D. *J. Phys. Chem.* **1996**, *100*, 16104
[26] Dateo, C.E.; Clary, D.C. *J.Chem.Phys.* **1989**, *90*, 7216
[27] CRC Handbook of Chemistry and Physics, Lide D.R. ed,. CRC Press, 1998
[28] González-Lafont, A.; Truong, T.; Truhlar, D.G. *J.Phys.Chem.* **1991**, *95*, 4618.
[29] Doubleday, Jr. C; Bolton, K.; Peslherbe, G.H.; Hase, W.H. *J.Am.Chem.Soc.* **1996**, *118*, 9922.
[30] Doubleday, Jr. C.; Bolton, K.; Peslherbe, G.H.; Hase, W.H. *J.Phys.Chem.A.* **1998**, *102*, 3648
[31] Bolton, K.; Hase, W.L.; Peslherbe, G.H. in: Modern Methods for Multidimensional Dynamics Computation in Chemistry, D. L. Thompson (Ed.), World Scientific, Singapore, 1998, p. 143.
[32] Hase, W.L.; Buckowski, D.G. *Chem. Phys. Lett.* **1980**, *74*, 284
[33] The initial phase of the S–C1 stretching normal mode has been selected so that the S and C atoms approach each other. However, one computation has been made with an initial random phase and the results were found to be essentially identical.
[34] Nyman, G.; Nordholm, S.; Schranz, H.W. *J.Chem.Phys.* **1990**, *93*, 6767
[35] Thompson, D.L., GENDYN program
[36] Stewart, J.P.P. MOPAC 7.0, a General Molecular Orbital Package QCPE, 1993, p455

This page is intentionally left blank

2. Liquids

NEW THEORETICAL IDEAS FOR AN OLD PROBLEM: EXCITED STATE PROTON TRANSFER IN SOLUTION

GIOVANNI GRANUCCI

Dipartimento di Chimica e Chimica Industriale, Università di Pisa, via Risorgimento 35, 56126 Pisa, Italy

THU-HOA TRAN-THI, THOMAS GUSTAVSSON, PHILIPPE MILLIÉ

Laboratoire Francis Perrin CNRS FRE 2298, CEA Saclay, DSM/DRECAM/SPAM, 91191 Gif Sur Yvette Cedex, France

JAMES T. HYNES

Département de Chimie, CNRS UMR 8640 PASTEUR, Ecole Normale Supérieure, 24 rue Lhomond, 75231 Paris, France and Department of Chemistry and Biochemistry, University of Colorado, Boulder, Colorado 80309-0215 USA

Recent theoretical results on the title topic, informed by femtosecond and steady state spectroscopic studies, are reviewed. It is argued that current conceptions of excited state intermolecular proton transfer of the hydroxyarene acid-base type in solution need to be replaced, including aspects associated with the dynamics and with electronic structure descriptions. One example of the latter—the source of the greatly enhanced acidity in the excited state compared to the ground state—is given special attention.

1 INTRODUCTION

In this plenary lecture, we review some of our recent work [1–3] addressed to the title topic for acid-base proton transfers. Certainly, the intermolecular transfer of a proton in the excited electronic state of an aromatic (hydrodroxyarene) acid ROH in solution to a solvent molecule acting as a base has been an important theme in photochemistry for many years [4, 5], and is one that has been strongly invigorated in recent times by modern femtochemistry experimental studies [3, 6, 7]. However, it cannot be claimed that it is at all understood from a microscopic point of view; indeed, a central theme of this paper is that considerable revision of previous ideas is in order.

Perhaps the most striking feature of ESPT reactions is the greatly enhanced acidity—typically by a factor of 10^6 in the acidity constant—of the excited state acid ROH* compared to its ground state form ROH [8,9]. This feature has long been interpreted in terms of a charge transfer (CT) picture, in which upon excitation there is a direct CT from the non-bonding orbital of the oxygen of the OH group into a π^* orbital of the aromatic ring system R, creating a partial positive charge on the O. As a result of this n-π^* CT, the coulombic repulsion of the H is then thought to explain the enhanced acidity [8,9].

An alternate, more spectroscopic, perspective for ESPT has often been used [10,11]. Here the focus is on a ^1L_b-^1L_a description, invoking electronic states respectively polarized along the short and long axes of the acid molecule. In this view, the ESPT is viewed as a transition between the spectroscopically accessed state (^1L_b) and a state of different symmetry (^1L_a). The latter is thought to be more polar (although why this is so is not immediately clear), and most importantly for our present purposes, this $^1L_b \rightarrow {}^1L_a$ dynamical transition is thought to constitute the ESPT act itself.

The above two perspectives are largely orthogonal; for example, the $^1L_b/^1L_a$ scenario is silent concerning the cause of the enhanced acidity. Further, as discussed in Sec. 2, neither perspective addresses very important dynamical and electronic structure issues already present for the ground state reaction. Clearly, fundamental questions remain about ESPT and new perspectives are required both to guide and to be probed by femtochemistry experiments.

The outline of the remainder of this paper is as follows. In Sec. 2, we quickly review some crucial issues for ground electronic state PT in solution, focussing on those novel aspects that will also be important for ESPT. Section 3 discusses recent experimental results, both femtochemical and steady state, which have provided the impetus to formulate a new view of ESPT, while the initial theoretical results, focussed on electronic structure issues for ESPT, are presented in Sec. 4. Concluding remarks are offered in Sec. 5. We make no attempt to cover the details of the research involved or to provide detailed references, for which the reader is referred to the original papers [1–3].

2 GROUND ELECTRONIC STATE PROTON TRANSFER ISSUES

As indicated in the Introduction, there will be a number of issues in common for ESPT and ground state PT (GSPT). Since more detailed discussions are already available [12], we recount here the most important of these only briefly, focussed on the acid-base GSPT

$$AH \cdots B \longrightarrow A^- \cdots HB^+ \tag{1}$$

in a hydrogen-bonded complex surrounded by solvent molecules.

In the traditional view of GSPT in solution, the key focus is on classical motion of the proton over the proton barrier; further, to the extent that the solvent is considered, it is implied that the solvent is in equilibrium with the moving proton at each instant. In a quite different perspective [12], none of these visions is correct. Thus, the proton nuclear motion is always quantum. That motion is typically faster than that of the polar solvent molecules—which are strongly electrostatically coupled to the reactant pair in eq. 1. As a consequence, the solvent molecules must rearrange first in order for the PT to proceed, the reaction coordinate is determined by the solvent rather than the proton, and the reaction activation energy is governed by this solvent rearrangement rather than being given by the height of the proton barrier. A molecular level illustration of the sort of solvent rearrangement involved in the reaction coordinate and governing the activation energy is shown in Scheme 1 for the acid ionization of hydrochloric acid HCl in water [12c,d], which involves a change in the coordination number of the proton-accepting water from the value appropriate for water (four) to that appropriate for the hydronium ion H_3O^+ (three).

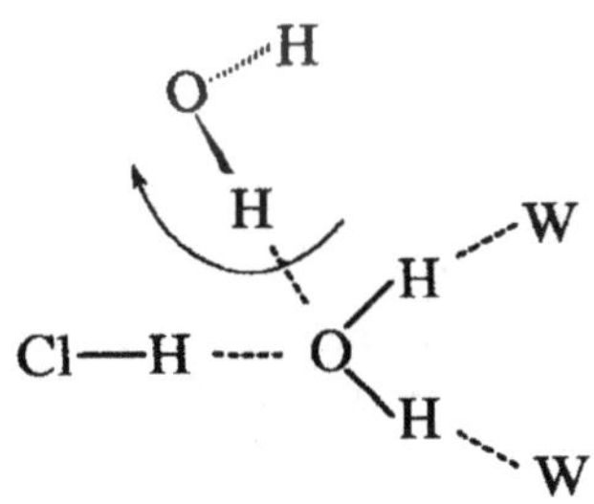

Scheme 1. Microscopic solvent motions involved in hydrochloric acid proton transfer in water.

The remarks made above apply both to the case where the proton tunnels and when it does not, the latter situation still remaining completely quantum, in strong contrast to the standard picture of classical motion of the proton over a barrier when tunneling does not occur. All the GSPT features just recounted will *remain* critical for ESPT dynamics. Indeed, as an early indication of this, it can be shown [13] that activation free energy–reaction free energy relations of just the sort successfully applied in ESPT ultrafast experiments [14] follow directly from the new perspective outlined above.

A final key aspect of GSPT reactions in solution—which is actually included in the description above, but which has not yet been emphasized—is the electronic rearrangements associated with the proton transfer act. Briefly, there is now considerable support [12] for the little used Mulliken picture [15], in which there is charge transfer from the non-bonding orbital of the base to the anti-bonding orbital of the acid. This CT, which we label n-σ^*, presumably must *also* occur in ESPT in solution, and is not referred to in either of the CT or $^1L_b/^1L_a$ perspectives mentioned in the Introduction. In the theoretical treatment of GSPT, this Mulliken n-σ^* CT is described by the resonance coupling mixing of two valence bond (VB) states, which have respectively the basic charge character of the reactants and products in eq. 1 [12].

In summary, the dynamical and electronic features applying for GSPT in solution should also be present for ESPT in solution. In the following, we take this for granted, and focus instead on the *additional* features that could be involved for ESPT.

3 EXCITED STATE PT: EXPERIMENTAL CLUES FROM PYRANINE

In the view of the present authors, the need for a complete reconsideration of both the n-π^* and $^1L_b/^1L_a$ (limited) perspectives for ESPT in solution was clearly demonstrated by femtosecond experimental results on the pyranine (3sPyOH) molecule (Scheme 2) in water [3].

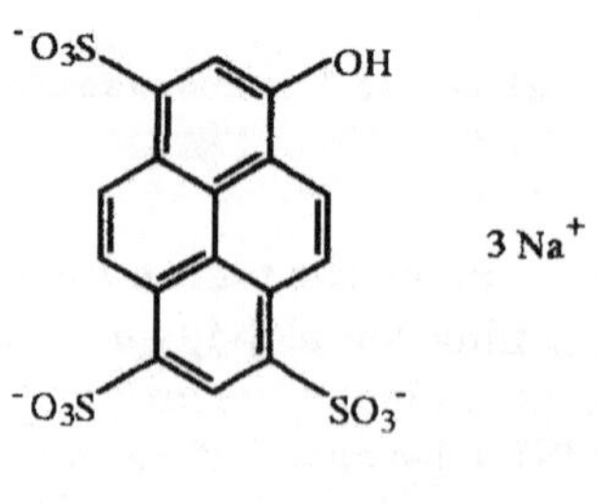

Scheme 2. Pyranine (8-hydroxy-1,3,6-pyrenetrisulfonic acid trisodium salt).

The results are summarized briefly in Scheme 3, whose first major feature is that three times were observed in the experiment, suggesting three successive steps, only one of which is the actual PT step. The last of these timescales (~ 87 ps) is consistent with earlier picosecond experimental results of Pines *et al* [16] and is evidently associated with the PT act itself, also influenced by some diffusion recombination of the produced ions $3sPyO^{-*}$ and H_3O^+. The first of these times (less than ~ 300 fs) can be attributed to ordinary solvation dynamics of the excited 3sPyOH* molecule produced in the initial Franck-Condon transition $S_0 \rightarrow S_1$. The intermediate step, with a time scale of ~ 2.2 ps, presented something of a mystery. The fact that some change in oscillator strength was associated with it—and not for the first or last steps—suggested a change in electronic state in the step.

$$\begin{array}{ccccccc} & h\nu & & (1) & & (2) & & (3) & \\ \mathrm{ROH} & \longrightarrow & \mathrm{ROH}^* & \longrightarrow & \mathrm{ROH}^{*\prime} & \longrightarrow & \mathrm{ROH}^{*\prime\prime} & \longrightarrow & \mathrm{RO}^{-*} + \mathrm{H_3O^+} \\ & & & \sim 300\ \mathrm{fs} & & \sim 2.2\ \mathrm{ps} & & \sim 87\ \mathrm{ps} & \\ & & & \delta\nu = -1380 & & \delta\nu = -1200 & & \delta\nu = -2800 & \\ & & & \delta f = 0 & & \delta f \simeq 0.4 & & \delta f = 0 & \end{array}$$

Scheme 3. Summary of some results of the pyranine ESPT femtosecond experiment at pH=4. Characteristic time scales, frequency shifts (in cm^{-1}) and oscillator strength changes are indicated for each step.

These results are in clear variance with what would be expected in either of the n-π^* CT or the $^1L_b/^1L_a$ scenarios. In the former perspective, a stronger Stokes shift should be observed; in the latter perspective, there should only be one step following any initial solvation dynamics, the $^1L_b \rightarrow {}^1L_a$ transition which is supposed to be the PT act itself. In view of this situation, a tentative alternate scenario was constructed [1, 17] , illustrated in Scheme 4.

The scenario in Scheme 4 is presented in two languages. In the first, valence bond (VB) language, the three steps are respectively: a) a first step consisting of the solvation dynamics for the locally excited (LE) state reached in absorption, a state with little charge transfer character of any kind; b) a second step associated with n-π^* CT from the oxygen of the OH into the ring system, producing a CT state, the second VB state; no PT is involved in this step; c) a third step involving n-σ^* CT, similar to that in the ground electronic state (see the Introduction) to produce a PT electronic state, which is the final VB state. This last step is the PT step proper.

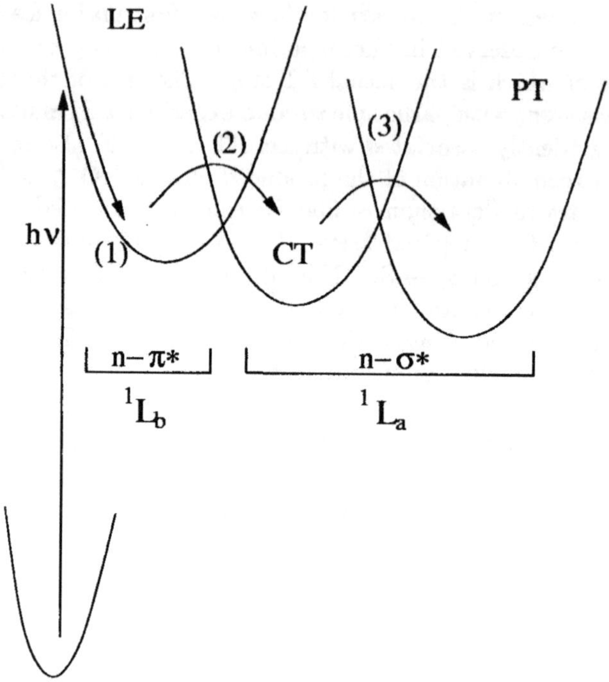

Scheme 4. Proposed scheme for the excited state dynamics of pyranine in water, both in the valence bond and ${}^1L_b/{}^1L_a$ perspectives. The electronic character involved in steps (2) and (3) are indicated, the first referring to the O to aromatic ring charge transfer and the second referring to the proton accepting base to OH bond charge transfer.

In the alternate ${}^1L_b/{}^1L_a$ language version depicted in Scheme 4, the first step is solvent relaxation in the ${}^1L_b$ state, equal to LE, while the second step is the ${}^1L_b \rightarrow {}^1L_a$ transition, but is not the PT step, since the ${}^1L_a$ state comprises two VB states, the conversion between them constituting the PT.

The validity of the new ESPT image of Scheme 4 was then pursued in a combined experimental/theoretical study involving analysis of the solvent dependence of absorption and steady state fluorescence of pyranine [1]. A very brief recounting of two important results of that extensive study is that, under experimental pH conditions where *no* proton transfer is possible, a) the absorption and fluorescence spectra are not mirror images of each other, consistent with the interconversion of an LE= ${}^1L_b$ state reached in absorption to a fluorescent ${}^1L_a$ state, ***without*** any PT having occurred, and b) a solvatochromic analysis of the absorption and fluoresence indicates the fluorescing state has significant n-π^* character and can be identified as ^1L_a character, as distinct from the LE state reached in absorption, which has ^1L_b character. All these features are completely consistent with the Scheme 4 scenario, and not with the previous conceptions.

4 EXCITED STATE PT: ELECTRONIC STRUCTURE FOR PHENOL AND SUBSITUTED PHENOLS AND THEIR ANIONS

The ideas and results outlined above indicated that it was important to carry out high level electronic structure calculations on simpler systems to examine in detail the electronic character of an acid and its anion in the excited state, a program carried out for phenol and cyano-phenols (ortho, meta, para) in vacuum [2]. Phenol was selected as the simplest aromatic acid (although upon excitation, electron transfer can occur [18]), while the cyanophenols are susceptible to experimental time dependent experiments in solution, the reaction medium of ultimate primary interest. It is to be stressed however, that these calculations do not include any proton accepting base partner, and so are silent on the presumably n-σ^* Mulliken CT aspect of the problem.

To address the fairly challenging electronic structure issues, the ground and first two excited electronic states were calculated for the acids and the anions. We performed ab initio calculations using the double-ζ cc-pVDZ basis set. To properly describe the anions, we added a set of s and p diffuse functions on three alternated carbon atoms of the cycle and on the oxygen. For the cyanophenols, diffuse functions were also added on the C and N atoms of the cyano group. Such a basis set is unsuitable to describe Rydberg states, but our major ultimate interest was focussed on the aqueous-phase behaviour, where contribution from Rydberg states is unlikely. The ground and the two first $\pi\pi^*$ excited states were obtained with CASSCF state averaged calculations, followed by CASPT2 perturbative calculations in order to take into account the repolarization of the σ system together with the dynamical correlation. For phenol and phenolate the active space consisted of eight electrons and eight π orbitals: the O p_π, the six ring π and one extravalence π orbital. For the cyano derivatives, the active space contained two more electrons and two more π orbitals. In the following, we focus only on the phenol results; those for the cyanophenols were qualitatively similar.

The first departure from accepted view is summarized in Table 1. The first important point to be seen there is that for the acid, there is in fact *very*

Table 1. Features of the first two valence excited singlets of phenol and phenolate. T_e is the computed transition energy, CT is the charge transfer from O to the ring (with respect to the ground state), estimated from the variation of the dipole moment (for PhO^-) or from the variation of the Lowdin charges (for PhOH).

Molecule	state	T_e (eV)	CT (e)
PhOH	S_0	0	0
	S_1 ^1L_b	4.6	0.02
	S_2 ^1L_a	6.3	0.09
PhO^-	S_0	0	0
	S_1 ^1L_a	3.8	0.45
	S_2 ^1L_b	4.1	~ 0

little n-σ^* CT from the O to the ring in the $S_0 = {}^1L_b$ state, with only a bit more in $S_1 = {}^1L_a$. On the other hand, for the anion, significant CT was found in $S_0 = {}^1L_a$ and much less in $S_1 = {}^1L_b$, an inverted pattern compared to the acid.

This inversion and significant charge transfer in S_0 of the anion was pursued in a calculation of the state energies of the anion as a point proton was approached to the O along two geometries, shown in Fig. 1, both for the absolute energies and the excitation energy from the ground state. The upper, absolute energy, panel displays the endothermic character of acid ionization for isolated phenol, with the trends we wish to stress are more clearly shown in the bottom, transition energy panel.

As the proton approaches the PhO^- anion along the CO axis, Fig. 1 shows that there is a crossing of the ^1L_a and ^1L_b curves. Since there is less negative charge on the O in the anion ^1L_a state, that curve rises more sharply than does the ^1L_b curve, where the charge on the anionic O is very similar to the ground state value. In the more physically relevant approach of the proton towards the anionic O along a lone pair axis, electronic coupling produces an avoided crossing between the ^1L_a and ^1L_b curves. Thus, there is a conical intersection in the problem, as a function of the OH angle. This should prove to be important in the excited state dynamics, although it does not seem likely that the conical intersection point itself would be easily reached.

We have already noted that significant n-π^* CT—expected in the S_0 state of phenol according to the standard explanation of enhanced acidity—was not observed. The question then is: what it the source of the enhanced excited

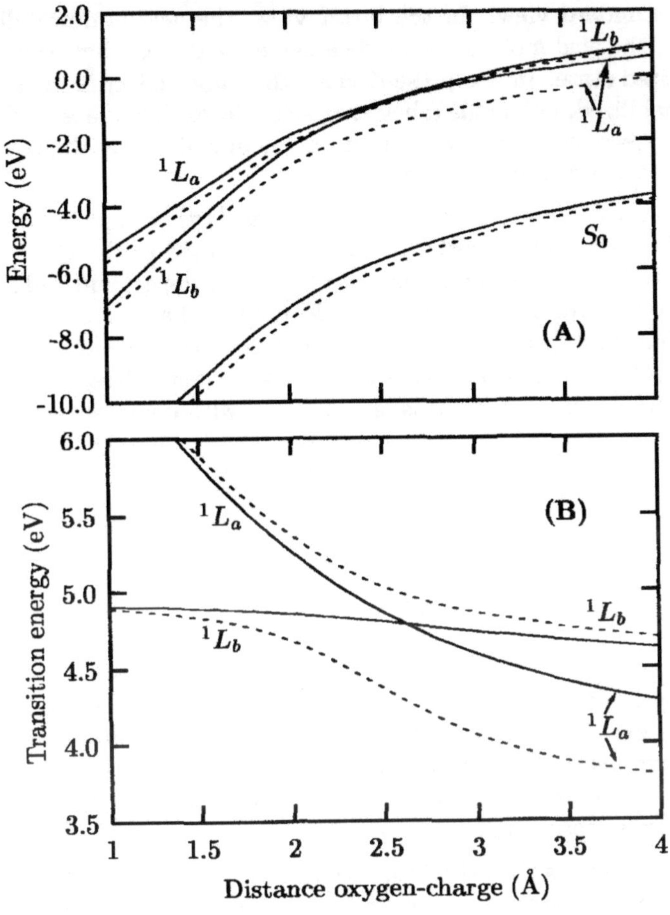

Figure 1. Interaction between the phenolate anion and the proton, modelled as a unit positive charge. Full lines: the charge is approached on the C-O axis. Dashed lines: the charge is approached following the O-H direction in phenol. (A) S_0, ^1L_a and ^1L_b energies with respect to the ground state of phenolate alone; (B) ^1L_a and ^1L_b transition energies.

state acidity? In fact, the answer is already contained in the results and the discussion presented above: it is primarily due to an effect on the anionic side of the ESPT reaction and not on the product acid side of the reaction, as in the standard view. In the latter view, the acid is destabilized in the OH bond—the acid molecule is made a better acid, i.e. a better proton donor in the excited state. In the present view, the major effect is instead that the anion is stabilized, i.e., made a better base in the excited state. Since negative charge has been significantly reduced on the anionic O by the n-π^* CT, it is a worse proton acceptor in the excited state.

The novel perspective just recounted was further supported by calculations of the electrical potential in the neighborhood of the O, which can be connected to the acidity, for the acid and anion in the ground and first excited states. Further, the picture above would predict that there should be a significant blue shift on absorption of the anion in solution—due to the reduced charge on O^- in the first excited state, the solvent stabilizes the ground state more than the excited state; this is consistent with observations.

5 CONCLUDING REMARKS

In this paper, we hope to have shown that new concepts are necessary to account for excited state proton transfer reactions in solution, and a number of them have been proposed and supported. These new concepts should be probed by future femtosecond experiments and should prove useful in interpreting those experiments.

We should stress that the basic scenario Scheme 4 provides, in our view, an important and useful framework for thinking about ESPT. One can certainly anticipate that there were be important variations in this scheme. For example, the second VB state, of ${}^1L_a$ character and with significant n-π^* character could, depending on the acid involved (as well as the solvent) be sufficiently lowered in energy that a ${}^1L_b \to {}^1L_a$ transition could lead without significant barrier to PT, provided that the barrier arising from the crossing of that VB state with the final proton transfer VB state were sufficiently low.

Clearly much remains to be done on both the experimental and theoretical sides. For the former, experiments on cyanophenol ESPT in solution are underway. For the latter, calculations for ESPT in solution aimed at clarifying, for example, the character of the ${}^1L_b \to {}^1L_a$ transition as well as the PT act itself are underway.

6 AKNOWLEDGMENT

Two of the authors (T.-H. T.-T. and JTH) acknowledge the support of an NSF-CNRS US-International Collaboration grant. JTH acknowledges support from NSF grants CHE- 9700419 and CHE-0108314. GG acknowledges support from EC-TMR grant ERB4001GT962151.

REFERENCES

1. T.-H. Tran-Thi, C. Prayer, P. Millié, P. Uznanski, J.T. Hynes J. Phys. Chem., accepted, preprint available.
2. G. Granucci, J.T. Hynes, P. Millié, T.-H. Tran-Thi J. Am. Chem. Soc. **122**, 12243 (2000).
3. (a) T.-H. Tran-Thi, T. Gustavsson, C. Prayer, S. Pommeret, J.T. Hynes Chem. Phys. Lett. **329**, 421 (2000). (b) C. Prayer, T.-H. Tran-Thi, T. Gustavsson *AIP conference Proceedings 364*, Ed. A. Tramer, p. 333 (1995).
4. (a) T. Förster Z. Elektrochem **54**, 142 (1950). (b) H.H. Jaffé, H. Lloyd Jones J. Org. Chem. **30**, 1965 (1965). (c) E.L. Wehry, L.B. Rogers J. Am. Chem. Soc. **87**, 4234 (1965). (d) Z.R. Grabowski, A. Grabowska Z. Phys. Chem. N.F. (Wiesbaden) **101**, 197 (1976). (e) Z.R. Grabowski, W. Rubaszewska J. Chem. Soc. Faraday Trans. I **73**, 11 (1977). (f) G.W. Robinson J. Phys. Chem. **95**, 10386 (1991).
5. For reviews, see P.F. Barbara, P.K. Walsh, L.E. Brus, J. Phys. Chem. **93**, 29 (1989). J. Syage J. Phys. Chem. **99**, 5772 (1995). See also the special issues: J. Phys. Chem. **95**, #25 (1991); Chem. Phys. **136**, #2 (1989); Ber. Bunsenges. Phys. Chem. **102**, #3 (1998).
6. S.K. Kim, J.J. Bren, D.M. Willberg, L.W. Peng, A. Heikal, J.A. Syage, A.H. Zewail J. Phys. Chem. **99**, 7421 (1995).
7. L. Genosar, B. Cohen, D. Huppert J. Chem. Phys. A **104**, 6689 (2000).
8. (a) E. Vander Donckt Progr. Reaction Kinet. **1**, 187 (1961). (b) G. Jackson, G. Porter Proc. Roy. Soc. **A200**, 13 (1961). (c) L. Salem *Electrons in Chemical Reactions* (Wiley, New York, 1982).
9. N. Mataga, T. Kubota *Molecular Interactions and Electronic Spectra* (Dekker, New York, 1970) Chap. 7.
10. A. Tramer, M. Zaborowska, Acta Physica Polonica **5**, 821 (1968).
11. (a) R. Knochenmuss, P.L. Muiño, C. Wickleder J. Phys. Chem. **100**, 11218 (1996). (a) R. Knochenmuss, I. Fischer, D. Luhus, Q. Lin Israel J. Chem. **39**, 221 (1999).

12. (a) J. Timoneda, J.T. Hynes J. Phys. Chem. **95**, 10431 (1991). (b) A. Staib, J.T. Hynes, D. Borgis J. Chem. Phys. **102**, 2487 (1995). (d) K. Ando, J.T. Hynes, J. Molec. Liquids **64**, 25 (1995). (d) K. Ando, J.T. Hynes, J. Phys. Chem. **101**, 10464 (1997). (e) K. Ando, J.T. Hynes, J. Phys. Chem. A **103**, 10398 (1999).
13. P. Keifer, J.T. Hynes J. Phys. Chem., in press.
14. E. Pines, B.-Z. Magnus, M.J. Lang, G. Fleming Chem. Phys. Lett. **281**, 413 (1997).
15. R.S. Mulliken J. Chim. Phys. **61**, 20 (1964).
16. (a) E. Pines, D. Huppert Chem. Phys. Lett. **126**, 88 (1986). (b) E. Pines, D. Huppert, N. Agmon J. Chem. Phys. **88**, 5620 (1988).
17. C. Prayer, T.-H. Tran-Thi, P. Millié, J.T. Hynes Abstract book of the 18th Intern. Conf. on Photochemistry, Warsaw, 1997.
18. O. Shimizu Photobiochem. Photobiophys. **4**, 347 (1982).

ULTRAFAST DYNAMICS OF HYDROGEN BONDS IN THE LIQUID PHASE: VIBRATIONAL QUANTUM BEATS AND DEPHASING

J. STENGER, D. MADSEN*, J. DREYER, E.T.J. NIBBERING, P. HAMM**, AND T. ELSAESSER

Max-Born-Institut für Nichtlineare Optik und Kurzzeitspektroskopie, Max-Born-Str. 2 A, D-12489 Berlin, Germany

**Current address: Department of Chemistry, Aarhus University, Langelandsgade 140, 8000 Aarhus C, Denmark*

***New address: Physikalisch-Chemisches Institut, Universität Zürich, Winterthurer Str. 190, CH-8057 Zürich, Switzerland*

Coherent vibrational motions of intramolecular hydrogen bonds (H-bonds) and dephasing of coherent O-H stretching excitations in intermolecular H-bonds were studied by femtosecond spectroscopy in the mid-infrared. The O-H stretching absorption of phthalic acid monomethyl ester (PMME-H) and the O-D absorption in its deuterated analog PMME-D, both forming an intramolecular H-bond, display oscillatory changes upon femtosecond excitation. Such signals provide the first evidence for a coherent wavepacket motion along a low-frequency coordinate in both PMME-H and PMME-D that couples anharmonically to the fast stretching motion and modulates the length of the hydrogen bond with a 100 cm^{-1} frequency. The oscillations display a damping time of approximately 0.5 ps, demonstrating the underdamped character of the low-frequency mode. Dephasing processes of coherent O-H stretching excitations in HDO dissolved in D_2O were investigated in 2- and 3-pulse photon echo experiments. We observe a very fast decay of the 2-pulse photon echo signal, suggesting dephasing rates on the order of 10^{13} s^{-1}, much higher than those of stretching vibrations of free O-H groups. The comparably fast dephasing is attributed to the large anharmonicity of the O-H oscillator in a hydrogen bond, enhancing its coupling to the fluctuating forces exerted by the liquid environment. 3-pulse photon echo peak shift experiments show that spectral diffusion within the O-H stretching band occurs in a wide temporal range, including dynamics on a 5 to 15 ps time scale.

1 Introduction

Hydrogen bonding represents an important type of local interaction which is relevant both for the structure and function of molecular systems like hydrogen-bonded liquids, proteins and enzymes. In a H-bond, a hydrogen atom is shared between two electronegative atoms of the donor and acceptor groups, thus linking donor and acceptor moieties. This geometry leads to drastic changes of the potential energy surface of the hydrogen and strongly affects vibrational motions of the coupled groups. As a result, vibrational spectroscopy has developed into one of the most important spectroscopic tools to analyze the properties of hydrogen bonds [1]. In most studies, the stationary infrared and/or Raman spectra have been investigated, in particular the stretching vibration of the hydrogen donor groups. Upon formation of a hydrogen bond, the O-H stretching band undergoes a red-shift of up to several hundred wavenumbers (cm^{-1}) corresponding to a reduction of the force constant of the O-H oscillator. The reduced force constant is mainly due to an enhanced anharmonicity of the hydrogen bond potential. This behavior is illustrated in Fig. 1 where the stretching band of the free O-H group of a phenol monomer is compared to the O-H band of HDO in D_2O being part of an intermolecular H-bond, and to the O-H band of PMME-H, forming an intramolecular O-H...O bond. The red-shift increases with increasing hydrogen bond strength, i.e., a decreasing distance between the heavy atoms in the H-bond. Systematic studies of H-bonds in solids have established a rather general relationship between the distance of the heavy atoms, i.e., the H-bond length, and the O-H stretching frequency [2].

In addition to the red-shift, a substantial broadening of the O-H stretching bands is found in the liquid phase, in some cases linked to a peculiar new substructure of the vibrational spectrum (c.f. Fig. 1). So far, this behavior is not well understood and different mechanisms have been invoked to explain it:

(i) In a disordered medium like a liquid, a distribution of H-bond lengths may exist. In such a case, a distribution of O-H stretching frequencies occurs, giving rise to an inhomogeneous broadening of the overall O-H stretching absorption.

(ii) The enhanced anharmonicity of the potential energy surfaces results in a coupling of different vibrational degrees of freedom. In particular, the O-H stretching motion can couple to low-frequency modes changing the length of the

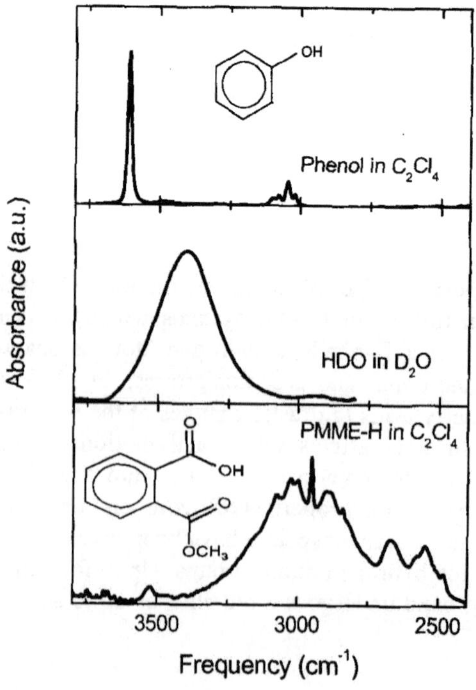

Figure 1. O-H stretching bands of phenol monomers in C_2Cl_4, HDO in D_2O, and PMME-H in C_2Cl_4. The O-H band of phenol at 3610 cm^{-1} has a width of 20 cm^{-1}. The O-H band in the weak hydrogen bond of HDO has a maximum at 3400 cm^{-1} and a width of 250 cm^{-1}. PMME-H with a medium-strong hydrogen bond shows an O-H vibrational band that peaks at 3000 cm^{-1} and extends over 700 cm^{-1} with marked substructure.

hydrogen bond [3,4]. In the O-H absorption band, such a coupling leads to the formation of sidebands which are shifted by plus/minus one or several low-frequency quanta relative to the (pure) O-H stretching excitation.

(iii) In liquids, the hydrogen bond is subject to fluctuating forces exerted by the environment which result in a dephasing of the O-H stretching mode and in a damping of low-frequency vibrations. In the literature, different theoretical models have been developed to account for such effects. In Ref. [5], direct dephasing of high-frequency stretching vibrations by fluctuating electric fields has been considered. Other models invoke coupling of low-frequency modes of the hydrogen bond to solvent fluctuations, resulting in a stochastic motion of the low-frequency H-bond vibrations [6-8]. Because of the coupling of low- and high-frequency motions in the hydrogen bond, a stochastic low-frequency motion results in a dephasing of the O-H stretching transitions and its sidebands. In a situation where the time scales related to inhomogeneous broadening are different from the time scale of phase relaxation, homogeneous broadening due to dephasing processes can be separated from inhomogeneous broadening. In liquids, however, such a separation of time scales represents an approximation and – in general – one has to consider the full frequency fluctuation correlation function to describe the system realistically.

Linear optical studies, e.g. of stationary absorption spectra, do not allow for a separation of the different broadening mechanisms and the underlying microscopic dynamics which occurs on ultrafast time scales. Here, nonlinear optical techniques in the ultrashort time domain are required. In recent years, a number of femtosecond infrared studies of hydrogen bonded systems has been reported, mainly concentrating on incoherent processes of vibrational relaxation and spectral diffusion which occur after excitation of O-H or O-D stretching vibrations [9-12]. In this article, we address the *coherent* vibrational dynamics of hydrogen bonds. We report the first observation of coherent low-frequency oscillations which are due to the propagation of vibrational wavepackets along intramolecular hydrogen bonds. Femtosecond pump-probe experiments in the mid-infrared give detailed insight into wavepacket generation, oscillation phases and damping of the coherent motions. In a second series of experiments based on photon echo techniques, we investigate the ultrafast dephasing of coherent O-H stretching excitations in water. We determine dephasing rates on the order of 10^{13} s^{-1} and demonstrate that spectral diffusion processes cover a broad range in time extending up to about 15 ps.

2 Experimental Techniques

Femtosecond mid-infrared pulses tunable in the range from about 3600 cm^{-1} (λ=2.77 μm) to 1670 cm^{-1} (λ=6.0 μm) were derived from the output of an regeneratively amplified Ti:sapphire laser (repetition rate 1 kHz, pulse energy 300 μJ, pulse duration 90 fs) by parametric frequency mixing in two different setups [13,14].

Pulses in the 3600 to 3000 cm^{-1} range were generated in a two-stage setup using a $KTiOPO_4$ (KTP) nonlinear crystal. The first stage is seeded by a single filament white light continuum and pumped by 3 μJ pulses at 800 nm. The output of the first stage was fed back into the same crystal and amplified by mixing with a 200 μJ pulse at 800 nm. The energy of the mid-infrared pulses of 130 fs duration varies from 5 μJ at low frequency to 3 μJ at high frequencies. Pulses between 3300 and 1670 cm^{-1} were generated by difference frequency mixing in a $AgGaS_2$ crystal seeded by tunable near-infrared pulses which are derived from the output of the Ti:sapphire laser. This source which has been described in detail in Ref. [14] provides mid-infrared pulses of 1 to 2 μJ energy and a 100 to 130 fs duration. The spectral bandwidth has a value of about 150 cm^{-1}. Both mid-infrared sources were optimized for low pulse-energy fluctuations being less than 1 %.

In the mid-infrared pump-probe experiments, the intense pump pulse induces a nonlinear absorption change in the range of the O-D/O-H stretching absorption of PMME-D/H which is monitored by a weak time-delayed probe pulse. The transmitted probe pulse was either detected by a spectrally integrating HgCdTe detector or spectrally dispersed and detected with a multichannel HgCdTe detector.

Mid-infrared photon echo experiments on HDO in D_2O were performed with the 3-pulse scheme depicted in Fig. 2. Three pulses traveling into the directions $\mathbf{k}_1$, $\mathbf{k}_2$, and $\mathbf{k}_3$ were derived from the output of the infrared source. The pulses 1 and 2 with a mutual delay τ, the coherence time, generate a transient grating in the sample.

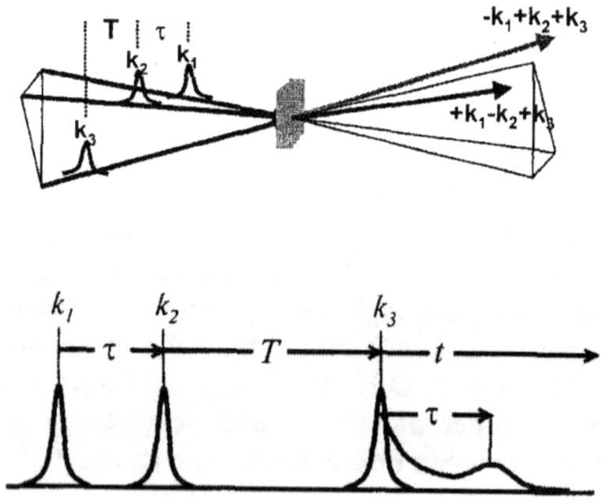

Figure 2. Schematic of the 3-pulse photon echo experiment. The pulses 1 and 2 with propagation direction $\mathbf{k}_1$ and $\mathbf{k}_2$ generate a transient grating in the sample from which pulse 3 induces coherent signals emitted into the phase-matched directions $-\mathbf{k}_1+\mathbf{k}_2+\mathbf{k}_3$ and $\mathbf{k}_1-\mathbf{k}_2+\mathbf{k}_3$. Such signals are recorded by time-integrating infrared detectors. Lower part: Sequence of the 3 pulses in time (τ: coherence time, T: population time). The signal emitted after pulse 3 displays features of a free induction decay and a photon echo.

The delayed pulse 3 (T: population time) induces the emission of coherent third-order signals into the phase-matched directions $-\mathbf{k}_1+\mathbf{k}_2+\mathbf{k}_3$ and $\mathbf{k}_1-\mathbf{k}_2+\mathbf{k}_3$. During the first time interval τ, the system evolves in a coherent polarization state whereas it evolves in a population state during the second time period T.

Different experiments were performed:

- 2-pulse photon echo: For T=0, i.e. the simultaneous arrival of pulses 2 and 3 in the sample, one measures the decay of the macroscopic nonlinear polarization corresponding to a 2-pulse photon echo.
- Photon-echo peak shift: For a fixed population time T, the signals emitted into the two phase-matched directions are mirror images of each other with maxima at different values of the coherence time τ. The so-called peakshift is defined as half of the time interval between the two maxima. In a peak-shift measurement, this quantity is recorded as a function of T.
- Transient grating decay: The decay of the population grating generated by the pulses 1 and 2 with τ=0 is monitored in a measurement of the diffracted signals as a function of the population time T.

PMME-D was prepared as described in Ref. [15]. PMME-H and PMME-D were dissolved in C_2Cl_4 (concentration c=30 mM). The solution of HDO in D_2O had a concentration of 0.2 M, corresponding to an optical density of 0.4 at 3400 cm^{-1}. The measurements were performed in sample cells of 0.3 to 0.5 mm thickness, the beam diameters on the sample were on the order of 100 μm.

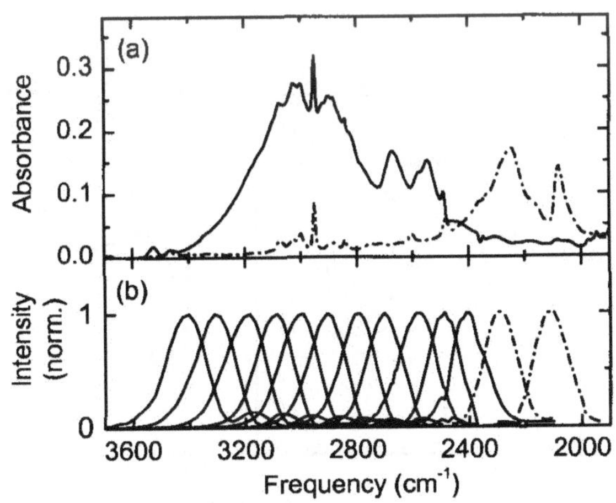

Figure 3. (a) Steady-state O-H stretching band of PMME-H (solid line) and O-D stretching band of PMME-D (dash-dotted line). The sharp line at 2950 cm^{-1} is due to C-H stretching absorption. (b) Spectra of femtosecond pulses used in the pump-probe experiments. Solid lines: Experiments with spectrally integrated probe pulses. Dash-dotted lines: Experiments with spectrally resolved probe.

3 Coherent Wavepacket Propagation along Intramolecular Hydrogen Bonds

The structure and the vibrational spectra of PMME were analyzed by quantum chemical calculations which are discussed in Ref. [16]. The lowest energy conformer of PMME consists in a quasi-planar structure in which the hydrogen bond is part of a seven-membered ring. This 7-ring does not entirely fit into the plane of the benzene ring so that the substituents are somewhat tilted out-of-plane. The heavy atom hydrogen bond O-O distance of 256 pm and the H...O distance of 160 pm indicate an intramolecular hydrogen bond of intermediate strength and a well-defined geometry.

The steady-state O-H and O-D stretching bands of PMME-H and PMME-D are shown in Fig. 3 (a). In the femtosecond pump-probe experiments with PMME-H, the infrared pulses were tuned throughout the O-H stretching band, as indicated by the pulse spectra in Fig. 3 (b), and the spectrally integrated probe pulses transmitted through the sample were detected. The time evolution of the nonlinear change of absorbance at the different spectral positions is plotted in Fig. 4. Around delay zero, i.e. during the temporal overlap of pump and probe pulses, there is a minor (<10 %) signal contribution due to the third order nonlinearity of the solvent and the windows of the sample cell. The dominant signal originates from changes of the O-H stretching absorption of PMME-H. For probe frequencies between 2500 and 3200 cm^{-1}, we observe a pronounced decrease of infrared absorption which is strongest

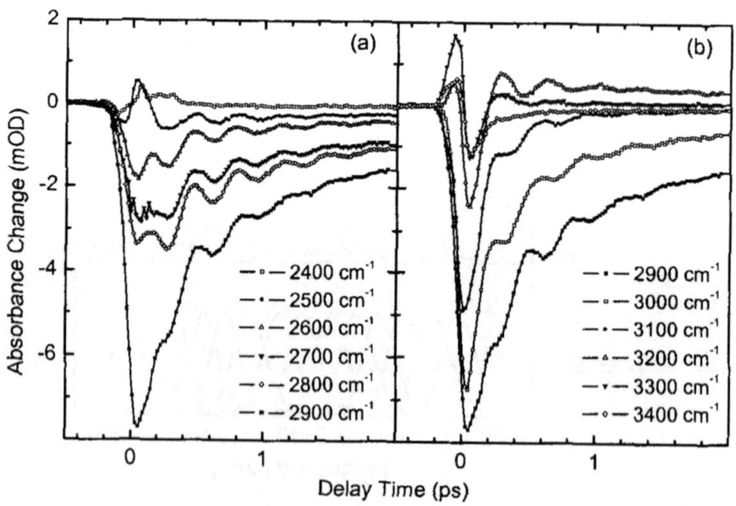

Figure 4. Pump-probe transients measured with PMME-H. The change of absorbance $\Delta A=-\log(T/T_0)$ is plotted as a function of the time delay between pump and probe pulses at the frequencies indicated (T_0, T: transmission of the sample before and after excitation). The spectrally integrated probe is detected.

around 2900 cm^{-1}, i.e. around the maximum of the steady-state infrared band. The bleaching transients display fast decays within the first 1 ps, followed by a slower relaxation on a time scale of 30 ps (not shown). At 3300 and 3400 cm^{-1}, an absorption increase is found which rises within the first 200-300 fs and shows a slow decay within 30 ps. All transients are superimposed by strong oscillatory signals with a frequency of 100 cm^{-1} and a damping time of approximately 500 fs.

The transient O-D stretching absorption of PMME-D reveals a similar behavior. In Fig. 5, we present results for pump pulses centered at (a) 2100 cm^{-1} and (b) 2300 cm^{-1} and spectrally dispersed probe pulses. Depending on the probe frequency, one observes either bleaching or enhanced absorption. At 2028 cm^{-1} (Fig. 5 a), the enhanced absorption decays with a time constant of about 400 fs. The bleaching transients (except for a probe frequency of 2271 cm^{-1}) show a fast decay with 400 fs, followed by a slower decay on a 20 to 30 ps time scale (not shown). Again, pronounced oscillations with a frequency of 100 cm^{-1} (Fig. 5 c), i.e. identical to that in PMME-H, are superimposed on the incoherent pump-probe signals.

For an analysis of the pump-probe data, we consider the level scheme and the potential energy diagram in Fig. 6. In addition to the vibrational transition from the v=0 to the v=1 state, the red-shifted v=1 to v=2 transition and the blue-shifted v'=0

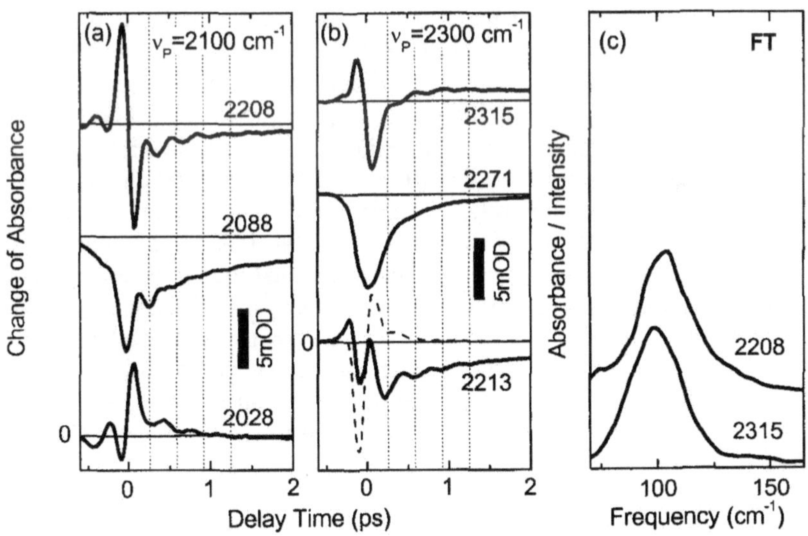

Figure 5. (a,b) Time resolved nonlinear change of vibrational absorption of PMME-D as measured with spectrally resolved probe pulses (ν_P: center frequency of the pump pulses, probe frequencies are given in cm^{-1}). The dashed line in (b) represents the response of the solvent. (c) Fourier transforms (FT) of the oscillatory signals.

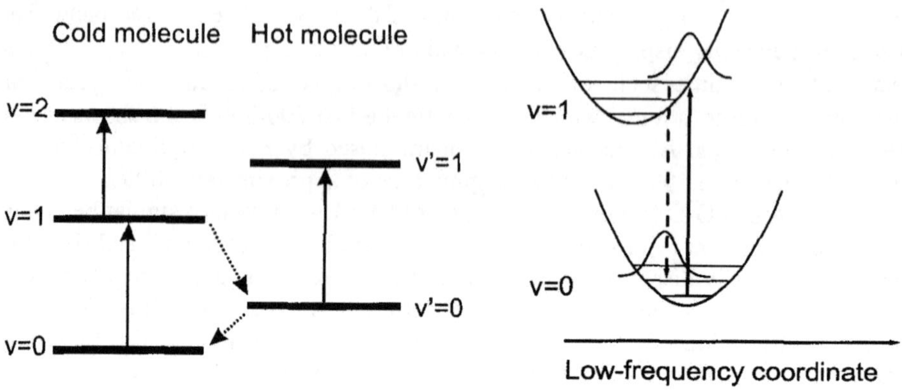

Figure 6. Left: Level structure implementing the v=0, v=1, and v=2 states of the O-H/O-D stretching mode and the two levels of the vibrationally hot molecules which are formed when the v=1 state has decayed. Right: Schematic potential energy diagram for the O-H/O-D stretching mode along the low-frequency coordinate. Femtosecond excitation creates wavepackets in both the v=0 and v=1 state of the O-H/O-D mode by superposition of several levels of the low-frequency mode.

to v'=1 transition of molecules with a vibrationally hot ground state are taken into account. The pump pulse promotes part of the molecules from the v=0 to the v=1 state, resulting in a bleaching of the v=0→1 transition due to the depletion of the v=0 state and stimulated emission from the v=1 state. Such two contributions dominate the bleaching signals observed at delay times <1ps in Figs. 4 and 5. The transient population of the v=1 state gives also rise to a red-shifted v=1→2 absorption which is evident from the PMME-D transient at 2028 cm^{-1} (Fig. 5 a) and – much less pronounced – from the PMME-H data taken at 2400 cm^{-1} (Fig. 4). The decay of excited state absorption reflects the depopulation of the v=1 level with time constants of ~400 fs for PMME-D and <250 fs for PMME-H. Population relaxation is due to a coupling of the v=1 state to other vibrational modes at lower frequency, transferring the energy of the O-H (O-D) excitation into the vibrational manifold. As a result, molecules with a hot vibrational system are created in which the O-H (O-D) stretching mode is in the v'=0 ground state. In the hot molecules, the H-bond is weakened and, thus, the O-H (O-D) stretching absorption is blue-shifted compared to the initial absorption. This mechanism gives rise to (i) the enhanced absorption of PMME-H at 3300 and 3400 cm^{-1} and of PMME-D at 2315 cm^{-1}, and (ii) to the bleaching signals extending to delay times longer than 1 ps. Dissipation of the vibrational excess energy into the surrounding solvent – a process occurring on a time scale of several tens of picoseconds – leads to the slow nonexponential decay of the absorption changes. It should be noted that the fast rise of the blue-shifted absorption of PMME-H at 3400 cm^{-1} gives independent support for a decay of the v=1 state with a sub-250 fs time constant.

We now discuss the oscillatory signals present in the pump-probe transients. In both PMME-H and PMME-D, the oscillations display a frequency of 100 cm^{-1}, much lower than the O-D/O-H stretching frequency. The oscillations are due to wavepacket motions along a low-frequency mode which couples anharmonically to the fast stretching modes. Wavepacket generation is visualized with the help of the potential energy scheme in Fig. 6 (b) which is based on an adiabatic separation of the time scales of the fast stretching and the slow low-frequency motion. Several vibrational levels of the low-frequency mode are drawn into the potential energy curves of the v=0 and v=1 state of the stretching oscillator. In our experiments, the spectral field envelope of the pump pulse covers several levels of the low-frequency mode, allowing for the excitation of a coherent superposition of low-frequency states, i.e. of wavepackets. Wavepackets in the v=1 state of the stretching oscillator are created via vibrational absorption whereas wavepacket motion in the v=0 state is induced by a Raman-like process within the bandwidth of the pump pulse. This Raman process is resonantly enhanced by the vibrational transition.

The enhanced absorption of PMME-D at 2028 cm^{-1} (Fig. 5 a) is due to the $v=1 \rightarrow 2$ transition and – thus – the oscillations on this transient reflect wavepacket motion in the v=1 state. The damping time of the oscillations of 400 fs is close to the v=1 lifetime, suggesting that vibrational population relaxation represents a major damping mechanism. In PMME-H, the v=1 lifetime of the O-H stretching mode has a substantially smaller value of 200 to 250 fs, i.e. close to the oscillation period of the 100 cm^{-1} mode. As a result, oscillations in the v=1 state are not observed.

Wavepacket motions in both the v=0 and v=1 states are relevant for the analysis of the oscillations found on top of the bleaching signals. As shown schematically in Fig. 6 (b), the $v=0 \rightarrow 1$ transition frequency of the O-D/O-H stretching mode depends on the elongation along the low-frequency mode. Wavepacket motion along this coordinate results in a periodic modulation of the O-D/O-H transition frequency. In PMME-D (Fig. 5), this mechanism leads to the oscillatory blue-shift of the O-D stretching band which is evident from the oscillatory enhancement of absorption at 2315 cm^{-1}. Consistent with this picture, there are no oscillations around the maximum of the O-D band at 2271 cm^{-1} and oscillations of opposite phase at 2213 cm^{-1}. PMME-H (Fig. 4) shows the same behavior. Here, however, only wavepacket motion in the v=0 state is relevant [16]. As frequency shifts of the O-D/O-H stretching bands are a measure of the length of the hydrogen bond, the oscillatory blue-shift is equivalent to a modulation of the O....O distance in PMME, determining the strength of the hydrogen bond.

A normal mode analysis of PMME provides insight into the microscopic nature of the low-frequency mode displaying a frequency of 100 cm^{-1}. As discussed in detail in Refs. [15,16], the periodic modulation of the O...O distance is attributed to an out-of plane torsional motion by which the two ring substituents in PMME move relative to each other, thus changing the O...O distance. The frequency of this mode remains unchanged upon H/D exchange and – thus – identical oscillation frequencies occur in the PMME-H and PMME-D data. This intramolecular mode displays a comparably long damping time on the order of 0.5 ps, resulting in an underdamped character of this mode. This is in contrast to many theoretical studies of H-bond dynamics in which overdamped low-frequency motions have been assumed. It should be noted that there are modes at higher frequencies also modulating the length of the hydrogen bond [17]. The spectral bandwidth of our present pump pulses and the time resolution of the experiment, however, are not sufficient to excite wavepackets along such modes and to follow their time evolution.

In conclusion, we have provided the first evidence for coherent low-frequency wavepacket motions along hydrogen bonds, modulating the H-bond length and strength. A similar behavior has been observed in hydroxydeuterated 2-(2'-hydroxyphenyl)benzothiazole where the intramolecular O-D…N bond is modulated by a 118 cm^{-1} in-plane deformation mode [18]. The generation and propagation of wavepackets along H-bonds with double minima potentials may allow for inducing and optically controlling hydrogen transfer reactions.

4 Vibrational Dephasing and Spectral Diffusion in Liquid Water

Dephasing processes in liquids which lead to a decay of coherent vibrational excitations occur - in most cases - in the time domain between 0.5 and several picoseconds and result in essentially Lorentzian vibrational bands dominated by homogeneous broadening. So far, homogeneous broadening of the stretching bands of hydrogen bonded O-H groups has remained nearly unexplored. A measurement of the homogeneous linewidth and/or the underlying dephasing dynamics requires nonlinear techniques like hole burning or photon echo methods, the latter allowing for an observation of the dephasing dynamics in real-time. In the following, we discuss the first femtosecond 2- and 3-pulse photon echo studies of O-H stretching bands in water. As a model system for water, HDO in D_2O is investigated.

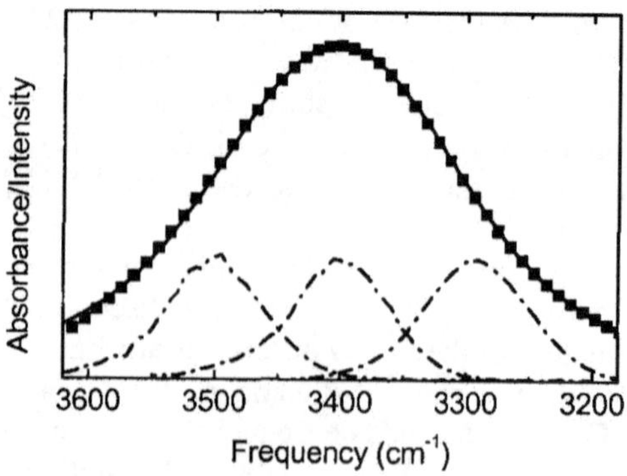

Figure 7. Absorption spectrum of the O-H stretching vibration of HDO in D_2O (symbols). The solid line gives the absorption spectrum obtained from the fit of the spectrum and the 2-pulse photon echo data. Dash-dotted lines: Pulse spectra.

In contrast to the sterically well-defined intramolecular hydrogen bonds discussed in section 3, there is an inhomogeneous distribution of H-bond geometries of HDO in D_2O contributing to the overall O-H stretching band shown in Fig. 7. A number of femto- and picosecond experiments in which nonlinear changes of the O-H stretching absorption were measured, have led to conflicting models for the O-H stretching band and the underlying H-bond geometries: In Refs. [10,11], a continuous distribution of transition frequencies and, thus, H-bond lengths has been assumed in order to explain spectral diffusion processes within the O-H band which display a characteristic time constant of 700 fs. In contrast, 3 different local substructures have been invoked in Ref. [9] to explain hole burning experiments with subpicosecond pulses. In the following, we report 2- and 3-pulse photon echo measurements at different spectral positions (c.f. Fig. 7) within the O-H band, providing direct insight into the ultrafast dephasing dynamics and spectral diffusion.

The wavevector geometry of our photon echo experiments and the different types of measurements perfomed have been explained in section 2. In Fig. 8, we present 2-pulse photon data for HDO in D_2O (solid symbols) and for a 1 mm thick CaF_2 window (open symbols) measured under the same experimental conditions.

The diffracted intensity recorded with a time integrating detector is plotted as a function of the coherence time τ between pulse 1 and the pulses 2 and 3 arriving

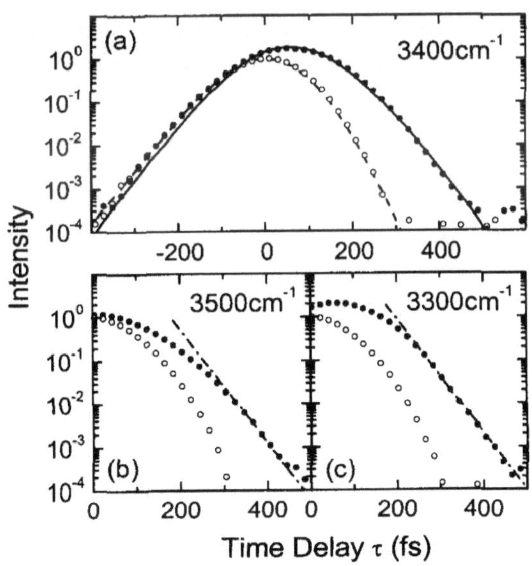

Figure 8. Results of the 2-pulse photon echo measurements at 3 spectral positions. The diffracted intensity is plotted on a logarithmic scale as a function of the coherence time τ for HDO in D_2O (solid symbols) and for CaF_2 which exhibits an instantaneous response. The solid lines represent the fit of the data (see text). Dash-dotted line: Monoexponential decay with a time constant of 33 fs.

simultaneously in the sample. The CaF_2 signal is caused by a nonresonant electronic third order nonlinearity displaying an instantaneous response. Thus, the measured time evolution is directly proportional to the third-order correlation function of the mid-infrared pulses from which delay zero and the time resolution of the experiment are derived. The asymmetry of the cross correlation function reflects a slight asymmetry of the pulse envelopes which originates from the parametric generation process of the mid-infrared pulses.

The signal from HDO measured at 3400 cm^{-1} shows a maximum at a delay time of 45 fs, i.e. a peak shift which is due to the inhomogeneous broadening of the vibrational transition. At longer delays, the signal decreases by several orders of magnitude within several hundred femtoseconds. For time delays longer than 250 fs, this decay is close to an exponential with a decay time of 33±5 fs, distinctly slower than the decrease of the third-order correlation function of the pulses. Similar transients with identical decay times are found at the other spectral positions.

Results of the 3-pulse photon echo experiments are presented in Fig. 9 [20]. The photon echo peak shift and the transient grating signal are plotted as a function of the population time T. In agreement with the data in Fig. 8, the peak shift at T=0 has a value of approximately 50 fs. The peak shift decays to about 20 fs within the first 200 fs, independent of the spectral position. At longer times, the peak shift in the low-frequency part of the O-H stretching band is very different from the nearly constant values of 10 to 20 fs at the two other spectral positions: At 3300 cm^{-1}, the peak shift rises again up to a value of 50 fs at T=2 ps and eventually decays to less than 20 fs at T=4 ps.

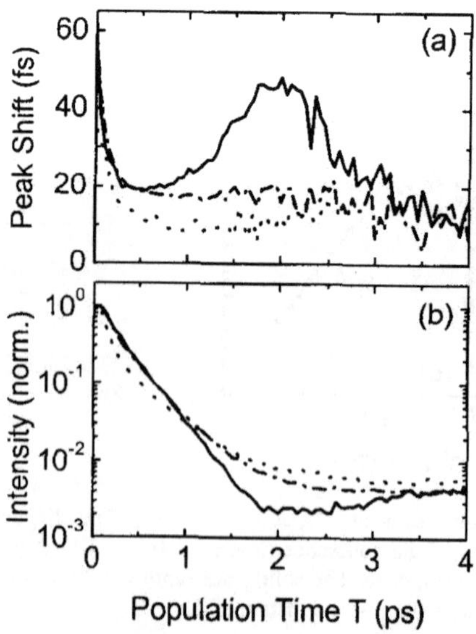

Figure 9. Results of the 3-pulse photon echo experiments. (a) Photon-echo peak shift as a function of the population time T for pulses at 3300 cm^{-1} (solid line), 3400 cm^{-1} (dash-dotted line) and 3500 cm^{-1} (dotted line). (b) Normalized transient grating signal for the same spectral positions.

The transient grating signals in Fig. 9 (b) display exponential decays with a time constant of approximately 300 fs at short population times T. For T>1 ps, the signals at 3400 and 3500 cm^{-1} merge into a small constant value. At 3300 cm^{-1}, the intensity has a minimum at 2 ps and rises again to a constant value which remains nonzero up to population times of 100 ps. The peak shift and the transient grating data remain unchanged for a variation of the pulse energies by a factor of five, excluding artifacts from higher order nonlinearities. Propagation effects distorting the envelopes and optical phase of the pulses can be neglected in our optically thin sample (optical density 0.4 at 3400 cm^{-1}).

The photon echo data reveal a very fast decay of phase coherence with identical decay times throughout the O-H stretching band of HDO in D_2O. The measured decay times are much shorter than the v=1 lifetime of HDO of about 1 ps and – thus – vibrational population relaxation plays a minor role for dephasing. The peak-shift measured in the 3-pulse photon echo experiments displays (i) a finite value of 50 fs for a population time T=0, giving evidence of inhomogeneous broadening of the O-H band, and (ii) a time dependence which depends on the spectral position. Most surprisingly, the peak shift at 3300 cm^{-1} shows a maximum around T=2 ps. A quantitative interpretation of the photon echo results requires a theoretical analysis of the third-order response of the system by considering the relevant nonlinear response functions which are directly related to the frequency fluctuation correlation function of the liquid [21].

In Ref. [19], we have presented an analysis of the 2-pulse photon echo data based on the simplifying assumption of a separation of the time scales of dephasing and inhomogeneous broadening. An exponential correlation term with a time constant τ_c was introduced for the fluctuations in the liquid which underlie dephasing and the inhomogeneous broadening was considered quasi-static. In the calculation of the nonlinear response, the v=0, v=1, and v=2 levels of the O-H oscillator and the respective optical transitions were taken into account and all 3-level Feynman diagrams were included (c.f. Fig. 6). Both the absorption spectrum and the time-dependent 2-pulse photon echo were fitted with this model (solid lines in Figs. 7 and 8). It was found that within this model the data can be well explained revealing a pure dephasing time $T_2^*=(\Delta_h^2\tau_c)^{-1}=90$ fs (Δ_h: fluctuation amplitude, $\tau_c<T_2$: correlation time). This $T_2^*=90$ fs corresponds to a homogeneous linewidth of 120 cm^{-1}. This result clearly demonstrates that a substantial part of the total width of the O-H stretching band of 235 cm^{-1} (FWHM) is caused by fast dephasing processes in the liquid. Including spectral diffusion processes with a time constant of 700 fs [11] or on longer time scales does not change the characteristics of the fast dephasing dynamics.

An analysis of the photon echo peak shift and transient grating data extending well into the picosecond regime requires a more sophisticated approach for the frequency fluctuation correlation function. In particular, models including only 3-level Feynman diagrams cannot account for the delayed maximum of the peak shift in the low-frequency part of the O-H stretching bands and for the nonzero transient grating signal at long population times T. As is discussed in detail in Ref. [20], the v'=0 and v'=1 levels of vibrationally hot molecules and the corresponding blue-shifted vibrational transition have to be taken into account, in addition to the 3 levels of the cold molecules (Fig. 6). The relaxation of excited molecules from the v=1 level to the v'=0 state which is connected with a loss of phase memory leads to a bleaching signal on the $v=0\rightarrow1$ transition which exists for much longer time than any contribution from the v=1 state and eventually decays by thermal relaxation. As a result, the corresponding frequency grating component survives for very long times and allows the observation of slow spectral diffusion processes. In addition, there are signal contributions from the blue-shifted $v'=0\rightarrow1$ transition which is uncorrelated to the $v=0\rightarrow1$ transition.

A detailed model calculation of the peak shift and the transient grating signals which is based on this 5-level scheme, has been presented in Ref. [20]. The experimental findings are well reproduced by calculations with a frequency fluctuation correlation function that – in addition to the fast component discussed above – contains two exponentially decaying (Kubo) terms with time constants of τ_1=700 fs and τ_2=15 ps. The slow τ_2 component of spectral diffusion implies a long-lived phase memory in the system, corresponding to structures in the liquid persisting in the picosecond domain. The delayed maximum in the peak shift originates from the interference of contributions decaying and/or being formed on a time scale of T_1, the relaxation time of the v=1 level and from the long-lived component in the frequency fluctuation correlation function.

Vibrational frequency fluctuation correlation function for the 3 vibrational modes of H_2O dissolved in D_2O have been derived from molecular dynamics simulations [22]. Such functions are very close to biexponential decays with time constants of 50 fs and 800 fs. A comparison with our photon echo results suggests that the fast fluctuation component underlies the fast dephasing as observed in the 2-pulse photon echo study, and the slower component is identical to the 700 fs dynamics of spectral diffusion. So far, simulations for the time domain of the slowest 15 ps component found in our measurements are not available.

The dephasing rates measured here for the stretching vibration of hydrogen-bonded O-H groups are about one order of magnitude higher than those of other stretching vibrations. The stretching band of free O-H groups of H_2O monomers in organic solvents shows a spectral width of ~35 cm^{-1}, corresponding to dephasing times >300 fs [23]. This suggests that the coupling between the vibrational oscillator and the fluctuating forces exerted by the solvent is strongly enhanced upon formation of hydrogen bonds. From a theoretical point of view, vibrational pure dephasing has been described by expanding the interaction potential V between the oscillator and the solvent along the vibrational coordinate Q of the oscillator [24]:

$$V=\left[\frac{dV}{dQ}\right]_{Q=0}Q+\frac{1}{2}\left[\frac{d^2V}{dQ^2}\right]_{Q=0}Q^2+..=F_1Q+F_2Q^2+..$$

This gives – up to the second order in Q – the (fluctuating) frequency shift $\delta\omega(t)$:

$$\hbar\delta\omega(t)=(Q_{11}-Q_{00})F_1(t)+\left[(Q^2)_{11}-(Q^2)_{00}\right]F_2(t)+..$$

Here, Q_{ii} and $(Q^2)_{ii}$ represent expectation values in the v=0 and v=1 states. For a harmonic oscillator, the difference $(Q_{11}$-$Q_{00})$ vanishes whereas $[(Q^2)_{11}$-$(Q^2)_{00}]$ has a nonzero value. For an anharmonic oscillator, $(Q_{11}$-$Q_{00})$ is nonzero and $[(Q^2)_{11}$-$(Q^2)_{00}]$ becomes much smaller than in the harmonic case. Thus, anharmonicity makes the oscillator sensitive to the fluctuating force $F_1(t)$ but reduces the influence of the second order term. It has been shown for different anharmonic systems that indeed the first-order term dominates the total pure dephasing rate [24,25]. Expressing $(Q_{11}$-$Q_{00})$ in terms of the intramolecular anharmonicity Δ which is the difference between the v=1→2 and the v=0→1 transition frequencies of the oscillator, one derives a pure dephasing rate Γ

$$\Gamma\propto\frac{\Delta}{\hbar^2\omega\mu}\int_0^\infty\langle F_1(t)F_1(0)\rangle dt$$

Here, ω and μ are the mean oscillation frequency and the reduced mass of the oscillator, respectively. The dephasing rate is proportional to the time integral of the correlation function of the fluctuating force $F_1(t)$ with a prefactor depending linearly on the anharmonicity. Upon formation of a hydrogen bond, the anharmonicity of the oscillator potential is strongly enhanced and makes the oscillator much more sensitive to the fluctuating force. In the present case, the anharmonicity of the O-H stretching oscillator has a value of about 270 cm^{-1}, much higher than that of the free O-H vibration of <100 cm^{-1}. Concomitantly, the hydrogen bonded system studied here displays dephasing rates which are about 3 times higher than for the free O-H vibration (dephasing time >300 fs). Compared to this strong effect, the changes in the second order term are of minor importance.

In conclusion, we have presented the first photon echo study of liquid water. Our results demonstrate that fluctuations in such a hydrogen bonded liquid occur over a broad range in time, extending from sub-100 fs dynamics up to about 15 ps. The frequency fluctuation correlation function derived from our experiments exhibits a fast sub-100 fs component dominating the vibrational dephasing dynamics and two slower components with time constants of 700 fs and 15 ps underlying the complex spectral diffusion dynamics within the O-H band.

5 Acknowledgements

Part of this work was supported by the Deutsche Forschungsgemeinschaft (Sonderforschungsbereich 450) and by the Fonds der Chemischen Industrie. We thank J. Manz, O. Kühn, H. Naundorf and G. Paramonov, Freie Universität Berlin, for valuable discussions.

References

1. Pimentel G.L., McClellan A.L., The Hydrogen Bond (W.H. Freemann, San Francisco 1960).
2. Hadzi, D. (Ed.), Theoretical Treatments of Hydrogen Bonding (John Wiley & Sons, Chichester 1997).
3. Henri-Rousseau, O., Blaise, P., The infrared spectral density of weak hydrogen bonds within the linear response theory, *Adv. Chem. Phys.* **103** (1998) pp. 1-186.
4. Maréchal Y., Witkowski A.J., Infrared spectra of hydrogen bonded systems, *J. Chem. Phys.* **48** (1968) pp. 3697-3705.
5. Rösch N., Ratner M.A., Model for the effects of a condensed phase on the infrared spectra of hydrogen-bonded systems, *J. Chem. Phys.* **61** (1974) pp. 3344-3351.
6. Bratos S., Profiles of hydrogen stretching IR bands of molecules with hydrogen bonds – stochastic theory. 1. Weak and medium strength hydrogen bonds, *J. Chem. Phys.* **63** (1975) pp. 3499-3509.
7. Robertson G.N., Yarwood J., Vibrational relaxation of hydrogen bonded species in solution, part 1 and 2, *Chem. Phys.* **32** (1978) pp. 267-299.
8. Staib A., Hynes J.T., Vibrational predissociation in hydrogen-bonded OH...O complexes via OH stretch OO stretch energy transfer, *Chem. Phys. Lett.* **204** (1993) pp. 197-205.
9. Laenen R., Rauscher C., Laubereau A., Dynamics of local substructures in water observed by ultrafast infrared hole burning, *Phys. Rev. Lett.* **80** (1998) pp. 2622-2625.
10. Woutersen S., Emmerichs U., Bakker H.J., Femtosecond mid-IR pump-probe spectroscopy of liquid water: Evidence for a two-component structure, *Science* **278** (1997) pp. 658-660.
11. Gale G.M., Gallot G., Hache F., Lascoux N., Bratos S., Leicknam J.C., Femtosecond dynamics of hydrogen bonds in liquid water, *Phys. Rev. Lett.* **82** (1999) pp. 1068-1071.
12. Deak, J.C., Rhea S.T., Iwaki L.K., Dlott D.D., Vibrational energy relaxation and spectral diffusion in water and deuterated water, *J. Phys. Chem. A* **104** (2000) pp. 4866-4875.

13. Kaindl R.A., Wurm M., Reimann K., Hamm P., Weiner A.M., Woerner M., Generation, shaping, and characterization of intense femtosecond pulses tunable from 3 to 10 μm, *J. Opt. Soc. Am. B* **17** (2000) pp. 2086-2094.
14. Hamm P., Kaindl R.A., Stenger J., Noise suppression in femtosecond mid-infrared light sources, *Opt. Lett.* **25** (2000) pp. 1798-1800.
15. Stenger J., Madsen D., Dreyer J., Nibbering E.T.J., Hamm P., Elsaesser T., Coherent response of hydrogen bonds in liquids probed by ultrafast vibrational spectroscopy. *J. Phys. Chem. A* **105** (2001) pp. 2929-32.
16. Madsen D., Stenger J., Dreyer J., Hamm P., Nibbering E.T.J., Elsaesser T., Femtosecond mid-infrared pump-probe study of wave packet motion in a medium-strong intramolecular hydrogen bond, *Bull. Chem. Soc. Jpn.,* submitted.
17. Paramonov G.K., Naundorf H., Kühn O., Ultrafast multidimensional dynamics of strong hydrogen bonds, *Eur. Phys. J. D* **14** (2001) pp. 205-215.
18. Madsen D., Stenger J., Dreyer J., Nibbering E.T.J., Hamm P., Elsaesser T., Coherent vibrational ground-state dynamics of an intramolecular hydrogen bond. *Chem. Phys. Lett.* **341** (2001) pp. 56-62.
19. Stenger J., Madsen D., Hamm P., Nibbering E.T.J., Elsaesser T., Ultrafast vibrational dephasing of liquid water. *Phys. Rev. Lett.* **87** (2001) 027401.
20. Stenger J., Madsen D., Hamm P., Nibbering E.T.J., Elsaesser T., A photon echo peak shift study of liquid water, *J. Phys. Chem. A*, submitted.
21. Mukamel S., *Principles of nonlinear optical spectroscopy* (Oxford University Press, New York 1995).
22. Diraison M., Guissani Y., Leicknam J.Cl., Bratos S., Femtosecond solvation dynamics of water: Solvent response to vibrational excitation of the solute, *Chem. Phys. Lett.* **258** (1996) pp. 348-351.
23. Graener H, Seifert G., Vibrational and orientational relaxation of monomeric water molecules in liquids, *J. Chem. Phys.* **98** (1993) pp. 36-45.
24. Rey R., Hynes J.T., Vibrational phase and energy relaxation of CN^- in water, *J. Chem. Phys.* **108** (1998) pp. 142-153 and references therein.
25. Oxtoby D.W., Levesque D., Weis J.J., Molecular dynamics simulation of dephasing in liquid nitrogen, *J. Chem. Phys.* **68** (1978) pp. 5528-5533.

Femtosecond Mid-infrared Spectroscopy of Aqueous Solutions

H. J. Bakker and M. F. Kropman
FOM Institute AMOLF, Kruislaan 407, 1098 SJ Amsterdam, The Netherlands

We report on the direct measurement of the dynamics of water molecules in the solvation shell of an ion in aqueous solution. We find that the lifetime of the O–H stretch vibration of HDO molecules in the solvation shell of the halogenic anions Cl^-, Br^-, and I^- is much longer than the lifetime of the O–H stretch vibration of the HDO molecule in bulk D_2O solution. This difference in lifetime allows for a clear separation of the response of the solvation shell from that of the bulk liquid. A detailed investigation of the spectral dynamics of the solvating HDO molecules reveals that the hydrogen-bond dynamics of aqueous solvation shells of ions are surprisingly slow compared to the hydrogen-bond dynamics of bulk liquid water.

1 Introduction

The chemical reactivity of (bio)molecules in an aqueous environment is largely determined by the strong and specific solvation (hydrogen-bond) interactions with the surrounding water molecules. Unfortunately, there is very little information on these interactions, mainly because most experimental techniques cannot distinguish the response of water molecules in the solvation shell of a molecule or ion from the response of bulk water.

Up to now, most of the information on the properties of aqueous solvation shells originates from studies on ionic solutions. For these systems, structural information on the solvation shell has been obtained with infrared spectroscopy on small gas-phase clusters,[1,2,3,4] cw (continuous wave) Raman spectroscopy,[5,6,7] nuclear magnetic resonance (NMR) studies,[8]ons and molecular dynamics simulations.[9,10,11,12,13] In many of these studies evidence was found that halogenic anions are solvated by water via directional $O\text{–}H\cdots Y^-$ hydrogen-bond interactions ($Y^- = F^-, Cl^-, Br^-, I^-$). In these studies also information was obtained on the number of water molecules in the first solvation shell of the ion and on the distance between these molecules and the ion.

An elegant method to get experimental information on the dynamics of aqueous solvation shells is to excite and detect a probe molecule dissolved in water.[14] In this type of experiment, the dynamical properties of the solvating water molecules are deduced from the time-dependent response of the probe molecule. A disadvantage of this technique is that it probes the dynamics of solvating water molecules indirectly.

Recently, it was demonstrated that the microscopic dynamics of water can be measured with nonlinear spectroscopic techniques that employ femtosecond mid-infrared laser pulses.[15,16,17,18,19,20,21,22,23] With this technique it was found that the reorientational motion of water shows a bimodal distribution that is strongly correlated to the strength of the hydrogen-bond interactions between the water molecules.[15] It was also observed that an excitation of the O–H stretching mode is extremely rapid transferred from one water molecule to another, leading to a rapid delocalization of this excitation over the liquid.[20] The techniques has also been used to investigate the hydrogen-bond dynamics of liquid water, and it was found that the hydrogen bond between two water molecules stretches and contracts on a very short time scale, with a characteristic time constant of approximately 500 femtoseconds.[18,19] In this contribution, we report on the use of femtosecond mid-infrared spectroscopy to separate the response of water molecules in the solvation shell of an anion from the response of the rest of the water molecules.

2 Experiment

The dynamics of aqueous solvation shells are studied with two-color femtosecond mid-infrared pump-probe experiments. The mid-infrared pulses are generated via parametric frequency-conversion processes that are pumped with the output of a Ti:sapphire regenerative amplifier. These pulses have a central wavelength of 800 nm, an energy per pulse of 3 mJ and a duration of approximately 100 femtoseconds. The repetition rate of the system is 1 kHz. In a first multi-pass parametric generation process in a BBO crystal (4 mm length), part of the pulse at 800 nm is converted into signal and idler pulses at wavelengths of approximately 1250 nm and 2200 nm, respectively. The idler pulse at 2200 nm is frequency doubled in a second BBO crystal (2 mm length) to generate a pulse at 1100 nm. This pulse is used as a seed pulse in a parametric amplification process in a KTP crystal (5 mm length) that is pumped by another part of the 800 nm pulse. This process results in the amplification of the pulse at 1100 nm and the generation of a pulse with a wavelength of approximately 3000 nm that is used in the experiment. The setup contains two chains of parametric generation and amplification processes to generate independently tunable pump and probe pulses. The wavelength of these pulses can be varied between 2.7 and 4 μm by rotating the BBO and KTP crystals. The pulses have a typical energy of 20/2 μJ (pump/probe) and a duration of approximately 200 femtoseconds.[19] The spectral bandwidths of the pump and probe pulses are 80 and 60 cm^{-1}, respectively.

We performed femtosecond mid-infrared pump-probe experiments on the O–H stretch vibration of HDO molecules in an aqueous solution consisting of HDO (0.5 M) in D_2O and different concentrations (1 M, 2 M, 3 M, 6 M) of KF, NaCl, NaBr, NaI, and $MgCl_2$. The intense mid-infrared pump pulse excites a

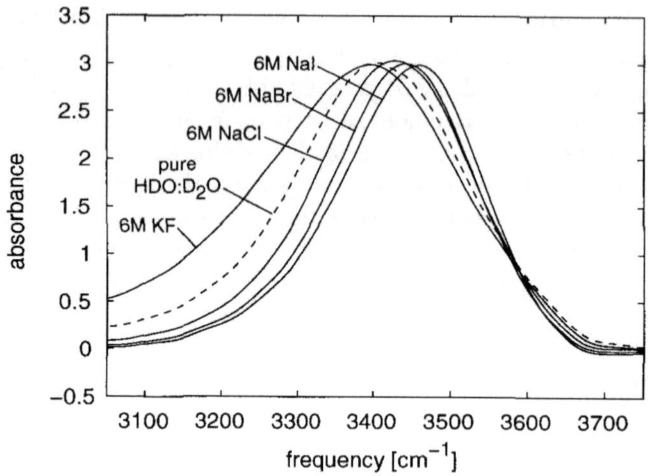

Figure 1: Absorption spectra of a solution of 0.5 M HDO in D_2O and of solutions of 0.5 M HDO and 6 M KF, NaCl, NaBr, or NaI in D_2O.

significant fraction of the HDO molecules to the first excited state ($v_{OH} = 1$) of the O–H stretch vibration. The energy relaxation and spectral dynamics of this excitation are probed with the weak, independently tunable probe pulse. The polarization of the probe pulse is at the magic angle (54.7^o) with respect to the polarization of the pumping pulse in order to avoid the measurements to be affected by reorientation of the HDO molecules. In the experiments, both the pump and probe frequencies are tuned through the absorption band of the O–H stretching mode of the HDO molecules.

3 Results

In Fig. 1, the absorption spectrum of the O–H stretch vibration of HDO is shown for solutions of 6 M KF, NaCl, NaBr, and NaI in HDO:D_2O. Within the halogenic series (F^-, Cl^-, Br^-, I^-), the absorption spectrum of the O–H stretch vibration is observed to shift to higher frequencies.

In Fig. 2, pump-probe transients are shown for a solution of 0.5 M HDO in D_2O and for a solution of 0.5 M HDO and 6 M NaI in D_2O. For both solutions, a bleaching signal ($\ln(T/T_0) > 0$, with T the transmission of the probe and T_0 the transmission of the probe in absence of the pump) is observed. This bleaching results from the depletion of the $v_{OH} = 0 \rightarrow 1$ transition, and decays due to population relaxation out of the $v_{OH} = 1$ state. For the HDO:D_2O solution,

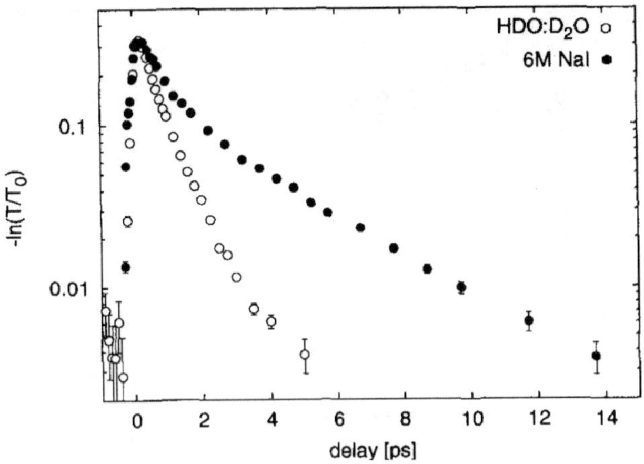

Figure 2: Pump-probe transients measured with a pump and probe frequency of 3400 cm^{-1} for a solution of 0.5 M HDO in D_2O and for a solution of 0.5 M HDO and 6 M NaI in D_2O.

the decay is single-exponential with a time constant of 740 femtoseconds, as has been observed before.[17] In contrast, for the solution containing NaI, the relaxation is non-exponential and evolves from a fast initial decay into a slow decay with a time constant of approximately 3.5 picoseconds. To investigate the origin of the second component, we varied the concentration, and we also performed experiments on solutions of KF, NaCl, NaBr, and $MgCl_2$.

In Fig. 3, pump-probe transients are shown that are measured for four different salt solutions. For all salts, the observed transients can be described as a sum of two components of which the first shows an exponential decay with a time constant of 0.8±0.1 ps. The second component decays exponentially with a time constant that strongly depends on the dissolved salt. For KF, NaCl, NaBr, and NaI we find time constants of 0.7±0.2 ps, 2.6±0.3 ps, 3.4±0.3 ps, and 3.9±0.3 ps, respectively. The amplitude of the second component scales linearly with the concentration of dissolved salt. Interestingly, changing the nature of the cation has no effect on the experimental observations: solutions of $MgCl_2$ and NaCl give exactly the same results if the concentration of Cl^- is the same. In contrast, the time constant does strongly depend on the anion, and increases within the halogenic series F^-, Cl^-, Br^- to I^-. In view of these observations, the second component is assigned to HDO molecules that solvate the anion via the formation of an $O\text{–}H\cdots Y^-$ hydrogen-bond (with $Y^- = F^-, Cl^-, Br^-, I^-$). The time constant of this component corresponds to the vibrational lifetime T_1 of these HDO molecules.

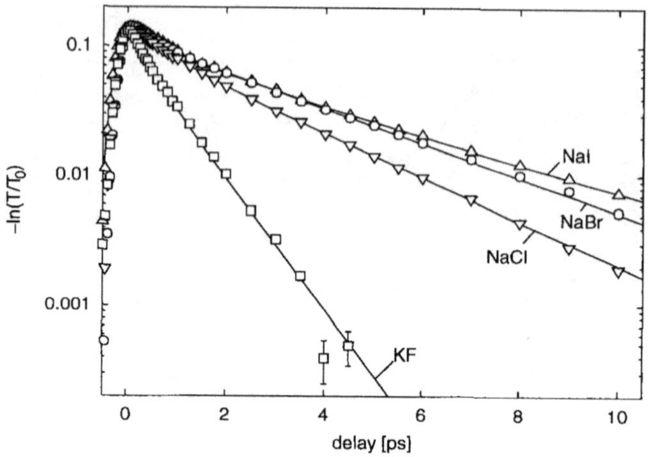

Figure 3: Pump-probe transients measured for aqueous solutions of different salts in HDO:D_2O. The transients were measured using a pump frequency of 3450 cm^{-1} and a probe frequency of 3200 cm^{-1}, resonant with the $v = 1 \rightarrow 2$ transition. For the solutions of NaCl, NaBr, and NaI, a biexponential decay is observed with time constants of 0.8±0.1 ps for the first component, and of 2.6±0.3 ps (NaCl), 3.4±0.3 ps (NaBr), and 3.9±0.3 ps (NaI), for the second component.

The amplitude and time constant of the first component do not depend on the concentration of dissolved salt, which suggests that this component mainly results from bulk HDO molecules. This interpretation is supported by the observation that this component has a time constant of 800 femtoseconds, which is very similar to the vibrational lifetime T_1 observed for a solution of HDO in D_2O.[17] The first component likely also contain the response of the HDO molecules in the first solvation shell of the cations, because it was observed that the cations have no effect on the amplitude and the time constant of the second component. For the HDO molecules near the cation, the O–H groups will point away from the ion and will form O–H···O hydrogen bonds with bulk D_2O molecules.[9,10,11] For O–H···O hydrogen-bonded systems, it has been found that the lifetime of an O–H stretch vibration strongly depends on the strength of the O–H···O hydrogen-bond interaction.[24,25] The O–H···O hydrogen bonds formed by HDO molecules in the first solvation shell of the cation with bulk D_2O molecules will be similar to the hydrogen bonds formed between bulk HDO and bulk D_2O. As a result, the O–H stretch vibration of an HDO molecule in the first solvation shell of a cation is expected to have a similar vibrational lifetime T_1 as the O–H stretch vibration of an HDO molecule in the bulk liquid, in agreement with the observations.

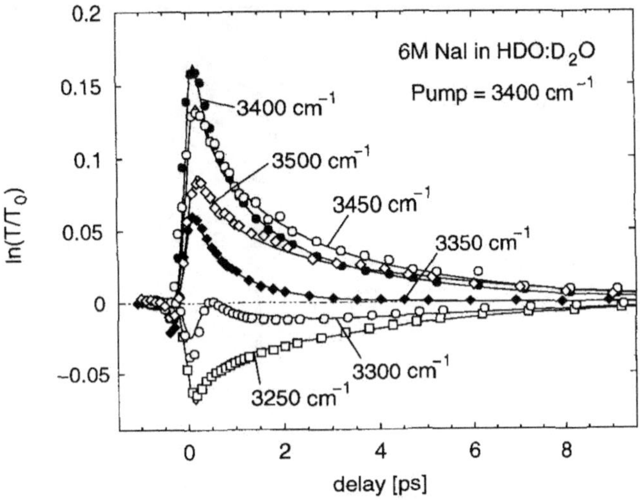

Figure 4: Pump-probe transients measured for an aqueous solution of 6 M NaI in HDO:D_2O. The transients are measured at six different probe frequencies and a pump frequency of 3400 cm^{-1}. The solid curves are calculated with the Brownian oscillator model described in the text.

In Fig. 4, pump-probe transients are presented for a solution of 6M NaI in HDO:D_2O. These transients are measured at a pump frequency of 3400 cm^{-1} and six different probe frequencies. At probe frequencies $\geq$3350 cm^{-1}, a transmission increase ($\ln(T/T_0) > 0$) is observed due to the bleaching of the $v_{OH} = 0 \rightarrow 1$ transition, whereas at frequencies $\leq$3300 cm^{-1} a transmission decrease ($\ln(T/T_0) < 0$) is observed due to the $v_{OH} = 1 \rightarrow 2$ induced absorption. At a probe frequeny of 3300 cm^{-1}, which is close to the frequency at which the induced absorption turns into a bleaching, the observed dynamics is very complicated, which indicates that the spectral response of the O–H stretch vibration of HDO molecules that solvate the I^- ion differs from that of HDO molecules dissolved in bulk D_2O.

In Fig. 5, transient spectra measured for a solution of 6M NaI in HDO:D_2O are presented at four different delays after excitation with an intense pump pulse at 3400 cm^{-1}. These spectra correspond to the pump-probe transients shown in Fig. 4. At short delays after the excitation, the maximum of the bleaching is near the central frequency of the pump (indicated by the arrow). This bleaching is significantly narrower ($\approx$ 120 cm^{-1}) than the width of the linear absorption spectrum ($\approx$ 250 cm^{-1}), shown as the dashed curve for comparison. Hence, the excitation leads to a spectral hole in the O–H stretch

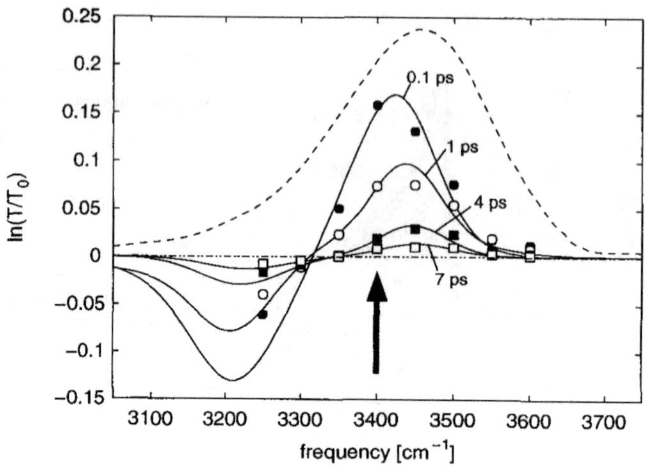

Figure 5: Transient spectra measured at four delays for an aqueous solution of 6 M NaI in HDO:D_2O. The spectra are measured with a pump frequency of 3400 cm^{-1}. The solid curves are calculated with the Brownian oscillator model described in the text.

absorption band, which shows that only a subset of the O–H oscillators is excited. The excitation of a subset of the O–H stretch oscillators implies that a subset of the distribution in hydrogen-bond lengths is selected, because the O–H stretch frequency is strongly correlated to the O–H···O and O–H···I^- hydrogen-bond lengths.[26,27]

At somewhat longer delay times (up to 2 ps), the transient spectrum shows a broadening due to the hydrogen-bond dynamics of liquid water. As a result of these dynamics, the hydrogen bonds are continuously stretched and contracted which, due to the correlation between the O–H stretch frequency and the hydrogen-bond length, leads to a broadening of the spectral hole. At even longer delay times (> 2 ps), the transient spectrum shows a clear blue shift and becomes narrower again. This effect is only observed if the HDO:D_2O liquid contains Cl^-, Br^-, or I^- ions. To explain this latter observation, it should be realized that after a few picoseconds only excited O–H···I^- are observed, because all excited O–H···O oscillators have relaxed. Hence, the transient spectra shown in Fig. 1 after 4 and 7 picoseconds are (transient) absorption spectra of O–H···I^- oscillators only. This observation constitutes an important advantage of time-resolved nonlinear spectroscopy over conventional cw spectroscopic techniques: the response of the water molecules that solvate the Cl^-, Br^- or I^- can be separated with high selectivity from the response of all the other water molecules by simply choosing the appropriate delay-time window.

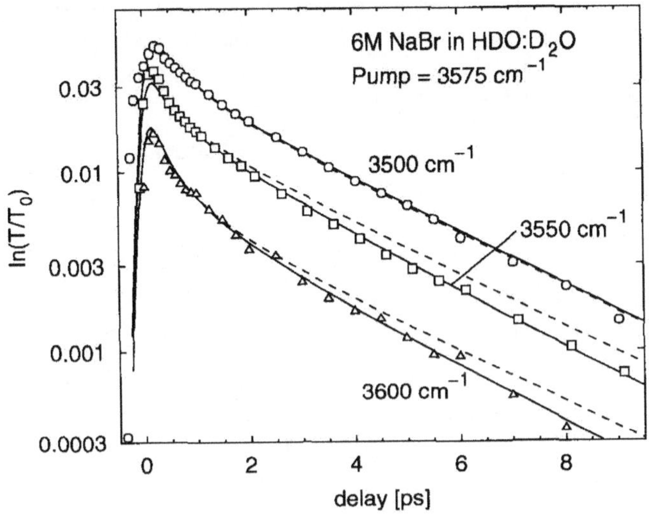

Figure 6: Pump-probe transients measured for an aqueous solution of 6 M NaBr in $HDO{:}D_2O$. The transients are scaled with respect to each other and are plotted on a logarithmic scale. The solid curves are calculated with a τ_c of of 25 picoseconds for the modulation of the $O\text{–}H\cdots Br^-$ hydrogen bond. The dashed curves are calculated with the same model using $\tau_c = \infty$.

For the $O\text{–}H\cdots Br^-$ and $O\text{–}H\cdots I^-$ components, the largest relative amplitude is observed for pump and probe frequencies of approximately 3500 cm^{-1}. For the $O\text{–}H\cdots Cl^-$ component, the largest relative amplitude is observed for pump and probe frequencies of approximately 3450 cm^{-1}.

In Figs. 6 and 7, pump-probe measurements of solutions of NaBr and NaCl in $HDO{:}D_2O$ are presented. These transients again consist of a fast decay, associated with bulk HDO molecules and HDO molecules that solvate the Na^+ ions, and a slow decay due to HDO molecules that solvate the halogenic anions. Interestingly, the time constant of the second component shows a small but significant dependence on the probe frequency. To illustrate this more clearly, Figs. 6 and 7 contain calculated dashed curves that run parallel at large delay times (> 3 ps). The time constant of the second component is observed to increase with increasing frequency difference between pump and probe. If the pump is at 3575 cm^{-1} (Fig. 6), the fastest decay is observed at 3600 cm^{-1} and the slowest decay at 3500 cm^{-1}; if the pump is at 3525 cm^{-1} (Fig. 7), the fastest decay is observed at 3500 cm^{-1} and the slowest decay at 3600 cm^{-1}. This shows that the decay time constant of the second component is affected by a slow spectral diffusion process with a time constant

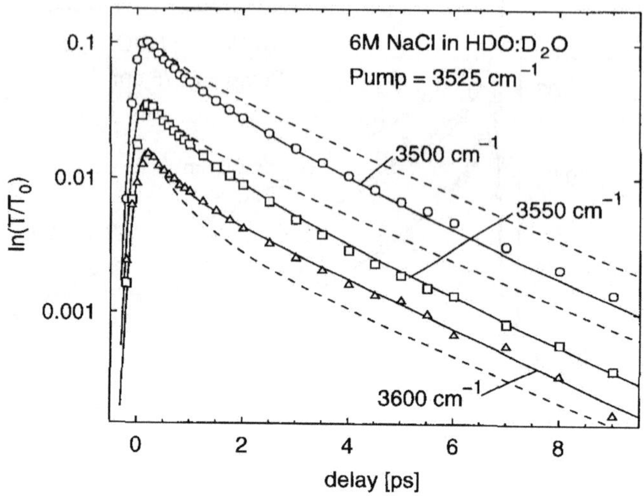

Figure 7: As Figure 6, but for a solution of 6 M NaCl in HDO:D_2O. The solid curves are calculated using a τ_c for the modulation of the O–H$\cdots$$Cl^-$ hydrogen bond of 12 picoseconds, the dashed curves are calculated with $\tau_c = \infty$.

of approximately 10 picoseconds. Due to this spectral diffusion process, excited molecules (spectrally) diffuse away from the excitation frequency, which leads to a faster decay at probe frequencies close to the pump frequency and to a slower decay at probe frequencies that differ from the pump frequency.

4 Discussion

4.1 Vibrational lifetimes of HDO in aqueous solvation shells

The vibrational lifetimes T_1 of the O–H stretch vibration of an HDO molecule in the first solvation shell of Cl^-, Br^-, and I^- are exceptionally long compared to the vibrational lifetime of approximately 800 femtoseconds of an HDO molecule dissolved in bulk D_2O. For the latter system, the O–H stretch vibrational energy was found to be accepted either by a combination of the HOD bending mode [21] and the O–H$\cdots$O hydrogen bond, or by the hydrogen bond alone.[28] The long vibrational lifetimes of the HDO molecules in the solvation shell of the larger anions can be understood if the O–H$\cdots$$Y^-$ hydrogen bond accepts at least part of the energy of the excited O–H stretch vibration. Then, there are two ways in which the long vibrational lifetimes can be explained from the properties of the O–H$\cdots$$Y^-$ hydrogen bond.

First, it is observed that the O–H stretch frequency increases (the redshift from the gas-phase value decreases) in the series F^-, Cl^-, Br^-, I^-(see Fig. 1 and Refs. 29 and 30), which indicates that the strength of the hydrogen-bond interaction decreases.[25] The weaker hydrogen-bond interaction in turn leads to a decrease of the anharmonic interaction terms between the O–H stretch vibration and the hydrogen-bond mode,[25] and thereby to a slow-down of the vibrational relaxation.

Second, the reduced mass of the O–H···Y^- hydrogen-bond stretch vibration, which determines the energy spacing of the levels of this mode, increases within the halogenic series F^-, Cl^-, Br^-, I^-. As a result, the level spacing of the hydrogen-bond vibration decreases. Therefore, the dissipation of the excitation energy of the O–H stretch vibration will involve a higher level of excitation of the hydrogen-bond vibration for I^- than for F^-. The anharmonic interaction term responsible for the energy relaxation of O–H···I^- will thus be of higher order in the hydrogen-bond coordinate than the anharmonic interaction term responsible for the relaxation of O–H···F^- . In general, the magnitude of the anharmonic interaction decreases with increasing order in the vibrational coordinates.[25,29] At this moment, it is not clear which of the two above-mentioned effects, strength of the anharmonic interaction with the hydrogen bond, or reduced mass of the hydrogen-bond vibration, is most important in slowing down the relaxation in the O–H···Y^- series.

The vibrational lifetime of an HDO molecule solvating F^- is comparable to that of HDO dissolved in bulk D_2O. For water molecules solvating this anion, the absorption spectrum is not blueshifted but slightly redshifted with respect to the absorption spectrum of HDO dissolved in D_2O. In addition, the reduced mass of the O–H···F^- hydrogen bond is similar to that of the O–H···O hydrogen bond. Hence, the T_1 of HDO molecules solvating F^- is expected to be similar to that of HDO dissolved in bulk D_2O, in perfect agreement with the experimental observations.

4.2 Hydrogen-bond dynamics of HDO in aqueous solvation shells

In the previous section it was shown that the spectral response of the O–H stretch absorption band of HDO molecules forming an O–H···Y^- hydrogen bond (with $Y^- = Cl^-, Br^-, I^-$), is affected by a slow spectral diffusion process with a time constant of approximately 10 picoseconds (Figs. 6 and 7). This spectral diffusion directly reflects the (stochastic) modulation of the hydrogen-bond length, because the transition frequency of the O–H stretch vibration and the length of the hydrogen bond are strongly correlated[26,27,30].

In order to determine the precise value of the correlation time constant τ_c of the stochastic modulation of the O–H···Y^- hydrogen bond, we modeled the data of Figs. 4-7 with the Brownian oscillator model.[31] The Brownian oscillator model assumes harmonic-potential energy functions for the low-frequency mode (hydrogen bond) that are displaced with respect to each other in the ground and excited states of the O–H stretch oscillator. In case the hydrogen-bond mode is strongly overdamped, a Gaussian absorption spectrum for the O–H stretch vibration results, in quite good agreement with the experimental observations. Within the Brownian oscillator model, the spectral dynamics of the transition of the O–H stretch vibrational transition is described with two parameters: the correlation time constant τ_c, representing the time scale at which the hydrogen-bond length is stochastically modulated, and the width Δ of the Gaussian absorption line shape $e^{-\omega^2/2\Delta^2}$ of the O–H stretch vibration. This width increases with increasing displacement of the harmonic-potential energy functions. The pump-probe signal $\ln(T/T_0)$ as a function of delay τ and pump and probe frequencies is given by:

$$\ln(T/T_0)(\omega_1,\omega_2,\tau) = \frac{2\pi e^{-\tau/T_1}}{\sqrt{(\Delta^2+\delta_1^2)\alpha^2(\tau)}} e^{(-\omega_1-\omega_{eg})^2/2(\Delta^2+\delta_1^2)} \times \left\{ e^{-[\omega_2-\omega_g(\tau)]^2/2\alpha^2(\tau)} + e^{-[\omega_2-\omega_e(\tau)]^2/2\alpha^2(\tau)} \right\}, \tag{1}$$

with

$$\omega_g(\tau) = \omega_{eg} + e^{-\tau/\tau_c}(\omega_0 - \omega_{eg}) \tag{2}$$

$$\omega_e(\tau) = \omega_{eg} - 2\lambda + e^{-\tau/\tau_c}(\omega_0 - \omega_{eg} + 2\lambda) \tag{3}$$

$$\lambda = \hbar\Delta^2/2k_BT \tag{4}$$

$$\omega_0 = \omega_1 \frac{\Delta^2}{\Delta^2+\delta_1^2} + \omega_{eg}\frac{\delta_1^2}{\Delta^2+\delta_1^2} \tag{5}$$

$$\alpha^2(\tau) = \Delta^2\left[1 - \frac{\Delta^2}{\Delta^2+\delta_1^2}e^{-\tau/\tau_c}\right] + \delta_2^2, \tag{6}$$

where ω_1, δ_1 and ω_2, δ_2 are the center frequencies and the spectral widths of pump and probe, respectively. The frequency ω_{eg} is the maximum of the linear absorption spectrum and T_1 is the vibrational lifetime. The parameter λ represents half the Stokes shift. The Stokes shift follows from the difference between the positions of the minima of the hydrogen-bond potentials of the ground and excited state of the O–H stretch vibration. Excitation of the O–H stretch vibration to the $v_{OH} = 1$ state thus results in a shift of the hydrogen-bond length to a new equilibrium position and an associated redshift of 2λ of the stimulated emission out of the $v_{OH} = 1$ potential to the $v_{OH} = 0$ potential. For harmonic potentials, the Stokes shift is directly related to the width Δ of

Table 1: Parameters of the two Brownian oscillators used to describe the data shown in Figs. 4-7. All values are determined for a concentration of anions of 6 M.

	ω_{eg}	$\Delta\omega$	T_1	τ_c
O–H···O	3420±10 cm^{-1}	280±20 cm^{-1}	0.8±0.1 ps	0.5±0.2 ps
O–H···Cl^-	3440±15 cm^{-1}	160±15 cm^{-1}	2.6±0.3 ps	12±3 ps
O–H···Br^-	3470±15 cm^{-1}	130±15 cm^{-1}	3.0±0.3 ps	25±5 ps
O–H···I^-	3490±15 cm^{-1}	105±15 cm^{-1}	3.7±0.3 ps	18±5 ps

the absorption band. The Stokes shift does not depend on temperature, since it is only determined by the shape and displacement of the harmonic potentials. Nevertheless, in expression (4) the temperature T enters, because the width Δ of the absorption band depends on the thermal occupation of the ground-state potential. The two terms between the braces in equation (1) correspond to the bleaching (depletion of the $v = 0$), and to the stimulated emission $v = 1 \rightarrow 0$ contribution to the transmission change $\ln(T/T_0)$.

All transients can be well fitted with two Brownian oscillators that represent the O–H···O and O–H···Y^- components, respectively. In Table 1, the central frequency ω_{eg}, the spectral width $\Delta\omega$ (full width at half maximum: $\Delta\omega = \sqrt{8\ln 2}\Delta$), the vibrational lifetime T_1, and the correlation time τ_c, that result from these fits, are presented.

The results of the calculations are shown in Figs. 4-7 by the solid curves. The O–H···O component has the same set of parameters for all different salt solutions. These parameters are quite similar to those of a pure solution of HDO in D_2O. Only the width of the absorption band is somewhat larger (280 cm^{-1} instead of 260 cm^{-1}), which possibly results from the contribution of HDO molecules solvating the cations. The calculated curves are sensitive to the value of the correlation time constant τ_c. We find τ_c=12±3 ps for Cl^-, τ_c=25±5 ps for Br^-, and τ_c=18±5 ps for I^-. For comparison, in Figs. 6 and 7 also results obtained with $\tau_c = \infty$ (dashed curves) are shown. These curves run parallel at large delays, because for $\tau_c = \infty$ the transients are no longer affected by spectral diffusion. This behavior is in clear contrast to the experimental observations.

The central frequencies ω_{eg} of the O–H···Cl^-, O–H···Br^- and O–H···I^- absorption components are in excellent agreement with the results from a study in which a double-difference spectroscopic technique was used to obtain the change in O–D stretch frequency due to salt addition.[32] After multiplying by 1.36 to convert O–D into O–H stretch frequencies,[33] the central absorption frequencies found in these studies are 3441 cm^{-1} (O–H···Cl^-), 3476 cm^{-1} (O–H···Br^-), and 3495 cm^{-1} (O–H···I^-).

The central frequency ω_{eg} of the absorption band of the O–H$\cdots$Y$^-$ component increases in the halogenic series Cl$^-$, Br$^-$, I$^-$, whereas the width of the absorption band decreases within this series. Both trends reflect the decrease of the strength of the hydrogen-bond interaction between the solvating HDO molecule and the anion in the halogenic series. The relation between the O–H stretch vibrational frequency and the length of the O–H$\cdots$X hydrogen bond has been measured for many hydrogen-bond acceptors X, including the halogenic anions.[27,30] This relation can be used to determine the distribution of O–H$\cdots$Y$^-$ hydrogen-bond lengths from the width of the O–H$\cdots$Y$^-$ absorption components. For the O–H$\cdots$Cl$^-$, O–H$\cdots$Br$^-$, and O–H$\cdots$I$^-$ hydrogen-bonds, we obtain widths of the length distribution of $0.20(0.05)\times10^{-10}$ m, $0.21(0.05)\times10^{-10}$ m, and $0.12(0.04)\times10^{-10}$ m, respectively. These widths are small compared to the width of the O–H$\cdots$O absorption band of HDO:D_2O of $0.36(0.02)\times10^{-10}$ m. It should be realized that the abovementioned widths form upper limits, because the O–H$\cdots$Y$^-$ absorption component may also contain contributions that do not result from the variation in hydrogen-bond length. Hence, the true O–H$\cdots$Y$^-$ hydrogen-bond distributions may be narrower.

The values of τ_c of the solvation shells are surprisingly long in comparison to the value of τ_c of the O–H$\cdots$O in bulk liquid water. For a pure solution of HDO in D_2O, the spectral diffusion of the O–H stretch absorption band was observed to have a correlation time constant τ_c of approximately 500 femtoseconds.[18,19] As a result, the excited spectrum (spectral hole) of the O–H stretch vibration of HDO dissolved in D_2O acquires the shape of the linear absorption spectrum within a few picoseconds.[18] The much longer time scale of the spectral diffusion of the solvating HDO molecules implies that the stochastic modulation of the O–H$\cdots$Y$^-$ hydrogen-bond length is much slower than the stochastic modulation of the O–H$\cdots$O hydrogen-bond length in bulk liquid water.

The small width and long τ_c of the O–H$\cdots$Y$^-$ absorption component imply that the water molecules that directly bind to the Y$^-$ halogenic anion form a relatively stable and well-defined structure. It should be noted that the solvation shells of F$^-$ and of the cations likely show similar dynamics as the solvation shells of Cl$^-$, Br$^-$ and I$^-$, but these dynamics could not be measured because the O–H stretch vibrational lifetime of the water molecules in these solvation shells is comparable to that of bulk HDO:D_2O.

The value of τ_c is observed to be somewhat longer for the solvation shells of Br$^-$ and I$^-$ than for the solvation shell of Cl$^-$. This observation can be well explained within the framework of the Brownian oscillator model. Following this model, the time constant τ_c is related to the frequency ω_{HB} of the hydrogen-bond stretch vibration via $\tau_c = \gamma/\omega_{HB}^2$,[34] with γ the damping. The dependence of τ_c on ω_{HB} can be understood from the fact that tight bonds (high ω_{HB}) show faster dynamics than weak bonds (low ω_{HB}), leading to a faster decay of $\langle R_{HB}(\tau)R_{HB}(0)\rangle \propto e^{-\tau/\tau_c}$.[35] Exchanging Cl^- for Br^- or I^- leads to a decrease of ω_{HB},[36] because the hydrogen bond becomes weaker and because the reduced mass of the hydrogen-bond vibration increases. Hence, if the damping γ of the hydrogen-bond stretch vibration is similar for O–H···Cl^-, O–H···Br^-, and O–H···I^- hydrogen bonds, the observed increase of τ_c going from Cl^- to Br^- and I^-, can be well explained from a decrease of the hydrogen-bond stretch frequency ω_{HB}.

5 Conclusions

We employed two-color femtosecond mid-infrared spectroscopy to study the dynamics of the aqueous solvation shells of the halogenic anions F^-, Cl^-, Br^-, and I^-. We find that the O–H stretch vibrational lifetimes of HDO molecules solvating Cl^-, Br^-, and I^- are much longer than the lifetime of 800 fs of the O–H stretch vibration of HDO dissolved in D_2O. As a result, a few picoseconds after excitation, only HDO molecules in the solvation shells of these anions remain excited, which enables a highly specific study of the spectral dynamics of these molecules.

The spectral dynamics of the HDO molecules solvating Cl^-, Br^-, and I^- are studied by measuring transients at different pump and probe frequencies. From this study, we find that the stochastic modulation of the length of the hydrogen bond between the water molecules and the halogenic anions has a characteristic time constant τ_c of 10–25 ps, which is 20–50 times longer than the value of τ_c of the O–H···O hydrogen bonds of bulk liquid water. In addition, the distribution of lengths of the hydrogen bond between the water molecules and the halogenic anions is observed to be much narrower than the distribution of hydrogen-bond lengths in pure liquid water. The long τ_c and the narrow distribution of hydrogen-bond lengths indicate that the aqueous solvation shell forms a relatively long-living and well-defined structure. The value of τ_c is observed to be larger for the aqueous solvation shells of Br^- and I^- than for the solvation shell of Cl^-, which can be explained from the dependence of τ_c on the frequency of the hydrogen-bond vibrations between the HDO molecules and the solvated anions. Finally, it should be noted that for F^- and the cations the hydrogen-bond dynamics likely will be as slow as for the large halogenic anions. However, these dynamics could not be measured with the present technique, because for these ions the vibrational lifetime of the solvating HDO molecules is as short as for HDO dissolved in D_2O.

Acknowledgment

This research is part of the research program of the Stichting Fundamenteel Onderzoek der Materie (Foundation for Fundamental Research on Matter) and was made possible by financial support from the Nederlandse Organisatie voor Wetenschappelijk Onderzoek (Netherlands Organization for the Advancement of Research).

References

1. C. G. Bailey, J. Kim, C. E. H. Dessent, and M. A. Johnson, Chem. Phys. Lett. **269**, 122 (1997)
2. J.-H. Choi, K. T. Kuwata, Y.-B. Cao, and M. Okumura, J. Phys. Chem. A **102**, 503 (1998)
3. O. M. Cabarcos, C. J. Weinheimer, J. M. Lisy, and S. S. Xantheas, J. Chem. Phys. **110**, 5 (1999)
4. P. Ayotte, C. G. Bailey, G. H. Weddle, and M. A. Johnson, J. Phys. Chem. A **102**, 3067 (1998)
5. G. E. Walrafen, J. Chem. Phys. **36**, 1035 (1962)
6. G. E. Walrafen, J. Chem. Phys. **40**, 3249 (1964)
7. G. E. Walrafen, J. Chem. Phys. **52**, 4176 (1970)
8. H. G. Hertz, in *The Chemical Physics of Solvation, Part B Spectroscopy of Solvation* ed. R. R. Dogonadze, E. Kálmán, A. A. Kornyschev, and J. Ulstrup (Elsevier, Amsterdam, 1986), Chap. 7
9. K. Hashimoto and K. Morokuma, Chem. Phys. Lett. **223**, 423 (1994)
10. T. Asada and K. Nishimoto, Chem. Phys. Lett. **232**, 518 (1995)
11. L. M. Ramaniah, M. Bernasconi, and M. Parrinello, J. Chem. Phys. **109**, 6839 (1998)
12. G. Peslherbe, B. M. Ladanyi, and J. T. Hynes, J. Phys. Chem. A **104**, 4533 (2000)
13. A. Chandra, Phys. Rev. Lett. **85**, 768 (2000)
14. R. Jimenez, G. R. Fleming, P. V. Kumar, and M. Maroncelli, Nature **369**, 471 (1994)
15. S. Woutersen, U. Emmerichs, and H. J. Bakker, Science **278**, 658 (1997)
16. R. Laenen, C. Rauscher, and A. Laubereau, Phys. Rev. Lett. **80**, 2622 (1998)
17. S. Woutersen, U. Emmerichs, H.-K. Nienhuys, and H. J. Bakker, Phys. Rev. Lett. **81**, 1106 (1998)
18. G. M. Gale, G. Gallot, F. Hache, N. Lascoux, S. Bratos, and J.-C. Leicknam, Phys. Rev. Lett. **82**, 1086 (1999)
19. S. Woutersen and H. J. Bakker, Phys. Rev. Lett. **83**, 2077 (1999)

20. S. Woutersen and H. J. Bakker, Nature **402**, 507 (1999)
21. J. Deak, S. Rhea, L. Iwaki, and D. Dlott, J. Phys. Chem. **A104**, 4866 (2000)
22. R. Laenen, T. Roth, and A. Laubereau, Phys. Rev. Lett. **85**, 503 (2000)
23. M. Kropman and H. J. Bakker, Science **291**, 2118 (2001)
24. R. E. Miller, Science **240**, 447 (1988)
25. A. Staib and J.T. Hynes, Chem. Phys. Lett. **204**, 197 (1993)
26. A. Novak, Struct. Bonding **18**, 177 (1974)
27. W. Mikenda, J. Mol. Struct. **147**, 1 (1986))
28. H.-K. Nienhuys, S. Woutersen, R. A. van Santen and H. J. Bakker, J. Chem. Phys. **111**, 1494 (1999)
29. A. Nitzan and J. Jortner, Mol. Phys. **25**, 713 (1973)
30. W. Mikenda, S. Steinböck, J. Mol. Struct. **384**, 159 (1996))
31. S. Mukamel, *Principles of Nonlinear Optical Spectroscopy* (Oxford University Press, New York, 1995)
32. P.-Å. Bergström and J. Lindgren, J. Phys. Chem. **95**, 8575 (1991)
33. M. J. Wójcik, J. Lindgren and J. Tegenfeldt, Chem. Phys. Lett. **99**, 112 (1983)
34. G. N. Robertson and J. Yarwood, Chem. Phys. **32**, 267 (1978)
35. M. Bonn, M. J. P. Brugmans, A. W. Kleyn, R. A. van Santen, and H. J. Bakker, Phys. Rev. Lett. **76**, 2440 (1996)
36. P. Ayotte, G. H. Weddle, J. Kim, and M. A. Johnson, J. Am. Chem. Soc. **120**, 12361 (1998)

ULTRAFAST EXCITED STATE PROTON TRANSFER: REACTIVE DYNAMICS BY MULTIDIMENSIONAL WAVEPACKET MOTION

S. LOCHBRUNNER, K. STOCK, V. DE WAELE, AND E. RIEDLE

Lehrstuhl für BioMolekulare Optik, Sektion Physik, Ludwig-Maximilians-Universität, Oettingenstr. 67, 80538 München, Germany
E-mail: stefan.lochbrunner@physik.uni-muenchen.de

The ultrafast excited state intramolecular proton transfer was investigated in a number of molecules with identical and differing heteroatoms in the H-chelate ring. Time resolved absorption measurements with 30 fs resolution were performed with a fully tunable UV-visible pump probe spectrometer. A very similar behavior of all systems is found. The strongly Stokes shifted product emission is delayed by a small but finite amount and all signals display strong oscillatory contributions. The step like emission rise indicates that the reactive process proceeds as a ballistic wavepacket motion from the Franck-Condon region to the product well. The delay in the order of 50 fs results from the inertia of the participating nuclei and not from rate determining coupling matrix elements. The signal oscillations are due to ringing in a finite number of product normal modes that are displaced by the proton transfer reaction. From the detailed analysis of the periodic wavepacket motion a clear microscopic picture of the initial dynamics and the molecular deformations is reconstructed.

1 Introduction

Ultrafast photoreactions proceed on the time scale of vibrational motions and have to be described as wavepacket dynamics on potential energy surfaces (PES). A rate description in a state to state picture can be quite misleading. With a temporal resolution better than the period of relevant skeletal vibrations one should be able to follow the motion of the wavepacket along the potential energy surface even in large molecules in solution and gain direct insight into the reaction mechanism. In particular the observation of coherently excited vibrations reveals the changes of the molecular geometry during the process.

We apply this approach to excited state intramolecular proton transfer (ESIPT). ESIPT allows to study reactive proton dynamics under well defined conditions. In the electronically excited state, the hydrogen atom of an hydroxy group that is part of a H-chelate ring is transferred to an acceptor atom on the other side of the ring (see Fig. 1) [1,2]. The ESIPT leads to a characteristic and pronounced Stokes shift between the absorption in the UV and the fluorescence appearing in the visible (see Fig. 2). For several ESIPT molecules a time scale of 60 fs for the process has been found [3,4] and a number of studies show indications for wavepacket dynamics and coherent excitation of vibrational modes [5-7]. By time resolved spectroscopy on 2-(2'-hydroxyphenyl)benzothiazole (HBT) we found very pronounced coherent vibrational motions after an ESIPT process continuing for many vibrational periods and related them to structural changes during the ESIPT [8,9].

Here we report on our recent transient absorption measurements with a temporal resolution of 30 fs on selected ESIPT molecules (see lower part of Fig. 1). Two of them, HBT and 2-(2'-hydroxyphenyl)benzoxazole (HBO) have a nitrogen atom as proton acceptor. In the case of the other two, o-hydroxybenzaldehyde (OHBA) and 1-hydroxy-2-acetonaphtone (HAN), the proton switches between two oxygen atoms.

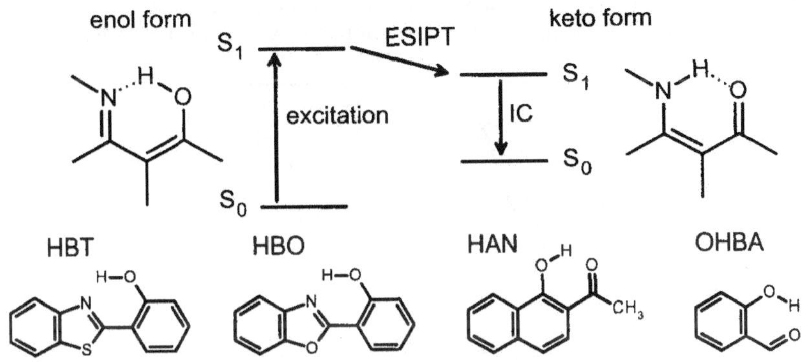

Figure 1. Enol- and keto-form of a H-chelate ring and the energetics for the excitation, ESIPT and internal conversion (IC). The molecules shown in the lower part are investigated in the presented study.

For all of them we find signatures of a ballistic wavepacket motion and characteristic oscillatory signal contributions. The vibrational modes involved in the ESIPT have been identified by comparing the observed frequencies with results from *ab-initio* calculations. From this we deduce the deformations of the molecular geometry during the process.

2 Experimental

Time resolved absorption measurements are performed with a fully tunable UV-visible pump probe spectrometer based on two noncollinearly phasematched optical parametric amplifiers providing a time resolution of 30 fs [8,10]. The ESIPT molecules are excited at their first absorption band by laser pulses with a center wavelength of about 340 nm (see Fig. 2). The pulses are generated by frequency doubling the output of one NOPA in a 100 µm thick BBO-crystal and subsequently compressed in a fused silica prism sequence leading to a pulse duration of about 25 fs. To avoid two photon absorption a low pump pulse energy of 60 nJ (focused to 100 µm diameter) is chosen and only about 5 % of the molecules are excited. As probe pulses with ten times less energy the reflection of a beam from a second

NOPA off a fused silica substrate is used. Various probe wavelengths between 450 and 680 nm are applied and the transmission change is probed from the blue to the red wing of the product emission (see Fig. 2). The crosscorrelation and time zero at the sample position is determined by difference frequency generation in a 25 μm thick BBO-crystal. The crosscorrelation has typically a duration of 35 fs (FWHM); the time zero is determined to better than ±5 fs.
The ESIPT molecules are solvated in cyclohexane with concentrations between 10^{-3} to 10^{-2} M to achieve an absorption of about 50 % at the excitation wavelength. As sample a 70 μm thick liquid jet or a flow cell with 200 μm thick windows and a channel thickness of 120 μm is used. The transmission change in the pure solvent is measured as well but no significant signal contributions from the solvent are found.

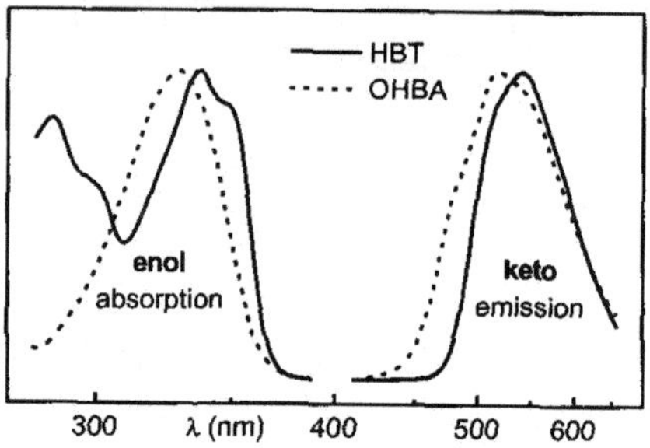

Figure 2. Absorption and fluorescence spectra of HBT and OHBA in cyclohexane. Both molecules show a large Stokes shift which is characteristic for ESIPT molecules [1].

3 Observations

The time dependent transmission change in the spectral region of the product fluorescence is shown in Figure 3 for excitation of HBT and HBO. After an initial decrease the transmission increases with some delay and pronounced oscillatory signal contributions appear. Since the probe wavelengths are far to the red with respect to the absorption spectrum bleaching and ground state dynamics do not contribute to the observed transmission changes. The positive changes are the result of the stimulated emission from the electronically excited state to the ground state of the product. The signal rise indicates the appearance of the product which is the keto form generated by the ESIPT. The delay between the excitation and the transmission rise is attributed to the duration of the ESIPT. The initial transmission decrease is due to excited state absorption (ESA) which appears as soon as the molecule is electronically excited and which does not change much during the ESIPT.

The Fourier transform spectra in Figure 3 show that two strong oscillations and two weaker ones contribute to the signal in the case of HBT. For HBO two strong oscillations with frequencies very similar to HBT are found. The role of these oscillations is discussed below.

To quantify the various signal contributions a model function is fitted to the data. It consists of a transmission decrease at time zero and a step like increase shifted by a variable delay with respect to time zero. The rise of both contributions is convoluted with the experimental crosscorrelation and their decay is modeled by a double exponential function accounting for vibrational redistribution and internal conversion. The oscillatory contributions are fitted by exponentially damped cosine functions. The results of the fitting procedure are shown as solid lines in Figure 3.

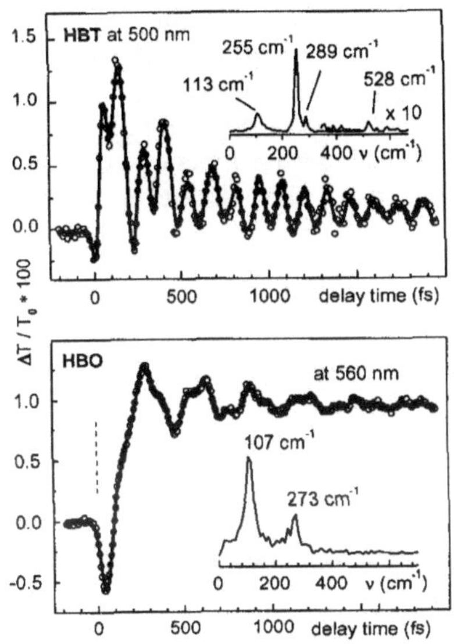

Figure 3. Time resolved transmission change after optical excitation of HBT and HBO (500 nm and 560 nm probe wavelength). The solid lines represent model functions fitted to the data (circles). For details see text. The insets show the Fourier transform spectra of the oscillatory signal components.

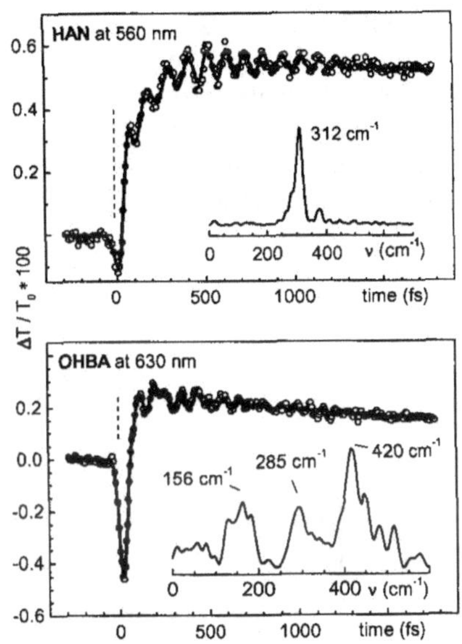

Figure 4. Transient transmission change of HAN and OHBA (560 nm and 630 nm probe wavelength). The solid lines represent model functions fitted to the data (circles). For details see text. The insets show the Fourier transform spectra of the oscillatory signal components.

For HBT the delay of the emission amounts to 30 fs in the blue wing and 50 fs in the red wing of the emission spectrum [9]. The oscillation frequencies of HBT are determined to 113 cm^{-1} and 255 cm^{-1} for the strong and 289 cm^{-1} and 528 cm^{-1} for the weak contributions. The data of HBO can be well fitted with an emission delay of 80 fs and oscillations at 107 cm^{-1} and 273 cm^{-1}.

Figure 4 shows the transient transmission of HAN and OHBA probed in the red wing of the product fluorescence at 560 nm and 630 nm. The experimental traces exhibit similar features as in the case of HBT and HBO, i.e. a transmission increase with a characteristic delay and strong oscillations. The fitting procedure as outlined above results in the solid traces. The delay of the emission is determined to about 30 fs for HAN and 45 fs for OHBA. The HAN trace has strong oscillations with a frequency of 312 cm^{-1} and weaker ones at 370 cm^{-1}. In the case of OHBA oscillations with three frequencies at about 156 cm^{-1}, 285 cm^{-1} and 420 cm^{-1} contribute to the signal [11].

4 Ballistic Wavepacket Motion

The ESIPT process is not associated with strong vibrational dephasing. Otherwise no oscillatory signal contributions would appear after completion of the proton transfer. The wavepacket prepared by the excitation stays well defined during the process and does not spread to a large extent. This is in agreement with the fact that the delayed transmission rise can be well modeled by a step function convoluted with the crosscorrelation instead of an exponential rise. This observation is demonstrated in Figure 5.

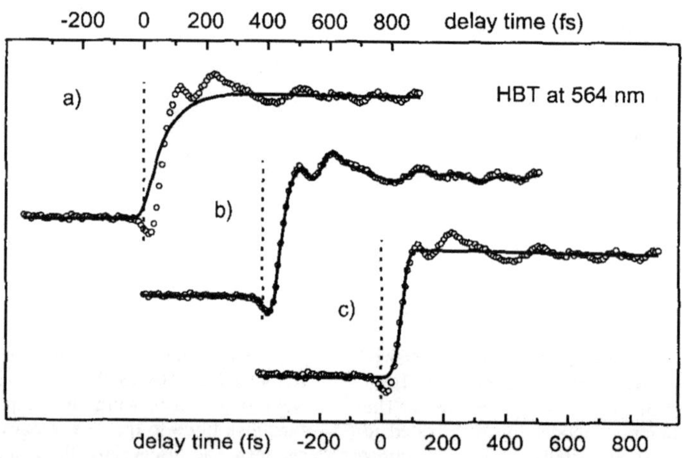

Figure 5. Fit of an exponential (a), solid line) and a delayed, step like (c), solid line) emission increase to the transient transmission of HBT at a center probe wavelength of 564 nm. The fit of the complete model function is shown in b).

It shows the transient transmission of HBT at 564 nm, i.e. near the maximum of the emission spectrum where the oscillations are much less pronounced as in the wings (see below). The data (circles) is modeled by a simple exponential increase ($\tau = 68$ fs, a), solid line) and a delayed step like rise ($\Delta t = 65$ fs, c), solid line). Both approaches give the same time scale but the exponential increase deviates at short delay times strongly from the data trace whereas the step like rise fits much better and resembles the characteristic form of the trace. The full fit shown in b) includes the ESA and the oscillatory contributions. By taking the ESA into account the fitted emission delay reduces to 50 fs.

An exponential time dependence is expected if the relevant process can be characterized by the depopulation of one state and the corresponding population of other states via a rate governed process. This situation exists for instance if there is a comparatively weak coupling between two states and Fermi's Golden Rule can be applied [12] or the system has to cross a significant energy barrier in order to arrive at the product form [13]. In contrast, a step like signal change indicates that the originally prepared wavepacket reaches the detection window within a short interval at a well defined time. The wavepacket has to be rather confined in this case and the process can be characterized as a movement of the complete wavepacket without pronounced spreading [8,9,14]. For the ballistic wavepacket motion the delay results from the inertia of the participating nuclei and not from rate determining coupling matrix elements.

The emission delay in the order of 50 fs is the time the wavepacket needs to move from the Franck-Condon region of the enol form to the product minimum, i.e. the keto form. It represents the duration of the proton transfer. This time scale is too long for a barrierless displacement of only the proton indicating that the ESIPT involves a deformation of the molecular skeleton [4,9]. The nature of this deformation can be deduced from the observed oscillations as discussed below.

5 Coherent Excitation of Vibrations

The observed signal oscillations are due to coherently excited vibrational modes in the product form of the molecule. Contributions from coherent motions in the ground state can be excluded because the spectral detection region is shifted far to the red of the ground state absorption.

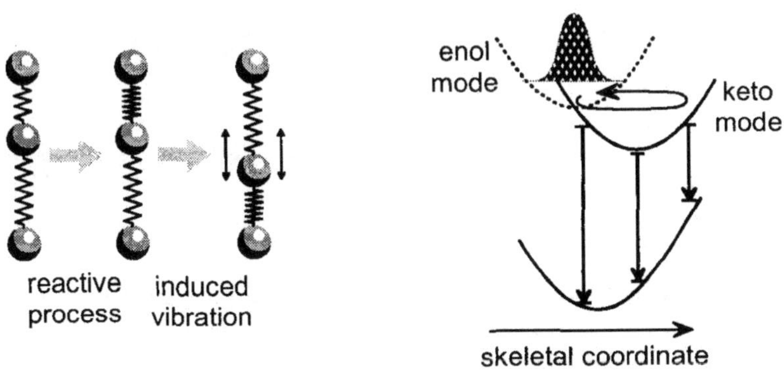

Figure 6. Ringing due to fast change of the electronic configuration: The reactive process leads to a fast change of the electronic configuration and the bond strengths. Thereafter the molecule oscillates around the equilibrium geometry associated with the new bond strengths leading to a periodic shift of the optical transition frequency.

Vibrations can be coherently excited by a sudden change of the electronic configuration leading to a redistribution of electron density. This situation is sketched in Figure 6.

If bond strengths change within a sufficiently short time interval the molecular geometry does not adopt adiabatically to the new equilibrium associated with the new bonds due to the inertia of the nuclei. At first the molecule is displaced with respect to the new equilibrium geometry along those coordinates where the potential minimum is shifted by the electronic configuration change and it starts to oscillate in the corresponding vibrational modes. A sudden electronic configuration change can appear for example due to an optical excitation or during a reactive process. Both is encountered in the case of ESIPT molecules. The coherent vibrational motion of the molecule leads to a periodic shift of the mean transition frequency and to oscillations of the spectrum along the frequency axis. In the case of HBT the strength of the ESA does not vary strongly in the visible [15] and spectral signal changes are mostly due to the shape of the emission spectrum. Therefore a very close correlation exists between the probe wavelength dependent amplitude of an oscillatory signal contribution and the slope of the emission spectrum. This is depicted in Figure 7 for the dominant contribution observed in the case of HBT (255 cm^{-1}).

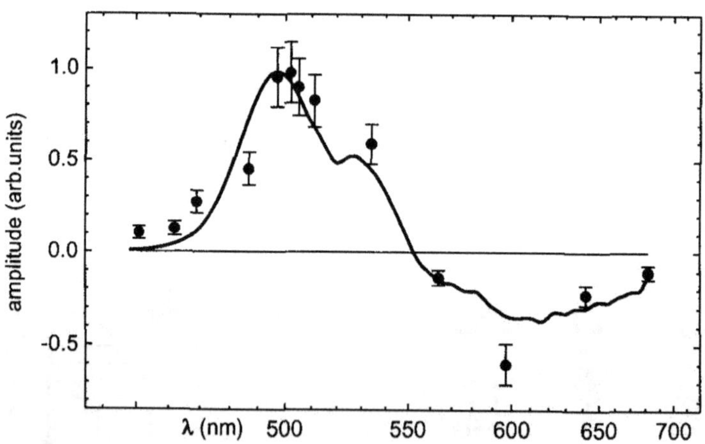

Figure 7. Amplitude of the experimentally observed oscillations at 255 cm^{-1} (circles) and slope of the emission cross section (solid line) of HBT.

The specific mechanism responsible for the coherent excitation of vibrational modes can be determined from the phase of the signal oscillations. At the time a vibrational motion starts the molecule is at the turning point and the total phase of a cosine function can be regarded to be zero. From the fitting result for the phase the starting time of an oscillation can be calculated modulo half of the vibrational period. This analysis was performed in the case of HBT [15]. It was found that the three higher frequency vibrations start at about 30 fs to 50 fs after the optical excitation, i.e. when the visible emission is observed for the first time. Since this emission is attributed to the keto form these vibrational motions are excited by the electronic configuration change associated with the change from the enol to the keto form. This points to a configuration change within a short time interval and not to a gradual variation during the ESIPT process. For the vibration with the lowest frequency the phase corresponds to a starting time shifted by a quarter of a vibrational period relative to the configuration change. The molecule is initially accelerated along this coordinate; this indicates that the nuclear motion is induced by the electronic configuration change associated with the optical excitation.

6 Deformation of the Molecular Geometry

The observed signal oscillations are due to a characteristic "ringing" of the product molecule in vibrational modes which are involved in the geometric changes during the reaction. The specific modes and the associated displacements of the nuclei are identified by comparing the frequencies with *ab-initio* results.

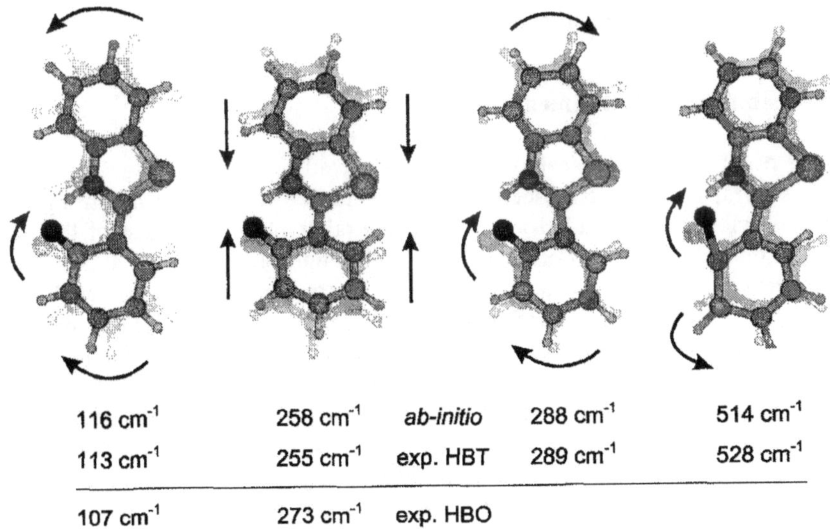

Figure 8. Coherently excited vibrations of keto-HBT. In HBO coherent excitation of vibrations is observed which correspond to the two lower frequency modes. The listed frequencies result from *ab-initio* calculations on HBT and from fits to HBT and HBO data.

For HBT the equilibrium geometry and the normal modes of the S_1-state were calculated using the CIS method with the 6-31G* basis set as implemented in the Gaussian98 program package [16]. The normal modes of OHBA in the electronic ground state have been calculated by the DFT method using the B3LYP functional and the 631G** basis set. In Figure 8 those vibrational modes of HBT are shown which contribute to the oscillations observed by the experiment. They involve in plane bending motions that shorten the distance between the oxygen and nitrogen atom. According to the analysis of the calculated reaction path a change along this coordinate drives the proton transfer. In HBO only the sulfur is replaced by an oxygen in comparison to HBT and it can be safely assumed that the vibrational modes of both molecules are very similar. Obviously HBO oscillates in modes corresponding to the two dominant ones of HBT. For OHBA we encounter a very similar situation [11].

For all ESIPT molecules investigated we observe the coherent excitation of only a few frequency vibrations which are in-plane deformations of the H-chelate ring. The analysis of the phases for the case of HBT reveals that the optical excitation leads to strong acceleration of the molecule along the coordinate belonging to the lower frequency mode of the two dominant ones. It is a bending motion of the entire molecule which reduces mostly the proton donor/acceptor distance (see Fig. 8). Then the electronic configuration and thereby the bonds change to those of the keto form. Subsequently coherent vibrational motions take place which move the nuclei of the chelate ring towards their new equilibrium positions. Several normal modes of the resulting keto configuration are needed to describe the intrinsic deformations and they are coherently excited.

7 Irreversibility by Multidimensionality

The ESIPT occurs on a time scale of 50 fs and a well defined wavepacket is observed for several hundred femtoseconds after excitation. Obviously no strong vibrational redistribution on a sub-100 fs time scale takes place and most of the energy stays in the contributing coordinates for much longer than the ESIPT duration. This raises the question why no back reaction occurs and the ESIPT is irreversible.

Our results indicate that several vibrational degrees of freedom participate in the ultrafast reaction. This multidimensional character is responsible for the irreversibility of the ESIPT without the need for a loss of coherence in individual modes on this time scale. For a reversal of the reaction a full recurrence of the wavepacket in all coordinates would be needed. This would take much longer than the period of a representative vibrational motion. Only on this strongly extended time scale the coupling to other coordinates leads to a flow of energy out of the reactive coordinate space or even into the surrounding solvent and the back reaction becomes impossible.

8 Conclusions

In conclusion, our transient absorption measurements show that the ESIPT proceeds as a ballistic wavepacket motion in the skeletal modes. The observed signal oscillations show that a well defined wavepacket exists for more than one picosecond despite the reactive process and the interaction with the solvent. The geometric changes are identified by the ringing of the molecule in the participating coordinates. In this way the microscopic mechanism of the ESIPT process is revealed. After the optical excitation a skeletal bending motion reduces the proton donor-acceptor distance. When it is sufficiently shortened a fast electronic configuration change occurs. Subsequently the molecule is accelerated along additional skeletal coordinates towards the equilibrium geometry associated with the new bond strengths. Besides the electronic degree of freedom several vibrational modes contribute and the essential features of the dynamics can only be understood by a multidimensional model. Nevertheless most of the energetically accessible vibrations do not play an important role and a realistic description of the ESIPT has to consider only a restricted number of vibrational degrees of freedom.

We observe the signature of these dynamics for all investigated molecules. They have reaction centers with different degrees of symmetry. In the case of HBT and HBO the proton donor is an oxygen and the acceptor a nitrogen atom. This is a much less symmetric situation than in HAN and OHBA where the donor and the acceptor are both oxygen atoms. Nevertheless very similar dynamics are observed. This indicates that the proposed mechanism for the ESIPT does not depend on the details of the reaction center. Accordingly, it should be generally valid for ultrafast ESIPT within a H-chelate ring and should also apply to compounds like methyl salicylate, salicylic acid and TINUVIN P.

The ballistic nature of the wavepacket dynamics can be distinguished from rate governed processes if sufficient time resolution is provided [8,9,14]. If the crosscorrelation is too long the differences between the two models are obscured. We believe that many short exponential time constants found by experiments with a time resolution in the order of 100 fs actually represent ballistic wavepacket motions. An exponential fit to such data results indeed in the correct time scale. But care should be taken in interpreting these results in terms of well defined states with a rate like population transfer from one to the other.

After the ESIPT the investigated molecules show a pronounced ringing in the involved vibrational modes. It appears since the electronic configuration changes due to the optical excitation and the ESIPT itself occur on a time scale much shorter than a representative vibrational period. Because of the small electron mass such configuration changes should be generally extremely fast. We expect therefore that ultrafast reactive processes lead very often to such a ringing. It should be observable for all situations where the prepared wavepacket is not split due to a limited transition probability at a significant energy barrier or electronic coupling region and if not a strongly dephasing interaction with the surrounding is involved.

The presented example shows that the analysis of the ringing gives direct insight into the structural changes during an ultrafast process. The normal modes of the product which have large projections on the occurring skeletal deformations are identified by the frequencies of oscillatory signal components. The role of these modes within the process can be determined from the phases. This kind of analysis should allow us to reveal the mechanism of numerous ultrafast processes.

9 Acknowledgement

We are indebted to J. Piel and A. J. Wurzer for their help with the experiment and the data evaluation and W. Zinth for continuous support. Funding by the Deutsche Forschungsgemeinschaft within the Schwerpunktprogramm "Femtosekunden-spektroskopie" is gratefully acknowledged.

References

1. Weller A., Z. Elektrochem. **60** (1956) pp. 1144.
2. Elsaesser T. and Kaiser W., Chem. Phys. Lett. **128** (1986) pp. 231.
3. Arthen-Engeland T., Bultmann T., Ernsting N. P., Rodriguez M. A. and Thiel W., Chem. Phys. **163** (1992) pp. 43.
4. Herek J. L., Pedersen S., Bañares L. and Zewail A. H., J. Chem. Phys. **97** (1992) pp. 9046.
5. Chudoba C., Riedle E., Pfeiffer M. and Elsaesser T., Chem. Phys. Lett. **263** (1996) pp. 622.
6. Jethwa J., Ouw D., Winkler K., Hartmann N. and Vöhringer P., Z. Phys. Chem. **214** (2000) pp. 1367.
7. Ernsting N. P., Kovalenko S. A., Senyushkina T., Saam J. and Farztdinov V., J. Phys. Chem. A **105** (2001) pp. 3443.
8. Wurzer A. J., Lochbrunner S. and Riedle E., Appl. Phys. B **71** (2000) pp. 405.
9. Lochbrunner S., Wurzer A. J. and Riedle E., J. Chem. Phys. **112** (2000) pp. 10699.
10. Riedle E., Beutter M., Lochbrunner S., Piel J., Schenkl S., Spörlein S. and Zinth W., Appl. Phys. B **71** (2000) pp. 457.
11. Stock K., Bizjak T. and Lochbrunner S., Chem. Phys. Lett., to be submitted.
12. Freed K. F., in *Radiationless Processes in Molecules and Condensed Phases*, F. K. Fong, Ed. (Springer, Berlin, 1976) pp. 23.
13. Steinfeld J. I., Francisco J. S. and Hase W. L., *Chemical Kinetics and Dynamics*, (Prentice Hall, Englewood Cliffs, New Jersey, 1991).
14. Cheng P. Y., Zhong D. and Zewail A. H., Chem. Phys. Lett. **237** (1995) pp. 399.
15. Lochbrunner S., Wurzer A. J. and Riedle E., J. Chem. Phys, to be submitted.
16. M. J. Frisch et al., Gaussian 98, Revision A.7.

ULTRAFAST DYNAMICS OF PHOTOACID-BASE REACTION

B. COHEN AND D. HUPPERT

Raymond and Beverly Sackler Faculty of Exact Sciences, School of Chemistry, Tel Aviv University, Tel Aviv 69978, Israel

We have measured the direct proton transfer rate between a photo-acid, 8-hydroxypyrene 1,3,6 - trisulfonate (HPTS) and a base, acetate anion in water using a pump probe technique with ~ 150 fs time resolution. Our data analysis of aqueous solutions of 0.5M-4M sodium-acetate shows a good agreement with the diffusive model. The intrinsic rate constant at contact of the reactive species was found to be $1.5 \cdot 10^{11}$ $M^{-1}s^{-1}$.

1 Introduction

The dynamics of fast bimolecular reactions in solution have, for a long time, been a subject of interest to both experimentalists and theoreticians. The kinetics of many physical, chemical and biological processes are influenced by diffusion [1,2,3,4,5,6,7,8,9,10]. Several of these processes occur in the excited state [9,10,11]. Bimolecular irreversible diffusion influenced reactions between donors and acceptors, in the pseudo-unimolecular limit when one reactant, the acceptor, is in excess, are the subject of the "Smoluchowski theory" [12,13,14,15,16]. The initial decay is faster due to the excess of acceptors at close proximity to a given donor. As this density approaches its steady-state limit, the reaction becomes exponential, with the diffusion-control rate coefficient, k_D.

Proton transfer to solvent (PTTS) from an excited organic photo-acid (ROH*), and the subsequent geminate recombination process can be treated by reversible diffusion formalism [17,18],

$$ROH^* \underset{k_a}{\overset{k_d}{\rightleftarrows}} RO^{-*} + H^+ (aq).$$

Good examples are provided by hydroxyarenes (ROH dyes), such as 1-naphthol, 2-naphthol and their derivatives that become strong acids in the first electronically excited state [2,3,4,9,10,11]. Light serves here as an ultrafast trigger for PTTS.

In PTTS the solvent serves as the base namely, the proton acceptor. Alternately, one may introduce a given concentration of a base, B, into the solution. Weller [3,4] was the first to monitor the kinetics between an excited photoacid (2-naphthol) and anionic bases

$$ROH^* + B^- \xrightarrow{k_{PT}} RO^{-*} + BH \tag{1}$$

Using steady-state fluorescence, he has found a rate constant of 2.9×10^9 $M^{-1}s^{-1}$ for the reaction between 2-naphthol and acetate in water. Laws and Brand [19] were the first to monitor it in the time-domain, fitting a bi-exponential to the decaying fluorescence signal. It was pointed out [20] that Weller's effective rate constant might be limited by diffusion, so that the "intrinsic" proton-transfer rate parameter which we call k_{PT} (to differentiate from Weller's overall rate constant, k_B) may be somewhat higher, $k_{PT} = 6.5 \times 10^9$ $M^{-1}s^{-1}$. Pines and coworkers [21], who have applied extremely large base concentrations in order to measure the direct proton transfer within a photoacid-base pair that has pre-formed already in the ground state,

$$ROH^* \cdots B^- \longrightarrow RO^{-*} \cdots BH$$

find (in 8M acetate) the same rate constant as Weller's.

For the reaction between excited 8-hydroxypyrene-1,3,6-trisulfonate (HPTS) and acetate they report $k_{PT} = 9 \times 10^9$ $M^{-1}s^{-1}$.

In this study we use a femtosecond pump-probe technique to follow the nonexponential dynamics of HPTS with an acetate anion base at intermediate concentrations.

2 Experimental

Pump-probe experiments reported employed a laser system based on the design of reference [22]. In brief, laser pulses (50 fs duration, centered near 800 nm with a pulse energy of ~600 μJ) at a 1 kHz repetition rate were generated by a Ti:Sapphire-based oscillator (Coherent Mira seed) amplified by a multipass Ti:Sapphire amplifier (Odin Quantronix). Each laser pulse was split into two parts: pump and probe. The probe portion was used to generate white-light continuum in a 2mm Sapphire disk. The probe light was obtained by wavelength selection of this continuum using interference filters producing approximately 3 nm of bandwidth. The pump portion of the light was chopped at 35 Hz and optically delayed by a translation stage. Samples were excited by the second harmonic of the amplified laser (~400 nm). In the course of this study we also used a 360 nm pulse, the fourth harmonic of an OPA at 1440 nm (TOPAS, Light Conversion) with an energy per pulse of ~5 μJ. Within the signal to noise of the experimental data we were not able to find a significant difference between the two-excitation wavelengths. The second harmonic at 400 nm or the fourth harmonic of 1440 nm were generated in a 0.5 mm β-BBO crystal. The cross correlation between the pump and the probe pulses is measured to be 250 fs FWHM indicating that after deconvolution our experimental time resolution amounts to about 150 fs.

3 The Smoluchowski Model

According to the Smoluchowski model, the survival probability of a single (static) donor (here, ROH*) due to its irreversible reaction with a concentration $c = [B^-]$ of acceptors is given by[12,14]

$$S(t) = \exp\left(-c\int_0^t k(t')dt'\right) \tag{2}$$

where $k(t)$ is the time-dependent rate coefficient (or, reactive flux at the contact distance, a) for the donor-acceptor pair

$$k(t) = k_{PT}\, p(a,t) \tag{3}$$

whose proton-transfer rate constant is k_{PT}. The pair (ROH*/B^-) density distribution, $p(r, t)$, is governed by a Smoluchowski equation [diffusion in a potential $U(r)$] in three dimensions[1]

$$\partial p(r,t)/\partial t = Dr^{-2}\frac{\partial}{\partial r}r^2 e^{-\beta U(r)}\frac{\partial}{\partial r}e^{\beta U(r)}p(r,t) \tag{4}$$

where D is the relative diffusion coefficient of the pair, $U(r)$ their interaction potential, $\beta = 1/k_B T$ and $k_B T$ is the thermal energy. To calculate $k(t)$, one imposes an initial equilibrium distribution

$$p(r,0) = \exp[-\beta U(r)] \tag{5}$$

and a "radiation" boundary condition at the contact distance ($r = a$), depicting irreversible recombination occurring upon the binary collision [13]

$$4\pi D a^2 e^{-\beta U(a)}\frac{\partial}{\partial r}e^{\beta U(r)}p(r,t)\Big|_{r=a} = k_{PT}\, p(a,t)\,. \tag{6}$$

One of the goals of our study is to determine the intrinsic bimolecular rate coefficient, k_{PT}.

The interaction potential, $U(r)$, was taken as the Debye-Hückel (DH) screened Coulomb potential [23],

$$\beta U(r) = (R_D/r)e^{-\kappa(r-a)}/(1+\kappa a) \tag{7}$$

for the electrostatic repulsion within the negatively charged HPTS/Ac^- pair, with screening due to the NaAc salt. Both the Debye length, R_D, and the inverse of the DH screening length, κ, may be calculated from the static dielectric constant of the solutions, ε, according to

$$\kappa^2 = 8\pi\beta e^2 c/\varepsilon, \quad R_D = \beta e^2/\varepsilon, \tag{8}$$

where e is the electronic charge. Thus $R_D = 7.1$Å for neat water.

Although the DH theory is inaccurate at high salt concentrations, for a 2M salt solution $1/\kappa \approx 2$ Å is so small that the effect of the potential on the results is only marginal. We have solved eq 4 numerically, using a user-friendly Windows application for Spherically-Symmetric Diffusion Problems (SSDP, ver. 2.61) [24] to yield $k(t)$ of eq 3.

For $U(r) \equiv 0$ it is possible to solve the above equations analytically for $k(t)$ [25]. This is no longer true when $U(r) \neq 0$. In this case, Szabo [14] has found an approximate expression for the time dependent rate constant:

$$k(t) = \frac{4\pi D a_e k_{PT} e^{-\beta U(a)}}{k_{PT} e^{-\beta U(a)} + 4\pi D a_e}\left\{1 + \frac{k_{PT} e^{-\beta U(a)}}{4\pi D a_e} e^{\gamma^2 Dt}\,\mathrm{erfc}\left[\left(\gamma^2 Dt\right)^{1/2}\right]\right\} \tag{9}$$

where γ is given by:

$$\gamma = a_e^{-1}\left(1 + \frac{k_{PT} e^{-\beta U(a)}}{4\pi D a_e}\right) \tag{10}$$

erfc is the complementary error function and a_e is an effective radius defined by

$$a_e^{-1} = \int_a^\infty e^{\beta U(r)} r^{-2} dr \tag{11}$$

Equation 9 is exact when the potential is zero, i.e., $U = 0$ and $a_e = a$. When a potential is introduced, it behaves correctly at both $t = 0$ and $t = \infty$

$$k(0) = k_{PT} e^{-\beta U(a)}, \quad k(\infty) = [k(0)^{-1} + k_D^{-1}]^{-1} \tag{12}$$

where $k_D = 4\pi Da_e$ is the diffusion-control rate constant. The non-exponentiality in $S(t)$ is a result of a non-constant $k(t)$.

4 Results

4.1 *Pump-Probe Measurements of HPTS in Neat Water*

The pump-probe transmission signal, ΔT, of HPTS in neat water solution, measured at 540 nm, changes signs from $\Delta T < 0$ to $\Delta T > 0$ at about 50ps. The signal consists of three distinct time regimes, a short time component of ~ 700 fs, a second component of ~3 ps and a long time component of about 100 ps. In basic solution (pH~10) the HPTS molecule is already in the ground state in its deprotonated form, RO^-. The pump-probe signal of RO^{-*} is negative at all times and assumes 70% of the absolute value within 300 fs, about twice the time response of the system. The rest of the signal has a growth time of ~700 fs. A 700 fs component is also seen in the signal of neutral solution where the ROH species are excited.

We attribute the positive part of the signal to absorption due to the optical transition $ROH(S_1) \rightarrow ROH(S_2)$. The negative component arises from the RO^{-*} stimulated emission due to transition to the ground state. The total signal at 540 nm is analyzed by:

$$-\Delta T \propto \sigma_{ROH^*} S(t) - \sigma_{RO^{-*}} (1 - S(t)) \qquad (13)$$

where σ_{ROH^*} denotes the absorption cross section of the S_1 to a higher excited state $S_1 \rightarrow S_n$ transition of ROH and $\sigma_{RO^{-*}}$ denotes the emission cross section of RO^{-*}. $S(t)$ and $(1-S(t))$ are the excited states survival probabilities of ROH* and RO^{-*} respectively. The short components arise from changes in σ_{ROH^*}. We checked whether the signal is linear with the power of the excitation pulse. We changed the laser power by a factor of seven by reducing the pump energy from 5 μJ to ~ 0.7 μJ. The power dependence measurements revealed that the signal shape is independent of the excitation power and the signal intensity is linear with the excitation power.

4.2 *The origin of the short components*

The time resolved fluorescence signal using upconversion technique measured by Tran-Thi and coworkers [26] also exhibited three time components with slightly shorter life times than we found in this study. The fastest component of 0.3 ps was attributed to solvation dynamics and the 2.2 ps component to a specific solute-solvent hydrogen bond formation. The long component of 87 ps fits approximately to the fluorescence decay of ROH* and to the growth of the fluorescence of RO^{-*}

measured by time correlated single photon counting technique [17,18]. The changes in the values of the decay time between the two studies may arise by the differences in the laser power used and the two methods of measurement.

4.3 The Direct Proton Transfer to the Acetate Base

At concentrations of > 0.1M the steady state diffusion controlled rate of the

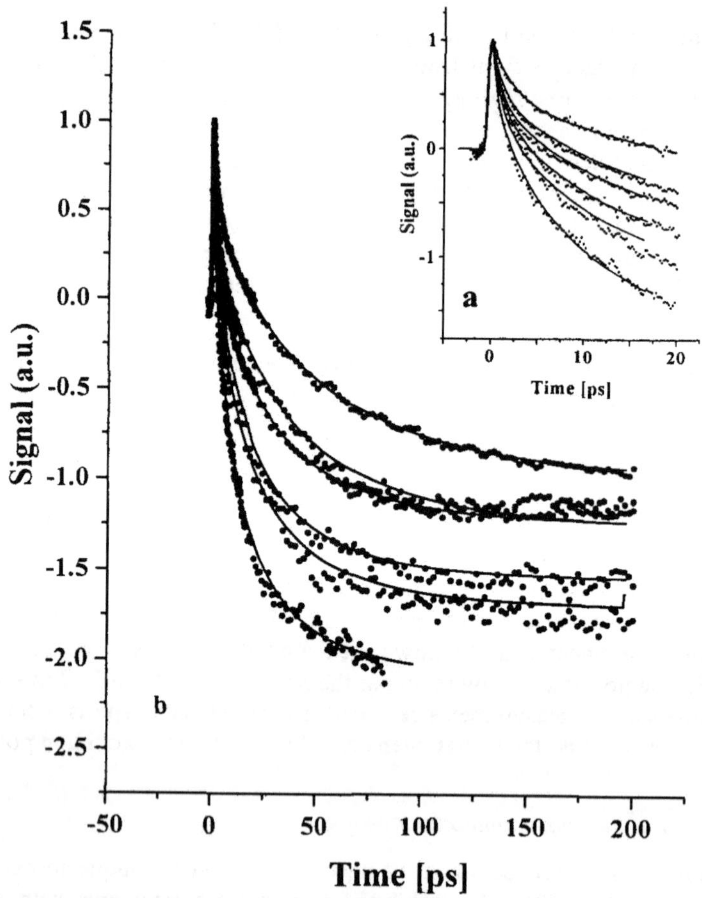

Figure 1 Pump-probe signal (dots) of HPTS aqueous solutions containing sodium acetate salt (NaAc) at several salt concentrations (Top to bottom: 0.5M, 1M, 2M, 3M and 4M) along with computer fits (solid line) at two time scales. The samples were pumped at 400 nm and probed at 540 nm. inset: a) short times b) long times;.

direct proton transfer is faster than the proton transfer rate to water $k_d \sim (120\text{ps})^{-1}$. At lower acetate concentrations the effect on the ROH* and RO$^-$* populations of direct proton transfer to a base is rather small. Figure 1 shows the effect of the base concentration on the pump-probe signal. The larger the base concentration, the faster the formation of the RO$^-$*.

We have used the diffusive reaction model at the Debye-Smoluchowski spherical-symmetric level to quantify the experimental data of direct proton transfer from excited HPTS ROH* to an acetate anion. Within the model there are several parameters that determine the actual dynamics *D, U(r), c,* k_{PT} and *a*. The mutual diffusion constant of the reactants *D*, was determined by Weller [3,4,11]: $D = D_{Ac^-} + D_{HPTS} \cong 2\cdot 10^{-5}\text{cm}^2\text{s}^{-1}$. *U(r)*, the repulsive potential is computed by the Debye-Hückel screening formula (Eqs. 7-8). For the contact radius *a* we used 7Å, the accepted literature value for proton transfer reaction [3,4,11]. It also resembles the actual size of HPTS and a water molecule in the first solvation shell surrounding it. k_{PT}, the intrinsic proton transfer reaction rate constant, is used as an adjustable parameter. In the fitting procedure of the experimental data its value was kept constant for all acetate concentrations 0.5 – 4 M. The pump-probe signal changes sign as time progresses. The relative amplitude of the positive and negative parts of the experimental signal is determined by the ratio of the cross-sections of absorption $\text{ROH}(S_1) \xrightarrow{h\nu} \text{ROH}(S_2)$, σ_{ROH}, and the emission cross section of a photon at 540 nm, σ_{RO^-}, by RO$^-$* which is formed by the direct proton transfer reaction. From the data simulations we find at 540 nm a ratio: $\sigma_{ROH}/\sigma_{RO^-} \cong 1/2.8$.

In general, the signal at 540 nm was analyzed as follows:

$$y = a_1 e^{-t/\tau_1} + a_2 e^{-t/\tau_2} + a_3 S'(t) - a_4(1 - S'(t)) \qquad (14)$$

where a_1, t_1 are the amplitude and the lifetime of the fastest component (0.7ps). a_2 and t_2 are the parameters of the slower component (for neat water 3.0ps). $S'(t) = S(t)e^{-k_d t}$, $S(t)$ is the survival probability computed from Eq. 2, k_d is the proton transfer rate of HPTS to water ($k_d = (120\text{ ps})^{-1}$). The population of ROH*, has a relatively short lifetime, since it also reacts with the water molecules and transfers the proton to the water in competition with the Ac$^-$ base reaction. The product $a_3 S'(t)$ gives the relative contribution of the ROH*(*t*) population to the signal. a_3 is related to the absorption cross section $\text{ROH}(S_1) \rightarrow \text{ROH}(S_n)$ (σ_{ROH}). $S'(t)$ the survival probability of the protonated form converts to ($1 - S'(t)$), the deprotonated form survival probability by the proton abstraction reaction. a_4 is related to the emission cross section of the transition $\text{RO}^-(S_1) \rightarrow \text{RO}^-(S_0)$, ($\sigma_{RO^-}$).

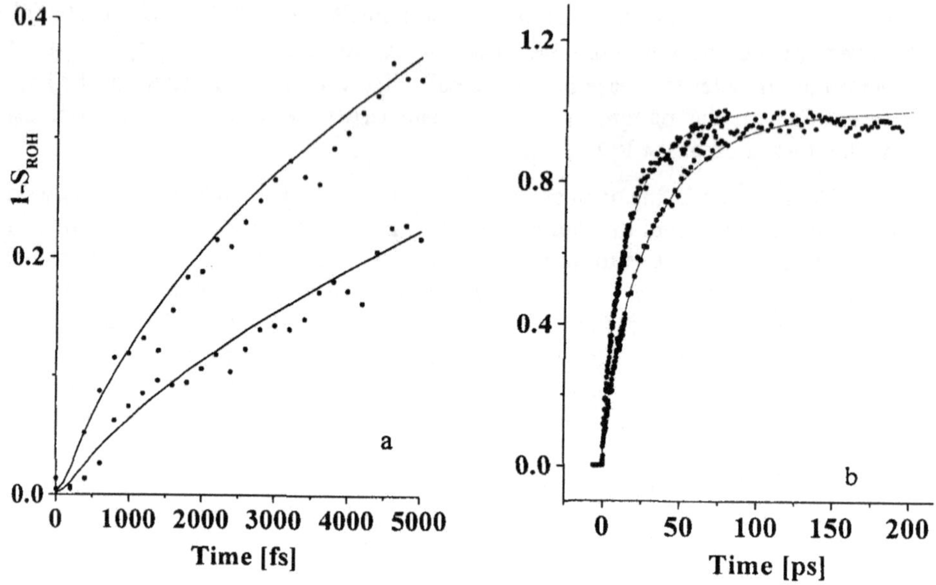

Figure 2. The population of RO^{-*} taken from the pump-probe signal (shown in fig. 1) of HPTS aqueous solutions containing sodium acetate salt (NaAc) at two salt concentrations (2M and 4M) along with computer fits (solid line) at two time scales. a) short times; b) long times.

Figure 2 shows the population of RO^{-*}, (1-S_{ROH}), as a function of time of HPTS in 2M and 4M sodium acetate aqueous solutions. The experimental signal (dots) was extracted from the pump-probe signal using Eq. 14 to subtract the non relevant components of the signal. We get good fits to the experimental results for the value of $k_{PT} = 1.5 \cdot 10^{11}\ M^{-1}s^{-1}$ at all acetate concentrations. The pseudo first order rate at 4 M sodium acetate is ~$(3ps)^{-1}$. The fastest signal component that was found for neat water solution as 0.7 ps does not depend on the acetate concentration while the second component, of 3 ps reduces to ~ 1.0 ps at acetate concentrations ≥ 2 M.

5. Discussion

The analysis of the pump-probe data is based on the diffusive model, given in Eqs. 2 – 12. The free parameter of the data fit is the intrinsic proton transfer rate constant, k_{PT}, from the photo-acid to the acetate anion. All other parameters are known from published literature values. For water we found an intrinsic second order rate constant, k_{PT} = $1.5 \cdot 10^{11}$ $M^{-1}s^{-1}$. It is about 20 times larger than the diffusion controlled rate constant $k_D = 4\pi D a_e N'$, where $D = 2 \cdot 10^{-5}$ cm^2/s, $a_e \sim 4.5$ Å and $N' = N_A/1000$, where N_A is the Avogadro's number. In such a case the time-dependent rate constant, Eq. 9, changes drastically with time. Figure 3 shows the time

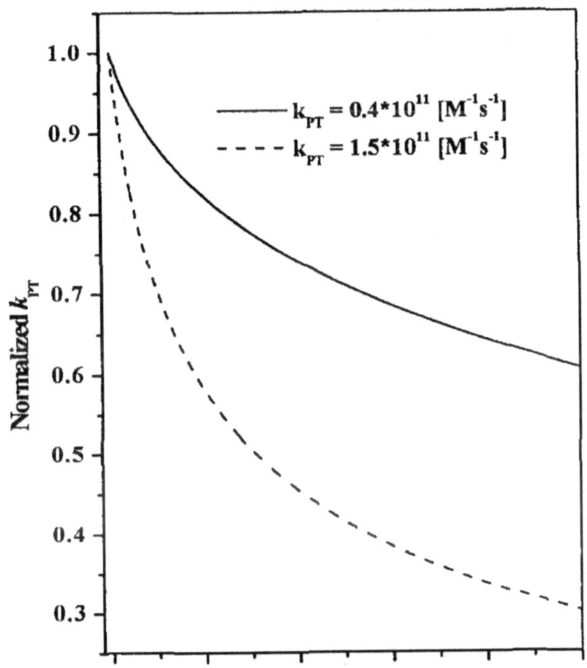

Figure 3 Comparison of $k(t)$ at two different values for k_{PT}, with $D = 2\text{x}10^{-9}$ cm^2/s and Z = 3

dependence of $k(t)$ calculated from Eq. 9 with the relevant parameters of the acid-base reaction. At $t = 0$ it assumes the value $k_{PT}e^{-\beta U(a)}$ (for 4M NaAc, $e^{-\beta U(a)} \approx 0.5$ and $k_{PT}e^{-\beta U(a)} = 8 \cdot 10^{10} M^{-1}s^{-1}$). The long time asymptotic expression of $k(t)$, is given by Eq. 15.

$$k = \underset{t \to \infty}{k(t)} = \frac{4\pi D a_e k'_{PT} e^{-\beta U(a)}}{k'_{PT} e^{-\beta U(a)} + 4\pi D a_e} \qquad (15)$$

where $k'_{PT} = k_{PT} / N'$. In our case $k_{PT}e^{-\beta U(a)}$ is much larger than k_D (for 4M NaAc $D \cong 0.7 \cdot 10^{-5}\,\text{cm}^2/\text{s}$) and therefore $k \approx k_D$, ($k_D = 3 \cdot 10^9$ $M^{-1}s^{-1}$, more than 20 times smaller than its initial value).

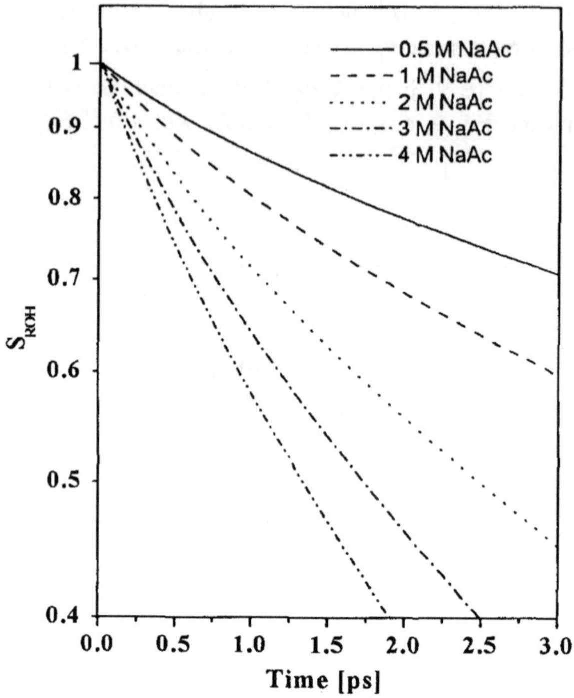

Figure 4. Calculated Survival Probabilities of ROH*, $S(t)$, at different NaAc concentrations (Top to bottom: 0.5M, 1M, 2M, 3M and 4M).

The survival probability of ROH, $S(t)$, is calculated from Eq. 2 and is shown in Figure 4. Since $k(t)$ is time-dependent, $S(t)$ decays nonexponentially and $S'(t)$ decays exponentially at very long times.

Using ordinary homogeneous chemical kinetic equation formalism cannot produce nonexponential decay rates.

6. Summary

The Smoluchowski approximation [12] is the most well known result in the kinetics of irreversible diffusion-influenced reactions.

Since it was first published in 1917, more than 50 years have elapsed before it was tested. In many cases other processes often obscured the results and the overall impression was that the transient phase is unimportant.

In the present study we tested the Smoluchowski theory in the ultrafast domain. The transient initial reaction rate can be observed even in the hundredth femtosecond time domain. Thus femtosecond spectroscopic methods can be utilized to test transient diffusion-influenced reactions and theories involving random walk of the reactants and products, as well as solvation dynamics and hydrogen bonding.

References

1. Rice, S. A. in: Bamford, C. H.; Tipper, C. F. H.; Compton, R. G. (Eds.), *Comprehensive Chemical Kinetics*, Vol. 25, Elsevier, Amsterdam, **1985**.
2. Förster, T. *Z. Elektrochem.* **1950**, *54*, 531.
3. Weller, A. *Z. Elektrochem.* **1954**, *58*, 849.
4. Weller, A. *Z. Phys. Chem. N. F.* **1958**, *17*, 224.
5. Schulman, E. F. *Fluorescence and Phosphorescence Spectroscopy*; Pergamon Press: New York, **1977**.
6. Nemzek, T. L.; Ware, W. R. *J. Chem. Phys.* **1975**, *62*, 477.
7. Lakowicz, J. R. in: Lakowicz, J. R. (Ed.), *Topics in Fluorescence Spectroscopy*, Vol. 4, Plenum, New York, **1994**.
8. Sikorski, M.; Krystkoviak, E.; Steer, R. P. *J. Photochem. Photobiol. A* **1998**, *117*, 1.
9. Eigen, M. *Angew. Chem., Int. Ed. Engl.* **1964**, *3*, 1.
10. Ireland, J. F.; Wyatt, P. A. H. *Adv. Phys. Org. Chem.* **1976**, *12*, 131.
11. Weller, A. *Prog. React. Kinet.* **1961**, *1*, 187.
12. Von Smoluchowski, M. *Z. Phys. Chem.* **1917**, *92*, 129.
13. Collins, F. C.; Kimball, G. E. *J. Coll. Sci.* **1949**, *4*, 425.
14. Szabo, A. *J. Phys. Chem.* **1989**, *93*, 6929.
15. Blumen A.; Zumofen, G.; Klafter, J. *Phys. Rev.* **1984**, *B30*, 5379; *J. Phys. (Paris)* **1985**, *46*, C7-3.
16. Szabo, A.; Zwanzig, R.; Agmon, N. *Phys. Rev. Lett.* **1988**, *61*, 2496.
17. Pines E.; Huppert, D.; Agmon, N. *J. Chem. Phys.* **1988**, *88*, 5620.
18. Agmon, N.; Pines E.; Huppert, D. *J. Chem. Phys.* **1988**, *88*, 5631.
19. Laws, W. R.; Brand, L. *J. Phys. Chem.* **1979**, *83*, 795.
20. Lawrence, M.; Marzzacco, C. J.; Morton, C.; Schwab, C.; Halpern, A. M. *J. Phys. Chem.* **1991**, *95*, 10294.
21. Pines, E.; Magnes, B-Z.; Lang, M. J.; Fleming, G. R. *Chem. Phys. Lett.* **1997**, *281*, 413.
22. S. Backus, J. Peatross, C. P. Huang, M. M. Murnane and H. C. Capteyn, Opt. Lett. **20**, 2000 (1995)

23. Robinson, R. A; Stokes, R. H. *Electrolyte Solutions*, 2nd ed., Butterworths, London, **1959**.
24. Krissinel', E. B.; Agmon, N. *J. Comput. Chem.* **1996**, *17*, 1085.
25. Gösele, U. M. *Prog. React. Kinet.* **1984**, *13*, 63.
26. C. Prayer, T. Gustavsson, T-H. Tran-Thi, In "Fast Elementary Processes in Chemical and Biological Systems" Vol. **364** AIP 333-339 (1996); Tran-Thi TH, Gustavsson T, Prayer C, Pommeret S, Hynes JT *Chem. Phys. Lett.* 329 (5-6): 421-430, 2000

BREAKING, MAKING AND ROTATING CHEMICAL BONDS OF 1′-HYDROXY-2′-ACETONAPHTHONE IN SOLUTIONS AND IN NANOCAVITIES

J.A.ORGANERO[1], L. SANTOS[2], AND A. DOUHAL[1]*

[1]*Departamento de Química Física, Sección de Químicas, Facultad de Ciencias del Medio Ambiente, Campus Tecnológico de Toledo, Avenida Carlos III, S.N., Universidad de Castilla-La Mancha, 45071 Toledo, Spain.*
[2]*Departamento de Química Física, Facultad de Químicas, UCLM, Ciudad Real, Spain.*
**Corresponding author: adouhal@amb-to.uclm.es, fax: + 34-925-268840.*

In this contribution, we report on time-resolved emission studies of 1'-hydroxy-2'-acetonaphthone (HAN) in solution and in nanocavities. HAN undergoes a photoinduced intramolecular proton-transfer (PT) reaction and a twisting motion of the protonated acetyl group giving birth to keto rotamers (KR*) in S_1. In water solutions, proton transfer in the enol (E*) structure and a subsequent internal rotation in the formed keto-type tautomer (K*) occur in few of tens of ps, to give birth to solvated rotamers (KR*), having lifetime in the ns-time scale. When caging HAN in nanocavities, the dynamics of twisting motion is greatly affected and the channel leading to KR* can be closed. To elucidate the experimental observation, we used the results of ab-initio calculations at the S_0 and S_1 states in gas phase.

Introduction

The dye 1′-hydroxy-2′-acetonaphthone (HAN), is an aromatic molecule with two groups able for forming an intermolecular H-bond (Scheme). This bond is comparable to that in methyl salicylate (MS) but with a larger pi-electronic molecular system. HAN has been studied in gas[1-4] and condensed phases[5,6]. In jet molecular beam, fluorescence data indicate a single conformer at the S_0 state[1]. The excitation of HAN at ~ 389 nm (0-0 transition) originates a structured emission which corresponds to the excited enol tautomer (E*) and two emission bands from the keto-type structures with maxima at 426 and 452 nm. The OH/OD isotopic substitution abnormally suppresses the emission of E* due to the change in the coupling between the vibrational modes of the involved structures.[1] No enol emission was observed at excess energies above ~300 cm^{-1}, providing experimental evidence of the existence of an energy barrier (~ 0.9 kcal/mol) for the proton transfer in the potential energy surface at S_1. The fluorescence lifetime of HAN in the molecular beam is 9.4 ns at zero point level, and it is independent on the observation wavelength, suggesting an equilibrium between E* and K* structures. On the basis of the structured and abnormally shifted emission band of the tautomer, the ground state potential energy surface has been suggested to contain additional minima. Cheng and co-workers[3], using femtosecond time resolved multiphoton ionisation spectroscopy studied the dynamics of HAN at S_1 pumping at different excitation wavelengths 385-410 nm. They detected an unresolved fast initial decay followed by a long-time non-exponential decay with two components: τ_1~ 61-86 ps related to the proton transfer process, and τ_2~ 610-1010 ps ascribed to the emission of K* structure The results are explained on the basis of a previously proposed mechanism[1,2] in which, the excited enol form is equilibrated with the keto tautomer crossing a low energy barrier at the S_1 state. Nanosecond-microsecond time-resolved experiments in solution and in rigid media were carried out by Tobita and co-workers[5]. They concluded that following the proton transfer process at S_1 a conformational change takes place in K* to produce a metastable state (keto-type rotamer, KR*). The twisting motion in the keto type structure, K*, has been also observed in the proton-transfer cycle of 3',4'-benzo-2'-hydroxychalcone, a derivative of HAN.[7] Recently, Stolow and co-workers[4] using time-resolved photoelectron spectroscopy observed two transient bands in the spectrum of excited HAN, and assigned them to the photionization of E* and K*. They also concluded that proton transfer occurs within 50 fs. Depletion of K* in S_1 state via internal conversion to the ground state occurs in 30 ps, and is enhanced by a close-lying n,π* state. Because of the large excess of energy of the pump (>2500 cm^{-1}), the experiment could not detect the small energy barrier for the proton transfer process (~ 0.9 kcal/mol).[1]

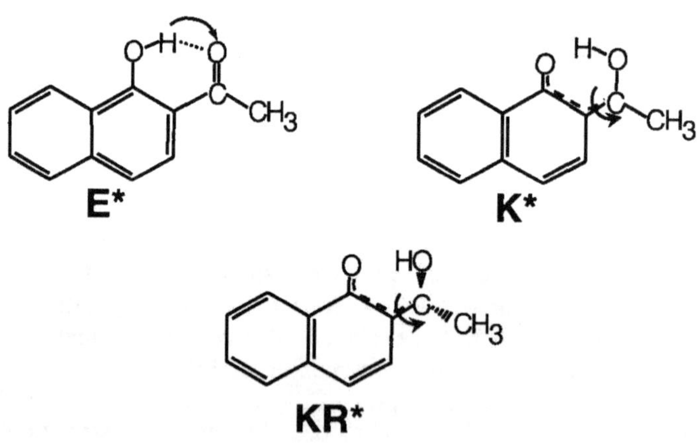

Scheme: Molecular structures of Enol (E*), Keto (K*) and Keto Rotamer (KR*) of 1′-hydroxy-2′-acetonaphthone at the S_1 state. The arrows indicate the involved motion.

Theoretical calculations

Ab initio calculations have been performed within the Gaussian 98 program.[8] The S_0 state has been studied through the restricted Hartree-Fock method (RHF), whereas a CI all-single-excitations with a spin-restricted Hartree-Fock reference ground state (CIS) has been used for the excited S_1 state[9]. The energies of the S_0 state have been recalculated through the Möler-Plesset perturbation theory up to second order (MP2)[10]. All calculations have been done with the split-valance 6-31G-(d,p)[11].

Our theoretical calculations have shown that at the S_0 state, E is the most stable structure and it has an intramolecular H-bond between OH and the CO groups. K structure does not have a local minimum in S_0. Other local minima exit as a result of rotation of OH and acetyl groups or protonated acetyl group[6]. These rotamers may be accessible through electronic or vibrational excitation.

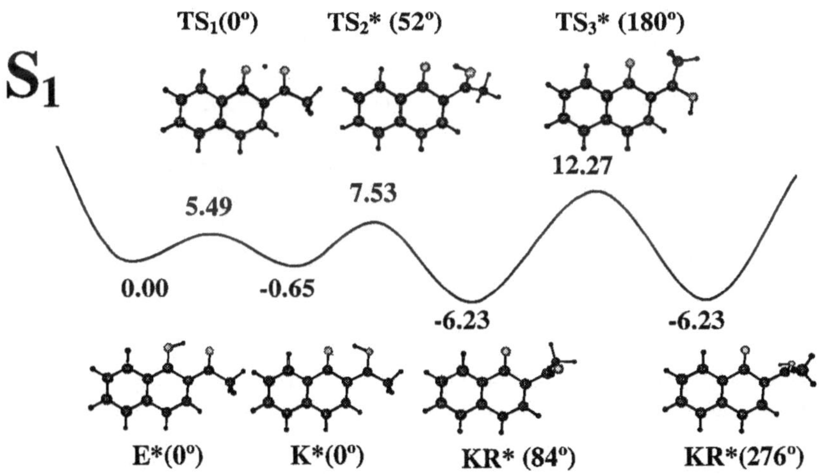

Figure 1: Schematic energy profile including the involved structures for proton transfer and twisting motion of HAN in S_1 state. Energies are given in kcal/mol relative to E*. It is also indicated between brackets the dihedral angle between the planes of acetyl group and naphtalenic frame.

Upon electronic excitation, the behaviour of HAN is completely different from that at S_0.[12] Figure 1 shows a schematic energy profile for the stationary points located at the S_1 state, and related to the proton-transfer reaction and twisting motion. Figure 1 shows the involved structures with the relevant rotational angle of the carbonyl group with respect the naphtalenic frame. K* is almost isoenergetic to E* and the transition state connecting E* and K* involves an energy barrier of only ~ 5.5 kcal/mol. Taking into account that HAN is a large organic molecule, and that the theoretical calculations are related to an electronic excited state, the obtained result is in agreement with the experimental observation in gas phase giving 0.9 kcal/mol as an energy barrier to the proton transfer, and the existence of an equilibrium between E* and K*.[1,2,3,6]

The calculations suggest also that a rotation of the acetyl group in K* leads to KR*, which now is the most stable structure in S_1 (6.23 kcal/mol relative to E*). The protonated acetyl group is almost perpendicular to the plane defined by the naphthalene ring. By continuing the rotation of the acetyl group we found another transition state (TS_3*) that connects KR* with an equivalent molecular structure. The calculations have shown that the origin of the energy barrier for the transfer in excited E might be in-plane and out-of-plane vibrational motions of OH and $CO(CH_3)$ groups involved in the H-bonded chelate ring. We have found active in plane vibrational modes at 325 and 381 cm^{-1} and out-of-plane motions at 380 cm^{1} correlated with the bands at 307, 365 and 370 cm^{-1}, respectively obtained in jet molecular beam[1] and which correspond to the bending deformations of the H-bond ring. In addition to that, the excitation spectrum in the molecular beam experiment[1] showed bands at 203 cm^{-1} and 677 cm^{-1} which might be correlated with the theoretically found modes at 189 cm^{-1} and 539 cm^{-1}, respectively, and assigned to C-O-H out-of-plane mode and in-plane bending motion of C=0 group, respectively. The high activity of the O-H out-of-plane motion (bending mode at 822 cm^{-1}) in E* may explain the suppression of emission signal in molecular beam beyond ~900 cm^{-1}.[1,2] All these data give a clear picture of how proton transfer occurs in excited HAN.[12]

Solution and nanocavities observations: Experimental evidences of proton transfer and twisting motions

Uv-visible absorption and emission spectra of 1-methoxy,2-acetonaphthone (a methylated derivative of HAN (OH/OMe) in which PT can not occur) in methyl cyclohexane (MC, an inert solvent) at 298 K shows a normal Stokes shift ~2300 cm^{-1}, while in the case of HAN the shift is ~6100 cm^{-1}. The difference in spectral shift is interpreted in terms of a intramolecular proton transfer in HAN at S_1 in accordance with previous studies.[1,2,3] The emission spectrum of HAN in MC at 298 K presents a broad band with a maximum at 480 and high shoulder at ~ 460 nm (figure 2A). Lowering the temperature to 80 K, the emission shows a structured band with clear maxima at 460 and 485 nm (figure 2A). The difference between both spectra may be caused by an energy barrier separating two different keto type phototautomers (K* and twisted rotamer KR*). Because the excitation spectra at both temperatures and gated at different wavelengths of the emission spectrum (not shown) are similar, and not different from the absorption one, the existence of two different conformers of HAN at S_0 is discarded.

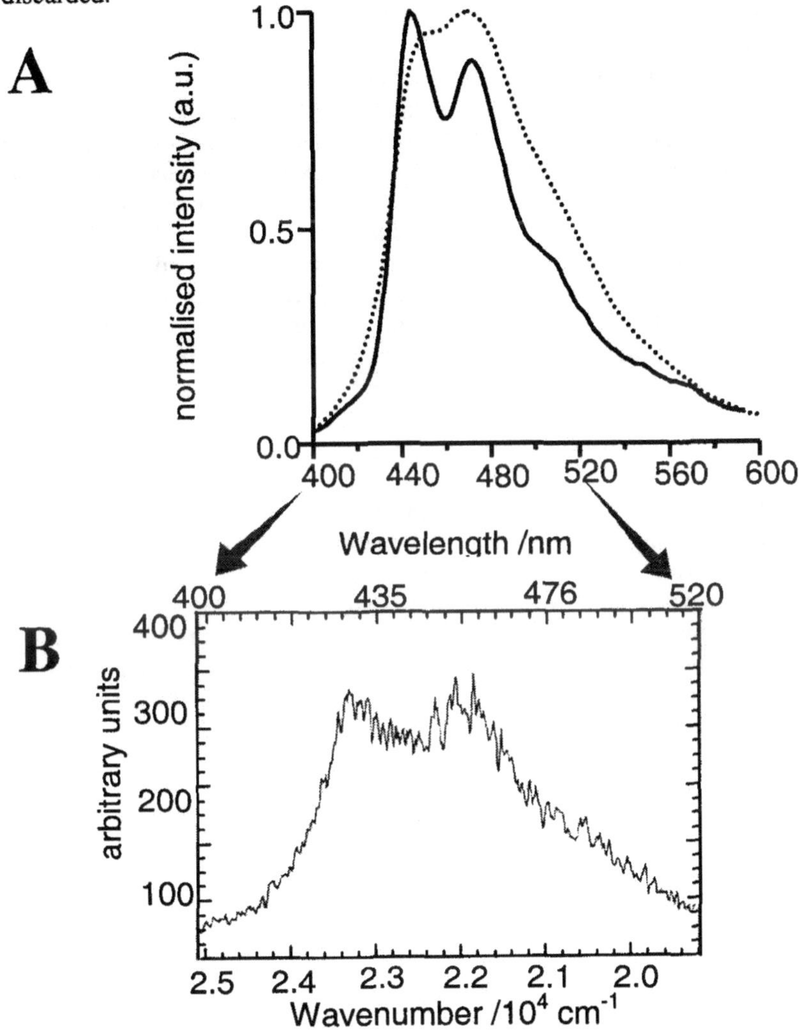

Figure 2: **A)** Steady state emission spectra of HAN in methylcyclohexane at 298 K (___) and at 80 K (....) and excited at 380 nm. **B)** Dispersed fluorescence spectrum of a jet cooled molecular beam of HAN excited at 380 nm ($0^0 + 836\ cm^{-1}$) (adapted from ref. 1).

Thus, to explain the structured PT fluorescence band, we suggest the emissions of K* and KR*. Interestingly these bands might be correlated to those observed in molecular beam at 426 and 452 nm, respectively (Fig. 2B). The coldest conditions in the supersonic molecular beam and the absence of solvation in these conditions might shift the emission maxima of these structures to the blue (higher energies).

To obtain some information about the dynamics of electronically excited HAN in solution, we gated the emission in MC at 80 and 298 K (figure 3). At 298 K, two time constants (± 10 ps) of 40 ps (22%) and 100 ps (78%) were used to obtain a biexponential fitting function of the decay at the PT band. The shortest component, assigned to the decay of K* and which might be considered as a limiting time for the twisting motion in this structure to produce KR*. The longest component is assigned to KR*. In this book Lochbrunner and co-workers show a rising component of the transmission signal of excited HAN in cyclohexane at the PT region using time-resolved absorption measurements. They found a time constant of 30 fs assigned to the duration of proton motion in E*. Lowering the temperature to 80 K (Fig. 3), the emission decay shows a remarkable change. The two components of the decay, assigned to K* and KR*, are 2.15 ns (4%) and 8.34 ns (96%) (we could not get a good fit using a single exponential function). The presence of efficient nonradiative processes in excited HAN at higher temperatures is clearly observed. We identify one of theses possible nonradiative channels as a twisting motion of the C-C bond linking the protonated acetyl group to the aromatic ring, and a torsion of the naphthalene molecular frame in K* and in KR*. In other solvents, we also found biexponential decays with lifetimes of 20-60 ps for the shortest component and 100-300 ps for the longest one. Our assignment agrees with the conclusions of Tobita and co-workers who proposed the existence of a metastable rotamer conformer following the proton transfer process in the S_1 state.[5]

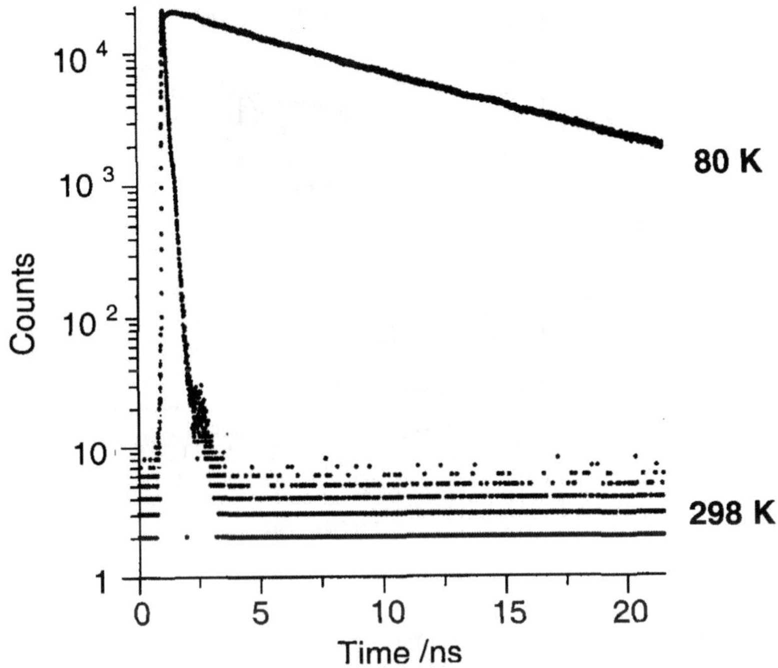

Figure 3: Emission decays of HAN in MC at 80 and 298 K, when gated at 480 nm and excited at 393 nm.

To further demonstrate the involvement of twisting motion in K* and formation of KR*, we studied the effect of the nanocavity size (β- and γ-CD's with interior diameter of ~ 8 and 9.5 Å, respectively) on the spectroscopy and dynamics of HAN in water solutions. In a larger cavity (γ-CD), the emission of HAN:γ-CD complex (maxima at 500 nm) shifts to the red when compared to the HAN:β-CD emission complex (maxima at 460 nm). This is due to a larger relaxation of the phototautomers into γ-CD. Emission decays of HAN in water, β-CD and γ-CD solutions (figure 4) clearly demonstrate that the size of the encapsulating nanocavities affects the dynamics of excited HAN. In pure neutral water we observed a biexponential decay with two components: ~90 ps (99%) and ~ 6 ns (1%), assigned to K* and KR* structures, respectively. Using β-CD as a caging medium, a triexponential function was needed to fit the decay. The lifetime components are ~90 ps (34%), ~200 ps (63%), and ~742 ps (3%). The 90 ps-component is assigned to the free HAN in water, while the two other components are assigned to the encapsulated tautomers K* and KR*, respectively. Because of confinement, these last times are longer than the first one (due to free HAN) .

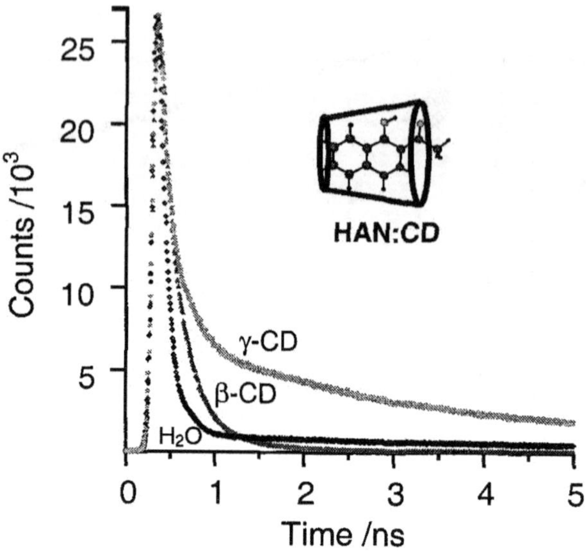

Figure 4: Emission decays of HAN in water and in presence of 6 mM of β- and γ-CD in a neutral water solution, gated at 480 nm and excited at 393 nm (T= 298 K).

The fluorescence decay using γ-CD is fitted with a triexponential function, having three components, ~100 ps (82%), similar to the lifetime of K* in water, ~650 ps (9%) and ~4.23 ns (9%), assigned to encapsulated K* and KR*, respectively. In this case, the protection of HAN by a larger cavity makes possible a longer lifetime of K* and KR*. Thus, the confinement governs the relaxation route of the formed tautomers.[13-15]

Conclusion

The experimental observation shows the involvement of twisting motion of the formed tautomer (K*) in the PES of HAN at S_1 and the formation of a keto rotamer (KR*) having emission at the red side of the spectrum. We demonstrated that these internal motions and related photophysics can be controlled by choosing the size of the caging nanocavity. A result in full agreement with our previous calculations and recently reported findings.[1-6, 12-15]

Acknowledgements

This work was supported by the Ministry of Science and Technology (Spain) through projects PB98-0310 and 1FD1997-1658, and by the USA-Spain cooperative program (project PHNCCT-20002).

References

1. Douhal, A.; Lahmani, F.; Zehnacker-Rentien, A. Chem. Phys. **178** (1993) 493.
2. Douhal, A.; Lahmani, F.; Zewail, A. H. Chem. Phys. **207** (1996) 477.
3. Lu, C.; Hsieh, R.-M. R.; Lee, I-R.; Cheng, P-Y. *Chem. Phys. Lett.* **310** (1999) 103.
4. Lochbrunner, S.; Schultz, T.; Shaffer, J. P.; Zgierski, M. Z.; Stolow, A. J. Chem. Phys. **114** (2001) 2519.
5. Tobita, S.; Yamamoto, M.; Kurahayashi, N.; Tsukagoshi, R.; Nakamura, Y.; Shizuka, H. J. Phys. Chem. A. **102** (1998) 5206.
6. Organero, J. A.; García-Ochoa, I.; Moreno, M; Lluch, J. M.; Santos L.; Douhal. A: Chem. Phys. Lett. **328** (2000) 83.
7. Tokumura, K.; Nagaosa, K.; Ohta, Y.; Matsushima, R. Chem. Phys. Lett. **295** (1998) 516.
8. Frisch, M. J.; et al. Gaussian 98; Gaussian Inc.: Pittsburgh, PA, (1998).
9. Foresman, J. B.; Head-Gordon, M.; Pople, J. A.; Frisch, M. J. J. Phys. Chem., **96** (1992) 135.
10. Møller, C.; Plesset, M.S.; Phys. Rev. **46** (1934) 618.
11. Ditchfield R.; Hehre, W.J.; Pople, J. A.; J. Chem. Phys., **54** (1971) 724, **56** (1972) 2257; Hariharan, P.C.; Pople, J. A. Mol. Phys. **27** (1974) 209; Gordon, M. S.; Chem. Phys. Lett. **76** (1980) 163; Clark, T.; Chandrasekar, J.; Spitznagel, G.W.; Schleyer, P.; J. Comput. Chem. **4** (1983) 294.
12. Organero, J. A.; Moreno, M.; Santos, L.; Lluch, J. M.; Douhal, A. J. Phys. Chem. A, **104** (2000) 8424.
13. Douhal, A. Science **276** (1997) 221.
14. Zhong, D.P.; Douhal, A.; Zewail, A.H. Proc. Natl. Acad. Sci. (USA) **97** (2000) 14052, and references therein.
15. Douhal, A. en Femtochemistry, De Schryver, F.C.; De Feyter, S. and Schweitzer, G. (Edits.), Wiley-VCH, Chapt. 15 (2001) 267.

VIBRATIONAL RELAXATION AND ENERGY EQUILIBRATION IN LIQUID WATER

A. J. LOCK, S. WOUTERSEN AND H. J. BAKKER
FOM Institute for Atomic and Molecular Physics
Kruislaan 407
1098 SJ Amsterdam
The Netherlands
E-mail: a.lock@amolf.nl

We have studied the vibrational relaxation and equilibration dynamics of liquid water (H_2O) using time-resolved femtosecond mid-infrared pump-probe spectroscopy. It is found that the energy equilibration leads to a shift of the absorption band of the OH–stretch vibration (ν_{OH}) and a decrease of the absorption cross section. For water, we find an equilibration time constant of 0.55 ± 0.05 ps. We observe that the equilibration time strongly increases going from water to alcohols like methanol, ethanol and propanol.

1 Introduction

The dynamics of chemical reactions in aqueous solution are governed both by the dynamics of the solvation interactions and the rate at which water molecules can accept the energy dissipated in the reaction. The importance of water as a solvent is illustrated in Ref.1, where the relaxation of vibrationally excited CH_3Cl molecules in aqueous solution is studied. Theoretical work [2] demonstrates that the excess vibrational energy is accepted by so-called low-frequency bath modes, like rotations and translations, but no time scale was given for this process of equilibration. Vodopyanov performed saturation studies in which water and ethanol layers were excited with intense picosecond laser radiation [3,4,5,6]. A major problem in these studies was that it was not possible to separate the time scale of the relaxation of the OH–stretch excitation from that of the full equilibration of the energy in the liquid.

Here we present a study on the equilibration dynamics of water and alcohols at room temperature following a local dissipation of energy. The equilibration dynamics are monitored by probing the spectral changes of the OH–stretch absorption band with femtosecond mid-infrared laser pulses.

2 Experimental Technique

In order to get mid-infrared pulses resonant with the OH–stretch vibrational modes of liquid water, we use a Ti:Sapphire based laser system including an optical parametrical amplification and generation (OPA/OPG) stage and non-linear crystals for sum- and difference frequency mixing [7,8]. The resulting infrared pulses can be tuned from 2.8 to 3.3 μm (3000-3600 cm^{-1}), have a FWHM duration of 180 fs and a spectral width of 80 cm^{-1}. The pump pulses have typical energies of 30 μJ, the probe pulses have energies of about 1 μJ.

The pump pulses are used to excite a fraction of the molecules from the vibrational ground state to the $\nu_{OH} = 1$ state. Thereby energy is put into a specific degree of freedom. Probe pulses are used to monitor the thermalization of the excitation of the pump pulse. The transmission of the probe and pump pulses through a 100 μm pinhole was typically 80% and 50%, respectively.

The transmission of the probe pulse with (T) and without (T_0) pump pulse is measured as a function of delay time between pump and probe pulse and the relative transmission change $\ln(T/T_0)$ is calculated. The probe polarization was set at the magic angle with respect to the polarization of the pump pulse. The samples consisted of a thin layer of liquid (≈ 1 μm for water) between two sapphire windows. In order to avoid accumulated heating, the sample was rotated. The temperature rise in the focus due to a single pump pulse is calculated to be approximately 50 K for water.

3 Results

The conventional absorption spectrum of liquid water is measured with an infrared spectrometer and is displayed in figure 1 for different temperatures. We observe a blue shift of the maximum of the absorption band and a decrease of the absorption cross section as a function of temperature.

In Figure 2, the transmission change $\ln(T/T_0)$ as a function of delay time is shown for pure water. The molecules are excited with a pump pulse with a central frequency of $\nu = 3250$ cm^{-1} and are probed at different frequencies. The two upper curves in the left panel (probe at 3000 cm^{-1} and 3100 cm^{-1}) show an induced $\nu_{OH} = 1 \rightarrow 2$ absorption signal around $t = 0$ that, with increasing delay, rapidly decays and rises to a high final level of transmission. This level of transmission remains constant on a very long time scale, which indicates that this level of transmission is associated with a completely thermalized system.

When probing in the blue wing of the absorption band, at 3460 cm^{-1} and 3500 cm^{-1}, a bleaching peak around $t = 0$ and a similar increase to a high final level of transmission for longer delay times are observed. For the probe frequencies in the centre and red wing of the absorption band (3200 cm^{-1} and 3400 cm^{-1}), the bleaching peak around $t = 0$ is hardly visible, because the long time transmission increase due to thermalization is very large and

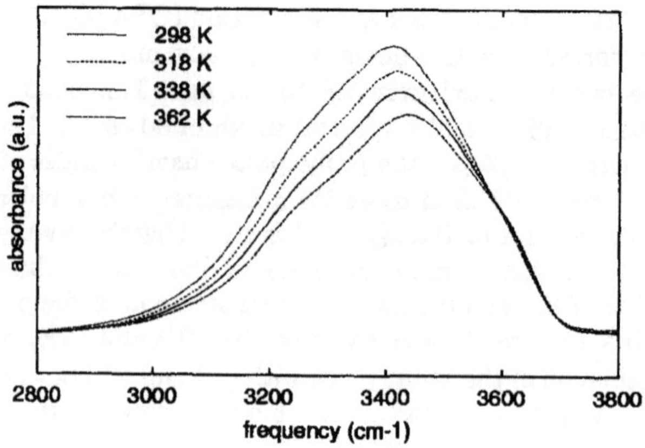

Figure 1: Absorption spectrum of the OH–stretch vibration of water for different temperatures

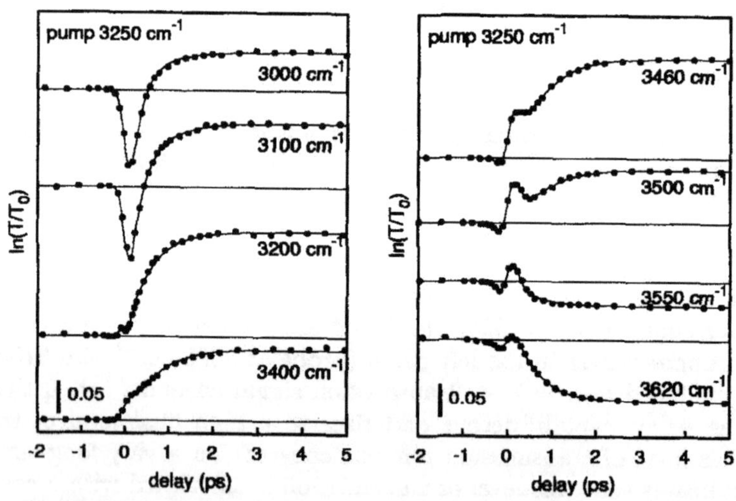

Figure 2: Transmission change $\ln(T/T_0)$ for water as a function of delay time between the probe and pump pulse. The frequency of the pump pulse is fixed at 3250 cm^{-1}. The probe frequency is indicated above each scan.

	H_2O	CH_3OH	C_2H_5OH	C_3H_7OH
τ_{eq} (ps)	0.55 ± 0.05	7.2 ± 0.5	13.5 ± 2.0	15 ± 5

Table 1: Measured equilibration time constants for different hydrogen-bonded liquids

dominates the bleaching around $t = 0$. The highest probe frequencies (3550 and 3620 cm^{-1}) show, after a small bleaching peak, a decrease to a lower (negative) final level of transmission.

We performed the same measurements on methanol (CH_3OH), ethanol (C_2H_5OH) and 1-propanol (C_3H_7OH). Qualitatively the same behaviour was observed: a bleaching peak or induced absorption depending on the probe frequency and an increase or decrease to a high or low final level of transmission (results not shown).

4 Model

To fit the data we use a model[8] that consists of the vibrational ground state (0), the first excited state (1), populated by the pump pulse, the second excited state (2) and the intermediate level 0* to which the population of state 1 relaxes with a time constant T_1. In this model, the decay of the intermediate 0*-state back to the ground state represents the thermalization of the system. The associated time scale τ_{eq} is the equilibration or thermalization time. The absorption cross sections for the $0 \to 1$ and $0^* \to 1^*$ transitions are $\sigma_{01}(\omega)$ and $\sigma_{01}^*(\omega)$, respectively.

The curves resulting from the fit are depicted in Figure 2 for water. Due to the overwhelming contribution of the thermalization effect in water, most measurements shown in Figure 2 are not very sensitive to the value of the vibrational lifetime T_1 of the OH–stretch mode. The best way to extract this first relaxation time is to use the value for the equilibration time τ_{eq}, which can be obtained from measurements in the red wing of the absorption band (3200 cm^{-1}), as a constant in the fit for the $1 \to 2$ induced absorption as measured in the upper two curves. This results in a value for T_1 of 250 ± 20 fs. For τ_{eq} we find: 550 ± 50 fs for all fitted curves. Applying the same model to the alcohols yield equilibration times τ_{eq}, as displayed in the table above.

5 Discussion

In our experiment we use the response of the OH–stretch vibration as a probe for the relaxation processes. The OH–stretch vibration is very sensitive for the positions and orientations of the H_2O molecules, in particular for the hydrogen bond length, as the exact resonance frequency depends linearly on the distance between two neighbouring water molecules[9,10]. Hence, the observed blue shift of the OH–stretch absorption band and the decrease of the cross-section at longer delay times directly reflect the repositioning of the molecules that follows from the dissipation of energy. It should be noted that the second step of the relaxation, i.e. the equilibration, involves two subsequent processes. The first process is formed by the energy redistribution in which the energy of the non-thermal intermediate level becomes thermally distributed over all degrees of freedom. After this redistribution, the equilibration is not yet complete, because putting energy into a particular degree of freedom does not imply that the coordinate corresponding to this degree of freedom instantaneously finds its new equilibrium position. It takes some time after the energy has been redistributed before the low-frequency translational and rotational degrees of freedom have acquired their new equilibrium positions. This repositioning of the molecules forms the second process in the equilibration and is what is observed in the spectral response of the OH–stretch.

If the repositioning of the molecules forms the rate-limiting step of the equilibration, the equilibration rate is expected to scale inversely with the mass of the molecules. This agrees quite well with the observation that the equilibration time constant increases going from water to methanol, ethanol and propanol. The repositioning shows the same trend as the time scales observed in dielectric relaxation and rotational correlation experiments. The Debye relaxation times are 8 ps for water, 50 ps for methanol, 160 ps for ethanol and 320 ps for 1-propanol[11,12]. Hence, the equilibration time constant appears to be dominated by the repositioning of the molecules, not by the energy redistribution. Only for the fastest liquid, liquid water, the energy redistribution may also play a significant role in the equilibration time constant.

6 Conclusions

Using two-color femtosecond mid-infrared pump-probe spectroscopy, we investigated the vibrational relaxation and equilibration dynamics of liquid water and alcohols. The dynamics of the latter equilibration process is monitored via the spectral response of the OH–stretch vibration. We observe that the equilibration leads to a blue shift of the OH–stretch absorption band and a decrease of the absorption cross section. For water, we observe a very short equilibration time constant of 0.55 ps, which is much shorter the time constants of than the alcohols. This means that water can adapt extremely fast, much faster than alcohols, to a deposition of energy.

Acknowledgment

The research presented in this paper is part of the research program of the Foundation for Fundamental Research on Matter (FOM) and was made possible by financial support of The Netherlands Organization for Scientific Research (NWO).

References

1. R. M. Whitnell, K. R. Wilson, and J. T. Hynes, *J. Chem. Phys.*, **96**, 5354 (1992).
2. R. Rey and J. T. Hynes, *J. Chem. Phys.*, **104**, 2356 (1996).
3. K. L. Vodopyanov, *Sov. Phys. JETP*, **55**, 1049 (1982).
4. K. L. Vodopyanov, *Sov. Phys. JETP*, **64**, 67 (1986).
5. K. L. Vodopyanov, *Sov. Phys. JETP*, **70**, 114 (1990).
6. K. L. Vodopyanov, *J. Chem. Phys.*, **94**, 5389 (1991).
7. U. Emmerichs, S. Woutersen, and H. J. Bakker, *J. Opt. Soc. Am. B*, **14**, 1478 (1997).
8. A. J. Lock, S. Woutersen, and H. J. Bakker, *J. Phys. Chem. A*, **105**, 1238 (2001).
9. A. Novak, *Struct. Bonding (Berlin)*, **18**, 177 (1974).
10. D. Hadži and S. Bratos, In *The Hydrogen Bond, volume II, chapter 12*, ed. P. Schuster, G. Zundel, and C. Sandorfy (Elsevier, Amsterdam, 1976).
11. J. Barthel, K. Bachhuber, R. Buchner, and H. Hetzenauer, *Chem. Phys. Lett*, **165**, 369 (1990).
12. J. T. Kindt and C. A. Schmuttenmaer, *J. Phys. Chem.*, **100**, 10373 (1996).

FIRST OBSERVATION OF THE WHOLE BELL-SHAPED ENERGY GAP LAW FOR PHOTOINDUCED INTRAMOLECULAR CHARGE SEPARATION REACTIONS

HAIK CHOSROWJAN, NOBORU MATAGA AND YUTAKA SHIBATA

Institute for Laser Technology, Utsubo-Hommachi 1-8-4, Nishi-ku, Osaka 550-0004, JAPAN
E-mail: haik@ile.osaka-u.ac.jp

NAOYA YOSHIDA AND ATSUHIRO OSUKA

Department of Chemistry, Graduate School of Science, Kyoto University, Kyoto 606-8502, JAPAN
E-mail: yoshida@kuchem.kyoto-u.ac.jp

Photoinduced ultrafast CS reaction from S_2 state of a series of electron donor (Zn-5,15-bis(3,5-di-tert-butylphenyl)-porphyrin) directly linked at meso-position with imide acceptors (phthalimide derivatives, pyromellitimide and napthylimide) were examined by fs fluorescence up-conversion technique in five solvents of different polarities and viscosities in 200K – 300K temperature range. The fluorescence dynamics in 100 fs regime could be accurately reproduced by single exponential functions in all solvents examined. The whole bell-shaped energy gap law k_{CS} vs. $-\Delta G_{CS}$ for photoinduced intramolecular CS reactions was unambiguously observed for the first time. The Marcus-Jortner model incorporating intramolecular *hfm* reproduced reasonably well the data in polar solvents. The nature and role of ultrafast solvation coupled with CS process are discussed.

Introduction

The dynamics and mechanisms of photoinduced electron transfer (ET) in condensed media are the important fundamental problems in chemistry and biology [1]. The ET reactions in solution are regulated by various factors including the magnitude of electronic interaction between donor (D) and acceptor (A), free energy gap, solvation dynamics as well as intramolecular vibrations coupled with ET process. When the nonadiabatic mechanism is applicable, the reaction could be well understood on the basis of known theories [1-3], however, to our knowledge, the expected whole bell-shaped energy gap low for the photoinduced charge separation (CS) reactions has not been clearly observed yet [4]. For this purpose, measurements of the ET from S_2 state of zinc (II)-porphyrin (D) to covalently linked acceptors may be very fruitful, in view of the relatively long lifetime of the S_2 state (~2.3 ps) as well as a wide range of the free energy gap ($-\Delta G_{CS}$) available for the CS reaction. Accordingly, CS reaction from S_2 state of a series of electron donor (Zn-5,15-bis(3,5-di-tert-butylphenyl)-porphyrin) directly linked at meso-position with imide acceptors (phthalimide derivatives, pyromellitimide and napthylimide) were examined by fs fluorescence up-conversion technique (IRF=110 fs) in five solvents of different polarities and viscosities: acetonitrile (ACN), glycerol triacetate (GTA), tetrahydrofuran (THF), toluene (TOL) and cyclohexane (CH). In the present study on the photoinduced CS processes of ZP-A supramolecular systems with fixed D-A distances in various solvents, we aim to elucidate the following problems: **(1)** clear-cut observation of the inverted effect in the energy gap dependence of the photoinduced CS reaction, **(2)** effects of the solvent polarity and viscosity on the photoinduced ET dynamics including the problem of its dependence on solvent relaxation time τ_L, **(3)** the nature of the ultrafast ET reaction from the higher excited electronic state (S_2) including the effects of the intramolecular high frequency modes and ultrafast solvation dynamics, **(4)** temperature and excitation wavelength effects on the ultrafast ET of the supramolecular systems.

Experimental

The measurements of the fluorescence dynamics were made by a fluorescence up-conversion apparatus similar to that described elsewhere [5] but the fwhm of the instrumental response was improved to 110 fs. Supramolecular samples of ZP directly linked at meso-position with A (imide compound) were synthesized and purified by standard methods and their structural formulas are indicated in Figure 1.

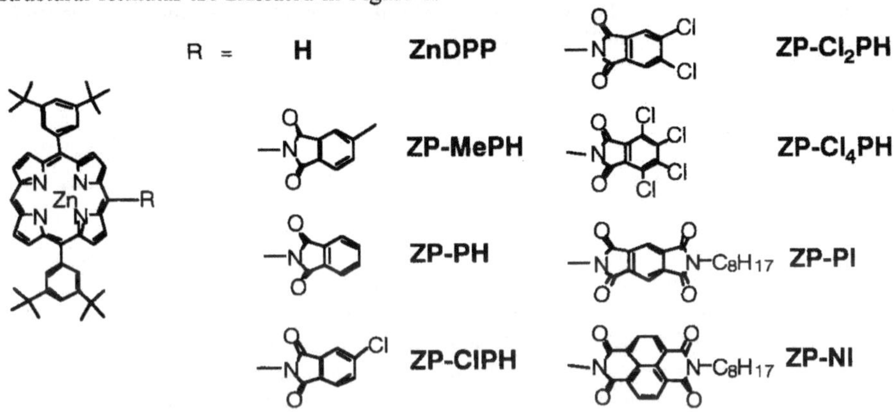

Figure 1. Structural formulas of the ZnDPP (ZnP) and ZP-A supramolecular compounds.

All solvents used were spectro or GC grade. Sample solutions in an optical cell of 2 mm path length were stirred by using micro magnetic stirring bar during the measurements. By comparison of the absorption spectra before and after the measurements it has been confirmed that the photo-decomposition of ZP-A systems are negligibly small during the measurements. For the measurements of temperature effects on fluorescence dynamics in THF and TOL solutions, an optical cell containing sample solution was set in a quartz dewer with optical flat windows and temperature was controlled with cold nitrogen gas stream.

Although the D and A are directly linked in the ZP-A systems, the Soret band at $\Delta\mathbf{E(S_2)}$ = 2.9 eV is practically the same throughout the series. The imide group shows no absorption band around the Soret band and the negligible D-A electronic coupling in the ground state (no charge transfer bands observed) can be ascribed to the nearly perpendicular conformation of the A moiety as revealed by X-ray crystallography. Moreover, we have confirmed by MO calculations (PM3 INDO/S using WinMOPAC), that the LUMO of the attached A moieties commonly have a node at the connecting meso-position, thereby minimizing the electronic interactions in the ground state.

To determine the free energy gap between reactant and product states, we have measured one electron oxidation potential ($\mathbf{E_{ox}}$) and reduction potential ($\mathbf{E_{red}}$) of ZP-A compounds in DMF (N,N-dimethylformamide). The obtained ($\mathbf{E_{ox}}$ - $\mathbf{E_{red}}$)/eV values are: 2.152 for ZP-

MePH, 2.140 for ZP-PH, 2.080 for ZP-ClPH, 2.034 for ZP-Cl_2PH, 1.836 for ZP-Cl_4PH, 1.592 for ZP-PI and 1.322 for ZP-NI. The free energy gaps ($-\Delta G_{CS}$) for different solvents used in our experiments were then calculated by using the equation:

$$-\Delta G_{CS} = \Delta E(S_2) - (E_{ox} - E_{red}) - \Delta G_S$$

where ΔG_S is the correction term given by

$$\Delta G_S = (e^2/2)(r_D^{-1} + r_A^{-1})(\varepsilon_s^{-1} - \varepsilon_r^{-1}) - (e^2/\varepsilon_s r)$$

Here r_D and r_A are effective radii of ZP^+ and A^- ions, **r** is the center-to-center distance between the ions, ε_r is the dielectric constant of DMF and ε_s is that of the solvent used for the fluorescence measurements. The values for r_D, r_A and **r** were taken from our previous studies on similar porphyrin-acceptor linked compounds [6,7].

Results

1.1 *Fluorescence decay dynamics and k_{CS} vs. $-\Delta G_{CS}$ relations in different solvents for ZP-A supramolecular systems*

Figure 2 shows ultrafast S_2 fluorescence decay curves of ZP and seven ZP-A dyads in THF solution. Quite similar decay curves with comparable S/N ratios were observed also in other solvents examined. The fluorescence dynamics of these systems undergoing ultrafast CS reactions in 100 fs regime could be accurately reproduced by single exponential functions in all solvents examined here. Using the obtained decay time constant τ, the rate constant for CS k_{CS} has been determined by $k_{CS} = \tau^{-1} - \tau_0^{-1}$ relation, where τ_0 (~2.3 ps) is the lifetime of the S_2 state of ZP monomer itself. If the strong S_2 fluorescence quenching of ZP-A dyads is due to the ultrafast CS, $ZP^*(S_2) - A \rightarrow ZP^+ - A^-$, it should be confirmed by fs-ps absorption spectroscopy on these transient species. Actually, such measurements have been made on ZP-NI, ZP-PI and ZP-MePH in THF providing k_{CS} values $9.5*10^{11}s^{-1}$, $5.1*10^{12}s^{-1}$ and $3.5*10^{12}s^{-1}$, respectively [8], which agree exactly with those obtained by the fluorescence studies. Moreover, the $S_2 \rightarrow S_1$ internal conversion of ZP in THF clearly showed the rise of rather strong S_1 fluorescence in accordance with the decay of the S_2 fluorescence [9], however, the rise of a very weak S_1 fluorescence in ZP-A dyads was an order of magnitude faster indicating that the internal conversion is strongly suppressed by competing CS process.

Figure 3 shows obtained k_{CS} values plotted against $-\Delta G_{CS}$. The inverted, top and normal regions in the energy gap law for the photoinduced CS reaction have been unambiguously observed here. In polar solvents ACN, THF and GTA, the ultrafast reorganization of the solvent in the course of CS plays an important role of the reaction coordinate and causes an activation barrier, giving rise to the normal region with relatively small $-\Delta G_{CS}$ values. However, such normal region was not recognized in CH and TOL solutions, which can be ascribed to the much smaller solvent reorganization energies, leading to the negligibly small activation barrier at relatively small $-\Delta G_{CS}$ values. The energy gap law of these supramolecular systems in less polar or nonpolar solvents becomes somewhat similar to that of the intramolecular radiationless transition (internal conversion) in the weak coupling limit [3].

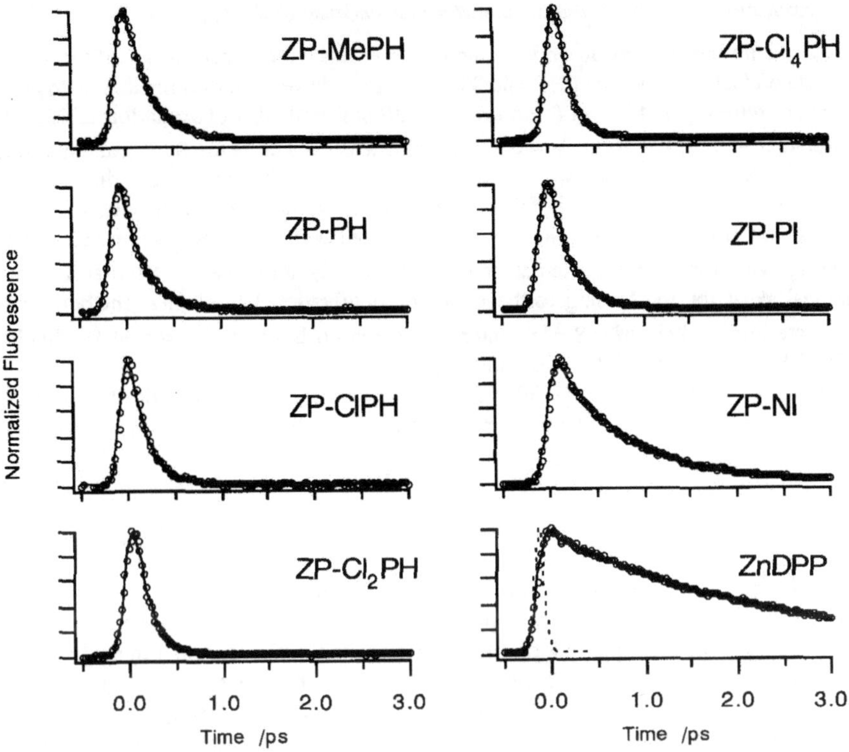

Figure 2. Normalized fluorescence decay curves of ZP and ZP-A dyads (open circles) excited at 405 nm and *monitored at 430 nm at 20°C in THF solution. The instrument response function (dashed line) and convoluted* single exponential function fittings (solid lines) are also given.

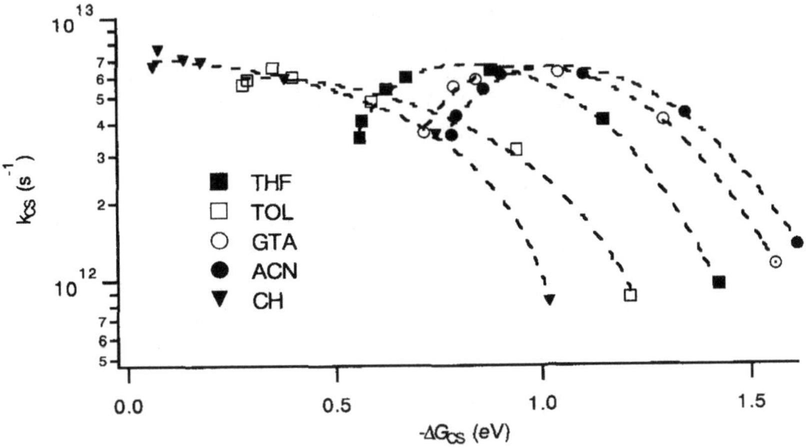

Figure 3. **k**$_{CS}$ vs. **-ΔG**$_{CS}$ plots for five different solvents. The dashed lines are polinomial extrapolations just for eye guidance.

1.2 Excitation wavelength and temperature dependence of the k_{CS}

The k_{CS} and its energy gap dependence upon the excess vibrational energy of the initial excited state of ZP-A dyads in THF and TOL solutions have been examined. We compared the case of excitation at 405 nm (monitored at 430 nm) with that of excitation at 425 nm (monitored at 455 nm) at 20°C, located in the blue and red side of the Soret band maximum, respectively. No changes in k_{CS} within the experimental error was observed indicating that the excitation to the hot S_2 state is followed by ultrafast vibrational redistribution (~10 fs) filling the vibrational states near the bottom of S_2, from which the CS reaction takes place competing with the $S_2 \rightarrow S_1$ internal conversion and S_2 fluorescence. Similar ultrafast vibrational redistribution leading to the formation of vibrational levels near the bottom of the S_1 state immediately after $S_2 \rightarrow S_1$ internal conversion have been observed for ZnTPP and ZnDPP in ethanol and THF solutions [9].

The k_{CS} values within 200K – 300K temperature range remained practically unchanged in THF and TOL solutions. The CS observed here seems to be rather close to barrierless process.

Modeling and Discussion

What is the reaction coordinate in extremely rapid ET reactions observed in this work? The concept of reaction coordinate in polar solvents has often been identified with the **diffusive** rotational reorientation of the permanent molecular dipole moments of the solvent (solvation coordinate). However, in our case, the k_{CS} rate of a given ZP-A supramolecular compound only slightly differs in five liquids with greatly different physical properties such as viscosity, dipole moment, molecular weight, and dielectric constant, effectively excluding the diffusive solvent response (slow solvation) being the reaction coordinate. Nevertheless, the observed k_{CS} vs. $-\Delta G_{CS}$ bell-shaped relations in polar solvents and absence of the normal region in nonpolar and slightly polar solvent indicate that the CS reactions are indeed solvent dependent, however coupled only to the ultrafast **nondiffusive** solvent response (fast solvation) playing the role of solvation coordinate. This fast solvation could arise from such nondiffusive intermolecular motions as librations, translations, interaction-induced effects, internal rotations or other ultrafast motions and is observed not only in polar nonviscous ACN [10] but also in viscous GTA. Y. J. Chang *et al.* [11] carried out Optical-heterodyne-detected Raman-induced Kerr-effect spectroscopy on GTA and predicted that about 25% of the solvation dynamics in GTA will occur in less than 100 fs regime. We have measured the dynamic Stokes shift (DSS) of coumarin 153 dye fluorescence in GTA and clearly recognized ultrafast DSS in 100 fs regime in addition to main components with very long relaxation times (~100 ps).

On the other hand, we can not exclude the important role of the coupling of the CS process with intramolecular high frequency modes (*hfm*) which facilitate the reaction [3] and act as an other reaction coordinate, because the k_{CS} in nonpolar CH, viscous GTA and polar ACN are very close. Summarized, there are at least two reaction coordinates and the photoinduced CS reaction in ZP-A dyads occurs on a multidimentional potential surface.

As we have shown, the fluorescene decay dynamics were accurately reproduced by single exponential functions in all solvents examined. Hence, the Sumi-Marcus model [12] can not be applied in our case because it predicts non-exponential decays due to the ET at various solvation states along the slow solvation coordinate. Our data in polar solvents can be reasonably well reproduced by Marcus-Jortner equation (Figure 4), incorporating both the solvent reorganization and the averaged intramolecular *hfm* coupled with the CS process:

$$k_{CS} = \left(\pi / \hbar^2 \lambda_s k_B T\right)^{1/2} V^2 \sum_n \left(e^{-S}\left(S^n / n!\right)\right) \exp\left\{-\left(\Delta G_{CS} + \lambda_s + n\hbar < \omega >\right)^2 / 4\lambda_s k_B T\right\}$$

where V is the electronic coupling matrix element, $S = \lambda_v / h<\omega>$ is the electron-vibration coupling constant, λ_v is the reorganization energy associated with the averaged angular frequency $<\omega>$ of *hfm* and λ_s is the solvent reorganization energy. The observed results in three polar solvents are reasonably well reproduced by using parameters close to those used in the related previous studies [4]. However, it becomes difficult to reproduce the observed results in TOL and impossible in CH solutions probably because of the oversimplification of the model incorporating only one averaged *hfm*.

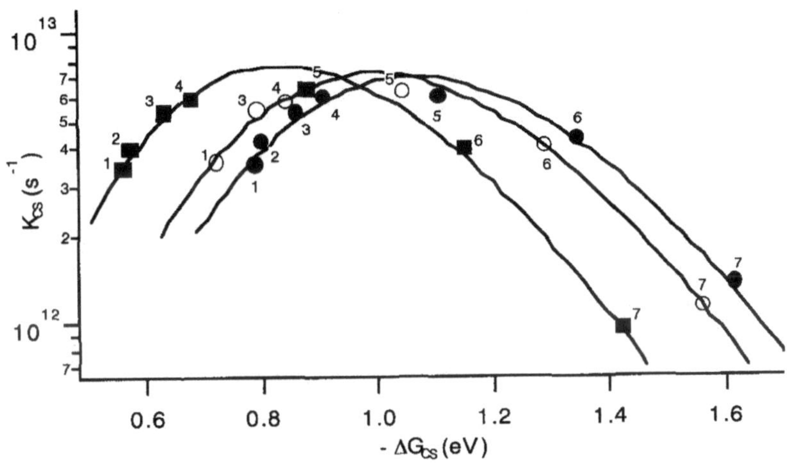

Figure 4. k_{CS} vs. $-\Delta G_{CS}$ relations in THF (■), GTA (○), ACN (●) at 20 °C and simulations (solid lines) with Marcus-Jortner equation. ZP-A compounds: 1. ZP-MePH, 2. ZP-PH, 3. ZP-ClPH, 4. ZP-Cl_2PH, 5. ZP-Cl_4PH, 6. ZP-PI, 7. ZP-NI. Parameter values in the global fitting are: $V = 0.024$ eV, $\lambda_v = 0.3$ eV, $<\omega> = 0.15$ eV throughout all three solvents, and λ_s for each solvent was determined to get the best fit between calculated and observed k_{CS} values. The obtained λ_s values are: (ACN) = 0.76 eV, (GTA) = 0.69 eV, and (THF) = 0.54 eV.

Investigations on photoinduced CS process from S_1 state of ZP-A compounds in various solvents are now going on in our laboratory to examine whether the CS reaction dynamics and mechanisms are similar to those from S_2 state discussed in this work. The results will be reported in a forthcoming detailed report.

Acknowledgements

H. C. thanks The Ogasawara Foundation for the Promotion of Science and Engineering, for financial support.

References

1. Marcus R. A., and Sutin N., *Biochem. Biophys. Acta* **811** (1985) p. 265.
2. Barbara P. F. and Jarzeba W., *Adv. Photochem.* **15** (1990) p. 1.
3. Bixon M. and Jortner J., *Adv. Chem. Phys.* **106** (1999) p. 35.
4. Mataga N. and Miyasaka H., *Adv. Chem. Phys.* **107** (1999) p. 431.
5. Chosrowjan H., Mataga N., Nakashima N., Imamoto Y. and Tokunaga F., *Chem. Phys. Lett.* **270** (1997) p. 267.
6. Asahi T., Ohkohchi M., Matsusaka R., Mataga N., Zhang R. P., Osuka, A. and Maruyama K., *J. Am. Chem. Soc.* **115** (1993) p. 5665.
7. Osuka A., Noya G., Taniguchi S., Okada T., Nishimura Y., Yamazaki I. and Mataga N., *Chem. Eur. J.* **6** (2000) p. 33.
8. Okada T., Osaka University, *private communication.*
9. Mataga N., Shibata Y., Chosrowjan H., Yoshida N. and Osuka A., *J. Phys. Chem. B.* **104**, (200) p. 4001.
10. Rosenthal, S. J., Xie, X., Du, M. and Fleming, G. R., *J. Chem. Phys.* **95** (1991) p. 4715.
11. Chang, Y. J. and Castner, Jr., E. W., *J. Chem. Phys.* **99** (1993) p. 7289.
12. Sumi H. and Marcus R. A., *J. Chem. Phys.* **84** (1986) p. 4894.

LIQUID PHASE PHOTOCHEMISTRY OF THE DI- AND POLYHALOGENATED ALKANES CONTAINING IODINE: FEMTOSECOND TRANSIENT ABSORPTION STUDY OF THE PHOTODISSOCIATION AND IN-CAGE ISOMERIZATION

ALEXANDER N. TARNOVSKY, MAGNUS WALL, MARCUS RASMUSSON, VILLY SUNDSTRÖM, AND EVA ÅKESSON

Department of Chemical Physics, Lund University, Box 124, SE-221 00 Lund, Sweden

Abstract: We present a femtosecond pump-probe study of the photodissociation and in-cage isomerization reactions of several dihalomethanes of the series CH_2XI (X=Cl, Br, I), a polyhalomethane (CHI_3) and several vicinal dihalides dissolved in acetonitrile upon UV excitation.

1 Introduction

The photodissociation of di- and polyhalogenated alkanes containing iodine in the gas phase has been extensively investigated over the years due to the general interest in molecular decomposition, control of intramolecular dynamics and due to implications for chemical lasers [1-13]. More recently, atmospheric chemistry evaluated the role of these molecules in ozone depletion as they represent a potentially significant source for halogen compounds emitted into the atmosphere [14,15]. Only a a few studies of these molecules have been performed in the condensed phase, focusing primarily on microscopic understanding of how solvent influences the outcome of chemical reactions and to attain a deeper understanding of the "cage effect [16-18]. General photochemistry of halomethylenes with specific interest towards their organic synthetic application as reagents in photocyclopropanation of olefins was also investigated [19-22].

The lowest energy transitions in di- and polyhalogenated alkanes containing iodine involve promotion of one of the n electrons of iodine to the σ^*orbital of the C-I bond. There is an agreement in the literature that the initial process following the photoexcitation is a highly efficient homolytic C-I bond cleavage in both gas phase and solution [20,22]. In solution, it was proposed that the resulting radicals undergo electron transfer within the caged radical pair that eventually gives a carbocation, which reacts with olefins in photocyclopropanation [20]. For diiodomethane, CH_2I_2, in particular, the α-iodo-cation (CH_2I^+) was conjectured to exist, although the identity of this reaction intermediate was not established by direct observations. When the dihalomethanes CH_2XI (X=Cl, Br, I), CH_2X_2 (X=Cl, Br) [24,25] and CCl_4 [26] embedded in cold matrices were photolyzed with UV light corresponding to the long-wavelength part of their absorption spectra, the matrices developed characteristic two band structured absorption spectra assigned to isomers H_2C-X-X, H_2C-X-I and Cl_2C-Cl-Cl. The assignment was based on IR-spectra and the fact that the process was at least partly reversible when the photoproduct bands were irradiated. An isomer production mechanism was suggested, in which the initially formed halomethyl-halogen radical pair cannot dissociate in the solid matrix and collapses into an isomer. The isomers disappeared quickly upon warming the polyethylene matrix above 100 K. Ab initio calculations demonstrated some stability of the isomers of CH_2Cl_2, CH_2Br_2, CH_2ClBr, and CH_2I_2 [25,27,28]. These matrix isolation studies turned out to be seminal in understanding the photochemistry of dihalomethanes because similar spectral features as those of the dihalomethane isomers had been observed for several polyhalomethanes, for example, CHI_3, $CHBr_3$ and CF_2Br_2, upon UV irradiation in hydrocarbon glasses [29,30].

The intriguing behavior of di- and polyhalomethanes upon UV excitation stimulated our interest in exploring the short time photodissociation dynamics of these molecules and subsequent photochemical changes of the reaction products in solution. In this paper, we show that the CH_2XI isomers are formed upon UV excitation of several dihalomethanes in room temperature liquids. We discuss how fast is the isomer formation and what is the isomerization mechanism in solution. Specifically, we continue to study the fast photoreactions of several dihalomethanes (CH_2I_2, CH_2BrI, CH_2ClI), one polyhalomethane (CHI_3) and several vicinal dihalides ($C_2H_4I_2$, $C_2F_4I_2$ C_2F_4BrI) in acetonitrile initiated by ~130 fs pulses at 266 nm (CH_2I_2, CH_2BrI [32], and CH_2ClI [31], vicinal dihalides [34]), at 310 nm (CH_2I_2, [18]) and 350 nm (CH_2I_2 and CHI_3 [33]).

2 Experimental

Our femtosecond transient absorption spectrometer is based on an amplified Ti:Sapphire laser/regenerative amplifier system. The amplified output at 800 nm, ~0.9 mJ per pulse, 100 fs pulse duration, 1 kHz repetition rate, was split into two beams to generate the pump light (at 310, 350 or 266 nm) and probe light, which was either a quasi-monochromatic light (300-1220 nm) produced via an OPA or a white-light continuum (330-1000 nm), see Ref [18,31,35]. for details. The 310 and 350 nm pump light pulses were generated by frequency quadrupling the signal output of the second OPA. The 266 nm pump light is produced in a frequency tripler. The details of detection system and method of data acquisition can be found in [18,31,35]. The shown time-resolved spectra were compensated for chirp. All measurements were performed at magic angle polarization conditions. The instrumental response function was from 180 to 280 fs (excitation at 310 and 350 nm) and ~220 fs (excitation at 266 nm).

3 Results and Discussion

3.1 Dihalomethanes

In the gas phase, for relatively low energy UV photons (<~5 eV) as in our experiments, 1) the excitation is to n(I)-σ*(C-I) states of CH_2XI (X=Cl, Br, I), [1,4,8-10], thought to be purely repulsive in the respective C-I bond [1,3,8-10]; 2) product distribution and energy disposal are influenced by curve crossings occurring in the low-lying dissociative channels, and are dependent on excitation wavelength [1-10]:

$$CH_2XI \xrightarrow{h\nu} CH_2XI^* \rightarrow CH_2X+[I(^2P_{3/2}), I^*(^2P_{1/2})] \qquad (1)$$

where the CH_2X radical is formed vibrationally and rotationally excited [2,3,7,9,10], $I(^2P_{3/2})$ and $I^*(^2P_{1/2})$ are the iodine-atom in the ground state or spin-orbit excited state. Resonance Raman studies in solution also suggest that the short-time photodissociation dynamics of these molecules upon excitation at a wavelength longer than 260 nm are consistent with C-I bond breaking [23]. We expect three competitive reactions to follow:

$$CH_2XI^* \rightarrow CH_2X+[I(^2P_{3/2}), I^*(^2P_{1/2})] \text{ in-cage} \rightarrow$$
$$\rightarrow H_2C\text{-}X\text{-}I, \text{ formation of isomer} \qquad (2)$$
$$CH_2XI^* \rightarrow CH_2X+[I(^2P_{3/2}), I^*(^2P_{1/2})] \text{ in-cage} \rightarrow$$
$$\rightarrow CH_2XI, \text{ recombination into parent,} \qquad (3)$$
$$CH_2XI^* \rightarrow CH_2X+[I(^2P_{3/2}), I^*(^2P_{1/2})] \text{ in-cage} \rightarrow$$
$$\rightarrow CH_2X_{\text{solvated}} + I, I^*_{\text{solvated}}, \text{ cage escape} \qquad (4)$$

We have measured the time-resolved spectra between 300-940 nm and kinetics at many probing wavelengths from 300 to 1220 nm to identify the processes that occur following the photodissociation of the dihalomethanes and the nature of the reaction products. For all the studied dihalomethanes we could distinguish the following characteristic stages of the transient absorption evolution (see the spectra of CH_2BrI as an example, Fig. 1). First, a transient absorption signal with an instantaneous rise is observed within the whole probe range followed by a fast 350 fs - sub-100 fs decay depending on the compound (for CH_2I_2 at 310 nm – 350 fs [18]). Second, following this initial fast phase, there is a broad and structureless transient absorption signal developing from the ultraviolet to the near infrared. Third, at longer times, the transient absorption grows and sharpens into characteristic absorption bands in the ultraviolet and visible spectral regions: this occurs with wavelength-dependent time constants of a few picoseconds (1-9 ps) and tens of picoseconds (30-50 ps). The bands attain equilibrium at about 50 ps, manifesting the formation of the product species, Fig. 1. The temporal profiles of the transient absorption measured at band maxima are identical. Finally, the product absorption decays with a time constant of ~120 ps and ~2.5 ns for CH_2ClI and CH_2BrI, respectively. For CH_2ClI, this decay is accompanied by formation of an absorption band at ~360 nm, and for CH_2BrI – very weak absorption around 300-315 nm. The absorption bands of the CH_2I_2 product were observed at all exciting wavelengths and were nearly stable out to 500 ps (a decay time of these bands was measured to be ~230 ns using nanosecond laser photolysis, Table 1).

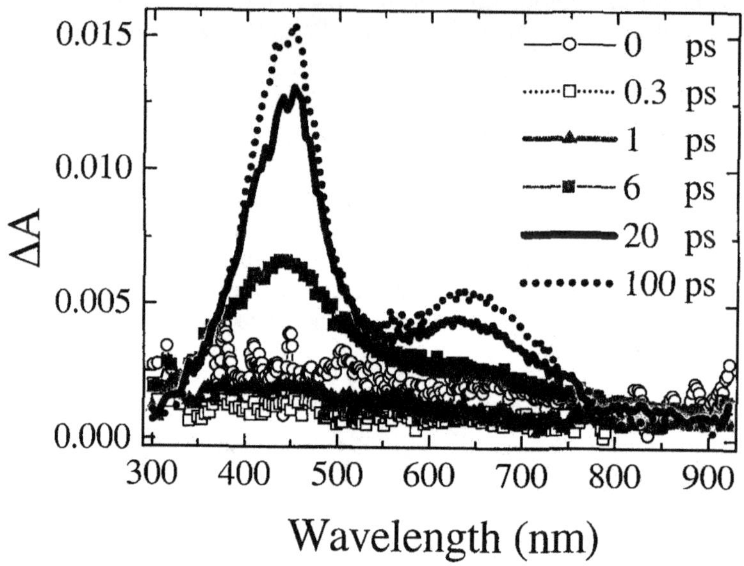

Figure 1. Measured transient absorption spectra of CH_2BrI (30 mM) in CH_3CN. The time delays between the 266 nm pump and the probe pulses are given in the inserts. Below 360 nm, the transient absorption spectra are constructed from kinetics measured at single wavelengths in a 110 μm thick jet and scaled to the main spectrum measured in the flow cell of 0.5 mm pathlength.

The dual absorption band spectra of the photoproduct species observed in the present study can be assigned readily to the H_2C-I-I, H_2C-Br-I and H_2C-Cl-I isomers. This is evident from comparison between the observed bands and the absorption spectra of the UV photolyzed cold matrices of the parent compounds, see Fig. 2. This assignment is supported by DFT calculations, which predict an intense ground to singlet state transition at 425, 422, 453 nm in the H_2C-I-I, H_2C-Br-I and H_2C-Cl-I species [36-38]. Recent transient resonance Raman studies confirmed that UV excitation of CH_2I_2, CH_2BrI and CH_2ClI in room temperature solutions yields the corresponding isomers [36-39]. Since the pump-probe kinetics recorded at the absorption peaks are similar, we attribute both the strong and weak bands to the isomer, most likely to the transitions from the ground state to two low-lying excited states (Fig. 3). This is supported by the observation that irradiation of low temperature matrices with light corresponding to one of the H_2C-X-I absorption bands depleted the absorption in both bands [24,25].

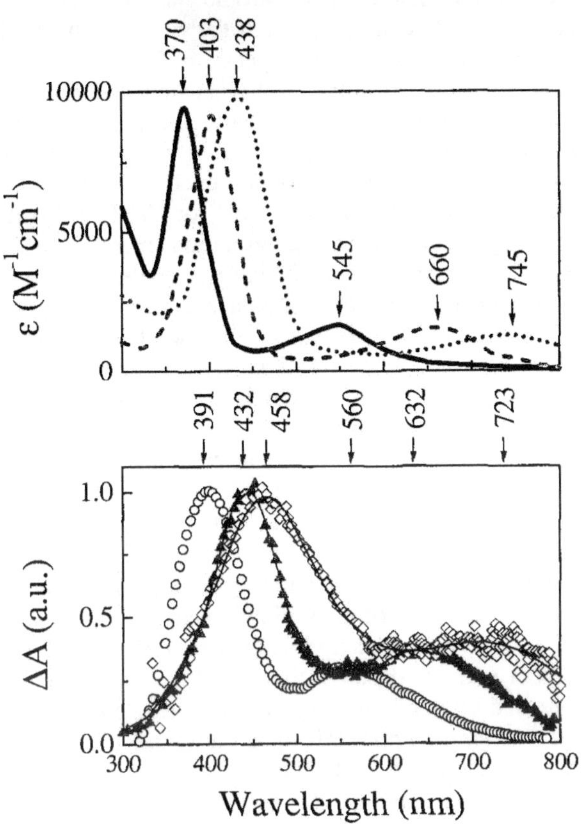

Figure 2. Top panel shows the absorption spectra of the isomers H_2C-I-I (solid line), H_2C-Br-I (dashed line), and H_2C-Cl-I (dotted line) in N_2 matrix at 12 K redrawn from Ref. [24,25]. Bottom panel shows the femtosecond pump-probe spectra of CH_2I_2 (open circles), CH_2BrI (solid triangles), and CH_2ClI (open diamonds) measured at room temperature in acetonitrile. The excitation wavelengths were 266 nm for CH_2BrI and CH_2ClI, and 310 nm for CH_2I_2. The spectra were recorded at the delays 200 ps (CH_2I_2), 100 ps (CH_2BrI) and 50 ps (CH_2ClI) and assigned to the photoisomers H_2C-I-I (Ref. [18]), H_2C-Br-I (Ref. [32]), and H_2C-Cl-I (Ref. [31]).

Alternative assignments for the absorption bands to I-atoms, CH_2X radicals, hot CH_2XI parent molecule formed due to primary geminate recombination and molecular cation, CH_2XI^+, can be ruled out. Solvated I-atom forms a charge-transfer complex with acetonitrile molecules and its absorption is known to peak at 275 nm [40]. The absorption spectrum of the CH_2I radical in the gas phase has been measured in the UV-region 200-400 nm [41], and it is very unlikely that CH_2I has any appreciable optical absorption above 400 nm in solution. The reddest absorption edge of the CH_2Br radical is agreed to lie below 335 nm [42]. The CH_2Cl radical absorbs below 230 nm [42]. Even though vibrationally hot parent molecules are very likely to absorb in the visible, the electronic spectrum will be flat, featureless and much reduced in strength relative to that of the vibrationally cool molecules, not like the observed structured and intense spectrum. In addition, vibrationally hot parent molecules are expected to attain a thermalized absorption spectrum faster than a few tens of picoseconds [43-44], in sharp contrast with the relatively long-lived absorption bands. Finally, direct production of CH_2XI^+ from CH_2XI in this study can be excluded since the energy of the excitation photon is much smaller than the ionization potential of CH_2XI (9.46 eV for CH_2I_2, [45]) and the transient absorption is linearly proportional to the excitation energy. An indirect route of formation of CH_2XI^+, in which dihalomethane aggregates exist prior to excitation and a formation of a charge-transfer complex between the I photofragment and CH_2I_2 is followed by full electron transfer to produce $CH_2I_2^+$ and I^- [17], strongly contradicts the lack of solute concentration effects in the transient absorption dynamics within the observed bands for CH_2I_2 [18] and two other dihalomethanes [31,32].

How are the isomers formed? To answer this question we turn to the transient absorption spectra at times < 50 ps. The instantaneous rise of the transient absorption represents population of the electronically excited state (n,σ*) of the parent molecule by the excitation pulse (excited state absorption ESA in Figure 3). The fast decay of this signal corresponds to the wave-packet motion on the excited state potential energy surface out of the initially excited Franck-Condon region, and can be associated with C-I bond lengthening and CH_2X moiety adopting more planar configuration or with the decay of the initially excited state to the lower lying excited state(s) followed by bond breaking. The resulting radical pair recombines to yield the H_2C-X-I isomer in noticeable quantities and the spectral evolution at longer times is associated mainly with the dynamics of the isomer, namely with growth of the isomer concentration (a part of the few picoseconds spectral evolution) and with vibrational relaxation of the isomer (the few picoseconds evolution and tens of picoseconds evolution). First, we discuss the vibrational relaxation. The few tens of picoseconds component reflects narrowing of the two absorption bands of the isomers. As to the faster few picoseconds process, faster absorption rise is observed at the wings than at the band peak. Outside of the thermalized isomer spectrum, the rise is even faster and is followed by a fast decay. These observations are typical for vibrational relaxation of small molecules in solution [46]. Further to the red of the weaker band, probe light samples the upper region of the isomer ground state potential surface (Fig. 3). This region is populated by the hottest and fastest formed isomers, as a part of internal energy stored in the radicals is yet not significantly dissipated due to interactions with the cage. This suggests that the sub-picosecond absorption rise times at the reddest probe wavelength in these experiments (1220 nm) correspond to the earliest observation of the isomers (Table 1). For shorter probe wavelength, the rise time becomes slower because vibrational relaxation in the lower lying vibrational states of the isomer ground state is probed. Despite sharpening and narrowing of the isomer spectra, the overall area under the bands is growing with time, suggesting a growth of isomer concentration. The isomer formation time is estimated by applying the conservation rule of the dipole strength for the electronic transition with time [47]. This rule implies that the value of the integral $\int_{band} \Delta A(\nu,t) d\ln\nu$ is linearly proportional to isomer concentration, where ν is the wavenumber, ΔA the isomer absorption at different times, and the integration is performed within the isomer absorption band. This approach yields an isomer formation time constants of several picosecond (Table 1) for all the studied CH_2XI molecules.

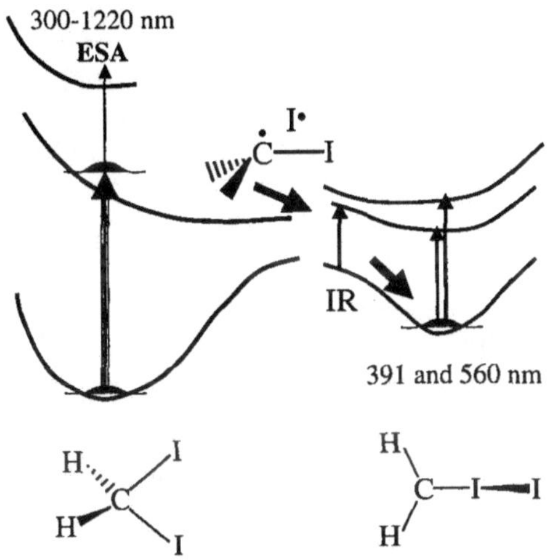

Figure 3. Isomerization scheme for diiodomethane. See text for details.

We propose that the H_2C-X-I isomers are formed in the solvent cage. Our experiments suggest that it takes about 1 ps for the hot CH_2X radical and I atom to start recombining at the energy level that corresponds to the highly excited vibrational states of the ground state of the H_2C-X-I isomer. The major fraction of the H_2C-X-I population appears on a time scale of several picoseconds, very likely, in parallel with vibrational cooling of the nascent radicals. This would be consistent with ab initio calculations of isolated isomers of the dihalomethanes CH_2Cl_2, CH_2Br_2, CH_2ClBr, and CH_2I_2 that suggest a shallow ground state potential energy minimum of the isomers with respect to the potential energy of the planar radical and iodine [25,27]. Still, the formed isomers are able to house some excess energy as indicated by spectral evolution of the isomer absorption. The isomers are formed in appreciable amount, the isomerization of CH_2I_2 into H_2C-I-I is most efficient: the quantum yield of isomerization is estimated to be 70% (Table 1).

The thermal stabilities of isomers differ, H_2C-I-I is most and H_2C-Cl-I the least stable. Our nanosecond laser photolysis experiments show that following 266 nm excitation, the transient spectra are dominated by the absorption of the isomers H_2C-I-I and H_2C-Br-I immediately after excitation (H_2C-Cl-I is too short-lived to be observed in the nanosecond experiments), and on the microsecond time scale - by well known absorption of I_3^- for all three dihalomethanes. The formation of I_3^- takes place following the diffusion-controlled dark reactions between I and I^-, where the isomers H_2C-X-I undergoing heterolytic cleavage of the X-I bond are the very likely precursors for I^-, and naturally CH_2X^+. Recent modelling of the photocyclopropanation reaction with diiodomethane as a reagent on a DFT level of theory suggested that the isomer H_2C-I-I is a more likely methylene transfer agent than CH_2I^+ [21]. This would be consistent with our results since H_2C-I-I is formed in noticeable quantities and long-lived enough to react with olefin.

	H_2C-I-I	H_2C-Br-I	H_2C-Cl-I
ε (Mol/l cm) [1)]	5 600	6 520 [7)]	5 280 [6)]
Yield (%) [2)]	70	25 [7)]	9 [6)]
Isomer appearance time (ps) [3)]	0.95 [5,8)]	1 [7)]	<1.5 [9)]
Isomer formation time (ps) [4)]	5.5 [5)]	7 [7)]	9 [6)]
Lifetime (ns)	230	2.5 [7)]	0.12

Table 1. Summary of photophysical and photochemical properties of the isomers H_2C-I-I I, H_2C-Br-I, and H_2C-Cl-I in acetonitrile.

[1)] The value at the maximum of the strong band based on the extinction value in the cold N_2 matrix (ε=10 000 l/Mol cm, [24,25]) and assumption that the transition dipole strength of the strong band is conserved upon going from the matrix to the solution. [2)] Based on the ε values of isomers and the absolute measurements of the amount of energy absorbed (femtosecond experiments). [3)] The transient absorption rise time at the probe wavelength of 1220 nm is taken to represent the hottest and the fastest isomers appearing after excitation. [4)] The isomer formation time found when applying the transition dipole strength conservation rule. [5)] Upon 310 nm excitation. [6)] Preliminary results. [7)] Ref. [32]. [8)] Ref. [18]. [9)] Ref. [31].

3.2 *Other molecules: Polyhalomethanes and Vicinal Dihalides*

Experiments on iodoform, CHI_3, may potentially reveal to what extent the above discussed caging mechanism applies to polyhalomethanes. Excitation of iodoform in the gas phase leads to effective cleavage of the C-I bond [1,4]:

$$CHI_3 \xrightarrow{h\nu} CHI_3^* \rightarrow CHI_2 + [I(^2P_{3/2}), I^*(^2P_{1/2})] \quad (5)$$

Following excitation at 350 nm of CHI_3 in acetonitrile, the transient absorption at early times after excitation (0.5-1 ps) exhibits a broad absorption in the 370-730 nm range, which is interpreted to be due the hot CHI_2 radical around 400 nm and the hot photoproduct, identified as the isomer, HIC-I-I [33] (Fig. 4).

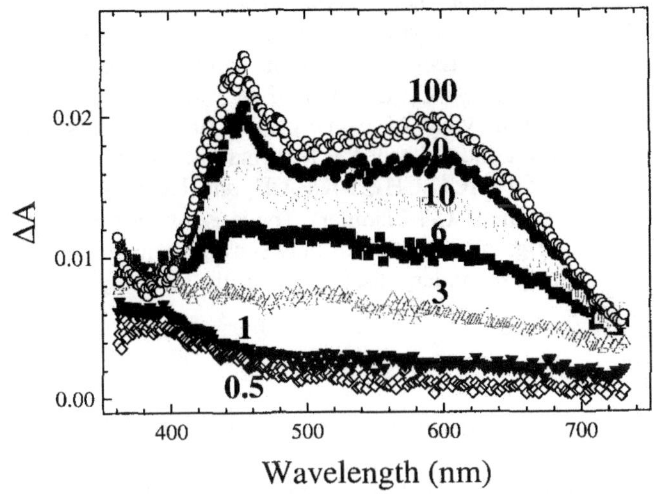

Figure 4. *Measured transient absorption spectra of CHI_3 (10 mM, 0.5 mm pathlength flow cell) in* CH_3CN. The time delays (in picoseconds) between the 350 nm pump and the probe pulses are given next to spectral curve.

As time progresses, the radical cools as evidenced by a blue shift of the radical absorption, while the isomer absorption bands at 450 and 600 nm grow in intensity and sharpen. DFT calculations confirm that HIC-I-I is responsible at least for the band at 450 nm [48]. The absorption rise time at the wings of the isomer bands is faster than at the wavelengths corresponding to the band maxima, at which a rise time of 7 ps is measured. This suggests that HIC-I-I is formed vibrationally excited, similarly to the dihalomethane isomers. The time constant of 7 ps is likely to represent the formation of HIC-I-I via in-cage recombination of the iodine atom and the moderately hot CHI_2 fragment [33].

In the examples given above, the primarily released iodine atoms are, due to their large size and mass, likely to remain in the solvent cage and interact with the C-X moiety of the primary radicals. Will photoisomer formation occur in the vicinal dihalides such as ICH_2CH_2I, ICF_2CF_2I and ICF_2CF_2Br? In these molecules, the coupling between halogen atoms is much smaller than that for CH_2XI because of their predominant anti-conformation. Fast dissociation of vicinal dihaloethanes in the gas phase upon $n \rightarrow \sigma^*$ excitation may be followed by a secondary carbon-halogen cleavage as a consequence of vibrational energy redistribution [10,11].

After excitation at 266 nm of ICH_2CH_2I, ICF_2CF_2I and ICF_2CF_2Br in acetonitrile, formation of I_2 within tens of picoseconds was observed for ICH_2CH_2I

and ICF_2CF_2I, while IBr was found for ICF_2CF_2Br [34]. The predominant mechanism for I_2 formation is secondary dissociation of the haloethyl radicals followed by geminate combination of iodine atoms. Early after excitation, weak absorption bands are observed around 400 nm for ICH_2CH_2I and ICF_2CF_2I and around 410-440 nm for ICF_2CF_2Br, which decayed on a time scale of 10-100 ps depending on the dihaloethane. The spectral position of these bands agrees with the band of H_2C-I-I at 390 nm, and with the red shift when a C-Br is substituted for the C-I bond. We tentatively assign the minor bands to the isomers of vicinal dihaloethanes, I-I-CH_2CH_2, I-I-CF_2CF_2 and I-Br-CF_2CF_2. Based on the intensity of these bands, the quantum yield of isomerization of the vicinal dihaloethanes is estimated to be less than one tenth of the I_2 formation quantum yield (assuming an extinction coefficient for I-ICH_2CH_2, I-ICF_2CF_2 and I-$BrCF_2CF_2$ of ~6 000 l/Mol cm, i.e. the same as for the H_2C-I-I isomer). The reasons for the small isomerization quantum yield are the greater separation between the halogen atoms in the parent molecule and competing secondary dissociation of the haloethyl radicals.

4 Conclusion

Femtosecond pump-probe experiments demonstrate that the dihalomethanes CH_2XI, X=Cl, Br, I) and iodoform upon UV excitation in solution at room temperature form the isomers H_2C-I-I, H_2C-Br-I, H_2C-Cl-I and HIC-I-I in appreciable quantities. An in-cage isomerization mechanism is suggested, in which the isomerization is preceded by C-I bond breaking producing radical fragments. Due to interaction of the fragments with the solvent cage on a time scale of several picoseconds, fragments dissipate a part of their excess energy and form the moderately vibrationally excited isomer. This isomer undergoes vibrational relaxation on a time scale of several and tens of picoseconds to yield the vibrationally relaxed isomer.

5 Acknowledgements

Financial support from the Swedish Natural Science Research Council, Magnus Bergwall Foundation, Crafoord Foundation and Knut and Alice Wallenberg Foundation is gratefully acknowledged.

References

1. Kawasaki M., Lee S. J. and Bersohn R., Photodissociation of molecular beams of methylene iodide and iodoform. *J. Chem. Phys.* **63** (1975) 809-814.
2. Baughcum S. L. and Leone S. R. Photofragmentation infrared emission studies of vibrationally excited free radicals CH_3 and CH_2I, *J. Chem. Phys.* **72** (1980) 6531-6545.

3. Kroger P. M., Demou P. C. and Riley S. J., Polyhalide photofragment spectra. I. Two-photon two-step photodissociation of methylene iodide, *J. Chem. Phys.* **65** (1976) 1823-1834.
4. Gedanken A. and Rowe M. D., Magnetic circular dichroism of iodoform and methylene iodide, *Chem. Phys.* **36** (1979) 181-186.
5. Hunter T. F. and Kristjansson K. S., Yield of $I(^2P_{3/2})$ in the photodissociation of CH_2I_2, *Chem. Phys. Lett.* **90** (1982) 35-40.
6. Marvet U., Brown E. J. and Dantus M., Femtosecond concerted elimination of halogen molecules from halogenated alkanes, *Phys. Chem. Chem. Phys.* **2** (2000) 885-891.
7. Jung K.-W., Ahmadi T. S. and El-Sayed M. A., Photofragment translational spectroscopy of CH_2I_2 at 304 nm: Polarization dependence and energy partitioning, *Bull. Korean Chem. Soc.* **18** (1997) 1274-1280.
8. Lee S. J. and Bersohn, R., Photodissociation of a molecule with two chromophores. CH_2IBr, *J. Phys. Chem.* **86** (1982) 728-730.
9. Butler L. J., Hintsa E. J., Shane S. F. and Lee Y. T., The electronic state-selective photodissociation of CH_2BrI at 248, 210, and 193 nm, *J. Chem. Phys.* **86** (1987) 2051-2074.
10. Schmitt G. and Comes F. J., Competitive photodecomposition reactions of chloroiodomethane, *J. Photochem. Photobiol. A* **41** (1987) 13-30.
11. Zhong D., Ahmad S. and Zewail A. H., Femtosecond elimination reaction dynamics *J. Am. Chem. Soc.* **119** (1997) 5978-5979.
12. Nathanson G. M., Minton T. K., Shane S. F. and Lee Y. T. Ultraviolet photodissociation and thermochemistry of CH_2BrCH_2I, CF_2BrCF_2I, and CF_2ICF_2I, *J. Chem. Phys.* **90** (1989) 6157-6170.
13. Dantus M., Coherent nonlinear spectroscopy: From femtosecond dynamics to control, *Ann. Rev. Phys. Chem.* **52** (2001) 639-679.
14. Mössinger J. C., Shallcross D. E. and Cox R. A., UV-VIS absorption cross-sections and atmospheric lifetimes of CH_2Br_2, CH_2I_2 and CH_2BrI, *J. Chem. Soc. Far. Trans.* **94** (1998) 1391-1396.
15. Roehl C. M., Burkholder J. B., Moortgat G. K., Ravishankara A. R. and Crutzen P. J., Temperature dependence of UV absorption cross sections and atmospheric implications of several alkyl iodides, *J. Geophys. Res.* **102** (1997) 12819-12829.
16. Schwartz B. J., King J. C., Zhang J. Z. and Harris C. B., Direct femtosecond measurements of single collision dominated geminate recombination times of small molecules in liquids, *Chem. Phys. Lett.* **203** (1993) 503-508.
17. Saitow K., Naitoh Y., Tominaga K. and Yoshihara K., Photo-induced reactions of CH_2I_2 in solution studied by the ultrafast transient absorption spectroscopy, *Chem. Phys. Lett.* **262** (1996) 621-626.
18. Tarnovsky A. N., Alvarez J.-L., Yartsev A. P., Sundström V. and Åkesson E., Photodissociation Dynamics of Diiodomethane in Solution. *Chem. Phys. Lett.* **312** (1999) 121-130.

19. Blomstrom D. C., Herbig K. and Simmons H. E., Photolysis of methylene iodide in the presence of olefins, *J. Org. Chem.* **30** (1965) 959-964.
20. Kropp P. J., Photobehavior of alkyl halides in solution: Radical, carbocation, and carbene intermediates, *Acc. Chem. Res.* **17** (1984) 131-137.
21. Phillips D. L. and Fang W. H., Density functional theory investigation of the reactions of isodihalomethanes (CH_2X-X where X =Cl, Br, or I) with ethylene: Substituent effects on the carbenoid behavior of the CH_2X-X species. *J. Org. Chem.* **66** (2001) 5890-5896.
22. Miranda M. A., Perez-Prieto J., Font-Sanchis E. and Scaiano J. C., One- vs two-photon processes in the photochemistry of 1,n-dihaloalkanes, *Acc. Chem. Res.* **34** (2001) 717-726.
23. Phillips D. L. Resonance Raman intensity analysis of directly photodissociating haloalkane and dihaloalkane molecules. *Progr. Reac. Kin. Mech.* **24** (1999) 223-286.
24. Maier G. and Reisenauer H. P., Photoisomerization of dihalomethanes, *Angew. Chem. Int. Ed. Engl.* **25** (1986) 819.
25. Maier G., Reisenauer H. P., Hu J., Schaad L. J. and Hess B. A., Jr., Photochemical isomerization of dihalomethanes in argon matrices, *J. Am. Chem. Soc.* **112** (1990) 5117.
26. Maier G. and Reisenauer H. P., Photoisomerisierung von tetraclormethan in einer argon-matrix, *Tetrahedron Lett.* **30** (1989) 4105-4108.
27. Glukhovtsev M. N. and Bach R. D., Methylene-iodonium ylide: An isomer of diiodomethane. *Chem. Phys. Lett.* **269** (1997) 145-150.
28. Orel A. E. and Kühn O., Cartesian reaction surface analysis of the CH_2I_2 ground state isomerization. *Chem. Phys. Lett.* **304** (1999) 285-292.
29. Simons J. P. and Tatham P. E. R., Photochemical production of trapped electrons in solutions of polyhalogenomethanes in hydrocarbon glasses, *J. Chem. Soc. A* (1966) 854-859.
30. Brown G. P. and Simons J. P., Spectroscopy and photochemistry at low temperatures. Part 3. - Photochemical production of ions and molecular complexes in dilute solutions of polyhalomethanes, *Trans. Far. Soc.* **65** (1966) 3245-3257.
31. Tarnovsky A. N., Wall M., Rasmusson M., Pascher T. and Åkesson E. Photodissociation dynamics of chloroiodomethane in acetonitrile studied by ultrafast pump-probe spectroscopy. *J. Chin. Chem. Soc.* **47** (2000) 769-772. Special Issue.
32. Tarnovsky A. N., Wall M., Gustafsson M., Lascoux N., Sundström V. and Åkesson E., Ultrafast study of the photodissociation of bromoiodomethane in acetonitrile upon 266 nm excitation. Submitted to *J. Phys. Chem. A.*
33. Wall M., Tarnovsky A. N., Sundström V. and Åkesson E., Photodissociation dynamics of iodoform in solution. Manuscript in preparation.

34. Rasmusson M., Tarnovsky A. N, Pascher T., Sundström V. and Åkesson E., Photodissociation of ICH_2CH_2I, ICF_2CF_2I and ICF_2CF_2Br in solution. Manuscript in preparation.
35. Rasmusson M., Tarnovsky A. N., Åkesson E. and Sundström V., On the use of two-photon absorption for determination of femtosecond pump-probe cross-correlation functions, *Chem. Phys. Lett.* **335** (2001) 201-208.
36. Zheng X. and Phillips D. L., Solvation can open the photoisomerization pathway for the direct photodissociation reaction of diiodomethane: Transient resonance Raman observation of the isodiiodomethane photoproduct from ultraviolet excitation of diiodomethane in the solution phase, *J. Phys. Chem. A* **104** (2000) 6880-6886.
37. Kwok W. M., Ma C., Phillips D., Parker A. W., Towrie M., Matousek P. and Phillips, D. L., Picosecond time-resolved resonance Raman observation of iso-CH_2Br-I following A-band photodissociation of CH_2BrI in the solution phase, *Chem. Phys. Lett.* **341** (2001) 292-298.
38. Kwok W. M., Ma C., Parker A. W., Phillips D., Towrie M., Matousek P., Zheng X. and Phillips, D. L., Picosecond time-resolved resonance Raman observation of the iso-CH_2Cl-I and iso-CH_2I-Cl photoproducts from the "photoisomerization" reactions of CH_2ICl in the solution phase, *J. Chem. Phys.* **114** (2001) 7536-7543.
39. Kwok W. M., Ma C., Parker A. W., Phillips D., Towrie M., Matousek P. and Phillips, D. L., Picosecond time-resolved resonance Raman observation of the iso-CH_2I-I photoproduct from the "photoisomerization" reaction of diiodomethane in the solution phase, *J. Chem. Phys.* **113** (2000) 7471-7478.
40. Treinin A. and Hayon E., Laser photolysis of iodide ions: spectroscopic evidence for iodine atoms and kinetics of their reaction with iodine in various solvents, *Int. J. Radiat. Phys. Chem.* **7** (1975) 387.
41. Sehested J., Ellermann T. and Nielsen O. J., A spectrokinetic study of CH_2I and CH_2IO_2 radicals, *Int. J. Chem. Kinet.* **26** (1994) 259-272.
42. Villenave E. and Lesclaux R., The UV absorption spectra of CH_2Br and CH_2BrO_2 and the reaction kinetics of CH_2BrO_2 with itself and with HO_2 at 298 K. *Chem. Phys. Lett.* **236** (1995) 376-384.
43. Bingemann D., King A. M. and Crim F. F., Transient electronic absorption of vibrationally excited CH_2I_2: Watching energy flow in solution. *J. Chem. Phys.* **113** (2000) 5018-5025.
44. Charvat A., Aßmann J., Abel B. and Schwarzer D., Real-time probing of intramolecular vibrational energy redistribution and intermolecular vibrational energy transfer of selectively excited CH_2I_2 molecules in solution, *J. Phys. Chem. A* **105** (2001) 5071-5080.
45. *CRC Handbook of Chemistry and Physics*, 80th ed.; Ed. Lide, D. R. (CRC Press LLC: Boca Raton, FL, 1999) Section 10.

46. Owrutsky J. C. Raftery D. and Hochstrasser R. M., Vibrational relaxation dynamics in solution. *Annu. Rev. Phys. Chem.* **45** (1994) 519-555.
47. Bultmann T. and Ernsting N. P., Competition between geminate recombination and solvation of polar radicals following ultrafast photodissociation of bis(p-aminophenyl) disulfide, *J. Phys. Chem.* **100** (1996) 19417-19424.
48. Zheng X. M. and Phillips D. L. Solvation effects on the iodoform ultraviolet direct photodissociation reaction: opening the photoisomerization channel, *Chem. Phys. Lett.* **324** (2000) 175-182.

ULTRAFAST DYNAMICS OF SOLVATED ELECTRONS IN LIQUID AMMONIA SOLUTIONS

ANDREAS-NEIL UNTERREINER

Lehrstuhl für Molekulare Physikalische Chemie, Institut für Physikalische Chemie
Universität Karlsruhe, Kaiserstr. 12, 76128 Karlsruhe, Germany
E-mail: andreas.unterreiner@chemie.uni-karlsruhe.de

JÖRG LINDNER AND PETER VÖHRINGER

Max-Planck-Institut für Biophysikalische Chemie, Biomolekulare und Chemische Dynamik,
Am Fassberg 11, 37077 Göttingen, Germany
E-mail: jlindne@gwdg.de, pvoehri@gwdg.de

We have performed the first femtosecond near-infrared pump-probe experiments of equilibrated solvated electrons in liquid ammonia. Using pump-pulses centered at 1280 nm and concentrations of (2±1) 10^{-4} mol/l, transients were recorded for probe wavelengths between 1280 and 1840 nm with a time resolution of 100 fs. Immediately after photoexcitation, the absorption spectrum of the solvated electron is red-shifted with respect to the steady-state spectrum. Subsequently, a dynamic spectral blue-shift is observed that occurs under shape stability with a time constant of (200±50) fs. Our analysis clearly demonstrates that these pump probe investigations can be understood in terms of a simple temperature-jump technique. Furthermore, the concentration dependence of the relaxation dynamics has been investigated. For concentrations in excess of 10^{-3} mol/l, long lived transient signal contributions are observed which imply the formation of nascent paramagnetic species from spin-paired electrons.

1 Introduction

Solvated electrons with a well-defined concentration can be synthesized chemically by dissolving sodium metal in NH_3 [1, 2], and are stable for up to several days. Although this lifetime easily exceeds that of its aqueous analogue, femtosecond time-resolved studies on the relaxation processes of the electron in NH_3 have not been reported so far. Stationary properties, such as the susceptibility and conductivity, critically depend on the concentration of solvated electrons: Below 10^{-3} mol/l, only paramagnetic electrons exist. With increasing concentration, a distinct transition from paramagnetic to diamagnetic behavior can be observed [3]. For concentrations above 10^{-1} mol/l, even an insulator-metal transition occurs [4, 5] which is accompanied by a change of color from deep blue to bronze. The equilibrated absorption spectrum of solvated electrons in ammonia is broad, featureless and asymmetric with a maximum located at about 1520 nm at a temperature of –50 °C. A number of features common to solvated electrons in polar media are worth highlighting:

1. Shape stability of the absorption spectrum over a large temperature range extending into the critical region of the solvent [6].
2. A linear continuous shift of the spectrum with temperature characterized by a large temperature coefficient of $\alpha = -18\ cm^{-1}/K$ [7] ($-19.4\ cm^{-1}/K$ in water [8]).
3. For constant temperature, the spectral properties remain independent of the concentration. Only for concentrations larger than 10^{-3} mol/l, a small spectral shift of the stationary absorption spectrum is observed [3].

For the hydrated electron, we previously showed unambiguously that following photoexcitation a time-dependent spectral shift of the linear absorption is correlated with the temporal evolution of the electron temperature [9, 10]. This has been interpreted as a cooling process causing the optical absorption spectrum to dynamically shift to higher frequencies with a time constant of (300 ± 50) fs [10]. Additional transient states (such as the wet electron) were not necessary to invoke. Furthermore, a simple correlation between optical sum rules and the cooling dynamics explains the optical response of aqueous electrons in water sufficiently. Therefore, photoexcitation can be understood as a simple temperature-jump experiment leading immediately to a canonical ensemble of solvated electrons with an elevated temperature, T. Non-adiabatic transitions through which the electron returns to its ground state occur within the time-resolution of the experiment. Here, we demonstrate first femtosecond near-infrared pump-probe experiments aimed at unraveling the cooling dynamics in liquid NH_3 solutions at the paramagnetic-diamagnetic transition [11] at $-50\ ^0C$.

2 Methods

Pump-probe experiments were carried out with two independently tunable femtosecond optical parametric amplifiers (OPA) driven by a standard Ti:sapphire oscillator / regenerative amplifier system (800 nm, 70 fs, 650 μJ). One OPA was capable of generating 100-fs duration pulses in the near-infrared spectral region between 1200 and 2400 nm. Both its signal and idler outputs were used for probing. Pump pulses with a duration of 100 fs were derived from the signal output of the other OPA which was tuned to a wavelength of 1280 nm

The relative timing between the two OPA outputs was adjusted by computer-controlled optical delay lines. Pump and probe pulses were spatially overlapped in the sample with an Al-mirror (rcc = 1 m). The pump and probe energies were 10 μJ and 150 nJ, respectively. The intensities of the probe pulse in front of and behind the sample were monitored with PbS diodes whose output was directed to gated integrators, normalized with an analog divider and digitized in a PC computer by a data acquisition card.

Details of the synthesis of solvated electrons in liquid ammonia are described in great detail in Ref. [12]. Briefly, liquid ammonia was dried over sodium for several weeks at $-78\ ^0C$. Fresh sodium ammonia solutions with specified concentrations were prepared in a mixing apparatus attached to a low-temperature fused silica sample cell with an optical thickness of 0.1 mm. All measurements were performed at –50 °C. The electron concentration was determined from the absorbed power of the pump beam.

3 Results and Discussion

The transient response of solvated electrons in liquid ammonia to 1280 nm near-infrared excitation was investigated for eight different probe wavelengths ranging between 1280 and 1840 nm and for concentrations between $4\cdot10^{-3}$ and $2\cdot10^{-4}$ mol/l. For the analysis, we focus on the data obtained at the lowest concentrations where spin-pairing under formation of diamagnetic electrons can be neglected [3]. The concentration dependence will only be discussed qualitatively further below.

3.1 Spectral Dependence of the Transients

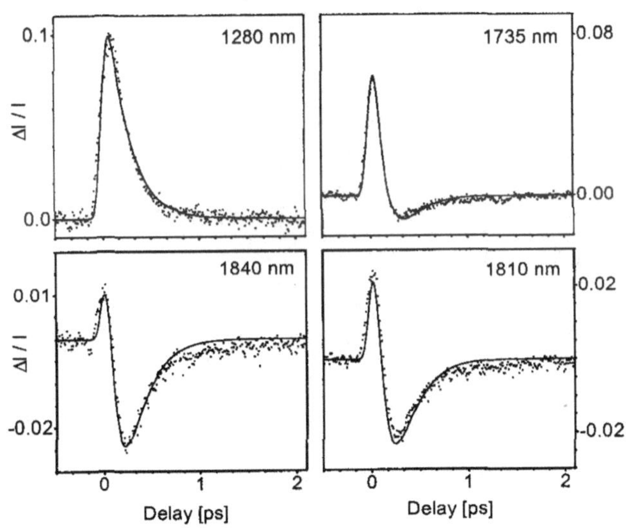

Figure 1. Pump-probe transients with detection at 1280, 1735, 1810 and 1840 nm. Experimental data are indicated as dots. The pump was centered at 1280 nm and the sample temperature was –50°C. The solid curves correspond to model simulations.

Four representative pump-probe signals for an electron concentration of $(2\pm1)\cdot10^{-4}$ mol/l and a sample temperature of –50 °C are displayed in **Figure 1**. On the high-frequency edge of the stationary absorption spectrum (1280 nm), the signals are governed by an immediate bleach that recovers on a time scale of about 1 ps.

Concurrently, on the low frequency edge (1735, 1810 and 1840 nm), an immediate bleach is followed by a transient absorption of which the latter decays within 1 ps.

As we found previously for the hydrated electron [9], this behavior is reminiscent of a dynamic blue-shift of the resonance originating from an instantaneous laser-induced temperature jump and subsequent dynamics of electron cooling. The spectral implications of such a scenario are qualitatively sketched in **Figure 2**, which compares the equilibrium absorption spectra of the solvated electron for two different temperatures. Note that raising the temperature by an amount ΔT causes the spectrum to shift by an amount $\Delta\nu_0$ to lower frequencies according to the temperature coefficient, α. The power spectra of the pump and probe lasers centered at 1280 and 1810 nm are also shown for reference. The optical response of electrons in ammonia was simulated assuming:

1. an instantaneous red-shift of the stationary electron absorption spectrum brought about by a pump-induced temperature jump at zero time delay. In **Figure 2**, the amount of the shift $\Delta\nu_0$ and the hot spectrum are labeled accordingly.
2. a subsequent dynamic blue-shift of the resonance due to relaxation of the system back into complete thermal equilibrium, i.e. cooling. In the figure, the dynamic evolution of the resonance due to cooling is indicated by an arrow.

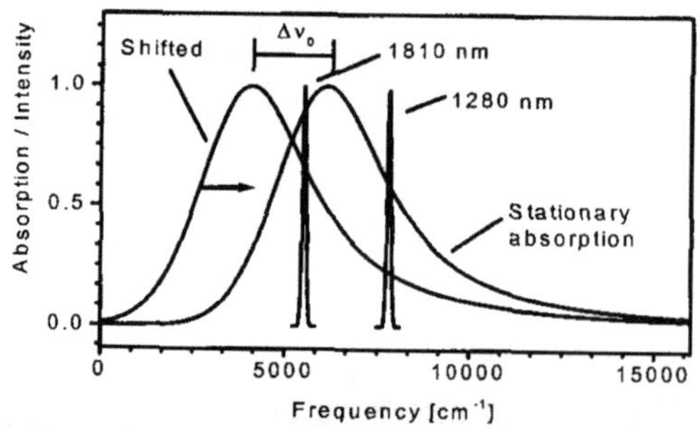

Figure 2. Schematic for the simulation of the transients. Prior to optical excitation, the electron absorption spectrum (denoted stationary spectrum) is located around 6750 cm^{-1} corresponding to a temperature of –50°C. The pump-pulse at $t = 0$ causes the spectrum to instantaneously shift to the red by an amount $\Delta\nu_0$ while the spectral shape is preserved. Subsequent cooling results in a gradual red-shift under shape stability as indicated by the horizontal arrow. For comparison, the pump and probe spectral widths are also displayed.

It was further assumed that during thermalization, the spectral shape of the electron resonance remains unchanged. From the figure, it can be deduced that the instantaneous pump-induced red-shift results in an immediate bleaching for probe wavelengths originally located on the high frequency edge of the stationary absorption spectrum (e.g. 1280 nm in **Figure 2**). Subsequently, this bleaching is expected to fully decay as the resonance returns to its thermalized spectral position.

Inspection of **Figure 1** demonstrates that the experimental pump-probe transient recorded at 1280 nm is indeed in quantitative agreement with this behavior. In contrast, for probe wavelengths originally located on the low frequency edge of the stationary absorption spectrum (e.g. 1810 nm in **Figure 2**), a distinctly different dynamic response is expected. Again, an initial bleach can expected at zero time delay provided the pump-induced red-shift is sufficiently large. In this instance, the probe window is located on the high-frequency tail of the resonance where the absorption cross section can be considerably smaller as compared to its equilibrium value probed at negative time delays. Subsequently, cooling drives the resonance back to its thermalized spectral position, thereby sweeping the absorption spectrum through the detection window at 1810 nm. As a result, the initial bleach turns into a transient absorption. The latter reaches a maximum when the dynamically shifting peak absorbance coincides exactly with the detection window. More quantitatively, the ratio between maximum transient bleach and maximum transient absorption depends critically on the precise location of the detection window with respect to the equilibrated resonance. Finally, while the spectrum fully approaches its original thermalized position, this absorbance decays to zero. Such a complicated temporal behavior can indeed be observed as evidenced in **Figure 1** for three different detection wavelengths originally located on the low frequency edge of the thermalized electron resonance.

For the simulations, the stationary absorption spectrum of solvated electrons in liquid ammonia was measured and was approximated by a Lorentzian (Gaussian) line shape on the high (low) frequency side of the absorption spectrum. The dynamic frequency shift $\Delta\nu(t)$, of the absorption spectrum due to excitation and electron cooling was assumed to be single exponential:

$$\Delta\nu(t) = \theta(t)\, \Delta\nu_0 \exp(-t/\tau) \qquad (1)$$

where $\theta(t)$ is the Heavyside function, $\Delta\nu_0$ is the amplitude of the dynamic frequency shift and τ is the relaxation time. The instrument response function of the pump-probe spectrometer was about 100 fs (FWHM) and was taken into account by convoluting the simulated transients with a Gaussian of the same duration. As can be seen from **Figure 2**, the spectral width of pump and probe laser was roughly 30 times smaller than that of the electron absorption spectrum and was therefore neglected. The temporal response for a given probe wavelength was then calculated using Beer's law.

Model simulations are reproduced in **Figure 1** as solid curves. Using a relaxation time of $\tau = (200 \pm 50)$ fs and an initial shift $\Delta\nu_0 = (3000 \pm 900)\ \mathrm{cm}^{-1}$, an excellent agreement between simulation and experiment is obtained. Minor deviations can only be seen at the low frequency edge of the originally thermalized stationary absorption spectrum for delays around 1 ps.

The model applied above describes the following mechanism for electron relaxation: Prior to optical excitation, solvated electrons are in thermal equilibrium with the solvent NH_3 at the temperature $T(t=-\infty) = -50°C$ where t is the pump-probe time delay. The pump pulse defining $t = 0$ excites the electrons to an optically allowed excited state from which they decay immediately back to the ground state. As judged by the high quality of the model simulations, this internal conversion effectively generates a canonical distribution of hot electrons. Hence, within our time-resolution, a temperature can be assigned throughout the entire relaxation dynamics. In essence, photoexcitation of the solvated electron is interpreted as a simple temperature-jump (T-jump) such that at $t = 0$, the temperature is quasi-impulsively raised to $T(t = 0) = T(t=-\infty) + \Delta\nu_0/\alpha$. Consequently, this T-jump results in an instantaneous red-shift of the stationary absorption spectrum as exemplified in **Figure 2**. Subsequent to heating, the excess energy initially delivered by the pump pulse is dissipated into the solvent. Cooling of electrons results in a spectral blue-shift of their resonance which can be quantified in the time-dependence of $\Delta\nu_0$. From our analysis, a value $\Delta\nu_0$ of $(3000 \pm 900)\ cm^{-1}$ is obtained corresponding to a T-jump of (167 ± 50) °C. Therefore, the pump pulse creates a hot electron distribution, and therefore, a transient absorption spectrum at $t = 0$ which is equivalent to a temperature of 117 °C. Due to the finite time resolution this value corresponds to a lower limit for the initial temperature. Since the relation $\Delta\nu_0 = \alpha\,\Delta T$ holds for all pump-probe time delays, the excess energy decays exponentially as given by Eq. (1). Furthermore, since the spectral shape remains constant, the canonical nature of the electron distribution is preserved during thermalization.

To emphasize, directly after excitation and during the entire process of cooling, the absorption spectrum of the electron is equal to an *equilibrium* spectrum at a temperature $T(t) > T(t=-\infty)$. This implies that at each instant in time (i.e. independent of the pump-probe time delay), the electron is in *thermal equilibrium* with its immediate microscopic surroundings. The relevant microscopic vicinity of the electron is then of course defined only by those fluctuations and those nuclear degrees of freedom of the solvent which are capable of coupling to the electronic transition dipole operator. These may very well be solvent coordinates which also facilitate the initial non-adiabatic transition which must follow optical excitation and which force the electron to return to its ground state. Therefore, with the electron occupying its excited state upon illumination at $t = 0$, some solvent nuclear degree of freedom might approach a critical configuration upon which the ground and excited electronic states come close enough in energy for the non-adiabatic transition to occur. The electron returns to the ground state and the excess energy initially deposited by the pump field is effectively converted into kinetic energy in the critical solvent degree of freedom. If this excess kinetic energy is rapidly redistributed (or equilibrated) among all the other nuclear coordinates which can couple to the electron transition dipole moment, the absorption spectrum merely serves as a reporter for the kinetic energy and thus for the temperature of this subset of solvent degrees of freedom.

3.2 Concentration dependence

The simulations shown in **Figure 1** refer to experiments conducted at a concentration of about (2±1)· 10^{-4} mol/l. This is in the dilute concentration regime where paramagnetic electrons dominate. As mentioned above, susceptibility and conductivity experiments indicate that at concentrations of the order of 10^{-3} mol/l a transition from paramagnetic to diamagnetic electrons occurs. To gain first insights as to how this transition influences the cooling dynamics, the concentration dependence was studied using probe-wavelengths located on the low and high frequency edge of the stationary absorption spectrum. Representative results are shown in **Figure 3**. At 1420 nm, a residual long-time transient absorption can be seen for concentrations above 10^{-3} mol/l. Simultaneously, on the low frequency edge at 1810 nm, a long-time transient bleaching is detected. Both of these long-time signal contributions are strongly concentration dependent and persist without a significant decay on time scales as long as 100 ps.

A concentration dependence of the optical response of solvated electrons has been reported neither in aqueous solution nor in any other polar solvent studied so far. Within the T-jump model outlined above, these residual bleaching and absorption components indicate that the electron absorption spectrum at infinite delays is blue-shifted with respect to the thermalized spectrum at $t = -\infty$. Therefore, these components indicate that relaxation ultimately establishes a temperature which is below –50 °C. Obviously, such a finding is entirely unrealistic. It is however known that upon going through the paramagnetic-diamagnetic phase transition, the absorption spectrum of the metal-ammonia solution shifts by about 400 cm^{-1} to lower frequencies [3]. Therefore, the absorption spectrum of diamagnetic electrons is located slightly to the red of the paramagnetic resonance. Assuming that spin-paired electrons can undergo photo-induced dissociation, the effective concentration of paramagnetic electrons will temporarily increase upon optical excitation. Such a pump-induced un-pairing will then result in an absorption spectrum that is indeed blue-shifted with respect to that of the thermodynamically stable paramagnetic-diamagnetic mixture defined at negative time delays. This interpretation, however, requires the geminate (or nascent) electrons pairs to recombine to spin-pairs on time scales significantly larger than 100 ps. We are currently in the process of studying the excitation energy dependence of these long-time dynamics in more detail.

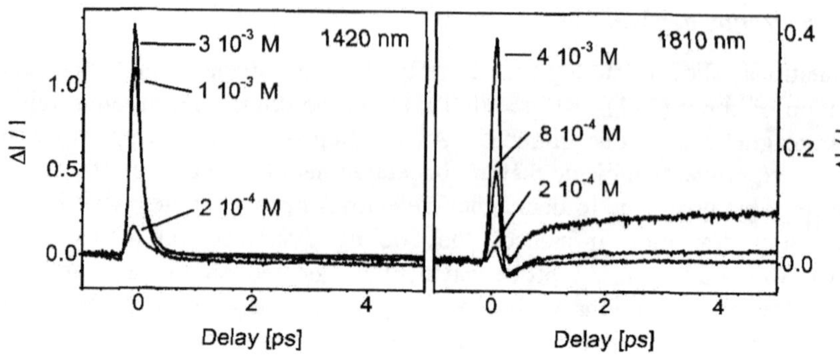

Figure 3. Concentration dependence of the transients at 1420 and 1810 nm. The pump pulse was centered at 1280 nm.

4 Conclusions

In summary, we have performed the first femtosecond near-infrared pump-probe experiments on solvated electrons in liquid ammonia. Our results indicate that optical excitation corresponds to a temperature jump on a time scale of 100 fs. The time-dependent electron absorption spectrum is very well described by an equilibrium absorption spectrum characterized by a time-dependent canonical temperature T(t). A dynamic spectral blue-shift under shape stability reflects the dynamics of electron cooling on a time scale of about 1 ps. For concentrations near the paramagnetic-diamagnetic phase transition, residual transient bleach and transient absorption contributions give first experimental indications for a photo-induced breakage of diamagnetic (i.e spin-paired) electrons. Finally, this work demonstrates that studying the dynamics of solvated electron in liquid ammonia solutions is highly advantageous to aqueous or other polar solutions because the electron concentration can be controlled in a well-defined fashion without photo-ionizing the neat liquid solvent or additional species dissolved in the liquid.

5 Acknowledgements

This work has been financially supported by the SFB 195 "Lokalisierung von Elektronen in makroskopischen und mikroskopischen Systemen" and the SFB 357 "Molekulare Mechanismen unimolekularer Reaktionen". We thank Prof. H. Hippler for helpful discussions and financial support.

References

1. M. Gold and W.L. Jolly, Absorption spectra of metal-ammonia solutions, *Inorg. Chem.* **1** (1962) pp. 818-827.
2. P.P. Edwards, The electronic properties of metal solutions in liquid ammonia and related solvents, *Adv. Inorg. Chem. Radiochem.* **25**, (1982) pp. 135-185.
3. U. Schindewolf, Paramagnetische, diamagnetische und metallähnliche Zustände gelöster Elektronen in Metall-Ammoniaklösungen, *Z. Phys. Chem. Neue Folge*, **112** (1978) pp. 153-166.
4. P.P Edwards, The electronic properties of metal solutions, *Phys. Chem. Lett.* **10** (1981) pp. 189-228.
5. U. Schindewolf, Thermodynamics and electrochemistry of sodium-ammonia solutions and of solvated electrons in ammonia, Ed. J. V. Acrivos, *Physics and Chemistry of Electrons and Ions in Condensed Matter*, (1984) pp. 361-383.
6. V. Herrmann and P. Krebs, Temperature Dependence of Optical Absorption Spectra of Solvated Electrons in CD_3OD for $T \leq T_c$, *J. Phys. Chem.* **99** (1995) pp. 6794-6800.
7. T.R. Tuttle and S. Golden, Solvated Electrons: What is Solvated?, *J. Phys. Chem.* **95** (1991) pp. 5725-5736.
8. F.-Y. Jou and G.R. Freeman, Temperature and Isotope Effects on the Shape of the Optical Absorption Spectrum of Solvated Electrons in Water, *J. Phys. Chem.* **83** (1979) pp. 2383-2387.
9. A. Hertwig, H. Hippler, A.N. Unterreiner and P. Vöhringer, Ultrafast Relaxation Dynamics of Solvated Electrons in Water, *Ber. Bunsenges. Phys. Chem.* **102** (1998) pp. 805-810.
10. A. Hertwig, H. Hippler and A.N. Unterreiner, Transient spectra, formation, and geminate recombination of solvated electrons in pure water UV-photolysis: an alternative view, *Phys. Chem. Chem. Phys.* **1** (1999) pp. 5633-5642.
11. U. Schindewolf, Excess Electrons in Nonaqueous Solvents, *Pure & Appl. Chem.* **53** (1981) pp. 1329-1344.
12 A.-N. Unterreiner, Ultraschnelle Relaxationsdynamik solvatisierter Elektronen in flüssigem Ammoniak und Wasser, Ph.D. Thesis, Universität Karlsruhe, Verlag Mainz, Aachen, (1998).

VIBRATIONAL COHERENCE AND SOLVENT-INDUCED NON-ADIABATIC COUPLINGS

C. R. GONZALEZ, S. FERNANDEZ-ALBERTI, J. ECHAVE
Centro de Estudios e Investigaciones, Universidad Nacional de Quilmes, 1876 Bernal, Argentina
E-mail: *seba@unq.edu.ar*

J. HELBING AND M. CHERGUI
Institut de Physique de la Matière Condensée, BSP, Université de Lausanne, CH-1015 Lausanne-Dorigny, Swizerland
E-mail: Majed.chergui@ipmc.unil.ch

The ultrafast non-radiative relaxation processes between electronic states of Hg_2 trapped in a cryogenic argon matrix after photon excitation at 266 nm is studied by molecular dynamics *with quantum transitions (MDQT). Population transfer by matrix-induced couplings and* vibrational energy relaxation compete on the same time scale during the dynamics. In spite of multiple non-adiabatic crossing events between the 14 different electronic excited states involved in the dynamics, our simulations show that the vibrational coherence of the Hg_2 oscillator is retained during the first few picoseconds. These results demonstrate that vibrational coherence of a solute can be preserved even after several solvent-induced curve-crossing events.

1 Introduction

The possibility to create and visualize vibrational wave packets by femtosecond spectroscopy techniques allows us to follow nuclear motion in molecular systems [1]. After femtosecond laser excitation, a wave packet evolves on the initially populated potential energy surface or surfaces (PES). As the molecular states interact, the wave packet can cross over to initially unexcited PES, their population depending on the strength of the corresponding couplings matrix elements between PES's. It was demonstrated in the case of isolated NaI molecules [2], that the splitting of the initial wave packet between tow PES's does not lead to a loss of vibrational coherence. This point is crucial in view of applications in coherent control experiments, where one wants to drive the molecular system to a specific channel at the expense of all others [3].

In the condensed phase, one has to deal with additional possibilities of loosing the phase coherence initially created in the solute by a femtosecond laser pulse. First, the fluctuations of the solvent molecules can destroy the phase coherence of the solute as a result of collisions between solute and solvent molecules. In spite of this, many examples now exist, in which vibrational coherences were observed for several oscillation cycles within a single PES in the case of molecules in crystalline solids [4,5], in ordered van der Waals clusters [6], in liquids [7,8] and even of chromophores in proteins such as the photosynthetic reaction centre [9], rhodopsin [10] and Myoglobins [11]. Vibrational coherence has also been reported for intermolecular modes in van der Waals solids [12]. Second, non-adiabatic couplings between solute PES's can be induced by the solvent. Due to the solvent fluctuations, these couplings will not be confined to a specific solute configuration, but will be modulated by the latter.

Over the past ten years, several theoretical works have dealt with non-adiabatic dynamics in the condensed phase [13,14,15,16,17,18]. However, the emphasis was usually not on the issue of vibrational coherence and its survival after multiple curve-crossing events, except for Ref. 13. This is due to the still ambiguous experimental identification of such events in the condensed phase. One of the reasons behind this may be the lack of adequate model systems that could allow a clear cut identification of coherence in curve crossing events, analogous to the NaI molecule in the gas phase [2]. The model system *par excellence* for condensed phase dynamics is the I_2 molecule in rare gas matrices for which vibrational coherence was demonstrated after dissociation and cage-induced recombination on the A-state [4,5], and for direct excitation into the bound B-state [8]. Nevertheless, this molecule does not easily lend itself for an identification of a coherent curve-crossing event, due to inefficient non-adiabatic couplings occurring in ordered rare gas matrix environments and to the lack of adequate probe windows.

The Hg_2 dimer avoids these limitations. Its gas phase potential curves are shown in Figure 1. Vibrational coherence of Hg_2 in the D 1_u excited state has been predicted by simulations based on a quantum mechanical wave packet description [19]. These simulations were compared to the femtosecond measurements of Hg_2 photo association in the gas phase by Dantus and co-workers [20,21]. In these experiments, a UV femtosecond pulse transfers population from a collision pair of Hg atoms in the very shallow ground state to the bound D-state, thus leading to association. As demonstrated in ref. 19, vibrational coherence could not be observed in these experiments due to the distribution of impact parameters at the elevated temperatures of mercury vapour. In low temperature rare gas matrices on the other hand, pairs of Hg atoms are locked at fixed locations of the lattice and cannot rotate. Even though there is a distribution of configurations for the dimer pair, a specific inter nuclear ground state configuration can be selected by the choice of the excitation wavelength. Experiments have shown that excitation of one of the ungerade states (D or F) of Hg_2 in argon matrices [22,23,24] leads to emission from only the lowest A $0^{\pm}_{g}$ state. As can be seen in Figure 1, this implies that intramolecular relaxation processes occur through multiple curve-crossing events, funnelling population down to the A-state. Thus, Hg_2 appears as an interesting model system both for investigating the vibrational coherence in condensed phase curve-crossing events and for a detailed understanding of non-adiabatic couplings in the population relaxation of a chromophore. As a prospective study for future time-resolved relaxation dynamics experiments, in this contribution we demonstrate the conservation of vibrational coherence in multiple curve-crossing events induced by the solvent.

2 Results and Discussion

We have performed molecular dynamic (MD) simulations of the non-radiative relaxation process after photon excitation of the Hg_2 molecule embedded in a cryogenic argon matrix. The system was composed of a Hg_2 molecule surrounded by 498 Ar atoms with minimum image convention. The initial conditions were taken from ground state equilibrated configurations at an effective temperature T'=48K given by the Thermal Harmonic Quantum Correction [25] applied for the experimental temperature of T_0= 4 K [23].

The initial configurations fulfil the classical Franck Condon (FC) condition for the electronic transition at 266 nm from the electronic ground state to one of the 14 possible adiabatic electronic excited states. Excited state trajectories were started from the initial configurations by putting into the system an energy corresponding to a 266 nm photon. In that way, a projection onto the Hg_2 diabatic electronic basis, defined by the states of the Hg_2 molecule in the gas phase, shows an average initial condition with most of population in the diabatic $D1_u$ state.

The Molecular Dynamics with Quantum Transitions (MDQT) method developed by Tully [26,27] is used in order to simulate the dynamics on the several electronic coupled potential energy surfaces (PES) involved in the process. Briefly, the method treats the electronic degrees of freedom ($\boldsymbol{r}$) quantum mechanically, while the motions of the nuclei ($\boldsymbol{R}$) are treated classically. The nuclei evolve on a PES, which is defined by a single electronic state at a given time. Hops from one electronic state to another are governed by the coefficients of the electronic wave function. More details can be found elsewhere [26,27]. In the present work, the electronic wave packet is developed on eigenstates of the electronic Hamiltonian $H_{el}(\boldsymbol{r};\boldsymbol{R})$ defined as the sum of the electrostatic Hg-Hg and Hg-Ar interactions.

The electronic PES of the Hg_2 molecule considered in the simulations of the non-radiative relaxation process [28,29,30,31,32] are shown in Figure 1. We assume that none of them are coupled in the gas phase, but couplings are induced in the condensed phase by the interaction of the diatomic molecule with the environment. In the present work, the interstate couplings are obtained from a diatomic-in-molecules (DIM) [16] treatment of the mixing between different states of Hg_2 induced by the argon atoms. The states, written as $|J\,\Omega\rangle$ vectors, are developed in the L-S or uncoupled representation as

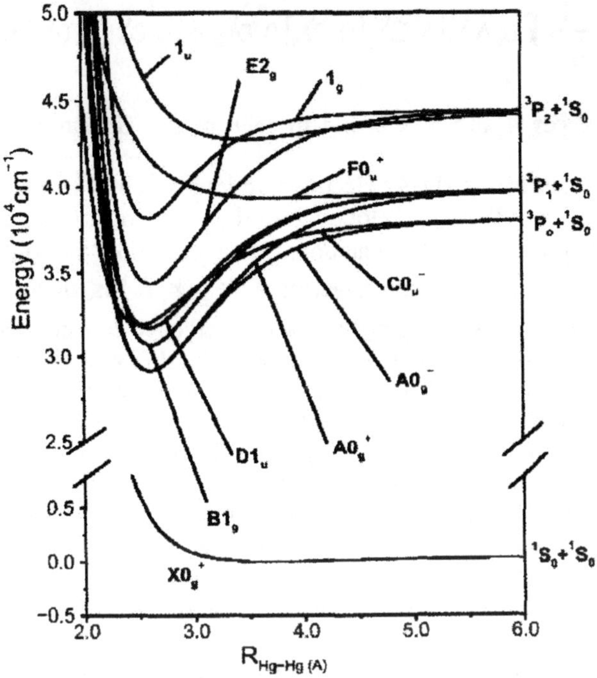

Figure 1. Potential energy surfaces of Hg_2 as functions of the Hg-Hg distance ($R_{Hg\text{-}Hg}$). The ground state $X0_g^+$ is plotted according to Ref. 28 using 3.69 Å as equilibrium distance. The $F0_u^+$, $D1_u$ and upper 1_u state parameters were taken from Refs 29 and 30. The parameters of $A0_g^{\pm}$, $B1_g$, $E2_g$ and 1_g were reported in Ref. 31. The $C0_u^-$ state has not yet been observed in the gas phase and was taken from theoretical calculation of Ref. 32.

$$|J,\Omega\rangle = \sum_{\Omega=\Lambda+\Sigma} C(L=1,\Lambda;S=1,\Sigma;J,\Omega) \times |L=1,\Lambda\rangle|S=1,\Sigma\rangle \qquad (1)$$

where C are the Clebsh-Gordan coefficients and

$$|L,\Lambda\rangle|S,\Sigma\rangle = \frac{1}{\sqrt{2}}\left[|L,\Lambda\rangle_A|S,\Sigma\rangle_A|0,0\rangle_B|0,0\rangle_B \pm |0,0\rangle_A|0,0\rangle_A|L,\Lambda\rangle_B|S,\Sigma\rangle_B\right] \tag{2}$$

with the sub indices *A* and *B* referring to each of the Hg atoms in the diatomic molecule.

The Hamiltonian of the Hg-Ar electrostatic interaction is written as the sum of electrostatic Hg(A or B)-Ar(j) diatomic Hamiltonians constructed from empirical potentials of the usual Σ and Π orientations. We then follow the same procedure as Batista and Coker [16]. The resulting matrix of $H_{el}(\boldsymbol{r};\boldsymbol{R})$ is finally diagonalised in order to obtain the corresponding eigenvalues and eigenvectors at a given time.

Figure 2 represents the average population of the different electronic states as a function of time up to 10 ps. They were obtained as the weighted average over all trajectories after projecting them onto the diabatic electronic basis that correspond to the states of the Hg_2 molecule in the gas phase. The population distribution at 10 ps is: $A0_g^-$ (~75%), $A0_g^+$ (15%), $C0_u^-$ (~ 8%) and $B1_g$ (~2%). Since the $A0_g^+$ and the $A0_g^-$ states are almost degenerate at low energies, the fact that about 90% of population is in these states, is consistent with the $A0_g^+ \rightarrow X0_g^+$ assignment to the only emission band experimentally observed in solid Ar [22,24]. In agreement with the experimental inferences, the $B1_g$ state is not observed in Ar since it is not populated. Indeed, from figure 2 we can see that, in our simulations, the $B1_g$ state is populated during the non-radiative relaxation dynamics but there is a further $B1_g \rightarrow A0_g^\pm$ step which empties it in the first 2 ps of the dynamics. The $C0_u^-$ state is also populated at early times, and its population then remains constant for the time of the simulation. However, the $C0_u^-$ emission is not observed in Ar [24], which may be due to the fact that its radiative lifetime is too long to compete with non-radiative relaxation on a μs-ms time scale. Even if it is quenched on the nanosecond time scale, it would be difficult to detect its contribution as a rise time of the A-state emission, given that the latter is populated to more than 90% in the first picoseconds. From the analysis of the spectroscopic data, it was suggested that the branching of population between the different electronic states takes place primarily close to the Hg-Hg asymptotic limit and not at a later stage after relaxation within a given electronic state [24]. This conjecture is confirmed by the present simulations.

Figure 3 shows the average, over all trajectories, of the Hg-Hg distance as a function of time. Immediately after excitation, at t=0, the molecule experiences a sudden contraction, typical of the photo association process [19-21]. It is interesting to note that the first oscillation brings the average inter nuclear distance near its initial value despite that the Hg_2 molecule looses ~ 0.1 eV of its initial internal energy. This is the result of a sudden non-adiabatic jump at still large Hg-Hg distances during the very first Hg_2 contraction. There is an important population transfer to the B-state (fig. 2) that has a larger oscillation amplitude than the D-state (fig. 1). Non-adiabatic population transfer due to the matrix-induced couplings occur on the same timescale as the relaxation of vibrational energy. Therefore, vibrational and electronic relaxation in the manifold of electronic states compete at the same level in the dynamics of the Hg_2 molecule in the condensed phase.

Given that Figure 3 represents the Hg-Hg distance averaged over all excited states, the mere fact that an oscillatory motion shows up is already a clear indication that vibrational coherence is preserved, independent of when and where the curve-crossing events take place. In order to distinguish between the loss of vibrational coherence due to collisions with Ar atoms and the loss due to non-adiabatic jumps,

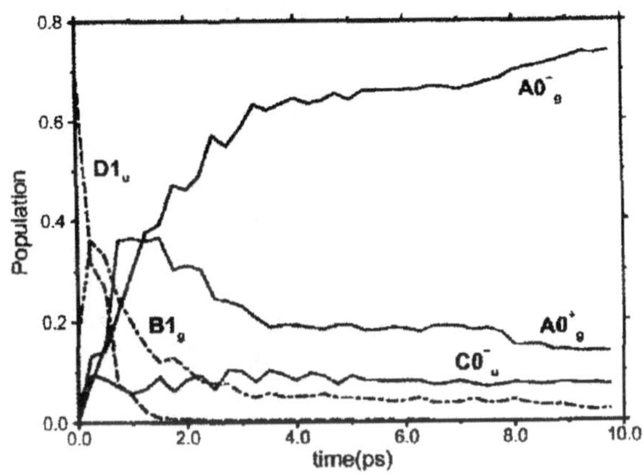

Figure 2: Population of the different electronic surfaces as a function of time. This is the population projected from the adiabatic curves onto the diabatic electronic basis that corresponds to the states of Hg_2 molecule in the gas phase.

we present in Figure 4a, trajectories resulting from a simulation where we only consider the $D1_u$ state. Couplings to other states are not taken into account in these trajectories and population relaxation to other electronic states is therefore not possible. In that way, we observe the loss of vibrational coherence due to the interaction with the matrix environment. The trajectories retain a reasonable degree of coherence up to about ten cycles. The vibrational relaxation in the D state is evident from the smoothly decreasing amplitude. These trajectories are to be compared with those of Figure 4b, which were run using the 14 coupled electronic states and MDQT to simulate the effects of non-adiabatic couplings. These trajectories yield the average trajectory of Figure 3.

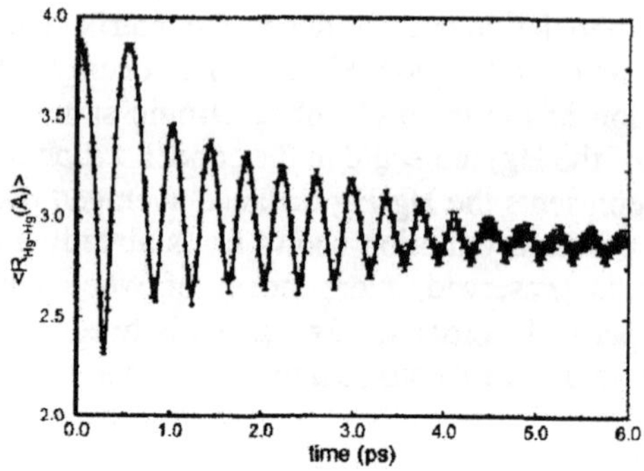

Figure 3: Average over all trajectories of the Hg-Hg distance ($R_{Hg\text{-}Hg}$) as a function of time.

We can see a more rapid loss of vibrational coherence than in Figure 4a. Even in this case however, vibrational coherence is preserved over many cycles. Therefore, the loss of vibrational coherence of the Hg_2 oscillator is due to both the non-adiabatic couplings induced by the environment and to collisions of the Hg atoms with the Ar matrix atoms. Although Figure 2 shows that the initially created wave packet branches in and out of different electronic states, Figures 3 and 4b show that the trajectories retain their vibrational coherence during several oscillations. The dynamics of matrix atoms has also been investigated and we found that they move on the time scale of the curve-crossing events [33].

3 Conclusions

We have performed mixed quantum/classical simulations using the MDQT method of Tully [26,27] for the non-radiative relaxation process after excitation at 266 nm of Hg_2 trapped in a cryogenic argon matrix. Fourteen electronic surfaces and their couplings were included in the calculation.

The results indicate that the transfer of population between adiabatic electronic states occurs in a few picoseconds after photo excitation. The simulations show that within 10 ps, most of the final population is in the lowest $A0_g^{\pm}$ state. Most important in this work is the demonstration that the vibrational coherence is conserved even after several curve-crossing events that are induced by the matrix. The vibrational coherence is gradually lost due to both the non-adiabatic transitions and the collisions of the Hg atoms with the Ar matrix atoms. A detailed account of this work

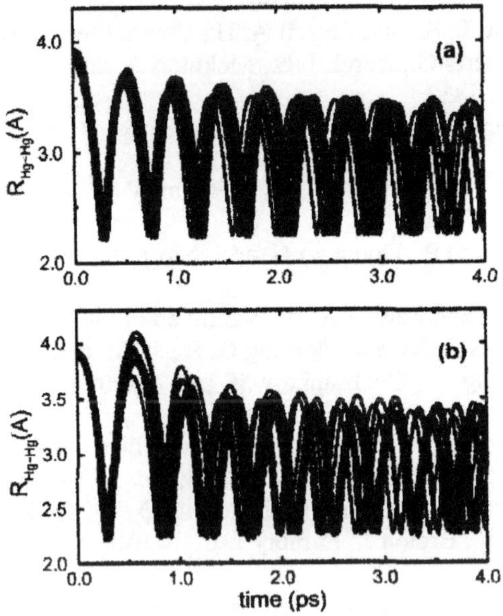

Figure 4. Hg-Hg distance ($R_{Hg\text{-}Hg}$) as a function of time. a) Ten trajectories corresponding to a dynamical simulation where we only consider the $D1_u$ state. b) Ten trajectories using the 14 coupled electronic states and MDQT to simulate the effects due to non-adiabatic couplings.

is presented in ref. 33 .

4 Acknowledgments

This work was supported by the Swiss NSF via grant 61897.00/1, the Universidad Nacional de Quilmes (UNQ) and the Fundacion Antorchas (Argentina).

References

1. Zewail A. H., J. Phys. Chem. A **104** (2000) 5660

2. Rosker M. J., Rose T. S. and Zewail A. H., Chem. Phys. Letters **146** (1988) 175; Cong P., Roberts G., Herek J. L., Mokhtari A. and Zewail A. H., J. Phys. Chem. **100** (1996) 7832
3. Assion A., Baumert T., Bergt M., Brixner T., Kiefer B., Seyfried V., Strehle M., and Gerber G, Science **282** (1998) 919.
4. Zadoyan R., Li Z., Martens C. C. and Apkarian V. A., J. Chem. Phys. **101** (1994) 6648
5. Bargheer M., Dietrich P., Donovan G. and Schwentner N., J. Chem. Phys. **111** (1999) 8556.
6. Liu G., Wang J. K. and Zewail A. H., Nature **364** (1993) 427
7. Scherer N. F., Jonas D. M. and Fleming G. R., J. Chem. Phys. **99** (1992) 153.
8. Zadoyan R., Sterling M., Ovchinnikov M. and Apkarian V. A., J. Chem. Phys. **107** (1997) 8446.
9. Vos M. H., Lambry J. C. , Robles St. J., Youvan D. C., Breton J., Martin J. L., Proc. Natl. Acad. Sci. U.S.A. **88** (1991) 8885; Vos M. H., Rappaport F., Lambry J. C., Breton J., Martin J. L., Nature **363** (1993) 320; Vos M. H., Jones M. R., Hunter C. N., Breton J., Lambry J. C., Martin J. L., Biochemistry **33** (1994) 6750.
10. Wang Q., Schoenlein R. W., Peteanu L. A., Mathies R. A. and Shank C. V., Science **266** (1994) 422.
11. Liebl U., Lipowski G., Négrerie M., Lambry J. C., Martin J. L. and Vos M. H., Nature **410** (1999) 181.
12. Jeannin C., Portella-Oberli M. T., Jimenez S., Vigliotti F, Lang B. and Chergui M., Chemical Physics Lett. **316** (2000) 51.
13. Ben-Nun M., Levine R. D. and Fleming G. R., J. Chem. Phys. **105** (1996) 3035
14. Gersonde I. and Gabriel H., J. Chem. Phys. **98** (1993) 2094.
15. Krylov A. I. and Gerber R. B., J. Chem. Phys. **106** (1997) 6574.
16. Batista V. S. and Coker D. F., J. Chem. Phys. **105** (1996) 4033.
17. Yu N., Margulis C. J. and Coker D. F., submitted to J. Chem. Phys.
18. Fernandez-Alberti S., Halberstadt N., Engel V., Beswick J. A.and Echave J., J. Chem. Phys. **113** (2000) 1027.
19. Backhaus P. and Schmidt B., Chem. Phys. **217** (1997) 131
20. Marvet U. and Dantus M., Chem. Phys. Lett. **245** (1995) 393.
21. Gross P. and Dantus M., J. Chem. Phys. **106** (1997) 8013
22. Crépin C. and Tramer A. , J. Chem. Phys. **100** (1994) 5475.
23. Helbing J., Haydar A. and Chergui M., Chem. Phys. Lett. **310** (1999) 43
24. Helbing J., Chergui M. and Haydar A., J. Chem. Phys. **113** (2000) 3621.
25. Bergsma J. P. , Berens D. R., Wilson K. R., Fredkin D. R., Heller E. J., J. Phys. Chem., **88** (1984) 612
26. Tully J. C., J. Chem. Phys. **93** (1990) 1061
27. Hammes-Schiffer S. and Tully J. C., J. Chem. Phys., **101** (1994) 4657

28. Kopersky J., Atkinson J. B. and Krause L., J. Mol. Spec. **184** (1997) 300.
29. Kopersky J., Atkinson J. B. and Krause L., Chem. Phys. Lett. **219** (1994) 161
30. Kopersky J., Atkinson J. B. and Krause L., Can. J. Phys. **72** (1994) 1070
31. Niefer R. J., Supronowicz W., Atkinson J. B. and Krause L., Phys. Rev. A, **35** (1987) 4629.
32. Czuchaj E., Rebentrost F., Stoll H. and Preuss H., Chem. Phys., **214** (1997) 277
33. Gonzalez C. R., Fernandez-Alberti S., Echave J. and Chergui M., J. Chem. Phys. (in press, Feb. 2002)

A NEW ANALYSIS OF THE SOLVENT-INDUCED CHARGE TRANSFER IN ADMA BY SUBPICOSECOND SPECTROSCOPY

M. M. MARTIN, P. PLAZA AND P. CHANGENET-BARRET

UMR CNRS-ENS 8640 PASTEUR, Département de Chimie,
Ecole Normale Supérieure, 24 rue Lhomond 75231 Paris Cedex 05, France
E-mail: Monique.Martin@ens.fr

A. SIEMIARCZUK

Photon Technology International, 347 Consortium Court
London, Ontario, N6E 2S8, Canada

The photodynamics of 4-(9-anthryl)–N,N'-dimethylaniline (ADMA) was reexamined in acetonitrile and benzonitrile by transient absorption spectroscopy. In acetonitrile, the solvated charge-transfer state (CT) is observed immediately after excitation. In benzonitrile the CT is observed after a 20-ps delay and the initial excited state exhibits spectral characteristics of the locally-excited state (LE). A slightly activated solvent-induced charge-transfer in the normal region following an ultrafast torsional relaxation in the LE is proposed to explain the kinetics. In order to try to unify our and other's observations, we propose that the charge transfer proceeds on an excited-state surface defined by the internal torsion and solvent-motion coordinates.

1 Introduction

Among electron donor-acceptor anthracene derivatives, 4-(9-anthryl)–N,N'-dimethylaniline (ADMA) was reported to exhibit complex solvent-dependent photodynamics. Okada et al [1] showed that ADMA fluorescence is due to a highly polar charge-transfer state (CT), involving electron transfer from dimethylaniline to anthracene. They proposed that the photodynamics involves multiple excited states with different degrees of charge transfer, solvation and twist angles [2,3] whereas Tominaga et al [4,5] found it to be well described by a strongly adiabatic reaction involving solvation as the single coordinate.

We reexamined the excited-state dynamics of ADMA upon excitation with a 500-fs laser pulse at 370 nm by pump-probe transient absorption spectroscopy. The continuum probe was optimised down to 330 nm in order to detect the intense anthracenic transient absorption band [6]. The experiments were carried out in acetonitrile and benzonitrile. Acetonitrile is a more polar solvent than benzonitrile and its average solvation time [7] is 0.26 ps whereas that of benzonitrile is 5.1 ps. The transient spectra are compared to those we obtained for bianthryl [6] and the photoinduced charge transfer is analysed within a two-coordinate (torsion and solvation) reaction picture.

2 Experimental

500-fs, 30-40-μJ, 370-nm pulses were used as the pump and 500-fs, 300-μJ, 573-nm pulses for generating the continuum probe. The differential absorption spectra (ΔD) were measured in the 330-780-nm range with a polychromator and a double-diode-array detector and corrected from the probe group velocity dispersion. The pump-probe response function, was fitted to a 1.0 ps (FWHM) Gaussian. ADMA solutions were prepared with spectroscopy-grade acetonitrile and benzonitrile and used in a 1-mm non-recirculating cell. The samples had an ≈0.8 optical density at the pump wavelength. They were not deaerated. The corrected steady-state fluorescence spectra were measured with dilute samples.

3 Results

The time-resolved and steady-state spectra of ADMA in acetonitrile and benzonitrile are compared in Fig. 1. The ground-state absorption spectra, appearing as a broadened and red-shifted anthracene-like absorption, do not show much difference in these two solvents except for a broader red-wing in benzonitrile. In both cases, the fluorescence Stokes shift is large, even 40 nm larger in acetonitrile.

In acetonitrile, the spectra show transient absorption in the whole probed range with maxima around 340 nm, 480 nm and 670 nm. An ultrafast spectral evolution occurs between 0.5 and 1.5 ps characterised by a decay of ΔD between 340 and 360 nm together with a rise of the 340-nm and 480-nm bands. A tiny negative dip is seen at 390 nm, the maximum of the bleaching band. There is no further change for larger picosecond delays.

In benzonitrile, similar spectra are reached 20 ps after excitation. The initial transient spectrum is however different with two broad absorption bands at 357 nm and 583 nm. As time increases, a fast decrease of the initial UV band accompanies a red shift of the minimum of ΔD until about 530 nm, close to the maximum of the fluorescence spectrum. At 1.5 ps the minimum of ΔD is slightly negative, indicating that stimulated emission is involved in the observed shift. Concomitantly, the 583-nm band decays, the broad and flat absorption band above 650 nm slightly increases and a peak appears at 485 nm. The final UV band lies at 340 nm, i.e. 18 nm blue-shifted from the initial band. Kinetics at 357 nm, 486 nm and 593 nm are compared in Fig. 2. A detailed study of the solvent polarity effects on these kinetics is reported elsewhere [8].

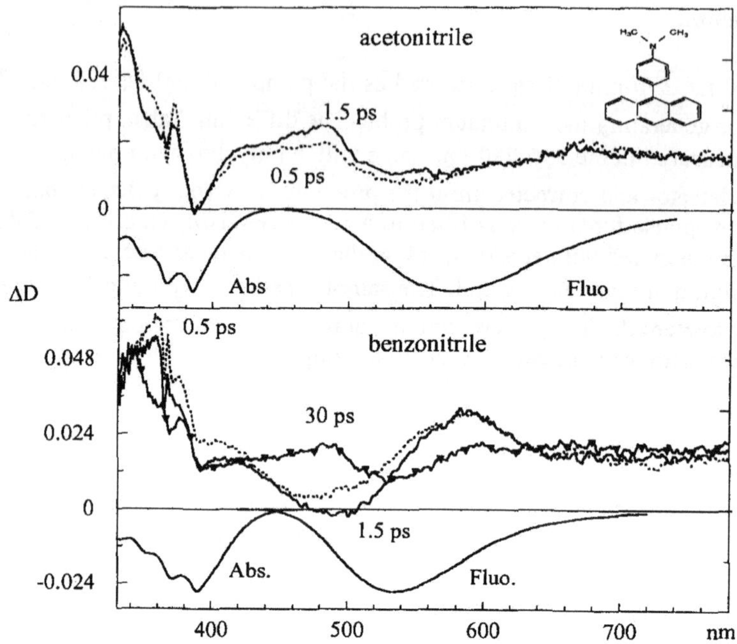

Figure 1. Differential absorption spectra (ΔD) of ADMA after 500-fs excitation at 370 nm in acetonitrile and benzonitrile at different delays. Upside down : normalised ground state absorption (Abs.) and fluorescence (Fluo.) spectra.

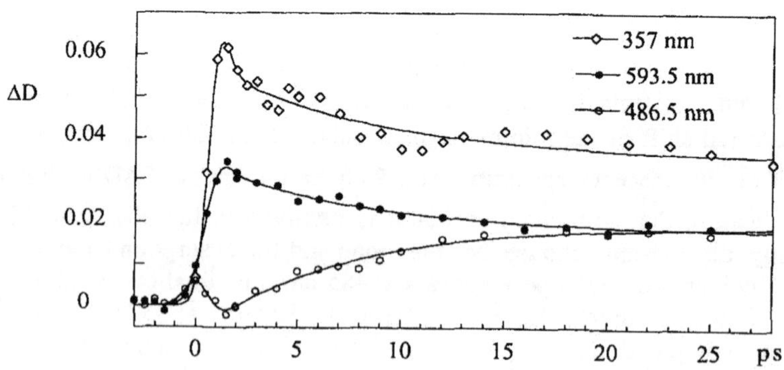

Figure 2. Kinetics and fit curves of the time-resolved differential absorption at selected wavelengths for ADMA in benzonitrile.

4 Discussion

4.1 Interpretation Of the transient spectra

In acetonitrile, the transient spectrum observed immediately after excitation can be interpreted by the superposition of the dimethylaniline cation radical (330 and 550 nm bands) and anthracene anion radical (480 nm band) absorptions [9], resulting from the photoinduced charge separation [1]. Since the solvation time is subpicosecond, one can attribute the 1.5 ps spectrum to the solvated CT.

In benzonitrile, the two initial absorption bands (357 nm and 583 nm) are similar to the ones we previously found for bianthryl (BA) [6]. Fig. 3 shows the 1.5-ps spectra of BA in ethanol and anthracene in cyclohexane. Anthracene exhibits a strong transient absorption band around 370 nm and a much weaker one around 600 nm.

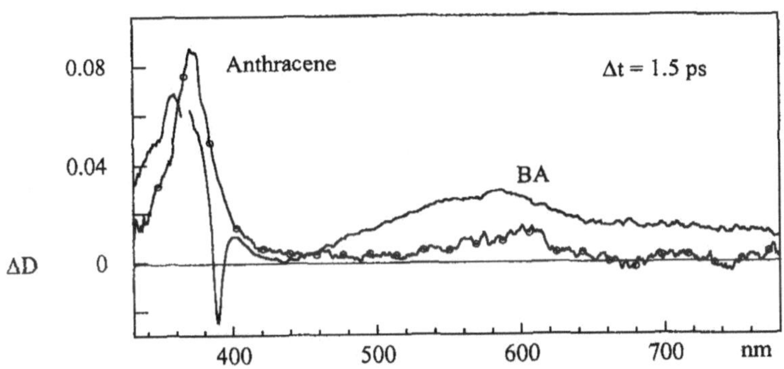

Figure 3. Transient absorption spectra of bianthryl in ethanol and anthracene in cyclohexane [6] for a 1.5 ps delay after 500-fs excitation at 370 nm.

One concludes that the initial bands observed for ADMA in benzonitrile are anthracene-like bands and that the initially probed state is the LE. This is in agreement with previous observations in butyronitrile and hexanenitrile [3]. Moreover, we reported an initial fast loss of structure (≤1 ps) in the gain spectra of BA that we attributed to the inertial component of the conformational relaxation within the LE [6] (the two anthracene moieties go from orthogonal to a ≈60° equilibrium angle in the LE [10]). From the similarity of the BA and ADMA initial spectra, one concludes that, for ADMA at 1.5 ps in benzonitrile, the coupling between the anthracene and dimethylaniline moieties is similar to that in the partially torsionally relaxed LE of BA. Thus, we propose that an ultrafast inertial torsion occurs also in the LE of ADMA prior to the charge transfer process. The final spectrum observed at 30 ps (Fig. 1b) is very similar to that in acetonitrile (Fig. 1a) and can be attributed to the solvated CT.

4.2 Reaction Kinetics and reaction path

The kinetics observed at selected wavelengths (Fig. 2) could be fitted by one or two exponentials, plus a nanosecond decay [1,11]. In addition to an ultrashort time constant (within our set-up resolution), a component of 8 ± 2 ps is found for the LE decay and CT rise. A fit with a single exponential leads to an average 6 ps. At 486 nm, the initial fast change in sign of ΔD is attributed to the red-shift of the gain spectrum as solvation proceeds. Following Tominaga et al's model [4,5], the ultrafast component observed in benzonitrile may correspond to the escape from the LE-character region of the S_1 adiabatic potential along the solvent coordinate. However, the LE character is expected to decay faster than solvation dynamics while the excited-state population propagates on the S_1 barrierless potential [4,5]. In the present study, the LE has a 6-8 ps decay component in benzonitrile, which is larger than the average solvation time. In contrast to dimethylformamide where the reaction was shown to be barrierless in the inverted regime [4,5], the 6-8 ps component found here in benzonitrile indicates a small barrier on the S_1 potential between the LE and CT minima. The reaction can be viewed as a precursor-successor reaction in the normal regime along the solvent coordinate. Such a process is known for BA in various polar solvents [6,12].

4.3 A Two-dimensional view of the photodynamics

Along the lines of the previously proposed solvation mechanism involving a gradual change in the electronic configuration and geometrical structure [2,3], we propose that the probed reaction occurs on an excited-state surface defined by both the internal twisting and the solvent motion, following a solvent-dependent path. Such a two-dimensional perspective was theoretically developed by Kim and Hynes [13] for the activated charge transfer reaction in DMABN and successfully described our own experimental observations [14]. The model foresees that the barrier crossing is mainly along the torsional coordinate in acetonitrile whereas in methanol the solvent coordinate is predominant [13].

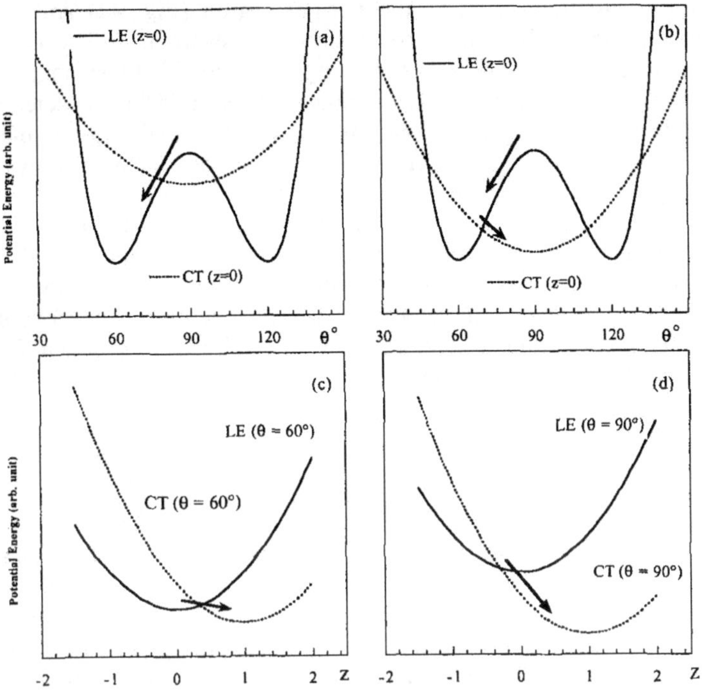

Figure 4. Schematic view of the photoinduced reaction involving both torsion and solvent coordinates: (**up**) torsional potentials of the diabatic LE and CT for a ground-state equilibrium solvent configuration in a polar solvent - (a) the crossing angle is close to 90° and the electronic coupling is weak - (b) the crossing angle is close to 60° and the electronic coupling is large. (**down**) potentials of the diabatic LE and CT along the solvent coordinate - (c) for a 60° torsion angle, the crossing is in the normal region - (d) for a 90° torsion angle, the crossing is in the inverted region.

In order to try to unify our and others' observations for ADMA we propose the mechanism illustrated by Fig. 4. Considering that there is no electronic coupling between the diabatic LE and CT for a 90° twist angle, the initial LE population projected from the orthogonal ground-state population by the excitation pulse evolves quickly towards the 60° equilibrium twist angle on a steep torsion potential while the solvent is still in the equilibrium configuration around the ground state. This statement implies that the CT and LE torsional potentials cross for a twist angle close to 90°. This condition is fulfilled when the CT potential exhibits a single minimum at 90° and is just slightly below the LE as previously suggested [15] and schematically illustrated in Fig. 4a. For a 60° twist angle, the electronic coupling is nonzero and the CT is above the LE so that along the solvent-coordinate they cross in the normal region (Fig. 4c) and the charge transfer is expected to be a precurssor-successor reaction. This would describe ADMA in benzonitrile and BA. Assuming that the CT twist angle is 90° at equilibrium, a change in the angular distribution back to 90° is expected while the reaction proceeds along the solvent-coordinate; the final angular distribution depending on the torsional potential shape. Let us consider now the case illustrated by Fig. 4b where the CT torsional potential is lowered in energy, for example in a more polar solvent. The CT crosses the LE for an angle close to 60° where the electronic coupling is nonzero. After the initial torsion in the LE, one may thus expect the population to cross to the CT which lies below the LE so that, for a broad range of angles, the diabatic LE and CT cross in the inverted region along the solvation coordinate (Fig. 4d) as proposed by Tominaga et al [4,5]. Such a general picture may reconcile the existing interpretations of ADMA photodynamics.

5 Conclusion

The UV-visible transient absorption spectra of 4-(9-anthryl)–N,N'-dimethylaniline were measured in acetonitrile and benzonitrile after excitation with 500-fs pulse at 370 nm. In both solvents, the spectral characteristics of the solvated charge-transfer state, i.e. the superposition of the dimethylaniline cation radical and the anthracene anion radical absorption bands, are observed. The photoinduced charge transfer kinetics is found to be within the experimental resolution in acetonitrile and slighly slower or comparable to that of solvation in benzonitrile. In the latter solvent, the initially excited state could be identified as the LE by comparison with bianthryl and anthracene. We conclude that in benzonitrile the photoinduced reaction is very similar to that in bianthryl and can be described by a slightly activated solvent-induced charge separation in the normal region following an ultrafast inertial torsional relaxation in the LE. To reconcile the existing interpretations of ADMA photodynamics, we propose that the charge-transfer reaction proceeds on an excited-state surface defined by the internal twisting and solvent coordinates, in a similar manner to what was previously theoretically demonstrated for DMABN [13], with a solvent-dependent reaction path.

6 Acknowledgments

MMM, PP and PCB thank Prof. J.T. Hynes (UMR CNRS 8640, ENS-Paris France, Univ. Boulder, Colorado, USA) for very helpful discussions. MMM thanks Prof. A. Douhal (Universidad de Castilla-La Mancha, Toledo, Spain.) for having invited her to communicate these results at the Femtochemistry V Conference.

References

1. Okada T., Fujita T., Kubota M., Masaki S., Mataga N., and, Ide R., Sakata Y.and Misumi S., Intramolecular electron donor-acceptor interactions in the excited state of (anthracene)-$(CH_2)_n$-(N,N-dimethylaniline) systems, *Chem. Phys. Lett.* **14** (1972) pp. 563-568.
2. Okada T., Mataga N., Baumann W.and Siemiarczuk A., Picosecond laser spectroscopy of 4-(9-anthryl)-N,N-dimethylaniline and related compounds, *J. Phys. Chem.* **91** (1987) pp. 4490-4495.
3. Mataga N., Nishikawa S., Asahi T.and Okada T., Femtosecond-picosecond laser photolysis studies on the photoinduced charge separation and charge recombination of a produced ion pair state of some typical intramolecular exciplex compounds in alkanenitrile, *J. Phys. Chem.* **94** (1990) pp. 1443-1447.
4. Tominaga K., Walker G. C., Jarzeba W.and Barbara P. F., Ultrafast charge separation in ADMA : experiment and theoretical issues, *J. Phys. Chem.* **95** (1991) pp. 10475-10485.
5. Tominaga K., Walker G. C., Kang T. J., Barbara P. F.and Fonseca T., Reaction rates in the phenomenological adiabatic excited-state electron-transfer theory, *J. Phys. Chem.* **95** (1991) pp. 10485-10492.
6. Jurczok M., Plaza P., Martin M. M., Meyer Y. H.and Rettig W., Excited state relaxation paths in 9,9'-bianthryl and 9-carbazolyl-anthracene: a sub-ps transient absorption study, *Chem. Phys.* **253** (2000) pp. 339-349.
7. Horng M. L., Gardecki J. A., Papazyan A.and Maroncelli M., Subpicosecond measurements of polar solvation dynamics: Coumarin 153 revisited, *J. Phys. Chem.* **99** (1995) pp. 17311-17337.
8. Martin M. M., Plaza P., Changenet-Barret P.and Siemiarczuk A., UV-visible subpicosecond spectroscopy of 4-(9-anthryl)-N,N'-dimethylaniline in polar and nonpolar solvents : a two-dimensional view of the photodynamics, *J. Phys. Chem. A, Mataga Festschrift, in press* pp. .
9. Shida T., In *Electronic absorption spectra of radical ions* , Elsevier (Amsterdam, 1988) .
10. Khundkar L. R.and Zewail A. H., Exciton and vibronic effects in the spectroscopy of bianthracene in supersonic beams, *J. Chem. Phys.* **84** (1986) pp. 1302-1311.
11. Siemiarczuk A.and Ware W. R., Complex excited-state relaxation in p-(9-Anthryl)-N,N-dimethylaniline derivatives evidenced by fluorescence lifetime distributions, *J. Phys. Chem.* **91** (1987) pp. 3677-3682.
12. Kang T. J., Jarzeba W., Barbara P. F.and Fonseca T., A photodynamical model for the excited state electron transfer of bianthryl and related molecules, *Chem. Phys.* **149** (1990) pp. 81-95.
13. Kim H. J.and Hynes J. T., Excited state intramolecular charge transfer rates of p-dimethylaminobenzonitrile (DMABN) in solution: A two-dimensional

dynamics perspective, *J. Photochem. Photobiol. A: Chem.* **105** (1997) pp. 337-343.

14. Changenet P., Plaza P., Martin M. M.and Meyer Y. H., Role of intramolecular torsion and solvent dynamics in the charge transfer kinetics in triphenylphosphine oxide derivatives and DMABN, *J. Phys. Chem. A* **101** (1997) pp. 8186-8194.
15. Siemiarczuk A., Koput J.and Pohorille A., Structural requirements for intramolecular charge-transfer in excited state of 4-(9-Anthryl)-N,N-dimethylaniline, *Z. Naturforsch* **37a** (1982) pp. 598-606.

BREAKDOWN OF HYDRODYNAMIC BEHAVIOR: SOLUTE ROTATIONAL DYNAMICS FROM THE SMALL-MOLECULE TO THE POLYMER LIMIT

MARK M. SOMOZA, MIKHAIL I. SLUCH AND MARK A. BERG

Department of Chemistry and Biochemistry,
University of South Carolina, Columbia SC 29208, USA
e-mail:berg@mail.chem.sc.edu

Solute motion in small-molecule solvents is governed by macroscopic hydrodynamics, as in the Stokes-Einstein-Debye (SED) models of rotation and diffusion. In polymers on the other hand, the same transport properties for small solutes are disconnected from the macroscopic viscosity. We have used rotation of anthracene as a probe of local friction and examined the transition from small-molecule to polymeric solvent. Two polymers were tested: polyisobutylene (PIB) and polydimethylsiloxane (PDMS). These two polymers have similar structures, but exhibit very different transport properties. PDMS has one of the highest solute diffusivities among common elastomers, while PIB has one of the lowest. Several new results are found: (1) The SED behavior of small-molecule solvents is lost when the solvent length exceeds a dynamic correlation length for torsional motion. Other characteristic polymer lengths, e.g. the entanglement length or persistence length, are not important. The solute-to-solvent size ratio is not the main factor in the breakdown of hydrodynamic behavior. (2) The rotational friction on small solutes can be dramatically different from the macroscopic viscosity in high polymers. In PDMS with a viscosity of 1×10^5 cP, anthracene rotates as if its local viscosity is 5 cP. The rotational friction mirrors the translational friction governing small-molecule diffusion. (3) The polymer chain length retains a weak effect on the local friction, even after the main decoupling of macro and nanoviscosity occurs. (4) The single-exponential relaxation of small-molecule solvents is lost for even very short oligomers, but further lengthening of the polymer does not increase the non-exponentiality.

1 Introduction

The dynamics of solutes in small-molecule liquids are typically discussed in terms of simple hydrodynamics. If the solute is larger, or of comparable size to the solvent molecules, the hydrodynamic model is invoked and the dynamics are related to the macroscopic viscosity. The Stokes-Einstein-Debye (SED) model predicts solute rotation inversely proportional to solvent viscosity.

The dynamics of polymer melts are viewed quite differently, as the macroscopic viscosity must be quite different from the effective viscosity experienced by a monomer-scaled solute molecule. This phenomena is due to the two-dimensional nature of polymers. As the polymer lengthens, the frictional forces restraining the translation of the entire molecule can originate far away from any given segment of the chain, while the interactions on the length scale of the solute remain local and are thus largely unchanged. Bulk viscosity scales with chain length while local viscosity must depend largely on interactions on the length scale of the solute molecule.

Our experiments have bridged the gap between these two regimes, small-molecule and polymeric solvent, by comparing the rotational friction experienced by a small solute with the macroscopic viscosity as the solvent is lengthened from the small-molecule limit to the entangled polymer. We attempt to answer the question of what length distinguishes polymers from small-molecule liquids when viewed on a short length scale.

2 Experimental Section

Because of the large range of rotation times covered in this study (6 ps to 180 ns), two experimental techniques were used in different ranges. All low viscosity samples (<100 cP) were measured by heterodyne-detected dichroism decay in a standard polarization-spectroscopy setup [1] using <100 fs frequency doubled pulses (375 nm) from a Ti:Sapphire laser. A flowing sample eliminated thermal effects and artifacts from photo-byproducts. The two lowest viscosity samples of each polymer, 0.65 cSt PDMS and isooctane, were pumped through a flow cell with a 1mm path length.

The longer rotational times were obtained by fluorescence anisotropy decay using standard time-correlated single-photon-counting (TCSPC) techniques. The typical instrument response function was 50 ps FWHM. In all experiments, the sample temperature was 25±0.5°C.

3 Calibration of a Nanoviscometer

The effective hydrodynamic volume of the solute molecule, anthracene, was determined from rotational decay measurements in a variety of small-molecule solvents [2]. These solvents were chosen to have molecular volumes equal to or smaller than that of anthracene. The data follow the modified SED equation [3]

$$\tau_r = \frac{\lambda V_h}{6kT}\eta + \tau_0 \tag{1}$$

very well over a broad range of viscosities from 0.3 – 60 cP (Fig. 1). Using these data as well as slip boundary conditions [4], the hydrodynamic shape of anthracene was modeled as an ellipsoid (11.8×7.5×4.2 Å) very close to the van der Waals shape (11.8×7.8×4.0 Å). This shape has an accidental degeneracy in rotational constants, and as a result, the model predicts a single exponential anisotropy decay. Single-exponential decays were observed. Thus, as long as the solvent molecules are smaller than those of the solute, anthracene is a nearly ideal hydrodynamic rotor [2].

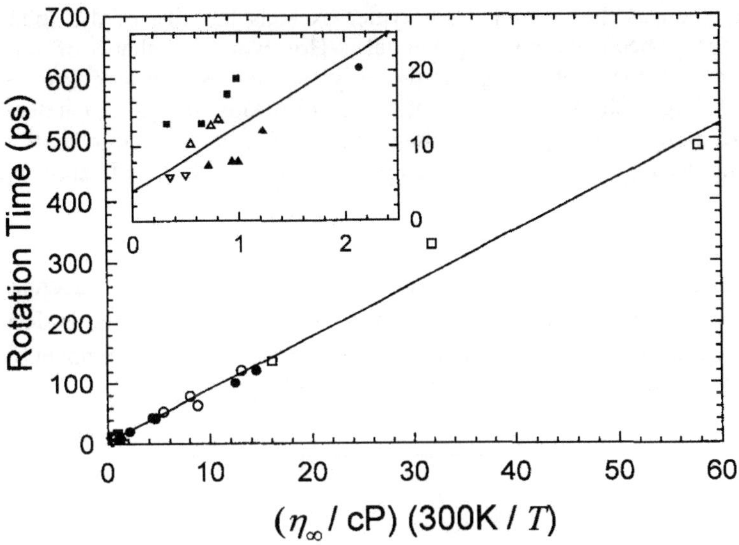

Figure 1. Rotation times τ_r for anthracene in small-molecule solvents versus solvent viscosity. The line is a fit to a hydrodynamic model (Eq. 1, $\tau_0 = 3.95$ ps, slope = 8.68 ps/cP). (∇) – from our dichroism measurements in isooctane and M = 162 g/mol PDMS. (■) – from UV-IR double resonance measurements in CCl_4, C_2Cl_4, C_6D_6, and C_3D_6O by Lettenberger et al [5]. (▲) – from fluorescence depolarization measurements in cyclohexane by Jas et al [6]. (Δ) – from our dichroism experiments in toluene [7]. (●) – from our previous dichroism experiments in benzyl alcohol [2]. (○) – from our TCSPC experiments in benzyl alcohol. (□) – from our TCSPC experiments in ethylene glycol and cyclohexanol.

Anisotropy decay data for the two polymers were combined with the small-molecule calibration, using Eq.1 to calculate an apparent viscosity for rotation or "nanoviscosity". This definition of viscosity is essentially identical to a "macro" viscosity, but applies more accurately to the local environment in which the anthracene molecules are located, and will reflect the dynamics at this length scale.

4 Results and Discussion

4.1 Polyisobutylene (PIB)

We measured rotational diffusion of anthracene in eleven PIB samples ranging in molecular weight from M = 114 to 85,000 g/mol, corresponding to a ten-order-of-magnitude range in bulk viscosities (0.5 – 5×10^9 cP) and a chain length (ℓ) range between 5 and 3040 atoms. Three major results are seen in the data.

First, as soon as the size of the polymer exceeds the size of the solute, the anisotropy decay becomes non-exponential. However, the shape of the non-exponential decay does not change as the polymer increases in size from 350 to 85,000 g/mol. Fig. 2 shows a master plot of the anisotropy decays. Each decay has been shifted along the logarithmic time axis to adjust for the change in its corresponding decay time. The common shape is fit is by a stretched exponential:

$$r(t) = r_0 \exp[-(t/\tau)^{\beta}], \tag{2}$$

with $r_0 = 0.27$ and $\beta = 0.52$. The transition is abrupt. The decay in isooctane (114 g/mol), a low M_W analog of PIB with a chain length (ℓ) of 5 atoms, is definitely single exponential, while in 350 g/mol PIB ($\ell = 13$) it is fully non-exponential and can be fit with the same stretched exponential as the longer PIBs.

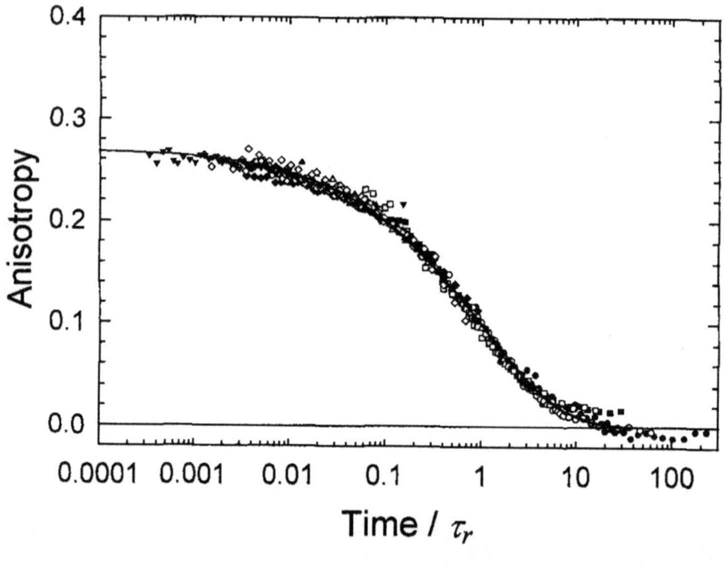

Figure 2. Master plot of anisotropy decays in PIB ($M \geq 224$ g/mol). The time scale for each data set is divided by a scaling time τ_r. The solid curve is a stretched exponential fit (Eq. 2) with $r_0 = 0.27$ and $\beta = 0.52$.

The second result is that the initial anisotropy is lower than in small-molecule solvents. The theoretical maximum anisotropy is 0.40, although the full maximum is rarely observed. In the case of anthracene in small-molecule solvents, we have measured a value of $r_0 = 0.34 \pm 0.02$, with the missing anisotropy assigned to a subpicosecond inertial decay component [2]. However, the initial value of $r_0 = 0.27$ observed in PIB is much smaller than the small-molecule value. Similarly low values of r_0 have been seen in other polymers [8-10]. The unusually large anisotropy deficit could represent a fast process in polymers that allows partial reorientation of the solute.

The third result is that the rotation times follow SED behavior for short polymers, but then show a sharp transition to a weak dependence on macroscopic viscosity as the polymer length increases (Fig. 3). For PIB, the hydrodynamic relationship between bulk viscosity and solute rotation is preserved to viscosities of 2000 cP. At these viscosities, the extended solvent molecules are about three times longer than the maximum dimension of the solute. Thus, PIB provides an example showing that hydrodynamic SED behavior can occur even when the solvent is significantly larger than the solute. This result directly contradicts the most common theories for the breakdown of hydrodynamics, which indicate that hydrodynamics will only be valid when the solvent is smaller than the solute.

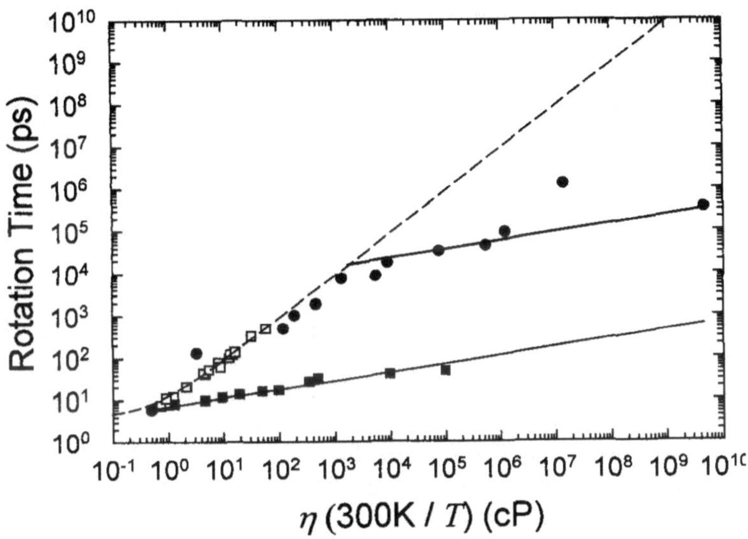

Figure 3. Rotation times of anthracene in PIB (●) and PDMS (■) with superimposed small-molecule solvent rotation data (□). The fit through the latter (dashed line) is the SED model of Eq. 1 and Fig. 1. The power law fits to the polymer data have a common exponent of 0.2.

Although SED behavior persists for unexpectedly large solvents, it break down dramatically for sufficiently long polymers. For the longest PIB, there is a greater than five orders-of-magnitude difference between the observed rotation time and the time predicted by the SED model. The simple, empirical straight line fits in Fig. 3 show that the transition from small-molecule behavior to polymer-like behavior occurs abruptly. They also highlight the fact that the rotation time does continue to increase, even after the breakdown of the SED model. We believe that this slow increase is due to end effects and that the rotation time will level off in the limit of an infinitely long polymer.

4.2 *Polydimethylsiloxane (PDMS)*

PDMS and PIB are very similar at the level of chain structure. However, both neutron scattering in the bulk polymers [11] and diffusivity of solutes indicate that the dynamics of the two polymers at short length scales are quite different. To help understand the origin of these differences, we measured the rotational anisotropy of anthracene in eleven PDMS samples with viscosities ranging from 0.5 cP to 1×10^5 cP. This viscosity range is smaller than that of the PIB samples, however, because PDMS has a different relationship between chain length and viscosity, the actual size range was greater, $\ell = 5$ to $\ell = 3760$. The PDMS data is presented alongside the PIB data in Figs. 3 and 4. The rotational dynamics of PDMS are completely different from those in PIB, despite their similarity in static structure. In particular, three differences stand out.

First, rotational decay times are much faster in PDMS, even for the same macroviscosity. At the short end of the size scale, both PIB and PDMS show rotation times consistent with the SED behavior of small-molecule solvents. The rotational decay times for anthracene are 5.9 and 6.2 ps, respectively, in the shortest PIB and PDMS. However, while the rotation times in PIB increases rapidly with increasing viscosity, rotation times in PDMS stay almost constant, increasing by only a factor of eight over the entire range. Most interestingly, the rotation time in the highest molecular weight PDMS sample (140,000 g/mol, $\ell = 3760$) is only about one third of that in second lightest PIB sample (224 g/mol, $\ell = 9$).

Second, while the rotational motion in PIB follows the small-molecule hydrodynamic model until intermediate molecular weights, increasing linearly with viscosity, the PDMS data is immediately decoupled from this relationship. The rotational dynamics of anthracene in PDMS are almost completely independent of the size or viscosity of the polymer.

Third, the (weak) power law dependence between the size of the PDMS molecule and the rotation time is constant, while PIB shows a distinct transition between strong and weak dependence of rotational dynamics on polymer size.

In one regard, PDMS and PIB are similar. PDMS also exhibits non-exponential decay once the chain lengthens beyond the size of the solute. The anisotropy decays with a single exponential in the smallest PDMS sample (162 g/mole, $\ell = 5$), but the next polymer in the series (340 g/mole) fits to the same stretched exponential decay ($\beta = 0.73$) as the higher weight samples.

5 *Breakdown of Hydrodynamics in PIB and PDMS*

The breakdown of hydrodynamics for molecular-sized objects is often interpreted in relation to the ratio of the volumes of compact solvents and solutes. However, compact solvents of increasing size are difficult to achieve experimentally and are replaced by polymers of increasing length. Because the solvent is increasing in size in only one dimension, the dynamics can not be easily viewed in terms of free-space models and packing inefficiencies. In our interpretation, the primary factor for the breakdown of hydrodynamics is the torsional rigidity of the polymer chain. So long as the solvent molecules are short, or torsionally rigid, the interactions with the solutes obey rigid body dynamics and follow the DSE hydrodynamic relationship. If the solvent molecule is long enough to be torsionally flexible on the time scale of the interaction with the solute, an additional mechanism arises to accommodate solute motion—displacements of only subsections of the polymer chain. This dynamic channel is never available on a macroscopic length scale and thus macro and nanoviscosities diverge.

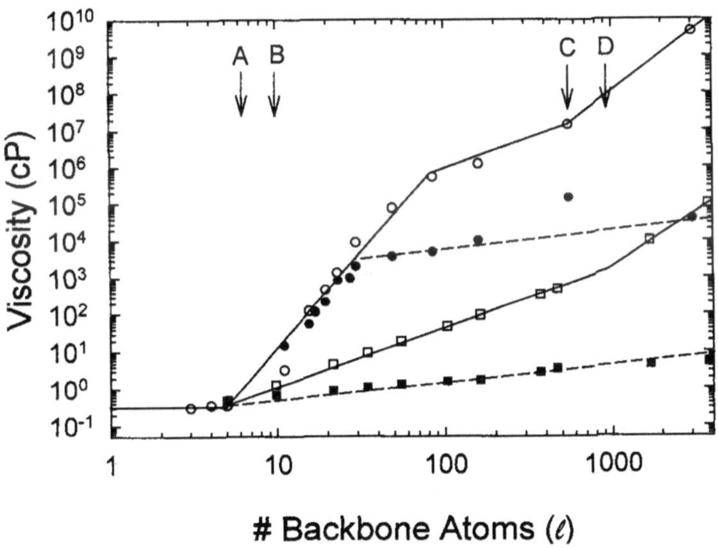

Figure 4. Bulk and nanoviscosity of PIB and PDMS as a function of polymer chain length. The solid symbols represent the values for the nanoviscosities [PIB(●) and PDMS (■)] while the empty symbols represent the bulk viscosities [PIB(○) and PDMS (□)]. The lines are empirical fits through the data. The arrows are significant length scales associated with the data: **A**-solute size, **B**-persistence length (similar for both polymers), **C** and **D**-entanglement lengths of PIB and PDMS, respectively.

PIB has a high rotational barrier, the methyl groups prevent easy torsion of the carbon-carbon backbone. Thus, short and intermediate length chains exhibit rigid body behavior and the nanoviscosity scales with bulk viscosity. As the polymer length increases, torsional mechanisms (crankshaft motion and end-flipping) start to displace rigid body motions as the lowest energy response to the motion of the solute molecules. However, these transitions can not accommodate the motions of very large objects, and thus, the macroviscosity continues to increase as the nanoviscosity levels off.

PDMS is structurally similar to PIB, but the longer Silicon-Oxygen bonds separate the methyl groups sufficiently to allow easy torsion of the backbone. This additional flexibility accounts for the observed behavior. The smallest PDMS molecule, (162 g/mole) fits the hydrodynamic model, see Figs. (1 and 4), but even slightly longer chains are too flexible on the picosecond time scale to allow rigid body motions, and the lowest energy response to accommodate anthracene rotation is a displacement of a segment of the polymer through rotation of Si-O bonds. Again, macroscopic relaxation is only sensitive to translation of the whole polymer molecule and therefore macro and nanoviscosity diverge.

6 Conclusions

The breakdown of the simple hydrodynamic model of a liquid described by the DSE equation is clearly observed. We have found that DSE behavior can extend to systems were the solvent is larger than the solute if the solvent is rigid on the time scale of solute rotation.

These studies have bridged the two regimes of small-molecule and polymeric solvent. The transition occurs with the development of a dynamic flexibility of the polymer chain. The mechanism that triggers the transition is not a simple ratio between the size of the solvent and that of the solute, but rather the height of the torsional barriers of the polymer.

7 References

1. Alvari, D. S., R. S. Harman and D. H. Waldeck, Optically heterodyned polarization spectroscopy. Measurement of the orientational correlation function. *J. Chem. Phys.* **92** (1990) p. 4055.
2. Zhang, Yunhan, Mikhail I. Sluch, Mark M. Somoza and Mark A. Berg, Ultrafast Dichroism Spectroscopy of Anthracene in Solution. I. Inertial versus Diffusive Rotation in Benzyl Alcohol. *J. Chem. Phys.* **115** (2001) pp. 4212-4222.
3. Alms, G. R., D. R. Bauer, J. I. Brauman and R. Pecora, Depolarized Rayleigh Scattering and Orientational Relaxation of Molecules in Solution. I. Benzene, Toluene and para-Xylene. *J. Chem. Phys.* **58** (1973) p. 5570.

4. Sension, R.J. and R.M. Hochstrasser, Comment on: Rotational friction coefficents for ellipsoids and chemical molecules with slip boundary conditions. *J. Chem. Phys.* **98** (1992) p. 2490.
5. Lettenberger, M., F. Emmerling, N. H. Gottfried and A. Laubereau, Orientational motion of anthracene in liquid solution studied by IR/UV double-resonance spectroscopy. *Chem. Phys. Lett.* **240** (1995) pp. 324-329.
6. Jas, G. S., Y. Wang, S. W. Pauls, C. K. Johnson and K. Kuczera, Influence of temperature and viscosity on anthracene rotational diffusion in organic solvents: Molecular dynamics simulations and fluorescence anisotropy study. *J. Chem. Phys.* **107** (1997) pp. 8800-8812.
7. Zhang, Yunhan and Mark A. Berg, Ultrafast dichroism spectroscopy of anthracene in solution. IV. Merging of inertial and diffusive motions in toluene. In preparation.
8. Stein, A. D., D. A. Hoffmann, A. H. Marcus, P. B. Leezenberg, C. W. Frank and M. D. Fayer, Dynamics in Poly(dimethylsiloxane) Melts: Fluorescence Depolarization Measurements of Probe Chromophore Orientational Relaxation. *J. Phys. Chem.* **96** (1992) pp. 5255-5263.
9. Inoue, T., M. T. Cicerone and M. D. Ediger, Molecular Motions and Viscoelasticity of Amorphous Polymers Near T-G. *Macromolecules.* **28** (1995) pp. 3425-3433.
10. Hyde, P. D., M. D. Ediger, T. Kitano and K. Ito, Local Segmental Dynamics of Polyisoprene in Concentrated Solutions and in the Bulk. *Macromolecules.* **22** (1989) pp. 2253-2259.
11. Arbe, A., M. Monkenbusch, J. Stellbrink, D. Richter, B. Farago, K. Almdal and R. Faust, Origin of internal viscosity effects in flexible polymers: A comparative neutron spin-echo and light scattering study on poly(dimethylsiloxane) and polyisobutylene. *Macromolecules.* **34** (2001) pp. 1281-1290.

FEMTO- AND PICOSECOND FLUORESCENCE DYNAMICS OF MEROCYANINE 540: EXPERIMENTS AND MODELING

I. BUGÁR AND D. CHORVÁT JR.

International Laser Center Bratislava, Ilkovičova 3, 812 19 Bratislava, Slovakia
E-mail: bugar@ilc.sk

T. PÁLSZEGI

Institute of Physical Chemistry, Vienna University, Währingerstrasse 42, A-1090 Wien, Austria
E-mail: palszegi@ap.univie.ac.at

P. MACH AND J. URBAN

Department of Chemical Physics, Faculty of Mathematics, Physics and Informatics, Comenius University, Mlynská dolina F1, 842 15, Bratislava, Slovakia
E-mail: mach@fmph.uniba.sk

V. SZÖCS

Institute of Chemistry, Faculty of Natural Sciences, Comenius University, Mlynská dolina CH2, 842 15, Bratislava, Slovakia
E-mail: szocs@fns.uniba.sk

Time- and spectral-resolved fluorescence of merocyanine 540 (MC540) was measured by up-conversion method on femto- and picosecond time scales in water. A spectral narrowing behaviour was observed on the obtained curves, concerned with fluorescence decay on the both wings of the steady-state spectra. Three decay components were found analyzing the MC540 fluorescence time evolution curves. The slowest decay component was found to be not wavelength dependent, with average value 84.4 ± 3.3 ps. The two remaining components were wavelength dependent with the respective lifetimes of several hundred femtoseconds and several picoseconds. The interpretation of the obtained results on the several picosecond time-scale is based on quantum chemical computations of the structural changes of the MC540 chromophore and its vibrational spectra. Utilizing the results of the computation a theoretical model was developed in order to fit the experimental curves. Further an ethanolic solution of MC540 was examined by the same method, in order to show the different short time behaviour in different environments. The used Brownian oscillator model provides a suitable theoretical basis for explanation of the observed differences in the fluorescence behaviour.

1 Introduction

Merocyanine 540 (MC540) is a lipophilic polymethine dye. Its conjugated system involves donor and acceptor parts resulting high polarity. The value of its dipole moment is further affected by conformational motions of the SO_3^- terminated flexible said chain [1]. The aqueous environment supports the formation of the zwitterionic form of molecules, which easily create dimers [2]. An important property of MC540 is that its preferentially binding to the leukemia, lymphoma and viral affected cells, what allows its use in photochemotherapeutic treatment [3]. Fluorescence properties of this compound are highly sensitive to the environmental factors. Recently, the increase of nonradiative deactivation by an increase of hydrogen bonding ability has been shown [4]. In the same work a high variety of MC540 lifetime and quantum efficiency in organic solvents in the 100 – 1000 ps and 0.04 – 0.5 ranges has been presented. Remarkable results concerning the environmental effect have been published, namely the fluorescence lifetime dependence on the ratio of a water-dioxane mixture in solvent [5].

The molecular structure of the MC540 in the ground electronic state has been analyzed on the quantum chemical level using the *ab initio* Hartree-Fock method and the density functional theory [6]. The explanation of the fluorescence properties of MC540 requires a detailed theoretical and experimental study of the excited state and the solvent interaction dynamics of this material. The first steps of this study are presented in this work, first and foremost an experimental femto- and early picosecond time-scale analysis of the fluorescence. Further a quantum chemical computations based fluorescence dynamic model is presented providing a suitable interpretation of the obtained results.

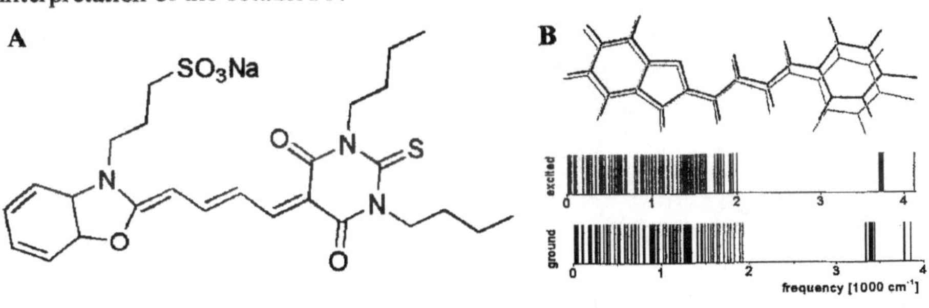

Figure 1. (**A**) The molecular structure of the MC540. (**B**) The geometry and vibrational frequencies change of the restricted molecular structure upon excitation according the computations (gray - excited state).

2 Experimental study

The time- and spectral-resolved fluorescence of MC540 was measured by an up-conversion technique, using a femtosecond Ti:sapphire laser (CDP TiF50) and the commercial fluorescence up-conversion apparatus (CDP FOG100). The detailed description of the measurement system was published elsewhere [7]. The majority of the measurements were realized using bidistilled water as a solvent, further ethanol p. a. was used to compare the short time fluorescence behaviour in these two environments.

The laser generated 65 - 95 fs pulses at 800 - 808 nm and the output power was in the range of 240 – 275 mW at 80 MHz pulse repetition rate. The cross-correlation function FWHM between the excitation and gate field was 210 fs for water samples. Water solutions were prepared in 10^{-3} - 10^{-5} M concentration range in order to examine the influence of the change of the dimer-monomer ratio on the temporal behavior of fluorescence. The fluorescence of MC540 in the water was examined in the wavelength range of 480 – 700 nm, which is approximately three times broader as the FWHM of the fluorescence spectra (λ_{max} = 585 nm).

All measurements were accomplished at magic angle adjustment of the polarization control. In the case of the water samples two different time scales were used: 3 ps and 300 ps. All measurements were realized at room temperature. Experimental data with sub-picosecond resolution were processed using the multiexponential deconvolution analysis [8]. For data processing the measured cross-correlation functions were used as the instrumental response. Measurements performed on the longer time scale were approximated by the same functions without the deconvolution procedure. The wavelength dependence of acquired fitting parameters was analyzed on the particular time scales.

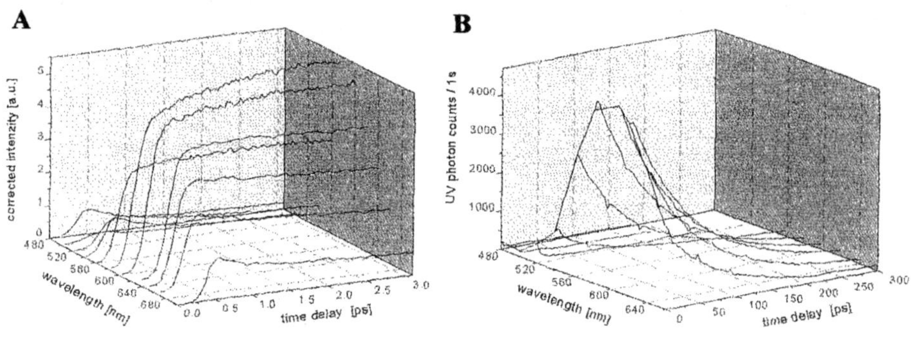

Figure 2. Wavelength and time-resolved fluorescence of MC540 in water at concentration 0.268 M (**A**) on the 3 ps and (**B**) on the 300 ps time-scale.

The typical wavelength resolved fluorescence behaviour of the MC540 in the water on the 3 ps and 300 ps time scales is illustrated in figure 2. Obtained results in the case of the water solution can be summarized as follows: Three different time constants τ_1, τ_2, τ_3 of the fluorescence relaxation were recognized in the most of analyzed wavelengths. The short-time fluorescence behaviour was a fast decay on the wings and a slight increase in the central part of the steady-state spectra. The results of the data processing are presented in the table 1. The mentioned increase was not possible properly determine by deconvolution procedure, therefore the τ_1 value is lacking for wavelengths 560 – 620 nm. The whole analysis was accomplished for three different concentrations (1.34 mM, 0.268 mM, 53.6 μM), and no significant differences were observed. Due to this fact the presented results in the table was obtained by averaging values found out for three different samples.

Table 1. The results of the fluorescence data processing in the case of aqueous solution of MC540.

λ [nm]	Two – exponential deconvolution method			Simple fit	
	τ_1 [fs]	τ_2 [ps]	a_1/a_2	τ_2 [ps]	τ_3 [ps]
480	264	4.67	3.76	-	12.2
500	277	3.82	4.16	-	21.3
520	403	5.12	2.68	5.2	72.2
540	623	18.6	0.52	4.7	81.1
560	-	-	-	-	82.9
580	-	-	-	-	87.7
600	-	-	-	-	87.5
620	-	-	-	-	86.5
640	2865	42.6	0.11	-	82.6
660	1872	32.1	0.36	6.2	83.2
700	518	19.1	0.61	4.8	81.9

The complex view on the spectral resolved fluorescence relaxation on the several picosecond time-scales showed a spectral narrowing behaviour. The typical manifestation of this behaviour was the decrease of the shortest time constant value τ_1 with an increasing wavelength difference from λ_{max} on the both wings. The two exponential deconvolution analysis gave for all curves two pre-exponential factors a_1, a_2 which ratio shows the significance of the τ_1 component. The values of the a_1/a_2 were increasing by increasing wavelength difference from λ_{max}, which is a sign of the spectral narrowing character, too. Analyzing the 300 ps time-scale curves in the range of 540 – 700 nm the slowest time constant was found out with a stable value 84.4 ± 3.3 ps. In order to compare the data obtunded for water time-resolved fluorescence of the MC540 in ethanol on the 3 – 100 ps time scale were performed too. The obtained results showed a constant value of fluorescence intensity on the examined time-scales.

3 Quantum chemical computations

Some parameters characterizing the potential energy hypersurface of ground and first excited states was calculated in order to give a base for the model of the fluorescence dynamics. The minimal basis set was used due to the size of the

MC540 molecule, as well as the model approximation for the study of dynamic properties. Further simplification concerned cutting off three side chains and replacing them with hydrogen atoms, illustrated in the figure 1., thus dealing only with the chromophore part of the molecule.

The STO-3G basis set [9] was used for the HF calculations of the ground state. The excited state was calculated using CIS approach [10,11]. The Hessian and vibrational frequencies for both ground and excited states were calculated at a fixed geometry, optimized for a ground state at the HF/STO-3G level. All calculations were done using the GAUSSIAN98 program [12].

The calculation of the molecular geometry (Fig.1.B) showed that the relative orientation of the heterocyclic aromatic groups is changing by the electronic excitation. This was caused by the changes in the dihedral angles and bond lengths of the polymethine chain, connecting the aromatic substituents, in a trans conformation. The geometrical reorganization of MC540 after the excitation was used for the development of the theoretical model of the fluorescence up-conversion signal. Namely, the excited state intramolecular potential displacement relatively to the ground state minimum was applied. The comparison of the results of the vibrational analysis in ground and in excited states showed a small difference of the vibrational normal mode distributions in the stable geometry's (Fig.1.B), therefore a Duschinsky effect was not involved in the model [13].

4 Model of the fluorescence dynamics

The frequency ω and the time-delay t_d dependent fluorescence up-conversion signal was modeled using results of Ungar and Cina [14], as follows:

$$F(\omega,t_d)=\int_{-\infty}^{\infty} dT' G(t_d-T')\tilde{F}(\omega,T'),$$

where $\tilde{F}(\omega,T')$ represents the theoretical formula for the deconvolved fluorescence spectrum, $G(t_d-T')$ is the intensity cross-correlation function of the exciting and gate laser pulses and T' is the average time interval over which excited state dynamics occurs.

The deconvolved fluorescence spectrum $\tilde{F}(\omega,T')$ is dependent on the excitation and gate pulse intensities and frequencies, and on the material response function. The latter one correlates operators of the type $U(t)=\exp(-iH_g t)\exp(iH_e t)$ and of their hermitian conjugates. Here, g and e denote the two lowest singlet electronic states of the chromophore. The $U(t)$ expresses the simultaneous time-evolution of chromophoric and solvent vibrational, rotational, etc. degrees of freedom, modulating the electronic transition energy.

The material response function is calculated averaging the variables of H_g and H_e. In this study, modifying the method [14], for H_g and H_e the second cumulant approximation, and the multi-mode Brownian oscillator (BO) description of Mukamel [15] was used. These oscillators represent harmonic intramolecular and solvent vibrations of atomic cores, determining the electronic transition frequency $\omega(t)$. Then, in the case of temporary well-separated Gaussian pulses:

$$\tilde{F}(\omega,T') = \int_{-\infty}^{\infty} dt \int_{-\infty}^{\infty} dt' e^{i\omega t} e^{i\Omega_{exc} t'} I_{gt}(t/2) I_{exc}(t'/2) \exp(-f_1(t,t',T')) \cos(f_2(t,t',T')),$$

where

$$f_1(t,t',T') = g_{Re}(-t) + g_{Re}(t') - g_{Re}(\frac{t'-t}{2} - T') + g_{Re}(\frac{t'+t}{2} - T') - g_{Re}(\frac{t'-t}{2} + T') + g_{Re}(\frac{t'+t}{2} + T')$$

and

$$f_2(t,t',T') = g_{Im}(-t) + g_{Im}(t') - g_{Im}(\frac{t'-t}{2} - T') + g_{Im}(\frac{t'+t}{2} - T') - g_{Im}(\frac{t'-t}{2} + T') + g_{Im}(\frac{t'+t}{2} + T')$$

The exponent $f_1(t,t',T')$ determines the spectral line broadening and $f_2(t,t',T')$ is causing the dynamic Stokes-shift of the fluorescence peak intensity. The time variables t and t' are combinations of pulse delays [15]. Further, $g_{Re}(t)$ and $g_{Im}(t)$ are the real and imaginary parts of the line-shape function $g(t)$, which is given by the auto-correlation function of $\omega(t)$.

In our model we assumed for $g(t)$:

- the same intramolecular vibrational modes for the displaced g and e state harmonic potentials
- no interdependence between intramolecular and solvent-dependent potential fluctuations and reorganizations, i.e. $g(t) = g_{intra}(t) + g_{sol}(t)$, which are additive for different oscillator frequencies, too
- that $g_{intra}(t)$ is determined by underdumped $C = C$ stretching dominated BO
- that $g_{sol}(t)$ is determined by overdumped BOs.

Explicitly, $g_{intra}(t)$ are oscillatory functions of frequencies ω_i^{BO}, slightly exponentially dumped by rate γ_i, and dependent on the dimensionless potential displacements d_i. These BOs generate the photoinitiated intramolecular reorganization related Stokes-shift λ_{intra} with contributions $\lambda_{intra,i} = d_i^2 \omega_i^{BO}/2$. The components of $g_{sol,i}(t)$ are defined both by the root-mean-squared amplitude Δ_i and by the correlation time Λ_i^{-1} of the solvent determined transition frequency fluctuations. Explicitly,

$$g_{sol,i}(t) = (\Delta_i^2/\Lambda_i^2)(\Lambda_i t - 1 + e^{-\Lambda_i t}) + i(\lambda_{sol,i}/\Lambda_i)(1 - e^{-\Lambda_i t}),$$

where $\Lambda_i = (\omega_i^{BO})^2/\gamma_i$ and $\lambda_{sol,i} = \Delta_i^2/2k_BT$, if $T >> 0$ K. λ_{sol} is usually composed of ultrafast (fs) and ps time-scale components, representing the inertial and diffusive dynamical Stokes-shifts, respectively [15].

Table 2. involves the parameters used for the calculation of the theoretical up-conversion signal of MC540 in water, having one underdumped and two overdumped BOs for the photoinitiated reorganization processes. The parameter values were originated from the displacement and from the Stokes-shift equal to 1400 cm^{-1}. The actual parameters of the underdumped and overdumped BOs were interrelated to cause stationary Stokes-shifts in an equal ratio.

Table 2. The BO parameters for the line-shape functions, used in the theoretical up-conversion signal

Underdumped	$d \in \langle 0.9; 1.2 \rangle$	$\gamma \in \langle 20; 80 \rangle cm^{-1}$	$\omega^{BO} = 1600 cm^{-1}$
1. overdumped	$\Lambda_1 \in \langle 0.0025; 0.005 \rangle fs^{-1}$	$\Delta_1 \in \langle 170; 225 \rangle cm^{-1}$	----
2. overdumped	$\Lambda_2 \in \langle 0.00014; 0.00033 \rangle fs^{-1}$	$\Delta_2 \in \langle 170; 225 \rangle cm^{-1}$	----

The signal calculation was realized by a self-made Fortran program. The program utilized a simplex method for the fitting of the BO and pulse intensity parameters and a 2D fast Fourier transform procedure [16]. The fitting procedure was based on the intervals of parameters presented in the table 2. Assuming comparable intramolecular and solvent related dynamical Stokes-shifts the fitting results had not a good quality. In order to reach better fit both the root-mean-squared transition frequency amplitude fluctuations and their correlation times, i.e. Δ_i and Λ_i, respectively, were increased by one order.

The frequency and time dependent 3D theoretical signal presented in figure 3A was reached using parameter values: $\Lambda_1 = 0.01$ fs^{-1}, $\Lambda_2 = 0.001$ fs^{-1}, $\gamma = 70$ cm^{-1}, $\Delta_1 = 5000$ cm^{-1}, $\Delta_2 = 11000$ cm^{-1} and $d = 1$. The slices of this fit, as well as of signals fitted by lower number of BOs are shown in figure 3B. The first fit was made with one underdumped BO, the second one with one underdumped and one overdumped and the third one with parameters showed above. The increasing number of BOs resulted better correspondence with the experimental signal for long and medium wavelengths. The trend was opposite for the short wavelengths.

The typical fast decay of the signal was possible to reproduce theoretically for the short wavelengths, corresponding to the fs-scale inertial solvent reorganization. The untypical behaviour in the experimental signal was the spectral narrowing, i.e. the fast decay also in the long wavelength wing. For this phenomenon our theoretical model should have to be modified, e.g. by including some another non-radiative relaxation channel (intersystem crossing, intramolecular charge-transfer, etc.). These effects would be possible to incorporate into the model including extra relaxation terms in the correlation function of the transition frequency fluctuation. Finally, the observed differences between the aqueous and ethanolic solutions could be discussed in the terms of BO model. The vibrational spectra of liquid water differentiated it from other polar solvents [17]. The ultrafast dynamics of the aqueous system resulted from large amplitude, high-frequency intermolecular librational and vibrational motions, which were expressed in the presented model with used overdamped oscillators.

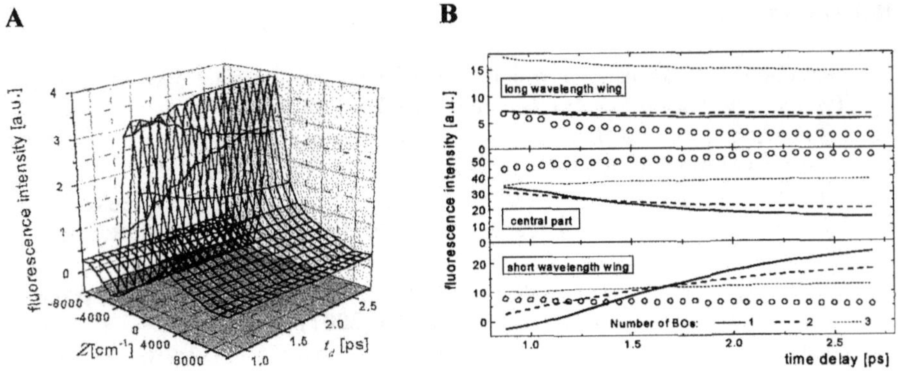

Figure 3. The optimal fitting surface of the time- and spectral-resolved fluorescence data (**A**) for the all examined and (**B**) slices for some different parts of steady-state spectra (circles – experimental data).

5 Conclusions

The time- and spectral-resolved fluorescence of MC540 was measured on femto- and picosecond time scales in water and in ethanol. In the case of the aqueous solution a spectral narrowing behaviour was observed on the 3 ps time-scale and an exponential decay on the 300 ps time-scale with time constant 84.4 ± 3.3 ps. Between the three examined concentrations were observed no significant differencies indicating, that the changes in the ratio of dimer-monomer did not effect the fluorescence properties of MC540. In the case of ethanolic solution was no time-dependence of the fluorescence on the examined time scales was found. Further quantum chemical computations of the structural changes of the MC540 chromophore and its vibrational spectra were realized. The negligible differences between the ground and excited state vibrational spectra allowed modeling of the time-resolved fluorescence without Duschinsky effect. The calculations were based on the BO model, by which the spectral-resolved curves was possible to fit on the central part and on the short wavelength wing of the steady state spectra.

Acknowledgements

T.P. and V.Sz. thank the Austrian Ministry of Science and Transport for financial support in the framework of Ost-West-Kooperation-H, bm:wv, GZ 45.452/2-III/2/99. The support of the Science Grant Agency of Slovak Republic (Vega 1/7355/20) is also acknowledged.

References

1. Harriman A., Photoisomerization dynamics of merocyanine dyes in solution, *J. Photochem. Photobiol. A: Chem.* **65** (1992) pp. 79-93.
2. Sikurova L., Haban I. and Chorvat D., Dimers of merocyanine 540 I. in aqueous solutions, *Studia Biophysica* **125** (1988) pp. 197–201.
3. Sieber F., Spivak J. L. and Stucliffe A. M., Selective killing of leukemic cells by merocyanine 540-mediated photosensitization, *Proc. Natl. Acad. Sci. USA* **81** (1984) pp. 7584-7587.
4. Cunderlikova B. and Sikurova L., Solvent effects on photophysical properties of merocyanine 540, *Chemical Physics* **263** (2001) pp. 415–422.
5. Mandal D., Pal S. K., Sukul D. and Bhattacharyya K., Photophysical processes of merocyanine 540 in solutions and in organized media, *J. Phys. Chem. A* **103** (1999) pp. 8156–8159.
6. Mach P., Urban J. and Leszczynski J., Quantum chemical study of ground state merocyanine 540 model compounds, *Int. J. Quant. Chem.* **75** (1999) pp. 741–750.
7. Bugar I., Kovac J., Lengyel O. and Lee S. T., Time-resolved fluorescence study of organics for light emitting and laser devices, *Laser Physics*, in edition
8. Chorvat D. Jr., Chorvatova Z., Multiexponential analysis of fluorescence data, *Acta Physica Univ. Comen.* **36** (1995) pp. 105-110.
9. Hehre W. J., Stewart R. F. and Pople J. A., Self-consistent molecular-orbital methods I. Use of Gaussian expansions of slater-type atomic orbitals, *J. Chem. Phys.* **51** (1969) pp. 2657-2664.
10. Foresman J. B., Head-Gordon M., Pople J. A. and Frisch M. J., Toward a systematic molecular orbital theory for excited states, *J. Phys. Chem.* **96** (1992) pp. 135-149.
11. Trucks G. W. and Frisch M. J., Analytic second derivatives of excited states: configuration interaction singles theory and application, in prep. (1998)
12. Frisch M. J. at el, Gaussian, Inc., Pittsburgh PA, (1998)
13. Gribov L. A., Orville-Thomas W. J., Theory and methods of calculation of molecular spectra (John Wiley and Sons, New York, 1988)
14. Ungar L. W. and Cina J. A., Short-time fluorescence stokes shift dynamics, *Advances in Chemical Physics* **100** (1997) pp. 171–228.
15. Mukamel S., Principles of nonlinear optical spectroscopy (Oxford University Press, New York, 1995)
16. Press W. H., Teukolsky S. A., Vetterling W. T. and Flannery B. P., Numerical recepies in Fortran (University Press, Cambridge, 1992)
17. Lang M. J., Jordanides X. J., Song X. and Fleming G. R., Aqueous solvation dynamics studied by photon echo spectroscopy, *J.Chem.Phys.* **110** (1999) pp. 5884-5891.

PHOTODISSOCIATION OF HALOBENZENES IN SOLUTION

N. LASCOUX[a], M. RASMUSSON[a], R. LINDH[a], A. N. TARNOVSKY[a], M. KADI[b], O. KÜHN[c], V. SUNDSTRÖM[a], E. ÅKESSON[a,*]

[a]*Department of Chemical Physics, Lund University, Box 124, 221 00 Lund, Sweden*

[b]*Department of Physical Chemistry, Uppsala University, Box 532, 751 21 Uppsala, Sweden*

[c]*Institut für Physikalische und Theoretische Chemie, Freie Universität, Takustr. 3, 141 95 Berlin, Germany*

Abstract: We have studied the photodissociation of bromobenzene in solution after excitation at 266 nm. The kinetics in different solvents exhibit the following features: a pulse limited rise due to excited state absorption, a single exponential decay of 9±1 ps which we assign mainly to predissociation, and a long lived signal. Additional features in acetonitrile suggest the involvement of lower triplet states or a charge transfer complex between the phenyl ring and the bromine atom. *Ab initio* calculations of the lower singlet and triplet states of bromobenzene have guided us throughout the interpretations as well as complementary measurements in iodobenzene.

1 Introduction

The usefulness of aryl halides and particularly halobenzenes in chemical synthesis has made them the objects of numerous studies in molecular beams [1,2]. It is now known that excitation at 266 nm produces a bound (π,π^*) state. Dissociation can then occur by a predissociative mechanism due to spin-orbit coupling between the initially excited state and a repulsive (n,σ^*) triplet state. Therefore halobenzenes are good model compounds for comparing gas phase and solution and mostly for studying the influence of the solvent molecules on photodissociation. Besides, choosing the halide atom gives you the possibility to change the spin-orbit coupling and the dissociation time. Indeed, in the case of iodobenzene, two dissociation channels both leading to subpicosecond bond cleavage have been found [1]. In contrast, the weak spin-orbit coupling of chlorobenzene leads to very slow predissociation, probably in competition with other deactivation channels [2,3]. For bromobenzene, which has an intermediate spin-orbit coupling strength, the average dissociation time has been measured around 30 ps in a molecular beam [2]. To our knowledge, no previous time-resolved investigations of predissociation occurring on a similar time scale have been carried out in the condensed phase. In performing such a study on bromobenzene, we first aim at comparing the dissociation times in different environments but also at elucidating the processes that follow or compete with predissociation in solution. For example, the involvement of lower triplet states remains unclear and needs further investigation. In cold matrices, it had previously been shown that excitation of bromobenzene at 266 nm resulted in weak phosphorescence from two low-lying triplet states [4]. The exact mechanisms producing these triplets are unknown. Another motivation to our study is to obtain further information about the absorption spectrum of the phenyl radical from the transient absorption. Several previous investigations on the phenyl radical spectrum have yielded rather different results [5-10].

In this work, we present femtosecond transient absorption measurements of bromobenzene after excitation at 266 nm, mainly in acetonitrile (λprobe=300-930 nm), but also at two different probe wavelengths in other solvents. Transient spectra of iodobenzene in acetonitrile have also been measured for comparison, since the phenyl radical is formed upon dissociation of both halobenzenes after excitation at 266 nm in gas phase. Finally all the results have been related to ab initio calculations of the potential energy surfaces and oscillator strengths.

2 Experimental

The optical set-up for our femtosecond transient absorption measurements has been described elsewhere [11]. Briefly, our spectrometer is based on an amplified Ti:Sapphire laser and regenerative amplifier system (800 nm, 1kHz, 100 fs, 0.9 W) whose beam is split into two. The pump light at 266 nm is then generated through a frequency tripler. For the probe beam we have two possibilities depending on the kind of experiments performed: for kinetics measurements the probe light is quasi-monochromatic and generated through an OPA (300-1200 nm); for spectra measurements, we use a white-light continuum (330-1000 nm). In the first case, the probe and reference signals are measured on two photodiodes placed behind a spectrometer. In the second case, the probe and reference spectra are measured on a diode array after diffraction by a grating. The detection scheme used has been previously described [12] and recent refinements can be found in reference [13]. The plane of polarization of the pump beam could be varied with a Berek polarization compensator and that of the probe beam was defined using a Glan-Thomson polarizer. The excitation light was focused using a f=150 mm lens and overlapped with the focused probe beam (f=75 mm) in the sample. The angle between the beams was ~30°. The sample was contained in a 1 mm flow cuvette (Spectrosil quartz) with 1.25 mm thick walls. We ensured that the cuvette itself gave no contribution to the measured signals.

The solvents were purchased from Merck, iodobenzene from Fluka and bromobenzene from Janssen Chimica. All chemicals were used without further purification. In our pump-probe measurements, the concentration of bromobenzene was typically 40 mM.

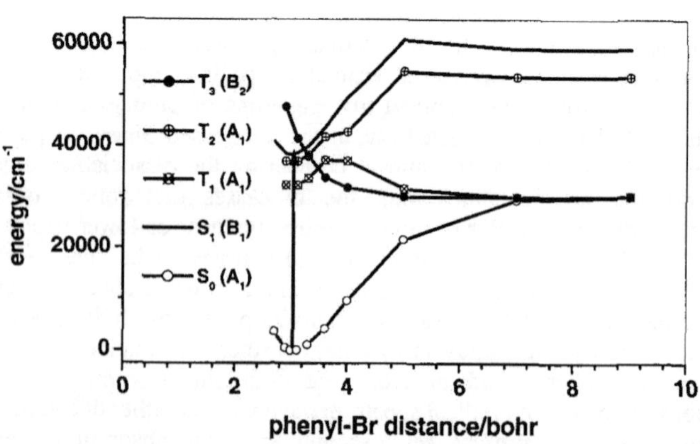

Figure 1. *Ab initio* potential energy curves along the dissociation coordinate for the lower electronic states of bromobenzene.

3 Results and discussion

The results of our *Ab initio* calculations are shown in fig.1 (for more details see ref. 11). The excitation (vertical arrow) promotes the molecules from the ground singlet state S_0 to the first bound singlet state S_1; conversion to the dissociative (π,π^*) triplet state T_3 can then occur via spin-orbit coupling. The calculations have also shown the presence of two lower bound triplet states T_1 and T_2, which are probably responsible for the dual phosphorescence observed by Okutsu *et al.* in glassy matrices at 77 K after excitation of bromobenzene at 266 nm [4].

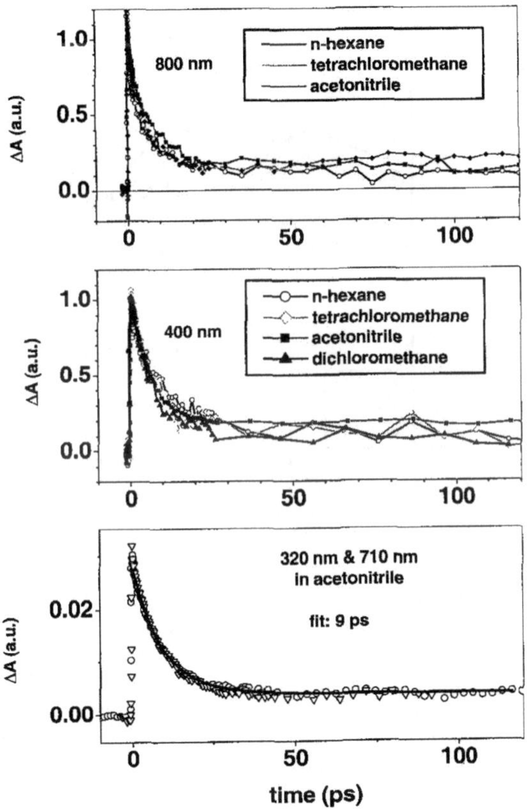

Figure 2. Transient absorption of bromobenzene in solution after excitation at 266 nm. For comparison, the kinetic traces in each panel were scaled to coincide at t=0 after removal of the solvent spike at t=0. Upper panel: kinetics in different solvents at λ_{probe}= 800 nm. Middle panel: kinetics in different solvents at λ_{probe}= 400 nm. Lower panel: kinetics in acetonitrile at 320 nm (triangles) and 710 nm (circles) and a corresponding fit (solid line) consisting of a constant contribution and a single exponential decay (9±1 ps).

Kinetics of bromobenzene in different solvents (n-hexane, dichloromethane, tetrachloromethane and acetonitrile) have been measured at 400 nm (Fig 2 middle panel) and 800 nm (Fig 2 upper panel) and our first observation is the absence of major solvent dependence. Indeed, all kinetics exhibit a first instantaneous rise due to the excitation of the molecules to the first singlet state S_1, then a single exponential decay (9±1 ps measured in acetonitrile at 320, 400, 710 and 800 nm) due to the conversion to the triplet states T_{1-3}, most probably to the dissociative triplet state T_3. This time is shorter than in the gas phase where the dissociation takes place in 28 ps [2], which could be expected because the solvent molecules have a strong influence on the spin-orbit coupling and the solvent environment might facilitates intersystem crossing from S_1 to one or several of the three triplet states. Finally, the last feature of these kinetics is a long lived signal which survives after 200 ps.

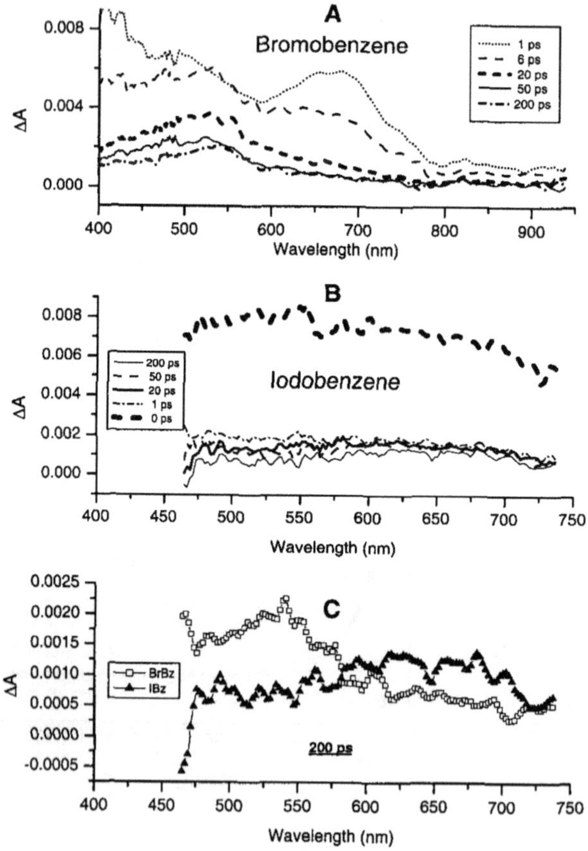

Figure 3. A- Transient spectra of Bromobenzene in acetonitrile measured with a diode array. Separate measurements in different spectral regions were joined together. Time delays: 1 ps, 6 ps, 20 ps, 50 ps and 200 ps. B- Transient spectra of Iodobenzene in acetonitrile measured in the same conditions. Time delays: 1 ps, 20 ps, 50 ps and 200 ps. C- Comparison between the spectra of Bromobenzene and Iodobenzene in acetonitrile at delay time 200 ps.

In acetonitrile we also measured spectra on the wavelength range 290-930 nm. They are partly given in fig. 3, panel A. Again we can see the global evolution consisting of a pulse limited rise followed by a 9 ps exponential decay and a long lived signal with a maximum at 530 nm, but at intermediate wavelengths, the dynamics look more complicated and the spectral shape is changing. This particular behavior will be studied in more detail later after evoking the case of the long lived signal.

The first candidate for explaining this long lived contribution is the phenyl radical. In an early work by Porter and Ward, the phenyl radical was reported to have an absorption band in the visible spectral region (425-525 nm) in the gas phase [5]. Absorption was also observed at 488 nm in another gas phase study [6]. Later studies, both in the gas-phase [7,9] (examining wavelengths below 360 nm and 570 nm, respectively) and in a matrix [10] (examining the interval between 190 nm and 2 500 nm), reported only major absorption bands in the UV spectral region; the first maximum was found at ~235 nm, but the extinction coefficients in refs. 9 and 10 differed by several orders of magnitude. However, if the assumption that the phenyl radical is responsible for the long lived signal is true, we should observe the same behavior in iodobenzene. Indeed, iodobenzene is known to dissociate very fast in the gas phase [1,2]. Therefore we have performed transient spectra measurements of iodobenzene in acetonitrile. The results are shown in fig 3, panel B. We observed a pulse limited rise and decay followed by a long lived signal. All spectra are very flat over the whole measured wavelength region (only the region of 350-740 nm is shown), which differs considerably from the observations in bromobenzene. For a better comparison between bromobenzene and iodobenzene, we have plotted both of these long lived signals together (Fig 3, panel C). The signal amplitudes after 5 ps are of similar magnitude for both molecules in the wavelength interval 380-450 nm, but considerably higher for bromobenzene around 520 nm, cf. Figure 3, panels A and B. Besides, the signal is almost negligible between 290-350 nm for iodobenzene but not for bromobenzene (not shown), whereas the amplitude of the long lived signal around 650 nm is bigger for iodobenzene, cf. Figure 3, panel C. From these results, we conclude that the phenyl radical does not give rise to a major contribution to the long-lived signals for neither iodobenzene nor bromobenzene between 290 and 930 nm and we assign the long lived signal in bromobenzene to most probably one of the lower bound triplet states.

Taking a closer look at the spectra of bromobenzene in acetonitrile (Fig. 3-A) we noticed an additional feature only present in this solvent: a band between 400 and 600 nm with a maximum at 530 nm that rises, reaches its maximum at 10 ps and then decays with a time constant of 10-20 ps. The presence of this band is also clearly manifested in the kinetics by a hump at intermediate wavelengths (Fig 4). Several candidates have to be considered for explaining this evolution. It could for instance be due to absorption from the long-lived triplet state during its spectral relaxation or the other low-lying bound (π,π^*) triplet state. Population of the two lower triplet states by intersystem crossing from S_1 is symmetry allowed. Internal conversion between the three lower triplet states and predissociation followed by recombination to the two lower triplet states is also possible. Another candidate for explaining the spectral features observed on the 10-20 ps time scale is a charge-transfer complex between the phenyl radical and a bromine atom, since charge-transfer complexes between bromine atoms and various substituted benzenes generally exhibit absorption around 530 nm [14]. In performing preliminary *ab initio* calculations on a complex between a bromine atom and a phenyl radical, we found a weak energy minimum for the triplet state complex with the bromine atom positioned above the π-system of the phenyl ring. If the 10-20 ps time scale dynamics are assigned to such a complex, the decay could be explained as being due to recombination.

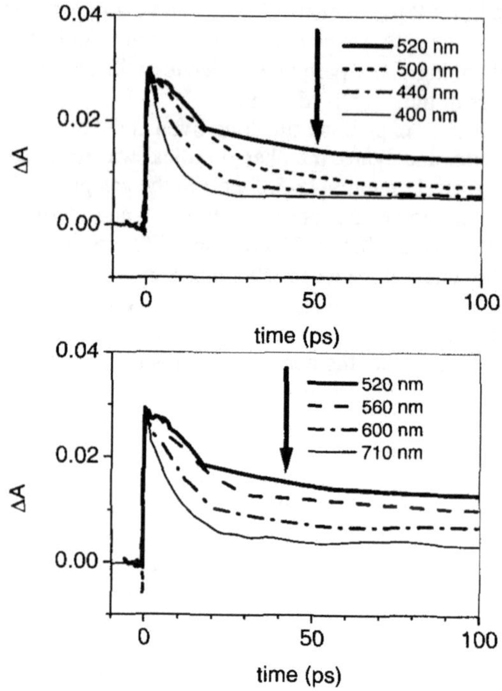

Figure 4. Transient absorption kinetics in acetonitrile. For comparison, the kinetics are scaled to coincide at t=0 (after removal of minor solvent contributions close to t=0 from some of the traces. For clarity, smoothed data are shown. Upper panel: kinetics in the region 400-520 nm. Lower panel b: kinetics in the region 520-710 nm.

4 Conclusion

We have performed femtosecond transient absorption measurements of bromobenzene in solution between 290 and 930 nm, upon excitation at 266 nm. The features common to all the measured kinetics are: a pulse limited rise due to the promotion of the molecules to the first excited bound singlet state, a single exponential decay of 9±1 ps which we assign mainly to predissociation via spin-orbit coupling to a dissociative triplet state, and a long lived signal with a maximum at 530 nm due to one of the low lying triplet states found by our calculations. In acetonitrile however, we observed the rise and the decay of an additional band between 400 and 600 nm within 10-20 ps, which suggests the involvement of the lower triplet states or a charge transfer complex between the phenyl ring and the bromine atom. Similar experiments in iodobenzene in acetonitrile have shown a pulse limited rise and decay followed by a flat long lived signal, which ruled out the involvement of the phenyl radical in the long lived signal.

5 Acknowledgements

This work was supported by the Swedish Natural Science Research Council, Trygger Foundation, Magnus Bergwall Foundation, Knut and Alice Wallenberg Foundation, Crafoord Foundation and the Swedish Institute.

References

1. P. Y. Cheng, D. Zhong, A. H. Zewail, Chem. Phys. Lett. 237 (1995) 399
2. M. Kadi, A. N. Tarnovsky, M. Rasmusson, J. Davidsson, E. Åkesson, accepted by Chem. Phys. Lett.
3. T. Ichimura, Y. Mori, H. Shinohara, N. Nishi; Chem. Phys. 189 (1994) 117
4. T. Okutsu, T. Kageyama, N. Kounose, J. Tsuchiya, H. Hiratsuka, Chem. Phys. Lett. 299 (1999) 597
5. G. Porter, B. Ward, Proc. Roy. Soc. A 287 (1965) 457
6. M. Preidel, R. Zellner, Ber. Bunsenges. Phys. Chem. 93 (1989) 1417
7. N. Ikeda, N. Nakashima, K. Yoshihara, J. Am. Chem. Soc. 107 (1985) 3381
8. B. Cercek, M. Kongshaug, J. Phys. Chem. 74 (1970) 4319
9. T. J. Wallington, H. Egsgaard, O. J. Nielsen, J. Platz, J. Sehested, T. Stein, Chem. Phys. Lett. 290 (1998) 333
10. G. Radziszewski, Chem. Phys. Lett. 301 (1999) 565
11. M. Rasmusson, R. Lindh, N. Lascoux, A. N. Tarnovsky, M. Kadi, O. Kühn, V. Sundström, E. Akesson, in preparation
12. U. Åberg, E. Åkesson, J.-L. Alvarez, I. Fedchenia, V. Sundström, Chem. Phys. Lett. 183 (1994) 269
13. A. N. Tarnovsky, M. Wall, M. Rasmusson, T. Pascher, E. Åkesson, J. Chin. Chem. Soc. 47 (2000) 769
14. K. D. Raner, J. Lusztyk, K. U. Ingold, J. Phys. Chem., 93 (1989) 564

DYNAMICS OF S_2 FLUORESCENCE QUENCHING AND OF CHARGE RECOMBINATION IN THE RESULTING ION PAIR

LAURINE ENGELI, ANA MORANDEIRA, PIERRE-ALAIN MULLER AND ERIC VAUTHEY

Department of Physical Chemistry of the University of Geneva, 30 quai Ernest Ansermet, CH-1211 Genève 4, Switzerland

The charge recombination (CR) dynamics of geminate ion pairs formed upon electron transfer quenching of benz[a]azulene, xanthione and Zn tetraphenylporphine in the S_2 state has been investigated. The ion pairs have two possible CR pathways: A) a highly exergonic CR to the neutral ground state and B) a moderately exergonic CR leading to the formation of neutral product with one of the components in the S_1 state. This investigation shows clear evidences of the predominance of path B. The relevance of these results to account for the absence of the inverted region in highly exergonic bimolecular electron transfer reactions is discussed.

1 Introduction

One of the remaining problems in bimolecular photoinduced electron transfer (ET) reactions is the absence of the inverted region, predicted by Marcus theory [1], in the high exergonicity regime [2]. The hypothesis of an increase of the ET distance with increasing free energy discrepancy can no longer be invoked for very large exergonicity ($\Delta G < -2$ eV). It has been proposed that, for these processes, the product is formed in an electronic excited state. In this case, the reaction is no longer highly exergonic and thus does not occur in the inverted regime. The product of charge separation reaction is a geminate pair of open-shell radical ions. These ions are known to have low energy electronic states. However, excited radical ions are difficult to detect in the condensed phase. They are in general very short-lived and do not, apart from a few exceptions, exhibit emission [3,4]. The situation is more favourable for charge recombination (CR) of geminate ion pairs (GIP), because the product molecules are closed shell species. However, in order to investigate CR to an electronically excited state, the GIP has to have a high energy and to lie above an electronic excited state of one of the neutral partners. One possibility is to generate the ions electrochemically, but in this case ionic recombination is homogenous and leads to the generation of triplet and singlet excited states, due to spin statistics. Moreover, the reaction being diffusion controlled, information on the actual CR dynamics is not accessible. The GIP can also be formed by ET quenching of a molecule in a triplet excited state. If the resulting triplet GIP is above the triplet state of the quencher, there are two CR pathways. However, only the pathway leading to the formation of the neutral state with the quencher in the triplet state is spin allowed. We have recently shown that only this spin allowed CR occurs [5].

We have investigated the dynamics of singlet GIPs that have two spin allowed CR pathways (see Figure 1): A) a highly exergonic CR to the neutral ground state and B) a moderately exergonic CR leading to the formation of the neutral product with one of the components in the S_1 state. Contrarily to the above-mentioned system, the competition between both CR pathways is mainly controlled by their relative exergonicity.

These GIPs are formed by ET quenching of benz[a]azulene (BA), xanthione (XA) and Zn tetraphenylporphine (ZnTPP) in the second singlet excited state by electron donors (with BA and XA) and electron acceptors (with ZnTPP).

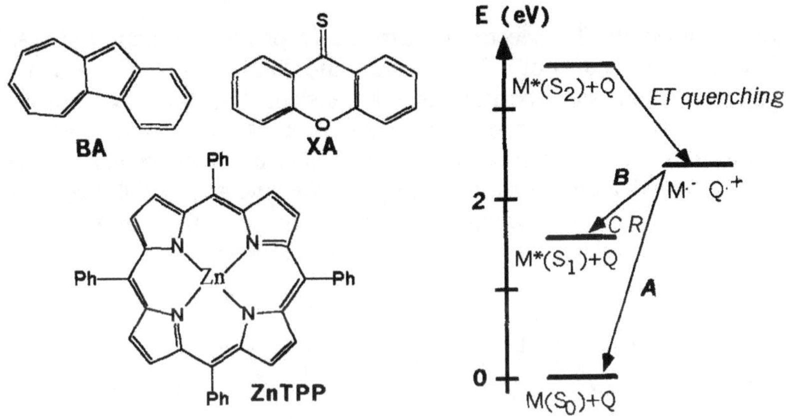

Figure 1: Energy diagram of the states involved in the S_2 quenching of M = BA, XA and ZnTPP (With ZnTPP, Q is the acceptor).

2 Experimental

The S_2 ET quenching of BA and XA and the resulting CR dynamics were investigated using the multiplex transient grating (TG) technique [6]. The TG experimental set-up has been described in detail in ref.[7]. Excitation was performed around 400 nm using the frequency-doubled output of a conventional 1 kHz Ti:Sapphire amplifier. The instrument response function (IRF) had a width of about 250 fs.

The S_2 quenching of ZnTPP has been investigated by fluorescence up-conversion. Excitation was performed with the frequency-doubled output of a Ti:Sapphire oscillator at 82 MHz. The IRF was around 200 fs with 0.4 mm thick samples and around 350 fs with 1 mm thick samples.

The quenchers were anisol (ANI), veratrol (VER), N,N-dimethylaniline (DMA), 1,4-dimethoxybenzene (DMB), 1,4-bicyclo[2,2,2]octane (DABCO) and acetophenone (ACP). ACP was used as electron acceptor (A) and the other quenchers as electron donors (D). All measurements were performed using acetonitrile (ACN) as solvent.

3 Results and discussion

3.1 *Benza[a]azulene (BA)*

Figure 2A shows the TG spectrum measured 15 ps after excitation of BA in ACN [8]. The spectral features of a TG spectrum are dominated by absorption, the band maxima being at the same position as in a transient absorption spectrum [6]. Figure 2B shows the time profile of the square root of the TG intensity, which is proportional to the concentration of the transient. This signal decays mono-exponentially with a time constant of 74 ps. BA is known to exhibit S_2 fluorescence, but with a much smaller quantum yield than azulene. This is in agreement with a shorter lifetime and thus this transient is ascribed to BA*(S_2).

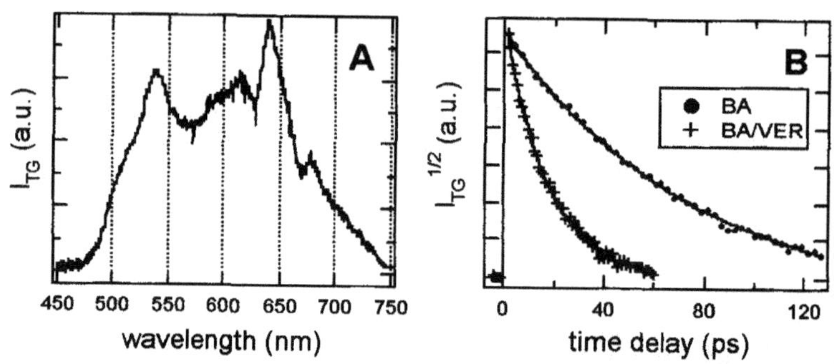

Figure 2: A) TG spectrum measured 15 ps after excitation of BA in ACN; B) decays of the square root of the TG intensity at 538 nm and best fits (from ref. [8]).

Upon addition of high concentrations of quencher (VER and DMA), the TG spectra remain the same but the band decays more rapidly (see Figure 2B) with time constants of 16 and 14 ps with 1 M VER and DMA, respectively. No other TG band was observed that could be ascribed to another species such as $VER^{\cdot+}$, $DMA^{\cdot+}$ and especially $BA^{\cdot-}$, which has a broad absorption band in the 500-550 nm region. However, ET as the quenching mechanism of BA*(S_2) is supported by the fact that the quenching efficiency increases with decreasing oxidation potential of the donor. With ANI, ET is endergonic and no quenching can be observed. Therefore, the absence of ion pair TG band can be explained by an ultrafast CR with a time constant much shorter than 14-16 ps. This is surprisingly fast if the exergonicity of

CR to the neutral ground state (path A) is considered. Indeed, CR of the pair perylene^{-}/DMA^{+}, which as a very similar exergonicity, occurs with a time constant of 4 ns, as expected for an ET in the inverted regime [7].

On the other hand, the exergonicity for CR according to path B is substantially smaller and is close to the barrierless region. In this case, CR with sub-picosecond time constants have been measured [7]. Consequently, these results strongly suggest that CR occurs predominantly via path B, i.e. results to BA*(S_1)+D. The lifetime of BA*(S_1) is not known, but it can be expected to be as short as that of azulene, which varies between 400 fs and 1.7 ps depending on the extent of vibrational energy [8]. This could explain why this state is not observed after the CR of BA^{-}/D^{+} pairs.

3.2 Xanthione (XA)

The TG spectrum measured with XA in ACN exhibits a broad band with a maximum at 610 nm, whose square root intensity decays with a time constant of 13 ps (see Figure 3A) [8]. The same lifetime has been reported for the S_2 fluorescence of XA in ACN [10] and therefore this transient can be ascribed to XA*(S_2). Upon addition of 1 M electron donor, its lifetime drops to 5.5 ps with VER and DABCO and to 3.5 ps with DMB and DMA.

With the weaker donors, VER and DMB (D_W), no new band was observed in the TG spectrum, although XA^{-} exhibits a broad absorption band in the 450-600 nm region [8].

The situation is different with the strongest donors, DMA and DABCO (D_S). In this case, a broad and weak TG band can be observed after the complete decay of the XA*(S_2) band. Its intensity remains constant over the whole time window of the TG experiment (up to 200 ps). This band can be reasonably ascribed to XA$^{\cdot-}$. At short time delay, it is completely hidden by that of XA*(S_2) and therefore its early dynamics cannot be observed.

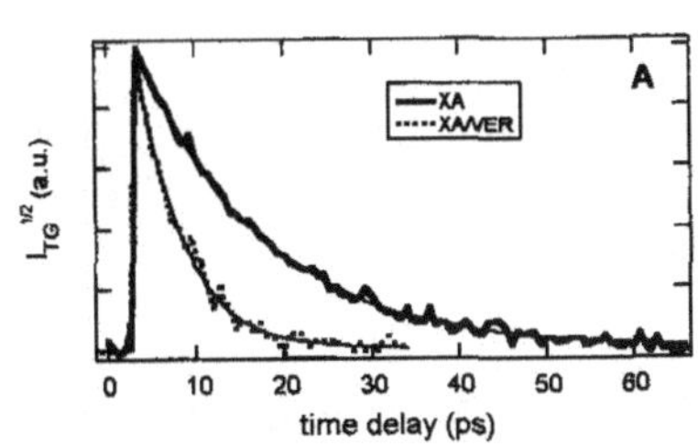

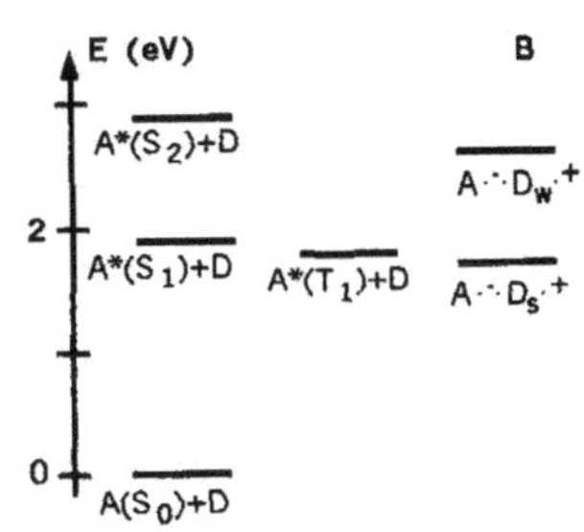

Figure 3: A) decays of the square root of the TG intensity with XA and with XA + 1M VER in ACN; B) Energy diagram of the states involved in the ET reaction of XA with weak and strong donors (from ref. [8]).

The free ion yield, determined by photoconductivity, depends strongly on the concentration of D_S. With DABCO, it first increases from 0.20 at 1 mM to 0.27 at 0.13 M and then decreases down to 0.11 at 1 M. A similar concentration dependence is observed with DMA. This effect can be understood by considering the energy diagram in Figure 3B. With D_S as quencher, the resulting GIPs are located below XA*(T_1)+D, and thus D_S can quench both XA*(S_1) and XA*(T_1). Both XA*(S_2) and XA*(S_1) being very short lived, they can only be quenched at high donor concentrations. This results to a singlet GIP, where CR is spin-allowed and moderately exergonic. Therefore, CR is much faster than dissociation into free ions and the free ion yield is small.

As XA*(T_1) is long lived, it can be efficiently quenched at low donor concentrations. This results to a triplet GIP, where CR is spin forbidden and thus slower than dissociation in to free ions. In this case, the free ion yield is high.

The weak donors can only quench XA*(S_2) and the ensuing GIPs are located above XA*(S_1)+D. As XA*(S_1) is very short lived and as XA*(T_1) cannot be observed in the spectral region of the TG set-up, the CR pathway of the $XA^{\cdot-}/D_W^{\cdot+}$ GIPs was determined by measuring the free ion yield with the system XA + 1M D_W + 1mM DABCO. The low DABCO concentration ensures that it only interacts with XA*(T_1). Therefore, the free ion yield is a measure of the triplet yield. This yield was measured to be independent on the quenching of XA*(S_2) by D_W, This indicates that ET quenching of XA*(S_2) generates a GIP that populates XA*(S_1) and XA*(T_1), i.e. that CR occurs along path B. The whole process can be summarised by the following scheme:

$$XA^*(S_2) \xrightarrow{+D_W} XA^{\cdot-}D_W^{\cdot+} \xrightarrow{CR} XA^*(S_1) \xrightarrow{ISC} XA^*(T_1) \xrightarrow{+D_S} XA^{\cdot-}D_S^{\cdot+} \xrightarrow{diss} XA^{\cdot-} + D_S^{\cdot+}$$

If CR was following path A, the XA*(S_1) and XA*(T_1) states would not be populated and therefore the free ion yield would be much smaller.

3.3 *Zn tetraphenylporphine (ZnTPP)*

Figure 4A shows the fluorescence spectrum of ZnTPP in ACN excited at 410 nm in the Soret band. The band centred at 430 nm is S_2 fluorescence, while the band in the 600-650 nm region is fluorescence from S_1. Figure 4B shows the time profile of the S_2 fluorescence. The ZnTPP*(S_2) state has a lifetime of 2.4 ps in ACN, the main deactivation channel being the internal conversion to S_1. Indeed, the time profile of the S_1 fluorescence at 600 nm (see Figure 4C) exhibits a rise with the same time constant.

Upon addition of large concentrations of acetophenone (ACP), the intensity of the S_2 fluorescence band decreases significantly, while that of S_1 fluorescence remains essentially the same. This S_2 quenching is due to the ET from ZnTPP*(S_2)

to ACP. The fact that the S_1 band is not affected by S_2 quenching is an indication that the quenching product, the GIP, undergoes CR to ZnTPP*(S_1) via path B.

This is further confirmed by up-conversion measurements of S_2 and S_1 fluorescence dynamics. Figure 4B shows the decay of the S_2 fluorescence with 4 M ACP in ACN. It is biexponential with time constants of 810 fs (60% of the total amplitude) and 2.35 ps. Considering the high donor concentration and the very short lifetime of ZnTPP*(S_2), the quenching is essentially static. Therefore, the larger decay time probably corresponds to the fraction of excited molecules that do not undergo ET. At higher ACP concentrations (6 M), both decay components are faster, around 250 fs (55% of the total amplitude) and 1.36 ps. The observed shortening of the fluorescence lifetime can account only partially for the large decrease of the S_2 fluorescence yield measured in the steady state spectra. This could be due to the presence of an important decay component of S_2 that is too fast to be resolved. For these measurements, the width of the IRF was around 350 fs (1mm thick samples). Moreover, the S_2 emission partially overlaps with a Raman band of ACP, whose contribution to the up-converted signal makes the extraction of fast fluorescence components difficult.

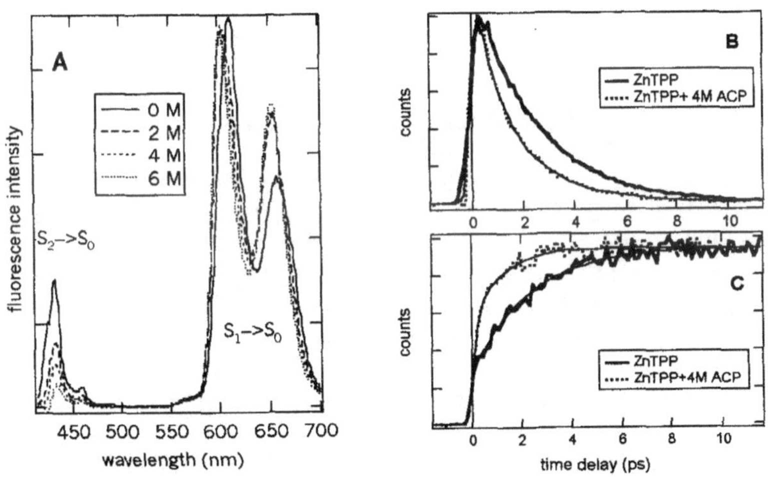

Figure 4: A) Steady-state fluorescence spectra of ZnTPP with various concentrations of ACP; Time profiles of the fluorescence at B) 430 nm and C) 610 nm of ZnTPP with and without ACP.

Figure 4C shows the fast dynamics of the S_1 fluorescence with 4 M ACP in ACN. The presence of ACP leads to a clear acceleration of S_1 formation. More than half of the total amplitude (55%) is reached with a time constant that is too fast to be time-resolved, the remaining intensity rises with a time constant of about 900 fs. At 6 M ACP, only the ultrafast rise is observed.

The acceleration of S_2 decay and the S_1 rise upon addition of electron acceptor clearly confirms that CR of the GIP formed upon ET quenching of ZnTPP*(S_2) occurs via path B. The close similarity of S_2 decay times and S_1 rise times indicates that CR is substantially faster than the quenching process. With an energy of about 2.6 eV, the GIP state is located almost exactly halfway between the S_2 and S_1 states of ZnTPP and thus acts as an efficient ladder for the non-radiative transition from S_2 to S_1.

4 Concluding remarks

We have presented clear evidences of the formation of an electronically excited product in donor/acceptor systems where ET to the ground state is highly exergonic. The presence of a low lying excited product state shifts the CR from the inverted to the barrierless regime.

This observation strongly suggests that this could also occur for highly exergonic charge separation processes, when low lying excited states of the radical ions are available. As most radical ions used in ET quenching experiments have excited states below 2 eV, the formation of electronically excited radical ions in highly exergonic ET processes might be a rather general process.

1. R. A. Marcus and N. Sutin, Biochim. Biophys. Acta *811* (1985) 265.
2. D. Rehm and A. Weller, Isr. J. Chem. *8* (1970) 259.
3. J.-C. Gumy and E. Vauthey, J. Phys. Chem. A *101* (1997) 8575.
4. P. Brodard, A. Sarbach, J.-C. Gumy, T. Bally and E. Vauthey, J. Phys. Chem. A *105* (2001) 6594
5. C. Högemann and E. Vauthey, J. Phys. Chem. A *102* (1998) 10051.
6. C. Högeman, M. Pauchard and E. Vauthey, Rev. Sci. Instrum. *67* (1996) 3449.
7. E. Vauthey, J. Phys. Chem. A *105* (2001) 340.
8. P.-A. Muller and E. Vauthey, J. Phys. Chem. A 105 (2001) 5994.
9. A. J. Wurzer, T. Wilhelm, J. Piel and E. Riedle, Chem. Phys. Lett. *299* (1999) 296.
10. C. J. Ho, A. L. Motyka and M. R. Topp, Chem. Phys. Lett. *158* (1989) 559.

ANALYSIS AND CONTROL OF CONDENSED-PHASE PHOTOCHEMICAL REACTIONS: ClF IN AR

M. BARGHEER, J. PIETZNER, P. DIETRICH AND N. SCHWENTNER
Institut für Experimentalphysik, FU-Berlin, Arnimallee 14, 14195 Berlin
Tel.: +49 30 83853530, Fax.: +49 30 83853050
E-mail: bargheer@physik.fu-berlin.de

We establish chlorine monofluoride (ClF) molecules isolated in rare-gas solids as a model system to study condensed phase photochemistry. It is a reactive system with a multidimensional potential energy surface typical for condensed phases which is now amenable to detailed theoretical description. We characterize the system spectroscopically. Results on the vibrational wave packet dynamics above the dissociation limit in the excited electronic state including cage motion are presented and interpreted consistently with experimentally measured recombination dynamics and energy dissipation. Several schemes of controlling photochemical reactions of ClF are discussed.

1 Introduction

The reactive ClF molecule is prepared in the inert environment of the solid rare gas argon, where it occupies a substitutional site in the fcc lattice. Fig. 1 shows the potential surfaces relevant for the discussion of the spectroscopy and dynamics of ClF in argon. The left panel reproduces the results of the calculation by Alekseyev et al. [1] for ClF in the gas phase. The right panel shows qualitatively, how the potential is changed due to the argon crystal. At large elongations of the ClF bond, the potential is bent up due to the repulsive interaction of the F-fragment with the first argon shell. The potential surface is multidimensional and will depend on the orientation of ClF in the crystal. However, there is a barrier for dissociation of the molecule induced by the first solvent shell and the dissociated F-fragment will move on the periodic potential in the argon lattice.

2 Spectroscopic characterization

Excitation of ClF to the valence manifold leads to emission from the lowest valence state $A' \rightarrow X$ with well-resolved vibronic structure[2]. This band is used to measure the solvational red-shift of the valence states (100 cm^{-1} in Ar) as well as the vibration frequency $\omega_e = 380\ cm^{-1}$ and anharmonicity $\omega_e x_e = 5.8$ of the ground (X) state [2]. More importantly, the intensity of the $A' \rightarrow X$ band is a measure for the remaining concentration of ClF in photo-

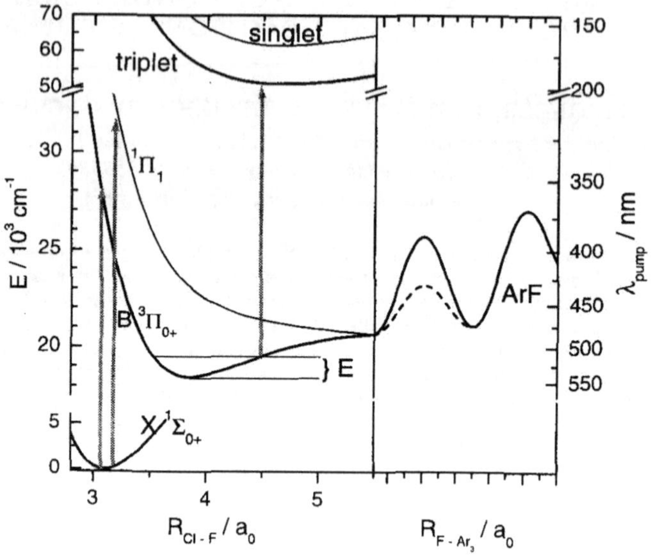

Figure 1. The left part of the diagram reproduces the gas phase potentials [1] for the X, B ($^3\Pi_0$), $^1\Pi_1$ and the ion pair triplet and singlet states to which transitions from B and $^1\Pi_1$ are allowed. The singlet ion-pair states are well sperated from the triplet states, so we are sure to probe only the B state with the probe wavelengths used in this experiment. The right part of the diagram is a sketch of the barriers imposed by the first and second argon shell. The dashed line shows the reduced barrier height for an open argon cage. (see text)

chemical experiments. Excitation to the dissociative part of one of the valence states, we observe a bleaching of this emission and derive a quantum yield for permanent dissociation of approx. 5% . This result must be contrasted with the dissociation efficiencies of 10^{-6} for Cl_2 in Ar and of 12 to 36% for F_2 in Ar, depending on the temperature. The comparison shows that the F fragment leaves the cage of argon atoms in the photo-dissociation of ClF, while Cl is trapped at the original lattice site. Two-pulse-excitation promotes the molecule to the ionic E state of ClF which leads to laser induced fluorescence (LIF) from the D' state after relaxation within the ionic manifold. Analyzing the dependence of the LIF on the wavelengths of the two pulses, we determine a solvation-redshift of the ionic states of 4500 cm^{-1} in absorption. From the fluorescence band a further redshift of 5500 cm^{-1} due to lattice rearrangement is calculated. This large stokes shift indicates a very strong coupling of the molecule to the argon lattice.

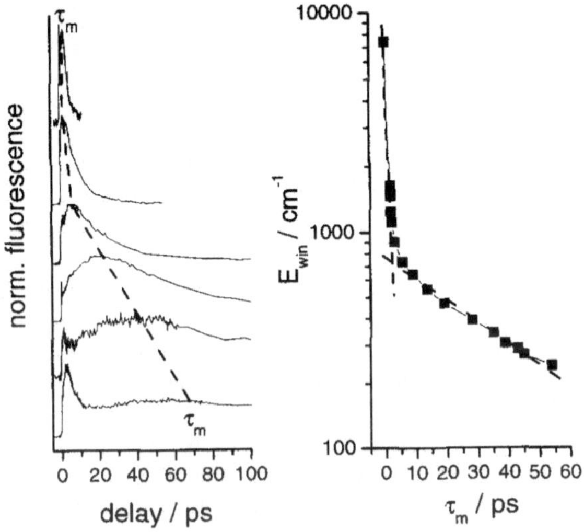

Figure 2. a) Pump-probe spectra for λ_{pump} = 387 nm and λ_{probe} = 319, 314, 306, 295, 286 and 282 nm. b) Mean energy of the wavepacket or vibrational population after time τ_m for λ_{pump} = 387 nm as obtained from the maxima in the spectra shown in Fig. 2.

3 Recombination dynamics

Since 95% of the photoexcited ClF molecules recombine, we are able to examine the corresponding dynamics by femtosecond pump-probe spectroscopy using LIF from the D' state as the detected signal. Exciting at 387 nm,

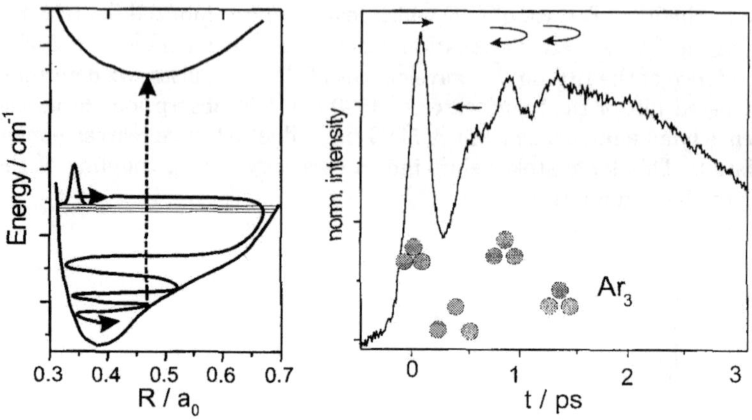

Figure 3. a) Sketch of the recombination dynamics as deduced from the pump-probe spectrum in b) $\lambda_{pump} = 387nm$ and $\lambda_{probe} = 319nm$.

i.e. 4900 cm^{-1} above the gas-phase dissociation limit and varying the probe-wavelength from 320 to 284 nm, we measure the vibrational energy relaxation in the bound valence state (B). Fig. 2a shows pump-probe spectra for selected probe wavelengths. From the difference potential $\Delta V = V_E - V_B$ by Alekseyev et al. [1] we calculate the energy E of the probe-window (Fig. 1), i.e. the vibrational levels in the B state which are probed by each of the wavelengths. The maximum of the pump-probe-envelope in the spectra of Fig. 2a occurs at the time delay τ_m, when the population of vibrational states is centered around the probe window. From the spectra we construct Fig. 2b, where the mean energy E of the vibrational population is plotted as a function time τ_m.

We observe two clearly distinct regimes of vibrational relaxation: A fast initial relaxation at a rate of 3000 cm^{-1}/ps for energies above the dissociation limit, when the vibrational wavepacket is bound by the solvent cage only. It is followed by a slow rate of 30 cm^{-1}/ps after the Cl and F fragments have formed the molecular Cl-F bond. Note that the high rate reflects the very fast energy dissipation during the first ps. It depends on the initial conditions of the impulsive energy transfer of the molecular fragments to the argon atoms and a different rate will be measured for other pump-wavelengths. However, as long as the excitation is above the dissociation limit of the molecule, the rate for the first collisions with the solvent clearly exceeds the rate for energy dissipation when the molecular Cl-F bond is re-established. Fig. 3b shows that the language used in the previous paragraph is justified: In the first ps we observe oscillations in the pump-probe spectrum which reflect the wavepacket dynamics of the ClF molecule, whereas later it is more meaningful to speak of vibrational population. The arrows in Fig 3 indicate how the wavepacket moves through the probe window for each peak.

The first peak occurrs after 50 fs, when the wave packet moves outwards through the probe window. The Cl and F fragments smash into the cage and lose most of their energy to the argon atoms. After 500 fs the wave packet returns to the probe-window, giving rise to the shoulder in the pump-probe spectrum. It reaches the inner turning point and passes the probe window a third time about 850 fs after the first peak. Due to the strong collision in the first excursion it has already lost most of the kinetic energy, the turning point lies close to the probe window in the third passage, and we observe only one peak corresponding to both outward and inward motion. Finally after 1250 fs we see it once more at the turning point. The oscillations ride on an increasing background due to the dispersion of the wavepacket on the anharmonic potential[3] and the statistics of the scattering by the matrix. After a few ps the regime of vibrational population relaxation is reached. The resulting scheme of the relaxation dynamics is presented in Fig. 3a, which is consistent with the rates obtained from Fig. 2b. The pump-probe spectrum in Fig 3b also carries information on the solvent motion. At time $t = 0$ the first argon shell is relaxed around the ClF molecule according to its size in the electronic ground state. In the excited (B-) state the ClF bond is on average 0.1 nm longer. Therefore the argon cage must adjust to accomodate the larger impurity, performing a strongly damped breathing motion with a period of 800 fs.[4,5] A larger cage shifts the energy in the ionic ClF states up, closer to the gas phase value. The probe-wavlelength $\lambda = 319$ nm corresponds to a probe window at the minimum of the difference potential. This makes the probe process very sensitive to shifts of the ionic potential surface resulting from solvent breathing. The resonance for the probe-transition is lost when the argon shell moves far out, causing the dip after 300 fs. At $t = 0$ the cage is tight and the first peak appears very strong. The solvent motion is only visible at a probe-wavelength close to the threshold for the probe transition.

4 Control

The above analysis shows that ClF isolated in rare gases is a good candidate to test new control schemes for photochemical reachtions in the condensed phase. Pulsed excitation of the ClF impurity dissociates the ClF bond and leads to vibrational motion of the argon cage. After excitation with a single pulse 95% of the molecules recombine. We attempt to increase the relatively low quantum efficiency for dissociation of 5% by the following double-pulse scheme: Exciting the molecule with a second pulse 100 - 300 fs after the first pulse will create a double-wavepacket of dissociating F-fragments approaching the argon cage. The first part of the wavepacket will induce the breathing motion of the argon cage as has been shown above. Then, the second part of the wavepacket should see a modified potential induced by the argon cage, which is now breathing. If the second wavepacket arrives at the argon shell at its biggest expansion, the barrier for cage exit will be lower. For this scenario the lifetime of a coherent superposition of vibrational states on a single molecule is crucial and will depend on the scattering processes of the molecule with the solvent. A related calculation [6,7] shows that pre-excitation of the argon lattice greatly increases the dissociation yield in the case of F_2 in argon. However in this scenario the argon lattice is excited already when the F_2 molecule is in its electronic ground state. This could be achieved by a pump-dump scheme or by increasing the temperature.

5 Conclusions

The recombination dynamics of the ClF molecule has been studied by femtosecond pump-probe spectroscopy. Two distinct regimes of energy loss are observed, one above and one below the gas-phase dissociation limit, respectively. The system will be used in future to study a new control scheme for photo-dissociation in the condensed phase, involving a superposition state of the reactant which actively influences the potential induced by the solvent environment.

Acknowledgments

This work has been funded by the Deutsche Forschungsgemeinschaft via SfB450.

References

1. A. B. Alekseyev, H.P. Liebermann, R. J. Buenker, and D. B. Kokh, J. Chem. Phys. **112**, 2274 (2000)
2. M. Bargheer, P. Dietrich, and N. Schwentner, J. Chem. Phys. **115**, 149 (2001)

3. M. Bargheer, K. Donovang, P. Dietrich, and N. Schwentner, J. Chem. Phys. **111**, 8556 (1999)
4. A. Goldberg and J. Jortner, J. Chem. Phys. **107**, 8994 (1997)
5. A. Heidenreich and J. Jortner, J. Electr. Spectr. **106**, 187 (2000)
6. G. Chaban, R.B. Gerber, M.V. Korolkov, J. Manz, M.Y. Niv, and B. Schmidt, J. Phys. Chem. A **105**, 2770 (2001)
7. R.B. Gerber, M.V. Korolkov, J. Manz, M.Y. Niv, and B. Schmidt, Chem. Phys. Lett. **327**, 76 (2000)

ULTRAFAST FEMTOSECOND CHARGE TRANSFER: SWITCHING FROM ADIABATIC TO NONADIABATIC DYNAMICS BY SMALL STRUCTURAL CHANGES

S. LOCHBRUNNER[1], K. STOCK[1], T. FIEBIG[2], E. PANDURSKI[2], AND E. RIEDLE[1]

[1) Lehrstuhl für BioMolekulare Optik, Sektion Physik, Ludwig-Maximilians-Universität, Oettingenstr. 67, 80538 München, Germany
E-mail: stefan.lochbrunner@physik.uni-muenchen.de

[2) Institut f. Physikalische u. Theoretische Chemie, TU München, 85748 Garching, Germany
E-mail: Torsten.Fiebig@ch.tum.de

The intramolecular charge transfer (CT) dynamics of three pyrenyl biphenyl ester isomers have been studied by femtosecond absorption spectroscopy. Two of them show an ultrafast CT and associated relaxation processes whereas the CT of the third one is slower by several orders of magnitude. The differing topology of the acceptor orbitals at the biphenyl ester moieties results in strong changes of the electronic coupling between the locally excited and the CT state. This leads to dynamics on adiabatic potential energy surfaces for the former two compounds and to a weak coupling regime for the latter one. It indicates that the charge flow in molecular systems can be designed in a deterministic manner with minor changes of other system properties by utilizing the topology of the acceptor orbitals.

1 Introduction

The dynamics of intramolecular charge transfer (CT) - one of the fundamental processes in chemistry and biology - are determined by the energetic location of the charge transfer state with respect to the locally excited (LE) state and the electronic coupling between these states [1-3]. The first factor has been extensively studied [4-6] while systematic studies of the second factor are sparse. This is due to the fact that variations of the electronic coupling through changed electronic wavefunctions usually lead to major changes in the geometry and the energetics of the molecules. Nevertheless the influence of the electronic coupling on the CT attracts more and more interest since it plays an important role in photosynthesis [7,8] and molecular electronics [9].

We carried out time resolved absorption studies on the three 4'-pyren-1-yl-biphenyl-i,j-dicarboxylic dimethyl ester isomers (with i,j = 2,3, 2,4 and 3,4) depicted in Figure 1 - abbreviated as **1e23**, **1e24** and **1e34**. Steady state spectroscopy and molecular orbital calculations showed that the energetic situation is very similar in all three compounds but the LUMO of the biphenyl ester moiety is very sensitive to the substitution positions of the esters [10]. This should lead to a small overlap between the electron donating and the accepting orbital in the case of **1e23** whereas the overlap should be large for **1e24** and **1e34**. By varying the overlap, the strength of the electronic coupling should also be strongly varied. Our time resolved experiments at a selected number of probe wavelengths have indeed shown that the compounds exhibit vastly different kinetics [11]. Additional time-resolved measurements with broad spectral resolution provide a deeper insight into the dynamics and allow the identification of CT products and intermediates.

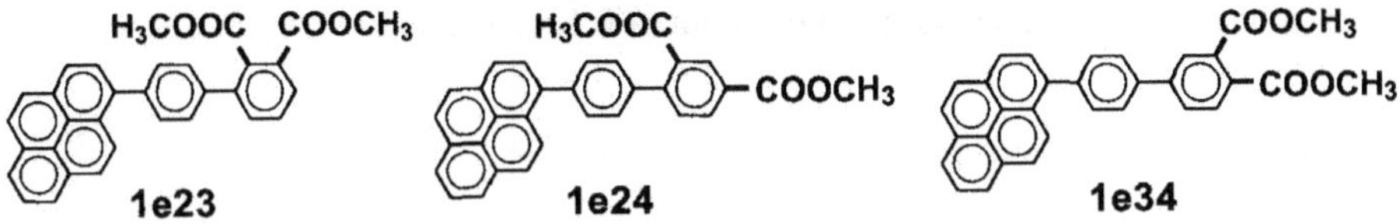

Figure 1. Structures of pyrene derivatives investigated in the presented study.

2 Experimental

The photodynamics of the compounds (see Figure 1) dissolved in acetonitrile were studied by femtosecond transient absorption spectroscopy with a fully tunable UV-visible pump probe spectrometer [12]. It is based on two noncollinearly phase-matched optical parametric amplifiers (NOPAs) pumped by a home built Ti:sapphire laser system [13]. The molecules were excited at their S_2-S_0 absorption band by laser pulses with a center wavelength of 340 nm generated by frequency doubling the output of the first NOPA. Induced transmission changes were probed by light pulses from the second NOPA at various wavelengths between 480 and 560 nm, covering the blue as well as the red wing of the CT emission. The crosscorrelation and time zero at the sample position is determined by difference frequency generation in a 25 μm thick BBO crystal. The crosscorrelation has typically a duration of 40 fs (FWHM). The results presented below were obtained with a magic angle geometry for the polarization of the pump and probe pulses to avoid transmission changes due to orientational relaxation.

The compounds were synthesized, characterized and purified by means of HPLC by Dr. W. Kühnle and his coworkers at the Max-Planck-Institute for Biophysical Chemistry in Göttingen [14]. To avoid dimerisation in the samples the experiments were carried out at low concentrations ($< 10^{-4}$ M). Therefore we had to use a cell with a large sample thickness of 1 mm to get an absorption for the pump pulses on the order of 50 % and reasonable transmission changes on the order of 2 %. Due to the group velocity dispersion in the solvent our time resolution was thereby smeared out to 100 fs. Due to the entrance window and half of the solvent layer time zero in the sample is shifted with respect to the BBO crystal. To determine time zero in the sample, the cell filled with acetonitrile was positioned in front of the BBO crystal and the change of the temporal overlap of pump and probe pulses with and without cell was divided by two.

The transmission change in the pure solvent was measured as well but no significant signal contributions from the solvent were found.

In a second set of experiments the molecules were excited at 260 nm by the third harmonic of the Ti:Sapphire laser pulses and the transient absorption was probed by a white light continuum (WLC) in the wavelength range between 460 and 720 nm. After passing the sample the WLC was dispersed by a spectrograph and spectrally resolved detected by a CCD camera. After time zero correction for the chirp of the WLC a time resolution of about 500 fs was achieved.

3 Results and discussion

The pump-probe transients of all isomers excited at 340 nm are shown in Figure 2 from –2 to 9 ps delay time for the probe wavelengths 480 and 530 nm. The curves for **1e24** and **1e34** are very similar while the ones for **1e23** deviate quite strongly. All curves show an instrument limited increase in optical density at time zero and an ultrafast initial change. For **1e23** the signal does not vary significantly after the first few 100 fs while the signals for **1e24** and **1e34** show large changes for the whole time interval depicted in Figure 2. It will be argued below that this behavior is due to ultrafast CT in **1e24** and **1e34** and extremely slow CT (on our time scale) in **1e23**.

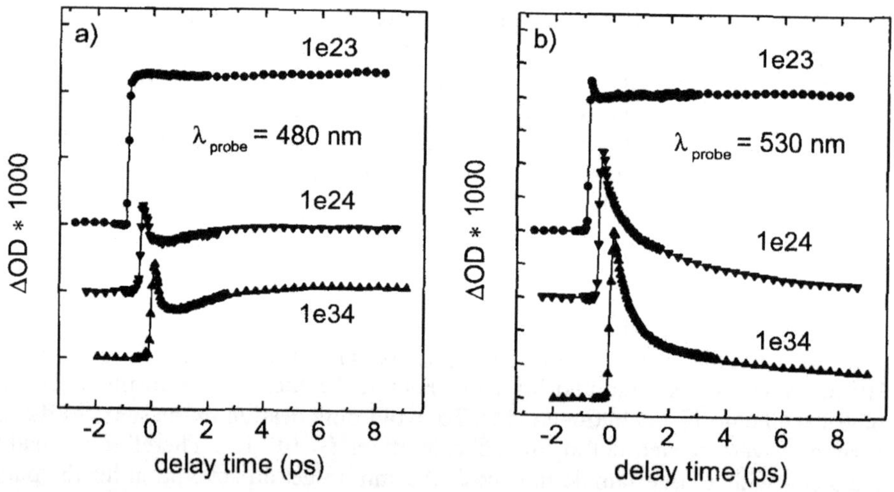

Figure 2. Time resolved transmission changes at a) 480 nm and b) 530 nm of the pyrene derivatives solvated in acetonitrile after photo excitation at 340 nm.

The initial component of about 100 fs appearing in the traces of all three compounds (see Figure 2) is due to the internal conversion from the S_2 to the S_1 state, both located on the pyrene moiety [11]. This interpretation is guided by the ultrafast dynamics of unsubstituted pyrene after excitation to its S_2 state. In this case a transient absorption band around 585 nm attributed to the S_{14}-S_2 transition appears immediately with the excitation and decays on a 150 fs time scale [15]. Simultaneously with its decay a new, slightly structured band at 460 - 500 nm emerges. It is due to excited state absorption originating from the S_1 state [16]. By analogy we attribute the absorption change within the first 200 fs observed for the pyrenyl biphenyl esters to the decay of a S_n-S_2 absorption (n = 10) of the pyrene moiety. At 480 nm the excited state absorption of the S_2 and the S_1 state are of similar strength. Depending on the details of the spectra a slight transmission increase results for **1e23** and a decrease for **1e24** and **1e34**.

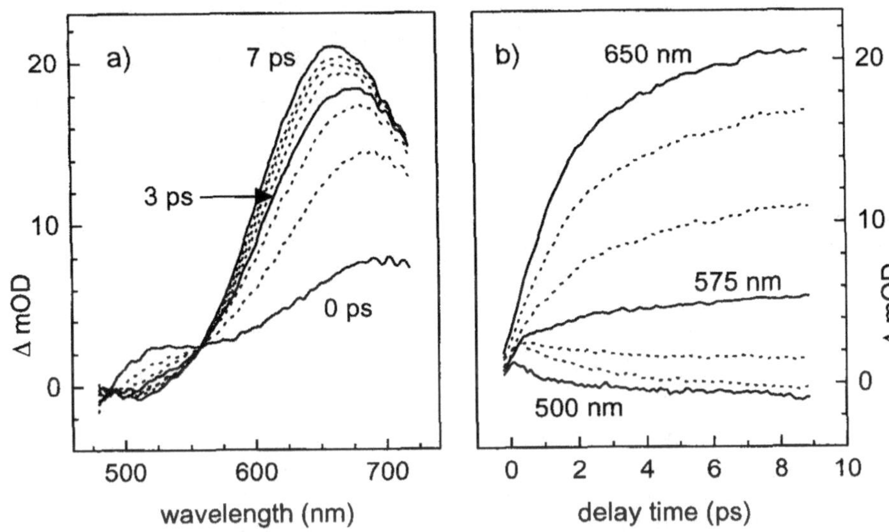

Figure 3. Time resolved absorption after excitation of **1e34** at 260 nm. a) Transient spectra between 0 and 7 ps are shown at increments of 1 ps. b) Time dependent absorption change at probe wavelengths between 500 and 650 nm with 25 nm increments.

After the initial relaxation from the S_2 to the S_1 state **1e24** and **1e34** show large subsequent changes in the transmission signal which can be described by two exponential components, one on the order of 400 fs and one of 3 ps [11]. The temporal evolution of the transient absorption spectrum is depicted in Figure 3 for the case of **1e34**. A quite similar time dependence of the absorption spectrum is expected for **1e24** because of the similarity between the traces of **1e24** and **1e34** in Figure 2.

According to Figure 2b) and Figure 3 the optical density decreases double exponentially between 500 nm and 550 nm. This is partially due to the decay of the transient absorption from the LE state and partially to the onset of stimulated emission from the CT state. The spontaneous emission of the CT state is strongly Stokes shifted for **1e24** and **1e34** [10], due to the transfer of the electron from the pyrene to the biphenyl ester moiety.

The transient absorption of **1e24** and **1e34** at 480 nm shows a decrease on the 400 fs time scale and an increase for longer times (see Figure 2a)). This indicates that the 400 fs time corresponds to the rise and the 3 ps time to a red shift of the CT emission band. We assign the fast component to the intramolecular CT itself and the slower one to geometric relaxation processes of the molecule as well as the solvent, both induced by the new charge distribution of the molecule. These relaxation processes lead to a further energy reduction of the CT state and to a red shift of the emission.

The strong transient absorption band at 660 nm observed by the spectrally resolved measurements on **1e34** shows a double exponential rise (see Figure 3b)). On the fast time scale the band strength and the transition dipole moment increase strongly indicating that electron density is redistributed as it is expected for a CT process. At longer times the absorption change is governed by a narrowing and slight blue shift of the band (see Figure 3a)). This supports the picture that intra- and intermolecular relaxation processes are responsible for the spectral evolution on the picosecond time scale.

In the case of **1e23** no significant changes of the transient absorption are observed after the initial contribution due to the S_2-S_1 relaxation (see Figure 2). In particular, the transient absorption does not decrease on the picosecond time scale as it were expected if emission from a CT state builds up or the transient absorption to higher electronic states is strongly changed. This lack of subsequent dynamics indicates that no CT occurs during the time span of hundred picoseconds which is scanned in our experiment [11]. This is in agreement with the cw fluorescence spectrum of **1e23** which is largely dominated by an emission that shows a small Stokes shift with respect to the absorption and is attributed to fluorescence from the LE state [10]. A much smaller second contribution appears as tail of the first band extending to the red. Its spectral position coincides with the CT emission of **1e24** and **1e34** [10]. The dominating LE fluorescence of **1e23** indicates that during the excited state lifetime of **1e23** only a small fraction of the molecules experiences CT.

4 Conclusions

Our transient absorption measurements show that after optical excitation the two pyrenyl biphenyl esters **1e24** and **1e34** undergo ultrafast CT and associated relaxation processes. In contrast, the CT process of the isomer **1e23** is found to be several orders of magnitude slower. Since the absorption spectra of all three compounds are very similar and their CT emissions are all in the same spectral region this difference in the dynamics can not be explained by assuming different energetics [10]. Because the isomers vary only in the positions of the methyl ester groups the density of states is for all of them quite similar and can not account for the vastly different CT dynamics either.

The only factor which is significantly varied is the electronic coupling between the LE and the CT state. The LUMO orbital of the biphenyl ester moiety accepts the electron which moves from the pyrene to the biphenyl ester moiety. According to semi-empirical calculations it extends over the two phenyl rings in the case of **1e24** and **1e34** but is localized on the outer phenyl ring in **1e23** [10,11]. Thus the spatial overlap between pyrene orbitals and the acceptor orbital is larger in **1e24** and **1e34** than in **1e23** giving rise to a much stronger electronic coupling for the former two.

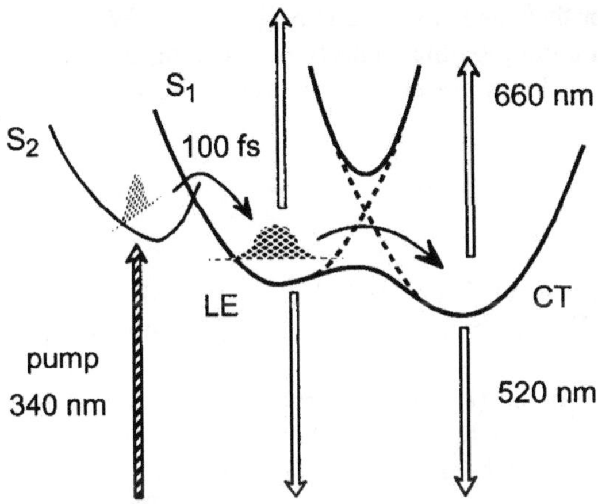

Figure 4. Model for the photo induced dynamics of the pyrenyl biphenyl esters. All of them show an ultrafast electronic relaxation from the optically excited S_2 to the S_1 state. In the case of **1e23** the transition probability from the locally excited (LE) to the charge transfer (CT) state is low and the CT rate is very small. In the case of **1e24** and **1e34** strong coupling between the LE and the CT state leads to an adiabatic PES which connects the LE with the CT minimum with a flat path. White arrows pointing up indicate excited state absorption contributing to the time resolved signals while white arrows pointing down symbolize stimulated emission.

A model of the involved potential energy surfaces is sketched in Figure 4. The diabatic surfaces of the LE as well as the CT state are similar for all three isomers (broken lines). But in the case of **1e24** and **1e34** the strong electronic coupling leads to a strong mixing of both states and to an adiabatic surface with a flat transition region. The process can then be described as a wavepacket motion on this adiabatic surface which has LE character at the excitation region and CT character at its minimum. The weak coupling observed for **1e23** fits in a nonadiabatic picture where the population of the LE state only slowly leaks out into the CT state.

These results clearly show that small yet specific modifications in the substitution scheme can lead to alternations of the relevant donator acceptor orbitals and thus to drastic changes in the excited state CT dynamics. It demonstrates that the topological properties of molecular orbitals can be utilized in a predictable manner to guide and design the charge flow in molecular systems. The presented example also shows that switching from an ultrafast CT to an almost forbidden one can be achieved just by varying the position of a substituent leading to extremely similar isomers. Changing the charge flow by this technique would barely influence other physical and chemical properties of the system. This is a great advantage for technical applications.

5 Acknowledgement

We warmly thank Dr. W. Kühnle (MPI Göttingen) for the synthesis, characterization and purification of the samples. We are indebted to Tanja Bizjak and Uli Schmidhammer for their help with the experiment and W. Zinth for continuous support of our work. Funding by the Deutsche Forschungsgemeinschaft (DFG) is gratefully acknowledged. TF is grateful for financial support from the Emmy-Noether-Program by the DFG.

References

1. Marcus R. A., Angew. Chem. **105** (1993) pp. 1161.
2. Verhoeven J. W., Scherer T., Willemse R. J., Pure & Appl. Chem. **65** (1993) pp. 1717.
3. Jortner J., Bixon M. (Eds.), "Electron Transfer: From Isolated Molecules to Biomolecules", Adv. Chem. Phys. **107** Part I & II (1999).
4. Wasielewski M. R., Chem. Rev. **92** (1992) pp. 435.
5. Il'ichev Y. V., Kühnle W., Zachariasse K. A., J. Phys. Chem. A **102** (1998) pp. 5670.
6. Mataga N., Miyasaka H. in "Electron Transfer: From Isolated Molecules to Biomolecules", eds.: Jortner J., Bixon. M., Adv. Chem. Phys. **107** Part II (1999) pp. 431.
7. Beratan D. N., Onuchic J. N., Betts J. N., Bowler B. E., Gray H. B., J. Am. Chem. Soc. **112** (1990) pp. 7915.
8. Zinth W., Huppmann P., Arlt T., Wachtveitl J., Phil. Trans. R. Soc. Lond. A **356** (1998) pp. 465.
9. Aviram A., Ratner M. (Eds.), Molecular Electronics: Science and Technology. New York Acadamy of Sciences, New York, 1998.
10. Fiebig T., Kühnle W., Staerk H., Chem. Phys. Lett. **282** (1998) pp. 7.
11. Fiebig T., Stock K., Lochbrunner S., Riedle E., Chem. Phys. Lett. **345** (2001) pp. 81.
12. Wurzer A. J., Lochbrunner S., Riedle E., Appl. Phys. B **71** (2000) pp. 405.
13. Riedle E., Beutter M., Lochbrunner S., Piel J., Schenkl S., Spörlein S., Zinth W., Appl. Phys. B **71** (2000) pp. 457.
14. The synthetic procedure used by Dr. Kühnle has never been published. Recently, several related compounds were synthesized as described in: Beinhoff M., Weigel W., Jurczok M., Rettig W., Modrakowski C., Brüdgam I., Hartl H., Schlüter A. D., Eur. J. Org. Chem. **2001** (2001) pp. 3819.
15. Neuwahl F. V. R., Foggi P., Laser Chem. **19** (1999) pp. 375.
16. Foggi P., Pettini L., Sànta I., Righini R., Califano S., J. Phys. Chem. **99** (1995) pp. 7439.

3. Coherent Control and New Techniques

This page is intentionally left blank

FROM AN ORIENTED RACEMATE VIA TORSIONALLY OR ELECTRONICALLY EXCITED STATES TO PURE ENANTIOMERS: DESIGN OF PUMP-DUMP LASER PULSES

KUNIHITO HOKI[†], DOMINIK KRÖNER[‡], JÖRN MANZ[‡]

[†]*Department of Chemistry, Graduate School of Science, Tohoku University, Sendai 980-8578, Japan*
E-mail: hoki@mcl.chem.tohoku.ac.jp

[‡]*Institut für Chemie – Physikalische und Theoretische Chemie, Freie Universität Berlin, Takustrasse 3, D-14195 Berlin, Germany*
E-mail: dkroener@chemie.fu-berlin.de

We design a series of two selective intense polarized laser pulses which transform an oriented, non-rotating racemate into pure enantiomers. The first "pump" pulse excites exclusively the undesired enantiomer into an excited, delocalized state. The second "dump" pulse transforms this excited state into a localized excited state of the target enantiomer. The net result is a mixture of the target enantiomer in the original non-excited state, and in an additional excited state. We compare two implementations of this pump-dump mechanism, using either IR or UV laser pulses for transitions via torsionally or electronically excited delocalized states, respectively. Quantum simulations of the laser-driven molecular dynamics are carried out in terms of density matrices representing the torsional motions and electronic transitions of the model system, H_2POSH. The results are explained by means of a restricted state representation.

1 Introduction

The transformation of a racemate into pure enantiomers by means of laser pulses is a challenging problem. Possible solutions are implied, in principle, by experimental demonstrations of partial enantiomeric enrichment using intense circularly polarized continuous-wave lasers, see e.g. Refs.[1,2] and the review.[3] The chiral information inherent to circularly polarized light may thus be transmitted to racemic photoresponsive materials, followed by even macroscopic expression of the chirality.[4] Laser pulses should achieve this goal on a ps or even fs time scale.

On the theoretical side, Salam and Meath have suggested to use circularly polarized pulses for selective excitations of left or right enantiomers.[5] Using a model two-level chiral molecular system, they could show that the relative populations of the excited enantiomeric states can be controlled within 100% for fixed oriented racemates, but only within 6% for randomly oriented samples.[6] Their approach does not change the overall 50%:50% populations of the two enantiomers directly, but sequel enantiomeric purification should be possible e.g. by photodestruction of the preferentially excited, but undesired enantiomer.[3]

In contrast, Shapiro, Brumer and coworkers have suggested a method for increasing the enantiomeric excess of a chiral enantiomer in a racemic mixture directly, starting from molecules in selected rotational states and using achiral i.e. linearly polarized IR plus UV laser pulses.[7,8] We have designed various alternative direct approaches in order to transform racemates into pure enantiomers, either starting from non-rotating, oriented molecules and using linearly polarized laser pulses,[9,10,11,12,13] or starting from rotating oriented molecules and using circularly polarized laser pulses.[14] Our approaches use either IR or UV laser pulses inducing transitions between torsional molecular states in the electronic ground states,[9,10,13,14] or between electronic ground and excited states,[11,12] respectively.

The purpose of this paper is to compare the efficiencies of IR versus UV laser pulses for our simplest method of enantiomeric purification of an oriented, non-rotating racemate, i.e. the pump-dump mechanism which has been introduced in Ref [12] for UV laser pulses. For this purpose, we present the first quantum model simulation for this mechanism using IR transitions in the electronic ground state, together with new applications to UV transitions via electronic excited states. The underlying model, adapted from Ref [11], is summarized in section 2, together with an explanation of the mechanism in terms of a restricted state representation, adapted from Ref [12]. The numerical results, which are calculated using a sufficiently large basis set for convergence, are presented and discussed in section 3. Our conclusions are in section 4.

2 Model and techniques

We consider the model of oriented, non-rotating, non-vibrating H_2POSH molecules, with their PS bond along the z-axis, and the POS fragment in the x/z plane, as indicated in Figure 1. The torsional states and energies are evaluated as eigenstates $|v\rangle$ and energies E_v of the molecular Hamiltonian H_{mol} ,

$$H_{mol}\,|v\rangle = E_v\,|v\rangle\,, \tag{1}$$

by means of quantum ab initio methods, see Ref [11]. Here $|v\rangle$ denotes e.g. the torsional states $|0A'\rangle$, $|0A''\rangle$, $|1A'\rangle$, $|1A''\rangle$ of the two lowest near-degenerate doublets in the electronic ground state, or a more excited torsional state $|eA'\rangle$ or $|eA''\rangle$ in the electronic ground or excited state; these eigenstates have *even* or *odd* symmetries, indicated by A' or A'', respectively. The levels E_v embedded in the torsional potential energy surfaces of the electronic ground and excited states (adapted from Ref [11]) are also shown in Figure 1.

The + and − superpositions of the doublet states yield ground (0) or first excited (1) torsional states representing left (L) and right (R) enantiomers, respectively,

$$|0L\rangle = \frac{1}{\sqrt{2}}(|0A'\rangle + |0A''\rangle), \qquad |0R\rangle = \frac{1}{\sqrt{2}}(|0A'\rangle - |0A''\rangle)$$
$$|1L\rangle = \frac{1}{\sqrt{2}}(|1A'\rangle + |1A''\rangle), \qquad |1R\rangle = \frac{1}{\sqrt{2}}(|1A'\rangle - |1A''\rangle) \qquad (2)$$

Initially, we assume that the system is prepared at low temperature such that only the two states of the lowest doublet are occupied with equal probablities, corresponding to a racemic mixture of the left and right enantiomers in their ground torsional states

$$\begin{aligned} \rho_0 &= \frac{1}{2}\left(|0A'\rangle\langle 0A'| + |0A''\rangle\langle 0A''|\right) \\ &= \frac{1}{2}\left(|0L\rangle\langle 0L| + |0R\rangle\langle 0R|\right) \\ &= \frac{1}{2}\rho_L + \frac{1}{2}\rho_R \end{aligned} \qquad (3)$$

The interaction with the laser pulses is given, in semiclassical dipole approximation, by

$$-\vec{\varepsilon}\cdot\vec{\mu} = -\varepsilon_x\cdot\mu_x - \varepsilon_y\cdot\mu_y - \varepsilon_z\cdot\mu_z, \qquad (4)$$

where $\vec{\varepsilon}$ is the electric field strength, and $\vec{\mu}$ is the molecular dipole, or the transition dipole operator, for IR or UV transitions, respectively. For convenience, we consider two cases with either polarized IR lasers propagating along the z-direction ($\varepsilon_z = 0$), or UV-pulses propagating along the x-direction ($\varepsilon_x = 0$) . The corresponding x and y components of the dipole interacting with the ε_x and ε_y components of the IR electric field , as well as the y and z components of the transition dipole operators interacting with the ε_y and ε_z components of the UV lasers, have A' and A'' symmetries, respectively. The laser-dipole interaction (4) can, therefore, be decomposed into two contributions which account for the laser interactions with the A' and A'' components of the dipole, or transition dipole operators, respectively,

$$-\vec{\varepsilon}\cdot\vec{\mu} = -\varepsilon_{A'}\cdot\mu_{A'} - \varepsilon_{A''}\cdot\mu_{A''} \qquad (5)$$

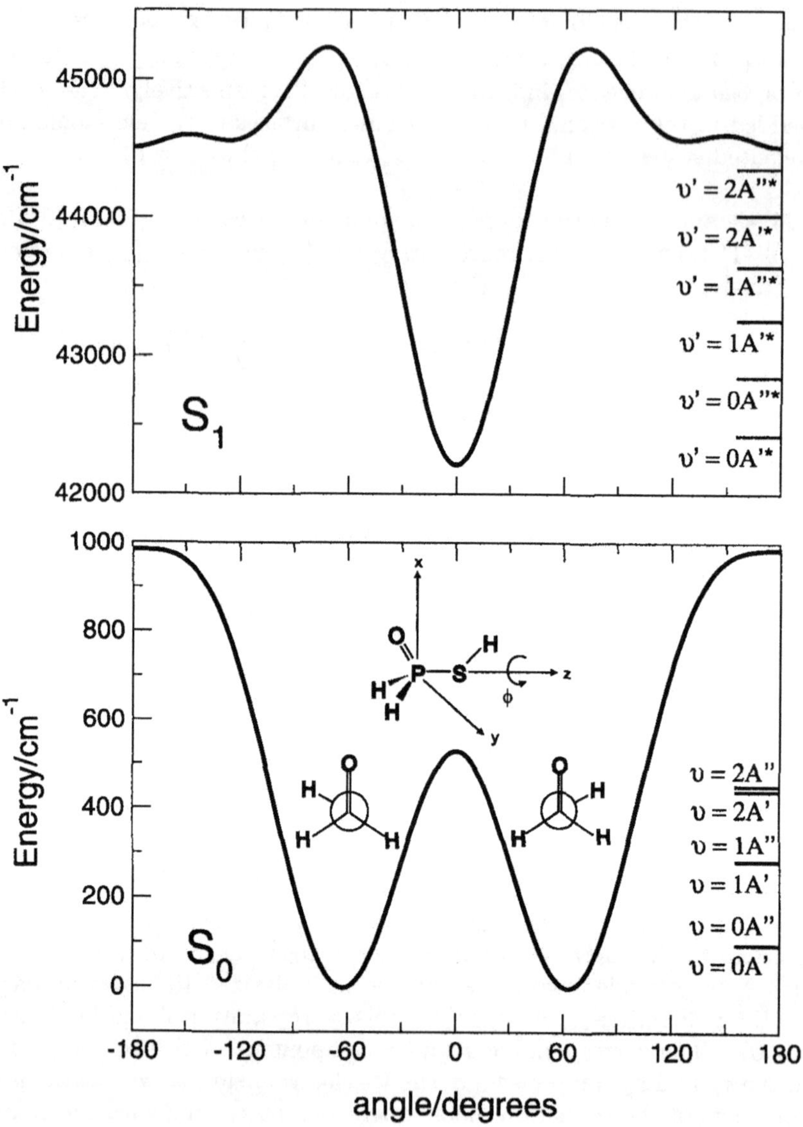

Figure 1. The model H_2POSH and its torsional levels embedded in the torsional potential energy surfaces of the electronic ground S_0 and excited states S_1, adapted from Ref[11]. The torsional states of the electronic excited state are labeled in analogy to the electronic ground state as $v' = nA'^*$ and $v' = nA''^*$ (instead of using $v' = 0, 1, 2, ...$ as in Ref[11]).

The Hamiltonian of the total, i.e. molecular plus electric (transition) dipole interaction system is thus

$$H = H_{mol} - \vec{\varepsilon} \cdot \vec{\mu} = H_{mol} - \varepsilon_{A'} \cdot \mu_{A'} - \varepsilon_{A''} \cdot \mu_{A''} \tag{6}$$

or in matrix notation

$$\langle v|H|w\rangle = E_v \cdot \delta_{vw} - \varepsilon_{A'} \cdot \langle v|\mu_{A'}|w\rangle - \varepsilon_{A''} \cdot \langle v|\mu_{A''}|w\rangle \,. \tag{7}$$

The corresponding A'- and A''- (transition) dipole matrix elements, adapted from Ref [11], couple the torsional states $|v\rangle$ and $|w\rangle$ with the same, or with opposite symmetries, respectively. The laser pulses are designed as two sequential (labeled $k = 1, 2$ or $k = p$ for *pump* and $k = d$ for *dump*) sub-pulses

$$\vec{\varepsilon} = \vec{\varepsilon}_1 + \vec{\varepsilon}_2 = \vec{\varepsilon}_p + \vec{\varepsilon}_d \tag{8}$$

starting at initial times $t_1 = 0$ and $t_2 = t_{p1}$, and with durations t_{p1} and t_{p2}, respectively. Specifically,

$$\begin{aligned} \varepsilon_{k,A'/A''} &= \varepsilon_{k,A'/A''}(t - t_k) \cdot \cos\left(\omega_k \cdot (t - t_k) + \eta_{k,A'/A''}\right) \\ &= \varepsilon^0_{k,A'/A''} \cdot s_k(t - t_k) \cdot \cos\left(\omega_k \cdot (t - t_k) + \eta_{k,A'/A''}\right) \end{aligned} \tag{9}$$

with amplitudes $\varepsilon^0_{k,A'/A''}$, carrier frequencies $\omega_{k,A'/A''}$, phases $\eta_{k,A'/A''}$, and the same shape function $s_k(t)$ for both components A' and A'' of the laser field. We use

$$s_k(t) = \sin^2\left(\frac{(t - t_k) \cdot \pi}{t_{pk}}\right), \tag{10}$$

during the times of the subpulses, $t_k < t < t_k + t_{pk}$. The sequential pump and dump pulses are designed essentially as π-pulses which transfer the undesired (say the right) enantiomer via the delocalized excited state into the first excited torsional state of the target (say the left) enantiomer. For this purpose the frequencies ω_p and ω_d are chosen such that the first *pump* pulse induces selective transition from states $|0A'\rangle$, $|0A''\rangle$ preferably to the excited state $|eA'/A''\rangle$, whereas the second *dump* pulse induces transitions from $|eA'/A''\rangle$ preferably into states $|1A'\rangle$, $|1A''\rangle$, thus

$$\hbar\omega_p = E_{eA'/A''} - E_0 \tag{11}$$

$$\hbar\omega_d = E_{eA'/A''} - E_1 \tag{12}$$

where E_0, E_1 denote average energies of the A' and A'' levels of the ground state and first excited torsional doublets, and $E_{eA'/A''}$ is the energy of the excited (e) state which may have either A' or A'' symmetry (in the eqns.

below both possibilities are indicated by A'/A''). Moreover, the amplitudes of the A' and A'' components of the pulses are chosen such that

$$\begin{aligned} 2\hbar\Omega_p(t) &= -\varepsilon_{pA'/A''}(t)\,\langle eA'/A''|\mu_{A'/A''}|0A'\rangle \\ &= +\varepsilon_{pA''/A'}(t)\,\langle eA'/A''|\mu_{A''/A'}|0A''\rangle \end{aligned} \tag{13}$$

$$\begin{aligned} 2\hbar\Omega_d(t) &= -\varepsilon_{dA'/A''}(t)\,\langle 1A'|\mu_{A'/A''}|eA'/A''\rangle \\ &= -\varepsilon_{dA''/A'}(t)\,\langle 1A''|\mu_{A''/A'}|eA'/A''\rangle \end{aligned} \tag{14}$$

with corresponding Rabi frequencies $\Omega_k(t)$ and Rabi areas

$$\phi_p(t) = \int_0^t 2\Omega_p(t)dt \quad \text{for } 0 \leq t \leq t_{p1} \tag{15}$$

$$\phi_d(t) = \int_{t_{p1}}^t 2\Omega_d(t)dt \quad \text{for } t_{p1} \leq t \leq t_{p1} + t_{p2} \tag{16}$$

for the individual transitions, mediated by $\mu_{A'}$ and $\mu_{A''}$. Finally, the pulse durations for then so-called π-pump and -dump pulses are chosen such that

$$\phi_p(t_{p1}) = \pi/\sqrt{2} \tag{17}$$

$$\phi_d(t_{p1} + t_{p2}) = \pi/\sqrt{2} \tag{18}$$

The factor $\sqrt{2}$ in eqns. (17), (18) accounts for the fact that both the A' and A'' components of the pump and dump π-pulses operate simultaneously. Alternatively, one may describe the corresponding processes also in terms of laser-π-pulse induced transitions from the initial, undesired state $|0R\rangle$ via the delocalized state $|eA'/A''\rangle$ to the target state $|1L\rangle$, see Ref.[12]. In that representation, the factor $\sqrt{2}$ is taken into account by using the total field strength and polarization angles,

$$\begin{aligned} \varepsilon_p &= \sqrt{\varepsilon_{pA'}^2 + \varepsilon_{pA''}^2} \\ \varepsilon_d &= \sqrt{\varepsilon_{dA'}^2 + \varepsilon_{dA''}^2} \end{aligned} \tag{19}$$

$$\begin{aligned} \tan\alpha_p &= \frac{\varepsilon_{pA''}}{\varepsilon_{pA'}} = -\frac{\langle 0A'/A''|\mu_{A'}|eA'/A''\rangle}{\langle 0A'/A''|\mu_{A''}|eA''/A'\rangle} \\ \tan\alpha_d &= \frac{\varepsilon_{dA''}}{\varepsilon_{dA'}} = +\frac{\langle 1A'/A''|\mu_{A'}|eA'/A''\rangle}{\langle 1A'/A''|\mu_{A''}|eA''/A'\rangle} \end{aligned} \tag{20}$$

Starting from the initial racemic mixture ρ_0, see eqn. (3), the time evolution of the density operator ρ subject to the pump and dump laser pulses , and neglecting any competing dissipative processes, is determined by the Liouville-von Neumann equation

$$i\hbar\dot{\rho} = [H, \rho] \tag{21}$$

which is solved numerically, using a sufficiently large set of molecular eigenstates, see section 3. In order to explain the underlying *pump-dump* mechanism, it is convenient to express the results of eqns. (3), (21) as

$$\rho(t) = U\rho_0 U^\dagger, \tag{22}$$

with unitary time evolution operator

$$U = U_d \cdot U_p \quad \text{for } 0 \le t \le t_{pp} + t_{pd} \tag{23}$$

corresponding to the two sequential pump and dump processes. The individual operators U_p and U_d may be expressed by means of the interaction (i) picture as

$$U_k = U_{k,ph} \cdot U_{k,i} \quad \text{for } t_k \le t \le t_k + t_{pk} \tag{24}$$

where the $U_{k,ph}$ account for rapid changes of the phases,

$$U_{k,ph} = \exp\left(\frac{-iE_v(t - t_{kp})}{\hbar}\right) \cdot \delta_{vw} \tag{25}$$

The $U_{k,i}$ exhibit much slower changes. They may be approximated by means of the rotating wave approximation RWA, provided that the pulse durations satisfy the relations

$$\frac{h}{(E_{eA'/A''} - E_0)} \ll t_p \ll \frac{h}{(E_{0A''} - E_{0A'})} \tag{26}$$

$$\frac{h}{(E_{eA'/A''} - E_1)} \ll t_d \ll \frac{h}{(E_{1A''} - E_{1A'})} \tag{27}$$

implying the time averages during field cycles

$$\overline{\cos(\omega_k \cdot t)} = 0, \quad \overline{\cos(\omega_k \cdot t) \cdot e^{(i\omega t)}} = \frac{1}{2} \tag{28}$$

In this approximation, the RWA matrix representations of $U_{p,i}$ may be restricted to just the five most important states $|0A'\rangle$, $|0A''\rangle$, $|1A'\rangle$, $|1A''\rangle$, $|eA'/A''\rangle$ (see the labels $0A'$ etc. below) which participate in the *pump-dump* mechanism,

$$U_{p,i,\mathrm{RWA}} \approx \frac{1}{2}\begin{pmatrix} 1+\cos\phi_p & 1-\cos\phi_p & 0 & 0 & -i\sqrt{2}\sin\phi_p \\ 1-\cos\phi_p & 1+\cos\phi_p & 0 & 0 & i\sqrt{2}\sin\phi_p \\ 0 & 0 & 2 & 0 & 0 \\ 0 & 0 & 0 & 2 & 0 \\ -i\sqrt{2}\sin\phi_p & i\sqrt{2}\sin\phi_p & 0 & 0 & 2\cos\phi_p \end{pmatrix}$$

$$\longrightarrow \frac{1}{2}\begin{pmatrix} 1 & 1 & 0 & 0 & -i\sqrt{2} \\ 1 & 1 & 0 & 0 & i\sqrt{2} \\ 0 & 0 & 2 & 0 & 0 \\ 0 & 0 & 0 & 2 & 0 \\ -i\sqrt{2} & i\sqrt{2} & 0 & 0 & 0 \end{pmatrix} \begin{matrix} 0A' \\ 0A'' \\ 1A' \\ 1A'' \\ eA'/A'' \end{matrix} \qquad \text{for } t \to t_{pp} \tag{29}$$

Likewise,

$$U_{d,i,\mathrm{RWA}} \approx \frac{1}{2}\begin{pmatrix} 2 & 0 & 0 & 0 & 0 \\ 0 & 2 & 0 & 0 & 0 \\ 0 & 0 & 1+\cos\phi_d & -1+\cos\phi_d & -i\sqrt{2}\sin\phi_d \\ 0 & 0 & -1+\cos\phi_d & 1+\cos\phi_d & -i\sqrt{2}\sin\phi_d \\ 0 & 0 & -i\sqrt{2}\sin\phi_d & -i\sqrt{2}\sin\phi_d & 2\cos\phi_d \end{pmatrix}$$

$$\longrightarrow \frac{1}{2}\begin{pmatrix} 2 & 0 & 0 & 0 & 0 \\ 0 & 2 & 0 & 0 & 0 \\ 0 & 0 & 1 & -1 & -i\sqrt{2} \\ 0 & 0 & -1 & 1 & -i\sqrt{2} \\ 0 & 0 & -i\sqrt{2} & -i\sqrt{2} & 0 \end{pmatrix} \begin{matrix} 0A' \\ 0A'' \\ 1A' \\ 1A'' \\ eA'/A'' \end{matrix} \qquad \text{for } t \to t_{pp} + t_{pd} \tag{30}$$

As a consequence, the density matrix $\langle v|\rho|w\rangle$ changes from the initial time ($t = 0$) to times t_{pp} and $t_{pp} + t_{pd}$ after the pump and dump pulses

$$\rho \approx \begin{pmatrix} \frac{1}{2} & & & & \\ & \frac{1}{2} & & & \\ & & 0 & & \\ & & & 0 & \\ & & & & 0 \end{pmatrix} \xrightarrow{\Omega_p} \begin{pmatrix} \frac{1}{4} & \frac{1}{4} & & & \\ \frac{1}{4} & \frac{1}{4} & & & \\ & & 0 & & \\ & & & 0 & \\ & & & & \frac{1}{2} \end{pmatrix} \xrightarrow{\Omega_d} \begin{pmatrix} \frac{1}{4} & \frac{1}{4} & & & \\ \frac{1}{4} & \frac{1}{4} & & & \\ & & \frac{1}{4} & \frac{1}{4} & \\ & & \frac{1}{4} & \frac{1}{4} & \\ & & & & 0 \end{pmatrix} \begin{matrix} 0A' \\ 0A'' \\ 1A' \\ 1A'' \\ eA'/A'' \end{matrix} \tag{31}$$

Using a mixed representation of localized ($|0L\rangle$, $|0R\rangle$, $|1L\rangle$, $|1R\rangle$) doublet states together with the delocalized excited state $|eA'/A''\rangle$, the corresponding changes of the density matrix are written as (see Ref. [12])

$$\rho \approx \begin{pmatrix} \frac{1}{2} & & & & \\ & \frac{1}{2} & & & \\ & & 0 & & \\ & & & 0 & \\ & & & & 0 \end{pmatrix} \xrightarrow{\Omega_p} \begin{pmatrix} \frac{1}{2} & & & & \\ & 0 & & & \\ & & 0 & & \\ & & & 0 & \\ & & & & \frac{1}{2} \end{pmatrix} \xrightarrow{\Omega_d} \begin{pmatrix} \frac{1}{2} & & & & \\ & 0 & & & \\ & & \frac{1}{2} & & \\ & & & 0 & \\ & & & & 0 \end{pmatrix} \begin{matrix} 0L \\ 0R \\ 1L \\ 1R \\ eA'/A'' \end{matrix} \tag{32}$$

The initial racemate of oriented molecules is driven, therefore, almost perfectly to a single enantiomer in the ground and first excited torsional states. These results are based on the RWA, without taking into account the competing effects of (multiphoton) transitions to other states, or tunnelling. Those effects are, indeed, negligible during the pump pulse, due to high selectivity of the laser pulses, and due to the much longer tunneling time of the corresponding enantiomers in the torsional ground state. In contrast, the tunneling time in the first excited torsional state is much shorter, i.e. possibly of the order of the duration of the dump pulse, so that it may become important to study the effects of competing tunneling during the dump pulse. This will be investigated elsewhere.

The results (22)-(32) are in near quantitative agreement with the accurate solutions of the Liouville-von Neumann equation presented in the next section.

3 Results and Discussions

The results of our quantum simulations of the laser driven transformation of the oriented, non-rotating model H_2POSH into almost pure enantiomers are shown in Figures 2 and 3, and in Table 1. These results have been obtained by solving the Liouville-von Neumann equation numerically (21), using a sufficiently large basis to ensure convergence. The propagations start from the initial density (3) which represents the racemate, and then they use the pump-dump laser pulses (8)-(10) in order to drive the system towards the density of the target enantiomer (31), (32). Specifically, Figures 2 and 3 show the results for the pump-dump IR and UV pulses, inducing transitions between torsional states in the electronic ground state, or between the electronic ground and excited states (indicated by *), respectively. The corresponding top, middle and bottom panels of Figures 2 and 3 show the pulse envelopes (see eqn. (10)), the populations of the most important eigenstates $|0A'\rangle$, $|0A"\rangle$, $|1A'\rangle$,$|1A"\rangle$, $|eA'/A"\rangle$, and the populations of the corresponding local states $|0R\rangle$, $|0L\rangle$, $|1R\rangle$, $|1L\rangle$ as well the the delocalized excited state $|eA'/A"\rangle$, respectively. The laser parameters and relevant (transition) dipole matrix elements are listed in Table 1.

Close inspection of Figures 2 and 3 shows that the populations of the five states shown in the middle and bottom panels add up to 1 almost perfectly, i.e. all other states are nearly irrelevant. This confirms the *pump-dump* mechanism as explained by the restricted (five) state representation in Section 2, except for marginal deviations caused by competing excitations of slightly interfering states (see e.g. the oscillatory patterns superimposed on the overall smooth time evolution of the probabilities). Our discussion of the results will, therefore, be based on the restricted state presentation, see section 2, in accord with Ref [12] but now for IR versus UV laser pulses.

ominant transitions, (transition) dipole matrix elements, and pump-dump laser pulses for the transformatio[n] [m]odel racemate into pure enantiomers.

perty/parameter	IR pump	IR dump	UV pump	UV dump
ninant transition	$0A' \to 2A'$	$2A' \to 1A'$	$0A' \to 5^* = 2A''^*$	$2A''^* \to 1A'$
ansition) dipole matrix ment/ea_0	$\langle 0A' \vert \mu_{A'} \vert 2A' \rangle$ $-6.7582 \cdot 10^{-3}$	$\langle 2A' \vert \mu_{A'} \vert 1A' \rangle$ $-89.581 \cdot 10^{-3}$	$\langle 0A' \vert \mu_{A''*} \vert 2A''^* \rangle$ $+29.929 \cdot 10^{-3}$	$\langle 2A''^* \vert \mu_{A''*} \vert 1A' \rangle$ $-42.244 \cdot 10^{-3}$
pled to laser field	$\varepsilon_{pA'}$(IR,x)	$\varepsilon_{dA'}$(IR,x)	$\varepsilon_{pA''}$(UV,z)	$\varepsilon_{dA''}$(UV,z)
d amplitude $\varepsilon^0_{kA'/A''}/GVm^{-1}$	1.25	0.10	3.82	−2.62
quency $\omega_k/2\pi c \cdot cm^{-1}$	343.90	156.51	44252.35	44066.66
ase $\eta_{k,A'/A''}/°$	0	0	0	0
ninant transition	$0A'' \to 2A'$	$2A' \to 1A''$	$0A'' \to 2A''^*$	$2A''^* \to 1A''$
ansition) dipole matrix ment/ea_0	$\langle 0A'' \vert \mu_{A''} \vert 2A' \rangle$ $+7.1303 \cdot 10^{-3}$	$\langle 2A' \vert \mu_{A''} \vert 1A'' \rangle$ $-55.517 \cdot 10^{-3}$	$\langle 0A'' \vert \mu_{A'*} \vert 2A''^* \rangle$ $-29.945 \cdot 10^{-3}$	$\langle 2A''^* \vert \mu_{A'*} \vert 1A'' \rangle$ $+43.854 \cdot 10^{-3}$
pled to laser field	$\varepsilon_{pA''}$(IR,y)	$\varepsilon_{dA''}$(IR,y)	$\varepsilon_{pA'}$(UV,y)	$\varepsilon_{dA'}$(UV,y)
d amplitude $\varepsilon^0_{kA''/A'}/GVm^{-1}$	1.185	0.145	3.82	2.62
quency $\omega_k/2\pi c \cdot cm^{-1}$	343.90	156.51	44252.35	44066.66
ase $\eta_{k,A'/A''}/°$	0	41	0	0
ation t_{pk}/ps	7.0	7.0	0.5	0.5
al field amplitude ε^0_k/GVm^{-1}	1.7224	0.1761	5.4023	3.7052
ensity $I_k/TWcm^{-2}$	0.78748	0.0082317	7.7469	3.6441
arization angle $\alpha/°$	+43.47	+55.41	+45	-45

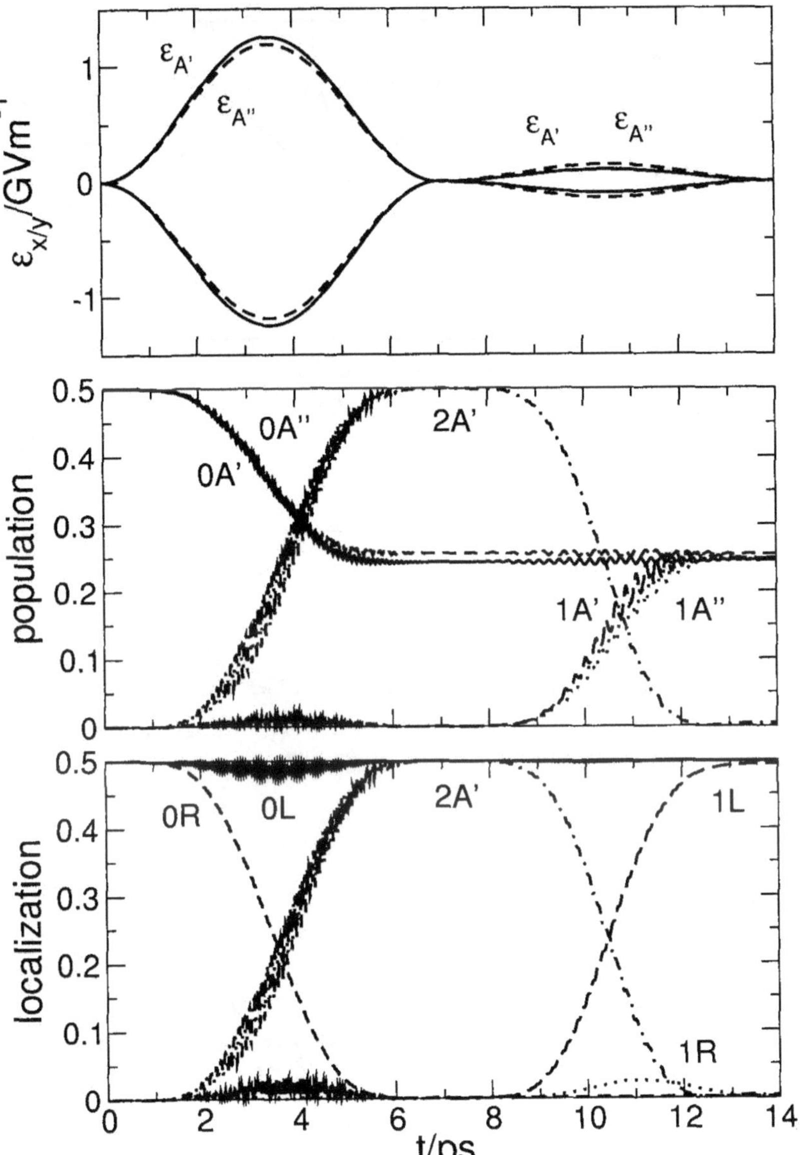

Figure 2. Transformation of the model H_2POSH racemate into pure left (L) enantiomers by means of IR pump dump laser pulses. Top panel: Envelopes of the laser pulses ($\varepsilon_x = \varepsilon_{A'}$ and $\varepsilon_y = \varepsilon_{A''}$). Middle panel: Populations of the five most important torsional states. Bottom panel: populations of the local states $|0R\rangle$, $|0L\rangle$, $|1R\rangle$, $|1L\rangle$ and of the intermediate excited delocalized state.

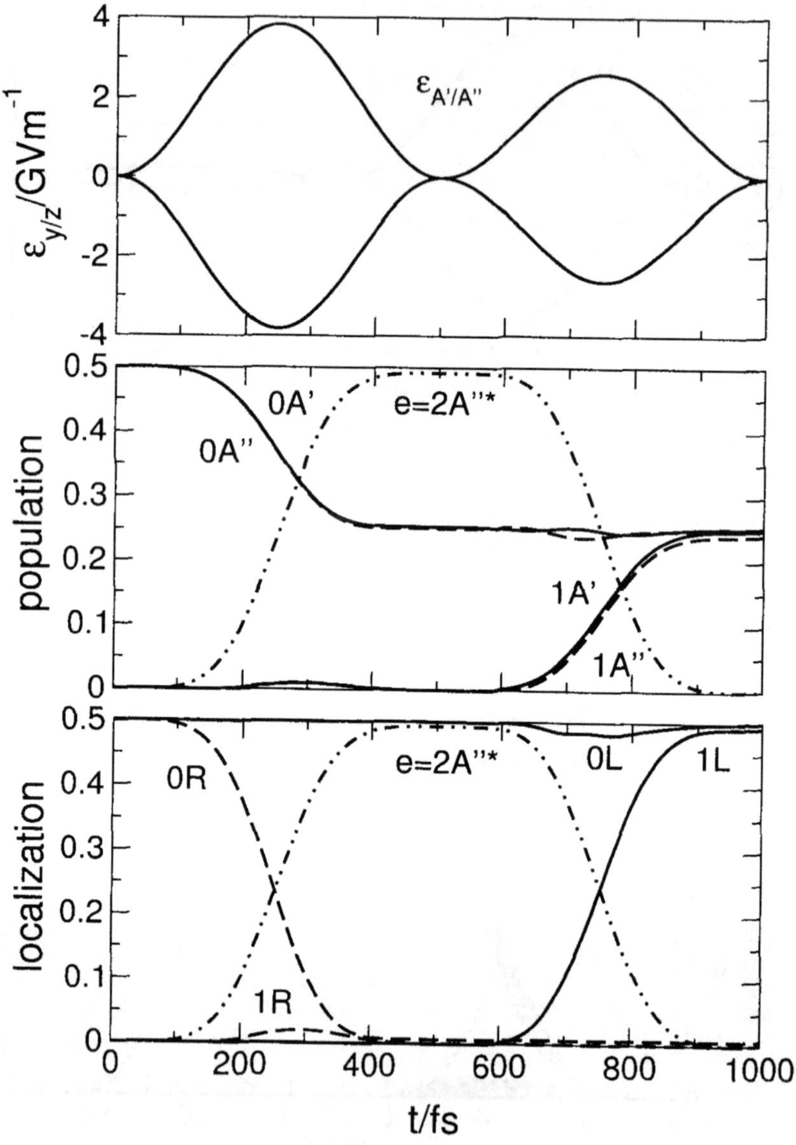

Figure 3. Transformation of the model H_2POSH racemate into pure (left L) enantiomers by means of UV pump dump laser pulses. The notations are as in Figure 2 (but $\varepsilon_y = \varepsilon_{A'}$, $\varepsilon_z = \varepsilon_{A''}$).

Both IR (Figure 2) and UV (Figure 3) laser pulses induce the same overall *pump-dump* mechanism, i.e. the first pulse transfers exclusively the undesired enantiomer (here: the right one, i.e. $|0R\rangle$) into the excited delocalized state, and the second pulse dumps it into the excited local state $|1L\rangle$ of the target enantiomer, without any significant effect on the target enantiomers $|0L\rangle$ which are already present in the initial racemate. The net effect is, therefore, the transformation of the initial 50%:50% mixture of enantiomers in the ground state, $|0L\rangle$ and $|0R\rangle$, into another 50%:50% mixture i.e. for the target enantiomer in its ground and excited states $|0L\rangle$ and $|1L\rangle$.

Irrespective of this overall similarity, there are also several significant differences between the IR and UV *pump-dump* mechanisms, which should have important consequences for the experimental implementations. These concern the frequencies, the choice of the torsionally/electronically excited intermediate states, the laser pulse durations and field strength, and the polarizations. For convenience, our comparison will be based on a choice of equal durations of the pump and dump pulses (different from Ref [12]). Moreover, we shall use rather short durations, implying rather large field strength, but restricted to the domain of the present mechanism where competing multiphoton transitions to other levels are negligible.

First, the laser frequencies are obviously different for the IR versus UV pump-dump mechanisms.

Second, the choice of the intermediate excited states $|eA'/A"\rangle$ for IR transitions in the electronic ground state e.g. $|0A'\rangle \rightarrow |eA'/A"\rangle \rightarrow |1A'\rangle$, or $|eA'/A"^*\rangle$ for UV transitions via electronic excited state e.g. $|0A'\rangle \rightarrow |eA'/A"^*\rangle \rightarrow |1A'\rangle$ is based on conditions (13-18) for π pulses. Accordingly, the most efficient sequential IR or UV pulses should be associated with the largest products of (transition) dipole matrix elements, $\langle 0A'|\mu_{A'}|eA'\rangle \times \langle eA'|\mu_{A'}|1A'\rangle$ or $\langle 0A'|\mu_{A"*}|eA"^*\rangle \times \langle eA"^*|\mu_{A"}|1A'\rangle$, respectively. This criterion yields $|2A'\rangle$ as intermediate state for IR transitions i.e. the pump pulse excites the first torsional overtone, followed by the dump transition to the fundamental excited torsional state. The corresponding matrix elements differ by approximately one order of magnitude (see Table 1), calling for much stronger field strengths of the pump pulses than for the dump pulses, see Figure 2 and Table 1. Alternatively, one may also use $|2A"\rangle$ as intermediate state, whereas higher excited states are not recommended because they would require even higher overtone excitations with higher intensities due to smaller transition dipole matrix elements. The same criterion yields $|5^*\rangle = |2A"*\rangle$ as intermediate state for the UV transition. This may be interpreted as a consequence of preferably near-vertical Franck-Condon type transitions between the initial, intermediate, and final states (Note, however, that we do not invoke the Franck-Condon approximation in our rigorous quantum calculations). An important difference between the IR and UV transitions is, therefore, that the intermediate state in UV transitions carries more torsional quanta (five) than for the IR transition (two). As a consequence, the level density close to the intermediate state is much higher for UV than for IR transitions. We anticipate, therefore, that the selectivity of the IR laser pulses will be better than for the UV transitions which could excite also near resonant, non-torsional states.

Third, the required laser pulse durations and field strength turn out to be longer and higher for IR laser pulses, than for UV pulses. This is again a consequence of the corresponding products of (transition) dipole matrix elements which are smaller for IR than for UV pulses. For the selected excited intermediate states, we have e.g., (see Table 1)

$$\left|\langle 0A'|\mu_{A'}|2A'\rangle \times \langle 2A'|\mu_{A'}|1A'\rangle\right| < \left|\langle 0A'|\mu_{A"*}|2A"^*\rangle \times \langle 2A"^*|\mu_{A"*}|1A'\rangle\right| \tag{33}$$

The conditions (13) - (18) for π-pulses then imply that the corresponding products of laser field durations times field strengths should be larger for IR than for UV pulses. For example, the applications shown in Figures 2 and 3 (see also Table 1) use 7 ps versus 0.5 ps pulse durations for IR versus UV pump pulses with similar field strengths.

Forth, the polarizations of the IR versus UV pump and dump pulses are also different. In most cases, linearly polarized pulses achieve high enantiomer selectivities, but for the IR dump pulse a phase shift of 41° (see Table 1) implying an elliptically polarization is found to be more efficient. The problem of elliptically, circulary, or linearly polarized pulese will be investigated elsewhere. For the present linearly polarized pulses ($\eta = 0$ in Table 1), the polarization angles are determined by eqn (19) as arctan of the ratios of the A' and $A"$ (transition) dipole matrix elements which are associated with the transitions between two doublet states $|0A'\rangle$ or $|0A"\rangle$ (and like wise $|1A'\rangle$ and $|1A"\rangle$) and the excited state. From Table 1, the resulting polarization angles are close to $-45°$ and $+45°$ for the UV pump and dump pulses, i.e. the UV pump and dump laser pulses have near orthogonal polarizations, see also the discussion in Ref [12]. A systematic investigation of UV transitions via other, less efficient intermediate states reveals, however, that this near orthogonal polarization is not a general rule. Instead it may be interpreted as a consequence of the incidental near co-incidence of the transition dipole functions close to the near vertical UV-transitions involved, see the Figures of the transition dipole functions shown in Ref [11]. In contrast, the polarization angles for the most efficient IR pump and dump pulses are +43.47 and +55.41 (see Table 1), i.e. they are non-orthogonal.

4 Conclusions

The present comparison of UV versus IR pump-dump laser pulses for the transformation of the model H_2POSH racemate into pure enantiomers confirms the overall mechanism which has been suggested in Ref [12], but it yields several differences which should be relevant for experimental implementations. By extrapolation of the exemplary results presented in Section 3, it should be advantageous, on the one hand, to use UV laser pulses, because the frequency domain is more easily accessible to present laser technology, and the pulses are also shorter and/or less intense, than IR laser pulses. Moreover, it may be easier to use orthogonal linear polarizations of the pump and dump pulses, as required for the present most efficient UV pulses, in contrast with non-orthogonal linear and elliptical polarizations of the IR pulses. The possibility to prepare UV pulses with arbitrary polarizations has been demonstrated recently.[16] On the other hand, IR laser pulses may be more advantageous than UV pulses because they allow transitions via low excited torsional states, in contrast with higher torsional and electronic excitations by UV pulses. Therefore, the corresponding level densities are lower for IR than for UV pulses, suggesting high selectivities of the IR pulses, whereas the UV pulses may excite competing non-torsional levels. These effects will be investigated elsewhere.

The present model assumes that the racemate has been preoriented e.g. by the method developed in Ref [14,15], see also Refs [17,18,19,20]. The combined effect of the laser pulses for orientation, plus those for the transformation of the oriented racemate into pure enantiomers, implies the transmission of chiral information from the laser pulses, i.e. from the laboratory, to the molecules. This approach may serve as an alternative method to enantioselective synthesis where the chiral information of some (e.g. organo-metallic) catalyst, or reaction partner, is imposed on the target enantiomer. While finishing this article, the 2001 Nobel prize in chemistry is awarded to W. Knowles, R. Noyori and B. Sharpless, and it is certainly adequate to give supreme credit to their,[21] and others[22] syntheses of enantiomers. The present approach (see. also Refs 5-14) may add new aspects, in particular to the laser-synthesis of enantiomers on ultrashort (fs-ps) time scales.

Acknowledgement

We whould like to express our thanks to Professor Y. Fujimura and Dr. L. González for helpful and stimulating discussions and support. Financial support by the Deutsche Forschungsgemeinschaft (D.K. and J.M., project Ma 515-18/1,2) and the Research Fellowship of the Japan Society for the Promotion of Science (K.H., project No. 6254) is also gratefully acknowledged. The computations have been carried out on our HP J5000 (Berlin) and Compaq Alpha (Sendai) computers.

References

1. J.J. Flores, W.A. Bonner and G.A. Massey, *J. Am . Chem. Soc.* **99**, 3622 (1977).
2. Y. Shimizu, J. Chem. Soc., *Perkin Trans* **1**, 1275 (1987).
3. Y. Inoue, *Chem. Rev.* **92**, 741 (1992).
4. N. P. M. Huck, W. F. Jager, B. de Lange and B. L. Feringa, *Science* **273**, 1686 (1996).
5. A. Salam and W. J. Meath, *J. Chem. Phys.* **106**, 7865 (1997).
6. A. Salam and W. J. Meath, *Chem. Phys.* **228**, 115 (1998).
7. M. Shapiro, E. Frishman and P. Brumer, *Phys. Rev. Letters* **84**, 1669 (2000).
8. D. Gerbasi, M. Shapiro and P. Brumer, *J. Chem. Phys.* **115**, 5349 (2001).
9. Y. Fujimura, L. González, K. Hoki, D. Kröner, J. Manz and Y. Ohtsuki, *Angew. Chem. Int. Ed* **39**, 4586 (2000).
10. K. Hoki , Y. Ohtsuki and Y. Fujimura, *J. Chem Phys* **114**, 1575 (2001).
11. L. González, D. Kröner and I.R. Solá, *J. Chem. Phys.* **115**, 2519 (2001).
12. K. Hoki, L. Gónzalez and Y. Fujimura, *J. Chem. Phys.* , (to be published).
13. K. Hoki and Y. Fujimura in ACS Books "Laser Control and Manipulation of Molecules", edited by A.D. Bandrauk, R.J. Gordon and Y. Fujimura, American Chemical Society, 2001, in press.
14. K. Hoki, D. Kröner and J. Manz, *Chem. Phys.* **267**, 59 (2001).
15. K. Hoki, Y. Fujimura, *Chem. Phys.* **267**, 187 (2001).
16. G. Gerber and T. Brixner, *Optics Lett.* **26**, 557 (2001).

17. B. Friedrich, D. R. Herschbach, *Phys. Rev. Lett.* **74**, 4623 (1995).
18. T. Seideman, *J. Chem. Phys.* **103**, 7887 (1995).
19. N. E. Henriksen, *Chem. Phys. Lett.* **312**, 196 (1999).
20. H. Sakai, C. P. Safvan, J. J. Larsen, K. M. Hilligsøe, K. Held, H. Stapelfeldt, *J. Chem. Phys.* **110**, 10235 (1999)
21. The Nobel lectures of W. Knowles, R. Noyori and B. Sharpless will be published in *Angew. Chem. Int. Ed.* **41** (2002).
22. B. Seelig, S. Keiper, F. Stuhlmann, A. Jäschke, *Angew. Chem. Int. Ed.* **39**, 4391 (2000).

ADAPTIVE QUANTUM CONTROL OF FEMTOCHEMISTRY IN THE GAS AND LIQUID PHASE

T. BRIXNER AND G. GERBER

Physikalisches Institut, Universität Würzburg, Am Hubland, 97074 Würzburg, Germany

E-mail: gerber@physik.uni-wuerzburg.de

Obtaining active control over quantum-mechanical systems is a fascinating prospect in modern physics. The basic principles of adaptive femtosecond quantum control are discussed. Specifically shaped femtosecond laser pulses are used to direct the dynamical evolution of quantum wavefunctions into desired target channels. This is achieved in a "closed-loop" implementation wherein experimental feedback signals are evaluated by a learning algorithm. In this iterative optimization scheme, knowledge of the Hamiltonian is not required. Specific experimental examples are automated femtosecond laser pulse compression, control of gas-phase molecular photodissociation, and selective control of molecular excitation under liquid-phase conditions. The complexity of these control problems is made visible. As further technological development, the technique of femtosecond polarization pulse shaping is discussed briefly, which makes it possible to vary the polarization state of light, intensity and oscillation frequency in a flexible manner on an ultrashort timescale.

1 Introduction

Controlling the outcome of chemical reactions—i.e., guiding the formation of desired products at the expense of starting materials and unwanted products—is a central theme in modern chemistry. Traditionally, chemists use macroscopic variables (such as temperature, pressure, and concentration) to influence the result of a certain reaction. Within the statistical framework of intermolecular collisions, one tries to shift the equilibrium between initial reactants and final products into the desired direction. However, the microscopic reaction path is not within the reach of the experimenter, thus limiting the realm of attainable control. In this context, light has the potential to be a very powerful tool. Making use of suitable photoexcitations, it should be possible to overcome reaction barriers inhibiting desired reaction pathways or to change the thermodynamic starting place of reactant species. Unfortunately, success of this "traditional" way of photochemistry is limited because of the complexity of the molecular potential energy surfaces (PES). In general, these potentials depend on the coordinates of all contributing atoms and thus constitute a complex multidimensional "landscape." It is intuitively clear that the quantum-mechanical temporal evolution of the system after simple photoexcitation does not necessarily lead to the desired end product.

For an accurate quantum-mechanical description of the system, it is necessary to account also for intramolecular coupling terms. To state it generally, the full molecular Hamiltonian is responsible for time evolution. The problem of control over photochemical reactions is therefore a difficult one. Nevertheless, substantial progress has been made both theoretically and experimentally by exploiting the broad range of quantum interference effects. The emerging discipline can be termed "quantum control" or "coherent control," wherein full utilization of the relevant features of the molecular Hamiltonian upon light–matter interaction is made.[1,2,3,4,5,6] Drawing from the coherence properties of lasers, the spectral–temporal characteristics of the applied electric fields are adjusted in such a way that a given initial quantum-mechanical wavefunction is transformed into a desired target wavefunction. This quantum-mechanically coherent (i.e., phase-preserving) manipulation of the system makes it possible to "select" out of the many possible trajectories through the complex potential energy surfaces those which lead to the desired goal. Constructive interference of reaction pathways is used to enhance desired products, whereas destructive interference reduces contributions from unwanted channels.

The beginning of quantum control focused on three types of single-parameter control schemes. These were mainly the "frequency-domain phase control" suggested by Brumer and Shapiro[7,8] and experimentally carried out on atoms[9] and small molecules,[10,11] the "time-domain pump–dump control" proposed by Tannor, Kosloff, and Rice[12,13] and experimentally applied in small molecular systems,[14,15,16,17,18,19] and "stimulated Raman adiabatic passage (STIRAP)" suggested and experimentally carried out by Bergmann and coworkers.[20,21,22] At first, these schemes appeared to be rather different. But nowadays it is clear that they merely emphasize different aspects of a general control methodology.

In the further development of experimental techniques, the use of ultrashort laser pulses with time-varying photon frequencies was also investigated. The control parameter was "linear chirp," corresponding to a linear increase or decrease of the momentary frequency as a function of time under the pulse envelope. Linearly chirped pulses were used to control electronic excitation in molecular systems.[23,24,25,26,27,28] Frequency-swept infrared pulses were also applied in "ladder-climbing" processes to control vibrational excitation.[29,30] These experiments represent a first step toward using "tailored" femtosecond pulses with specifically designed spectral–temporal characteristics.

In the examples described above, one variable (i.e., the phase difference between two CW lasers, or the time delay between two laser pulses, or the linear-chirp parameter) is changed and the effect noted in experimental observables. Therefore, these methods are referred to as single-parameter control schemes. The variable is also called the "control knob," since it represents the experimenter's "handle" on the modulations that are induced. For some systems with simple and predictable quantum dynamics, this can be an effective means. However, in the case of larger systems with complex multidimensional potential energy surfaces, these simple schemes may not be sufficient. One can envision that electric field shapes are needed which are precisely matched to a specific control problem.

The theory of specifically shaped femtosecond laser pulses in the context of coherent reaction product control was already discussed by Tannor and Rice.[12,31] They used a variational approach to find optimal electric fields. Rabitz and coworkers,[32,33] and independently Kosloff and coworkers,[34] have then applied optimal control theory to solve the problem of many-parameter quantum control theoretically. In chemical reaction control, the following question is asked: With a given target distribution of photoproducts and the quantum-mechanical equations of motion, what is the shape of the electric field required to guide the temporal evolution of the system appropriately?

An intuitive picture of the optimal many-parameter control process is that the wavepacket motion on the complex potential energy surfaces is controlled by "offering" those wavelengths at each specific point in the temporal evolution, which are needed to transfer the wavepacket onto some other surface which is better suited for its further evolution. Many of these transient Franck–Condon transitions might be present throughout the control process, and the coherence relations between the different vibrational modes are constantly being adjusted, such that finally an optimal steering of the wavepacket leads to a specific product channel.

Due to the short timescale of vibrational motion within molecules, this scheme requires correspondingly short (femtosecond) laser pulses whose spectral–temporal characteristics can be manipulated in a flexible manner. These laser pulses are called shaped or tailored laser pulses. Of course, the required modulations of the laser fields cannot be achieved by turning a single control knob, and in the early stages of many-parameter quantum control theory, it was not clear how such modulations could be achieved. However, technological progress in the area of ultrafast optics offered new possibilities. Femtosecond pulse shapers,[35,36,37,38,39,40] developed primarily for telecommunication purposes, allow convenient and sophisticated manipulation of femtosecond laser pulses.

An important observation to be made in this context is that, at least in gas-phase experiments on isolated molecules, coherence-destroying processes such as intermolecular collisions or spontaneous emission from excited states take place on a timescale much longer than the process to be controlled. The time evolution of the quantum system—governed by the Schrödinger equation—therefore remains completely deterministic and phase-preserving.

The procedure where optimal electric fields are first calculated and subsequently tested experimentally is also called open-loop control. Despite the theoretical success of optimal control theory, there remain problems in implementing this type of experiment.[1,41] First, and most important, theory is not able to predict the optimal laser fields because the required potential energy surfaces of more complex molecules are, in general, not accurately known. Numerical simulations are currently not possible for systems involving more than just a few atoms. And second, even if theory could predict such fields, it is not clear that they could be reproduced accurately enough under laboratory conditions at the location of interaction with the molecule. The obvious brute-force method of testing all possible laser pulse shapes within the experimental constraints is not feasible either, because the search space, given by all possible combinations of control knob settings, is simply too large to be scanned completely.

A general method to circumvent these problems and to find optimal laser fields under laboratory conditions was proposed by Judson and Rabitz in their pioneering work "Teaching lasers to control molecules."[41] They suggested that a "closed-loop" setup be employed in which the experimental output is included in the optimization process (Fig. 1). The quantum system should be excited with shaped femtosecond laser pulses, and the experimentally determined outcome of the light–matter interaction (i.e., the photoproduct yields) be evaluated in an optimization algorithm which uses the information to iteratively improve the applied laser pulse shape.

Step by step, an optimal laser field should be found without the need for a-priori knowledge of molecular Hamiltonians or any other detail about the investigated system or the initiated reaction processes. Instead, the experimental apparatus is used as an analog computer which solves the Schrödinger equation exactly with the true laboratory field. Successive steps of the control-field optimization are sequentially performed on "new" but identical quantum systems. This type of loop is also called learning loop or adaptive control loop, because only after completion of the control process on one molecule, the results are evaluated with the goal of "learning" how to perform better on the next molecule under identical initial conditions. The electric field iteratively "adapts" to the demands of the quantum system and the control objective designed by the experimenter.

Applications of the closed-loop learning scheme to molecular systems were first demonstrated in an experiment by Bardeen et al.[42] where the excited-

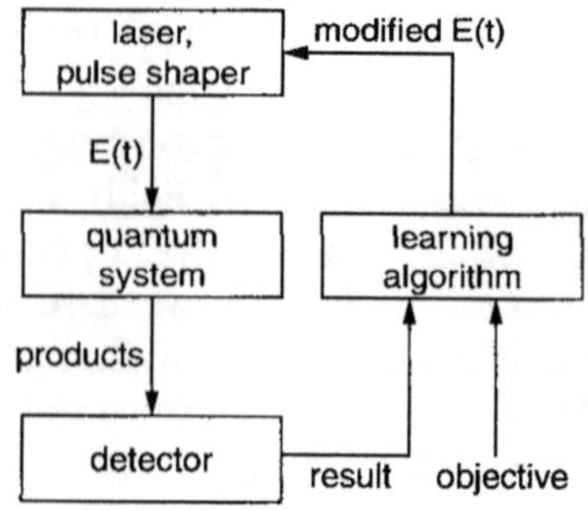

Figure 1. Closed-loop learning control. The combination of a femtosecond laser system and a computer-controlled pulse shaper is used to generate specific electric fields $E(t)$, which are irradiated onto a given quantum system, initiating photophysical or photochemical processes. After detection of the resulting photoproducts, a learning algorithm calculates modified electric fields based on the information from the experimental feedback signal and the user-defined control objective. The improved laser pulse shapes are tested and evaluated in the same manner. Cycling through this loop many times results in iteratively optimized laser pulse shapes and the desired control over the quantum system.

state population of a laser dye in solution was optimized, and in experiments performed in our group where photodissociation reactions of complex molecules were controlled.[43,44] Other examples include the excitation of different vibrational modes in a molecular liquid[45] and the control of vibrational dynamics in a four-wave mixing experiment.[46]

But optimal quantum control is not limited to molecular systems. Femtosecond laser pulse shapers and learning loops have been used for automated pulse compression[47,48,49,50,51] and optimized generation of arbitrary laser pulse shapes,[52,53] control of two-photon transitions in atoms,[54,55] shaping of Rydberg wavepackets,[56] optimization of high-harmonic generation,[57] and control of ultrafast semiconductor nonlinearities.[58] Shaped electric fields have also been suggested to be of use in the context of laser cooling.[59,60,61]

The adaptive many-parameter control approach represents the most general realization of the coherent manipulation of light–matter interaction. This is indicated by calling this method "optimal" control (literally meaning that there is no better way of doing it because all possibilities are already included). It automatically encompasses "other" cases of control scenarios provided that the spectral bandwidth of the shaped laser pulse contains all those frequencies with adequate intensities that would be required in the corresponding (single-parameter) schemes. Specifically, adaptive control includes pump–dump scenarios, linear chirp control, or phase control. Additionally, it should be noted that with repeated Rabi cycling, states can be reached which are not accessible with a single one-photon transition. This occurs when within a single Rabi cycle a photon from the blue edge of the broadband laser spectrum is absorbed while emission is stimulated by a photon from the red edge. This leaves the molecule in a higher ground-state vibrational level. Depending on where the last Rabi cyle of this sequence is stopped, the population is left in an energetically high-lying vibrational level of the ground or the excited state. This provides a mechanism for "simulating" the absorption of a photon whose wavelength may not even be included in the original laser spectrum. Another strong-field effect is the Stark shifting of levels, which may be exploited to advantage by moving levels in and out of certain Franck–Condon windows. Although with present-day technology it is not yet possible to also generate shaped femtosecond laser pulses fulfilling the STIRAP requirements (strong enough fields and slow "adiabatic" time evolution), this scheme is, in principle at least, nonetheless included in optimal control as well.

Considering the generality of the adaptive approach outlined here, it is clear that the separation of control methods into different categories by assigning distinct names ("coherent control," "quantum control," "optimal control," "pump–dump control," etc.) is only a matter of "historical" practice and convenience. This should not conceal the fact that their physical background is the same. Especially with the frequently made distinction between "coherent control" and "optimal control" it should be kept in mind that the adaptive optimal control approach is of course also a method of coherent control because it relies on the coherent manipulation of quantum systems.

2 Experimental Setup

The experimental setup of adaptive femtosecond quantum control is shown in Fig. 2. A femtosecond laser pulse shaper (right side of the figure) is used to impose specific spectral phase modulations. Details of our setup have been previously published.[48,50] Briefly, the device consists of a zero-dispersion compressor in a 4f-geometry, which is used to spatially disperse and recollimate the femtosecond laser pulse spectrum. Insertion of a liquid-crystal display (LCD) in the Fourier plane of the compressor provides a mechanism for convenient manipulation of the individual wavelength components. By applying voltages, the refractive indices at 128 separate pixels across the laser spectrum can be changed, and upon transmission of the laser beam through the LCD, a frequency-dependent phase is acquired due to the individual pixel voltage values. In this way, an immensely large number of different spectrally phase-modulated femtosecond laser pulses can be produced. In all experiments de-

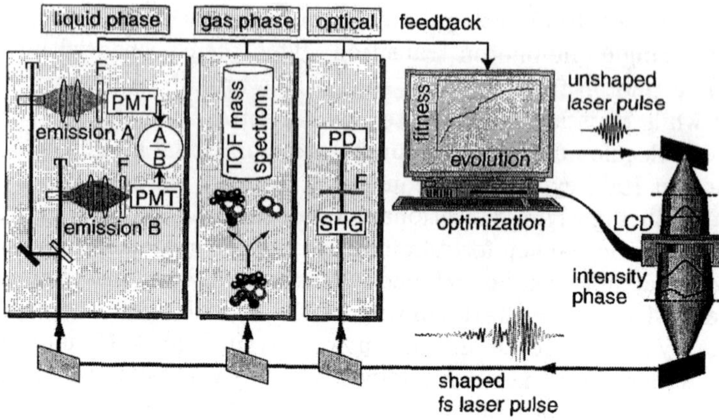

Figure 2. Experimental Setup. A femtosecond pulse shaper (right side) is used to generate phase-modulated laser pulses for closed-loop adaptive quantum control. Depending on the type of experiment, experimental feedback signals may be collected from purely optical signal processing (second-harmonic generation, SHG, detected in a photodiode PD), from gas-phase time-of-flight (TOF) mass spectrometry, or from liquid-phase emission spectroscopy (where the signals are detected after filtering with photomultiplier tubes, PMT). These signals are processed in a computer learning algorithm (based on an evolution strategy approach) which iteratively improves the applied laser pulse shape until an optimum is reached.

scribed below, the spectral amplitudes have not been changed, and therefore the integrated pulse energy remains constant for different pulse shapes. On the other hand, by virtue of the Fourier transform properties, spectral phase changes result in phase- and amplitude-modulated laser pulse profiles as a function of time.

The shaped laser pulses are then used in different types of adaptive quantum control experiments wherein experimental feedback signals guide an automated search for optimal electric fields within a learning algorithm. We have employed purely optical feedback from second-harmonic generation (SHG) for purposes of automated pulse compression, feedback from time-of-flight (TOF) mass spectrometry to control photodissociation reactions of complex molecules in the gas phase, and feedback from emission spectroscopy to control molecular excitation processes under liquid-phase conditions. In a general control experiment, the optimal pulse is not known in advance. And since the variational space of possible pulse shapes (given by the different combinations of the 128 LCD pixel voltage settings) is so huge, scanning the complete parameter space is impossible. In order to overcome this problem,

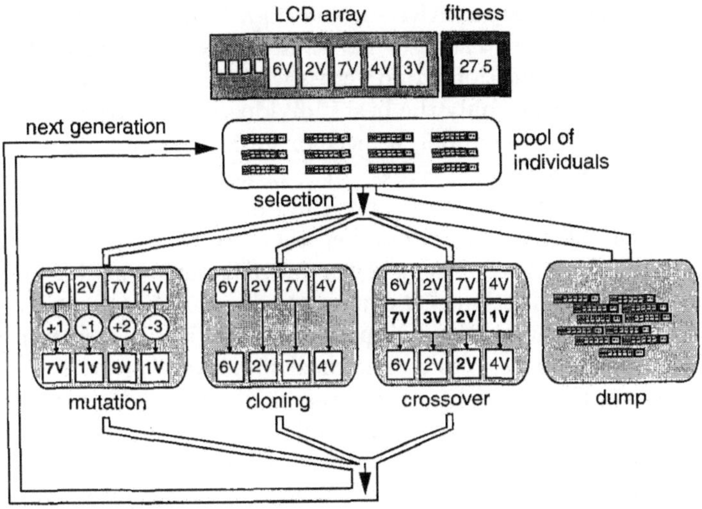

Figure 3. Evolutionary algorithm. This basic schematic shows one cycle of the optimization procedure. Each individual within the population pool represents a specific laser pulse shape (i.e., specific LCD pixel voltage values). Depending on the fitness, an individual is either selected for reproduction or it is rejected. Reproduction operations include mutation, cloning, and crossover of the LCD voltage arrays. The children thus produced constitute the next generation, which is then evaluated in the same way. Iterating through this loop many times finally leads to optimized laser pulse shapes.

we use a learning loop to find optimized electric fields taking into account the experimental outcome. In an iterative process, the electric laser fields are improved with a computer algorithm until the particular optimization objective is reached and the feedback signals approach the user-specified goal.

The global search method we use for this purpose is an evolutionary algorithm.[62,63] Our implementation has been described in detail elsewhere,[48] but we briefly outline the technique here. Evolutionary algorithms have working schemes inspired by biological evolution (Fig. 3). At the beginning of the optimization, a pool of randomly initialized "individuals" (i.e., different laser pulse shapes having a "genetic code" comprised of a linear array of LCD pixel voltages), are experimentally tested for their control performance. The success of a pulse is judged with respect to an optimization task determined by the experimenter at the outset. Since more than one experimental observable may be optimized, a "fitness function" is employed which takes into account all required signals from the experimental output. The better the outcome of the quantum control experiment, the higher is the fitness assigned to the individual laser pulse shape. After evaluating all individuals from the pool, some percentage of the fittest laser pulses are selected for reproduction. We then use "cloning," "mutation," and "crossover" operations to produce a new pool of laser pulses. Since only the best individuals of the previous generation serve as parents of the new pool, a mechanism is provided wherein offspring can inherit good "genetic properties." It is often the case that—by virtue of crossover and mutation—some of these children are "better adapted" to the environment than their predecessors, i.e., they are better suited to achieve a given quantum control task. The new generation is then tested again, followed by selection and reproduction. By cycling this evolutionary loop a number of times, the fitness of the best individuals increases from generation to generation until an optimum is found. The most important property of this type of optimization is the generality of the search. In principle, it works for any type of system, and the user does not need to provide any type of model for the system's response. The user who has an optimization objective simply has to choose the appropriate fitness function for purposes of evaluating the individual pulses.

3 Experimental Examples

3.1 Automated Pulse Compression

As a first implementation of adaptive closed-loop learning control, we describe briefly the technique of automated pulse compression in a purely optical realization of adaptive quantum control. Even with nowadays rapidly improving ultrashort laser technology, bandwidth-limited amplified femtosecond laser pulses are often not routinely available. In the commonly employed chirped-pulse amplification (CPA) scheme, the temporal pulse profile is stretched in time, and recompressed after amplification. In the ideal case, all spectral phase modulations should be removed at the output of the laser system. In practice, this is not easily achieved. Often the pulses have acquired higher-order phase terms upon propagation through the various optical components of the stretcher–amplifier–compressor setup leading to imperfect temporal pulse characteristics. Conventional compression of these pulses relies on the sophisticated design and careful alignment of all relevant optical parts. In this context, a computer-controlled laser pulse shaper represents an ideal tool. Complex spectral phase profiles can be easily generated, and dispersive phase terms can be compensated for without mechanical realignment. In combination with a learning loop, it is even possible to avoid the need for prior pulse-shape characterization.

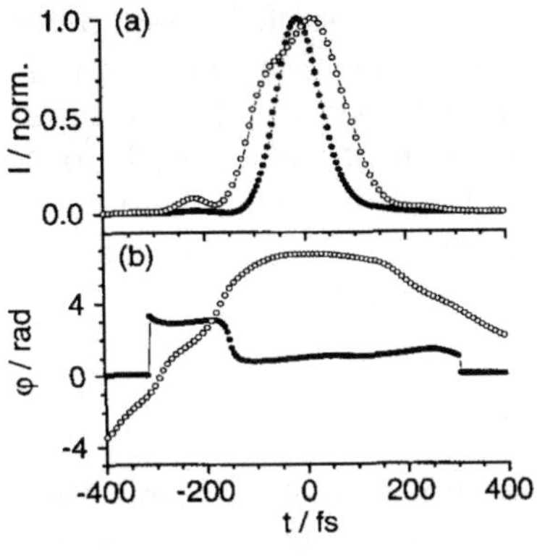

Figure 4. Pulse shapes before (open circles) and after (solid circles) SHG optimization. Obtained by analyzing the FROG traces, this graph shows (a) the temporal intensity and (b) the temporal phase.

The automated generation of bandwidth-limited (i.e., optimally short) amplified laser pulses is achieved by maximizing the nonresonant type-I SHG yield produced in a 100-μm BBO crsytal. After removal of the fundamental by a shortpass filter, the signal is recorded by a photodiode and used as feedback in the evolutionary algorithm. The SHG signal is proportional to $\int I^2(t)\,dt$. Therefore it is intuitively clear that the shorter the pulse at a given energy, the higher the second-harmonic yield because of the second-order nonlinearity. Maximizing SHG therefore leads to bandwidth-limited laser pulses.[47,48,49,50,51]

We analyzed the laser pulse shape achieved in this optimization procedure[50] (after 100 generations of the evolutionary algorithm) by SHG-FROG (Fig. 4). The original pulse before optimization (open circles) clearly displays an asymmetric distribution of the temporal intensity, with significant prepulse and "shoulder" contributions. After optimization (solid circles), a very smooth, symmetric, and temporally shortened pulse emerges which reaches the bandwidth limit. The variation of the phase (chirp) over the relevant temporal range is also reduced in the optimized profile.

This experiment shows that adaptive femtosecond pulse shaping can remove higher-order spectral phase modulations which often arise in femtosecond amplifier laser systems. If optical elements between the output of the pulse shaper and the experiment further modify the pulse by material dispersion, the adaptive technique can also be used to precompensate for these effects. The experimenter simply has to define the feedback signal—i.e., second-harmonic production—at the spot of the experiment, and the pulses will be optimally short at the desired location. This simple-to-use technology is likely to be included in future generations of femtosecond amplification systems. The successful demonstration of the closed-loop pulse-shaping technique is promising for the realization of quantum control experiments where the feedback signal is taken directly from the investigated quantum system. This is the topic of the following sections.

3.2 Gas-Phase Control

In this realization of quantum control in the gas phase, the optimization objective is to maximize (or minimize) branching ratios within photodissociation reactions in complex molecules. The question is if the evolutionary algorithm can find laser fields which preferably generate one photoproduct at the expense of another one. As one particular example, we have investigated $CpFe(CO)_2Cl$, where Cp=cyclopentadienyl=C_5H_5. Its structure is such that the central Fe atom is bonded directly to a C_5H_5 ring, two carbonyl ligands, and one Cl ligand. Upon irradiation with femtosecond laser pulses, a number of fragmentation pathways are observed, with dissociation times in the subpicosecond range. Thus it can be expected that adaptive femtosecond control may be possible. Out of the many possible dissociation channels of $CpFe(CO)_2Cl$, two were selected for optimizing a branching ratio. In the first channel, single carbonyl loss and ionization leads to $CpFeCOCl^+$ while in the second channel, more complete fragmentation leads to $FeCl^+$. The two optimization goals explored were maximization and minimization of the $CpFeCOCl^+/FeCl^+$ product ratio.[44]

The resulting product ratios, after optimization with the evolutionary algorithm, are shown in Fig. 5 together with the corresponding electric laser fields.[44,64] It is seen that the distributions indeed can be changed in the way requested by the fitness function. The $CpFeCOCl^+/FeCl^+$ ratios can be switched between about 5:1 (left) and 1:1 (right). Bandwidth-limited laser pulses (middle) are not suitable for achieving optimal conditions, and only an intermediate value for the branching ratio is observed. Likewise, arbitrarily phase-shaped long laser pulses—the starting place for both optimization processes—are not sufficient. Instead, specifically shaped laser pulses are required. As a conclusion of this experiment it is noted that reaction product control by adaptive femtosecond pulse shaping is obviously possible in complex systems. No information about the sample substance is needed in the optimization procedure. The closed-loop setup works in a completely automated fashion.

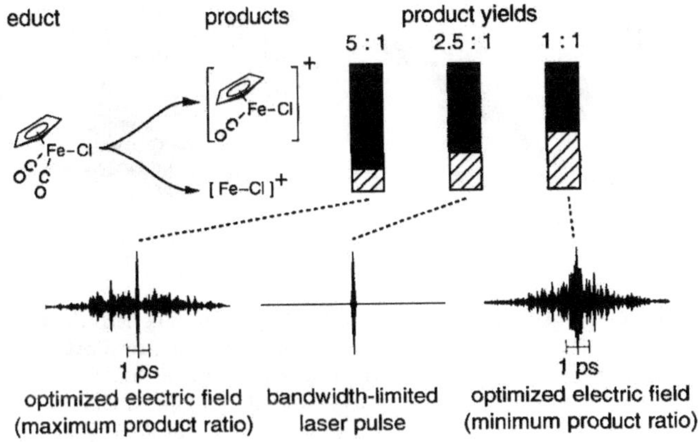

Figure 5. Quantum control of the $CpFeCOCl^+/FeCl^+$ product ratio. The top part displays relative yields of the two investigated product channels leading to $CpFeCOCl^+$ (solid blocks) and $FeCl^+$ (shaded blocks). The bottom part shows the temporal electric fields leading to a maximum (left) and minimum (right) $CpFeCOCl^+/FeCl^+$ ratio, as well as the electric field of a bandwidth-limited laser pulse (middle) leading to an intermediate branching ratio. The electric fields are calculated by Fourier-transforming the measured laser spectrum and the spectral phase applied by the pulse shaper.

It is also possible to analyze the significance of the resulting complex-looking electric field structures. This is done by performing what we call "correlation experiments."[64] These are used to "map out" specific features of the search-space topography within quantum control experiments, thus making visible their complexity directly. Specifically, the question is addressed if the results shown in Fig. 5 could be explained by a trivial "intensity" dependence. For this purpose, we employed again the SHG detection scheme used in the previous section on pulse compression. The SHG signal is recorded simultaneously with the photoproduct yields of the investigated molecular process (for each individual during the branching-ratio optimization).

It was already shown that SHG can be used as a measure for pulse intensity and that its maximization leads to optimally short pulses. Hence, if the outcome of the molecular product yield or product ratio in a quantum control experiment depends on the laser intensity in the same simple way as SHG does—where the SHG signal is directly related to laser intensity—then a correlation should be visible between the two signals. If, on the other hand, the two signals (SHG and feedback from the photochemical process) are not correlated, then the molecular control experiment requires a specific phase function and not simply the correct laser pulse intensity.

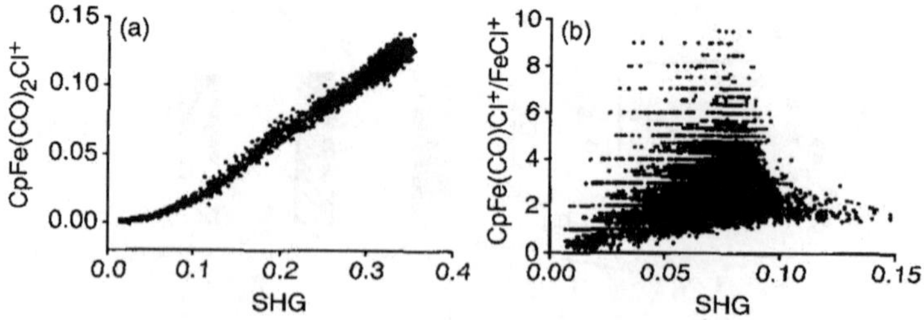

Figure 6. Correlation diagrams for $CpFe(CO)_2Cl$ quantum control. (a) In the case of $CpFe(CO)_2Cl^+$ maximization, the observed ion parent yields are plotted versus SHG efficiency for all individual laser pulse shapes during the optimization. (b) The product ratios are plotted versus SHG efficiency for all individuals of the $CpFeCOCl^+/FeCl^+$ maximization.

As a first test of this technique, direct ionization of $CpFe(CO)_2Cl$ was investigated, without being interested in bond cleavage. It is known from previous femtosecond experiments on organometallic molecules that the ionization process is strongly dependent on the pulse duration. Since the excitation process has a high-order multiphoton nature, it can be expected that higher laser pulse intensities increase the ionization yield. (Note, however, that the dependence of molecular ionization and fragmentation on laser pulse intensity may show different behavior if the higher intensity is either achieved by a higher laser pulse energy or by a shorter pulse duration.[65]) In the experiment undertaken here, the optimization goal of the evolutionary algorithm was to maximize the $CpFe(CO)_2Cl^+$ yield. For each individual pulse shape during this optimization, the SHG yield was measured as well. Figure 6a reveals a clear correlation between these two observables. The ion yield increases for shorter laser pulses, as does the SHG efficiency. Since there is a one-to-one correspondence, it is possible to map the unknown control problem onto the much simpler and well-understood SHG process.

In the next step, the initial and apparently complex problem of $CpFeCOCl^+/FeCl^+$ maximization was considered. This time a completely different behavior is observed (Fig. 6b). The correlation diagram reveals a broad distribution and therefore no clear correlation between the optimization goal and the SHG efficiency. This is most clearly seen by considering one specific SHG intensity and then observing the large range of different $CpFeCOCl^+/FeCl^+$ values achieved which all depend on the exact pulse shape. Therefore, the mechanism of control active in this molecule cannot be ascribed to multiphoton absorption probabilities alone, and the $CpFe(CO)_2Cl$

optimization results cannot be explained by a simple "intensity" variation of the applied laser fields. Specifically shaped electric fields of corresponding high complexity are required to reach the desired outcome with optimum efficiency.

It is believed that generalized concepts of correlation diagrams can be further exploited to shed more light on the fundamental processes involved in quantum control experiments. For example, one could think of correlations between different molecular fragments, SHG, other nonlinear signals, and also the structure of the electric field itself.

3.3 Liquid-Phase Control

By now it should have become evident that adaptive femtosecond pulse shaping is a very valuable methodology for controlling light–matter interactions in complex systems. One important potential use of such a methodology is in chemical synthesis where light could be exploited as a controllable reagent for the production of reactive intermediates or the formation of specific products. However, if adaptive pulse shaping can have promise as a general synthetic method it is undoubtedly the case that it must work in the solution-phase medium, where most preparative chemistry is done and where high enough particle densities are available to allow the processing of macroscopic amounts of chemical substances. This presents certain challenges that are relaxed in gas-phase control problems.

For most molecular systems, introducing a solvent environment strongly increases the complexity of the time-dependent distribution of energy following photoexcitation.[66] Furthermore, it is generally the case that the timescale in which vibrational coherence remains in a molecular system is decreased as the coupling between the solute and its environment increases. Since quantum optical control relies on the manipulation of interference phenomena, its viability at controlling solution-phase photochemistry requires that the challenges brought forth by rapid dephasing can be overcome. Adaptive pulse shaping, because of the implementation of feedback as a learning mechanism and the use of extremely large parameter spaces, is ideally suited for approaching this difficult physical problem. Successful demonstrations of control in the solution phase should begin to shed light on the question of what types of excited-state dynamics are amenable to control in the presence of solute–solvent interactions. Furthermore, as the techniques and associated theoretical methods develop, it may one day be possible to extract from optimal electric fields information about the time-dependent response of the solvent throughout the evolution of the molecule.

The experiments of the previous sections have laid the groundwork for adaptive control in the liquid phase. However, those experiments as well as all other literature demonstrations have focused on controlling the outcome of unimolecular phenomena, and little attention has been paid to selectively controlling mixtures of different molecules. The experiments discussed now consider simultaneous but selective multiphoton excitation of two distinct electronically and structurally complex dye molecules in solution.[67] The question is asked if and how coherent light fields can be used to selectively excite one specific molecule (but not another one) within liquid-phase solute–solvent mixtures.

The experimental setup was already shown schematically in Fig. 2. Single-parameter quantum control schemes (variation of wavelength, intensity, and linear chirp) and many-parameter adaptive control are investigated by use of the pulse shaper. The modified laser pulses irradiate methanol solutions of DCM (4-dicyanomethylene-2-methyl-6-*p*-dimethylaminostyryl-4H-pyran) and $[Ru(dpb)_3](PF_6)_2$, where dpb=4,4'-diphenyl-2,2'-bipyridine. Following two-photon absorption (as there is no intermediate resonance reached with the 800-nm fundamental wavelength) and excited-state dynamics, charge-separated configurations are reached in both molecular species. Spontaneous emission yields are then collected at approximate right angles, spectrally filtered at 650 nm for DCM (A) and at 630 nm for $[Ru(dpb)_3]^{2+}$ (B), detected by photomultiplier tubes (PMT), and temporally integrated by boxcar averagers. The emission ratio of molecule A versus molecule B is used as feedback signal in the evolutionary learning loop as it reflects the excitation ratio between the two molecules.

An obvious starting place for selective excitation of molecules using light is to adequately choose the corresponding excitation wavelength. However, in large molecules (especially in the liquid phase), absorption spectra are often broad and unstructured. This is also the case here. Since broadband 800-nm femtosecond laser pulses are employed, the second-order power spectrum for a bandwidth-limited laser pulse (Fig. 7a, dotted line) indicates which transition frequencies are possible in two-photon absorption processes. The DCM/$[Ru(dpb)_3]^{2+}$ linear absorption ratio within this spectral range (Fig. 7a, solid line) does not vary significantly, and therefore selective excitation of one molecule with respect to the other is not possible by varying the corresponding (one-photon) excitation wavelength in the relevant range.

Similarly, it is also not possible to vary the excitation ratio by changing the laser power through attenuation (Fig. 7b). This so happens because both molecules are excited in a two-photon process, and the intensity dependencies in the numerator and the denominator of the ratio cancel each other.

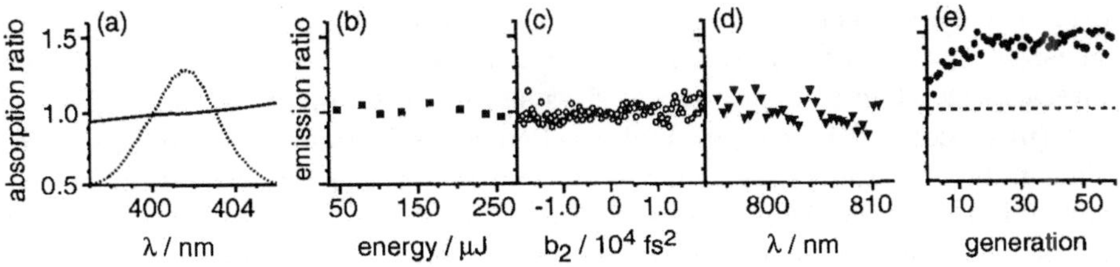

Figure 7. Control of molecular excitation. (a) The relative DCM/$[Ru(dpb)_3]^{2+}$ linear absorption ratio (solid line) is shown as a function of wavelength. The second-order power spectrum (dotted line) of a bandwidth-limited laser pulse is shown to illustrate possible two-photon transition frequencies. The DCM/$[Ru(dpb)_3]^{2+}$ emission ratio is plotted (b) for varying pulse energies of unshaped laser pulses, (c) for varying second-order spectral phase (i.e., linear chirp), (d) for scanning a symmetric and rectangular window of 5 nm width over the laser spectrum, and (e) for many-parameter phase-shaping as a function of generation number within the evolutionary algorithm.

Varying linear chirp is another possibility which has been exploited successfully in previous experiments on quantum control. However, under the conditions here the DCM/$[Ru(dpb)_3]^{2+}$ emission ratio does not change as a function of linear chirp (Fig. 7c). Qualitatively, this happens because the two molecules DCM and $[Ru(dpb)_3]^{2+}$ respond to linear chirp exactly in the same way: Introduction of second-order spectral phase increases the laser pulse duration, reduces the temporal peak intensity, and therefore reduces the transition probability for the induced nonresonant two-photon excitation processes.

Still another possibility of selective control in two-photon absorption would be given if the two-photon absorption cross sections were different for the two molecules. This was investigated by scanning a rectangular transmission window over the laser spectrum while recording the emission signals (Fig. 7d). The shown wavelength region is determined by the available spectral bandwidth (compare with the doubled frequencies of Fig. 7a). Within the statistical uncertainties, the DCM/$[Ru(dpb)_3]^{2+}$ emission ratio does not change and selective excitation is not possible.

All these results seem to indicate that the excitation processes for each molecule in the case of spectrally broad absorption bands cannot be separately controlled. However, this assumption is wrong, and selective excitation can be achieved if many-parameter adaptive quantum control is employed.

Using the DCM/$[Ru(dpb)_3]^{2+}$ emission ratio as a feedback signal, the evolutionary algorithm iteratively improves the signal by approximately 50 % (Fig. 7e). This demonstrates that one molecule (DCM) can be selectively excited with respect to the other molecule ($[Ru(dpb)_3]^{2+}$). It should be emphasized that control is possible in the presence of complex solute–solvent interactions. The feedback signal rises significantly above the level given by the (unsuccessful) single-parameter schemes (indicated by the dashed line). This failure of the other schemes is helpful for the interpretation of relevant control mechanisms. Since the initial excitation step is not controllable (Figs. 7a–d), nontrivial control of excited-state wavepacket dynamics must be involved which goes beyond the initial excitation step. This in turn shows that molecular phase coherences must exist which are not destroyed by interaction with the surrounding solvent on the time scale of the optimized pulse shape (100 fs to 1 ps). This is an important—and surprising—inference from the experimental data (which could not be obtained from the linear spectroscopy or the single-parameter control techniques alone), demonstrating the usefulness of adaptive quantum control as a sophisticated spectroscopic technique.

Applications are envisioned in chemical analysis where specific molecular properties could be made visible, or in bimolecular reaction control where specific educt molecules could selectively be prepared in reactive states ("activated") for purposes of chemical synthesis. For example, consider a mixture of three distinct chemical species in solution where photoexcited species A^* or B^* undergo rapid irreversible bimolecular reactions with C to form AC or BC, respectively. In the absence of selectivity, light would produce at the very least a mixture of AC and BC. Photoselective excitation, on the other hand, would allow for controlled production of individual bimolecular product channels (either AC or BC). Another application could be found in multiphoton fluorescence microscopy. Here, it would be possible to obtain emission signals preferentially from a specific fluorophore, and thus spatial (microscopic) resolution could be combined with "chemical resolution."

4 Femtosecond Polarization Pulse Shaping

The previous sections have illustrated how adaptively phase-shaped femtosecond laser pulses can be used to control complex quantum systems. An intuitive classical picture of the fundamental photoprocesses taking place in these molecular examples can be drawn in the following way: The electric field "pulls" at chemical bonds within the molecule at certain optimal times with specific frequencies. However, physical systems such as molecules are three-dimensional. The question therefore arises if it were not possible to make use of the vectorial properties of light and to additionally "pull" along the correct spatial directions. This requires controlling the polarization state of femtosecond light pulses as a function of time. We recently developed such

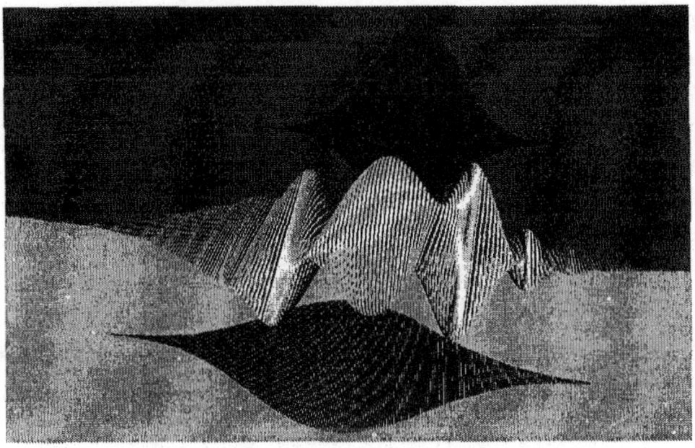

Figure 8. Quasi-three-dimensional electric field representation for a simulated polarization-shaped femtosecond laser pulse. Time evolves from left to right, and electric field amplitudes are indicated by the sizes of the corresponding ellipses. The momentary frequency can be indicated by colors or gray-shading, and the shadows represent the amplitude envelopes of component E_1 (bottom) and component E_2 (top) separately.

a technique which we call femtosecond polarization pulse shaping.[68] It introduces a fundamentally new way of manipulating electromagnetic waves. For the first time, temporal intensity, momentary frequency, degree of ellipticity, and orientation of elliptical principal axes can be varied in a complex manner within a single laser pulse.

The polarization pulse shaper is designed similar to the phase-only shaper of Fig. 2. However, the one-layer LCD is replaced by a two-layer LCD such that each layer manipulates one of two orthogonal polarization components. It is then possible to introduce spectral phases independently, and the resulting interference between the two polarization components leads to complex phase-and-polarization modulated pulse shapes in the time domain. Care must be taken in the experimental pulse-shape characterization as each optical element (such as mirrors, beam-splitters etc.) can further change the polarization state of light. Complete characterization is possible by dual-channel spectral interferometry.[69]

An example of a polarization-shaped laser pulse is shown in Fig. 8. In this quasi-three-dimensional representation, the sizes of the ellipses represent the electric field amplitude. Time evolves from left to right, and the shadows represent the electric field amplitudes of either polarization component alone. In a color plot, the momentary oscillation frequency can also easily be visualized. In the example here, it is seen that in the beginning and in the end of the laser pulse, linear polarization extends over a certain time period. In the intermediate time regime the ellipticity, elliptical orientation, intensity, and oscillation period vary, and also transient linear polarization states with different orientations are reached. This graph illustrates nicely the possibilities of polarization pulse shaping but also the difficulties in a correct characterization: From the linear amplitude components alone (shadows), it would not be possible to guess correctly the complex temporal evolution of the polarization state. Rather the complete amplitude and phase structures are relevant.[69]

Polarization femtosecond pulse shaping can be considered a novel spectroscopic technique, because the temporal as well as three-dimensional spatial properties of quantum wavefunctions can potentially be addressed and controlled. Immediate applications are found in obvious extensions of the experiments discussed in Section 3. But there are also fundamentally new prospects which rely on the vectorial manipulation of light–matter interactions. The door to many experimental possibilities has just been opened.

5 Conclusion

Obtaining active control over quantum-mechanical phenomena is a fascinating perspective of modern physics. Especially the optical control of chemical reactions was in the focus of research since the invention of the laser in the 1960s. However, most of the early attempts at achieving chemical selectivity by the application of light fields were unsuccessful because of the complexity of the underlying photoprocesses. Following the theoretical suggestion of Judson and Rabitz, a "closed-loop" approach was developed by which quantum-mechanical processes can be controlled even in complex systems. The experimental outcome of light–matter interactions are used as feedback within a learning algorithm which iteratively improves femtosecond pulse-shaper settings until an optimum is reached. This global optimization strategy does not require prior knowledge of the quantum system's Hamiltonian.

Technological applications have been demonstrated in the case of automated pulse compression. In molecular systems it is now possible to control gas-phase dissociation reactions and to assess the complexity of the control mechanisms. In many cases, the complex electric laser field structures are indeed significant for the optimal control of photoproduct branching ratios. A significant advance was also made in liquid-phase photoexcitation where solute–solvent interaction introduces additional complications. Nevertheless, it was possible to achieve selective excitation of specific molecular species by adaptive femtosecond pulse shaping, despite the failure of single-parameter schemes. This is promising for the future development of quantum control of chemical synthesis where different reactants have to be controlled simultaneously. Further advances in this research area are expected when use is made of vectorial electric field properties. This is possible with the novel method of femtosecond polarization pulse shaping, by which quantum dynamics in three-dimensional systems can be influenced in a very general way on an ultrashort timescale.

Acknowledgments

We are indebted to many dedicated coworkers who were involved in the experiments described herein. In this context, we would like to thank T. Baumert for his ideas and guidance he provided during the initial stages of this research project, and the following workers for their essential contributions: A. Assion, M. Bergt, N. H. Damrauer, B. Kiefer, G. Krampert, P. Niklaus, V. Seyfried, and M. Strehle. We also gratefully acknowledge financial support from the European Coherent Control Network (COCOMO): HPRN-CT-1999-00129, the German–Israeli Cooperation in Ultrafast Laser Technologies (GILCULT): FKZ-13N7966, the Sonderforschungsbereich 347 (Universität Würzburg): "Selektive Reaktionen Metall-aktivierter Moleküle," and the "Fonds der chemischen Industrie."

References

1. W.S. Warren, H. Rabitz, and M. Dahleh, *Science* **259**, 1581 (1993).
2. R.J. Gordon and S.A. Rice, *Annu. Rev. Phys. Chem.* **48**, 601 (1997).
3. S.A. Rice and M. Zhao, *Optical Control of Molecular Dynamics* (Wiley, New York, 2000).
4. H. Rabitz, R. de Vivie-Riedle, M. Motzkus, and K. Kompa, *Science* **288**, 824 (2000).
5. M. Shapiro and P. Brumer, in *Advances in Atomic, Molecular, and Optical Physics*, eds. B. Bederson and H. Walther, volume 42 of *Advances in Atomic, Molecular, and Optical Physics Series*, pp. 287–345 (Academic Press, London, 1999).
6. T. Brixner, N.H. Damrauer, and G. Gerber, in *Advances in Atomic, Molecular, and Optical Physics*, eds. B. Bederson and H. Walther, volume 46 of *Advances in Atomic, Molecular, and Optical Physics Series*, pp. 1–54 (Academic Press, London, 2001).
7. P. Brumer and M. Shapiro, *Chem. Phys. Lett.* **126**, 541 (1986).
8. M. Shapiro, J.W. Hepburn, and P. Brumer, *Chem. Phys. Lett.* **149**, 451 (1988).

9. C. Chen, Y. Yin, and D.S. Elliott, *Phys. Rev. Lett.* **64**, 507 (1990).
10. S.M. Park, S.P. Lu, and R.J. Gordon, *J. Chem. Phys.* **94**, 8622 (1991).
11. L.C. Zhu, V.D. Kleiman, X.N. Li, S.P. Lu, K. Trentelman, and R.J. Gordon, *Science* **270**, 77 (1995).
12. D.J. Tannor and S.A. Rice, *J. Chem. Phys.* **83**, 5013 (1985).
13. D.J. Tannor, R. Kosloff, and S.A. Rice, *J. Chem. Phys.* **85**, 5805 (1986).
14. T. Baumert, M. Grosser, R. Thalweiser, and G. Gerber, *Phys. Rev. Lett.* **67**, 3753 (1991).
15. T. Baumert and G. Gerber, *Isr. J. Chem.* **34**, 103 (1994).
16. E.D. Potter, J.L. Herek, S. Pedersen, O. Liu, and A.H. Zewail, *Nature* **355**, 66 (1992).
17. J.L. Herek, A. Materny, and A.H. Zewail, *Chem. Phys. Lett.* **228**, 15 (1994).
18. A. Assion, T. Baumert, V. Seyfried, V. Weiss, E. Wiedenmann, and G. Gerber, *Z. Phys. D* **36**, 265 (1996).
19. T. Baumert, J. Helbing, and G. Gerber, in *Chemical Reactions and Their Control on the Femtosecond Time Scale. XXth Solvay Conference on Chemistry*, eds. P. Gaspard and I. Burghardt, pp. 47–82 (Wiley, New York, 1997).
20. U. Gaubatz, P. Rudecki, M. Becker, S. Schiemann, M. Külz, and K. Bergmann, *Chem. Phys. Lett.* **149**, 463 (1988).
21. U. Gaubatz, P. Rudecki, S. Schiemann, and K. Bergmann, *J. Chem. Phys.* **92**, 5363 (1990).
22. K. Bergmann, H. Theuer, and B.W. Shore, *Rev. Mod. Phys.* **70**, 1003 (1998).
23. C.J. Bardeen, Q. Wang, and C.V. Shank, *Phys. Rev. Lett.* **75**, 3410 (1995).
24. B. Kohler, V.V. Yakovlev, J. Che, J.L. Krause, M. Messina, K.R. Wilson, N. Schwentner, R.M. Whitnell, and Y.J. Yan, *Phys. Rev. Lett.* **74**, 3360 (1995).
25. A. Assion, T. Baumert, J. Helbing, V. Seyfried, and G. Gerber, *Chem. Phys. Lett.* **259**, 488 (1996).
26. G. Cerullo, C.J. Bardeen, Q. Wang, and C.V. Shank, *Chem. Phys. Lett.* **262**, 362 (1996).
27. C.J. Bardeen, Q. Wang, and C.V. Shank, *J. Phys. Chem. A* **102**, 2759 (1998).
28. C.J. Bardeen, V.V. Yakovlev, J.A. Squier, and K.R. Wilson, *J. Am. Chem. Soc.* **120**, 13027 (1998).
29. B. Broers, H.B. van Linden van den Heuvell, and L.D. Noordam, *Phys. Rev. Lett.* **69**, 2062 (1992).

30. P. Balling, D.J. Maas, and L.D. Noordam, *Phys. Rev. A* **50**, 4276 (1994).
31. D.J. Tannor and S.A. Rice, in volume 70 of *Advances in Chemical Physics*, pp. 441–524 (Wiley, New York, 1988).
32. A. Shi, A. Woody, and H. Rabitz, *J. Chem. Phys.* **88**, 6870 (1988).
33. A.P. Peirce, M. Dahleh, and H. Rabitz, *Phys. Rev. A* **37**, 4950 (1988).
34. R. Kosloff, S.A. Rice, P. Gaspard, S. Tersigni, and D.J. Tannor, *Chem. Phys.* **139**, 201 (1989).
35. J.P. Heritage, A.M. Weiner, and R.N. Thurston, *Opt. Lett.* **10**, 609 (1985).
36. A.M. Weiner, D.E. Leaird, J.S. Patel, and J.R. Wullert II, *Opt. Lett.* **15**, 326 (1990).
37. A.M. Weiner, D.E. Leaird, A. Patel, and J.R. Wullert II, *IEEE J. Quantum Electron.* **28**, 908 (1992).
38. M.M. Wefers and K.A. Nelson, *Opt. Lett.* **20**, 1047 (1995).
39. M.M. Wefers and K.A. Nelson, *J. Opt. Soc. Am. B* **12**, 1343 (1995).
40. A.M. Weiner, *Rev. Sci. Instrum.* **71**, 1929 (2000).
41. R.S. Judson and H. Rabitz, *Phys. Rev. Lett.* **68**, 1500 (1992).
42. C.J. Bardeen, V.V. Yakovlev, K.R. Wilson, S.D. Carpenter, P.M. Weber, and W.S. Warren, *Chem. Phys. Lett.* **280**, 151 (1997).
43. A. Assion, T. Baumert, M. Bergt, T. Brixner, B. Kiefer, V. Seyfried, M. Strehle, and G. Gerber, in *Ultrafast Phenomena XI*, eds. T. Elsaesser, J.G. Fujimoto, D.A. Wiersma, and W. Zinth, volume 63 of *Springer Series in Chemical Physics*, pp. 471–473 (Springer, Berlin, 1998).
44. A. Assion, T. Baumert, M. Bergt, T. Brixner, B. Kiefer, V. Seyfried, M. Strehle, and G. Gerber, *Science* **282**, 919 (1998).
45. T.C. Weinacht, J.L. White, and P.H. Bucksbaum, *J. Phys. Chem. A* **103**, 10166 (1999).
46. T. Hornung, R. Meier, and M. Motzkus, *Chem. Phys. Lett.* **326**, 445 (2000).
47. D. Yelin, D. Meshulach, and Y. Silberberg, *Opt. Lett.* **22**, 1793 (1997).
48. T. Baumert, T. Brixner, V. Seyfried, M. Strehle, and G. Gerber, *Appl. Phys. B* **65**, 779 (1997).
49. A. Efimov, M.D. Moores, N.M. Beach, J.L. Krause, and D.H. Reitze, *Opt. Lett.* **23**, 1915 (1998).
50. T. Brixner, M. Strehle, and G. Gerber, *Appl. Phys. B* **68**, 281 (1999).
51. D. Zeidler, T. Hornung, D. Proch, and M. Motzkus, *Appl. Phys. B* **70** *[Suppl.]*, S125 (2000).
52. D. Meshulach, D. Yelin, and Y. Silberberg, *J. Opt. Soc. Am. B* **15**, 1615 (1998).
53. T. Brixner, A. Oehrlein, M. Strehle, and G. Gerber, *Appl. Phys. B* **70**

[Suppl.], S119 (2000).
54. D. Meshulach and Y. Silberberg, *Nature* **396**, 239 (1998).
55. T. Hornung, R. Meier, D. Zeidler, K.L. Kompa, D. Proch, and M. Motzkus, *Appl. Phys. B* **71**, 277 (2000).
56. T.C. Weinacht, J. Ahn, and P.H. Bucksbaum, *Nature* **397**, 233 (1999).
57. R. Bartels, S. Backus, E. Zeek, L. Misoguti, G. Vdovin, I.P. Christov, M.M. Murnane, and H.C. Kapteyn, *Nature* **406**, 164 (2000).
58. J. Kunde, B. Baumann, S. Arlt, F. Morier-Genoud, U. Siegner, and U. Keller, *Appl. Phys. Lett.* **77**, 924 (2000).
59. D.J. Tannor, R. Kosloff, and A. Bartana, *Faraday Discuss. Chem. Soc.* **113**, 365 (1999).
60. D.J. Tannor and A. Bartana, *J. Phys. Chem. A* **103**, 10359 (1999).
61. A. Bartana, R. Kosloff, and D.J. Tannor, *Chem. Phys.* **267**, 195 (2001).
62. D.E. Goldberg, *Genetic Algorithms in Search, Optimization, and Machine Learning* (Addison-Wesley, Reading, 1993).
63. H.-P. Schwefel, *Evolution and Optimum Seeking* (Wiley, New York, 1995).
64. T. Brixner, B. Kiefer, and G. Gerber, *Chem. Phys.* **267**, 241 (2001).
65. M. Bergt, T. Brixner, B. Kiefer, M. Strehle, and G. Gerber, *J. Phys. Chem. A* **103**, 10381 (1999).
66. G.R. Fleming, *Chemical Applications of Ultrafast Spectroscopy* (Oxford University Press, New York, 1986).
67. T. Brixner, N.H. Damrauer, P. Niklaus, and G. Gerber, *Nature* **414**, 57 (2001).
68. T. Brixner and G. Gerber, *Opt. Lett.* **26**, 557 (2001).
69. T. Brixner, G. Krampert, P. Niklaus, and G. Gerber, *Appl. Phys. B [Suppl.]*, in press (2002).

PHONON-POLARITONS: CONTROLLED PROPAGATION AND AMPLIFICATION

T. FEURER, NIKOLAY S. STOYANOV, JOSHUA C. VAUGHAN, DAVID W. WARD, AND KEITH A. NELSON

Department of Chemistry, Massachusetts Institute of Technology room 6-235, 77 Massachusetts Avenue, MA02139 Cambridge, USA

We report the coherent manipulation of mixed lattice vibrational and electromagnetic modes, phonon-polaritons, as they propagate through their crystalline hosts at light-like speeds, through the use of ultrashort excitation pulses with specified spatial and temporal excitation profiles. Terahertz (THz) *polaritonics* has applications in THz spectroscopy, experiments exploring linear and nonlinear lattice dynamics, and ultrahigh-bandwidth signal processing.

1 Introduction

Coherent optical control has generally been applied to spatially localized excitations including lattice and molecular vibrations and electronic states. [1] In such cases the primary control element is the temporal profile of the optical excitation waveform. However, some collective responses propagate across macroscopic distances. Both spatial and temporal excitation profiles may be specified to achieve coherent control over the dynamical evolution of such responses. In addition, spatially varying structures fabricated in the host material may be constructed to further control the behavior of propagating modes within it. A familiar example is the case of surface acoustic wave (SAW) devices, in which spatially and temporally varying electrical inputs in the MHz–GHz frequency range and fabricated structures including acoustic waveguides, resonators, and filters are used to manipulate acoustic wave propagation. A less familiar example is that of mixed polar lattice vibrational and electromagnetic modes, called phonon-polaritons, which have THz frequencies and which propagate at light-like speeds through their host crystals. [3] Their manipulation requires inputs with THz bandwidths, but at present there exists no family of polaritonic devices analogous to the SAW devices routinely used for signal processing at MHz–GHz bandwidths. However, capabilities for generation of spatially and temporally specified, ultrafast optical waveforms and the use of these waveforms for polaritonic manipulation have recently been demonstrated. [3] In addition, direct real-space imaging of polariton propagation [4] has enabled facile assessment of control experiments and refinement of spatiotemporally shaped waveforms used for them. Finally, experience in the acoustic and optical regimes provides extensive guidance for the fabrication of structures that are optimized for enhanced polaritonic control. Thus, the time is ripe for advances in THz polaritonics, with promise for fundamental objectives, including THz spectroscopy and coherent control over lattice dynamics, and for applications including THz-bandwidth signal processing. We present preliminary results in spatial and spatiotemporal control over phonon-polaritons in the ferroelectric crystal, lithium tantalate ($LiTaO_3$). Spatial control is achieved through projection of excitation light in a circular pattern at the sample, resulting in a polaritonic ring which propagates inward toward a focal point at the center. This focusing yields a substantial increase in polariton amplitude and intensity. Spatiotemporal control is achieved through the use of automated spatiotemporal femtosecond pulse shaping to generate many excitation beams with specified temporal profiles. Polariton focusing is again demonstrated, in this case with controllable focal length and propagation direction.

2 Phonon-Polaritons and Their Properties

The uncoupled electromagnetic mode whose dispersion relation

$$\omega = \frac{c}{n(\omega)}k = \frac{c}{\varepsilon^2(\omega)}k \tag{1}$$

is essentially linear, where n is only weakly frequency dependent, and the uncoupled transverse optical phonon, essentially dispersionless, become strongly mixed when their frequencies and wavevectors are both similar. Phenomenologically, the coupling can be described by the following pair of equations:

$$\vec{Q} = b_{11}\frac{d^2\vec{Q}}{dt^2} + \Gamma\frac{d\vec{Q}}{dt} + b_{12}\vec{E} \tag{2}$$

$$\vec{P} = b_{21}\vec{Q} + b_{22}\vec{E} \tag{3}$$

Solution of the coupled system results in a macroscopic dielectric constant given by

$$\epsilon(\omega) = \epsilon_\infty + \frac{\omega_T^2(\epsilon_0 - \epsilon_\infty)}{\omega_T^2 - \omega^2 - i\omega\Gamma} \tag{4}$$

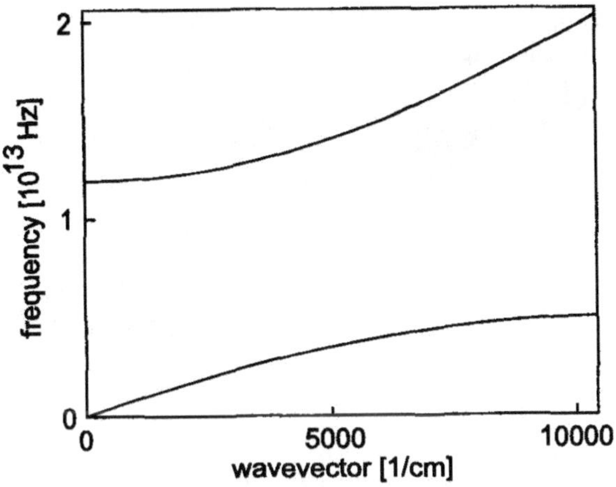

Figure 1. Polariton dispersion curve.

where ϵ_0 and ϵ_∞ represent the low and high frequency limits of the dielectric response, ω_T is the frequency of the transverse optical phonon, and Γ is the phenomenological damping rate. Equations (1) and (4) yield the dispersion curves for the upper and lower phonon-polariton branches, as shown in fig. 1 for $LiTaO_3$.

In our experiments we excite polaritons in the lower branch, with low wavevectors mostly in the linear part of the curve. For anisotropic crystals the dielectric response is a tensor. Lithium tantalate is a uniaxial crystal, and equation (4) describes the dielectric response for a phonon-polariton mode polarized along its optic axis, with wavevector perpendicular to the optic axis.

3 Polariton Focusing

3.1 Experimental Setup

We have performed spatiotemporal imaging of phonon-polaritons in an X-cut lithium tantalate crystal with dimensions 5 mm × 5 mm × 4 mm (depth). The polar axis of the crystal was aligned with the polarization of the incoming pump and probe beams. The experimental setup is shown in fig. 2. An amplified Ti:Sapphire laser was used to generate pulses of 50 fs duration and average power of 450 mW at 1 kHz repetition rate. The output was divided

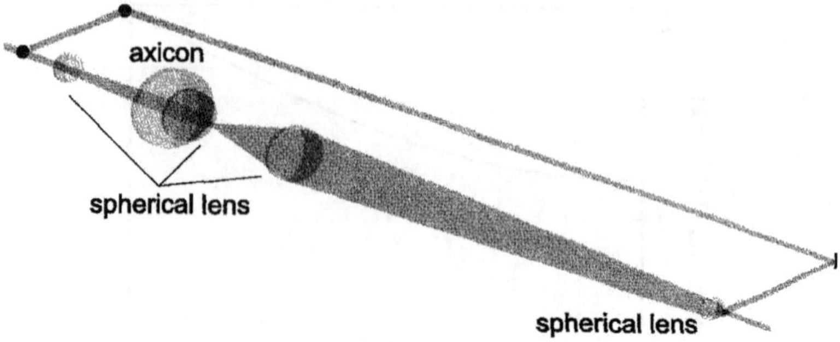

Figure 2. Experimental setup.

by a 90/10 beamsplitter into pump and probe beams. The pump light was directed through an axicon, which is a plano-conical optic used in combination with a lens to produce a circular beam profile. The circle was then imaged to a smaller ring, 3.1 mm in diameter, at the sample. The probe beam was directed to a computer-controlled translation stage and expanded to an area larger than the sample, to enable spatiotemporal imaging of the sample response. The transmitted probe light was projected by a camera objective onto a CCD. Phonon-polaritons modulate the phase, but not the amplitude, of the probe light. The phase information imparted to the probe beam was converted to amplitude information through Talbot self-imaging, in which images are recorded slightly out of focus of the image plane by the CCD camera. A subsequent long-pass filter removed 400 nm light produced by second harmonic generation in the sample. The CCD integrated the light and was read at a rate of 30 frames/sec. In order to achieve optimal signal contrast, a 'background' CCD image was recorded with the probe pulse preceding the excitation. The background was subtracted from every image recorded during the experiment. To further improve the signal contrast in some of the experiments, a number of images recorded at each delay were averaged. The same number of background images in these cases was also averaged.

3.2 Full-Circle Excitation

The images in figures 3 depict the polariton ring traveling inwards as it propagates in time. Both the 2D and 3D representations of the data at any delay clearly show the signal to be at its maximum at the horizontal extremes of the ring. No signal is observed at the vertical extremes since there is no

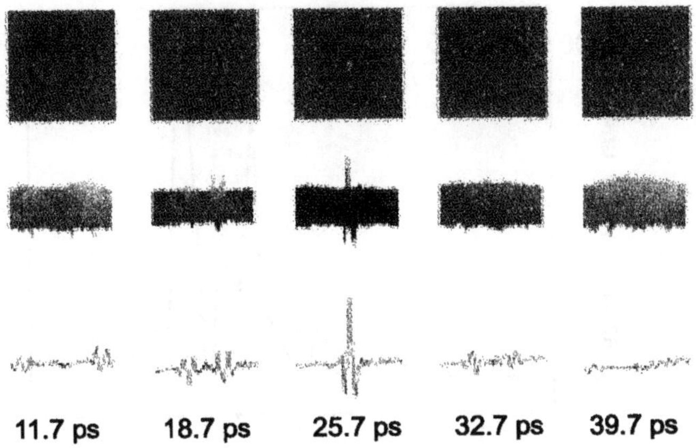

Figure 3. Full ring excitation. The top of the figure shows the CCD images from a 1.6 mm $\times$ 1.6 mm crystal area at various delays. The collapse towards the focal point occurs at 28.7 ps. The middle figures show the polariton amplitude derived from the CCD readings, represented by height along the vertical axis. The bottom figures show polariton amplitudes averaged over an angular range of $2\pi/5$ about the horizontal axis at various delays.

propagation along the vertical polariton polarization direction. The maximum polariton amplification occurs at the focus, most clearly shown by the one dimensional plots. To reduce the noise contributions, the 1D plots were constructed out of each image by averaging the signal across chords of $2\pi/5$ angular spread around the horizontal axis that passes through the focal point.

One way to assess the increase in polariton amplitude is to compare the absolute maxima of the signals from images recorded during focusing. Figure 4, constructed by plotting the maximum from each of the one-dimensional plots at the bottom of fig. 3, shows the results. As evident, the most substantial increase in amplitude magnitude is achieved around the focus. The overall increase in amplitude is about a factor of six, which corresponds to a factor of 36 increase in polariton intensity.

3.3 Half-Circle Excitation Peprendicular to the Crystal Axis

An experiment was performed by blocking half of the excitation light so that a semi-circle of pump light creates a polariton response that focuses along a direction perpendicular to the crystal axis. A notable feature of electromag-

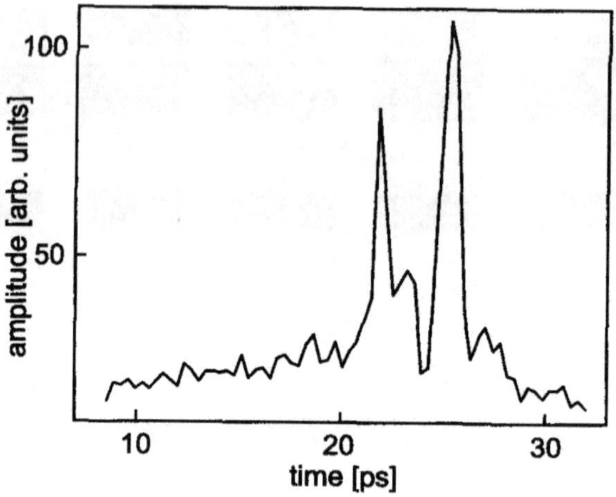

Figure 4. Maximum amplitude intensity of the polariton response as it propagates in time.

netic radiation that travels through focus is that it undergoes what is known as the Gouy phase shift. [5,6,7] This effect is illustrated in fig. 5. The figure was constructed by subtracting the mean value from each of the one-dimensional graphs, analogous to those at the bottom of fig. 3, color-coding the resulting plots and displaying them one below the other in temporal order.

The slight curve in the lines of constant wavefront around the focus corresponds to the speeding up of the phase fronts near the focus, i.e. to the Gouy phase shift. For a beam with Gaussian transverse distribution going through the focus, the phase shift depends on the number of transverse dimensions of focusing. The shift is π for a spherical wave, and $\pi/2$ for a cylindrical one. [5,6] In our case the focusing occurs only along one transverse dimension, so the expected phase shift is $\pi/2$. For the cylindrical case, the phase of a monochromatic wave is given as a function of the direction of propagation, z, along the beam axis by [5,6,8]

$$\phi(z) = -\frac{1}{2}\arctan\left(\frac{\lambda z}{\pi w_0^2}\right) \qquad (5)$$

where w_0 is the beam waist at the focus and λ is the wavelength. Though the polariton response in the present case is a nearly single cycle 'ripple'

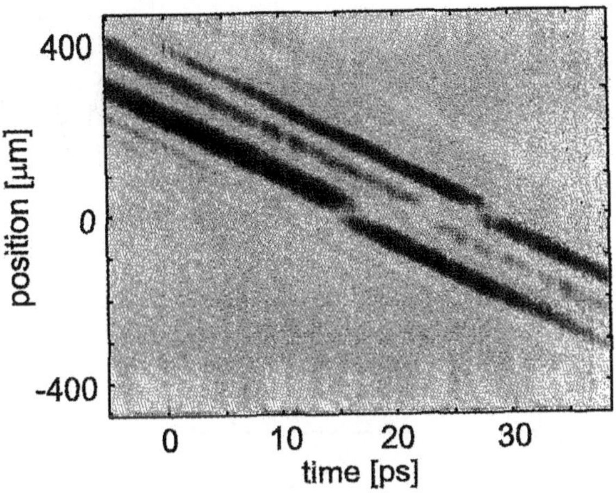

Figure 5. Semi-circular polariton ripple propagating through its focus. The figure shows the amplitude averaged around the direction of propagation over a $2\pi/5$ angle as a function of time.

with a broad frequency bandwidth, a useful comparison of the calculated phase shift of the central frequency can be made with the experimental data. To estimate the beam waist one can use the polariton group velocity, which for low wavevectors is $\sim 5 \times 10^7$ m/s, coupled with the frames at different time steps to calibrate for distance scale on the CCD images. Note that the calibration of the distance scale by using the size of the image of the excitation ring is difficult since the CCD camera is placed out of focus in order to image the polariton phase modulation. Thus, the beam waist, w_0(FWHM)$\approx$ 68 μm. We may estimate the center wavelength from the images in similar fashion to obtain $\lambda \approx 130$ μm (note that the spectrum peaks at 0.42 THz which corresponds to $\lambda = 120$ μm, consistent with the estimate). By tracing the propagation of one of the fringes in fig. 5 we can extract the empirical phase shift, and we can compare it with the calculated phase shift as a function of the propagation distance from equation (5) using the estimated parameters. The results are shown in fig. 6.

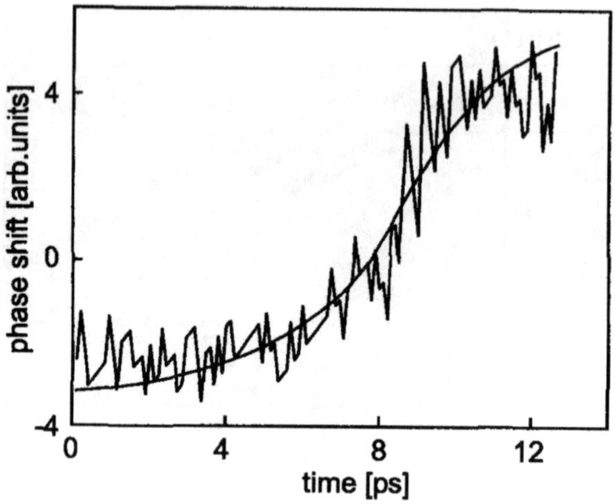

Figure 6. Experimental and simulated Gouy phase shift for semi-circular polariton focusing.

4 Spatiotemporal Pulse Shaping

4.1 Experimental Setup

Our phase-only pulse shaping arrangement is similar to conventional $4f$ spectral filtering pulse shaping arrangements [9], with the differences that the incoming beam was expanded in one dimension and a two-dimensional spatial light modulator (SLM) in a reflection geometry is employed. Extension of programmable femtosecond pulse shaping to include a spatial dimension through the use of a transmissive 2D SLM has been demonstrated [10], but poor fill factor and incomplete phase modulation severely restricted the use of the method. In the present case the reflective SLM (Hamamatsu model SLMM X7550-800) is optically addressed by backside illumination of a photosensitive layer. The diode laser light used for backside illumination is spatially filtered by a 2D transmissive SLM which consists of an array of 480 $\times$ 480 pixels. Thus, the reflective SLM consists of an array of 480 $\times$ 480 virtual pixels which provide a phase modulation between 0 and 2π radians with 8 bit resolution. The frequency components of a horizontally polarized 785 nm, 60 fs, 160 μJ optical laser pulse were dispersed in the horizontal direction by a 1200 lines/mm diffraction grating. A 10 cm cylindrical lens focused each of

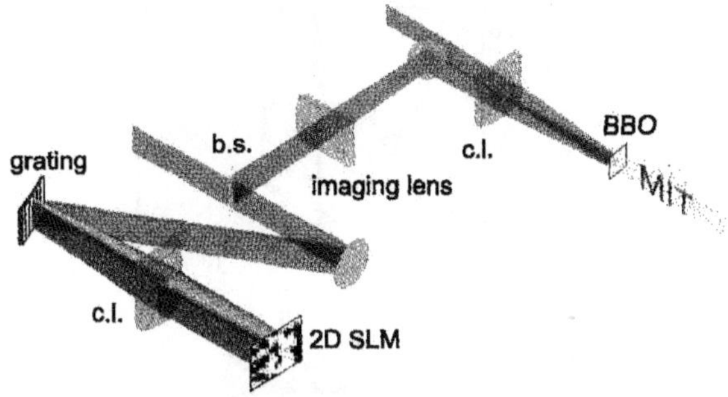

Figure 7. Setup for real-space shaping. The shaped pulse was cross-correlated with a reference pulse and the signal was recorded by a CCD camera.

the separated frequency components to a vertical line at the SLM. The phase-modulated frequency components were reflected back through the cylindrical lens and recombined at the diffraction grating. Finally, a beamsplitter separated out the retroreflected beam. The reduced number of optical elements in this $4f$ design simplified its alignment, although its reliance on a beamsplitter to separate the shaped pulse from the incoming beam lowered the overall throughput. In future experiments we plan to employ a different geometry such that the shaped pulse will be slightly displaced from the incoming beam. We expect that the throughput of the pulse shaper will be as high as 70%.

4.2 Real-Space Shaping

In real-space shaping, each of the 480 independently shaped regions was imaged to a different location on the sample, making it possible to excite and control spatiotemporal events in a coherent fashion. Real-space shaping allows the manipulation of polaritons as they propagate with light-like speed through the sample. It also allows for the creation of user-defined spatially and temporally varying responses.

To demonstrate real-space shaping, a horizontally-aligned 30 cm cylindrical lens was used to image the face of the SLM onto the BBO crystal with approximately 1:1 imaging. The shaped pulse was measured by cross-correlation with an expanded unshaped reference pulse on a motorized delay. The second harmonic radiation created by each of the incident beams alone was blocked, and the intensity of the cross-correlation signal was recorded by imaging (1:1) a BBO crystal onto a CCD camera for various delays of the

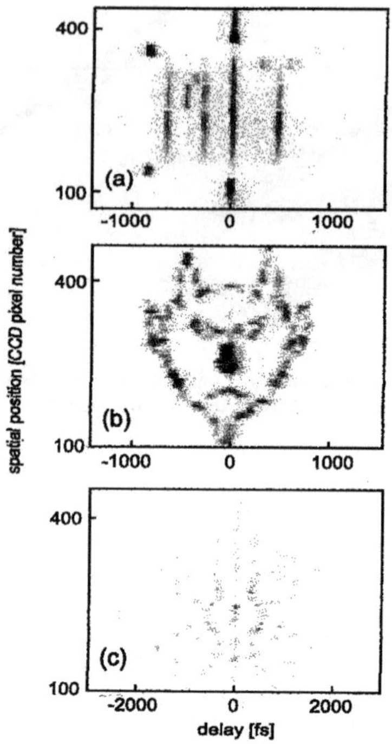

Figure 8. (a), (b) *Complex user defined two-dimensional waveforms.* (c) *Quasi random* waveform.

reference pulse. A blue colored-glass filter was used to eliminate scattered 785 nm light from the incoming pulses. Note that this arrangement allows for a simultaneous cross-correlation measurement of all shaped pulses in the vertical direction. The setup is shown in fig. 7

Three examples of complex 2D waveforms possible in real-space shaping are illustrated in fig. 8. The pulse shapes are about 4 mm high and 1.5 ps in duration. A new, fast iterative Fourier transform algorithm developed by Hacker et al. [11] was used to calculate the phase mask patterns necessary to generate the user-defined 2D pulse shapes presented in fig. 8(a) and (b). The benefit of this algorithm is especially significant for 2D phase-only pulse shaping because the iterative Fourier transform technique calculates a complicated SLM pattern consisting of over 100 vertical regions in a few seconds while standard simulated annealing or genetic algorithms would take many hours to perform the same task. It is expected that closed-loop experiments may produce even more complicated waveforms, as suggested by the pseudo random shape indicated in fig. 8(c).

4.3 Polariton Control

An immediate application of spatiotemporal pulse shaping is the controlled generation of phonon-polaritons in ferroelectric crystals. As an example a series of adjacent spots can serve as a phased array signal source for polaritons. If the spots are distributed along the ('horizontal') polariton propagation direction, then traveling wave amplification can be carried out. This has been reported recently and a greatly improved demonstration will be reported in the near future. In the present experiments adjacent spots were equally spaced and distributed along the vertical direction one above the other. If a pulse arrives at each spot simultaneously for example the net result is a vertical 'line source' to which the response is a part of counterpropagating polariton ripples that travel in the horizontal direction. On the other hand, if a pulse reaches the bottom spot first, then the one immediately above it, then the next and so on, with constant delay between two pulses, then the polariton phase front is tilted away from the vertical and the propagation direction correspondingly tilted away from the horizontal. Figures 9(a)-(d) show images of tilted polariton phase fronts produced in this manner, with different signs and magnitudes of the slope of temporal delay vs. spot position. Similarly, if both the top and the bottom spot are irradiated first then the two adjacent spots, and so on, with progressively shorter temporal delays between successive pulses, then curved wavefronts are produced. Polariton focusing may be achieved by creating a parabolic temporal delay along the line of excitation spots. Figures 9(e)-(g) show examples of simultaneous focusing and steering by applying a combination of linear and parabolic temporal delays to the excitation spots.

5 Conclusions

We have demonstrated polaritronic coherent control through the use of spatial and spatiotemporal control elements. The focusing of polaritons generated by a ring excitation pattern was shown. Applications of polariton focusing to linear and especially nonlinear THz spectroscopy of the crystalline hosts themselves and of externally situated samples are under current development. The objectives include study of nonlinear lattice dynamics and the dynamics

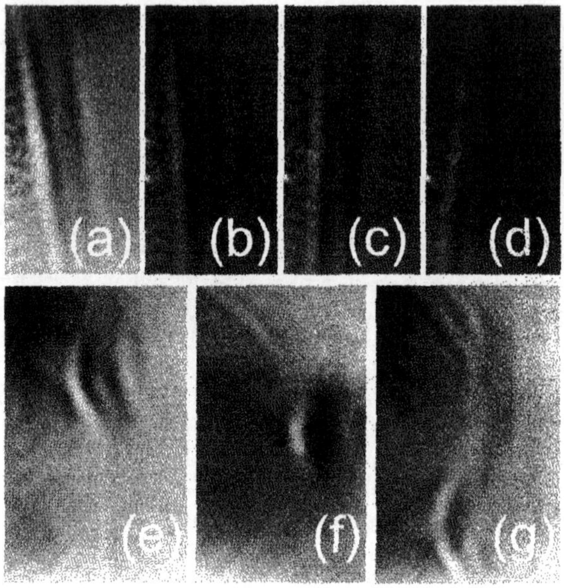

Figure 9. Polariton responses to a series of about 20 spots, generated through spatiotemporal femtosecond pulse shaping, situated one above the other in the vertical dimension. The excitation spots are located to the left of the images shown. (a)-(d) Tilted polariton wavefronts resulting from pulses arriving at the spots at progressively later times, with a linear relation between temporal delay and spatial location of the spot, resulting in a linear polariton phase delay. The wavefront and propagation direction depend on the slope of temporal delay vs. position. (e)-(g) Polariton steering and focusing specified through a combination of linear and parabolic phase delays.

of polar liquids through THz pump — THz probe and related measurements.

We have also demonstrated reconfigurable 2D femtosecond phase-only pulse shaping with high-fidelity output waveforms suitable for use in multiple-pulse spectroscopy and signal processing. Spatiotemporal pulse shaping has been employed for automated coherent control over polariton wavepacket steering and focusing. A wide range of applications in ultrafast spectroscopy and in control over propagating excitations and signals is anticipated.

Acknowledgments

This work was supported in part by NSF Grant no. CHE97-13388 and Army Research Office Grant no. DAAD19-01-1-0674. T. Feurer acknowledges financial support by the Max Kade Foundation.

References

1. S. Rice, M. Zhao, "Optical Control of Molecular Dynamics", Wiley, John & Sons, Incorporated (2000)
2. M. Artoni, J.L. Birman, Phys. Rev. B **44**, 3736 (1991)
3. R.M. Koehl, K.A. Nelson, "Coherent optical control over collective vibrations traveling at lightlike speeds," J. Chem. Phys. B **114**, 1443-1446 (2001)
4. R.M. Koehl, S. Adachi, K.A. Nelson, "Direct Visualization of Collective Wavepacket Dynamics", Phys. Chem. A **103**, 10260-10267 (1999)
5. A. E. Siegman, "Lasers", (University Science Books, Sausalito, 1986), pp. 682-685. Note that Gouy's name is mispelled as "Guoy" in this reference.
6. S. Feng, and Herbert G. Winful, "Physical origin of the Gouy phase shift", Optics Letters, **26**, 485-487 (2001)
7. A.B. Ruffin, J.V. Rudd, J.F. Whitaker, S. Feng, and H.G. Winful, "Direct Observation of the Gouy Phase Shift with Single-Cicle Terahertz Pulses", Phys. Rev. Letters, **83**, 3410-3413 (1999)
8. B. H. Saleh and M. C. Teich, "Fundamentals of Photonics", (John Willey & Sons, Inc., New York, 1991)
9. A.M.Weiner, "Femtosecond pulse shaping using spatial light modulators," Review of Scientific Instruments **71**, 1929-1960 (2000)
10. R.M. Koehl, T. Hattori, K.A. Nelson, "Automated spatial and temporal shaping of femtosecond pulses," Opt. Commun. **157**, 57-61 (1998)
11. M.Hacker, G.Stobrawa, T.Feurer, "Iterative Fourier transform algorithm for phase-only pulse shaping," Opt. Express **9**, 191-199 (2001)

ANALYSIS AND FEEDBACK-CONTROL OF ULTRAFAST FRAGMENTATION PROCESSES IN $CpMn(CO)_3$

C. LUPULESCU, P. ROSENDO-FRANCISCO, Š. VAJDA* AND L. WÖSTE

Insitut für Experimentalphsik, Freie Universität Berlin, Arnimallee 14, D-14195 Berlin, Germany

**E-mail: vajda@physik.fu-berlin.de*

Our contribution consists of real-time analysis and feedback control experiments of fragmentation/ionization ratio in the photo-excited cyclopentadienyl manganese tricarbonyl ($CpMn(CO)_3$) by using shaped femtosecond laser pulses. First, the dynamics of ultrafast fragmentation processes which occur via resonant multiphoton ionization were investigated. The observed exponential decay times obtained from the two-color pump and probe spectra indicate an ultrafast loss of the first ligand. Further we performed an adaptive feedback control optimization of the individual fragmentation pathways in the system in order to control the $CpMn(CO)^+/CpMn(CO)_3^+$ intensity ratio only by optimizing the phase of the applied laser field. The resulting optimized pulse forms reflect the capability of the optimization method to find tailored pulses which correlate with the ultrafast molecular dynamics during photofragmentation process.

1 Introduction

Metal-carbonyls containing CO ligands are adequate systems for the investigation and control of ultrafast split of ligands. When they are ionized by an electron gun, the recorded mass spectra reflect fragmentation patterns, strongly depending on the electron impact energy. Pump and probe experiments can be performed on these molecules, due to the chromophoric character of their metallic core. The pump pulse electronically excites the core, which transfers the energy to different bonds, causing the cleavage of the involved ligands. The temporal evolution is monitored by the probe pulse which ionizes the photo-excited system from its electronically excited state. The development of the pulse shaping techniques [1] combined with Rabitz' idea [2] to employ a self-learning algorithm in a feedback loop experiment has become a fascinating topic for a selective manipulation of the reaction dynamics by driving the molecule towards the desired fragmentation path. The first experiment was performed by the Gerber group [3] who optimized the shape of a femtosecond laser pulse in order to drive the multiphotonic process into the desired reaction channel by using a computer controlled feedback optimization loop. The possibility to use optimized laser pulses in order to control the reaction dynamics of organometallic molecules was already shown by theoretical calculations [4] and experiments [5]. It has been also demonstrated that from the optimized laser pulses it is possible to extract information about the dynamics which occur in the electronically excited state [6].

2 Analysis

The experiments were carried out in a Ti:sapphire oscillator-amplifier system and a molecular beam apparatus equiped with a quadrupole mass spectrometer used for the mass selection of the resulting photoions. The experimental set-up has been described elsewhere [5-6]. First, we investigated the dependence of different fragment ions coming from $CpMn(CO)_3$ as a function of the electron impact energy and laser pulse energy.

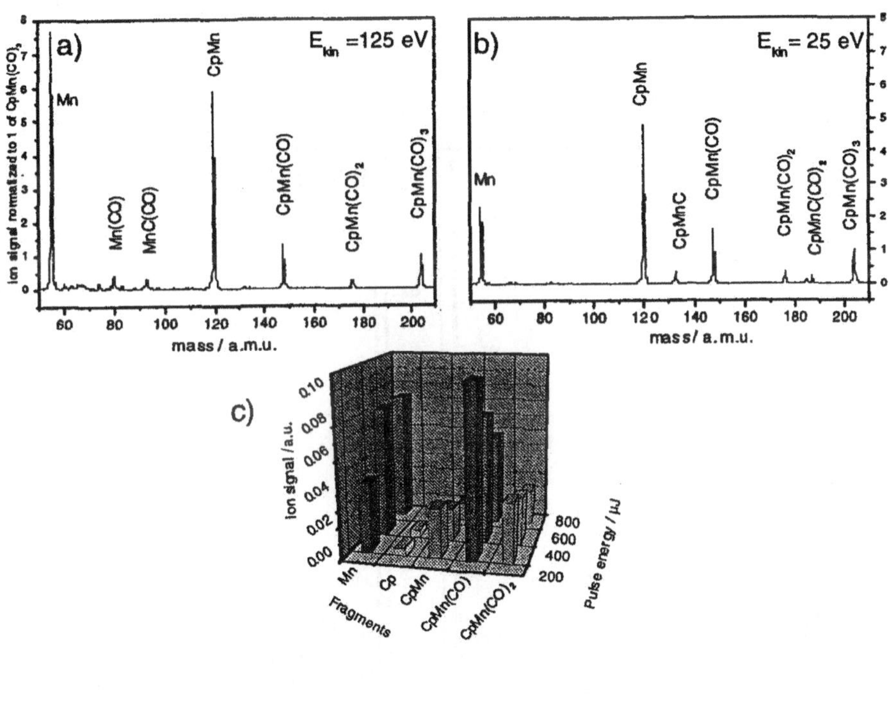

Figure 1. Mass spectra showing the dependence of the fragment intensities on the electron kinetic energy by applying 125 V (a) and 25 V (b) on the cathode and with a single 100 fs laser pulse centered at 800 nm at different pulse energies (c). The peaks were normalized to 1 for the $CpMn(CO)_3^+$ ion signal.

Fig. 1 shows the intensities of all fragments measured with an electron gun at two different electron kinetic energies (Fig.1a and Fig1b) and using a 100 fs single pulse centered at 800 nm (Fig.1c). The $CpMn(CO)^+$, $CpMn^+$ and Mn^+ ion peak intesities change dramatically while decreasing the kinetic energy of the electrons. At 800 nm the mass spectra feature a strong dependency of the Mn fragment peak intensity whilst the pulse energy was increased. Increasing the pulse energy, a decrease of the $CpMnCO^+$ fragment intensity can also be observed. The mass spectra recorded at 400 nm, not shown here, feature a saturation at lower laser pulse energy, which is the first indication of a resonant step in the multiphoton ionization process. The time-resolved mass scans, recorded at different time delays between the pump and the probe pulse, which are not presented here, indicate that the fragmentation process of $CpMn(CO)_3$ completes within 2 ps.

By employing a simple grating pair arrangement, linearly downchirped pulses can be created at a central wavelength of 800 nm. The distribution of the carbonyl compounds contained in the molecular beam, measured with a ~100 fs unchirped pulse is presented in Fig. 2a.

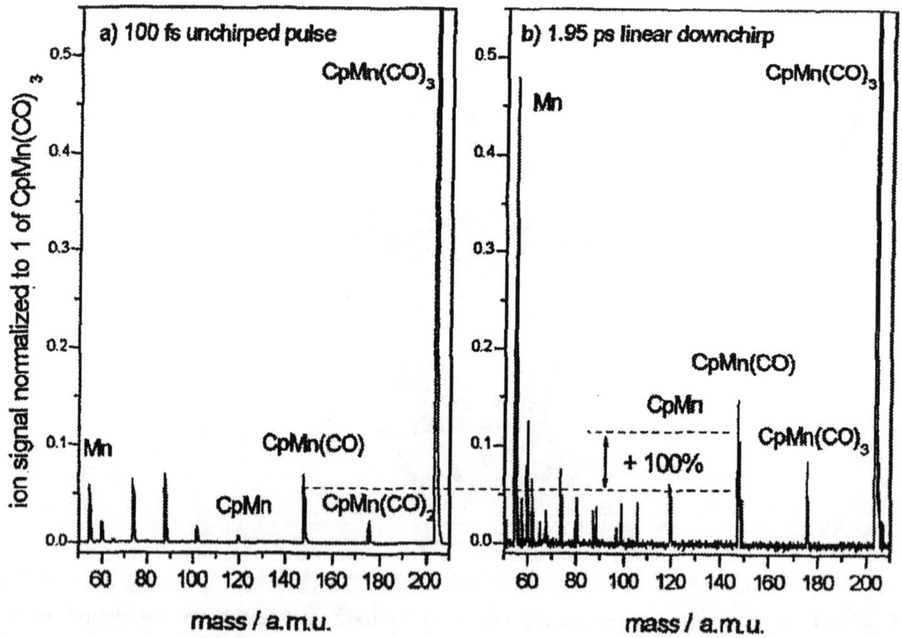

Figure 2. Mass spectra recorded with an ultrashort laser pulse of 100 fs time duration (a) and with a linearly downchirped pulse of about 1.95 ps. The peaks were normalized to 1 for the $CpMn(CO)_3^+$ ion intensity.

By intersecting the supersonic jet with a chirp pulse, the time separation of the different spectral components in the laser pulse (red wavelengths faster than blue wavelengths) favorizes an increase of different fragment ion yields emerging from the parent molecule. An increase of approx. 100% of CpMn(CO) ion intensity (relative to the mother ion) was observed for a 1.95 ps downchirp pulse (Fig. 2b).

Based on these findings, we performed the first pump and probe experiments on this organometallic compound, by employing laser pulses with a full-width at half maximum of about 110 fs at 400 nm for the excitation and ~90 fs laser pulses centered at 800 nm for the ionization process. Fig. 3 shows a set of time-resolved spectra of the parent $CpMn(CO)_3^+$ ion and the resulting fragment ions $CpMn(CO)_2^+$, $CpMnCO^+$, and $CpMn^+$. In order to extract the decay times from the experimental data, a deconvolution algorithm was used.

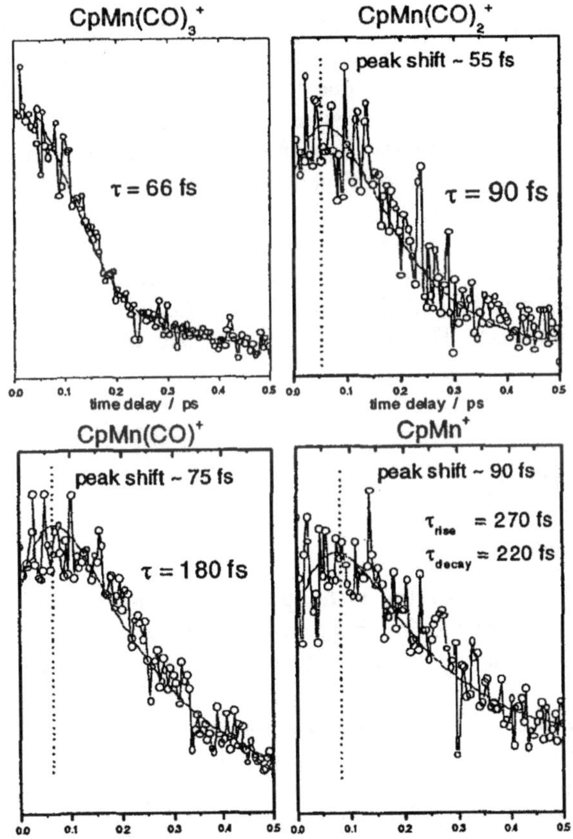

Figure 2. Pump and probe spectra reflecting the temporal evolution of the parent molecule $CpMn(CO)_3^+$ ion signal and all of its fragment ions. The lifetimes of each species and the time shifts of the maximum ion signal in the recorded transients are shown as well.

The experimentally observed deçay time (66 fs) of the photofragmenting $CpMn(CO)_3^+$ is in good agreement with the calculated time (~63 fs) for the cleavage of the first metal-CO bond [5]. Experimental investigations of other metal-carbonyl molecules indicated that the photodissociation may compete against internal conversion via conical intersections [7]. Hence, the $CpMn(CO)_2^+$ fragment may have enough energy in the ground state to break off more than one CO ligand. The other possibility (on purely energetical considerations) would be that the excitation photon (3.1 eV) has sufficient energy to photodissociate two CO ligands. Thus, the measured effective lifetime of $CpMn(CO)_2^+$ fragment ion (90 fs) may indicate the time while the CO ligand breaks off from the ground state of this molecule following fast relaxation, or the time for cleavage of the second Mn-CO bond from the electronically excited $CpMn(CO)_2^+$. The two time constants 220 and 270 fs, which describe the rise and decay time of the $CpMn^+$ transient ion signal, may be the contribution of different precursors, which are created on other reaction channels. An intuitive assignment of these time constants cannot be made at this point, but in the multiphotonic process, we consider several species with different intermediate states which can be consecutively populated. These alternative reaction channels can lead to the formation of the ion fragments $CpMn(CO)^+$, $CpMn^+$ and Mn^+ [8]. To summarize, the ultrafast decay times and the time shift observed in the transient ion signal indicate two possible processes: 1. in the excited state of $CpMn(CO)_3$, a direct break of the two CO bonds can take place, or 2. one CO ligand breaks off in the excited state and the second CO bond is lost after the returning of the molecule back to the ground state. This may lead to the formation of the individual ion.

3 Feedback Control

3.1 Experimental setup

Based on the features observed in the pump and probe spectra, the $CpMn(CO)_3^+$ molecule and its longest living fragment, $CpMn(CO)^+$, have been chosen for the feedback optimization experiment in order to control the ionization/fragmentation pathways.

A single 87 fs laser pulse centered at a wavelength of 800 nm passes through a pulse shaper, which consists of 2 gratings (2 x 1200 l/mm), 2 plano-convex lenses (f = 200 mm) in a 4f arrangement and a liquid crystal spatial light modulator (Cambridge Research and Instrumentation). Placed in the Fourier plane of this zero-dispersion compressor, the liquid crystal mask allows simultaneous phase- and amplitude modulation of the incoming femtosecond laser pulses by independently applying voltages on both arrays of 128 pixels. In this way, shaped pulses can be created, e.g. pulse trains, linearly chirped pulses, etc. The modulated pulses are focussed on the supersonic molecular beam of $CpMn(CO)_3$ and the resulted ions are mass-selected by a quadrupole mass spectrometer. The detected current is then taken as a feedback signal for the optimization algorithm based on evolutionary stategies [9]. The self-learning method begins with randomly generated shaped pulses created from originally transform-limited pulses and changes iteratively the modulator settings in order to find the pulse shape that yields the highest feedback ion signal. A detailed description of the optimization algorithm used in this experiment can be found in [5,6].

3.2 Results and Discussion

In order to avoid any undesired pulse energy effects, in the feedback-control experiment only the phase of the laser pulse was optimized. In Fig. 4a and Fig. 4b, instead of the absolute ion signal, which can vary with the detector instabilities and with fluctuations possible in a molecular beam, the relative ratios of the peak intensities are presented. The resulting $CpMn(CO)^+/CpMn(CO)_3^+$ ion yield ratios are shown in Table 1, indicating that the control experiment leads to a substantial increase of the desired products. After the optimization, the ratio of fragment/mother ion yield was improved by 150%.

$CpMn(CO)^+/CpMn(CO)_3^+$ mass peak ratio	Pulse form
1:13	Reference pulse before optimization (87 fs, unchirped)
1:16	After $CpMn(CO)_3^+$ yield optimization
1:6	After $CpMn(CO)^+$ yield optimization

Table 1. $CpMn(CO)^+/CpMn(CO)_3^+$ mass peak ratios obtained before and after the optimization experiments.

The three dimensional FROG traces of the optimized pulses in the case of $CpMn(CO)_3^+$ and $CpMn(CO)^+$ optimization are shown in Fig. 4c and Fig. 4d, respectively. The two optimized pulse shapes obtained from the SHG FROG analysis are presented in Fig. 4e for $CpMn(CO)_3^+$ and in Fig. 4f for $CpMn(CO)^+$ maximum ion yield.

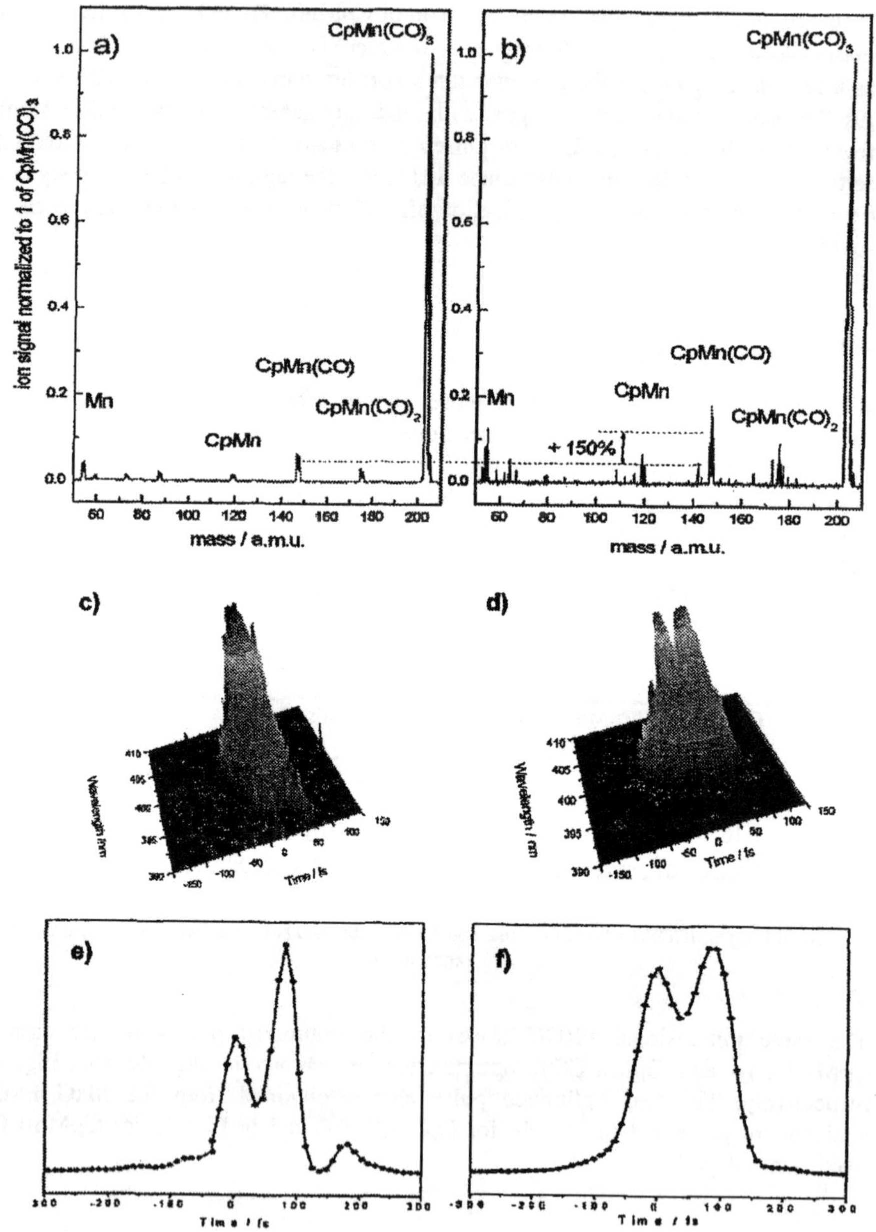

Figure 4. Comparison of mass spectra recorded with optimized pulse shapes for a maximum ion yield of $CpMn(CO)_3^+$ (a) and $CpMn(CO)^+$ (b). The SHG FROG traces in a three-dimensional representation of the optimized pulses for $CpMn(CO)_3^+$ (c) and $CpMn(CO)^+$ (d) and the optimized pulse forms obtained from the FROG analysis after the optimization experiment for $CpMn(CO)_3^+$ (e) and $CpMn(CO)^+$ (f) maximum ion signal.

The pulse sequence which produced the highest amount of $CpMn(CO)_3^+$ consists of three pulses, each of 36-40 fs time duration (Fig. 4e). The effective duration of the two leading pulses is approximately 100 fs, similar to the duration of the initially unchirped laser pulse (87 fs). The pulse form can be explained taking into account a two-photon excitation and 3-4-photonic ionization process together with the ultrafast 66 fs time decay of the $CpMn(CO)_3^+$ ion signal. The third following pulse can represent the second ionization step: a fraction of the electronically excited mother molecules, later excited by the second pulse will be ionized by the third pulse. We have to mention that the self-learning algorithm found the previously described triple pulse form and not a single short pulse. Because of relatively long time duration of the laser pulses used in the two-color pump and probe experiments (110 fs for the excitation and 90 fs for the ionization) the approx. 36-40 fs time shifts observed in Fig. 4e could not be resolved in the time-resolved spectrum. In order to gain more detailed information about the real-time process combined with the optimized pulse forms for each investigated compound, further experiments with sub-50 fs laser pulses are currently under progress. The pulse train acquired for the maximum of $CpMn(CO)^+$ fragment ion yield consists of a double pulse sequence (Fig. 4f). The first pulse excites the $CpMn(CO)_3$ molecule. The second pulse arrives at about 80 fs after the first pulse. (The effective time duration of the double pulse is about 145 fs.) The second pulse ionizes the actual population of the involved state(s) of the neutral CpMn(CO), which later gives rise to a maximum $CpMn(CO)^+$ fragment ion signal. The arrival time of the second pulse correlates very well with the peak shift of the $CpMn(CO)^+$ time-resolved pump and probe spectrum with respect to the excitation process – 80 fs and 75 fs, respectively.

4 Summary

In this paper we have presented the ultrafast dynamics in the electronically excited $CpMn(CO)_3$ molecule using femtosecond laser pulses. We have extracted the lifetimes of several involved electronic states of the parent and fragment molecules. According to the experimental observations and the theoretical calculations, the time when the first Mn-CO bond breaks off is about 66 fs. The peak shift of the transient pump and probe spectra indicates the evidence for sequential population of the intermediate states of the excited $CpMn(CO)_3$ and $CpMn(CO)_2$, giving later rise to ion signals of $CpMn(CO)_3^+$, $CpMn(CO)_2^+$, $CpMn(CO)^+$, $CpMn^+$ and Mn^+. We have performed feedback optimization experiments on this photofragmenting system. The experiment increased the relative ion yields of the chosen reaction channels, implying the control of populations of the corresponding neutral molecular state(s). The optimized pulse shapes reflect the consecutive population of the precursors of the involved states already observed in the time-resolved spectra.

5 Acknowledgements

The authors thank Mr. Andreas Bartelt for providing the evolutionary algorithm used in the optimization experiment. We are grateful to Prof. Dr. Jörn Manz, Dr. Werner Fuß and Dr. Serghej Trushin for the stimulating discussions and to Deutsche Forschungsgemeinschaft for the financial support within the SFB 450. Cosmin Lupulescu and Porfirio Rosendo-Francisco gratefully acknowledge the financing of their PhD studies at Freie Universität Berlin by Gottlieb Daimler- und Karl Benz-Stiftung and Deutscher Akademischer Austauschdienst, respectively.

References

1. A.M. Weiner et al., IEEE J. Quantum Electronics **28** (1992) 908.
2. R.S. Judson and H. Rabitz, Phys. Rev. Lett **68** (1992) 1500.
3. G. Gerber at al., Science **282** (1998) 919.
4. J.E. Combariza, B. Just, J. Manz, G.K. Paramonov, J. Phys. Chem. **95** (1991) 10351.
5. C. Daniel, J. Full, L. Gonzales, C. Kaposta, M. Krenz, C. Lupulescu, J. Manz, S. Minemoto, M. Oppel, P. Rosendo-Francisco, Š. Vajda, L. Wöste., Chem. Phys. **267** (2001) 247.
6. Š. Vajda, A. Bartelt, E.-C. Kaposta, T. Leisner, C. Lupulescu, S. Minemoto, P. Rosendo-Francisco, L. Wöste, Chem. Phys. **267** (2001) 231.
7. S.A. Trushin, W. Fuss, K.L. Kompa, W.E. Schmid, Chem. Phys. **259** (2000) 313.
8. W. Fuss, W.E. Schmid, S.A. Trushin, J. Chem. Phys. **112** (2000) 8347.
9. H.-P. Schwefel, Evolution and Optimum Seeking, Wiley, New York, 1995.

STRONG FIELD COHERENT CONTROL IN MOLECULES

B. J. PEARSON, D. S. MORRIS, AND P. H. BUCKSBAUM

Physics Department, University of Michigan, Ann Arbor, MI, 48109, USA
E-mail: phb@umich.edu

T. C. WEINACHT

JILA, University of Colorado, Boulder, CO, 80309, USA

We describe our recent results using an automated learning apparatus to control vibrational Raman scattering in liquid phase. By directing intense, shaped ultrafast laser pulses into a liquid sample and using a measurement of the system as a feedback signal to a learning algorithm, we are able to reshape the laser pulses to enhance the Raman excitation. In addition to describing the achieved control, we discuss possible methods of using the results in order to learn about the system under investigation.

1 Introduction

The number of successful implementations of coherent control has grown tremendously since it was suggested to incorporate a feedback loop into the experimental apparatus.[1] Despite these accomplishments, feedback experiments have had limited success in utilizing the control to learn about the system being studied. This paper describes work using our adaptive learning machine to examine feedback control in liquid phase alcohol. We also discuss some of the methods we use to try and extract information about the molecules under investigation.

2 Experimental Method

Our experiments make use of the recent developments in ultrafast lasers and pulse shaping technologies. We use an amplified titanium sapphire laser system that produces pulses with about 3 THz bandwidth at a central frequency of 790 nm and a repetition rate of 10 Hz. The pulses are shaped in a acousto-optic pulse shaper which allows a computer to specify the amplitude and phase at any color within the spectral bandwidth of our laser pulse.[2] The shaped laser pulses are generated by our adaptive learning algorithm running on a desktop computer. The feedback signal collected from the experiment is returned to the algorithm in order to generate new shaped laser pulses based on the success of previous pulses in achieving a predetermined goal.

The details of the learning algorithm have been described elsewhere.[3]

Briefly, the algorithm starts with a collection of shaped laser pulses (a population) specified by the phase and amplitude at each frequency in the laser pulse. These pulses interact with the liquid sample, and the feedback measurement allows the algorithm to assign a fitness to each pulse shape which is a measure of how successful the pulse shape was at producing the desired effect. Our algorithm goes a step further than traditional genetic algorithms[4] by running many different search strategies concurrently. The algorithm adapts itself throughout each run, finding the most effective way to search the phase space.

In most of the experiments, we allow the algorithm to vary only the phases of the incident colors in the laser pulse. This maintains a constant energy in each laser pulse (around 800 μJ), and the algorithm works to determine how to best distribute this energy in time. The pulses are focused into a liquid cell, and the stimulated Raman spectrum in the forward direction is collected. This spectrum is used to calculate the fitness value for each pulse shape.

3 Results and Discussion

Following our previous work in methanol,[3] we have extended the vibrational control experiments to liquid phase ethanol (CH_3CH_2OH). We concentrate on excitation of the three C-H stretch Raman modes with energies of 2877 cm^{-1}, 2928 cm^{-1}, and 2972 cm^{-1}.[5] Figure 1 shows the resultant Raman spectra after excitation with the laser pulses. The dotted curve at the bottom is the spectrum from an unshaped laser pulse, while the three solid curves are the spectra after the algorithm has converged on each of the three vibrational modes by varying only the phases of the colors in the laser pulse. As Fig. 1 shows, the algorithm is quite successful at selective excitation of the different Raman modes.

We would now like to go further by examining the results of the learning process in an attempt to learn why the control was possible. One possible method of analysis involves examining the optimal pulse shapes which led to the vibrational excitation. However, it is not entirely obvious how the pulses should be viewed. We begin by looking at the intensity profile $I(t)$ of the incident laser pulse which led to excitation of the central mode in Fig. 1. Figure 2a demonstrates how the algorithm finds that a transform-limited pulse is not optimal, and instead converges to a structured pulse with a total duration of approximately 4 ps. However, as we often discover, the complexity of $I(t)$ in the time domain prevents any immediate conclusions to be drawn.

Since we are exciting vibrational modes with different energies, it may be more instructive to look at the pulse in a representation where we can see both

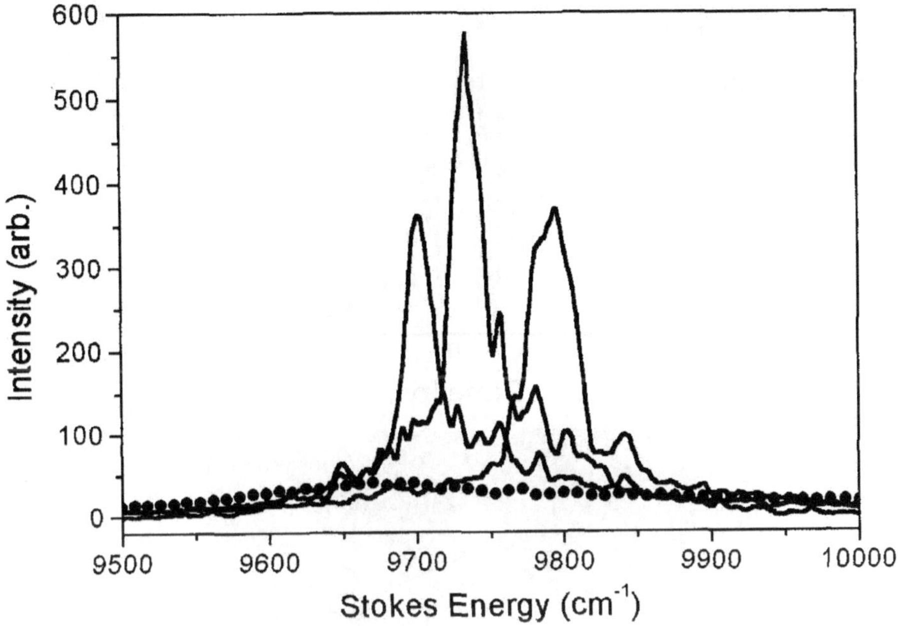

Figure 1. Stimulated Stokes spectrum in ethanol. The dotted curve is the Stokes spectrum after excitation with a transform-limited laser pulse. The three solid curves are the optimal spectra after the learning algorithm has converged on the three separate C-H stretch modes.

frequency and time together. Figure 2b shows a more complete description of the same laser pulse. This Wigner distribution is a two-dimensional (time and frequency) representation of the same pulse shape.[3] The Wigner plot is interesting since it allows us to see not only the areas where the intensity in the pulse lies, but also the coherences between these regions. In fact, this Wigner representation may actually provide too much information if what we are really after is a simple view of the pulse structure in both time and frequency. This can be accomplished by smearing the Wigner function with a two-dimensional Gaussian, resulting in the Husimi representation of the pulse shape.[3] The Husimi function is positive definite and depicts where the energy in the pulse is contained. One particular Husimi distribution for this pulse

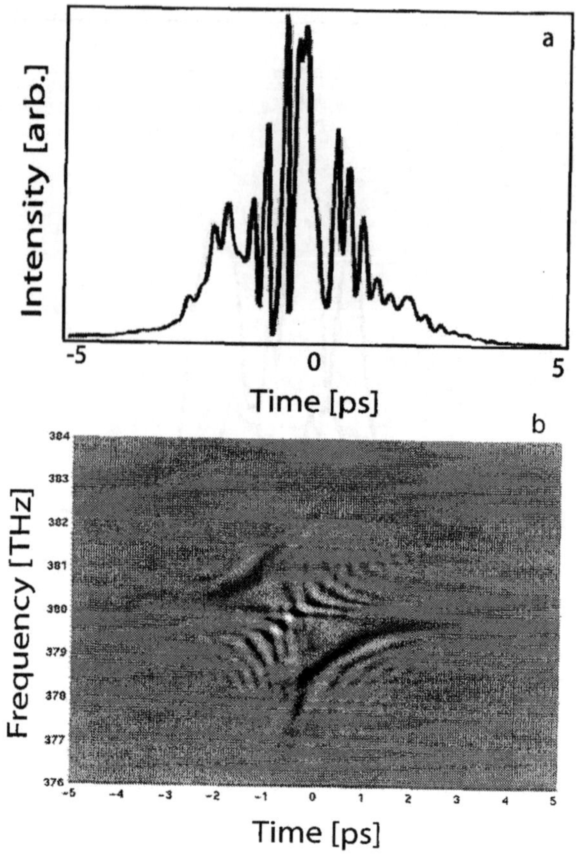

Figure 2. Two different views of the pulse shape that led to excitation of the central C-H vibrational mode. Panel a plots the intensity profile $I(t)$. Panel b is the two-dimensional Wigner representation of this same pulse shape.

shape is shown in Fig. 3a. In this plot, the two-dimensional Gaussian function used in the convolution has equal width in both time and frequency. The energy in the pulse can clearly be seen in two distinct regions, while the coherences between these two regions have been washed out by the convolution.

We are not limited to using a symmetric Gaussian for the convolution of the Wigner function. It may be more instructive to maintain better resolution in the time domain, as temporal correlations may be more important in our

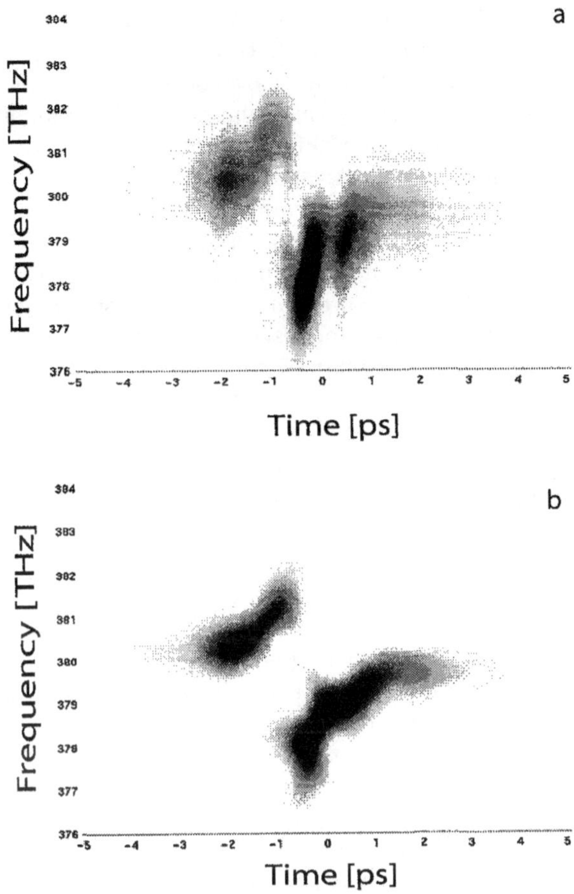

Figure 3. Two Husimi representations of the optimal pulse shape for exciting the central mode in ethanol. Panel a was generated using a symmetric Gaussian function to smooth the Wigner function. Panel b used an asymmetric Gaussian with twice the resolution along the time axis.

non-resonant excitation. Figure 3b shows the Husimi plot generated when the convolution is performed with an asymmetric smoothing Gaussian (the temporal width is halved while frequency width is doubled to preserve the total area). This plot preserves some features of the pulse shape, such as chirp, much better than in the case of the symmetric Gaussian. In experiments where the amplitudes of the incident colors are varied as well as the phases, the resolution along the frequency axis may become more important. However for these experiments where only the phases are controlled, it appears that more detailed information is conveyed by the asymmetric Gaussian smoothing.

We have been able to combine different views of the pulse to investigate the nature of the control. In contrast to ethanol, methanol (CH_3OH) has only two modes in the C-H region: a symmetric (2834 cm^{-1}) and asymmetric (2946 cm^{-1}) stretch.[6] Our learning algorithm was able to selectively excite either mode, and structure in the Husimi plots of the optimal pulse shapes from our previous experiments showed energy separations which matched the energy splitting between the two modes.[3] This is similar to true impulsive Raman scattering where the laser bandwidth is large compared to the energy separation of the vibrational levels. Meshulach and Silberberg showed that in this case the transition amplitude is proportional to $\int_{-\infty}^{\infty} \epsilon^2(t)e^{\imath\omega_o t}dt$, or the component of $I(t)$ at the resonance frequency between the two levels.[7] Of course our laser bandwidth is much smaller than the vibrational level spacing, but it is large compared to the splitting of the two modes. This led us to consider a "quasi-impulsive" model of the excitation, where instead of looking at the component of $I(t)$ at the resonance frequency of the Raman transition, we focus on coupling frequency between the symmetric and asymmetric stretches.

These results are summarized in Fig. 4. Figure 4a and 4c show the intensity profiles, $I(t)$, of the optimal laser pulse for exciting the symmetric and asymmetric stretch modes, respectively. The Fourier transforms of $I(t)$ for these two pulses are plotted in the lower curves in Fig. 4b and 4d. The upper curves are what the Fourier transform would be if the input pulse were transform-limited with our spectral bandwidth. The interesting feature to note is that the optimal pulse for exciting the asymmetric stretch mode contains a strong component at the coupling frequency between the two modes (3.3 THz), while the symmetric stretch pulse does not.

To further explore this idea we repeated the experiment using deuterated methanol (CD_3OD) where the C-H stretch mode energies are 2082 cm^{-1} and 2162 cm^{-1}.[8] The results from deuterated methanol are shown in Fig. 5 (same panel descriptions as Fig. 4). Here the behavior appears to be opposite from methanol: the optimal pulse for exciting the symmetric stretch mode contains the strong component at the coupling frequency of 2.4 THz.

Our analysis need not be limited to examination of the optimal pulse shapes. Since the algorithm adapts itself during each experimental run, we can investigate which search methods are most effective for each particular system. This is accomplished by observing the evolution of the operators. Just as each individual pulse shape has a fitness, each operator is assigned its own fitness

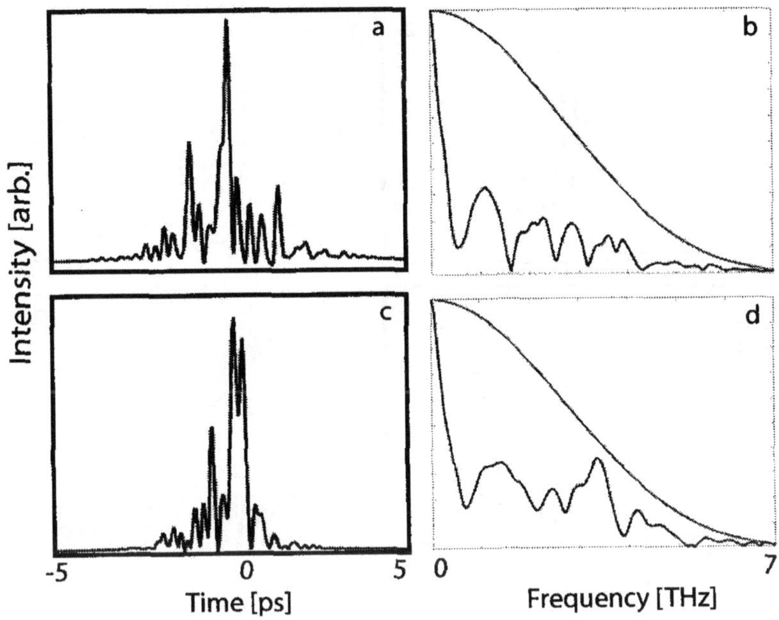

Figure 4. Quasi-impulsive analysis in methanol. Panel a (c) shows $I(t)$ for the optimal pulse shape which led to excitation of the symmetric (asymmetric) C-H stretch mode. The lower curve in panel b (d) plots the intensity of the Fourier transform of $I(t)$ for the pulse in panel a (c). The upper curve plots what the Fourier transform would be if the input pulse were transform-limited. The optimal pulse for exciting the asymmetric stretch mode contains a strong component at the coupling frequency (3.3 THz) of the two modes.

which is a measure of how well the operator is doing at producing fit new pulse shapes. The algorithm has a pool of operators available to choose from during an experimental run, and it selects the operators for use in reproduction in proportion to their fitnesses. This accomplishes two functions. First, less time is wasted on those operators which are not producing fit pulse shapes, thereby making the algorithm more efficient. Second, we are able to go back and examine which operators are best suited to a particular goal. Since some of the operators induce physically meaningful changes to the pulses, we have the potential to see which processes may be an important part of the control

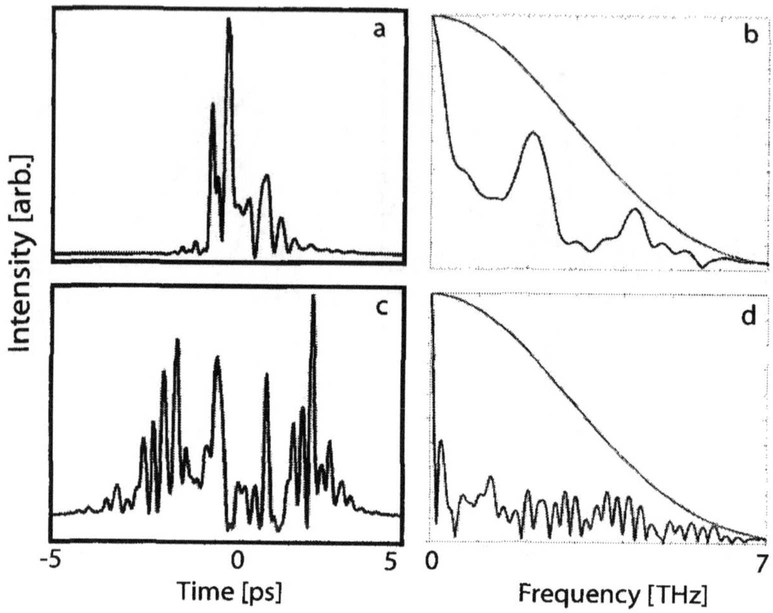

Figure 5. Quasi-impulsive analysis in deuterated methanol. Panel a (c) shows $I(t)$ for the optimal pulse shape which led to excitation of the symmetric (asymmetric) C-H stretch mode. The lower curve in panel b (d) plots the intensity of the Fourier transform of $I(t)$ for the pulse in panel a (c). The upper curve plots what the Fourier transform would be if the input pulse were transform-limited. In this case, the optimal pulse for exciting the symmetric stretch mode contains a strong component at the coupling frequency (2.4 THz) of the two modes.

mechanism.

For the particular pulse shape which led to the control in ethanol (see Fig. 2 and Fig. 3), we can examine the methods the algorithm used to search the available phase space. Figure 6 shows the fitness of each operator as a function of generation for optimizing the central mode in ethanol. The operators are initialized in a typical genetic algorithm configuration with *two-point crossover* dominant. However as soon as the operators are allowed to evolve, the algorithm finds that two-point crossover is not the most efficient searching method. Instead, the *polynomial phase* operator does very well. This oper-

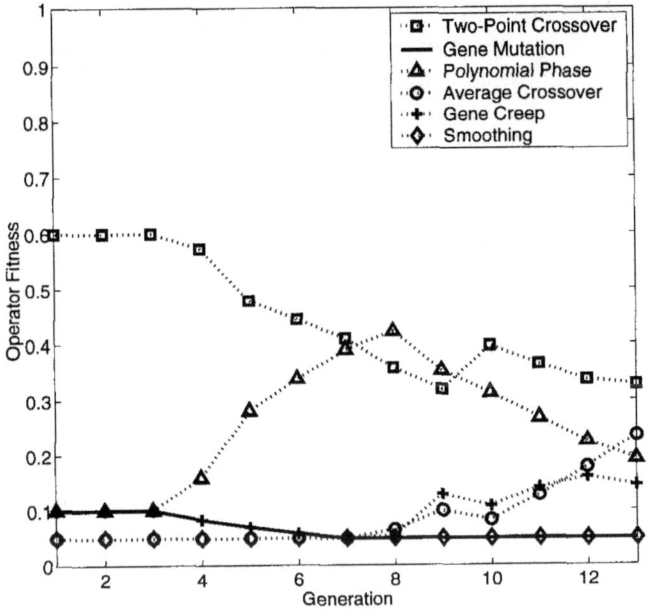

Figure 6. Fitness of each operator as a function of generation for excitation of the central C-H stretch mode in ethanol. The algorithm discovers that standard genetic algorithm operators (two-point crossover and mutation) are not the most efficient methods of searching the phase space. Instead the polynomial phase operator is optimal for creating fit new pulse shapes.

ator inserts sections in the pulse shape where the phase is proportional to a polynomial function of frequency ($\phi(\omega) \propto \pm(\omega)^n, n \in 0, 6$), effectively adding different orders of chirp across the spectrum. The higher order polynomial phase is evident in the Husimi distribution of Fig. 3b.

4 Conclusions

We have constructed an adaptive learning algorithm capable of controlling vibrational Raman excitation in liquid systems. Our feedback technique is general since no specific resonances are exploited in the excitation. We explore a variety of methods to extend the experiments to learn about the system being studied. We find that the details of the learning process and analysis of solutions are very important if one wishes to extract information as to why the control was possible. Our future plans include both extending the capabilities of the algorithm, such as incorporating an operator which works with a two-dimension representation of the pulse shape, as well as examining control in more complicated molecular systems.

Acknowledgments

The authors would like to thank Misha Ivanov, Robert Levis, Marcus Motzkus, and David Tannor for helpful discussions throughout the experiments. This work was supported by NSF grant 9987916.

References

1. R. Judson and H. Rabitz, *Phys. Rev. Lett.* **68**, 1500 (1992).
2. J. X. Tull, M. A. Dugan, and W. S. Warren, *Adv. Opt. Mag. Resonance* **20**, 1 (1997).
3. B. J. Pearson, J. L. White, T. C. Weinacht, and P. H. Bucksbaum, *Phys. Rev. A* **63**, 063412 (2001).
4. Davis, L. Ed. *Handbook of Genetic Algorithms* (Van Norstrand Reinhold, New York, 1991).
5. A. Lauberau, G. Kehl, and W. Kaiser, *Opt. Comm.* **11**, 74 (1974).
6. T. Shimanouchi, *Tables of Molecular Vibrational Frequencies: Consolidated Volume 1* (National Bureau of Standards, Washington, DC, 1972).
7. D. Meshulach and Y. Silberberg, *Phys. Rev. A* **60**, 1287 (1999).
8. H. D. Noether, *Infra-Red and Raman Spectra of Polyatomic Molecules* **10**, 693 (1942).

VIBRATIONAL POPULATION CONTROL IN LIQUID-PHASE METAL CARBONYLS

VALERIA D. KLEIMAN
Chemistry Department, University of Florida,
Gainesville, FL 32611-7200 USA
E-mail: kleiman@chem.ufl.edu

JOSEPH S. MELINGER
Electronics Science and Technology Division, Code 6812, Naval Research Laboratory,
Washington, DC 20375 USA
E-mail: melinger@ccf.nrl.navy.mil

EDWIN J. HEILWEIL
Optical Technology Division, National Institute of Standards and Technology,
Mail Stop 8441, Gaithersburg, MD 20899-8441 USA
E-mail: edwin.heilweil@nist.gov

Broadband picosecond and femtosecond mid-infrared chirped-pulse excitation with broadband probe techniques are being explored at NIST using multi-element, mid-infrared focal plane array detectors. Ultrafast infrared chirped IR pulse generation, detection and pump-probe spectroscopy schemes are reviewed. These methods are used to coherently control and probe vibrational overtone populations and the results modeled using the density matrix formalism. Coherent control of the T_{1u} CO-stretching vibrational overtone population distributions for tungsten-hexacarbonyl in room temperature n-hexane using deliberately chirped mid-IR excitation pulses is reported. Negatively chirped picosecond IR excitation pulses enhance v=2 population while intense femtosecond pulses produce only v=1 excitation.

1 Introduction

Excitation and control of ground electronic state vibrational population using specifically designed laser pulses is an increasingly active area of research [1-5]. This surge of interest has largely been inspired by the increased availability of commercial, solid-state, table-top ultrafast laser equipment and nonlinear frequency conversion methods for generating high power, broadly tunable infrared (IR) pulses in the near to mid-infrared wavelength regime (ca. 2 to 25 μm). Under routine situations, excitation is accomplished with near-transform limited (all frequency components of the pulse arrive at the sample simultaneously) infrared picosecond or femtosecond pulses. However, since polyatomic electronic ground state vibrational potential energy surfaces are generally anharmonic, it is possible that non-transform limited pulses may significantly alter final state distributions in multiphoton processes relative to their transform limited counterparts. Since ultrashort pulse-shaping and pulse-analysis techniques are now widely used, a whole new approach to studying and affecting vibrational dynamics on ground electronic state potential energy surfaces (PES) has become available. By appropriately controlling the arrival time of different frequency components of an IR excitation pulse, for example, one can in principle readily transfer and "control" excited state population in otherwise inaccessible regions of the PES. Comparing measured population dynamics (produced by using carefully chosen and characterized experimental pulse parameters) to theoretical calculations could then lead to better analytical models for potential surfaces of complex systems. This could be particularly valuable for studying condensed phase systems for which high resolution spectroscopic methods to determine PES are ineffective.

There are two broad-based schemes for accessing vibrationally excited ground electronic state surfaces via coherent control. One involves multi-photon excitation through an excited electronic state (i.e. Impulsive Raman-excitation with chirped visible pulses [6,7]). In this case, interpretation of the results involves detailed knowledge of the dynamics on excited electronic surfaces. Another possibility is through direct multi-photon "climbing" of a vibrational mode overtone ladder using phase-modulated IR excitation pulses (see below). This latter method requires prior knowledge of only ground electronic state parameters (vibrational state lifetimes, IR cross-sections, state-to-state couplings, etc.) for interpreting prepared population distributions.

Ultrafast pulses are advantageous for inducing and probing photochemical reactions on timescales shorter or comparable to those of competing relaxation or emission processes. In addition, when excitation schemes involving the coherent properties of the states are employed, interferences can be beneficially used to manipulate the final product state distribution [8,9]. In the case of vibrational populations, for example, state-selective distributions can be obtained by constructive or destructive interference between different components of the excited wavefunction. Therefore, it may be possible to affect efficient mode-specific chemical bond-breaking reactions in ground electronic state molecules using properly tailored ultrafast infrared pulses.

There are a number of theoretical predictions for phase-controlled multi-photon infrared excitation that have yet to be experimentally explored or verified. Just et. al. modeled the IR excitation of gas phase OH radicals and found that a linear chirp imposed on a $\sin^2$ pulse envelope could transfer population to higher vibrational levels if state selectivity was not required [2]. A similar result was found by Paramonov [5] where optimization of the pulse amplitude and pulse timing sequence improved the population selectivity on the final state. That work was extended by Korolkov et. al. to include preparation of high lying states that could lead to dissociation in diatomic molecules [3]. Melinger, et. al. explored the use of single pulses with amplitude and phase modulation to invert vibrational populations with high selectivity [4]. In that work, the robustness of the control for different pulse parameters was examined. While all of the calculations mentioned above were performed for model "isolated" diatomics or single mode oscillators, condensed-phase systems usually contain complex coupling mechanisms to the bath, as well as couplings of the vibrational modes within the molecule. For the work presented here, we consider these theoretical models as zeroeth-order approximations to the anharmonic vibrational overtone excitations that are possible in the condensed-phase when ultrashort IR pulses are employed.

Recently, an increasing number of experiments have applied coherent control methods. Control of the dynamics of an electronically excited wave-packet has been shown to occur in the gas phase [10] and in matrices [11]. Experiments aimed at modifying ground electronic state dynamics with chirped pulses have been performed using visible pulses interacting with electronic excited states [11,12]. Adiabatic rapid passage has also been used to control excitation of atomic systems. For example, both the chirp direction and field strength of an ultrashort pulse were used to alter the population transferred from the 5s to the 5d state in Rubidium atoms [13,14]. While negatively chirped pulses (red frequencies arriving after blue frequencies) increased the population transfer by 100 %, positively chirped pulses (red frequencies arriving before blue frequencies) inhibited the process. The specific use of unchirped IR pulses to climb a single oscillator vibrational ladder was recently reported for gas phase NO [15] and $W(CO)_6$ in the liquid phase [16].

Broadband infrared detection of controlled vibrational overtone state populations can be easily accomplished using multi-channel infrared array detectors, which has been employed by our group [17-19] and others [20,21]. In many respects, this approach is complementary to the accepted technique for performing ultraviolet-visible transient absorption spectroscopy. By generating broadband mid-infrared probe pulses a few picoseconds or less in duration and detecting a portion of an infrared spectral region for each laser pulse, data acquisition rates are vastly enhanced and system stability issues are diminished. It should also be pointed out that picosecond or femtosecond broadband probe pulses can be used to detect narrow-band transient absorption features as long as the transient absorber has lived on the order of the coherence (T_2) lifetime of the interrogated state [22]. These typically narrow-band features (3 cm^{-1} to 15 cm^{-1} FWHM) are spectrally resolved by an up-converter crystal (with CCD detection) or through the spectrograph resolution independent of the inherent time-bandwidth characteristics of the probe pulse. Demonstrations of the general broadband IR detection approach were first made using nonlinear IR frequency down and up-conversion with visible spectrograph-CCD detection [17,18]. With US Defense Department declassification and increased commercial availability of infrared focal-plane arrays for mid-infrared imaging use, direct broadband infrared spectroscopic detection covering the 1 μm to 12 μm range is now practical [23].

We present experimental results that demonstrate control of the vibrational state population distribution in a polyatomic condensed-phase system using tailored picosecond IR excitation pulses. Specifically, we investigate multi-photon vibrational ground electronic state T_{1u} mode excitation of $W(CO)_6$ in room temperature n-hexane solution [16] but we now use deliberately chirped IR pulses. We show how specific phase properties of an IR excitation pulse alter the vibrational population distribution within a molecular mode vibrational ladder. These results suggest the possibility of creating a desired excited vibrational state population distribution or potentially populating dissociative overtone states by proper choice of excitation pulse parameters. In the first section we explain how IR pulses are produced and analyzed. After discussing these techniques, results from studies on vibrational mode up-pumping performed using these broadband pump and probe schemes are presented.

2 Experimental Methods

The production of ultrafast mid-IR pulses begins with a fixed-frequency laser oscillator that is pulse amplified from nanojoule energies to microjoule (μJ) and millijoule (mJ) levels. The laser systems used include a 80 MHz Nd^{+3}:YAG picosecond oscillator (ca. 100 ps, 1.064 micron with second harmonic generation and pulse compression to 8 ps) and a Ti:Sapphire, 35 fs oscillator followed by regenerative and multiple-pass amplifiers, respectively. The amplified pulses are subsequently frequency-shifted or beat against another pulse (by frequency down-conversion steps) in an appropriate nonlinear crystal to reach the mid-IR range. Various frequency-shifting techniques include difference frequency generation between two synchronously pumped tunable dye lasers or optical parametric amplification (OPA, see below) of a weak infrared continuum which provides narrow-bandwidth tunability in the mid-IR region at moderate power levels (<20 μJ/pulse). Difference frequency mixing of amplified tunable visible dye oscillators beat against a fixed frequency pulse (e.g., the 532 nm second harmonic of the Nd^{+3}:YAG) in $LiIO_3$ crystals enables generation of independently tunable IR pump and probe pulses [24]. The above approaches produced tunable ~1 picosecond pulses with relatively narrow (3 cm^{-1} to 10 cm^{-1} FWHM) spectral bandwidths appropriate for frequency tuned up-pumping of condensed-phase systems.

Modifications to the above laser systems and frequency down-converting schemes are readily performed to generate broadband IR output. For example, synchronously-pumped picosecond visible or near-IR dye oscillators deliberately run without intracavity tuning elements permit the lasers to run broadband with output center frequency dictated by the dye gain spectrum [25]. Sub-picosecond dye laser pulses can also be fiber frequency chirped, amplified (to <1 mJ) in three-stage dye amplifiers, grating compressed to <150 fs pulse duration and then mixed against a 5 to 10 ps, 300 μJ dye pulses in $LiIO_3$ to yield 150 fs to 200 fs FWHM broadband (200 to 300 cm^{-1} FWHM) IR output [25]. When beat against the second harmonic of the Nd:YAG amplifier or another amplified picosecond (i.e., narrow-band) dye laser in $LiIO_3$ or $AgGaS_2$, this approach yields pulses in the 3 to 10 micron spectral region of up to several hundred wavenumbers bandwidth (FWHM). Difference frequency generation between a broadband seed pulse and narrow-band pump pulse ensures that the generated broadband IR pulse experiences minimal temporal dispersion or jitter while traversing the nonlinear crystal. Dye-based broadband sources (deliberately run without intracavity seeding) are inherently spectrally noisy and require shot-to-shot normalization and averaging during data collection (see below). However, there is no inherent reason why this general method cannot be extended to longer wavelengths for probing in the molecular "finger print" (6 μm to 12 μm) region. To date, many investigations of C=O stretching modes in the 5-7 μm range have been conducted [25,26] but little, if any known transient IR work has been performed in the 8 μm and longer wavelength range.

The other laser system employed for our work is based on a femtosecond Ti:sapphire oscillator (80 MHz, pulse duration <50 fs FWHM, ~800 nm, 0.35 Watt average power) and kilohertz multi-pass amplifier capable of delivering approximately 1 Watt of output power. This system possess the desirable characteristics of delivering ultrafast (but only slightly tunable) pulses which maintain their inherent near transform-limited broadband characteristics (>200 cm^{-1} FWHM) and high repetition rates for rapid data collection and signal averaging. Tunable frequency down-conversion employs a BBO-based (β-Barium Borate) OPA (with multiple BBO passes) that "split" the ~800 nm, amplified Ti:Sapphire input pulse into two longer-wavelength pulses (signal and idler tunable between 1.1 to 2.9 μm). These pulses are subsequently co-propagated and temporally recombined in a down-converting mixing crystal (e.g., Type I $AgGaS_2$,, 1 mm thick) to produce tunable, relatively broad-bandwidth (>150 cm^{-1} or >1.8 μm FWHM) mid-IR probing pulses. The salient features of producing IR pulses in this fashion are straight-forward tuning of the solid-state OPA to cover the desired mid-IR spectral range (accessible between 2 μm to 12 μm), natural short pulse duration for time- resolved measurements, the inherent energy stability (a few percent rms amplitude noise) and smooth nearly transform-limited broadband spectral output. Generated IR pulse energies of a few microjoules is more than sufficient for up-pumping experiments with direct IR focal plane array detection schemes discussed below.

Detection of transient population is accomplished with a liquid nitrogen cooled IR focal plane array (IRFPA) detector with InSb detector diodes (1 μm to 5.5 μm sensitivity and having 256x256 format with 65,536 total pixels on 30 μm centers) [27]. One requirement for pulsed spectroscopic applications is that an IR array be run in "snapshot" rather than normal imaging strobing read-out format. This characteristic is necessary because a constantly strobing read-out device is incapable of capturing a single ultrafast laser pulse spectrum over the width of the array. For a snapshot array, all diode electron capacitive wells accept charge during the same integration period. After pulse integration (typically ca. 10 μsec), the array is subsequently read out (via two analog outputs and 1 MHz A/D converters) by a clocking sequence and an image reconstructed from the analog-amplified and digitized chip output. Customized software was developed to vertically integrate the digitized spectral image to extract the two track spectra, and to perform normalization, averaging, and automated data collection with user display functions.

Our snapshot IRFPA supports 20 to 30 Hz frame rates by using dual outputs for the format described above. Typically, 4000 shots were accumulated and averaged to obtain a transient spectrum (giving type B, k=1 uncertainty of ca. 0.0025 OD units).

3 Vibrational Coherent Control with Chirped Picosecond Infrared Excitation [16, 28]

Investigations of "vibrational coherent control" of molecular vibrational populations are now attainable with specifically tailored high power, chirped infrared ultrafast pulses. Controlling the phase and amplitude of an infrared excitation pulse that is properly tuned to adiabatically sweep population up a molecular vibrational manifold has been theoretically shown to produce population in specific overtone states [4]. By choosing pulse properties for coherently exciting vibrational population above a bond dissociation energy, for example, it may be possible to "control" the dissociation process and hence improve chemical reaction yields over conventional thermally activated processes. Ultrafast (picosecond or shorter pulsewidths) chirped excitation naturally spans multiple anharmonic shifts (allowing access to higher energy vibrational states) as well as circumventing intrinsic population relaxation mechanisms. Broadband infrared probing is also ideal for monitoring population distributions as a function of time after coherent absorption of the excitation pump pulse.

We elected to study coherent up-pumping dynamics in solution-phase metal-hexacarbonyl systems because of their strong vibrational infrared absorption cross-sections, relatively simple ground-state vibrational spectra and small (ca. 15 cm^{-1}) anharmonic overtone shifts. It was felt that these systems are ideal candidates to demonstrate that population control could be achieved for polyatomic species in solution because the excited state population lifetimes (T_1~100 ps) were already known to be long-lived [16,24,25] and the coherence lifetimes for lower-lying states are also relatively long (T_2 > 4 ps). The first systematic investigation was conducted on the triply-degenerate CO-stretch T_{1u} mode (v=0->1 at 1986 cm^{-1}) of $W(CO)_6$ in n-hexane solution at room temperature [16]. It was known from previous work that the CO-stretch (v=1) state has a T_1 lifetime of ca. 120 ps and that the system in n-hexane exhibits narrow (ca. 3 cm^{-1} FWHM) features. Initial studies which used non-chirped, difference-frequency generated (near transform-limited) 2 picosecond, <10 μJ pump pulses tuned across the v=0->1 absorption feature revealed that population distributions spanning up to the v=3 overtone state could be altered by choice of center frequency. Generation of only CO(v=1) population was achieved when the pump pulse was tuned and overlapped with the high frequency side of the 1986 cm^{-1} absorption feature. However, up to v=3 excitation with nearly equal v=1 and v=0 populations (saturation) was achieved by tuning the pump pulse center frequency to the low frequency side of the same feature. Concomitant excited overtone state T_1 relaxation times were also deduced from the transient spectra, producing a monotonic lifetime reduction as one ascends the manifold.

3.1 Chirped Picosecond IR Pulse Excitation [28]

More advanced tests of vibrational "coherent control" theory were conducted by using deliberately chirped picosecond excitation pulses. By deliberately mixing positively or negatively chirped 2 picosecond visible pulses (at 589 nm) with much longer, narrowband dye laser pulses (8 ps FWHM at 650 nm) in $LiIO_3$ crystals, one generates chirped IR pulses which experience negligible group velocity delay distortion. The sign of the chirp was produced by either passing a sub-picosecond dye pulse through a short length of optical fiber (red-to-blue or positive chirp) or by deliberately lengthening a two-pass grating compressor to invert the phase (blue-to-red or negative chirp). Actual pulse chirp rates (ca. 10 cm^{-1}/ps) and signs were deduced from time versus directly obtained IR spectrum (via a spectrograph and InSb IR focal plane detector) from data-sets that were analyzed by the FROG (Frequency Resolved Optical Gating) iterative fitting algorithm [29].

Results for up-pumping $W(CO)_6$ with chirped pulse excitation were compared to excitation of the T_{1u} manifold using transform-limited pulses with center frequencies all tuned to the peak vibrational mode absorption frequency. Figure 1 shows transient broadband absorption spectra taken at 40 ps time delay for the three different pulse types. As depicted, one readily observes that the relative population amplitudes in the CO-stretch v=1 (1970 cm^{-1}) and v=2 (1955 cm^{-1}) levels are strongly affected by the

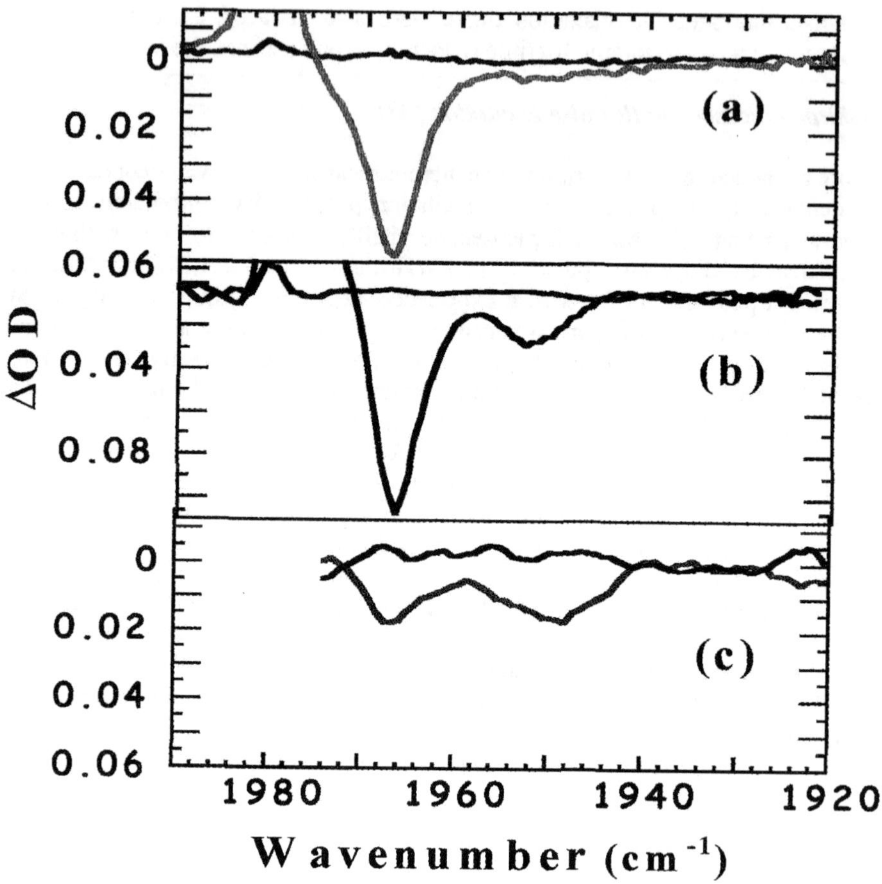

Figure 1. Transient infrared spectra obtained at 40 ps time delay for $W(CO)_6$ in n-hexane using (a) positively chirped, (b) no chirp and (c) negatively chirped IR excitation pulses centered at 1983 cm^{-1}. Note the suppression of v=1 population and increase in v=2->3 excited state absorption at 1950 cm^{-1}. The uncertainty in optical density (single value type B, k=1) is approximately 0.0025, as can be seen from the baseline noise obtained with the pump pulse blocked. Other values reported in this paper are derived from single measurements using standard evaluation methods.

chirp of the excitation pulse. Excitation with negatively chirped pulses and ten times lower pump pulse energy was found to produce higher level overtone populations than using non-chirped excitation. This general trend was confirmed by multi-level coherent up-pumping modeling using excitation pulse properties measured independently by the Frequency Resolved Optical Gating (FROG) technique [30].

3.2 Femtosecond IR Excitation and Probing

An attempt to extend the picosecond study to excitation of higher overtone levels using broader bandwidth femtosecond infrared excitation was made. Pulses generated by a tunable femtosecond mid-IR OPA (810 nm, 50 fs, 450 μJ 1 KHz input pulses) were analyzed by IR-VIS surface sum frequency cross-correlation. We used a fraction of the IR OPA output pulses together with 50 fs, 810 nm gating pulses at the surface of a z-cut quartz flat to generate sum frequency spectra (detected by a spectrograph/CCD). The ca. 2 μJ, 5 μm pulses were found to be slightly negatively chirped with an approximate pulse duration of 210 fs FWHM and 180 cm^{-1} FWHM bandwidth. The DFG output beam was propagated to the same experimental apparatus described earlier [16, 28] for performing up-pumping excitation except that ca. 10 % of the beam was split off with a 1 cm thick CaF_2 flat to generate a delayed set of probe and reference pulses.

Using the significantly larger bandwidth and shorter IR pulse duration, one might expect to up-pump vibrational population into much higher overtone levels than v=3 (as observed with picosecond IR pulses). However, as can be seen in Figure 2, only population in v=1 was produced (1970 cm^{-1} absorption) when using these excitation pulses (and about 5 % population was transferred from v=0 to v=1). The sequence of delayed transient spectra indicates that the v=1 population exhibits the expected T_1=110 ps lifetime for $W(CO)_6$ in n-hexane so the sample and probing conditions were proper for detecting v=2 and higher overtone populations. It remains to be explored (when using the measured IR pulse parameters in our adiabatic coherent pumping model) why excitation of only the v=1 level is observed. We are theoretically determining whether the high Rabi frequency used with picosecond T_2 excited state lifetimes or spatial frequency chirp can potentially produce only v=1 excitation under the applied excitation conditions.

4 Conclusions and Future Directions

A description of broadband ultrafast infrared pulse generation and multi-channel IR focal plane detection methods for coherent control of vibrational overtone states has been reviewed. We showed experimental results for overtone excitation of $W(CO)_6$ in n-hexane using near transform-limited picosecond pulses, positively and negatively chirped picosecond pulses of similar duration (10 cm^{-1}/ps chirp rates) and slightly negatively chirped 210 femtosecond IR pulses. Excitation with negatively chirped picosecond pulses does enhance v=2 population over non-chirped and positively chirped pulses, while the femtosecond excitation scheme only generates v=1 population.

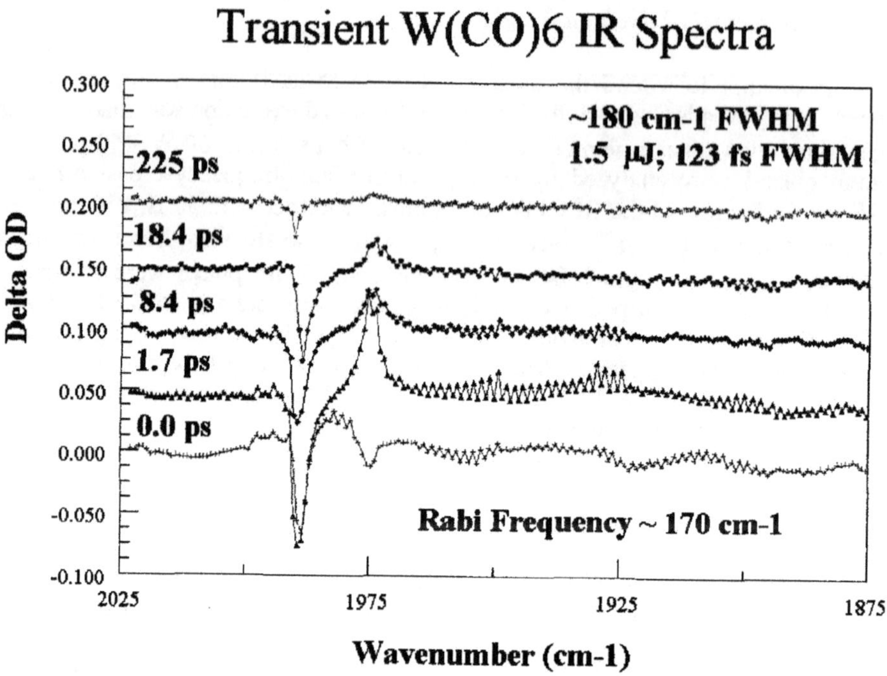

Figure 2. Transient IR spectra for $W(CO)_6$ in n-hexane at 273 K using 210 fs, slightly negatively chirped broadband excitation and probing pulses. These spectra indicate that pulses with these characteristics bleach the ground state T_{1u} v=0 level and generate only v=1 population which decays with the usual 110 ps lifetime. The low frequency oscillations (with high frequency is noise) for time delay of 0.0 ps and 1.7 ps arise from coherent interaction of the probe pulse with the radiating coherence of the v=0 and v=1 levels.

Theoretically predicted chirped-pulse excitation on the up-pumping dynamics of $W(CO)_6$ and related theories for diatomic rotational population distributions in the gas phase [4] suggest that overtone population distributions can be "controlled" with carefully generated chirped IR pulses. It remains to be seen if deliberately chirped broadband IR femtosecond and picosecond pulses with inherently larger bandwidths and varying chirp rates can produce pure populations in specific vibrational overtones that lead to chemical bond-breaking phenomena and ground state reactions [31-33].

5 Acknowledgements

The authors are indebted to the hard work and focussed experimental efforts of many of our colleagues without whom this research would never have been possible. These include former NIST/NRC postdoctoral research associates Drs. Steven Arrivo, Tom Dougherty (who passed away in 1997), Tandy Grubbs, Todd Heimer and Larry Iwaki. We are also indebted to Dr. John Stephenson, of the NIST Laser Applications Group, for his support of this research and valuable scientific discussions. Thanks also to Dr. Kimberly Briggman for help with the femtosecond mid-IR OPA and output pulse characterization measurements. Research funding for this work was provided by internal NIST STRS support.

References

1. Chelkowski S. and Bandrauk A.D., Raman Chirped Adiabatic Passage: a New Method for Selective Excitation of High Vibrational States J. Raman Spec., **28** (1997) pp. 459-466.
2. Just B., Manz, J. and Trisca, I., Series of Ultrashort Infrared Laser Pulses with Analytical Shapes for Selective Vibrational Excitation: Model Simulations for OH (v=0) -> OH (v<10), Chem. Phys. Lett, **193** (1992) pp. 423-434.
3. Korolkov, M. V., Manz, J. and Paramonov, G. K., State-Selective Control for Dissipative Vibrational Dynamics of HOD by Shaped Ultrashort Infrared Laser Pulses J. Phys. Chem., **100** (1996) pp. 13927-13940.
4. Melinger, J. S., McMorrow, D., Hillegas, C. and Warren, W. S., Selective Excitation of Vibrational Overtones in an Anharmonic Ladder with Frequency- and Amplitude-Modulated Laser Pulses. Phys Rev A **51** (1995) pp. 3366-3369.
5. Paramonov, G. K., Coherent Control of Linear and Nonlinear Multiphoton Excitation of Molecular Vibrations. Chem. Phys. **177** (1993), pp. 169-180.
6. Bardeen, C. J., Wang, Q and Shank, C. V., Selective Excitation of Vibrational Wave Packet Motion Using Chirped Pulses, Phys. Rev. Lett. **75** (1995) pp. 3410-3413.
7. Ruhman, S. and Kosloff, R., Application of Chirped Ultrashort Pulses for Generating Large-Amplitude Ground State Vibrational Coherences: A Computational Simulation. J. Opt. Soc. Am. B., **7** (1990) pp. 1748-1752.
8. Gaubatz, U., Rudecki, P., Schiemann, S. Bergmann, K., Population Transfer between Molecular Vibrational Levels by Stimulated Raman Scattering with Overlapping Laser Fields: A New Concept and Experimental Results. J. Chem. Phys, **92** (1990) pp. 5363-5376.
9. Gordon, R. J. and Rice, S. A., Active Control of the Dynamics of Atoms and Molecules, Annu. Rev. Phys. Chem., **48** (1997) pp. 601-641.

10. Kohler, B., Yakolve, V.V., Che, J. W., Krause, J. L., Messina, M., Wilson, K. R., Schwenter, N., Whitnell, R. M. and Yan, Y., Quantum Control of Wave Packet Evolution with Tailored Femtosecond Pulses, Phys. Rev. Lett. **74** (1995) pp. 3360-3363.
11. Bardeen, C. J., Che, J. W. Wilson, K. R., Yakovlev, V. V. Apkarian, V.A., Martens, C. C., Zadoyan, R., Kohler, B. and Messina, M., Quantum control of I_2 in the gas phase and in condensed phase solid Kr matrix, J. Chem. Phys., **106** (1997) pp. 8486-8503.
12. Ashworth, S. H., Hasche, T., Woener, M. , Riedle, E. and Elsaesser, R., Vibronic excitations of large molecules in solution studied by two-color pump–probe experiments on the 20 fs time scale J. Chem. Phys, **104** (1996) pp. 5761-5769.
13. Balling,, P., Maas and L. D., Noordam, Interference in climbing a quantum ladder system with frequency-chirped laser pulses, Phys. Rev. A., **50** (1994) pp. 4276-4285.
14. Borers, B., van Linden van den Heuvell, H. B. and Noordam, L. D., Efficient population transfer in a three-level ladder system by frequency-swept ultrashort laser pulses, Phys. Rev. Lett., **69** (1992) pp. 2062-2065.
15. Maas, D. J., Duncan, D. I., van der Meer, A. F. G., van der Zande, W. J. and Noordam, L., D., Vibrational Ladder Climbing in NO by Ultrashort Infrared Laser Pulses, Chem. Phys. Lett., **270** (1997) pp. 45-49.
16. Arrivo, S. M., Dougherty, T. P., Grubbs, W. T. and Heilweil, E. J., Ultrafast Infrared Spectroscopy of Vibrational CO-Stretch Up-Pumping and Relaxation Dynamics in Metal Hexacarbonyls. Chem Phys Lett **235** (1995) pp. 247-254.
17. Heilweil, E.J., Ultrashort-pulse Multichannel Infrared Spectroscopy using Broadband Frequency Conversion in $LiIO_3$. Optics Lett **14** (1989) pp. 551-553.
18. Dougherty T.P. and Heilweil, E. J., Dual Beam Subpicosecond Broadband Infrared Spectrometer, Optics Lett **19** (1994) pp. 129-131.
19. Arrivo, S. M., Kleiman, V. D., Dougherty, T. P. and Heilweil, E. J., Broadband Femtosecond Transient Infrared Spectroscopy using a 256x256 Element Indium Antimonide Focal-Plane Detector. Optics Lett **22** (1997) pp. 1488-1490.
20. Yang, H., Snee, P. T., Kotz, K. T., Payne, C. K., Frei, H. and Harris, C. B., Femtosecond Infrared Studies of a Prototypical One-Electron Oxidative-Addition Reaction: Chlorine Atom Abstraction by the $Re(CO)_5$ Radical. J. Am. Chem. Soc. **121** (1999) pp. 9227-9238.
21. Hamm, P., Wiemann, S., Zurek, M. and Zinth, W., A Dual-Beam Sub-Picosecond Spectrometer for the 5-11 μm Spectral Region and its Application to Vibrational Relaxation in the S_1 State. In *Ultrafast Phenomena IX,* ed. By Barbara P. F., Knox W. H., Mourou G. A., and Zewail A. H., (Springer-Verlag, Berlin; 1994) pp.152-153.
22. Beckerle, J. D., Cavanagh, R. R., Casassa, M. P., Heilweil, E. J. and Stephenson, J. C., Subpicosecond Transient Infrared Spectroscopy of Adsorbates: Vibrational Dynamics of CO/Pt(111), J Chem Phys **95** (1991) pp. 5403-5418.

23. Kidder, L. H., Levin, I. W., Lewis, E. N., Kleiman, V. D. and Heilweil, E. J., Mercury Cadmium Telluride Focal-Plane Array Detection for Mid-Infrared Fourier-Transform Spectroscopic Imaging. Optics Lett **22** (1997) pp. 742-744.
24. Beckerle, J. D., Cavanagh R. R., Casassa, M. P., Heilweil E. J. and Stephenson, J. C., Subpicosecond Study of Intramolecular Vibrational Energy Transfer in Solution-Phase $Rh(CO)_2$acac. Chemical Physics **160** (1992) pp. 487-496.
25. Dougherty, T. P., Grubbs, W. T. and Heilweil, E. J., Photochemistry of $Rh(CO)_2$(acetylacetonate) and Related Metal Dicarbonyls Studied by Ultrafast Infrared Spectroscopy. J Phys Chem **98** (1994) pp. 9396-9399.
26. Chudoba, C., Nibbering, E. T. J., Elsaesser, T., Dynamics of Site-Specific Excited-State Solute-Solvent Interactions as Probed by Femtosecond Vibrational Spectroscopy. In *Ultrafast Phenomena XI,* ed. by Elsaesser,T., Fujimoto, J. G., Wiersma, D. A. and Zinth, W. (Springer-Verlag, Berlin, 1998) pp. 535-537.
27. Heimer, T. A. and Heilweil, E. J., Sub-Picosecond Interfacial Electron Injection in Dye Sensitized Titanium Dioxide Films. In *Ultrafast Phenomena XI,* ed. by Elsaesser,T., Fujimoto, J. G., Wiersma, D. A. and Zinth, W. (Springer-Verlag, Berlin, 1998) pp. 505-507.
28. Kleiman, V. D., Arrivo, S. M., Melinger, J. S. and Heilweil, E. J., Controlling Condensed-Phase Vibrational Excitation with Tailored Infrared Pulses, Chem. Phys. **223** (1998) pp. 207-216.
29. Trebino, R., DeLong, K. W., Fittinghoff, D. N. Sweetser, J. N., Krumbugel, M. A., Richman, B. A. and Kane, D. J., Measuring ultrashort laser pulse in the time-frequency domain using frequency-resolved optical gating, Rev. Sci. Instrum., **68** (1997), pp. 3277-3295 .
30. Richman, B. A., Krumbugel, M. A. and Trebino, R., Optics Lett., **22** (1997), pp. 721-723.
31. Liu, W-K., Wu, B. and Yuan, J-M., Nonlinear Dynamics of Chirped Pulse Excitation and Dissociation of Diatomic Molecules. Phys Rev Lett **75** (1995) pp. 1292-1295.
32. Trushin, S. A., Sugawara, K. –I. And Takeo, H., Formation of Cr Atoms in the 5 μm Multiphoton Decomposition of $Cr(CO)_6$. Chem. Phys., **203** (1996) pp. 267-278.
33. Windhorn, L. and Kompa, K. L., Femtosecond Multiphoton Infrared Dissociation of $Cr(CO)_6$ in the Gas Phase. In *Femtochemistry and Femtobiology V Proceedings* (poster), ed, by Douhal, A. and Santamaria, J., (World Scientific Publishing Co, Singapore, 2002).

ADIABATIC REGIMES OF POPULATION TRANSFER IN MOLECULAR SYSTEMS

IGNACIO R. SOLÁ, BO Y. CHANG, AND JESÚS SANTAMARÍA
Dpt. de Química Física I, Universidad Complutense, 28040 Madrid, Spain

VLADIMIR S. MALINOVSKY
Michigan Center for Theoretical Physics & FOCUS Center, Dpt. of Physics, University of Michigan, Ann Arbor, MI 48109

In this contribution we review the dynamics of population transfer between vibrational levels of different electronic potentials. Based on adiabatic parameters and simple dynamical models we frame and further explore the connections between two imaginative schemes based on counterintuitive pulse sequences. Finally we present numerical results particularized for the Na_2 molecule that show how is possible to selectively excite the vibrational population along the bond coordinate in highly non Franck-Condon transitions.

1 Introduction

With the advent of femtosecond laser pulses, much progress has been achieved towards inducing and following the dynamics of molecules. A forward natural step aimed at controlling the dynamics so as to selectively break a certain bond of the molecule or populate a certain discrete quantum state.[1]

Succinctly, there have been two major approaches to treat this problem, that sometimes can be used in a complementary way.[2] One, that we may call mathematically oriented, was developed by using optimal control algorithms.[3] In this approach a functional of the goal is set up and its maxima are obtained by a variational scheme. The solution of the problem is then to find the so called optimal pulses, those that guarantee that the dynamics rends a value of the objective that is at least a local maximum of the functional. By means of optical control algorithms many different control scenarios were numerically tested with amazing success. However, the very intricate solutions often posed a challenge both from the theoretical point of view (analyzing the underlying mechanisms by which the pulses drive the system and trying to generalized them) and from the experimental point of view (reproducing in the laboratory the same kind of fields). After a proposal of Judson and Rabitz,[4] a major breakthrough was possible with respect to the former issue by bringing together advances in the technology of pulse shaping[5] and implementing in the laboratory evolutionary or genetic code algorithms.[6] The standard recipe to deal with the first issue was to deconstruct *a posteriori* the optimal pulses in order to unravel the role of the different components in the dynamics. However, due to the complexity of the fields, a complete understanding of the dynamics so as to being able to anticipate or generalize the underlying schemes, was always a difficult and expensive process, although current research along these lines may change the panorama.[7]

A second, physically oriented approach to control the dynamics comes from a careful analysis and generalization of the models of quantum dynamics of simple systems. In this case the solutions (*i.e.* the optimal pulses) are anticipated *a priori* based on the physical properties of a model, and are numerically tested afterwards for the molecular system. Usually the validity of such schemes rely on the validity of a reduced dimensional description of the molecular Hamiltonian, which accommodates better in the coordinate representation for short time dynamics (impulsive and perturbative models), and in the energy representation for long time dynamics (adiabatic and perturbative models).

Briefly there have been two most profitable schemes proposed in the energy representation. The first is described as a perturbative scheme implying amplitude and phase control, usually named coherent control.[8] The second enters into the category of adiabatic schemes, implying amplitude and either frequency chirp (Chirped Adiabatic Rapid Passage schemes[9]) or time-delay control (STIRAP type schemes[10]). In this contribution we will focus on the last one, addressing the selective preparation of molecules in vibrational states of excited potentials by means of STIRAP and higher energy related schemes.

The great advantage of adiabatic methods is that they assess a great stability on the dynamics and therefore rely on the control of very few parameters with ample margins. However this economy in the control is at the expense of energy in the field, since a typical requirement of adiabaticity implies that the effective pulse area must be much larger than π

$$\Omega_{eff}\tau \gg \pi \ , \tag{1}$$

(which can sometimes be as much as 100), where Ω_{eff} is the effective Rabi frequency or radiation coupling term of the process, and τ is the timescale or width of the pulses. This is to be compared with a minimum of $\Omega_{eff}\tau = \pi$ required for population inversion according to the Rabi formula. An obvious related consequence of adiabaticity is the requirement of using pulses with larger time-widths. If we have to use ultrashort pulses, then Eq. (1) would immediately lead us to employ very intense radiation sources, which can induce multi-photon processes or break the selectivity of the population transfer. The reduced dimensional Hamiltonian assumed in STIRAP is no longer valid. If, for instance, τ is around 1 picosecond, Ω_{eff} can be as large as 500 cm^{-1}, therefore larger than, or of the order of, the vibrational quantum ω_v. Due to the energy uncertainty, in this regime a wave packet can be formed and subsequently evolve, much in the same way as (or actually complementary to) the way the time uncertainty of femtosecond pulses is responsible for inducing the dynamics in ultra-short processes. Therefore a completely new model must be used to understand the dynamics in these conditions. Recently, Garraway and Suominen[11] and us[12] have proposed a model based on the effects of light modulation on the potentials to predict and control the outcome of the dynamics. A novel and very general scheme was proposed named Adiabatic Passage by Light Induced Potential shaping (APLIP).

In this paper we plan to review the connections between STIRAP and APLIP, so as to propose a different adiabatic condition for the APLIP method and therefore frame the different adiabatic scales ruling both schemes.

2 STIRAP in Na_2

We address the problem of selective preparation of single vibrational levels in excited potentials by means of STIRAP. To be specific consider the electronic states of Na_2 depicted in Fig. 1. Our initial state is the zeroth vibrational level in the ground electronic state ${}^1\Sigma_g$. We will use the nomenclature (X,0) for this state, with wavefunction $\phi_0^X(x)$. The goal is to invert the population to the zeroth level of the excited potential ${}^1\Pi_g$, that we name (B,0). Since the bond is more relaxed in the excited configuration, the transition that we are considering is highly non Franck-Condon (see Fig. 1).

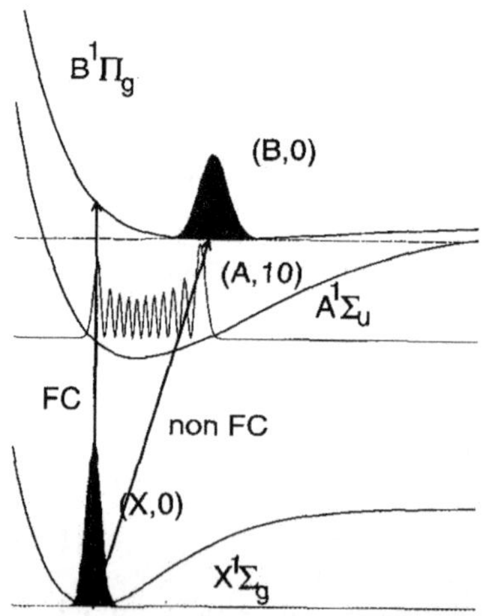

Figure 1. The Na_2 potentials considered in our system together with the initial and target wavefunctions. The proposed transition implies a large displacement of population along the bond coordinate, and therefore is highly non Franck-Condon. However a two-photon transition may be possible if we find a efficient "wavefunction bridge", as the (A,10) state in the figure.

Assuming the Franck-Condon approximation, the vertical transition would most naturally target highly excited vibrational levels in ${}^1\Pi_g$. Also the overlapping between the initial and final states is practically zero. The fact that the desired transition implies a considerable displacement of the population along the bond distance, a usual situation in molecular systems, certainly plays an important role in the problem.

However we can try to find an intermediate state that acts as a bridge between the initial and target wavefunctions (actually this is needed not only because of geometry reasons, but firstly because of symmetry constrains in our problem). This is always done in STIRAP implementations, which is a two-photon process. Therefore we must find a single intermediate vibrational state that couples efficiently with both the initial and final desired level. We show in Fig. 2 the Franck-Condon factors of the initial and target wavefunctions together with other wavefunctions of the same electronic potentials, with respect to different vibrational wavefunctions of the intermediate $^1\Sigma_u$ electronic state.

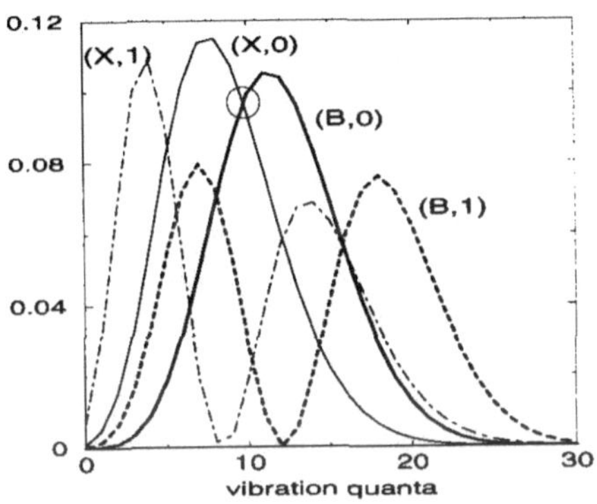

Figure 2. Franck-Condon factors of some wavefunctions of the initial and final electronic states with respect to different vibrational levels of the intermediate potential, $^1\Sigma_u$. The shape of the wavefunctions is reflected by $^1\Sigma_u$. For the overall two photon transition we must select a single intermediate state that maximizes the Franck-Condon factor for both transitions $(X,0) \rightarrow (A,v)$ and $(A,v) \rightarrow (B,0)$. This is achieved with the encircled state $(A,10)$.

Due to the geometry of the three electronic potentials, the repulsive part of the $^1\Sigma_u$ curve maps out in energy the coordinate dependence of the vibrational wavefunctions of the ground electronic state, by a reflection principle,[13] whereas the attractive part of $^1\Sigma_u$ serves as a map of the vibrational wavefunctions of the final electronic state. As shown by Fig. 2 the best compromise for having large Franck-Condon factors in both transitions is obtained for the tenth vibrational level (A,10). For this case the Franck-Condon factors are $p_{0,10}^{X,A} = |\langle\phi_{10}^A|\phi_0^X\rangle|^2 = 0.0937$, and $p_{10,0}^{A,B} = |\langle\phi_{10}^A|\phi_0^B\rangle|^2 = 0.0993$. Of course for different potentials it could be more difficult to find a good "wavefunction bridge". This happens for instance when either the ground or the target potential curve is not displaced with respect to the intermediate potential. Then the reflection principle fails for the wavefunctions of this potential and it may not be possible to find a single intermediate state with efficient couplings with both the initial and target levels.[14]

Once that an intermediate state is chosen and selected by tuning the lasers in resonance with its transitions, the STIRAP scheme is implemented by using a counterintuitive sequence of pulses, such that the laser that couples the intermediate (A,10) with the target state (B,0), $E_2(t)$, precedes the laser that couples the initial state (X,0) with (A,10), $E_1(t)$. Since the "second" laser is switched on first, with apparent no effect -all the population is at initial times in (X,0)- hence the name counterintuitive for the scheme. However this preparation stage has indeed an important role, most notably seen in the adiabatic representation.

If we define the Rabi frequencies of the transitions as the product of the field amplitude times the transition dipole moments (here approximated by the Franck-Condon factors) $\Omega_1(t) = E_1(t)p^{X,A}_{0,10}/\hbar$ and $\Omega_2(t) = E_2(t)p^{A,B}_{10,0}/\hbar$, then the Hamiltonian in the energy representation spanned by the three selected levels using the Rotating Wave Approximation[15] (RWA) is

$$H = \frac{1}{2}\begin{pmatrix} 0 & -\Omega_1(t) & 0 \\ -\Omega_1(t) & 0 & -\Omega_2(t) \\ 0 & -\Omega_2(t) & 0 \end{pmatrix} . \tag{2}$$

Diagonalising the Hamiltonian we obtain the time dependent eigenvectors (adiabatic dressed states) and eigenvalues (quasi-energies). Here it suffices to analyze the form of the dressed state with zero eigenvalue:

$$|\Phi_0(t)\rangle = \frac{1}{\Omega_0(t)}\left(\Omega_2(t)|\phi_0^X\rangle - \Omega_1(t)|\phi_0^B\rangle\right) , \tag{3}$$

where $\Omega_0 = \sqrt{\Omega_1^2 + \Omega_2^2}$, serves as a normalization factor. Now it is clear that if $\Omega_2(t) \gg \Omega_1(t)$ at initial times then the initial state (X,0) coincides with $|\Phi_0\rangle$. Also if at late times $\Omega_1(t) \gg \Omega_2(t)$, then $|\Phi_0\rangle$ will correlate with (B,0). Therefore, as long as the dynamics is adiabatic and we stay in the initially prepared dressed state, a counterintuitive sequence drives all the population from (X,0) to (B,0) via $|\Phi_0\rangle$. Finally, in order that the non-adiabatic couplings can be neglected, it has been proved[10] that a single requirement must be satisfied through all the time evolution. In its simpler form it can be reduced to the parameter

$$\xi_S = \Omega_0\,\sigma \gg 1 \tag{4}$$

as in Eq.(1). The Ω_0 here can be set as its value with maximum overlapping of the pulses, although it is usually evaluated introducing the peak amplitudes for both pulses.

There are two outstanding features of STIRAP. The first is that Eq. (4) is not upper bounded. The adiabatic requirements set only the minimum values for perfect population transfer from (X,0) to (B,0). This is to be contrasted with an intuitive sequence of lasers, which would lead to Rabi oscillations on the populations, the final probability in (B,0) ranging from 0 to 1 depending on whether $\Omega_0 \sigma$ is an even or odd multiple of π. Therefore STIRAP is a very robust procedure to transfer the population between specific quantum states. The second outstanding feature is that the evolution in the adiabatic state $|\Phi_0(t)\rangle$ has zero overlapping with the intermediate level (A,10). This means that the STIRAP dynamics drives all the population from the initial to the final level without exciting any intermediate state. Actually, the population in (A,10) during the time evolution can be used as a measure of the degree of accomplishment of the adiabaticity during the process.

In order to numerically test the validity of the STIRAP model in the molecular system, we solve the Schrödinger equation in the Born-Oppenheimer approximation, with the three potential curves of Fig. 1 (see Eq. (5) introduced later in the text). We use a Split-Operator propagator combined with Fast Fourier Transform techniques obtaining the wave packets $\psi_X(x,t)$, $\psi_A(x,t)$ and $\psi_B(x,t)$ (for details consult reference[16]). The vibrational populations are obtained projecting the wave packets on the vibrational eigenstates calculated by the Fourier Grid Hamiltonian method.[17]

In Fig. 3(a) we show the population histories for the STIRAP dynamics, using two Gaussian shaped pulses with maximum amplitude $E = E_1 = E_2 = 2 \cdot 10^{-4}$ a.u. (1 MV/cm) and time widths of $\sigma = 14.1$ ps. The time delay between the pulses is made equal to the time width. Full population passage takes place without barely populating the intermediate (A,10) level (the maximum population in this level is 0.023, therefore it hardly shows up in the figure). Using the field amplitudes, time widths and Franck-Condon factors, we estimate $\xi_S \sim 15.4 \sim 5\pi$, which seems enough for the perfect transfer.

Although according to Eq.(4) the population transfer can only improve increasing ξ_S, there is indeed an upper limit for the acceptable values of Ω_0, above which the starting assumptions of the model are no longer valid. The results of Figure 3(b) illustrate the case. Here we use the same laser parameters than in the previous case except that we make the field amplitude ten times stronger, $E = 2 \cdot 10^{-3}$ a.u. (10 MV/cm). Now instead of decreasing, the transient population in the intermediate state is larger and increases with time. Also the passage is not completed and part of the population returns to the initial state. Finally there is transient excitation of other levels, most notably (A,11) and (B,1). What happens is that the effective Rabi frequencies for other non resonant transitions are becoming of the order of the energy

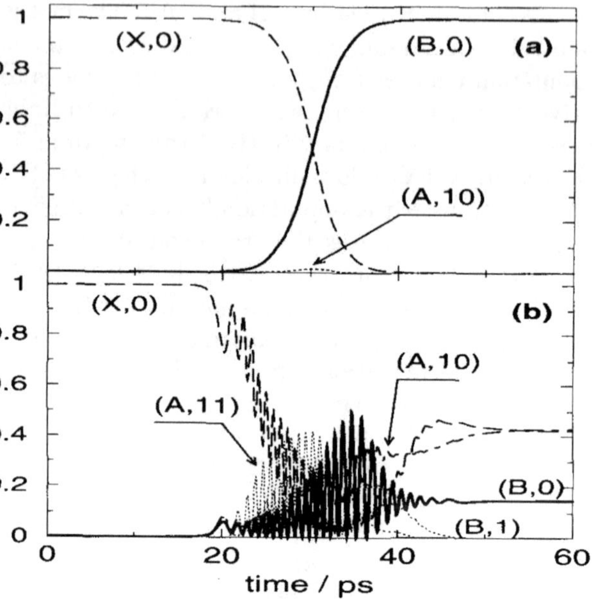

Figure 3. Population histories of the most important states participating in the dynamics of a STIRAP process. In (b) the laser amplitudes are ten times larger than in (a). Although the area of the pulses is larger, the STIRAP dynamics fails with increasing intensity because more states participate and the population cannot be locked anymore in the intermediate state. The parameters of the pulses are given in the text.

detuning, and therefore these transitions cannot be neglected in the Hamiltonian. The 3-level model breaks. Most notably the contribution of the path $(X,0) \rightarrow (A,11) \rightarrow (B,0)$ with effective Rabi frequency $\Omega_0^{0\rightarrow 11\rightarrow 0} \sim 2.6 \cdot 10^{-4}$ a.u. (comparable with the energy difference between (A,11) and (A,10) which is $\hbar\omega_{10,11}^{A} \sim 5.0 \cdot 10^{-4}$ a.u.) seems enough to disturb the locking of population in the intermediate level, and therefore not only there is substantial transient excitation of level (A,11) but also the population in (A,10) does not decay to zero at the end of the dynamics. Additionally, the non-resonant route $(X,0) \rightarrow (A,10) \rightarrow (B,1)$ with $\Omega_0^{0\rightarrow 10\rightarrow 1} \sim 1.9 \cdot 10^{-4}$ a.u. leads to transient excitation of (B,1). We could expect that the population histories can only become more complicated as the intensity of the fields is increased.

3 From STIRAP to APLIP

We have seen that STIRAP is a powerful method applicable to molecular systems. The limits of its success can be approximately framed by the conditions $\xi_S \geq 10$ (the exact number depending upon the system and pulse shapes) and $\Omega_0 < \omega_v$, where we identify ω_v as the vibrational quantum in any of the potentials implied. However, instead of stopping the search, here we continue to explore the behavior of the system at increasingly stronger intensities.

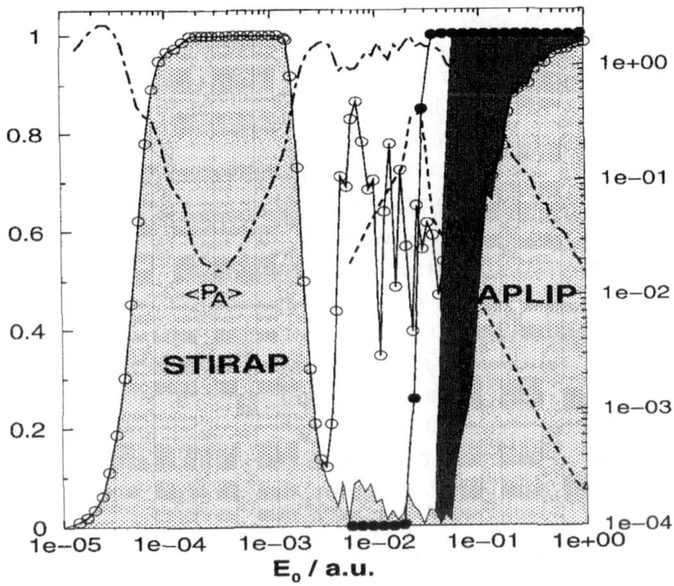

Figure 4. Efficiency and selectivity of the two-photon transfer as a function of the effective field amplitude E_0. The curve with open circles represents the final population on the $^1\Pi_g$ potential for the resonant STIRAP-type transition, and the light gray shaded curve gives the population on the $(B,0)$ state at final times. The time-averaged population on the $^1\Sigma_u$ potential, $\langle P_A \rangle$, is given by the dot-dashed curve, whose logarithmic scale is represented at the right side of the plot. For E_0 around $5 \cdot 10^{-5}$ a.u. the transfer is efficient and selective and the drop in $\langle P_A \rangle$ reveals the STIRAP signature. Increasing E_0 the efficiency first decays and then oscillates. Moreover the transition is no longer selective. However it recovers for very large values of E_0. However this is a different dynamic process called APLIP, whose performance can be improved by detuning the lasers with respect to any intermediate state. The solid circles and the dark-grey shaded curves provide respectively the overall population in $^1\Pi_g$ and $(B,0)$ at late times for detuning $\Delta = 0.014$ a.u. to the blue of the resonance with the intermediate level. The dashed line represents $\langle P_A \rangle$. All the parameters of the pulses are given in the text.

Figure 4 condenses all the results of efficiency and selectivity of the method as a function of the field amplitude parametrized by $E_0 = \sqrt{E_1^2 + E_2^2}$ evaluated using the peak amplitudes for both pulses. For all the calculations we used the same Gaussian shaped pulses with time width and time delay $\sigma = \tau = 14.1$ ps. The light-gray shaded area encloses the results for the final population on the target level (B,0), whereas the open circles represent the final population in the excited electronic state (B,all v). When both curves lie close together and take larger values the method is efficient and selective. The dot-dashed curve provides the time-averaged population in the intermediate electronic state, $\langle P_A \rangle = \int_0^{t_f} dt' |\langle \psi_A(x,t') | \psi_A(x,t') \rangle|^2 / \sigma$, whose values follow the logarithmic scale at the right side of the plot. This magnitude is proportional to the fluorescence signal coming from ${}^1\Sigma_u$.

Inspection of the results shows the expected behavior of STIRAP for lower amplitude fields. Once the adiabatic requirement is fulfilled, the method is highly efficient and selective and its signature is clearly noticed as the drop in $\langle P_A \rangle$ reveals a dark resonance process (no fluorescence). For larger values the population in the intermediate state is no longer locked and the population is no longer transferred to the target state, as Fig. 3(b) showed. However, as the field amplitude is further increased, the method seems to gain some efficiency, although it is still highly non selective and non robust, since the actual value of the final population is very sensitive with respect to the exact value of E_0. Finally for even stronger fields the method surprisingly recovers its previous efficiency and selectivity. Despite using huge fields tuned in resonance with transitions to intermediate states, the average population in the ${}^1\Sigma_u$ also decreases as if it were locked. Moreover the results can be improved by tuning both lasers off resonance with respect to ${}^1\Sigma_u$. The dark-grey shaded area encloses the final population in (B,0) and the filled circles represent all the population in the excited electronic state, now using lasers detuned 0.014 a.u. (3073 cm^{-1}) above the resonance with level (A,10) with all other parameters the same. Also the dashed line provides the corresponding results for $\langle P_A \rangle$. However, as we will see, in none of these cases the dynamics of the system can be considered of STIRAP type. It is actually a novel adiabatic passage process that has been named APLIP.[11]

In order to sample the system dynamics, Fig. 5 shows the population histories for two situations. In Fig. 5(a) we keep all the parameters of the Gaussian pulses the same except for the amplitude, which is made ten times larger ($E_1 = 0.02$ a.u. or 100 MV/cm). Now the Rabi frequencies for many different paths greatly exceed the energy difference from the resonance. Therefore there are many levels excited and the population passage can not proceed through a single adiabatic state. Moreover there are also Raman Stokes transitions

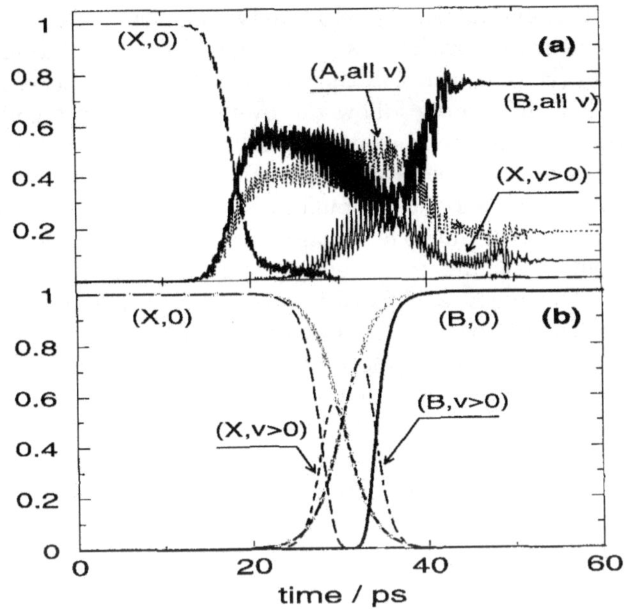

Figure 5. Population histories of the most important states participating in the dynamics. Although the amplitude of the pulses are increased by a factor of ten in (b), and more states participate, the signature of the process is again simplified revealing the APLIP dynamics. The parameters of the pulses are given in the text. In (b) the lasers are additionally detuned to the blue of the transition with the intermediate state. Although this feature is not strictly required to view the process it improves the quality of the transfer.

competing with two-photon excitation, leading to excitation in the ground potential. Despite the complexity of the problem, which is very dependent on the amplitude and duration of the fields, the main process is still the two-photon excitation to the $^1\Pi_g$ state, although negligible amount of population reaches the target level (B,0). The dynamics is not selective nor robust.

In Fig. 5(b) the pulse amplitudes are again ten times larger. For this case we have also changed the frequencies of the pulses, which are now detuned to the blue of the previous resonance by 0.014 a.u. Still, the same qualitative results are obtained without detuning. The overall dynamics is again simplified, and it follows the characteristic pattern of the so called APLIP dynamics: First the population is mainly excited to high vibrational levels in the ground potential by means of Raman Stokes processes; this is rapidly followed by two-photon absorption from these levels to high excited vibrational levels of the final potential, and finally, by selective two-photon de-excitation (at some point similar to Raman anti-Stokes processes) all the population is driven to the target state (B,0). During all the dynamics the population in $^1\Sigma_u$ never reaches the 0.1%. If we only follow the overall electronic populations (grey curves in the plot), then the behavior in adiabatic conditions is exactly like that of STIRAP. This can be proved by a relatively simple model.[14] However, the vibrational structure of the populations shows that this process is more complicated. It involves the creation and movement of vibrational wave packets in all the potentials, due to the energy uncertainty generated by the huge external fields. Obviously, as APLIP is revealed as a different process, a new set of adiabatic conditions must be found to adequately describe it. It is clear that ξ_S, which is almost 500 π and 5000 π for conditions in Fig. 5(a) and (b) respectively, fails to account for the success of the method. In order to do so we need to analyze a little bit further the APLIP dynamics.

The situation can be described like the following. In the RWA we consider the electronic potentials shifted by the energy of the photons (there are many equivalent ways of doing so while conserving the energy of the molecule plus field). This is shown if Fig. 6, where we chose the scenario with frequencies detuned to the blue of the resonance. The Born-Oppenheimer Hamiltonian is now

$$H = \begin{pmatrix} T + U_1(x) & -\Omega_1(t)/2 & 0 \\ -\Omega_1(t)/2 & T + U_2(x) & -\Omega_2(t)/2 \\ 0 & -\Omega_2(t)/2 & T + U_3(x) \end{pmatrix} , \qquad (5)$$

where T are the kinetic energy operators and $U_i(x)$ are the energy shifted potentials: $U_1(x) = V_{\Sigma_g}(x)$, $U_2(x) = V_{\Sigma_u}(x) - \hbar\omega_1$, $U_3(x) = V_{\Pi_g}(x) - \hbar(\omega_1 + \omega_2)$, with ω_1, ω_2 the laser carrier frequencies. The Rabi frequencies in the Franck-Condon approximation are exactly the laser amplitudes in frequency units $\Omega_1(t) = E_1(t)/\hbar$, $\Omega_2(t) = E_2(t)/\hbar$.

We want to displace the vibrational population along the bond distance, from the minimum in $U_1(x)$ to the minimum in $U_3(x)$. In order to do so, the two-photon coupling must provide a minimum of energy to the initial wavefunction that is larger than the energy barrier, or energy at the crossing point between the initial and final potential curves (see Fig. 6). This stage corresponds to the Raman Stokes processes that initialize the dynamics in Fig. 5(b). In our example $E_{bar} \sim 640$ cm^{-1}, which implies that the average energy must exceed that of level $v = 4$ (since this is actually an average energy, states with much more than four vibrational quanta are temporally excited). When both pulses fully overlap the effective two-pulse Rabi frequency Ω_{eff} must be larger than E_{bar}, allowing the wave packet to move from one potential

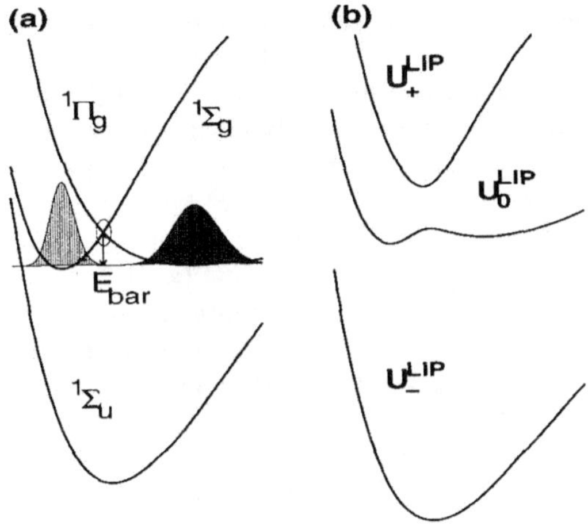

Figure 6. (a) Energy-shifted electronic potentials in the diabatic representation showing the energy barrier that the initial wavefunction must surmount to move along the bond distance before reaching the desired state; (b) Light-induced potentials (LIPs) that show the dynamics in the adiabatic representation using fully overlapping pulses of moderate intensity. In this representation the initial wavefunction moves as a wave packet in the initially prepared LIP, U_0^{LIP} in the figure. The energy of the pulses serves to modulate the LIP so as to reduce or eliminate the energy barrier allowing the wave packet to move from the left well (correlating with $U_1(x)$ at initial times) to the right well (correlating with $U_3(x)$ at late times) of the LIP, corresponding to the desired transition.

to the other, which is signaled as a two-photon absorption process. Finally, as the pulses are switched off, the transient energy disappears, and by two-photon resonance the wave packet ends up in the bottom of $U_3(x)$. This last step corresponds to Raman de-excitation and it is more difficult to explain fully by means of the former picture. It is the case that the selectivity of the transfer implies conservation of the vibrational quantum number, and therefore goes beyond energy resonance conditions.

The whole mechanism is better explained in terms of light-induced potentials.[12] In a quasi-adiabatic representation we diagonalize the part of the Hamiltonian consisting in the electronic potentials and laser couplings. The diagonal elements become the so-called light-induced potentials (LIPs). If the initial wavefunction is placed on a single LIP correlating at initial times with $U_1(x)$, then all the dynamical features can be related to the topological changes of this LIP. For instance, we may shape the pulses so that at late times this LIP correlates with $U_3(x)$. This implies both intensity and time-ordering requirements. The first one is already indicated in Fig. 6(b). In this figure we show the shape of the LIPs obtained with laser couplings $E_1 = E_2 = 0.02$ a.u. The LIP correlating at initial times with $U_1(x)$ is $U_0^{LIP}(x,t)$. It can be seen that the energy barrier has been already clearly reduced. Using larger intensities amounts to eliminating the energy barrier completely. Then, the time-ordering requirements in the pulse sequence serve to modulate the changes in the LIP so as to induce the movement of the wave packet in the proper direction, from the left to the right of $U_0^{LIP}(x,t)$.[18] Finally, the passage is completed if sufficient adiabatic requirements are satisfied. In this case non-adiabatic couplings arise both from the kinetic terms in the Hamiltonian, and from the time-dependence of the pulses. Both can induce transitions between different LIPs, leading to loss of efficiency, or changes in the shape of the wave packets, leading to non-selective excitation. A thorough examination of both allows to replace ξ_S for a single or multiple parameters ξ_A. However, there are still no easy models for a full treatment, that go beyond the scope of this paper.

Here we will just present a simple model that explains and reflects some of the requeriments of the method, allowing for a simple parametrization of ξ_A. In this model we assume that the wave packet is a point particle moving from $U_1(x)$ to $U_3(x)$ at speed v_c. The kinetic energy of the particle must be larger than the energy barrier E_{bar}, and we assume that this energy has been temporally given by the fields, through an effective two-photon coupling Ω_{eff}. Then a Landau–Zener model can provide the probability of crossing the barrier, which is given by[19]

$$P_{U_1 \to U_3} \approx 1 - \exp\left(-\pi \Omega_{eff}^2(t_c)/2\lambda\right), \tag{6}$$

with

$$\lambda = \left| \frac{v_c}{2\hbar} \frac{d(U_3 - U_1)}{dx} \right|_{x=x_c}, \tag{7}$$

where x_c is the crossing point. The adiabatic parameter is therefore proposed as $\xi_A = \Omega_{ef}^2(t_c)/2\lambda$.

Following this model, the detuning of the intermediate potential is highly recommendable to avoid the crossing of the particle moving from $U_1(x)$ or $U_3(x)$ to $U_2(x)$. At the region of interest, the potentials do not cross with $U_2(x)$ if the absolute value of the detuning $|\Delta|$ is made at least larger than E_{bar}. If additionally we assume that $|\Delta| \gg \Omega_1(t_c), \Omega_2(t_c)$, the effective Rabi frequency is approximately $\Omega_{eff} \approx \Omega_1(t_c)\Omega_2(t_c)/2\Delta$,[15] where t_c is the time of the crossing. For other cases we should substitute Δ by $\sqrt{\Delta^2 + \Omega_0^2(t_c)}$, although the present simple model does not give the right adiabatic properties in this limit. On the other hand, the parameter λ can be approximately evaluated by a number of means. For example, if we consider that $U_1(x)$ and $U_3(x)$ are displaced harmonic potentials with the same fundamental frequency, ω, then

$$\frac{d}{dx}|U_3(x) - U_1(x)| = m\omega^2 d, \tag{8}$$

where m is the reduced mass of the molecule and d the distance between the potential minima. Finally, v_c can be obtained by insisting that the kinetic energy of the wave packet must be larger than the energy barrier. Then $v_c > (2E_{bar}/m)^{1/2}$ and $\lambda \sim 2\omega E_{bar}/\hbar$. Summing up the pieces we obtain

$$\xi_A = \frac{\Omega_1^2(t_c)\Omega_2^2(t_c)\hbar}{16\Delta^2 E_{bar}\omega} . \tag{9}$$

Using this parameter and the results of Fig. 4 we find that a value of $\xi_A \sim 32 \sim 10\pi$ is needed so that APLIP can be efficient. The Landau–Zener model is too simple to explain the underlying mechanism of the dynamics (why the wave packet moves from $U_1(x)$ to $U_3(x)$) and of course, since it assumes a classical particle-like motion, it fails to account for the selectivity (or lack of) of the process. (In order to ensure selectivity, the model provides a very large value for $\xi_A \sim 240$.) However the simple form of ξ_A is enough to predict certain trends that have been observed. In particular the dependence with respect to the fields is of quartic order, which explains why the yield above the adiabatic threshold increases much faster than in the case of STIRAP (ξ_S is of first order in the field amplitude), a clear feature of Fig. 4. It depends also on E_{bar} in the expected way. Additionally it depends on the vibrational quantum number, so that for steepest potentials is more difficult to transfer the wave packet (even for fix barrier energy). Finally, a very notorious fact is the lack of dependence with respect to the time width of the pulses. Although it is clear that a minimum time is needed for APLIP to take place (for instance, selectivity demands that σ must be larger than ω^{-1}), the large amplitude requeriments always ensure that $\Omega_0\sigma \gg 1$, and therefore the minimum threshold in Ω_0 for the adiabatic transfer of the wave packet does not seem sensitive to σ, as has been revealed by numerical analysis.[14]

4 Summary and Discussion

In this paper we have considered the following problem: To invert the population between two electronic states in a selected vibrational transition that is highly non Franck-Condon. We have explored the efficiency of an adiabatic scheme based on a particular class of time delay control using a counterintuitive sequence of two lasers. For moderate intensities the scheme is generally termed STIRAP, and makes use of a resonant transition to an intermediary state that works as a "wavefunction bridge" between the initial and target states. The efficiency of the method is then ruled by a simple adiabatic condition that is no longer useful in the strong intensity regime. In this case another scheme, apparently similar, operates, named APLIP. In previous contributions we have studied the dynamics of APLIP and showed its performance using frequency detuned transitions, and both counterintuitive and intuitive sequences. Here we propose a very simple model for APLIP and we concentrate in the relations between both methods in terms of their different adiabatic properties, following the dynamics for increasing pulse intensities.

We have shown that both schemes are efficient and selective by numerical calculations particularized for a vibronic transition in Na_2. The range of efficiency of the STIRAP method is clearly revealed in our study and possible limitations are outlined. However our results do not show the limitations in APLIP. These appear when we consider the validity of the model (based on three-electronic potentials) adopted when using very strong laser fields. Although our results using APLIP may violate the model assumptions, in reality the threshold of selective transfer should appear for lower intensities, since the dipole moments in Na_2 are very large, whereas the calculations are performed using the Franck-Condon approximation. However, the utility of the method for more general systems will rely on our ability to extend the control in the modulation of the potentials beyond the Rotating Wave Approximation, and also in n-dimensional spaces. The use of frequency chirped pulses may become an important additional tool in the control process. Work along these lines is in prospect.

Acknowledgments

Financial support from the Dirección General de Investigación Científica y Técnica under Project No. PB98-0843 is gratefully acknowledged. I.R.S. also acknowledges financial support from the Ministerio de Educación y Ciencia for a post-doctoral fellowship and Prof. H. Rabitz for very stimulating conversations. Parts of this work were done during his stay at Princeton University.

References

1. S. A. Rice and M. Zhao, *Optimal control of molecular dynamics* (John Wiley & Sons, New York 2000).

2. V.S.Malinovsky and D.J.Tannor, *Phys. Rev.* A **56**, 4929 (1997); I. R. Sol, V. S. Malinovsky, and D. J. Tannor, *Phys. Rev.* A **60**, 3081 (1999).
3. A. P. Pierce, M. A. Dahleh, and H. Rabitz, *Phys. Rev.* A **37**, 4950 (1988).
4. R.S.Judson and H.Rabitz, *Phys. Rev. Lett.* **68**, 1500 (1992).
5. A.M.Weiner, J.P.Heritage, and R.N.Thurston, *Opt.Lett.* **11**, 153, (1986); A.M.Weiner, D.E.Leaird, G.P.Wiederecht, and K.A.Nelson, *Science* **247**, 1317, (1990); C.J.Bardeen, Q.Wang, and C.V.Shank, *Phys. Rev. Lett.* **75**, 3410 (1995).
6. C. J. Bardeen *et al*, *Chem. Phys. Lett.* **280**, 151 (1997); A.Assion *et al*, *Science* **282**, 919 (1998).
7. H.Rabitz, private communication.
8. M. Shapiro and P. Brumer, *Adv. At. Mol. Opt. Phys.* **42**, 287 (2000).
9. S.Chelkowski, A.D.Bandrauk, and P.B.Corkum, *Phys. Rev. Lett.* **65**, 2355 (1990); J. S. Melinger, S. R. Gandhi, A. Hariharan, J. X. Tull, and W. S. Warren, *Phys. Rev. Lett.* **68**, 2000 (1992); B. Y. Chang, I. R. Solá, V. S. Malinovsky, and J. Santamaría, *J. Chem. Phys.* **113**, 4901 (2000).
10. K. Bergmann, H. Theuer, and B. W. Shore, *Rev. Mod. Phys.* **70**, 1003 (1998).
11. B. M. Garraway and K.-A. Suominen, *Phys. Rev. Lett.* **80**, 932 (1998).
12. I. R. Solá, J. Santamaría, and V. S. Malinovsky, *Phys. Rev.* A **61**, 3413 (2000); I.R.Solá, B.Y.Chang, J.Santamaría, V.Malinovsky, and J.Krause, *Phys. Rev. Lett.* **85**, 4241 (2000); B.Y.Chang, I.R.Solá, J.Santamaría, V.Malinovsky, and J.Krause, *J. Chem. Phys.* **114**, 20 (2001); V. S. Malinovsky and J. L. Krause, *Chem. Phys.* **267**, 47 (2001).
13. R. Schinke, *Photodissociation Dynamics*, (Cambrigde University Press, 1993).
14. I. R. Solá *et. al.*, unpublished results.
15. B. W. Shore, *The Theory of Coherent Atomic Excitation* (Wiley, N.Y., 1990).
16. B. Y. Chang, I. R. Solá, V. S. Malinovsky, and J. Santamaría, *J. Chem. Phys.* **113**, 4901 (2000).
17. G. G. Balint-Kurti, R. N. Dixon, and C. C. Marston, *Faraday Trans. Chem. Soc.* **86**, 1741 (1990).
18. B.Y.Chang, I.R.Solá, J.Santamaría, V.Malinovsky, and J.Krause, *J. Chem. Phys.* **114**, 20 (2001).
19. B. M. Garraway and K. A. Suominen, *Rep.Prog.Phys.* **58**, 365 (1995).

ULTRAFAST LASER-DRIVEN HYDROGEN BOND DYNAMICS

H. NAUNDORF AND O. KÜHN

Institut für Chemie, Physikalische und Theoretische Chemie,
Freie Universität Berlin, Takustrasse 3, D-14195 Berlin
E-mail: ok@chemie.fu-berlin.de

The influence of anharmonic couplings on the dynamics of carboxy-deuterated phthalic acid monomethylester following laser excitation of the OD stretching vibration is investigated. Based on a Cartesian reaction surface Hamiltonian a 19-dimensional model is developed for which the vibrational quantum dynamics is studied using the multiconfiguration time-dependent Hartree method. In contrast to a mean field approach the multiconfiguration treatment leads to the onset of energy randomization on a time scale of a few picoseconds.

1 Introduction

The detailed knowledge about the structural and dynamical properties of hydrogen bonds is of central importance for the understanding of many processes in chemistry and biology [1]. Multidimensionality and the quantum nature of H-bond dynamics combine to a formidable theoretical challenge including the ab initio quantum chemical construction of a potential energy surface (PES) and the propagation of nuclear wave packets [2]. Since the determination of ab initio PES for more than a few degrees of freedom (DOF) is beyond reach one has to resort to approximate methods. The most powerful description is based on the partitioning of all coordinates into active and substrate ones [2,3,4,5,6,7]. This concept is particularly suited for H-bonded systems where the (active) reaction coordinates are closely related to the proton's motion which will cover the most anharmonic part of the PES. In contrast heavy atom motions of the molecular scaffold (the substrate) is usually confined to the harmonic region of the PES. In principle a formulation in terms of internal [3,4,5,6] or Cartesian [7] reaction coordinates is possible for the situation of a light proton moving with respect to the much heavier substrate. Below we will use the all-Cartesian reaction surface (CRS) Hamiltonian which has the numerical advantage that all couplings between the different DOFs are contained in the potential energy operator [7,8,9,10].

Given a high-dimensional CRS Hamiltonian straightforward solution of the time-dependent Schrödinger equation is impossible without further simplifications. The simplest approximation is a Hartree product ansatz for the total wave function. This yields a mean field description where each DOF moves

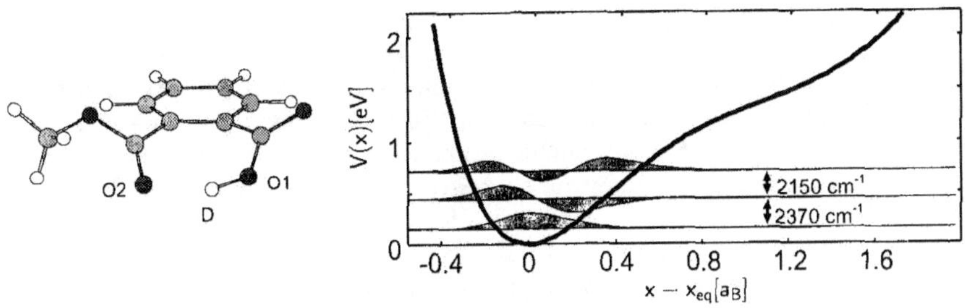

Figure 1. Left: Equilibrium structure of deuterated phthalic acid monomethyl ester with the H-bond O1-D···O2. Right: Potential $V(x)$ along the O1-D bond (reaction coordinate) with all other DOFs being frozen. Also shown are the wave functions for the lowest vibrational states in this potential. The equilibrium O1-D distance is $x_{\rm eq} = 0.98$ Å.

in the time-dependent average field due to the other DOFs (time-dependent self-consistent field or time-dependent Hartree (TDH)) [11,12]. As pointed out by Makri and Miller [13] this single Hartree product (configuration) form does not provide enough flexibility for the description of proton transfer processes where a multiconfiguration extension is mandatory. Meyer and coworkers [14] developed the multiconfiguration time-dependent Hartree (MCTDH) approach to a powerful tool for multidimensional wave packet propagations (for a recent review see Ref. [15]).

In the present contribution we combine the CRS Hamiltonian with a MCTDH nuclear wave packet calculation to investigate the influence of anharmonic couplings on the dynamics of carboxy-deuterated phthalic acid monomethylester (PMME-D) shown in Fig. 1. Our motivation derives from the recent experimental observation of ultrafast vibrational wave packet dynamics after excitation of the OD stretching vibration in this system [16]. In a previous study we had shown using the TDH approach that the oscillatory features in the IR pump-probe signal can be explained by the anharmonic coupling of the OD motion to a low-frequency substrate mode modulating the H-bond geometry [9]. While the TDH dynamics did not show any signatures of relaxation-type behavior, a 9-dimensional MCTDH study revealed partial energy randomization on a time scale of about 2 picoseconds [10]. Here we will combine the MCTDH and the TDH descriptions supplementing our previous 9D model by 10 additional DOFs. In Section 2 we summarize the theoretical background and give some details of our model for PMME-D. In Section 3 numerical results are discussed and the paper is summarized in Section 4.

2 Theoretical Model

The construction of the Cartesian reaction surface starts from the most stable configuration of PMME-D shown in Fig. 1. It has been obtained using second order Møller-Plesset perturbation theory (MP2) and a Gaussian 6-31+G(d,p) basis set (all quantum chemistry calculations have been performed using Gaussian98 [17]). A natural choice for the Cartesian reaction coordinate for the dynamics of the intramolecular H-bond O1-D···O2 would be the position of the deuterium within the molecular frame. However, since the O1-D stretching vibration is excited and monitored in the experiment [16], only its anharmonicity has to be considered and a one-dimensional reaction coordinate is chosen which points along the O1-D bond (here the x-direction). Following Refs. [7,9] the CRS-Hamiltonian will be expressed in terms of a zeroth-order part,

$$H_{\rm CRS}^{(0)} = H_{\rm sys} + \sum_n H_n = \frac{p_H^2}{2m_H} + V(x) + \frac{1}{2}\sum_n [p_n^2 + \omega_n^2 Q_n^2] \tag{1}$$

which describes the one-dimensional reaction coordinate x and the uncoupled set of harmonic mass-weighted substrate modes $\{Q_n\}$, and an interaction part

$$V_{\rm CRS} = -\sum_n f_n(x) Q_n + \frac{1}{2}\sum_{mn}(K_{mn}(x) - \delta_{mn}\omega_n^2) Q_n Q_m \tag{2}$$

Note that the zeroth-order substrate Hamiltonians H_n have been defined with respect to the frequencies at the reference configuration, i.e. $\omega_n^2 = K_{nn}(x_{\rm eq})$. Eq. (2) contains two types of couplings, linear and quadratic in the substrate coordinates. The quadratic term in Eq. (2) leads to a mixing between different substrate normal modes, if their character changes along the reaction coordinate. The linear term derives from the fact that on a CRS one does not necessarily move on a minimum energy path, i.e. the n-th mode is shifted by the reorganization energy (neglecting the mode-mode coupling)

$$\Delta E_n(x) = f_n^2(x)/2K_{nn}(x) \tag{3}$$

As discussed in Ref. [8] the reorganization energy gives a convenient measure for identification of those substrate modes which have the strongest impact on the reaction coordinate dynamics.

In Fig. 1 we show the potential $V(x)$ along the reaction coordinate for all substrate modes being frozen at their equilibrium configurations. The potential is rather anharmonic but does not show a double minimum, i.e. the H-bond is of medium strength and no proton transfer is possible. Note, that this conclusion also holds if the substrate coordinates are allowed to relax

Table 1. Parameters of the 19D model used in the MCTDH simulations. In the different rows the DOFs combined in each MCTDH particle are given; the TDH modes have not been combined. The grid dimensions are given in atomic units (mass-scaled for the substrate modes) and the last column shows the number of harmonic oscillator DVR functions used. (The grid for the reaction coordinate is given with respect to $x_{\rm eq} = 1.85 a_{\rm B}$. ΔE_n is the reorganization energy at $x = x_{\rm eq} + 0.2\AA$.) Note that in the dynamics simulations the potential energy terms in Eqs. (1) and (2) as well as the dipole moment have been fitted to polynomials and powers of exponentials.

mode	$\omega_i/2\pi c$ (cm^{-1})	ΔE_n (eV)	SPF	grid	DVR
$\nu_{\rm OD}$	2370	-	7	-0.8 / 1.1	40
ν_{23}	986	$1.273 \cdot 10^{-1}$		-100 / 100	40
ν_1	67	$9.5 \cdot 10^{-3}$	5	-250 / 250	40
ν_3	147	$8.6 \cdot 10^{-3}$	5	-205 / 225	32
ν_5	201	$8.9 \cdot 10^{-3}$		-175 / 200	32
ν_7	263	$1.76 \cdot 10^{-2}$	5	-175 / 175	32
ν_9	306	$9.1 \cdot 10^{-3}$		-175 / 175	32
ν_{13}	507	$1.22 \cdot 10^{-2}$	5	-150 / 150	32
ν_{24}	1011	$7.5 \cdot 10^{-3}$		-125 / 125	32
ν_6	227	$2.3 \cdot 10^{-3}$	1	-200 / 200	24
ν_8	283	$1.7 \cdot 10^{-3}$	1	-200 / 200	32
ν_{10}	340	$0.8 \cdot 10^{-3}$	1	-150 / 150	32
ν_{11}	412	$3.8 \cdot 10^{-3}$	1	-150 / 125	32
ν_{14}	563	$5.7 \cdot 10^{-3}$	1	-150 / 150	24
ν_{19}	754	$5.6 \cdot 10^{-3}$	1	-125 / 125	24
ν_{20}	846	$4.9 \cdot 10^{-3}$	1	-125 / 125	32
ν_{21}	861	$0.5 \cdot 10^{-3}$	1	-125 / 125	24
ν_{25}	1048	$1.8 \cdot 10^{-3}$	1	-100 / 100	24
ν_{47}	2169	$2.7 \cdot 10^{-3}$	1	-100 / 100	32

to their respective minimum energy values. As can be concluded from the wave functions corresponding to the lowest vibrational states shown in Fig. 1, the dynamics under low excitation conditions takes place in the vicinity of the equilibrium position. In order to judge the role of the substrate modes we have inspected their reorganization energy according to Eq. (3) at the representative position $x_{\rm eq} + 0.2\ \AA$ where $\sum_n \Delta E_n(x) = 0.24$ eV.

The most strongly coupled modes are listed in Table 1. In all cases displacement along the respective normal coordinates modulates the H-bond geometry. The 18 substrate modes in Table 1 comprise about 96 % of the total reorganization energy at $x_{\rm eq} + 0.2\AA$. In a previous study we had included the 8 modes with the largest reorganization energy (upper part of Table 1) in a numerical exact MCTDH study [10]. With the cut-off set at 6 meV they accounted for 85 % of the reorganization energy. By including 10 more modes (lower part of Table 1) we are in the position to give an almost complete description on the level of the CRS Hamiltonian.

One of the advantages of the CRS-Hamiltonian is its factorized structure with respect to the different DOFs. This makes it particularly useful for the implementation within MCTDH wave packet propagation [15]. The MCTDH method rests on the expansion of the total wave function according to

$$\Psi(x, Q_1, \dots Q_p, t) = \sum_{j_0=1}^{n_0} \dots \sum_{j_p=1}^{n_p} A_{j_0 \dots j_p}(t) \varphi_{j_0}^{(0)}(x,t) \prod_{\kappa=1}^{p} \varphi_{j_\kappa}^{(\kappa)}(Q_\kappa, t) \quad (4)$$

The time-dependent coefficients $A_{j_0 \dots j_p}(t)$ and single particle functions (SPFs) $\varphi_{j_\kappa}^{(\kappa)}$ are variationally optimized leading to a set of coupled nonlinear equations [15]. Each term in Eq. (4) constitutes a configuration; the case of a single configuration corresponds to the TDH limit. The SPFs have to be expanded in terms of a time-independent basis set. We have used a discrete variable representation (DVR) [18,19] with parameters given in Table 1. Further we have combined the more strongly coupled modes into multidimensional SPFs to facilitate efficient incorporation of correlations [20]. The MCTDH equations have been solved using the Heidelberg package [21]. The propagation with the laser field were made with the variable mean field method, using a Adams-Bashforth-Moulton predictor-corrector integrator with an initial step size of 0.01 fs. The following propagation without the field was performed with the constant mean field method with a step size of 0.005 fs. For the propagation of the SPFs a Bulirsch-Stoer extrapolation integrator is used while the coefficients are calculated with a short iterative Lanczos integrator (for details see Ref. [15]).

A total of 9 DOFs (upper part of Table 1) have been treated numerically exact by MCTDH, the weaker coupled modes (lower part of Table 1) are accounted for using only a single configuration (TDH modes). We will consider the dynamics of the O1-D stretching vibration as triggered by the interaction with an external laser field. Invoking the dipole approximation Eqs. (1) and (2) are supplemented by the time-dependent Hamiltonian

$$H_{\text{field}}(t) = -d(x)E(t) \quad (5)$$

where $d(x)$ is the dipole moment operator along the reaction coordinate. For the classical electric field we will use the form

$$E(t) = E_0 \Theta(\tau - t) \sin^2(\pi t/\tau) \cos(\Omega t) \quad (6)$$

with E_0, τ, and Ω being the amplitude, the duration, and the center frequency, respectively.

The quantum chemistry input for the CRS-Hamiltonian, Eqs. (1), (2), and (5), thus consists of the reference potential, the forces, the Hessian, and the dipole moment as a function of the reaction coordinate.

3 Numerical Results

In Fig. 2 we show the dynamics of the coordinate expectation values for the reaction coordinate and two substrate modes ν_1 and ν_{23}. These modes represent extreme cases, i.e. the lowest frequency (ν_1) and the strongest coupling (ν_{23}), cf. Table 1. The duration of the laser pulse has been set to $\tau = 300$ fs with $\Omega/2\pi c = 2425$ cm^{-1} being resonant to the $\nu_{\mathrm{OD}} = 0 - 1$ transition and $E_0 = 5 \times 10^{-4} E_{\mathrm{h}}/ea_{\mathrm{B}}$. The initial state was chosen as the vibrational ground state. Clearly, for these excitation conditions the dynamics takes place in the vicinity of the potential energy minimum as can also be seen from the variance Δx shown in the upper part of panel (A). The resonant driving causes $\langle x \rangle(t)$ to oscillate with a period given by the fundamental transition frequency. Note that the ground state expectation value of the reaction coordinate is different from x_{eq} due to the anharmonicity taken into account in the quantum mechanical averaging. Superimposed on the rapid oscillations are low-frequency modulations most notably with a period of about 500 fs. In order to rationalize this behavior let us consider the dynamics of the substrate modes which is solely due to the anharmonic coupling to the reaction coordinate. The origin of the 500 fs modulation of the $\langle x \rangle$ dynamics can be traced to the low-frequency mode ν_1 whose behavior is shown in Fig. 2C. Comparing panels (A) and (C) of Fig. 2 we notice that the maxima of $\langle Q_1 \rangle$ correlate with the minima in the envelope of $\langle x \rangle$. In other words, an elongation along Q_1 leads to a *compression* of the hydrogen bond and thus to a smaller amplitude of the OD vibration. Note that the variance Δx is modulated too. In Fig. 2B we show the dynamics of the most strongly coupled mode ν_{23}. It is driven by the interaction with the reaction coordinate which gives rise to rapid oscillations. The slower components cannot be correlated to the dynamics of the ν_1 mode only, since the coupling to the other modes via the off-diagonal elements of $\mathbf{K}(x)$ plays an important role.

A more detailed view of the energy flow dynamics is provided by looking at the expectation values for the uncoupled zero-order Hamiltonians defined in Eq. (1) and shown in Fig. 3. Similar to Fig. 2 we observe oscillations on different time scales. A comparison of panels (A) and (B) in these figures reveals that energy is rapidly exchanged between the reaction coordinate and

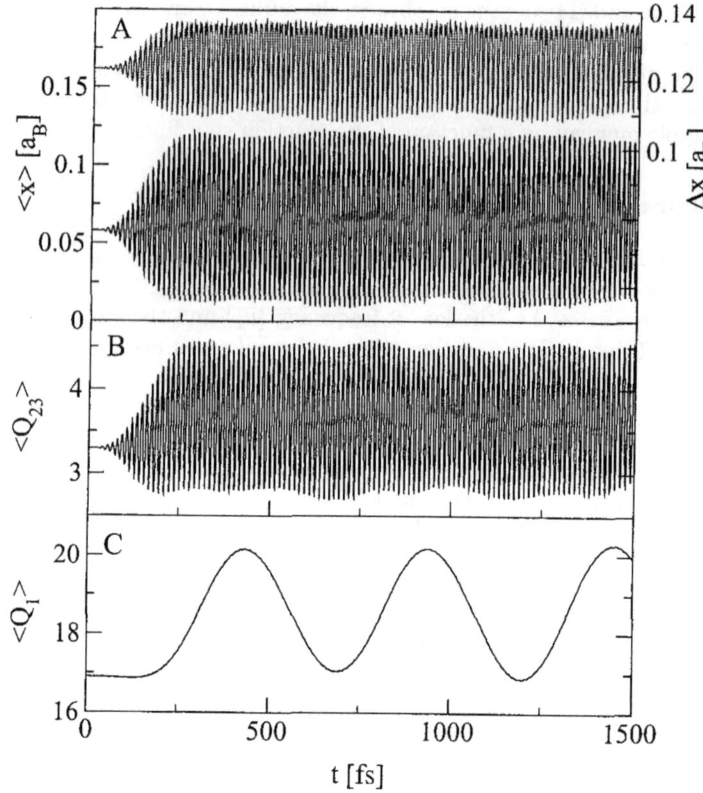

Figure 2. MCTDH coordinate expectation values:(A) reaction coordinate (left) and variance Δx (right), (B) ν_{23} mode, (C) ν_1 mode ($E_0 = 5 \times 10^{-4} E_h/ea_B$, $\Omega/2\pi c = 2425$ cm^{-1}, $\tau = 300$ fs).

the most strongly coupled mode ν_{23}. Comparing Figs. 2 and 3 we notice that energy flows into mode ν_{23} whenever $\langle x \rangle$ is approaching the inner turning point where the force $f_{23}(x)$ is largest. The overall excitation of mode ν_{23} is only modest due to the large energy mismatch with the reaction coordinate. An important point is that the behavior of $\langle E_x \rangle$ is characterized by a slow decrease which is superimposed on the oscillations. Estimating the time scale assuming a single exponential decay gives a relaxation time of about 20 ps.

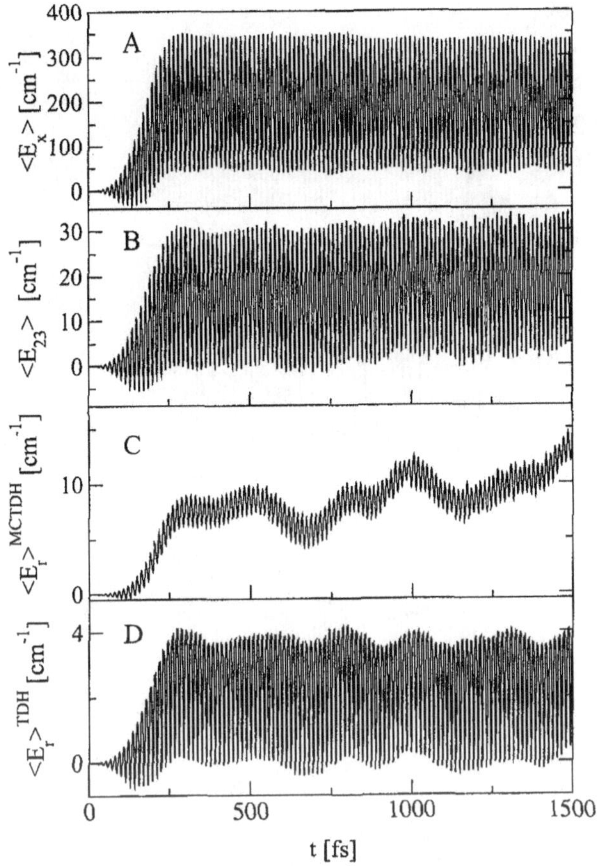

Figure 3. Change of expectation values of the zero-order Hamiltonians defined in Eq. (1) for excitation conditions specified in Fig. 2:(A) reaction coordinate, (B) ν_{23} mode, sum over all remaining MCTDH (C) and TDH (D) modes.

Simultaneously, $\langle E_{23} \rangle$ increases as does the energy of all remaining substrate modes, $\langle E_r \rangle^{\mathrm{MCTDH}}$, which are treated within the MCTDH approach as shown in Fig. 3C. The *TDH* modes are participating in the overall energy exchange as well as shown in Fig. 3D ($\langle E_r \rangle^{\mathrm{TDH}}$). However, their total excitation is rather small due to the weak anharmonic coupling (cf. Table 1). In addition

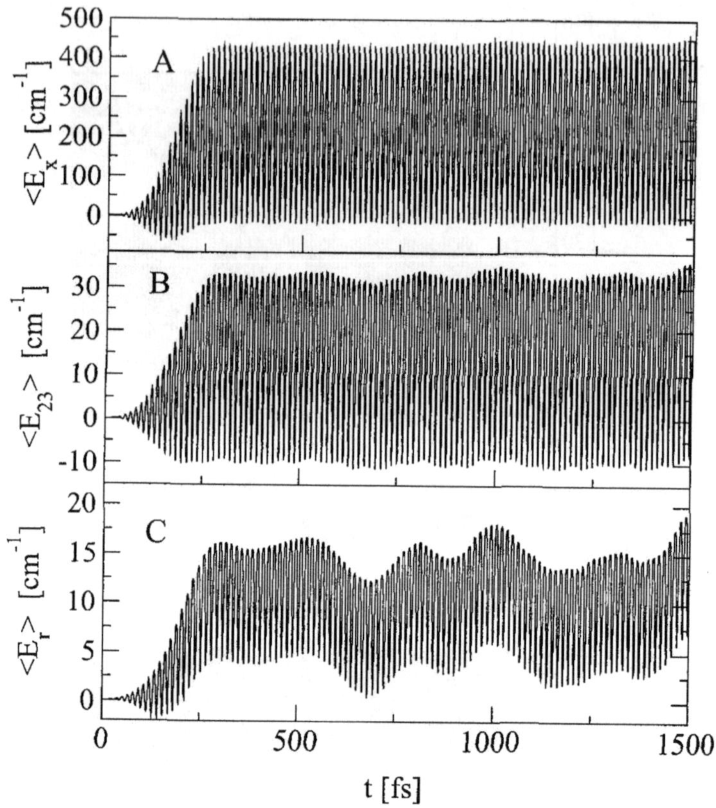

Figure 4. Change of expectation values of the zero-order Hamiltonians defined in Eq. (1) for excitation conditions specified in Fig. 2 and using the TDH approach:(A) reaction coordinate, (B) ν_{23} mode, (C) sum over all remaining modes.

no noticeable relaxation-type behavior is found for these modes.

It is interesting to note that despite the fact that the dynamics takes place only in the immediate neighborhood of the potential minimum, the mean field approach does not provide enough flexibility for the multidimensional vibrational wave function to describe the onset of energy randomization. This conclusion is supported by comparing Fig. 3 with the results of a full TDH simulation in Fig. 4. As far as it concerns the oscillation patterns TDH and MCTHD+TDH give rather similar results (cf. in particular panels (C)).

However, no relaxation-type behavior is found in the TDH limit (see also Ref.[9]).

4 Conclusions

The intramolecular hydrogen bond dynamics in a 19-dimensional model of PMME-D has been investigated by combining a reaction surface Hamiltonian with MCTDH/TDH vibrational wave packet simulations. Anharmonic coupling was found to lead to a modulation of the OD bond vibration dynamics after laser excitation, most notably on a time scale of about 500 fs. This dynamics corresponds to a low-frequency substrate mode which modifies the hydrogen bond geometry. Our calculations are in accord with the recent experimental demonstration of coherent vibrational dynamics in this system [16]. In the experiment, however, the OD excitation has a life time of about 500 fs in C_2Cl_4 solution. From our simulations we estimated the time scale for intramolecular energy randomization with respect to the ν_{OD} mode to be of the order of 20 ps. This highlights the importance of including the interaction with the solvent to establish the connection with the experiment. It is to be expected that the rather fast decay time of 500 fs will require a multi-quantum transition formulation of relaxation theory to model the interplay between intra- and intermolecular processes [22]. The CRS Hamiltonian should provide a good starting point for the development of an appropriate system-bath model [2].

Acknowledgments

This work has been supported by the Deutsche Forschungsgemeinschaft through the Sfb450. The authors gratefully acknowledge stimulating discussions with J. Dreyer, T. Elsaesser, P. Hamm, D. Madsen, E. T. J. Nibbering, and J. Stenger from the Max Born Institute Berlin as well as with H.-D. Meyer (Heidelberg) and G. A. Worth (London).

References

1. H. H. Limbach and J. Manz (eds), *Hydrogen Transfer: Experiment and Theory*, Ber. Bunsenges. Phys. Chem. **102**, (1998).
2. V. May and O. Kühn *Charge and Energy Transfer Dynamics in Molecular Systems*, (Wiley-VCH, Berlin, 2000).
3. W. H. Miller, N. C. Handy, and J. E. Adams, J. Chem. Phys. **72**, 99 (1980).

4. T. Carrington and W. H. Miller, J. Chem. Phys. **81**, 3941 (1984).
5. N. Shida, P. F. Barbara, and J. E. Almlöf, J. Chem. Phys. **91**, 4061 (1989).
6. D. Heidrich (ed), *The Reaction Path in Chemistry: Current Approaches and Perspectives*, (Kluwer Academic, Dordrecht, 1995).
7. B. A. Ruf and W. H. Miller, J. Chem. Soc. Faraday Trans. 2 **84**, 1523 (1988).
8. H. Naundorf, J. A. Organero, A. Douhal, and O. Kühn, J. Chem. Phys. **110**, 11286 (1999).
9. G. K. Paramonov, H. Naundorf, and O. Kühn, Eur. J. Phys. D **14**, 205 (2001).
10. H. Naundorf, G. A. Worth, H.-D. Meyer, and O. Kühn (submitted).
11. P. A. M. Dirac, Proc. Cambridge Philos. Soc. **26**, 376 (1930).
12. R. B. Gerber, V. Buch, M. A. Ratner, J. Chem. Phys. **77**, 3022 (1982).
13. N. Makri and W. H. Miller, J. Chem. Phys. **87**, 5781 (1987).
14. H.-D. Meyer, U. Manthe, and L. S. Cederbaum, Chem. Phys. Lett. **165**, 73 (1990).
15. M. H. Beck, A. Jäckle, G. A. Worth, and H.-D. Meyer, Phys. Rep. **324**, 1 (2000).
16. J. Stenger, D. Madsen, J. Dreyer, E. T. J. Nibbering, P. Hamm, and T. Elsaesser, J. Phys. Chem. A **105**, 2929 (2001).
17. M. J. Frisch *et al.*, *Gaussian 98*, Revision A.6, (Gaussian Inc., Pittsburgh, 1998).
18. D. O. Harris, G. G. Engerholm, and G. W. Gwinn, J. Chem. Phys. **43**, 1515 (1965).
19. J. C. Light, I. P. Hamilton, J. V. Lill, J. Chem. Phys. **82**, 1400 (1985).
20. G. A. Worth, H. D. Meyer, and L. S. Cederbaum, J. Chem. Phys. **109**, 3518 (1998).
21. G. A. Worth, M. H. Beck, A. Jäckle, H.-D. Meyer, *MCTDH package*, Version 8.2, (University of Heidelberg, Heidelberg, 2000).
22. V. M. Kenkre, A. Tokmakoff, and M. D. Fayer, J. Chem. Phys. **101**, 10618 (1994).

TOWARDS STRUCTURAL DYNAMICS RESEARCH WITH ULTRASHORT PULSE LASERS AND SYNCHROTRON RADIATION

C. BRESSLER, M. SAES, AND M. CHERGUI

Institut de Physique de la Matière Condensée, Université de Lausanne
CH-1015 Lausanne, Switzerland
E-mail: christian.bressler@ipmc.unil.ch

D. GROLIMUND AND R. ABELA

Swiss Light Source, Paul Scherrer Institut
CH-5232 Villigen PSI, Switzerland

We present a new technique to record single x-ray pulses in time-resolved XAS experiments at low frequencies around 1 kHz compared to the multibunch repetition rate at synchrotron radiation storage rings of several hundred MHz. This scheme of electronic gating of the multibunch train is applied to record high-quality EXAFS spectra. Further measurements show that this scheme permits us to record high-quality x-ray absorption spectra with a noise level below 3×10^{-5}. The achieved results imply the utility of this method with current synchrotron radiation sources.

1 Introduction

X-rays, neutrons and electrons are well-established tools to elucidate the structure of matter. X-ray diffraction [1] and x-ray absorption (XAS) methods [2,3] have proven to be powerful techniques for structural determination of biomolecules and of condensed matter. We are currently developing a scheme to exploit pulsed x-radiation for structural dynamics studies. X-ray absorption (XAS) techniques, such as extended x-ray absorption fine structure (EXAFS) and x-ray absorption near edge structure (XANES) spectroscopies, deliver local structural information [2,3]. While XANES is complicated to interpret from a structural point of view, it contains rather simple phenomena like the chemical shift, which describes the oxidation state of the atom [3]. On the other hand, EXAFS delivers a detailed picture of the local environment, and interatomic separations can nowadays be determined with an accuracy down to 100 fm [4]. In addition, EXAFS requires no periodic structures (as required in x-ray diffraction), and can be readily applied to disordered systems, e.g., liquids.

Time-resolved XAS aiming to visualize nuclear rearrangements on the femtosecond time scale is a young field. For this purpose ultrashort x-ray pulses are required. Furthermore, tunability of monochromatic x-rays over several hundred eV is a stringent prerequisite for XAS studies. Synchrotrons are today the most powerful source of x-rays. Synchrotron radiation (SR) is tunable over an extremely large region up to the very hard x-ray regime around 100 keV. However, its temporal resolution (pulse width) is limited to the 20-100 ps range. Furthermore, laser-pump SR-probe experiments depend on the available number of x-ray photons per single pulse at the repetition rate of the exciting laser, which is typically 1 kHz for commercial amplified fs-lasers. Therefore, such experiments can use only about 10^{-5} - 10^{-6} of the available photon flux at a synchrotron with its much higher pulse repetition rate of 100 - 500 MHz. Nevertheless, exploiting SR x-ray pulses in a pump-probe configuration has just recently become a reality in pioneering experiments by the Falcone group [5,6,7]. In such diffraction experiments, they were able to detect coherent phonons in InSb with the 30 ps pulses of the Advanced Light Source (ALS) in Berkeley, by exploiting an x-ray sensitive streak camera in jitter-free configuration thus obtaining a temporal resolution around 2-3 ps in averaging mode [7]. In another development, a recent breakthrough by Schoenlein *et al.* promises near-150 fs hard x-ray pulses to become soon available [8]. It is based on the slicing of an electron bunch in the SR storage ring by a femtosecond optical pulse. While the expected flux in the hard x-ray regime is rather low, this source represents the shortest x-ray source with existing technology.

To date the only ultrafast time resolved XAS experiment was performed by the Wilson and Kieffer groups [9]. In this experiment, gaseous SF_6 was photolysed with an intense optical laser pulse, and a subsequent continuum x-ray pulse from a laser-driven x-ray plasma source probed the XAS in the XANES region of the sulphur K edge. While the noise on the spectrum was still too large to observe the K edge jump itself, the disappearance of the SF_6 shape resonance (which occurs upon dissociation) was clearly observed. The absorption cross section of this shape resonance exceeds that of the sulphur K edge by nearly an order of magnitude [10], which is the main reason for the success of this experiment. Recently, a time-resolved EXAFS study with 14 ns resolution was reported at the Advanced Photon Source APS (Argonne National Laboratory, Chicago) [11]. Chen *et al.* measured the photoinduced structural changes associated with the ligands of Ni-TPP, a complex with a Ni center and octahedrically oriented ligands, via Ni K-edge EXAFS. However, data accumulation times of 300 sec per data point, adding up to ca. 20 hours for one entire spectrum, were necessary.

With the existing techniques for ultrashort x-ray pulse generation mentioned above, one has to evaluate the current capabilities to allow structural dynamical research using x-ray absorption. We have performed calculations of the expected pump-probe signal for a given condensed phase chemical system [12,13], which also underline the general utility of time-resolved XAS in condensed phase dynamics research. Large-scale facilities are currently in the planning stage to provide the scientific community with unprecedented intense bursts of hard x-ray pulses. X-Ray FELs will deliver *pulse* intensities, which are more than 10^6 times greater than at any existing source, however, their realization is still many years away [14]. At the same time, synchrotron radiation centers are evaluating the realization of so-called *Energy Recovery Linacs* (ERLs), which would benefit from all the advantages of conventional 3rd generation SR storage rings, but due to the single-cycle storage of electrons they would offer much more flexible arrangements, e.g., allowing decreased pulse widths down to the sub-ps range [15].

2 Experimental Methods

The experiments are carried out at beamline 5.3.1 of the *Advanced Light Source* (ALS), Berkeley. Bend magnet radiation is imaged 1:1 via a grazing incidence Pt-coated mirror (5 mrad deflection) inside the experimental hutch. A Ge (111) double monochromator provides us with monochromatic radiation in the 5 – 7 keV range with a bandwidth around 7 eV. The experimental chamber is filled with atmospheric helium gas (x-ray transmission > 99 % for > 4.5 keV radiation). The sample consists of a free-flowing liquid jet with a thickness of 0.1 mm. Excitation of the sample occurs with an amplified and frequency-tripled Ti-Sapphire laser at 1 kHz (266 nm,

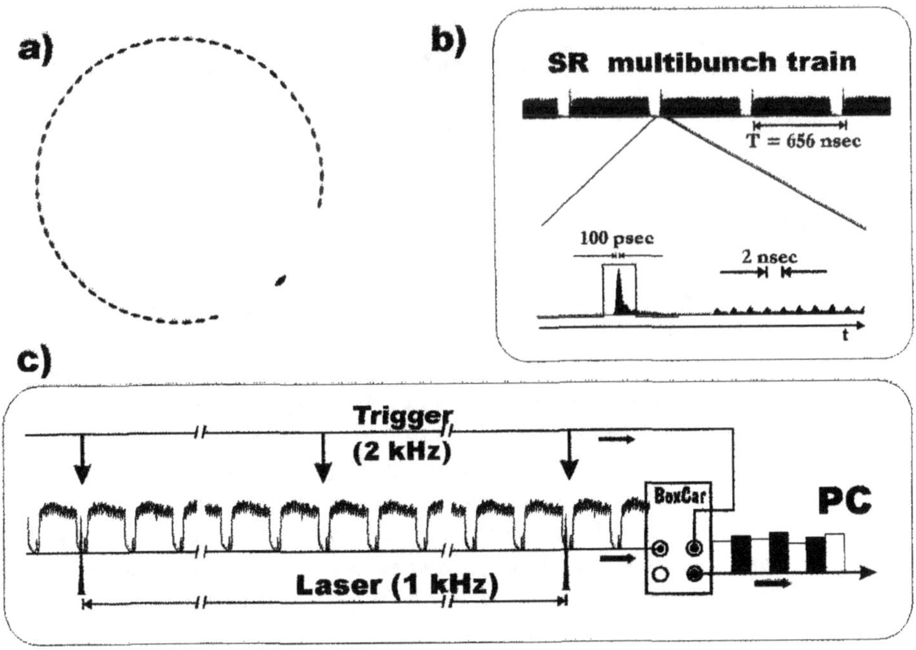

Figure 1. Data acquisition scheme for recording low-noise EXAFS in the pump-probe scheme. a) Special filling pattern of electron bunches at the ALS called *camshaft mode*, b) oscilloscope trace of the resulting time-dependent intensity. c) Recording the x-ray pulse train with a slower (smaller bandwidth) APD results

in an intensity pile-up of the multibunch trace (therefore, the single camshaft pulse appears smallier in intensity). The gated integrator records these single pulses at twice the laser repetition rate, so that its output signal train alternates between laser-pumped and laser-unpumped intensities (lower right in c).

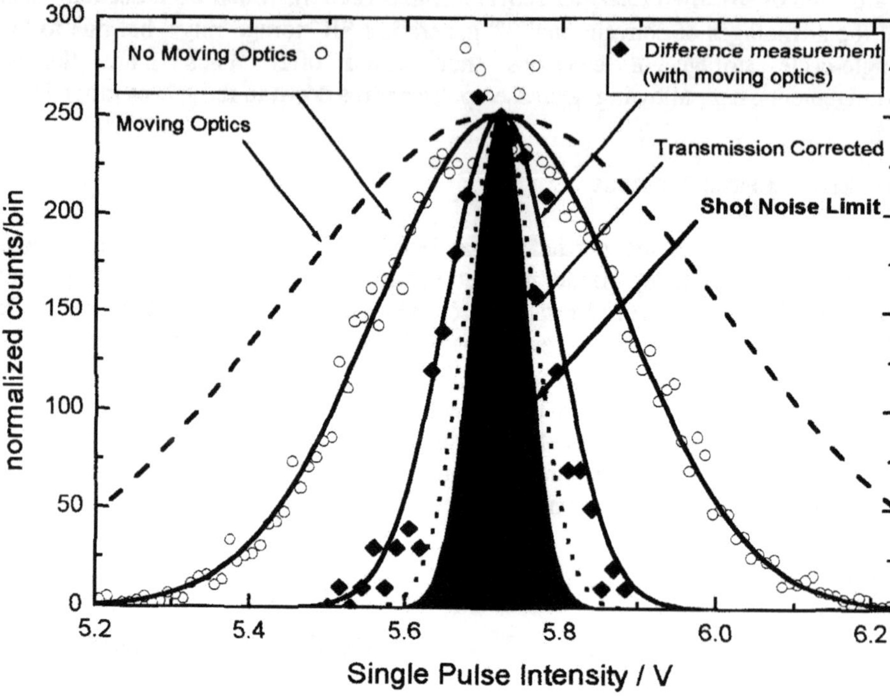

Figure 2. Experimentally measured deviations (normalized to the same max. intensities) of the recorded noise from the shot-noise limit (see text for details).

pulse energy > 100 μJ). The laser is focused on the sample with its beam size matching the x-ray focal size (ca. 0.3 mm). The measured jet velocity of 5 m/s indicates that after 60 μs the sample moves out of the probe focus. Therefore, pumping the sample at repetition rates below 15 kHz will always occur on a fresh sample. The x-ray pulse train transmitted through the sample is detected with an avalanche photodiode (APD), and processed with a gating integrator to select out the desired single x-ray pulse associated with the laser pump pulse.

The data acquisition scheme is shown in Fig. 1. A special storage ring electron filling pattern at the ALS, called the *camshaft mode* (Fig. 1 a), consists of a close-packed multibunch train followed by a 100 ns empty section. In this empty section a ten-fold denser electron bunch is placed, whose radiation we select with our gated integrator (see Fig. 1b with the indicated window gate around the camshaft pulse). In order to record high-quality spectra we have employed the following scheme

(Fig. 1c). The gated integrator is triggered at twice the laser repetition rate, so that every second pulse corresponds to the static (= unexcited) sample, while the other x-ray pulses monitor the transmission following photoexcitation at an adjustable time delay. Therefore, the gated integrator delivers an output of alternatively laser-pumped and unpumped intensities, which are then read into the computer and appropriately sorted. With this scheme we can potentially record high-quality EXAFS spectra of the photoinduced changes. Indeed, a careful analysis of the measured noise shows that we have nearly achieved the shot noise limit in this configuration. The measured noise is considerably reduced over other methods, which include additional systematic noise sources, as illustrated in Fig. 2. The step-scanning monochromator induces vibrations, which result in a rather constant noise source on the transmitted intensity. This is considerably reduced, when no optics move (the pulse height distribution is shown for 10^4 shots in Fig. 2), and the decreasing storage ring current is observed by a corresponding broadening of the pulse height distribution. In addition, using our latest development by measuring the difference spectra *during* a monochromator scan reduces the noise even more. After correcting the statistically derived effective flux for the limited x-ray absorption in the 30 μm thick APD, our measurements are very close to the specified flux of this beamline.

3 Results and Discussion

With our new difference-signal detection scheme we make best use of the x-ray source, since the recorded noise on these spectra correspond nearly entirely to the expected shot noise limit. But we can also record high-quality EXAFS without employing the difference measurement technique. Figure 3 illustrates this for the case of aqueous iodide. The energy spectrum has already been transformed into k-space (Fig. 3a), but the original data has not been treated (e.g., smoothed or Fourier-filtered). This spectrum represents a single scan accumulating 2500 single x-ray pulses per data point. The fit (and its Fourier transform, Fig. 3b) delivers a nice agreement with the literature, and the accuracy for determining the nearest neighbor shell of oxygen atoms is better than 10 pm [13].

Fig. 3 also illustrates the potential for measuring the nearest-neighbor distance in photogenerated iodine radicals. The I-O distance should change considerably upon photodetachment. Values for iodine are currently unknown, but in a theoretical study of aqueous Br radicals the predicted Br-O distance changed from 335 pm for the ion to ca. 280 pm for the neutral radical [16]. Assuming a similar reduction in the iodine-oxygen distances would result in a pronounced change in the EXAFS signal, as shown in the dashed curves in Fig. 3.

With this new setup we have attempted to perform a simple time-resolved XAS experiment on another model system, aqueous hexacyanoferrate. The UV photoionization of hexacyanoferrate is an interesting process, since both reactant and product are stable aqueous solutes. They can be easily identified both in optical absorption spectra [17], and by XAS through the well-known chemical shift at the Fe K edge [18]. We have investigated the photoionization process with 266 nm light (Fig. 4). Fig. 4a shows the XANES region of the Fe K edge for the pumped and unpumped sample, which have been offset for clarity. The inset enlarges

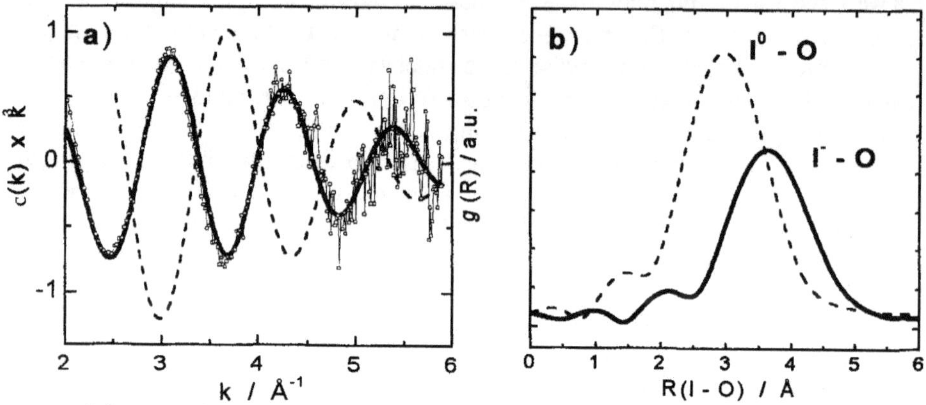

part of the pre edge

Figure 3. Static EXAFS above the L_3 edge of iodine of aqueous iodide (NaI concentration: 0.7 mol/l, thickness: 0.1 mm). Original data (open circles), and their fit (solid line) is shown in a), and the Fourier transform showing the nearest neighbor I-O distance in b) (solid line). For comparison, a simulation of the expected EXAFS of atomic iodine is shown (dashed curve in a) together with its Fourier transform (dashed curve in b).

edge region, which shows the correlated noise between two successive x-ray pulses. This noise cancels in the difference spectrum in Fig. 4b leading to an improved sensitivity. The pump-probe time delay was preset to ca. 200 ps in order to observe the chemical shift by the product species, whose difference signal should follow the solid curve in Fig. 4b.

Although the noise on the difference spectrum was < 0.1 % relative to the edge we did not observe a photoinduced signal, indicating that the photoexcitation yield was below 0.4 % within the probe beam. Extensive studies with intense pulsed UV beams show that the spatial intensity profile of the exciting laser is crucial to achieve maximum photolysis. Under ideal conditions we can excite 5 – 10 % of the sample. However, a poor laser beam profile with hot spots can decrease this value by more than an order of magnitude, which in this case remains below our sensitivity limit. Fig. 4 c and 4 d show the same experiment, but now with a pump-probe delay of 100 ns. Hereby we recorded the accumulated dose of 75 multibunches following the excitation laser pulse thus acquiring an improved sensitivity. On this time scale we do observe a pump-probe signal, but it results most likely from sample evaporation. Indeed, on the microsecond time scale we can easily measure the increased transmission through the jet due to laser evaporation, which also serves us as a convenient overlap monitor between the laser and x-ray beams.

The achieved sensitivity with our new technique is quite remarkable. The noise relative to the edge of ca. 0.1 % (for 100 ps temporal resolution), and 0.025 % (for 100 ns temporal resolution) correspond to an effective accumulated number of 10^8 to 10^9 x-ray photons, respectively. This implies that we can attempt to measure photo-

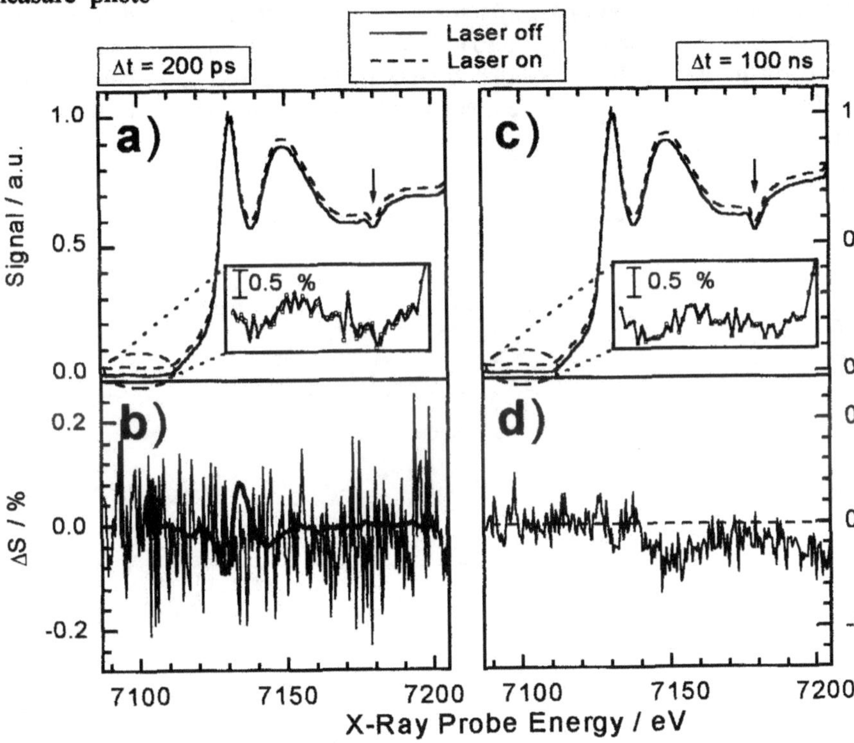

Figure 4. Time-resolved XAS on aqueous hexacyanoferrate (0.25 mol/l concentration and 0.1 mm thickness). a) Transmission spectra of the sample with (dashed) and without (solid) laser at a pump-probe time delay of 200 ps, inset enlarges the pre edge region, arrow denotes an experimental artifact (monochromator glitch). b) Difference spectrum between pumped and unpumped sample together with the expected difference spectrum due to the chemical shift. c) and d): same as a) and b), but with a larger pump-probe time delay (100 ns) due to gating 75 multibunches and acquiring better statistics exhibiting a laser-induced change (d).

induced structural kinetics on many condensed phase chemical systems. We have performed extensive model calculations of the expected pump-probe signal for a given experiment [12,13]. These calculations deliver a quantitative strategy of how to select the experimental conditions (e.g., optimum sample concentration, thickness, choice of solvent) as a function of laser pump parameters (pulse intensity and wavelength). With this simulation package we can assess the feasibility of virtually any kind of time-resolved EXAFS experiment, and Fig. 5 shows the range for some of these as a function of the degree of excitation within the probe beam. The lower boundary was calculated for one of the proposed chemical systems experiments, while the upper boundary was calculated for measuring high-quality time-resolved EXAFS in a dilute myoglobin system. Therefore, most condensed phase systems can be found in the shaded region, and the simple chemical systems can be found near the lower boundary. Our currently achieved sensitivity (ca. 10^9 accumulated x-ray photons) comfortably covers a wide class of possible experiments, if we assume that the laser can excite 1-10 % of the sample species (Fig. 5).

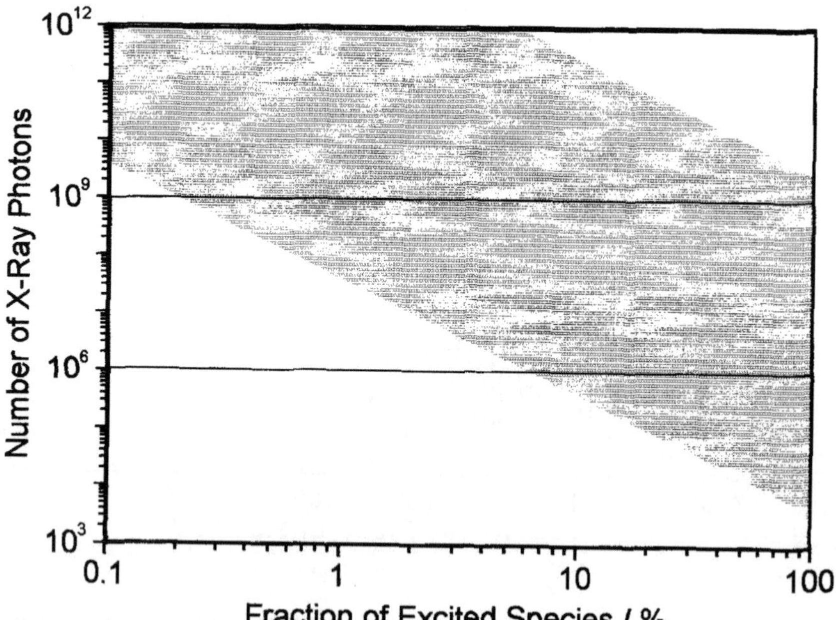

Figure 5: Feasibility range for time-resolved structural research exploiting x-ray absorption spectroscopy. The shaded region indicates the required x-ray flux as a function of the fraction of activated species for most condensed phase systems ranging from simple chemical systems (lower boundary) to ligand dynamics measured in dilute biological systems (upper boundary).

4 Conclusion

With the current progress in improving the experimental scheme we can envision the study of various photochemical reactions of potentially interesting systems including biomolecular functions. Advancing into these experimentally challenging systems is important also in view of the aforementioned evolving new ultrafast x-ray sources. While the time-slicing scheme is possible today at the ALS [8] and soon at the SLS [19], their anticipated flux is still far below those expected from future sources. Among those, X-FELs [14] and ERLs [15] promise to allow condensed phase dynamics research with x-rays to become a major field in the physical and life sciences.

5 Acknowledgements

This work was funded by the *Swiss National Science Foundation* (FNRS) via contract FN59146.99, by the *Paul Scherrer Institut* (Villigen, Switzerland), and by the ALS-Berkeley (USA). We thank Phil Heimann and Bob Schoenlein from the ALS for many fruitful discussions as well as for their support during the beamtime periods at the ALS.

References

[1] B. E. Warren, *X-Ray Diffraction*, Dover Publishers Inc., New York (1980)
[2] E. A. Stern, S. M. Heald, in: *Handbook on Synchrotron Radiation*, Vol. 1B, Ed: E. E. Koch, North Holland (1983), pp. 955 fff
[3] B. K. Agarwal, *X-Ray Spectroscopy*, Springer (1991)
[4] U. Buontempo, A. Filipponi, P. Postorino, R. Zaccari, J. Chem. Phys. **108**, 4131 (1998)
[5] J. Larsson, Z. Chang, E. Judd, P. J. Schluck, R. W. Falcone, P. A. Heimann, H. A. Padmore, H. C. Kapteyn, P. H. Bucksbaum, M. M. Murnane, R. W. Lee, A. Machacek, J. S. Wark, X. Liu, B. Shan, Optics letters **22**, 1012 (1997)
[6] J. Larsson, A. Lindenberg, P.A. Heimann, P.H. Bucksbaum, R.W. Lee, J.S. Wark, R.W. Falcone, Appl. Phys. A: Mater. Sci. Process. **66**, 587 (1998)

[7] A. M. Lindenberg, I. Kang, S. L. Johnson, T. Missalla, P. A. Heimann, Z. Chang, J. Larsson, P. H. Bucksbaum, H. C. Kapteyn, H. A. Padmore, R. W. Lee, J. S. Wark, R. W. Falcone, Phys. Rev. Lett. **84**, 111 (2000)
[8] R. W. Schoenlein, S. Chattopadhyay, H. H. W. Chong, T. E. Glover, P. A. Heimann, C. V. Shank, A. Zholents, M. Zolotorev, Science **287**, 2237 (2000)
[9] F. Ráksi, K. R. Wilson, Z. Jiang, A. Ikhlef, C. Y. Côté, J.-C- Kieffer, J. Chem. Phys. **104**, 6066 (1996)
[10] J. L. Dehmer, J. Chem. Phys. **56**, 4496 (1972)
[11] L. X. Chen, W. J. H. Jäger, G. Jennings, D. J. Gosztola, A. Munkholm, J. P. Hessler, Science **292**, 262 (2001)
[12] C. Bressler, M. Saes, M. Chergui, P. Pattison, R. Abela, Nucl. Instrum. Meth. **A 467-468**, 1444 (2001)
[13] C. Bressler, M. Saes, M. Chergui, D. Grolimund, R. Abela, P. Pattison, submitted to: J. Chem. Phys. (2001).
[14] J. Feldhaus, E. L. Saldin, J. R. Schneider, E. A. Schneidmiller, M. V. Yurkov, Opt. Comm. **140**, 341 (1997); see also: TESLA Technical Design Report, Part V, The X-Ray Free Electron Laser, Eds., G. Materlik, Th. Tschentscher (2001), downloadable at: http://tesla.desy.de/new_pages/TDR_CD/PartV/xfel.pdf
[15] see, e.g., the Cornell ERL website: http://erl.chess.cornell.edu/, and the online publication http://erl.chess.cornell.edu/Papers/WhitePaper_v41.pdf
[16] M. Roeselova, U. Kaldor, P. Jungwirth, J. Phys. Chem. A104, 6523 (2000)
[17] S. Pommeret, R. Naskrecki, P. van der Meulen, M. Ménard, G. Vigneron, T. Gustavson, Chem. Phys. Lett. **288**, 833 (1998)
[18] T. E. Westre, P. Kennepohl, J. G. DeWitt, B. Hedman, K. O. Hodgson, E. I. Solomon, J. Am. Chem. Soc. **119**, 6297 (1997)
[19] G. Ingold *et al.*, submitted to: Proc. Of the XXI ICFA Beam Dynamics Workshop and Laser Beam Interaction, Brookhaven National Laboratories (June 2001)

SOME PROSPECTS OF EUV-X-RAY FEMTOSECOND PULSES IN CHEMICAL PHYSICS

V.S. LETOKHOV
Institute of Spectroscopy of Russian Academy of Sciences, Troitsk, Moscow region, 142190, Russia,
E-mail: letokhov@isan.troitsk.ru

Recent impressive progress of high-order harmonics generation of high-power femtosecond laser pulses opened door to the femto/attosecond coherent VUV-X-ray pulses and their numerous potential applications in science and technology. I consider the possibility of photoelectron spectroscopy for chemical analysis with high spatial resolution and chemical selectivity (ESCA) using coherent femtosecond EUV-X-ray laser pulses. I discuss the trends of the current laser femtosecond instrumental revolution and extension of an electromagnetic oscillators to the X-γ -ray spectral range.

1 Introduction

The papers presented at this conference, especially the preceding papers on the *Techniques* section, show that we are witnessing a "femtosecond instrumental revolution" (Fig 1). This revolution has become possible thanks to the of important achievements: 1) radical simplification of the CW generation of stable femtosecond pulses [1] owing to the use of titanate sapphire crystals [2] and Kerr-lens mode-locking (KLM) effect therein [3] and 2) substantial simplification of the amplification of coherent fs-pulses by means of of their pre-amplification stretching and post-amplification compression (CPA regime) [4].

Femtosecond laser pulses of comparatively low energy made it possible to develop, in particular, femtosecond chemistry [5] that have been the subject matter of a series of conferences starting in 1993. Femtosecond pulses amplified to ultrahigh intensities enable one, first, to obtain high-order femtosecond harmonics in the EUV and soft X-ray regions [6, 7]. This is of principal importance, for this method makes it possible to transfer coherence from the optical to the X-ray region of the spectrum. In that case, there is no need to make radiation coherent by means of positive feedback, this being very difficult to implement in the X-ray region on a femtosecond time scale. What is more, it seems possible to produce sub-femtosecond (attosecond) radiation pulses along these lines [8]. Secondly, intense femtosecond laser pulses in the optical region allow one to produce and control ps–fs electron pulses [9-11]. This possibility has recently been demonstrated in the experiments on the generation of femtosecond X-ray synchrotron radiation pulses [12]. One can foresee tremendous prospects for femto/attosecond X (γ)-ray science and technology. In what follows, I will restrict myself to the qualitative discussion

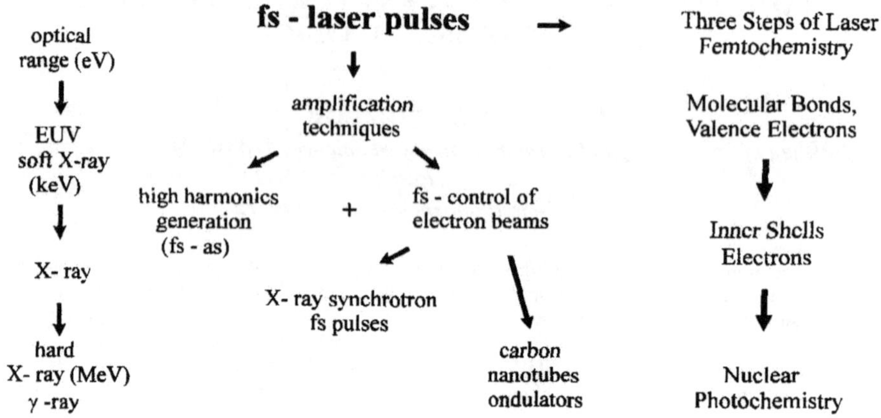

Figure 1. Laser femtosecond instrumental revolution.

of one such possibility based on the joint use of optical and X-ray femtosecond laser pulses.

2 The Problem and Possible Ways to Solve it

Gaining information on the molecular structure of individual biomolecules, especially such as DNA and proteins, is a very important problem facing the physicists developing new techniques and instruments for other domains of science (one can cite as an example X-ray crystallography, electron microscopy, etc.). I became interested in his problem a quarter of a century ago, when I suggested that a laser microscope be created using selective multiple-photon excitation and ionization by ultrashort laser pulses [13, 14]. Wet chemistry methods have since been developed using the works with many DNA replicas that led to the sequencing of the human genome. The original problem, however, still remains to be solved. First, many biomolecules cannot be crystallized, and so X-ray crystallographic techniques cannot be used with them. Secondly, many biomolecules are lacking so simple a structure of recurring molecular blocks (bases), and so the laborious wet chemistry techniques developed are inapplicable to them. This problem of the urgent need to create a "table-top DNA sequencer" and a "table-top protein 3D-microscope" has recently been stressed by Dyson [15]. Therefore, it makes sense to discuss the possibilities of solving this problem that have been opened up in connection with the "femtosecond instrumental revolution" (Fig. 1).

The requirements for the method of solving the above problem are exceptionally stringent: (1) atomic specificity, (2) ultrahigh sensitivity, and (3) nm–Å spatial (lateral, longitudinal) resolution. Combining all the above characteristics in one method would make it possible to determine the 3-D molecular structure of individual biomolecules.

The following methods can be regarded as potential. First, these include the well-known classical X-ray diffraction and electron diffraction techniques, each being capable of being extended to atomic-resolution holography (see, for example, [16–20]). These techniques, however, require that the specimens be either crystals or ensembles of oriented molecules. Moreover, for these techniques to be implemented, it is necessary that a many X-ray photons or electrons in individual shots be coherent. The latter requirement requires the intensity of the X-ray radiation that can cause damage to the structure of the specimens on a femtosecond time scale [21].

An alternative is to extend the well-known ESCA technique [22]. This technique is incoherent and can be used with individual biomolecules, and information can be accumulated in the course of many irradiation pulses of not very high intensity. Besides, the photoionization cross section of atoms exposed to X-rays is much greater than the scattering cross-section:

$$\sigma_{\text{ph.ion}} >> \sigma_{\text{X-ray scat.}} \tag{1}$$

Of course, to obtain a 2D image of the surface under study, one have to refuse the traditional electron spectroscopy techniques, i.e., one whould use the time-of-flight method for selecting photoelectrons. It is exactly for this purpose that the possibilities of using femtosecond optical and X-ray pulses jointly prove very suitable.

3 Femtosecond ESCA 2D-Microscopy

Figure 2 is a simplified illustration of the idea of ESCA microscopy with the atoms of the desired element in an individual molecule being ionized by focussed X-ray pulses having a energy of $h\nu_X$. The focal spot diameter d_{foc} is much greater than the wavelength λ_X of the X-ray radiation and the size a of the individual molecule of interest containing N ionizable atoms. The ionization probability of one such atom, $Y_{\text{ph.ion}}$, which depends on the fluence of the X-ray radiation pulse, $\Phi_X/h\nu_X$, and the photoionization cross section $\sigma_{\text{ph.ion}}$ ($Y_{\text{ph.ion}} \simeq \sigma_{\text{ph.ion}}\Phi_X$), is taken to be low ($< 1/N$). In that case, one X-ray pulse gives birth to no more than one photoelectron ($NY_{\text{ph.ion}} \lesssim 1$). This excludes the Coulomb repulsion of the photoelectrons produced with a low kinetic energy of $E_{\text{ph.el.}}$. An electron-optical system images these photoelectrons on the surface of a position-sensitive time-resolved (time-of-flight) electron detection system.

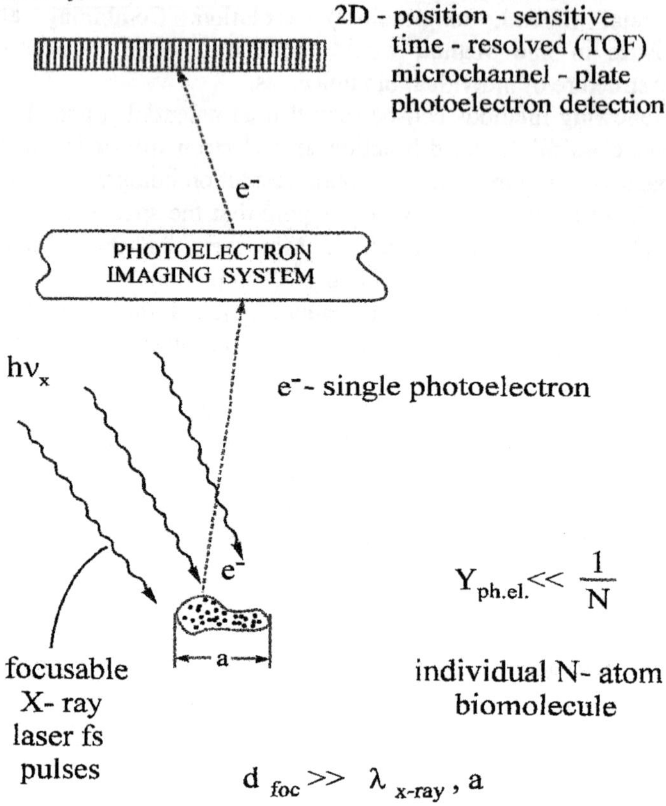

Figure 2. The simplified general idea of femtosecond ESCA 2D-microscopy.

By appropriately selecting the X–ray quantum energy $h\nu_X$ one can ensure the necessary atomic specificity of photoionization [22].

Ultrahigh sensitivity can be attained without damaging the biomolecule under study by the focused X-ray pulses by taking their energy fluence adequately low. Consider as an example the case where the photoionization cross section is $\sigma_{ph.ion} \simeq 10^{-20}$ cm^2 ($h\nu_X \lesssim 1$ keV). When an X-ray pulse with an energy of $E \simeq 1$ nJ in the spectral range $\Delta E \simeq 1$ eV is focused onto an area the size of $a \simeq 10$ μm (this provides for $h\nu_X\Phi_X \simeq 1$ mJ/cm^2·pulse·eV), the ionization probability of a single desired atom may come to some 10^{-7} photoelectron/atom·pulse, and for a molecule with $N \approx 10^5$ atoms, it will be $Y_{ph.ion} \approx 10^{-2}$ photoelectron/molecule·pulse. The energy of the X-ray pulse can probably be raised by a factor of 10^2 (up to a level of 0.1 J/cm^2·pulse·eV) in order to provide for the production of about one photoelectron per pulse in the biomolecule.

Photoelectrons are produced at the X-ray penetration depth that is much greater than the monolayer thickness. To achieve surface (monolayer) selectivity, it is necessary to eliminate the background noise due to the photoelectrons formed throughout the bulk of the specimen. To this end, use can be made of the effect of the photoelectron escape depth minimum $E^{min}_{ph.el.} \approx 30$ eV [23]. Selective detection of surface photoelectrons can be effected by at least two methods. First, use can be made of X-ray radiation with an energy of $h\nu_X = I + E^{min}_{ph.el.}$. This may be a monochromatized radiation of higher harmonics of laser pulses within a wide X-ray spectral range. Secondly, one can perform selective detection of photoelectrons with a kinetic energy of $E_{ph.el.} \approx 30$ eV. In our case of constructing a photoelectron image, this can be attained by means of time-of-flight (TOF) photoelectron detection. The selection of photoelectrons having the necessary energy $E_{ph.el.} \approx 30$ eV (velocity $v_e \approx 3\times10^8$ cm/s) in the energy range $\Delta E_{resol.}$ over the photoelectron flight path length $L_{el.fl.} \approx 1$mm requires that the TOF system used should have a very high time resolution, at a level of $\tau_{resol.} \approx 10^{-12}$ s. In principle, this is quite possible, provided that use is made of some of the energy of the initial high-intensity optical femtosecond pulses generating the X-ray pulses. This possibility is based in particular on the reflection of electrons from the strong light field of the femtosecond optical pulses [10].

4 Concept of Femtosecond Photoreflection of Electrons

Figure 3 schematically illustrates the idea of reflection of electrons by the strong field of an evanescent light wave, as suggested in [10]. The intense evanescent wave formed on the total internal reflection of a femtosecond light pulse penetrates into the vacuum to a depth of the order of the wavelength λ. The change in the electron velocity can be described in terms of the effective index of refraction, $n_{el} < 1$ [10]:

$$n_{el} = \left[1 - \left(\frac{\mu^2}{\beta^2}\right)\sqrt{1-\beta^2}\right]^{1/2}, \qquad (2)$$

where $\beta = v_e / c$, $\mu^2 = \frac{r_e}{mc}\lambda^2 I$, v_e is the electron velocity, r_e is classical electron radius, I is intensity, and λ is wavelength. Figure 4 shows n_{el} as a function of the intensity of laser radiation differing in wavelength for electrons with an energy of $E_{ph.el.} \approx 10$ eV. At the intensity in the range 10^{13}–10^{14} W/cm^2 n_{el} drops perceptibly.

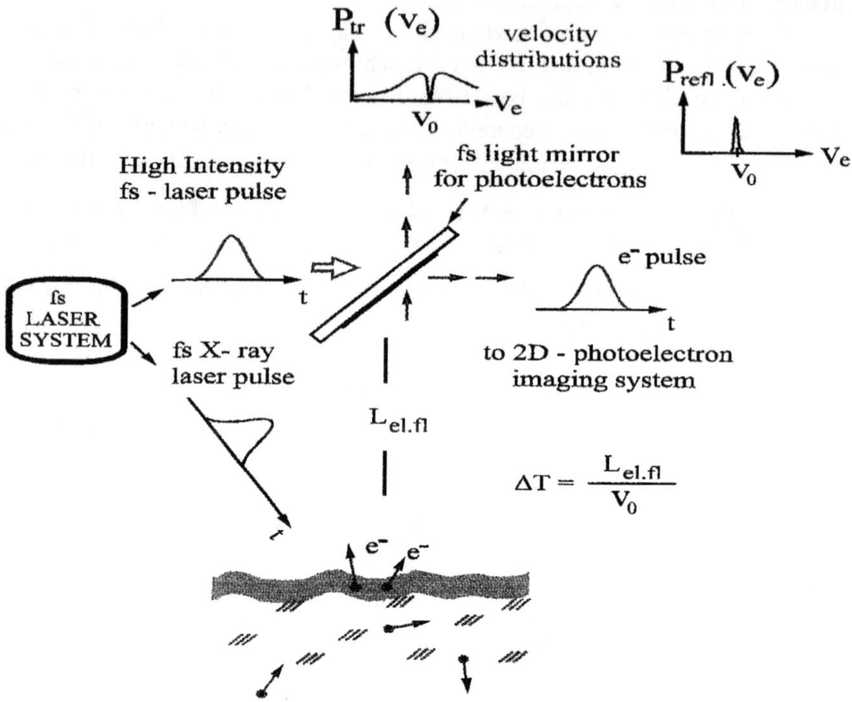

Figure 3. General idea of time-of-flight photoreflectron of photoelectrons by highly-intense femtosecond evanescent light wave.

Part of the powerful femtosecond pulse generating higher harmonics in the X-ray region is also capable of effectively reflecting electrons with the specified velocity v_0, whose time of flight from the surface, ΔT, is controlled by the time interval between the X-ray and the optical pulse. The reflected photoelectrons with a narrow velocity distribution $P_{refl.}(v_e)$ should be directed into the imaging system used for 2D-visualization purposes.

Naturally the photoreflection concept presented here is of rather qualitative character because this effect has not as yet been observed experimentally. However, it is of certain interest within the framework of "laser-induced electron optics" [9-11]. Moreover, this effect features some useful properties that are qualitatively illustrated in Fig. 5. Besides the possibility of velocity selection discussed above, it possesses the property of compensation of the velocity dispersion of the reflected electrons. Indeed the slower electrons (at the red edge of the peak of the distribution $P_{refl.}(v_{el.})$) must suffer reflection in a less intense light field of the molecules and cover a shorter distance. The opposite is true for the faster electrons. By using a curved evanescent wave, one can, in principle, attain electron beam

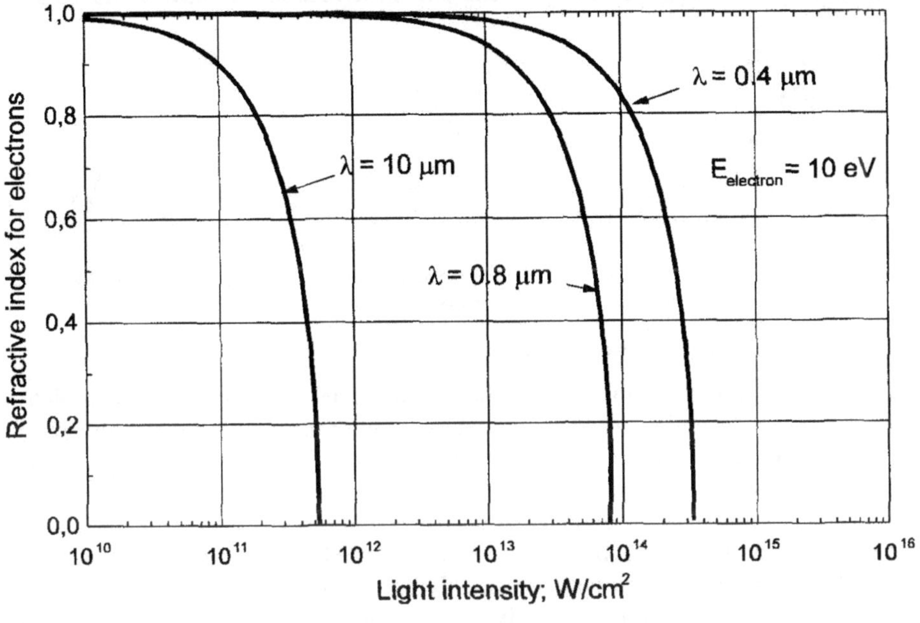

Figure 4. Change of effective refractive index n_e for the electrons with energy $E_{electrons}=10$ eV in presence of laser wave with intensity I (W/cm^2).

focusing simultaneously with reflection[1]. Of course, all these potentialities of "laser-induced" reflective electron optics should be the subject matter of future theoretical and experimental studies.

5 Discussion

The concept of the TOF-selection of ejected photoelectrons through femtosecond photoreflection allows one, in principle, to solve the problem of detection of photoelectrons coming from a thin surface layer. The problem of attaining the high longitudinal resolution necessary to realize 3D-microscopy with an atomic-scale resolution still remains to be solved for the task set in Sect. 2. It is

[1] In connection with the development of hollow waveguides with high-intensity femtosecond light pulses propagating therein [24], it is possible that waveguide propagation of electrons can also be observed in such waveguides.

quite possible that one will have to use the approach based on consecutive observation of thin surface layers, as demonstrated in [25].

As for the lateral spatial resolution, here again one is up against serious restrictions. In our experiments with one-photon ionization, we have achieved a

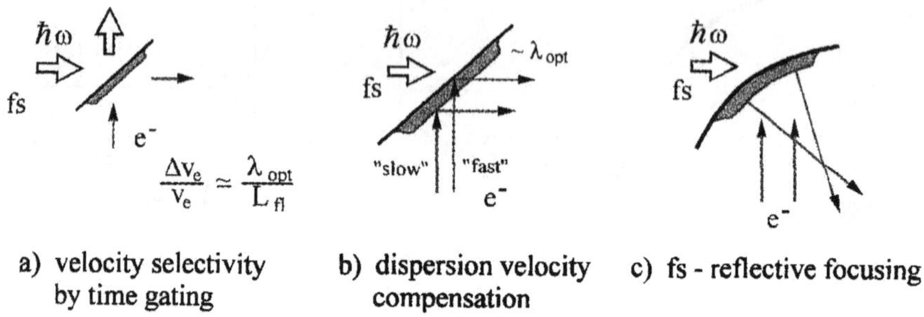

Figure 5. Valuable properties of photofrelection of electrons.

resolution of around 30 nm [26], and the use of a sharp needle tip and two-photon femtosecond-pulse excitation has made it possible to improve resolution to some 5 nm [27]. Progress in the development of photoelectron microscopy allows one to hope that a resolution of the order of 2 nm will be attained [28].

Therefore, the method considered here is potentially suitable for the solution of only the first part of the problem set in Sect. 2, namely, the table-top sequencing of DNA-like chain molecules with irregularly recurring molecular blocks spaced at some 1 nm on the average. To solve the entire problem, it would probably be necessary to combine several methods in a single setup.

6 Conclusion

It is interesting to consider the progress made in the production of femtosecond pulses in the EUV and X-ray regions of the spectrum within the framework of the more than century-long progress in obtaining electromagnetic oscillations of ever increasing frequency. Figure 6, which is an extension of the diagram presented in [29], illustrates advancement into the X-ray region owing to the generation of higher harmonics of femtosecond pulses [6, 7] and realization of the X-ray laser [30]. The further progress of X-ray lasers into the short-wavelength region is being effected by using a radiation generation regime wherein high-power ultrashort pulses producing population inversion in multiply-charged ions propagate synchronously with the X-ray pulses being generated. Such a traveling-wave excitaion mode was considered for the EUV region as far back as 1971 [31] and experimentally realized in [32]. This feedback-free coherent radiation generation regime is referred to as saturation amplification of spontaneous emission (SASE).

It is anticipated that the same regime will be implemented in the project on the generation of radiation in the 1–10-Å region with the free relativistic electron laser [33, 34].

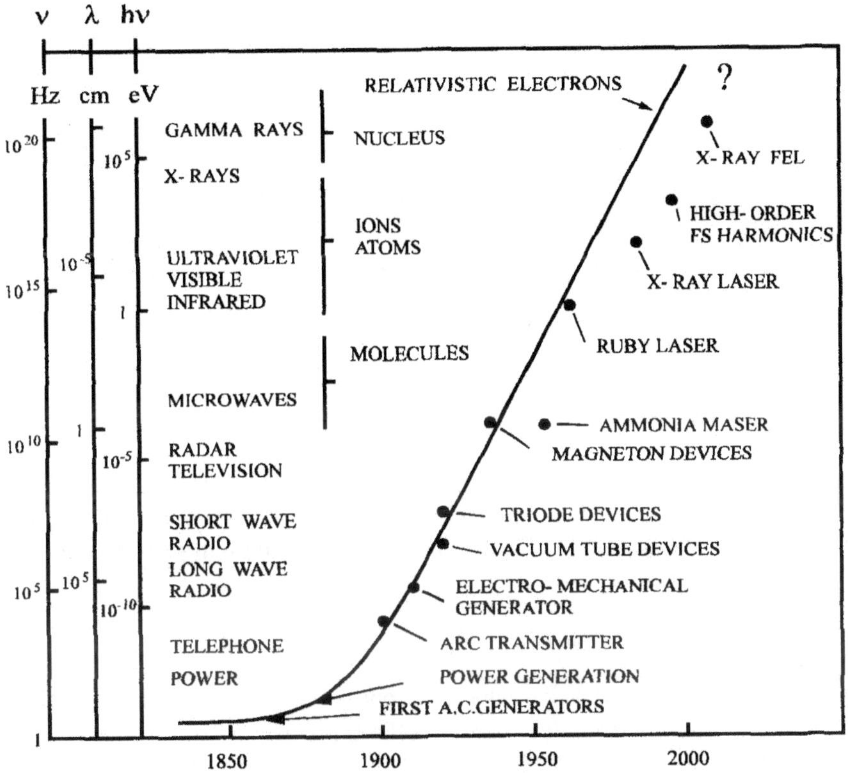

Figure 6. Progress in generation of electromagnetic coherent wave by various techniques.

In their turn, ultrashort relativistic electron pulses open up the principal possibility of using molecular-scale undulators (in the form of cable or crystal composed of carbon nanotubes) to obtain highly directional hard X (γ)-ray beams in the 100 keV-10 MeV energy range [35-37]. At the moment this possibility seems not very remote, for there have recently been produced self-assembling crystals of oriented single-wall carbon nanotubes [38] as shown on Fig. 7 electron pulses [11]. Such ultrashort electron buches in a low-density crystal (cable) of carbon nanotubes should be less damaging. It is not inconceivable that in the future femtosecond high-energy electron pulses will again be produced with the aid of ultrahigh-intensity femtosecond light pulses. Perhaps it will prove possible to extend the laser femtosecond instrumental revolution along this line to the hard X-

and γ-ray regions of the spectrum and extend the molecular femtochemistry to nuclear photochemistry (Fig. 1) [39].

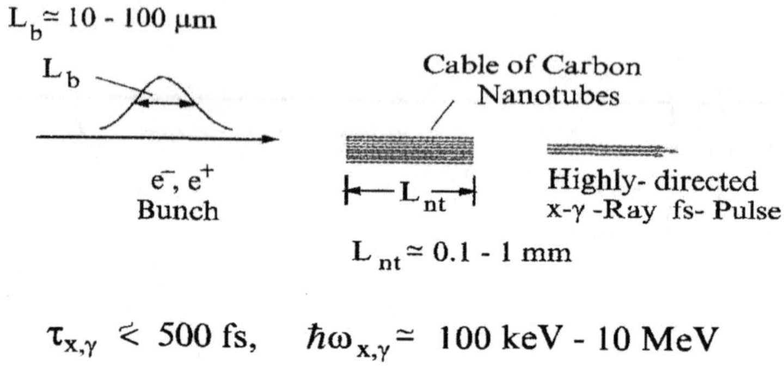

Figure 7. Potential possibility of production of hard X-ray and γ-ray cable of carbon nanotube as undulator.

7 Acknowledgments

I thank Prof. A. Douhal for his kind invitation for me to participate in this conference and give a talk on "Some Prospects" in the beautiful and inspiring Toledo. I am also grateful to Dr. M. Subbotin for his help in preparing Fig. 4.

References

1. Shank C.V., Ippen E.P., Subpicosecond kilowatt pulses from a mode-locked CW dye laser, *Appl. Phys. Lett.* **24** (1974) pp. 373–375.
2. Moulton P.F., Spectroscopic and laser characteristics of Ti:Al_2O_3, *J. Opt. Soc. Am.* **B3** (1986) pp.125–133.
3. Spence D.E., Kean P.N., Sibbett W., 60-fsec pulse generation from a self-mofe-locked Ti:sapphire laser, *Opt. Lett.* **16** (1991) pp. 42–44.
4. Strickland D. and Mourou G., Compression of amplified chirped optical pulses, *Opt. Comm.* **56** (1985) pp. 219–221.
5. Zewail A. Femtosecond Chemistry (World Scientific, Singapore, 1994).
6. L'Huillier A., Balcou Ph., High-order harmonic generation in rare gases with 1-ps 1053 nm laser, *Phys. Rev. Lett.* **70** (1993) pp. 774–777.
7. Guo T., Spielmann C., Barty C.P.F, Generation of hard X-ray by ultrashort terawatt lasers, *Rev. Sci. Instr.* **72** (2001) pp. 41–47.
8. Drescher M., Hentschel M., Kienberger R., Tempea G., Spielmann C., Reider G.A., Corkum P.B., Krausz F., X-ray approaching the attosecond frontier, *Science* **15** (2001) pp. 1923–1927.

9. Letokhov V.S., Focusing of electron beam by dipole force of ultrashort laser pulse, *JETP Lett.* **61** (1995) pp. 805–807.
10. Balykin V.I., Subbotin M.V., Letokhov V.S., Reflection of an electron beam by femtosecond light waves, *Optics Comm.* **129** (1996) pp. 177–183.
11. Minogin V.G., Fedorov M.V., Letokhov V.S., Formation of ultrashort electron pulses on scattering of an electron beam by a standing laser wave of ultrashort duration, *Opt. Comm.* **140** (1997) pp. 250–254.
12. Schoenlein R.W., Chattopadhyay S., Chong H.H.W., Glover T.E., Heimann P.A., Shank, S.V., Zholents A.A., Zolotarev M.S.,Generation of femtosecond pulses of synchrotron radiation, *Science* **287** (2000) pp. 2237–2340.
13. Letokhov V.S, The use of laser radiation in autoelectron and autoion microscopy for the observation of biomolecules, *Sov. J. Quant. Electr.* **5** (1975) pp. 506–510.
14. Letokhov V.S, Future applications of selective laser photophysics and photochemistry. In Proceed. of the Conference *Tunable Lasers and Their Applications*, ed. by Mooradian, T. Jaeger and P. Stokeseth (Springer, Berlin, 1976), pp. 122–139.
15. Dyson F. The Sun, Genome and Internet (Oxford Unuversity Press, 1999).
16. Szoke A., Time-resolved holographic diffraction at atomic resolution, *Chem. Phys. Lett.* **313** (1999) pp.777–788.
17. Rousse A., Rischel C., Gauthier J.-C., Femtosecond X-ray crystallography, *Rev. Mod. Phys.* **73** (2001) pp. 17–31.
18. Moffat K., Time-resolved biochemical crystallography: a mechanistic perspective, *Chem. Rev.* **101** (2001) pp. 1569–1581.
19. Fink H.-W., Stocker W., Schmid H., Holography with low-energy electrons, *Phys. Rev. Lett.* **65** (1990) pp. 1204–1206.
20. Len M., Denlinger J.D., Rotenberg E., Kevan S.D., Tonner B.P., Chen Y., Van Hove M.A., Fadley C.S., Holographic atomic images from surface and bulk W(110) photoelectron diffraction data, *Phys. Rev.* **59B** (1999) pp.5857–5870.
21. Doniach S., Studies of the structure of matter with photons from an X-ray free-electron laser, *J. Synchrotron Rad.* **3** (1996) pp. 260–267.
22. Siegbahn K., Nordling C., Fahlman A., Nordbeng R., Hamrin K., Hedman J., Johansson G., Bergmark T., Karlsson S.-E., Lindgren I., Lindberg B., *ESCA: atomic, molecular and solid structure studied by means of electron spectroscopy*, Nova Acta Regie Soc. Sci., Ups. **4** (1967) pp.20; K. Siegbahn, Nobel Lecture (1981).
23. Briggs D., Seah M.R., *Practical Surface Analysis* , 2-nd ed. (J. Wiley, Chichester, England, 1990).

24. Knight J.C., Birks T.A., Russel P.S.J., Atkin D.M., All-silica single-mode optical fiber with photonic crystal cladding, *Opt. Lett.* **19** (1996) pp. 1547–1549.
25. Magerle R., Nanotomography, *Phys. Rev. Lett.* **85** (2000) pp. 2749–2752.
26. Konopsky V.N., Sekatskii S.K., Letokhov V.S., Laser resonance photoelectron microscopy with a subwavelength-high spatial resolution: the first observation of a single color-centers on surface, *Optics Comm.* **132** (1996) pp.251–256.
27. Sekatskii S.K., Chekalin S.V., Ivanov A.L., Matveetz Yu.A., Stepanov A.G., Letokhov V.S., Femtosecond two-photon laser photoelectron microscopy, *Journ. Phys. Chem.* **A102** (1998) pp.4148–4153.
28. Smith N., Science with soft X-rays, *Physics Today* **29** (January 2001) pp. 29–34.
29. Baldwin G.C., Solem J.C., Comment of γ-laser prospects, *Laser Focus* **7** (June 1982) pp.8–9.
30. Matthews D.L., Hagelstein P.L., Rosen M.D., Eckart M.J., Ceglion N.M., Hazi A.U., Medecki H., MacGowan B.I., Trebes J.E., Whitten B.L., Campbell E.M., Hatcher C.W., Hawryluk A.M., Kauffman R.L., Pleasance L.D., Rambach G., Scofield J.H., Stone G., Weaver T.A., Demonstration of soft X-ray amplifier, *Phys. Rev. Lett.* **54** (1985) pp.110–113.
31. Knyazev I.N., Letokhov V.S., Excitation of far vacuum UV lasers by fast heating of plasma electrons in ultrashort pulsed optical field, *Optics Comm.* **3** (1971) pp. 332–334.
32. Nickles P.V., Shlyaptsev V.N., Kalachnikov M.P., Schnurer M., Will I., Sandner W., Short pulse X-ray laser at 32.6 nm based on transient gain in Ne-like titanium, *Phys. Rev. Lett.* **78** (1997) pp. 2748–2751.
33. Tatchyn R., Arthur J., Boyce R., Fasso A., Montgomery J., Vylet V., Walz D., Yotam R., Freund A.K., Howells M., X-ray optics design studies for the SLAC 1.5-15 Å Linac Light Source (LCLS), *Nucl. Instr. Meth.* **A429** (1999) pp. 397–406.
34. Elleaume P., Chavanne J., Faatz B., Design considerations for a 1 Å SASE undulator, *Nucl. Instr. Meth.* **A455** (2000) pp. 503–523.
35. Klimov V.V., Letokhov V.S., Hard X-radiation emitted by a charge particle moving in a carbon nanotube, *Phys. Lett.* **A222** (1996) pp. 424–428.
36. Klimov V.V., Letokhov V.S., Monochromatic γ-radiation emitted by a relativistic electron moving in a carbon nanotube, *Phys. Lett.* **A226** (1997) pp. 244–252.
37. Klimov V.V., Letokhov V.S., Hard highly directional X-radiation by a charged particle moving in a carbon nanotube. In *Atomic Physics Methods in Modern Research*, ed. by K. Jungman, J. Kowalski, I. Peinard, F. Trager (Berlin, Springer, 1997) pp. 167–184.

38. Schlitter R.R., Seo J.W., Gimzewski J.K., Durkan C., Saifullah M.S.M., Welland M.E., Single crystals of single-walled carbon nanotubes formed by self-assembly, *Science* **292** (2001) pp. 1136-1139.
39. Letokhov V.S., Ultrafast processes: from past to the future, In *Femtochemistry and Femtobiology: Ultrafast Reaction Dynamics at Atomic-Scal Resolution*, ed. by V. Sundström. Proceed. of Nobel Symposium, 1996, Bjorkborn (Imperial College Press, 1997) pp.755–763.

CONTROLLING THE VIBRATION AND DISSOCIATION DYNAMICS IN TRIATOMIC ALKALINE CLUSTERS

ŠTEFAN VAJDA*, COSMIN LUPULESCU, ANDREAS BARTELT, FILIP BUDZYN, PORFIRIO ROSENDO-FRANCISCO AND LUDGER WÖSTE

Institut für Experimentalphysik, Freie Universität Berlin, Arnimallee 14, D-14195 Berlin, Germany

E-Mail: vajda@physik.fu-berlin.de

This paper is devoted to the control of the nuclear dynamics in triatomic clusters by using tailored femtosecond laser pulses. First the control of the vibrational dynamics in bound electronic states of the homonuclear trimer Na_3 is presented, then the branching control of the predissociative 3-photonic ionization process in the heteronuclear trimer Na_2K is treated.

1 Introduction

The use of pulse-shapers in symbiosis with automized self-learning feedback loop algorithms [1] has opened new perspectives for controlling molecular dynamics in real time. Such an experiment allows a generation of tailor-made laser pulses, which can drive photoinduced -generally multiphotonic- processes at a maximum yield for a desired product along the most effective path on the potential energy surface(s). The first control experiment was performed by Gerber and his coworkers [2], who optimized the profile of an amplified laser pulse, either to maximize the parent ion or a chosen fragment ion yield in a highly dissociative multiphoton ionization process of metal-carbonyl compounds. An important issue remains the question concerning the information content coded in the optimized laser pulse form. In this regard it is advantageous to begin the series of systematic investigations on the simplest model systems, where the underlying molecular dynamics can be understood on the basis of already available experimental and theoretical data.

2 Choice of the molecular systems for control scenarii

Triatomic homo- and heteronuclear sodium-potassium clusters are suitable model systems for the investigation of optimal control scenarii. These clusters exhibit a large number of bound-bound and bound-free electronic transitions, allowing to study their vibrational dynamics and fragmentation kinetics by means of time-resolved multiphoton ionization spectroscopy.

The spectra of the sodium trimer reflect oscillatory wavepacket motions in its bound B-state when excited by a one-photonic transition with a femtosecond pump pulse at a central wavelength of 620 nm and probed in a subsequent one-photonic ionization step with a delayed fs pulse of the same wavelength [3-5]. When employing amplified laser pulses, the oscillatory structure of the pump-probe spectra reflects a motion of a wavepacket propagating in the ground state of this trimer as well (Gerber et al. [6]). Na_3 was chosen as a prototype model system *for controlling the population of different bound electronic states* by means of modulated femtosecond laser pulses of low intensity.

The heteronuclear K_2Na and Na_2K trimers undergo fast photofragmentation on the picosecond time scale when electronically excited in the wavelength region of 730-840 nm [7]. Moreover, in the case of Na_2K excited at 770 nm the residue of an oscillating wavepacket survives [8]. The transient Na_2K^+ signal - obtained in an one-color, one-photonic pump and two-photonic probe process - clearly exhibits an exponential decay with a time constant of 3.28 ps and superimposed oscillations with a period of about 500 fs. The pump-probe spectrum of the NaK fragment dimer recorded at 770 nm reveals an exponential risetime of 3.25 ps and superimposed oscillations with a period of about 440 fs, which corresponds to the known vibrational period of the excited NaK dimer [9]. Since coherence survives during the predissociation, the photodissociation and subsequent ionization process in the coupled Na_2K/NaK cluster system is well suited for an *optimal coherent control* experiment.

3 Experimental setup

Our molecular beam apparatus and the laser setup have been described in detail elsewhere [8]. Briefly, in the Na_3 experiment a commercial TiSa-OPO femtosecond laser system (Spectra Physics Tsunami and Opal) pumped by 15 W argon-ion laser is used, which produces 120 fs laser pulses at a central wavelength of 620 nm after frequency doubling. For the generation of downchirped laser pulses from unchirped 120 fs laser pulses centered at 620 nm a setup consisting of a grating pair with adjustable separation between the two gratings is used.
In the Na_2K experiment a commercial femtosecond laser (Spectra Physics Tsunami) pumped by a 10 W Nd:YVO4 laser (Spectra Physics Millenia X) is employed, which produced pulses of about 80 fs duration at a central wavelength of 770 nm. In the automized optimization experiment, a single input laser beam passes through a pulse shaper setup based on a double-mask liquid crystal spatial light modulator consisting of 2x128 pixels [10], which allows a simultaneous computer controlled phase- and amplitude modulation. The shaped pulses are then focussed on the neutral molecular beam and the current of the mass-selected product cation serves as feedback signal for the optimization algorithm. The optimization algorithm -

based on evolutionary strategies [11] - alters the pulse shaper settings until the ion signal levels off.

4 Experimental results and discussion

Na_3

A large set of Na_3^+ transients was recorded with differently chirped excitation pulses at a central laser wavelength of 620 nm and with an unchirped 120 fs ionization pulse of the same wavelength. Two typical spectra obtained by a 120 fs unchirped and a 400 fs downchirped excitation pulse are shown in Fig.1a and

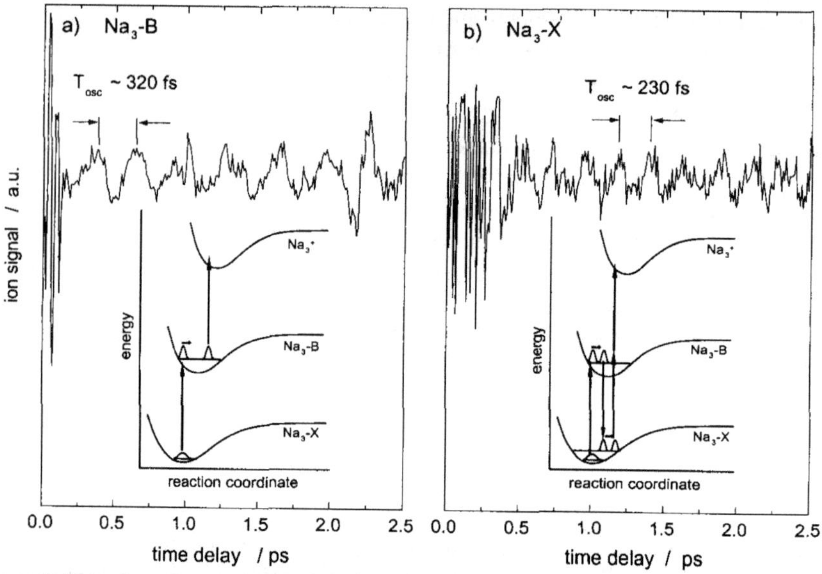

Figure 1. Time-dependent spectrum of Na_3^+ obtained by employing an unchirped 120 fs (a) and a 400 fs (b) downchirped pump pulse and an unchirped 120 fs probe pulse, both with central wavelength of 620 nm. The oscillation periods 320 fs and 230 fs represent the well known symmetric stretch ("breathing") mode of the B- and X-state of Na_3, respectively. The insets represent the schematic of the involved pump-probe (a) and pump-dump-probe (b) processes.

Fig.1b, respectively [4]. The Na_3^+ transient obtained with an unchirped 120 fs excitation pulse clearly reflects the 320 fs oscillatory component arising from the wavepacket propagation in the excited B-state of the trimer. The inset in Fig.1a schematically describes this pump-probe process. By employing a downchirped excitation pulse of 400 fs duration, the oscillation period suddenly changes to 230 fs – to a period characteristic for the wavepacket propagation in the ground state of Na_3 [5]. This observation can be explained by the Tannor-Kosloff-Rice scheme, which is shown in the inset in Fig.1b: The leading (shorter) wavelength components of the excitation pulse induce an evolving wavepacket in the B-state of Na_3, which is then dumped down into the ground state by the delayed longer wavelength components of the same pulse. The vibrational motion of the ground state is then monitored by the unchirped probe pulse via two photon ionization. The duration of the employed pulses was determined by intensity SFG cross-correlation with a 80 fs reference pulse centered at 775 nm, the sign and the order of the chirp from wavelength-resolved SFG cross-correlation measurements.

Na_2K/NaK

Our first series of feedback phase optimization experiments on the Na_2K/NaK system leading to the maximum ion yield of Na_2K^+ or NaK^+ showed, that the optimization algorithm is capable to increase the yield of the desired ion products very effectively [12]. The most important aspect of these experiments was the information content coded in the obtained pulse shapes. The analysis of the power spectra of the optimized pulses gave a detailed insight into the involved photoprocesses and underlying molecular dynamics. Briefly, the obtained pulse forms recovered the one-photon excitation and the two-photon ionization step, the time delay between the pulse elements matched exactly the oscillation period of the chosen cluster. Moreover, the ionization of the electronically excited species took place at the time of passage of the propagating wavepacket through the ionization Franck-Condon window located at the outer turning point of the involved potential energy surface. Though, the analysis of the power spectra in the time domain alone could not explain all features of the obtained pulses, for example, triple-pulse sequences obtained for NaK^+, where - intuitively - a double-pulse sequence could be expected. These observations reflected the high complexity of the involved processes in these even very small and simple molecular systems, and evoked the necessity to extend the searching space by an additional dimension: to study the role of individual frequency components of the laser pulses involved in the process.

We have performed a new series of feedback optimization experiments on the Na_2K/NaK system by modulating the phase of the laser pulses in order to obtain the maximum yield of Na_2K^+ or NaK^+. The obtained optimized pulse shapes were analyzed in the time- and frequency domain. Two typical X-FROG traces of pulses emerging from the optimization experiment for the maximum Na_2K and NaK ion yield, together with the pump-probe Na_2K^+ and NaK^+ signals are shown in Fig.2. The timing between the pulse elements matches exactly the "internal vibrational clock"- the oscillation period T_{OSC} - of Na_2K (500 fs) and NaK (440 fs) very well (Fig.2a and 2b, respectively). Moreover, the individual pulse elements exhibit different color and chirp structure: In the case of the pulse optimized for the maximum Na_2K^+ ion yield (Fig.2a), the first short and unchirped (excitation) pulse is followed by a more intense one, which exhibits a positive chirp in the frequency-time domain. Moreover, the first two pulses are red shifted with respect to the third pulse located at the central laser wavelength of 770 nm. The triple-pulse obtained for the optimization of the maximum NaK^+ ion yield (Fig.2b) is more structured. The first strongly red-shifted short pulse is followed by a longer intense one, which exhibits a complex substructure: The frontal part of this pulse exhibit upchirp, its

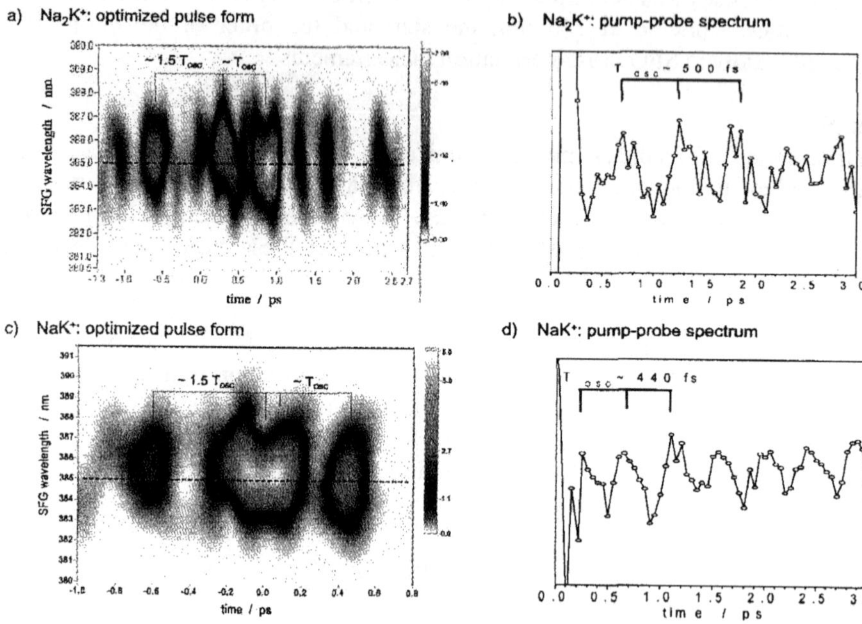

Figure 2a. X-FROG trace the acquired pulse shape for producing maximum yield of Na_2K^+. The dotted line indicates the central wavelength of the employed laser pulse, 770 nm.
Figure 2b. Oscillatory component of the time-resolved 3-photonic Na_2K^+ signal reflecting the asymmetric stretch mode of the electronically excited trimer.
Figure 2c. X-FROG trace of the acquired pulse shape for producing maximum yield of NaK^+.
Figure 2d. Time-resolved 3-photonic NaK^+ signal marked with the oscillation period of the excited dimer.

tail is downchirped. The third pulse is unchirped and located at the central laser wavelength. The substructure of the second pulse reflects the complex molecular dynamics taking place in the region of crossing of potential energy surfaces. One possible mechanism we can take into consideration is, that this pulse navigates the wavepacket around the crossing point of the involved potential energy surfaces so as to avoid the leakage into the Na_2K^+ channel.

In order to test the sensitivity of the optimization experiment to the degree of complexity of the investigated photofragmenting system and underlying processes, we changed the experimental conditions and enriched the beam of sodium-potassium clusters with larger Na_mK_n aggregates photofragmenting in the

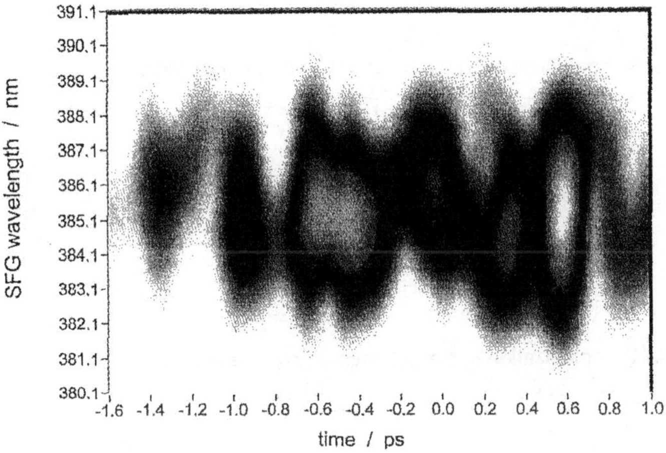

Figure 3. X-FROG trace the acquired pulse shape for producing maximum yield of Na_2K^+ in the presence of larger Na_mK_n aggregates in the molecular beam. The power spectrum of the pulse is shown in Fig. 4a.

wavelength region of the applied laser field [7]. The pulse shape obtained for the maximum ion yield of NaK^+ (see Fig. 3) underwent a dramatic change: A pulse train consisting of six pulses of rather different color, chirp and intensity structure was retrieved. The corresponding mass spectra and the power spectrum of the optimized pulse are shown in Fig.4a. In the presence of larger aggregates, the optimization led to an increase of the NaK^+ yield by about 80 %. The pulse profile reflects the broad spectrum of fragmentation processes leading to the formation of the NaK^+ photoproduct from a large number of precursors. Since the interpretation of such a pulse shape is presently not possible, we decided to decrease the complexity of the system even more by producing a molecular beam containing only NaK dimers but no larger aggregates. Such a system is fragmentation-free where only bound-bound transitions take place, thus only the excitation and subsequent ionization process has to be considered within the picture of the simplest approximation. The corresponding results of an optimization experiment on a NaK beam are shown in Fig.4b. The optimization led to an approximately 30 % increase of the NaK^+ yield. But, to a big surprise, the experiment did not end up with a simple pulse form! The power spectrum of the pulse reveals a triple-pulse structure: The first short pulse is followed by a second complex pulse delayed by 1.5 oscillation periods of the excited NaK dimer. The second pulse is followed by a third pulse after one oscillation period of NaK*. The timing of the pulses corresponds very well to the oscillation period of the excited dimer known from pump-probe data, the intensity ratio of the first and second pulse reflects the one-photon pump and two-photon probe step. Now new questions arose: What is the

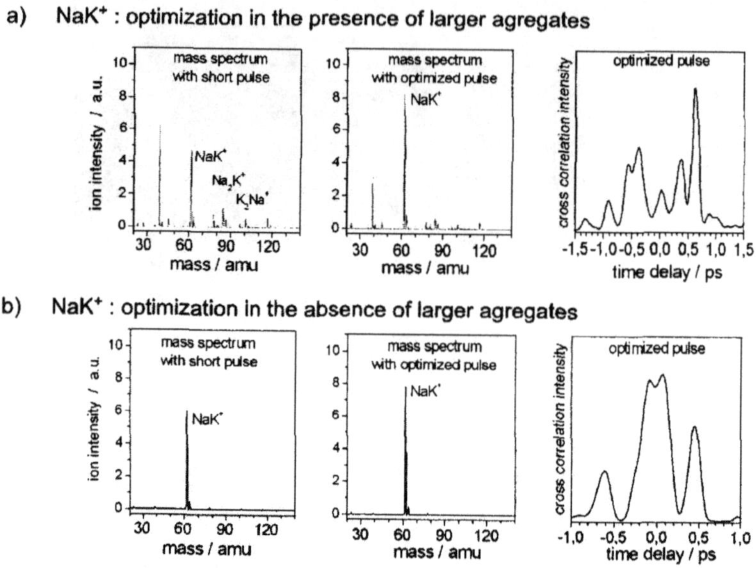

Figure 4a. Results of an optimization experiment for the maximum ion yield of NaK^+ in the presence of larger fragmenting Na_mK_n clusters. When employing the optimized pulse, a high increase (~ 80 %) of the NaK^+ mass peak intensity accompanied by the decrease of the $Na_mK_n^+$ peak intensities is observed. The power spectrum of the resulting pulse exhibits a very complex time sequence.

Figure 4b. Results of an optimization experiment for the maximum ion yield of NaK^+ in the absence of larger Na_mK_n clusters in the molecular beam. The optimized pulse produces an increase of the NaK^+ mass peak intensity by about 30 %. The power spectrum of the resulting pulse reveals a double-pulse substructrure for the central pulse.

origin of the complex substructure of the second pulse, and how can we explain the occurence of the third pulse? In the simplest case, a two step (or two pulse) process consisting of the excitation and ionization step by involving the ground-, excited- and ionic state of the NaK dimer should be fully sufficient. The complex pulse sequence retrieved from the experiment (Fig.4b) is very likely indicative of the involvement of more steps and/or states in the process. For example, one can consider a mechanism involving one or more additional excited intermediate states in the multiphoton process. Other possible mechanism could be the involvement of a dump process back to the ground state of the dimer followed by a transition of the evolving wave packet into an excited state(s) of the dimer before the final ionization step occurs. Since all optimization experiments ended up with pulse trains, it is straightforward to further simplify the experiment by reducing the searching space to pulse trains based on sine phase optimization as shown in [13]. In order to identify the involved physical processes and electronic states during the photoexcitation and photoinization of NaK, wavelength-dependent phase optimization experiments and experiments with simultaneous amplitude- and phase modulation are under preparation in close cooperation with theory [14].

5 Summary

This paper was devoted to the control of molecular dynamics in bound electronic states of the homonuclear sodium trimer Na_3 and in the photofragmenting heteronuclear triatomic Na_2K. In the case of Na_3, the propagation of the wavepacket in the B- and X-state of the trimer was controlled selectively with low intensity linearly chirped laser pulses: In the case of unchirped or slightly downchirped excitation pulse (up to ca. 300 fs duration) the wavepacket propagates in the B-state and the pump-probe spectrum reflects the symmetric stretch mode of this state. When employing a 400 fs downchirped excitation pulse, the leading blue spectral edge of the downchirped pulse creates a propagating wavepacket in the B-state of the homonucleus trimer, which is then dumped down by the following red spectral tail of the same pulse into the vibrationally excited region of the ground state. The recorded pump-probe spectrum reveals the symmetric stretch mode of the ground state, which is well known from earlier measurements with amplified laser pulses.

An active control of the multiphotonic process by means of feedback optimization was carried out by manipulating different branching ionization and fragmentation pathways of the photoexcited heteronuclear trimer Na_2K. By optimizing the applied laser field, the yield of the resulting parent and fragment ions could be controlled. The information content coded in the resulting pulse structures is very rich, and it reflects the intrinsic properties of the investigated multiphotonic process: In the Na_2K/NaK system, for example, the temporal structure of the pulses reflects exactly the vibration clock of the corresponding parent molecule or fragment, the intensity of the pulse elements is correlated to the number of photons involved in the process. The wavelength structure (colorness) and the chirp superimposed to the temporal structure of the retrieved pulse sequences indicate rather different pathways on the complex potential energy surfaces leading to comparable reaction product yields even in the case of the simplest investigated model system NaK (see [13] for more details, too).

6 Acknowledgements

The authors are grateful to Prof. Vlasta Bonacic-Koutecký and Prof. Herschel Rabitz for many stimulating discussions. C.L. acknowledges the financial support of the Gottlieb Daimler - und Karl Benz- Stiftung. This work is supported by the Deutsche Forschungsgemeinschaft in the frame of the Sonderforschungbereich 450 research project.

References

1. R. S. Judson, H. Rabitz, Physical Review Letters 68 (1992) 1500.
2. A. Assion, T. Baumert, M. Bergt, T. Brixner, B. Kiefer, V. Seyfried, M. Strehle, and G. Gerber, Science 282 (1998) 919.
3. S. Rutz, S. Greschik, E. Schreiber, and L. Wöste, Chem. Phys. Lett. 257 (1996) 365.
4. P. Rosendo-Francisco, PhD thesis, Berlin 2000
5. Š. Vajda, and L. Wöste, in „*Femtochemistry*",Edited by F.C. De Schryver, S. De Feyter and G. Schweitzer, John Wiley-VCH, 2001, ISBNs: 3-527-30259-X (Hardback); 3-527-60018-3 (Electronic), Chapter 10, pp. 199-216
6. T. Baumert, R. Thalweiser, and G. Gerber Chem. Phys. Lett. 209 (1993) 29.
7. P. Rosendo-Francisco, C. Lupulescu, B. Baptist, and Š. Vajda, *Journal of the Chinese Chemical Society* **47**, 705 (2000).
8. Š. Vajda, S. Rutz, J. Heufelder, P. Rosendo, H. Ruppe, P. Wetzel, and L. Wöste,J. Phys. Chem. A, 102 (1998) 4066.
9. L.-E. Berg, M. Beutter, T. Hansson, Chemical Physics Letters 253 (1996) 327.
10. A. Weiner et al., IEEE J. Quantum Electron. 28 (1992) 908.
11. H.-P. Schwefel, Evolution and Optimum Seeking, Wiley New York (1995).
12. Š. Vajda, A. Bartelt, E.C. Kaposta, T. Leisner, C. Lupulescu, S. Minemoto, P. Rosendo-Francisco , and L. Wöste, *Chem. Phys.* **267**, 253 (2001), Special Issue on „*Laser Control of Molecular Dynamics*", Guest Editors : R. de Vivie-Riedle, H. Rabitz, and K. Kompa
13. A. Bartelt, C. Lupulescu, Š. Vajda, and L. Wöste, in "Femtochemistry and Femtobiology" , Eds.: A. Douhal, World Scientific 2002, in press
14. V. Bonacic-Koutecký et al., private communications

FEEDBACK CONTROL OF ALKALI DIMERS WITH SINUSOIDAL PHASE MODULATED FS-PULSES: CAN WE LEARN FROM THE ACQUIRED PULSE SHAPES?

A. BARTELT, C. LUPULESCU, S. VAJDA, AND L. WÖSTE

Institut für Experimentalphysik, Freie Universität Berlin, Arnimallee 14, D-14195 Berlin

E-mail: bartelt@physik.fu-berlin.de

The controllability of three photon ionization pathways are investigated on model-like systems NaK and K_2. A closed learning loop for adaptive feedback control is used to find the optimal fs pulse form. Sinusoidal parameterizations of the spectral phase modulation are investigated in regard to the obtained optimal field. By reducing the number of parameters and thereby the complexity of the phase modulation, optimal pulse forms can be generated that carry fingerprints of the molecule's dynamical properties. This reduction of the searching space enables to find "understandable" optimal pulse forms and offers the possibility to gain insight into the photoinduced control process.

1 Introduction

In the field of controlling molecular dynamics with fs-laser pulses, experimental activities have increased enormously in the last few years. After Rabitz and Coworkers [1] postulated a closed loop experiment in 1992, various groups have successfully performed such optimization experiments . The objectives were the optimization of processes like e.g. population transfer [2], fragmentation [3] [4], ionization [5], and ISRS [6]. Without any knowledge of the molecular Hamiltonian, this method enables to find the optimal pulse form forcing the system into the target state, by using an iterative process. The methods combined programmable pulse shapers with optimization algorithms, latter being capable of solving high dimensional search problems within experimental conditions.

In many experiments the searching space was set as large as possible in order to find global solutions. In the case of liquid crystal pulse shapers, this is accomplished by taking all 128 phase and/or 128 amplitude values of the modulator pixels as optimization parameters ("free optimization"). Consequently the number of iterations needed for convergence is very large. In order to reduce the duration of the experiment, the modulation can be expressed by analytical functions (parameterization) [6]. The function parameters now serve as optimization parameters, the number of which is dramatically reduced in comparison to the free optimization. Due to the narrowed searching space, appropriate functions according to the optimization problem must be found in order to assure a possible solution.

Apart from the effectiveness and usefulness of the photochemical reaction control, one major goal of closed loop experiments is to gain information about the photoinduced control process [4] and molecular properties like potential energy surfaces [7]. This problem of inversion can be solved by analyzing the acquired optimal pulse form. In many cases free optimizations lead to optimal pulse forms which are very complicated and impossible to interpret. Some components of the pulse might not contribute to the effect and are therefore not considered by the optimization [8]. Hence, a very important step towards understanding the acquired pulses is to reduce the unimportant components of the light field. It was shown theoretically, that this is possible by adding "genetic pressure" to the cost functional [8], which tends to reduce the unimportant parameter values during the optimization.

In this contribution we use a sinusoidal spectral phase parameterization and investigate the effect of different parameter sets on the optimal pulse form. The constraints impinged by the parameterization reduce the solution space to certain subspaces or classes of solutions. The importance of single parameters are tested within the closed-loop experiment by reducing the number and range of values of the parameters. The optimized laser fields acquired for different parameter sets are compared.

A sinusoidal spectral phase parameterization is chosen for the following reason: We optimize three-photon transient ionization processes in model-like alkali dimers NaK and K_2. Fs-pump-probe spectroscopy on NaK and K_2 revealed wave packet dynamics in their excited states with oscillation periods of T_{osc}^{NaK}=440 fs and $T_{osc}^{K_2}$=500 fs, respectively [9] [10]. Free optimization experiments [5] carried out on these systems found optimal pulse trains consisting of several pulses with time separations of half and integer values of the corresponding wave packet oscillation times T_{osc}. Sinusoidal spectral phase modulations, however create pulse trains consisting of several pulses with controllable time separations. Therefore a sinusoidal parameterization is suitable for the ionization control of small alkali clusters.

2 Experimental Setup

The closed loop experiment consists of three components as depicted in Figure 1. The central part of the system is given by a programmable pulse shaper, which is able to create arbitrary fs pulse forms by computer control. In this experiment the pulse shaping is controlled by a self-learning optimization algorithm based on evolutionary strategies. The algorithm iteratively maximizes

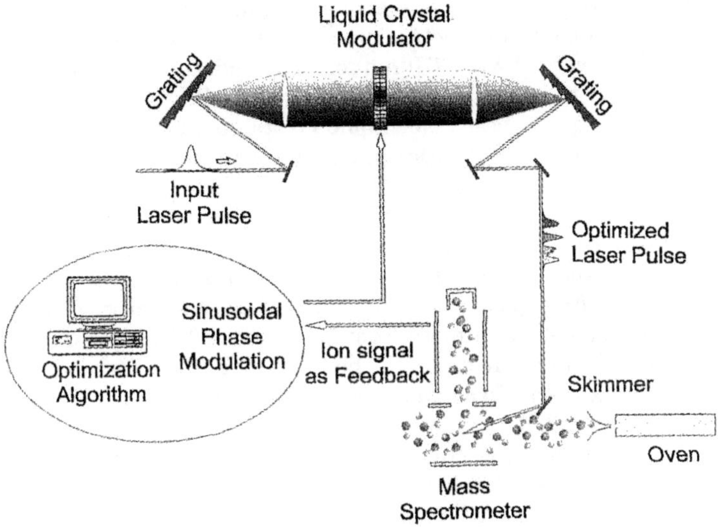

Figure 1. The closed loop experiment. A programmable pulse shaper is iteratively driven by a self-learning optimization algorithm. A molecular beam containing small alkali clusters is created by adiabatic expansion. The ion yield of the alkali dimers is taken as feedback. Parameter sets of a sinusoidal spectral phase modulation are optimized.

the ion yield of the alkali dimer under investigation. The optimal pulse form responsible for maximal ion yield is subsequently characterized.

The dimers are produced in an adiabatic coexpansion of alkali vapor and argon carrier gas through a nozzle of 70 μm diameter. In order to obtain mainly dimers, the temperature of the oven and the pressure of the argon were set to 650°C and to 2 bars, respectively. Further details of the alkali cluster production can be seen elsewhere[4]. The neutral alkali dimers are excited and ionized by a fs laser beam, which is focused onto the interaction region with the molecular beam. The cluster ions are mass selected by a quadrupole mass spectrometer and detected with a secondary electron multiplier.

The laser source consists of a Ti:Saphire oscillator which provides pulses of rather low energy, thus the experiments are carried out in the weak field regime. Pulses of 80 fs temporal length (fwhm) and an energy of 20 nJ per pulse are generated with a repetition rate of 80 MHz. Behind the pulse shaper the average power is about 800 mW. The intensity of the shaped pulses within the focus is estimated to be of the order of 1 GW cm^{-1}. The

MHz repetition rate assures a very high averaging rate, which overcomes laser pulse and molecular beam fluctuations. The experiments are carried out at the carrier frequency of $\lambda_0 = 770$ nm for NaK and $\lambda_0 = 834$ nm for K_2. The spectral width of the laser is about $\Delta\lambda/\lambda_0 \approx 1,2 \cdot 10^{-2}$.

The pulse shaper is based on a liquid crystal modulator mask (CRI-SLM-256) [11] [12] situated in the Fourier plane of a zero dispersion compressor [13], which consists of two gratings (1200 l/mm) and two lenses (f=200 mm) (see Fig. 1). The spectral phase is modulated by sinusoidal functions.

The closed loop experiment is driven by a non-deterministic optimization algorithm based on evolutionary strategies. The optimal pulse form is found by iteratively writing pulse form pattern on the modulator, until maximal ion yield is obtained. Each pulse form is determined by a set of parameters of the sinusoidal phase function. In the beginning, the parameter values are random. Modification of the values is accomplished in each iteration cycle by two operators called cross-over and mutation. After being written on the modulator, each pulse form creates an ion signal which represents its fitness. In the end of each iteration best individuals are selected according to their fitness, and serve as parent individuals for the next iteration. Details of the algorithm can be seen elsewhere [5].

The optimal pulse form is measured by intensity cross correlation and XFROG techniques. The reference pulse directly taken from the Titanium:Sapphire laser is characterized by SHG-FROG. In XFROG the cross correlation signal created by the convolution of shaped and reference pulse is frequency resolved. The cross correlation traces are recorded for many frequency components of the spectrum. The resulting spectrogram is a very intuitive picture of the time-frequency character of the pulse. For a complete characterization the temporal amplitude and phase can be calculated from the XFROG spectrogram by an iterative phase retrieval algorithm [14].

3 Sinusoidal spectral phase modulation

The spectral field distribution $E(\omega)$ in the fourier plane is modulated with a sinusoidal phase of the form

$$\psi(\Omega) = a \cdot \sin(\Omega + c). \tag{1}$$

The argument Ω is the frequency deviation from the carrier frequency ω_0 and is expressed in a Taylor expansion:

$$\Omega = \tau(\omega - \omega_0) + a_2(\omega - \omega_0)^2 + a_3(\omega - \omega_0)^3 \cdots \tag{2}$$

The shaped pulse form is given as the Fourier transform of the modulated spectral field $E_{mod}(\omega) = E_{in}(\omega) \cdot e^{i\psi(\omega)}$. In order to discuss the sinusoidal phase modulated pulses, first the non-linear Taylor coefficients are set to zero: $\psi(\Omega) = a \cdot \sin(\tau(\omega-\omega_0)+c)$. Due to the periodicity of the modulation a series of pulses is created. The time separation of the pulses within the pulse train increases linearly with the frequency of the phase modulation given by τ. Moreover, this pulse separation is equal to τ, so that τ is usually given in units of fs.

The amplitude of the modulation given by the parameter a can vary between 0 and 2π. Variation of a leads to different relative amplitudes of the single pulses. With small values of a a strong central pulse with small side peaks is observed, while increasing a leads to stronger outer pulses and the creation of new side pulses.

The modulation phase c governs the temporal phase shift $\Delta\varphi$ between the single pulses. For $c = 0$ (sinus) the phase shift amounts $\Delta\varphi = \pi$ for the first pulses up to the central pulse, and $\Delta\varphi = 0$ for the others. For negative sinus (c=π) the situation is opposite, e.g. $\Delta\varphi = \pi$ for the second half of the pulses and constant phase for the first ones. In the case of $c = \pi/2$ (cosinus) we find $\Delta\varphi = \pi/2$ for all pulses. Thus, a continuous variation of c between 0 and 2π allows classes of temporal phase shifts. While for each pair of pulses a phase shift between 0 or 2π is possible, this does not hold for the entire pulse train.

The introduction of second and third order Taylor coefficients a_2 and a_3 into the sinus argument leads to the chirping of the pulses. Whereas a pure Taylor expansion of the phase modulation leads to single chirped pulses, here the chirp is generated on the individual pulses of the pulse train. While a_2 causes linearly chirped pulses, a_3 introduces quadratic frequency chirp. The central pulse, however stays unchirped in both modulations.

4 Optimization of NaK^+

In contrast to the free optimization, where 128 discrete phase components of the spectral field are optimized, here only the parameters of the sinusoidal function 1 and 2 serve as optimization parameters. The searching space is dramatically reduced. Convergence is reached after the 5th to 8th generation in the case of three parameters, and after the 10th to 15th generation with five parameters. With a population size of 30 individuals one optimization experiment is completed within 2 and 5 minutes.

In the first experiment, all five parameters a, τ, c, a_2 and a_3 are optimized. The resulting optimal pulse form is shown in the upper row of Fig.

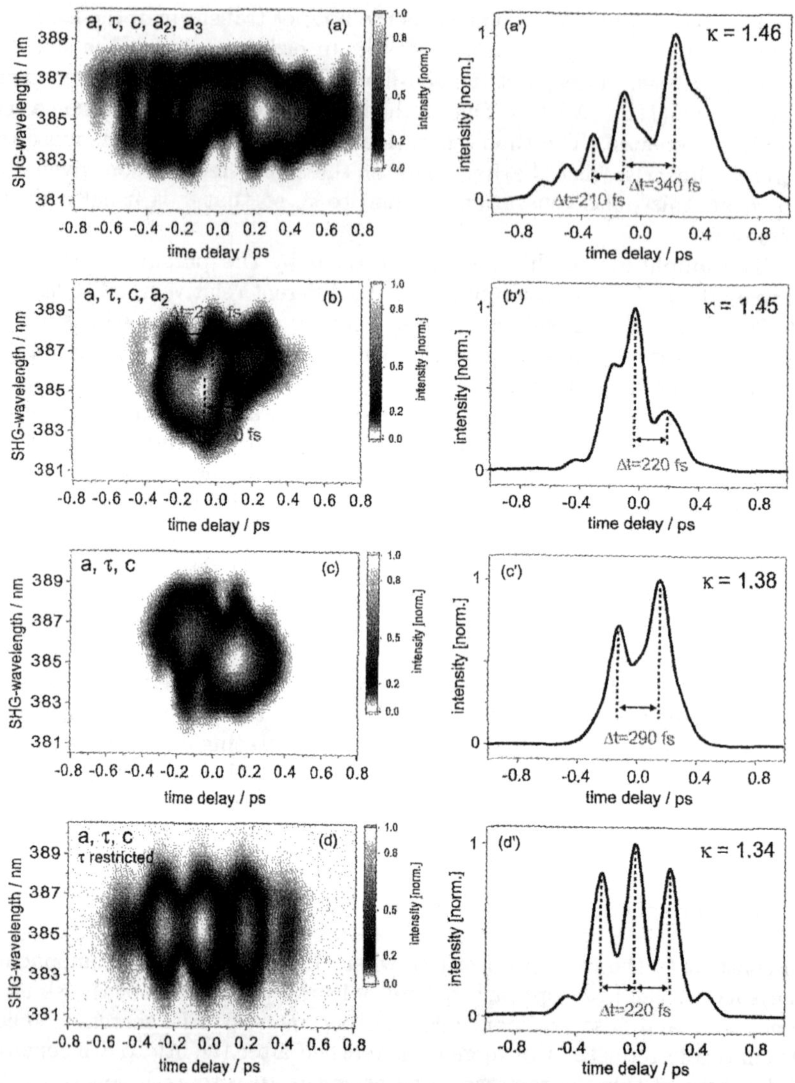

Figure 2. XFROG (left) and cross correlation (right) traces of optimized pulses for NaK^+ maximization with different parameter sets of a sinusoidal parameterization. (a) All five parameters are optimized. (b) Optimized result without third order Taylor coefficient a_3. (c) With three parameters a, τ and c a two-pulse sequence remains. (d) With the restriction $\tau > 180$ fs the three-parameter-optimized pulse form consists of three dominant pulses with a time separation of $\tau = 220$ fs each.

2 in its XFROG (a) and cross correlation (a′) representation. The ion yield I_{opt} obtained with the optimized pulse is compared with the ion yield I_{tl} obtained by applying a transform-limited pulse with the same energy. A significant increase of the ion yield is achieved with an optimization factor $\kappa = I_{opt}/I_{tl} = 1.46$. The XFROG trace reveals a highly structured pulse train consisting of several pulses which intensity is increasing with time. The temporal separations of the pulses as shown in the cross correlation trace (a′) vary between $\Delta\tau = 180$ fs and $\Delta\tau = 340$ fs, latter being the separation of the two main intense pulses. The pulses are strongly chirped with different chirp signs.

If we exclude the third order Taylor coefficient a_3, the 4-parameter optimization achieves an ion yield almost identical to the result obtained with all 5 parameters. The third order Taylor coefficient hence does not contribute to the effectiveness of the photoinduced ionization process NaK→NaK^+ and can be neglected. If we do so, the algorithm finds a more compact pulse form (Fig. 2b). The cross correlation (Fig. 2b′) reveals an intense middle pulse with a shoulder arriving 125 fs before the main pulse. It is followed by a less intense pulse arriving about 220 fs later. From the XFROG trace (b) we find the shouldered central pulse to be quadratically chirped in time. While the red wing of the pulse spectrum is divided into two pulses with a time separation of 220 fs, the blue components only show one pulse. Since $\frac{1}{2}T_{osc}^{NaK}$=220 fs, here first identifications of the pulse form with the wave packet oscillation period are found. However, the non-linear chirp of the pulse hinders an intuitive interpretation of the pulse form. The optimized parameters are listed in Tab. 1.

By neglecting also the second order parameter a_2, the chirp is excluded as a degree of freedom as well. The results of the optimization with only three parameters a, τ and c are shown in Fig. 2c. Now, only a double pulse is obtained with a time separation of $\Delta t = 290$ fs. As can be seen from Tab. 1 this separation is two times the corresponding parameter $\tau = 145$ fs. For small values of τ the sinusoidal phase modulation leads to "undeveloped" pulse train structures. For the used laser and pulse shaper setup, pulse trains with individually separated pulses are only generated if $\tau \geq 180$ fs. Smaller values of τ result in complex pulse structures that evolve from linearly chirped (cosinus) or quadratically chirped (sinus) pulses. The optimal pulse form in Fig. 2c is such a pulse structure, resulting from a negative sinus ($c \approx \pi$) with modulation frequency $\tau = 145$ fs and $a \approx \pi$. Its second pulse is more intense than the first one and exhibits a clear frequency separation (see the XFROG trace). The first pulse is red shifted, while the second pulse is slightly blue shifted. Although the pulse form is very clean, no intuitive interpretation of

Table 1. Optimized parameters for the NaK^+ optimizations with different sets of parameters of the sinusoidal phase parameterization. The results for the restricted optimization are also listed. The optimization yield κ in the last row is defined as the ratio of optimized ion yield with the ion yield obtained by a transform-limited pulse $\kappa = I_{opt}/I_{tl}$. The exclusion of a_3 only leads to a small change of κ. An additional exclusion of a_2, however decreases the ion yield significantly.

Parameter	a, τ, c, a_2, a_3	a, τ, c, a_2	a, τ, c	a, $\tau \geq 180$ fs, c
a [π]	1,64	0,92	1,03	0,9
τ [fs]	198	147	145	223
c [π]	1,07	0,07	1,13	1,56
a_2 [fs^2]	$-4{,}93 \cdot 10^3$	$2{,}24 \cdot 10^3$	0	0
a_3 [fs^3]	$-2{,}13 \cdot 10^5$	0	0	0
optimization yield κ	1,46	1,45	1,38	1,34

the pulse components is possible. The ion yield is reduced in comparison to the 4-parameter optimization.

As the result of the three-parameter optimization is an "undeveloped" pulse sequence, it is further necessary to introduce restrictions on the pulse parameters. In order to find the optimal pulse shape for fully developed pulse sequences, the restriction $\tau > 180$ fs is introduced. The restricted optimization using the three parameters a, τ and c was carried out by artificially labelling all those parameter sets with vanishing fitness, that include $\tau \leq 180$ fs.

The optimal pulse form is shown in Fig. 2d. The result is a pulse train consisting of three equally spaced pulses. The time separation of the pulses is exactly $\Delta\tau = 220$ fs. The efficiency of the pulse form compared to the unconstraint result (Fig. 2c) is only slightly reduced with an optimization factor of κ=1,34. As no Taylor coefficient was allowed, the pulses are unchirped (Fig. 2d). The central pulse is more intense than the other two pulses. Here, an optimal pulse form was found by the algorithm, which clearly describes and exploits the dynamical properties of the NaK molecule, given by the wave packet oscillation period [9] of T_{osc}^{NaK}=440 fs.

A "handwaving" explanation for the optimal pulse form can be given in the view of time-dependent Franck-Condon factors: The first pulse excites a wave packet in the excited $A^1\Sigma_u^+$ state of NaK. The second pulse arrives after $\Delta\tau = \frac{1}{2}T_{osc}^{NaK}$=220 fs which is identical to half the wave packet oscillation period in the excited state. From the phase of the oscillatory signal in the pump-probe spectrum, the creation of the wave packet could be localized at the inner turning point, whereas efficient ionization, leading to the high peak of the pump-probe signal, takes place at the outer turning point of the potential [9]. When the second pulse arrives $\Delta\tau = 220$ fs later, the wave packet is located at the outer turning point of the potential and effective ionization takes place. Meanwhile, the same pulse can also excite new molecules and create new wave packets, which can be transferred into the ion state by the third pulse arriving again $\Delta\tau = 220$ fs later.

5 Optimization of K_2^+

As a second example of pulse form complexity reduction the ionization control of $K_2 \rightarrow K_2^+$ is considered. The measurements are carried out at a central wavelength of $\lambda_0 = 834$ nm. At this wavelength, the most pronounced wave packet oscillations were found by fs pump-probe spectroscopy [15 10]. The oscillation period of the excited $(A)^1\Sigma_u^+$ state of K_2 was measured to be $T_{osc}^{K_2} =$ 500 fs. It was shown that like in the case of NaK, effective ionization takes place at the outer turning point of the potential, while the first pulse excites the wave packet in the inner turning point [15]. The contribution of the time dependent ion signal leads to a strong oscillatory amplitude. It was shown, that the ratio between the oscillatory amplitude and the integrated signal is about one magnitude smaller, when a UV probe pulse is used (two-color experiment) [16]. The efficient ionization pathway at the outer turning point of the potential could be explained by a resonant two photon process, exploiting an open Franck-Condon window for ionization via the intermediate $2^1\Pi_g$ state [17].

The optimization experiments of K_2 are also carried out with a Taylor expanded sinusoidal phase parameterization. The optimized pulses are compared for four different sets of parameters. The first optimization includes all five parameters. The results are shown in Fig. 3a. The cross correlation trace reveals an irregular sequence of pulses with separations between $\Delta t = 430$ and 900 fs. Some pulses are strongly chirped and frequency shifted in respect to each other (see XFROG trace Fig. 3a).

Different results are obtained, when the second order parameter a_2 is removed (Fig. 3b). The train consists basically of three pulses, one of which is very dominant. The distance of the most intense middle pulse to the last pulse is about $\Delta t_2 = 250$ fs, as can be seen from the cross correlation trace Fig. 3b'. As Δt_2 is identical with half of the oscillation period $\frac{1}{2}$ $T_{osc}^{K_2} = 250$ fs of K_2 in the excited $(A)^1\Sigma_u^+$ state, the separation Δt_2 is the first fingerprint of the

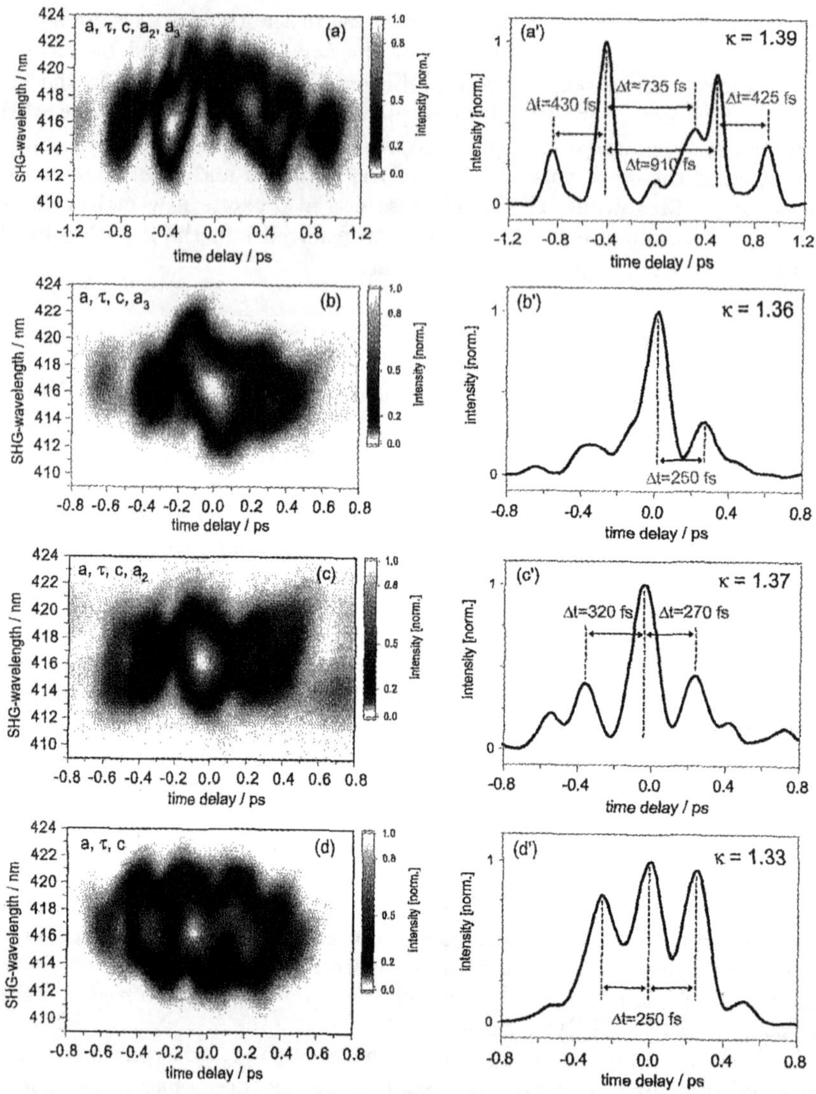

Figure 3. Optimized pulse forms for K_2. The optimized sets of parameters of the sinusoidal spectral phase modulation are shown inside the XFROG traces (left). The number of parameters is reduced from five (a) to three (d). The cross correlation traces are shown on the right. The optimization yields κ are given as well.

Table 2. Optimized sinusoidal phase parameters for K_2^+ optimization. The optimization yield κ in the last row is defined as the ratio of optimized ion yield with the ion yield obtained by a transform-limited pulse $\kappa = I_{opt}/I_{tl}$.

Parameter	a, τ, c, a_2, a_3	a, τ, c, a_3	a, τ, c, a_2	a, τ, c
a [π]	1,46	0,76	1,12	0,92
τ [fs]	347	398	210	250
c [π]	0,42	1,82	0,58	1,88
a_2 [fs^2]	-1,81$\cdot 10^3$	0	-7,33$\cdot 10^3$	0
a_3 [fs^3]	-2,49 $\cdot 10^4$	-2,35 $\cdot 10^5$	0	0
optimization yield κ	1,39	1,36	1,37	1,33

molecule's dynamics. The timing between first and second pulse, however does not match this criterion at first glance. On the other hand, the middle pulse is strongly up-chirped (see Fig. 3b), leading to different delay times for the spectral components. Although at some frequencies therefore the separation equals 220 fs, the role of the up-chirp in this case is unclear and makes the interpretation of the pulse form very difficult.

The optimization without third order parameter (Fig. 3c) results in a similar triple pulse sequence. Again the middle pulse is most intense compared to the first and third pulse. Also the XFROG trace (Fig. 3c) reveals an up-chirped middle pulse, but slightly less than the result with third order contribution (Fig. 3b). The timing between the pulses is $\Delta t_1 = 320$ fs and $\Delta t_2 = 270$ fs. Δt_1 does not coincide with the oscillation period of the molecule. The small deviation of Δt_2 however could be explained by the strong chirp of the middle pulse.

The situation seems to be cleaned up, when only the linear argument of the sinus is permitted, e.g. only the optimization parameters a, τ and c are used. An optimal pulse form is found (Fig. 3d), which again consists of three pulses. However, this time the pulse separation of all three pulses $\Delta t = 250$ fs matches exactly one half oscillation period $\frac{1}{2}T_{osc}^{K_2} = 250$ fs. Now the dynamical properties of the molecule are found directly in the intensity structure of the pulse. The interpretation of the control pulse form is similar to the one given above in case of the restricted optimization result of NaK.

In contrast to NaK, the three parameter optimization of K_2^+ leads directly to pulse trains consisting of three pulses with separations that match half the oscillation time of the molecule's wave packet. However, in the K_2^+ optimization also a second pulse form is found. Repetitive execution of the optimization experiment reveals two distinct optimal pulse forms with almost the same yield. The pulse form which was not presented here is a more compact and "undeveloped" pulse sequence. An intuitive approach can at this point not be given and will therefore be discussed in a further publication.

The reduction of the searching space leads to simple pulse forms, where "handwaving" explanations of a sequential 3-photon ionization process can be given. However, even the "simple" pulse structures must be analyzed by taking all electronic states and couplings into consideration. This is certainly true in the case of the more complex pulse forms, where the complexity may as well result from multi-photon sequential roundabout ways like stimulated raman scattering processes (ISRS). In order to decipher the richness of the different control scenarios an intense theoretical analysis is needed.

6 Conclusion

We investigated the adaptive feedback control results of parametrically phase shaped fs pulses. The spectral phase pattern of the pulse shaper was expressed by sinusoidal functions using second and third order Taylor expansion terms within the sinus argument. The goal was to optimize a three photon ionization pathway in model-like systems NaK and K_2. The obtained optimal pulse forms differed significantly, depending on the number and the range of values of the parameters. It could be shown that by choosing an appropriate parameterization and by reducing the complexity of the modulation, intrinsic properties of the molecule' s dynamical features could be found in the intensity profiles of the optimal pulse forms. For both molecules NaK and K_2 it was possible to extract a triple pulse train with pulse separations of half of the oscillation period $\frac{1}{2}T_{osc}$ of the respective molecular wave packet propagation. The interpretation of the photoinduced optimal pathway based on time dependent Franck-Condon windows implies the creation of the wave packet at the inner turning point and the effective ionization at the outer turning point of the potential.

Acknowledgments

The authors thank Prof. H. Rabitz for stimulating ideas and discussions and Dr. T. Feurer for his support. C. L. thanks the G. Daimler und C. Benz

Stiftung. The present work is supported by the Deutsche Forschungsgemeinschaft (SFB 450).

References

1. R. S. Judson and H. Rabitz *Phys. Rev. Lett.* **68**, 1500 (1992).
2. C. J. Bardeen, V.V. Yakovlev, K. R. Wilson, S. D. Carpenter, P. M. Weber and W. S. Warren *Chem. Phys. Lett.* **280**, 151 (1997).
3. A. Assion, T. Baumert, M. Bergt, T. Brixner, B. Kiefer, V. Seyfried, M. Strehle and G. Gerber *Science* **282**, 919 (1998).
4. S. Vajda, A. Bartelt, C. Kaposta, T. Leisner, C. Lupulescu, S. Minemoto, P. Rosendo-Francisco and L. Wöste *Chem. Phys.* **267**, 231 (2001).
5. A. Bartelt, S. Minemoto, C. Lupulescu, S. Vajda and L. Wöste *Eur. Phys. J.* D **16**, 127 (2001).
6. T. Hornung, R. Meier and M. Motzkus *Chem. Phys. Lett.* **326**, 445 (2000).
7. H. Rabitz and W. Zhu *Acc. Chem. Res.* **33**, 572 (2000).
8. J. M. Geremia, W. Zhu and H. Rabitz *J. Chem. Phys.* **113**, 10841 (2000).
9. J. Heufelder, H. Ruppe, S. Rutz, E. Schreiber and L. Wöste *Chem. Phys. Lett.* **269**, 1 (1997).
10. S. Rutz, R. DeVivie-Riedle and E. Schreiber *Phys. Rev. A* **54**, 306 (1996).
11. A. Weiner, D. Leaird, J. Patel and J. Wullert *IEEE J. Quant. Elec.* **28**, 908 (1992).
12. M. Wefers and K. Nelson *J. Opt. Soc. Am. B* **12**, 1343 (1995).
13. O.E. Martinez *IEEE J. Quant. Elec.* **23**, 59 (1987).
14. Femtosoft Technologies *XFROG 3.0* (2000).
15. S. Rutz and E. Schreiber in *Springer Series in Chemical Physics* ed. P.F. Barbara (Springer-Verlag , 1994).
16. S. Rutz and E. Schreiber *Chem. Phys. Lett.* **269**, 9 (1997).
17. R. De Vivie-Riedle, B. Reischl, S. Rutz and E. Schreiber *J. Phys. Chem.* **99**, 16829 (1995).

EXPERIMENTAL AND THEORETICAL STUDY OF QUANTUM CONTROL BY CHIRPED PULSE EXCITATION

D. HUPPERT[1], J. SEGAL[1] AND B. D. FAINBERG[1,2]

[1]*Raymond and Beverly Sackler Faculty of Exact Sciences, School of Chemsitry, Tel-Aviv University, Tel-Aviv 69978, Israel*

[2]*Holon Academic Institute of Technology, Department of Exact Sciences, 52 Golomb St.,Holon 58102, Israel*

The effects of the chirp and intensity of an ultrashort pulse on the transmission spectrum of Rhodamine 800 in several liquids are investigated experimentally and theoretically. High power negatively chirped pulses enhance the pump-dump process and the excited electronic state population is reduced. Excitation by positively chirped pulses diminishes this mechanism. We used a nonpertubative theoretical approach to calculate the absorption spectrum of the samples. The calculation results agree with experimental results for both negatively and positively chirped pulses.

1 Introduction

Optical control of chemical reactions [1] and quantum electronic states [2,3,4,5] has been a subject of great interest in recent years. The interaction of intense chirped ultra short pulses with molecular systems in liquid solutions has shown that it is possible to control the ratio of the ground state population to the excited population. The interaction of an intense pulse with a probe molecule in solution can be thought of in terms of sequential interactions with the field. In the nonlinear regime, more than one photon interacts with the molecular probe and both absorption and stimulated emission can occur.

The ground and the excited state potential energy surfaces of the molecule can be modeled by a two-dimensional surface. One coordinate is the generalized high frequency (HF) intramolecular coordinate and the second coordinate is the generalized solvent and low frequency (LF) intramolecular nuclear bath coordinate. The motion of the non-equilibrium initial population wave packet on the first excited electronic state results in a dynamic Stokes shift where the population moves from higher to lower energy. In a negative chirped pulse, the ordering of the frequency components is from high to low. The rate of the negatively chirp can be controlled to follow the motion of the excited state population and enhance the stimulated emission process and hence increase the ground state population. A positively chirped pulse cannot enhance the process and the end result is a higher population in the excited state.

The effects of varying the chirp and intensity of an ultrashort pulse, exciting probe molecules in liquid solutions, have been investigated experimentally by

Cerullo, Bardeen, Shank *et al.* [5] Bardeen, Wilson *et al.* [6], and Huppert *et al.* [7]. They measured the integrated fluorescence (which is directly proportional to the total excited state population) after the completion of the pulse action, as a function of pulse chirp. In addition, Shank *et al.* [5] and Huppert *et al.* [7] measured the absorption spectrum of chirped pulses. For low-power excitation, they found that the absorption and the size of excited state population were independent of chirp, while for high-power excitation the authors [5,6,7] observed strong chirp dependence.

One of the authors [8] has developed a nonpertubative analytic approach to the problem of the interaction of high-power chirped ultrashort pulses with molecular systems, which led to the picture of ``moving potentials". In the model, a strongly broadened vibronic molecular system of LF optically active (OA) intra- and intermolecular vibrations $\{\omega_s\}$ with dissipation interacts with a strong chirped pulse. Its field amplitude can be represented in the form:

$$\bar{E}(t) = \bar{\varepsilon}(t) exp(i\varphi(t)) \tag{1}$$

where $\varepsilon(t)$and $\varphi(t)$ are real functions of time, and $\varphi(t)$ describes the change of the pulse phase in a time t. The solution of the problem was based on the following assumptions.

1. The irreversible dephasing time of the electronic transition T' for such a system is much shorter than both the vibrational relaxation time τ_s and the pulse duration t_p.
2. The pulse frequency $\omega(t) = \omega - \dfrac{d\varphi}{dt}$ changes only slightly during T'.
3. The relaxation of the vibrational excitation in both the ground and excited electronic states can be described as diffusion with respect to the energy coordinate.

In a recent theoretical study [13] the effect of HF intramolecular modes on the chirped pulse excitation in condensed phase was investigated. The problem was solved using double-sided Feynman diagrams. It has been shown that the effect of the OAHF intramolecular modes strongly depends on the carrier pulse frequency. Incorporating the intramolecular modes increases the total excited state population, n_2 when a molecule is excited near $0 \rightarrow 1$ transition with respect to the OAHF vibration. If the molecule is excited near the $0 \rightarrow 0$ transition with respect to the OAHF intramolecular mode, the effect is the opposite. These results are explained by the picture of two-dimensional potentials corresponding to the OA HF and LF vibrations.

Whereas the information given by measuring the integrated fluorescence is completely clear (n_2), an accurate interpretation of the absorption spectrum of intense chirped pulses is a nontrivial problem [9,10,11]. It has been shown in Refs.[10,11] that the absorption spectrum of a strongly chirped pulse directly reflects the time evolution of the population wave packets in the ground and excited electronic states.

In this study we measured the time integrated absorption of Rhodamine 800 dye in three solvents: methanol, acetonitrile and butanol under intense positively and negatively chirped pulse. We fit the experimental data by calculating the time-integrated absorption using the theory developed in Refs.[10,11]. We found a good agreement between the experimental data and the theory.

2 Theoretical Background

Let us consider a molecule with two electronic states, $n = 1$ and 2, in a solvent described by the Hamiltonian

$$H_0 = \sum_{n=1}^{2} |n\rangle[E_n + W(Q)]\langle n| \tag{2}$$

where $E_2 > E_1$, E_n is the energy of state n, $W_n(Q)$ is the adiabatic Hamiltonian of reservoir R (the vibrational subsystems of a molecule and a solvent interacting with the two-level electron system under consideration in state n). The molecule is affected by electromagnetic radiation of frequency ω:

$$E(t) = \frac{1}{2}\vec{E}(t)\exp(-i\omega t) + c.c. \tag{3}$$

For phase modulated pulses the field amplitude $\vec{E}(t)$ can be presented by Eq.(1).

The absorption spectrum at a frequency Ω is given by [9,12]

$$\alpha(\Omega) = 2\omega\, Im[P(\Omega)/E(\Omega)] \tag{4}$$

where $P(\Omega)$ and $E(\Omega)$ are the Fourier transforms of the positive frequency components of the polarization, $P^+(t)$, and the field amplitude, $E(t)$, respectively.

Let us consider strongly chirped pulses when a chirped pulse duration, t_p, is much larger than that of the corresponding transform-limited one, t_{p0}[10]. Consider also one normal intramolecular vibration mode of frequency ω_0 whose equilibrium position is shifted by ΔQ under electronic transition. In the case under consideration, the absorption signal is given by [11]:

$$\alpha(\Omega) \sim \omega\frac{N\pi|D_{12}|^2}{\hbar}\exp(-S_0)\sum_{k=0}^{\infty}\frac{S_0^k}{k!}[\tilde{\rho}_{1s}(\omega_{21} + k\omega_0 - \omega(t), t) - \tilde{\rho}_{2s}(\omega_{21} - k\omega_0 - \omega(t), t)] \tag{5}$$

Here we assume that the quantum intramolecular modes are of a high frequency, so that $\hbar\omega_0 >> k_B T$.

The quantity $\tilde{\rho}_{1s}(\omega_{21} + k\omega_0 - \omega(t), t)$ in Eq.(5) is the population wave packet values in the ground electronic state 1 at the intersections of the moving "photonic replication" of state 1 with term 2 (for k=0) or with its vibrational "replications" (for k≠0). In a similar manner, the quantity $\widetilde{\rho}_{2s}(\omega_{21} - k\omega_0 - \omega(t), t)$ is the population wave packet values in the excited electronic state 2 at the intersections of term 2 with the moving "photonic replication" of state 1 (for k=0) or with its vibrational ``replications" (for k≠0).

The quantities $\widetilde{\rho}_{ns}\left(\omega_{21} - (-1)^n k\omega - \omega(t), t\right)$ satisfy differential equation (42) of Ref. [13]:

$$\frac{\partial}{\partial t}\widetilde{\rho}_{ns}(\alpha, t) = \widetilde{L}_{ns}\widetilde{\rho}_{ns}(\alpha, t) + (-1)^n \sqrt{2\pi\sigma_{2s}} \sum_{k=0}^{\infty} w_k(t)$$

$$\times\left[\delta(\omega_{21} + k\omega_0 - \omega(t) - \alpha)\widetilde{\rho}_{1s}(\alpha, t) - \delta(\omega_{21} - k\omega_0 - \omega(t) - \alpha)\widetilde{\rho}_{2s}(\alpha, t)\right] \quad (6)$$

where

$$\widetilde{L}_{ns} = \tau_s^{-1}\left[1 + (\alpha - \delta_{2n}\omega_{st})\frac{\partial}{\partial(\alpha - \delta_{2n}\omega_{st})} + \sigma_{2s}\frac{\partial^2}{\partial(\alpha - \delta_{2n}\omega_{st})^2}\right] \quad (7)$$

$$w_k(t) = exp(-S_0)\frac{S_0^k}{k!}\sigma_a(\omega_{21})J(t) \quad (8)$$

are the probabilities of light-induced transitions at $\alpha = \omega_{21} \pm k\omega_0 - \omega(t)$, τ_s is the correlation time, $\sigma_a(\omega_{21})$ is the value of the cross section corresponding to the maximum of the absorption band in the absence of HF intramolecular modes [8], $J(t)$ is the power density of the exciting radiation.

3 Experimental setup

The experimental setup consists of a femtoseconds laser source, optical parametric amplifier (OPA), a pulse shaping setup and a setup for absorption diagnostics. The laser source has two parts: a Ti:Sapphire based oscillator (Mira Seed Coherent) and a multipass based Ti:Sapphire amplifier (Quantronix Odin) which eventually produces laser pulses of 40 fs duration, centered near 800nm, with a pulse energy of ~600 μJ and a repetition rate of 1 KHz. These pulses are used to pump an OPA which generates tunable short intense and coherent pulses of 50 fs with ~20 nm bandwidth in the spectral range of 600-770 nm. The pulse shaping apparatus consists of a pair of 600 lines/mm gratings and two identical lenses with focal of f=15cm. By moving one of the gratings collinearly with the optical axis, the pulse gains positive or negative linear chirp [14]. The chirp sign depends on the direction of the movements. During the experiments, two chirp rates were chosen, -50000 and -50000 fs^2 and henceforth they will be noted as negative and positive chirp rate respectively. The pulse temporal shape was diagnosed by OKE (optical Kerr effect) autocorrelator setup for determining the pulse chirp [15] . Finally the sample is irradiated by 10 μJ pulses, which are focused by a 15 cm lens. The cell was placed about 2 cm off the focus (the spot size there was 360 μm) to avoid other nonlinear effects. Neutral density filters attenuated the pulse intensity.

In the experiments, we use Rhodamine 800 dye (R800), which was purchased from Exciton, in 1mm or 0.2mm quartz cell. The dye is dissolved in several solvents – methanol, acetonitrile and butanol. The optical density of the solution was 0.7. The absorption was measured by collecting the transmitted laser pulses by a fiber into a computerized spectrometer with a CCD array (CVI laser corporation).

4 Discussion

For solving this differential equation (eq. 6) numerically, we wrote a computer routine, which is based on Tal-Ezer - Kosloff time propogator in a Chebyshev polynomial expansion [16]. The routine calculates the populations as a function of the α coordinate and time and then calculates the absorption function by using eq. 5.

Simulations of the solutions are presented in figures 1-2. The relevant parameters are found in set I in table 1 and the pulse has a negative chirped pulse. In figure 1 we see a greater negative absorption, as the chirp is larger. When the chirp is larger, its duration is longer and the "red" part of the pulse, which dumps the excited population, comes after a longer time interval than the "blue" part of the pulse, which populates the excited state population. In figure 2, the negative absorption is greater for small relaxation time. The same logics explain the data in figure 2.

Table 1: Relevant parameters for calculation absorption lines regarding different sets. Set I consists of the parameters, which were used for the simulations shown in Fig.1-2. Set II is for the parameter used for fitting the results in Fig.3-4. λ_c is the carrier wavelength and λ_{12} is the wavelength at the peak of the absorption spectrum, q is a parameter which indicates the intensity of the incident pulse and the rest of the parameters are defined inside the article.

Set	t_{p0} (fs)	λ_{21} (nm)	λ_c (nm)	ω_0 (cm^{-1})	S_0	ω_{st} (cm^{-1})	q	$\Phi''(\nu)$(fs^2)	τ_s (fs)
I	30	684	700	1200	0.4	800	1.0	-50000	100
II	35	680	687	1370	0.4	800	2.5,0.8	±50000	100

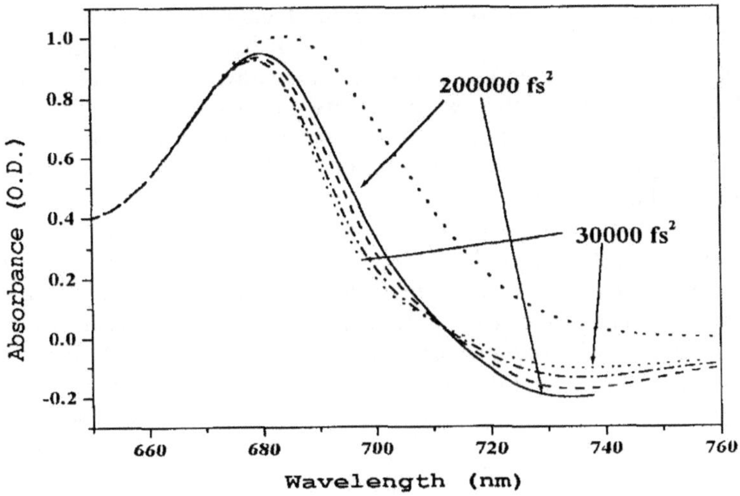

Figure 1: Simulation of the absorption of R800 in methanol for different chirps, see parameters in set *a* at table 1. Values of the chirp are 30000,50000,100000,200000 fs^2. The thick dotted line is the low power absorption spectrum.

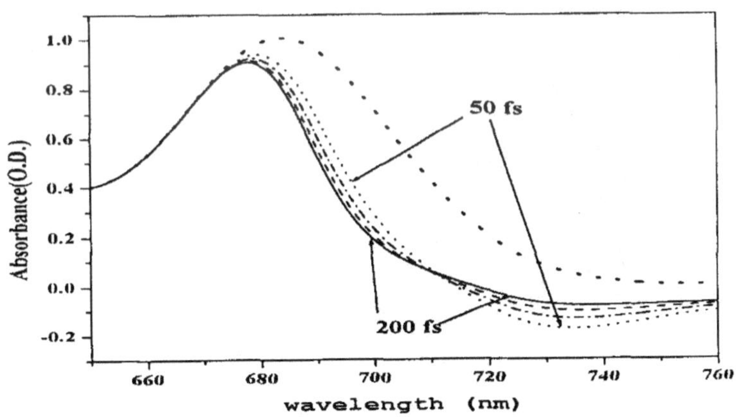

Figure 2: Simulation of the absorption of R800 in methanol for different relaxation times, see parameters in set *a* at table 1. Values of the relaxation time are 50,100,150,200 fs. The thick dotted line is the low power absorption spectrum.

Experimental results are presented in figures 3 and 4. For positive chirp, we see in figure 3, that a "hole" pattern was created in the absorption spectrum. This pattern is predicted by the theory and is due to the relatively narrow band (15-25nm) of the pulse. The "holes" in the experiment are narrower than those in the fits. A possible explanation is that the theory assumes that the pulse intensity is constant during the propagation through the sample. For achieving this condition we should use a dilute solution or a very thin sample. Using a dilute solution is impractical because the noise will be larger than the absorption changes. We also tried to use a thinner sample of 200 μm, and the results were the same. Therefore, a wave propagation analysis is required for more accurate fits. In figure 4, absorption spectra due to a negatively chirped pulse are presented for different intensities. A good correlation is achieved between the results and the theoretical fits. The experimental negative absorption in the spectra is smaller than the expected from the calculation. This discrepancy is due to the propagation effects and the spectral asymmetry of the pulse. This is the first time, to our knowledge, that theoretical calculated curves are fitted to absorption spectra, which are due to chirped pulse excitation.

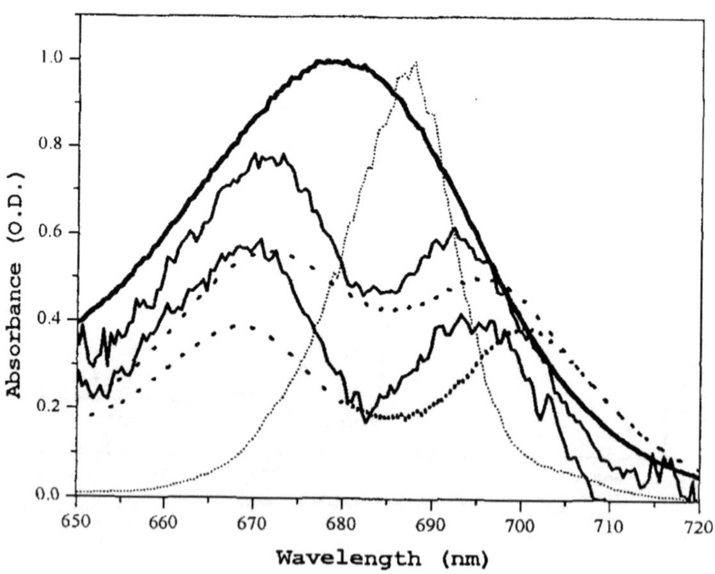

Figure 3: The absorption of R800 in methanol due to positive chirp pulse at 685nm ($\Phi''(\nu) = 50000$) and its theoretical fits (dotted lines). The lower lines for 2.5 μJ pulse. See parameters at set *b* in table 1. The bold line is the low power absorption spectrum. The thin dotted line is the pulse spectrum.

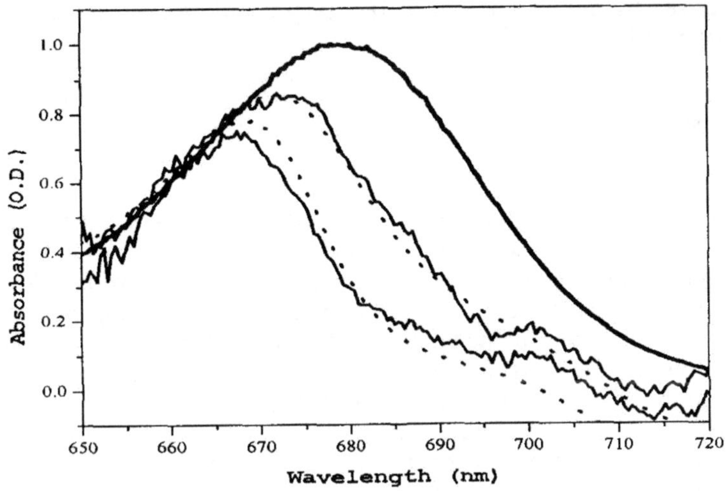

Figure 4: The absorption of R800 in methanol due to negative chirp pulse at 685nm ($\Phi''(\nu) = -50000$) and its theoretical fits (dotted lines). The lower lines for 2.5 μJ pulse. See parameters at set *b* in table 1. The bold line is the low power absorption spectrum.

We also simulated the effects of solvation energetic and dynamics on the absorption spectra. We compared the absorption spectra of a positive chirp pulse of R800 dissolved in fast relaxing solvents: methanol and acetonitrile. The relaxation time constant expresses the solvation effect in the solution. From table 3 in ref [17], we see that the solvation of coumarin 153 is nonexponential and could be described by a multiexponential decay. For acetonitrile about 70 percent of the Stokes shift is completed within the first 90 fs, but for methanol, the fastest substantial decay time component is rather long - 280 fs, which constitutes 34 percent of the Stokes shift. We could expect that effective decay time of R800 in acetonitrile will be faster than that of R800 in methanol. However, the solvents relaxation has rather a small effect on the absorption spectra. Simulations show that differences of the absorption spectra are more conspicuous at the wavelengths where negative absorption is seen. Therefore, a spectrally broader pulse should be applied for the absorption measurements in future experiments.

Another parameter that we should take into account more cautiously is the Stokes shift. Initially we substituted an empirical Stokes shift deduced from R800 fluorescence and absorption spectra. But theoretical fit based on this Stokes shift didn't match well the results. The reason for this discrepancy might be based on the complex solvation relaxation of R800. This means that the potential surfaces cannot be characterized by only a single low frequency coordinate but rather by a multi-dimensional coordinate.

5 Conclusions

In this paper, we examined the absorption spectra, which originated from chirped pulses as a possible diagnostic method for studying the solvation dynamics. We found that achieving an accurate fit requires a sample with a low optical density, at the cost of an increase of the signal to noise ratio, or alternatively improving the calculations by including the effects of the incident radiation through the sample propagation. However, the current theory provides rather good predictions of the change in the shape of the absorption band due to intense chirped pulse excitation. The strong chirp condition gives us a simple way to monitor the population difference dynamics through the absorption spectrum by the simple relation $\omega(t)=\omega-\mu(t-t_0)$. However, for this condition, the pulse duration in our experiments is too long to clearly distinguish between different solvation relaxation times (see Fig. 3). Using a shorter transform-limited (and spectrally wider) pulse better distinguishes between solvents. Another possibility for studying the solvation dynamics is to use chirped pulses that are not suited to the strong chirp condition and, therefore, their duration is not strongly different from that of a shorter transform-limited pulse. However, a simple relation between the absorption spectrum and the population wave packet dynamics is lost in this case [10,11]. The corresponding study concerning the non-strongly chirped pulses will be published elsewhere.

As for using strong linear chirp as a tool for controlling the molecule, the idea is to use the initial part of the pulse for excitation and the other part as a manipulator on the excited state population e.g. this part can induce stimulated emission. But to achieve substantial stimulated emission, we need a molecule, which has a larger Stokes shift and a fast relaxation time. For larger Stokes shift, a broader waveband is needed. Another possibility is to tailor a pulse by means of phase and amplitude modulator [18], the chirp of which is not necessarily a linear one, and the envelope of which is not necessarily gaussian, but it has other envelopes and chirps.

6 Acknowledgments

This work was supported by grants from the Israel Science Foundation and James-Franck German-Israel Program in Laser-Matter Interaction.

References

1. Gordon R. J. and Rice S. A., Active control of the dynamics of atoms and molecules, *Annu. Rev. Phys. Chem.* **48** (1997) pp. 601-641.
2. Paloviita A., Suominen K. A. and S. Stenholm, Molecular excitation by chirped pulses, *J. Phys. B* **28** (1995) pp. 1463-1474.
3. Ruhman S. and Kosloff R., Application of chirped ultrashort pulses for generating large-amplitude ground-state vibrational coherence: a computer simulation, *J. Opt. Soc. Am. B* **7** (1990) pp. 1748-1752.
4. Lin J. T., Hayashi M., Lin S. H. and Jiang T. F., Excitation of the NO molecule by chirped infrared laser pulses, *Phys. Rev. A* **60** (1999) pp. 3911-3915.
5. Cerullo G., Bardeen C. J., Wang Q. and Shank C. V., High-power femtosecond chirped pulse excitation of molecules in solution, *Chem. Phys. Lett.* **262** (1996) pp. 362-368.
6. Buist A. H. *et al.*, Probing microscopic chemical environments with high-intensity chirped pulses, *Opt. Lett.* **24**, (1999) pp. 244-246.
7. Fainberg B. D., Huppert D. and Segal J., Experimental and theoretical study of chirped pulse excitation, in *ultrafast phenomena XII. Springer series in chemical physics*, ed. by Elsaesser T., Mukamel S., Murnane M. M. and Scherer N. F. (Springer, Berlin Heidelberg, 2001),Vol. **66**, pp. 621-623.
8. Fainberg B. D., Nonperturbative analytic approach to the interaction of intense ultrashort chirped pulses with molecules in solution: Picture of "moving" potentials, *J. Chem. Phys.* **109** (1998) pp. 4523-4532.
9. Fain B. and Lin S. H., Phase-dependent absorption and stimulated emission in femtosecond pump-probe experiments, *Chem. Phys. Lett.* **207** (1993) pp. 287-293.
10. Fainberg B. D., Absorption spectrum of intense chirped pulse by molecules in solution and the time evolution of vibrationally non-equilibrium populations, *Chem. Phys. Lett.* **332** (2000) pp. 181-189.
11. Fainberg B. D. and Narbaev V., J. Chem. Phys., submitted.
12. Mukamel S., *Principles of nonlinear optical spectroscopy* (Oxford University Press, New York, 1995).
13. Fainberg B. D. and Narbaev V., Chirped pulse excitation in condensed phases involving intramolecular modes studied by double-sided Feynman diagrams for fast optical dephasing, *J. Chem. Phys.* **113** (2000) pp. 8113-8124.
14. Martinez O. E., 3000 times grating compressor with positive group velocity dispersion: application to fiber compensation in 1.3-1.6 μm region, *IEEE J. Quant. Elec.* **QE-23** (1987) pp. 59-64.
15. Kane D. J. and Trebino R., Single-shot measurement of the intensity and phase of an arbitrary ultrashort pulse by using frequency-resolved optical gating, *Opt. Lett.* **18** (1993) pp. 823-825.

16. Agmon N. and Kosloff R., Dynamics of two dimensional diffusional barrier crossing, *J. Phys. Chem.* **91** (1987) pp. 1988-1996.
17. Horng M. L., Gardecki J., Papazyan A. and Maroncelli M., Sub-picosecond measurements of polar solvation dynamics: Coumarin 153 revisited, *J. Phys. Chem.* **99** (1995) 17311-17337.
18. Weiner A. M., Heritage J. P. and Kirschner E. M., High-resolution femtosecond pulse shaping, *J. Opt. Soc. Am. B* **5** (1988) pp. 1563-1572.

ADAPTING OPTIMAL CONTROL THEORY TO A VARIETY OF MOLECULAR APPLICATIONS

REGINA DE VIVIE-RIEDLE*, THOMAS HORNUNG, ANGELIKA HOFMANN, CARMEN TESCH

Max-Planck-Institut für Quantenoptik, Hans-Kopfermann-Str.1, D-85748 Garching, Germany

E-mail: rdv@mpq.mpg.de

Microscopic understanding of ultrafast chemical and physical processes is a helpful basis to design control concepts using shaped laser pulses. We present applications of optimal control theory (OCT) to find the optimized laser fields for the selected objectives. Enclosure of the wavelength of the incident laser pulse in the OCT functional provides the best possible solution for the given molecule-laser system. Examples will be given for control of molecular reactions via conical intersections with the aim to manipulate velocity and final product formation and for control of vibrational population transfer as the basis for a new concept of a molecular quantum computer.

1 Introduction

Microscopic understanding of ultrafast chemical and physical processes is a helpful basis to design control concepts using shaped laser pulses. We present applications of optimal control theory (OCT) to find the optimized laser fields for the selected objectives. Enclosure of the wavelength of the incident laser pulse in the OCT functional provides the best possible solution for the given molecule-laser system. Examples will be given for control of molecular reactions via conical intersections with the aim to manipulate velocity and final product formation and for control of vibrational population transfer as the basis for a new concept of a molecular quantum computer.

2 Optimal Control Theory

The electric field of a femtosecond laser pulse is our optical tool to manipulate molecular processes by variation of its numerous pulse parameters. In general optimal laser fields for different control objectives can be found by variation of the functional K:

$$K\ [\Psi_i(t),\Psi_f(t),\varepsilon(t)] = \left|\langle\Psi_i(T)\,|\,\Phi_f\rangle\right|^2 - \alpha_0\int_0^T\frac{|\varepsilon(t)|^2}{s(t)}\cdot dt -$$

$$2\,\mathrm{Re}\left\{\langle\Psi_i(T)|\Phi_f\rangle\cdot\ \int_0^T\langle\Psi_f(t)|\frac{i}{\hbar}[H-\mu\varepsilon(t)]+\frac{\partial}{\partial t}|\Psi_i(t)\rangle dt\right\}. \qquad (1)$$

The resulting laser field ε (t) drives the system from an initial state $\Psi_i(0)=\phi_i$ at t=0 to the selected target state $\Psi_i(T)$ at time t=T. Variation of the functional *K* leads to a set of three coupled differential equations:

$$\begin{aligned}
&\varepsilon(t) = -\frac{s(t)}{\hbar\alpha_0}\cdot\mathrm{Im}\{\ \langle\Psi_i(t)|\Psi_f(t)\rangle\langle\Psi_f(t)|\mu|\Psi_i(t)\rangle\ \}\\
&i\hbar\frac{\partial}{\partial t}\Psi_i(t)=[H-\mu\varepsilon(t)]\Psi_i(t)\ ;\ \Psi_i(0)=\Phi_i\\
&i\hbar\frac{\partial}{\partial t}\Psi_f(t)=[H-\mu\varepsilon(t)]\Psi_f(t)\ ;\ \Psi_f(T)=\Phi_f.
\end{aligned} \qquad (2)$$

The first equation represents the optimal electric field in terms of the evolving wavepackets Ψ_i and Ψ_f. The two subsequent equations guarantee the compliance of the Schrödinger equation for Ψ_i and Ψ_f under the influence of the laser field ε (t). The overall shape function s(t) satisfies the demand of smooth switch on and off behavior[7]. The penalty factor α_0 limits the time averaged laser intensity and T denotes the overall pulse duration of the shaped pulse.

Eqs. (1-2) represent the standard optimal control equations for the iterative approach with insufficient constraints to obtain realizable pulses with a spectrum around a pre-specified wavelength. This is however of central importance to experiments, where a laser source together with a pulse shaping device can only be tuned over a specific wavelength range. Furthermore pulses with less than 5 femtoseconds are difficult to tailor. Optimal solutions beyond these experimental possibilities are not directly applicable and we therefore modified the functional given in eq. 1 to read:

$$\begin{aligned}
&L\ [\Psi_i(t),\Psi_f(t),\varepsilon(t)] =\\
&\sum_{k=1}^{2}\left(\left|\langle\Psi_i^k(T)\,|\,\Phi_f^k\rangle\right|^2 - 2\,\mathrm{Re}\left\{\langle\Psi_i^k(T)|\Phi_f^k\rangle\cdot\ \int_0^T\langle\Psi_f^k(t)|\frac{i}{\hbar}\left[H^k-\mu^k\varepsilon(t)\right]+\frac{\partial}{\partial t}|\Psi_i^k(t)\rangle dt\right\}\right)\\
&-\ \alpha_0\int_0^T\frac{|\varepsilon(t)|^2}{s(t)}\cdot dt\ .
\end{aligned} \qquad (3)$$

Here the sum runs over the molecular system with its free evolution Hamiltonian $H^{k=1}$ and its dipole transition moment $\mu^{k=1}$ and over the laser system represented as a two level system with a dipole transition moment $\mu^{k=2} >> \mu^{k=1}$. The additional objective $\left|\langle \Psi_i^{k=2}(T) | \Phi_f^{k=2} \rangle\right|^2$ represents the yield for the laser transition with given frequency ω_{12}. The present form of the functional assures the constant existence of the laser frequency ω_{12} in the optimized field that enables wavelength tuning and allows to find the best possible solution and control mechanism for a given laser system.

3 Optimization of a Molecule-Laser System

Exemplarily, we demonstrate the effect of the new functional (eq. 3) for the population transfer according to the vibrational transition v"=0 to v"=3 in the ground state ($X^1\Sigma_g^+$) of the potassium dimer which is mediated by the first excited state ($A^1\Sigma_u^+$). The equations for the OCT functionals (eqs. 1 and 3) are realized on *ab initio* potential energy and transition dipole surfaces. The Short Time Fourier Transform (STFT) of the various optimized laser pulses is shown in Fig. 1 together with their mask functions for transmission and phase. The mask function constitutes the direct interface with the experiment. In principle this function can be directly applied to the Liquid Crystal Spatial Light Modulator (LC-SLM) in the pulse shaper set up and serves as an "intelligent" initial guess for an experimental closed loop setup. The solution from the standard OCT (eq.1) results in a highly complex structured pulse in time and in frequency (Fig. 1, first column). The shaping process needs an initial Fourier limited pulse of nearly 5 fs (FWHM). The complexity of the pulse shape naturally translates into a complicated mask function for amplitude and phase resolved with more than 400 pixels. In summary, the solution offered by the standard OCT for this exemplary objective is hardly realizable in an control experiment.

Inclusion of the laser field parameter wavelength in (Fig.1, third column) or off resonance (Fig. 1, second column) with the main transitions for the molecular process allows to reduce the fields complexity. The intensity is decreased dramatically from $1.5 \cdot 10^{11}$ GW/cm^2 to $5.7 \cdot 10^{9}$ GW/cm^2; all non resonant transition frequencies (except the primary laser frequency) are eliminated and the resulting structure looses its unwanted complexity without significant loss of efficiency. The resulting mask functions show distinct frequency components and a less complex phase relation even for the detuned laser wavelength. The dots mark the relevant quantities for the phase function. When the laser system can be tuned in resonance with the dominant transition pathways, the resulting pulse structure becomes simple and adjustable with less then 50 pixels. An incident 50 fs Fourier limited pulse is now sufficient. These pulses are easy to generate experimentally with established techniques and can directly be used as initial guess delivered by theory. Further refinement in a closed loop experiment can be used to tackle with the imperfections in the molecular Hamiltonian and additional lab specific environmental effects. With this tool at hand we are able to design realistic and robust laser pulses for any given control objective.

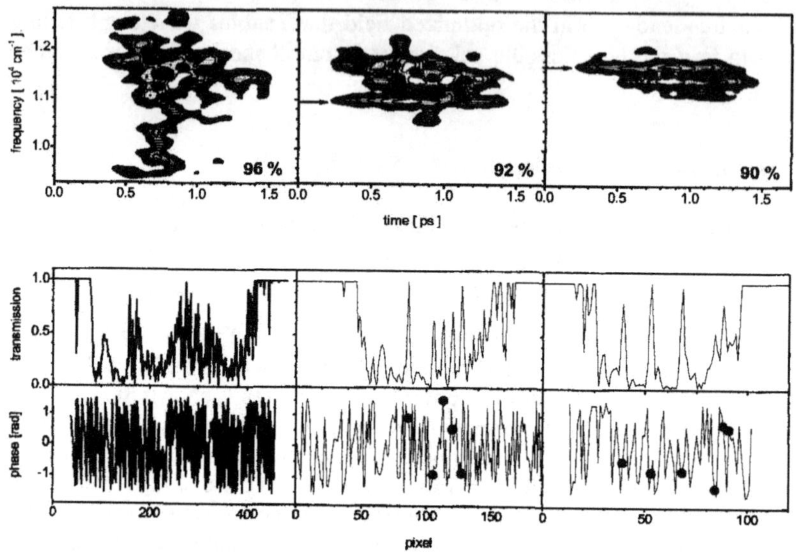

Figure 1. STFT (top) and extracted mask functions (bottom) for three different pulses, left: standard solution of OCT, middle: generalized solution of OCT including the incident laser system detuned from resonance, right: generalized solution including the incident laser system detuned from resonance.

4 Dynamics and Control Strategies for Ultrafast Reactions

Possible reactions of interest for coherent control are ultrafast reactions leading to different outcomes. Non-radiative transitions mediated by conical intersections (CoIns) fulfill these conditions. Conical intersections are regions of degeneracy between the electronic potential hypersurfaces involved which initiate barrierless and therefore ultrafast transitions between potential energy surfaces and act as forking points for different product channels. To define a control objective for a chemical reaction and to achieve its efficient control, it is helpful to study first, whenever possible, the systems spectroscopic properties and dynamical behavior.

We studied the photoinduced quantum dynamics of the electrocyclic ring opening of cyclohexadiene by means of wavepacket dynamics on non-adiabatically coupled potential energy surfaces[8,9]. The corresponding Schrödinger equation is given by:

$$i\hbar\frac{\partial}{\partial t}\Psi(q,t)=\left[\begin{pmatrix} T_{11} & K_{12} \\ K_{21} & T_{22} \end{pmatrix}+\begin{pmatrix} V_{11} & \mu_{12}\varepsilon(t) \\ \mu_{21}\varepsilon(t) & V_{22} \end{pmatrix}\right]\Psi(q,t) \tag{4}$$

with T the nuclear kinetic and V the potential energy part of the Hamiltonian, K_{12} the non-adiabatic coupling elements and $\mu_{12}\varepsilon(t)$ the time dependent laser-molecule interaction. The molecular system is treated in a two dimensional subspace, spanned by the reaction coordinates r and φ as defined in ref. 8. The corresponding potential energy surfaces, non-adiabatic coupling elements and transition dipole moments are calculated on the CASSCF level using the MOLPRO package. Implementation of the non-adiabatic coupling elements in the propagation routines is non-trivial, special care has to be taken to conserve the hermiticity of the molecular Hamiltonian[9]. Our first observation[8] was a branching of the wavepacket on the excited S_1 state towards two CoIns, labeled as $CoIn_{min}$ and C_2-CoIn in fig. 2, which are located at different positions in the reaction space.

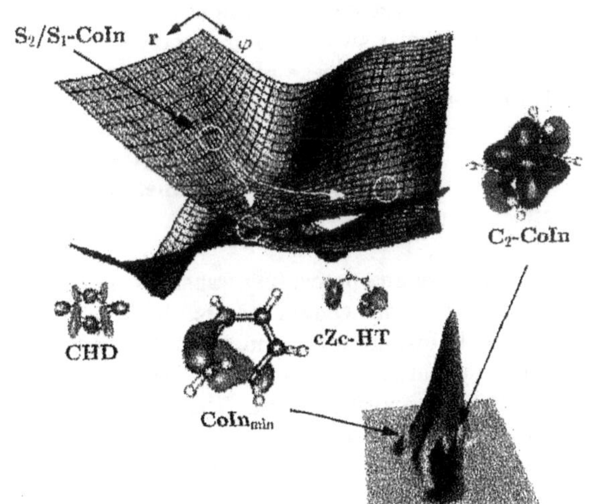

Figure 2. Ground- and excited state potential energy curves for the ring opening reaction of cyclohexadiene. Indicated are two conical intersections which allow the barrierless transfer to the groundstate, as well as their corresponding nuclear geometries forming a three center and four center bond, respectively.

In principle both CoIns allow the return to the S_0 ground state. From further quantum dynamical studies[9], we could deduce that the ring opening and product formation is indeed mediated by both conical intersections. Depending on which conical intersection is passed, different final product ratios for cyclohexadiene versus hexatriene are achieved. In addition the calculated nuclear geometries at the CoIns, a three center bond at $CoIn_{min}$ and a four center bond at C_2-CoIn (Fig. 2), initiate slow or fast reaction velocities. The transfer through C_2-CoIn is much more efficient since the four center bond geometry gives rise to a spatially broad coupling element and allows transfer to the ground state in a wide nuclear range. These features make the system an interesting prototype candidate to develop concepts for coherent control with respect to product distribution and reaction velocity.

Preliminary calculations varying the pulse width of the pump laser pulse[10] confirmed that the branching behavior of the prepared wavepacket can be influenced by the exciting laser light. Our future aim is to drive the reaction with optimized laser fields systematically through selected conical intersections and thereby to manipulate precisely the velocity and the product ratio. We could already demonstrate the focussing of wavepackets at conical intersections applying OCT for another three dimensional model system[10,11]. The transfer of these concepts to the ring opening reaction is subject of ongoing work. The microscopic control over the ring opening and product formation is of special interest to us because cyclohexadiene is found as the reactive center in larger molecules, such as fulgides, which are regarded as molecular switches with possible importance as nano-scale devices. Our future aim is to drive these molecules as ultrafast switches with the help of modulated laser pulses.

5 Molecular Quantum Computation: Implementation of a 2-qubit Quantum Gate

In the quantum world information processing translates into quantum information, with the quantum computer as one central research topic[12]. The quantal analogues of the classical "0 "and "1" bits are the quantum bits (qubits). They are realized by well defined two level quantum systems which in contrast to classical systems can also be prepared in their superposition states. The advent of this research field is dominated by concepts from the quantum optics community choosing atoms, ions, or quantum dots as an appropriate medium to realize a first quantum computer[13,14,15,16].

Recently, we proposed a new concept using molecules as the active device. We defined molecular normal modes, their overtones and combinations as basis states of a qubit system. Our molecular model system is acetylene, but in principle any molecular system can be used. For acetylene the two IR active modes are selected as basis for the qubit system. Each IR-active mode and its excitation define two separate Hilbert spaces, with a ground state $|0\rangle$ defined by n quanta and an excited state $|1\rangle$ defined by $n' \geq n+1$ quanta of the same vibrational mode. The complete two-qubit system is described as the direct product of the individual Hilbert spaces. Within the complete set of the five normal modes ($n_1n_2n_3n_4n_5$), the basis states read as:

$$|00\rangle = |00101\rangle$$
$$|10\rangle = |00201\rangle$$
$$|01\rangle = |00102\rangle$$
$$|11\rangle = |00202\rangle$$

The first entry refers to the asymmetric CH-stretch and the second entry to the cis-bending mode. The labeling is ambigious, any other allowed combination of quanta within either the IR or Raman active mode is also possible. We selected low lying vibrational levels to keep the interaction with the environment as low as possible and thus to avoid dephasing. The transition between two qubit states was selected to be a one photon process avoiding thereby enhancement of the necessary laser intensity. According to the two IR active modes we derived a two-dimensional nuclear model Hamiltonian. The corresponding *ab initio* potential energy and dipole moment surfaces are evaluated on the MP2 level and serve as input for the subsequent quantum dynamical calculations[17]. For state preparation and implementation of quantum logic operations according to Deutsch's requirements OCT is applied, i.e. the quantum gates are realized by modulated femtosecond pulses.

In our previous study[18], we could successfully demonstrate the implementation of 1-qubit quantum gates within each qubit separately. However, it became obvious that in a molecular two-qubit system, like in acetylene, it is important to differentiate between active and passive qubits. If the quantum gate operates on the asymmetric CH-stretch, the cis-bending mode, i.e. the second entry, is the passive qubit and vice versa. The influence of the passive qubit plays a relevant role for the construction of global gates. The pulse representing a global 1-bit quantum gate such as the Hadamard transformation:

$$H|0\rangle \rightarrow \frac{1}{\sqrt{2}}\left[|0\rangle + |1\rangle\right] \quad and \quad H|1\rangle \rightarrow \frac{1}{\sqrt{2}}\left[|0\rangle - |1\rangle\right]$$

must transform the two qubit basis state $|nn'\rangle$ into the corresponding superposition state regardless whether the passive qubit is in $|0\rangle$ or in $|1\rangle$. This requires flexible laserfields which can only be found with the help of OCT. The influence of the passive qubit is a typical molecular property and originates from the spatial variation of the nuclear wavefunctions representing the basis states. With increasing quantum numbers (e.g. compare input states in Fig. 3) the number of nodes increases.

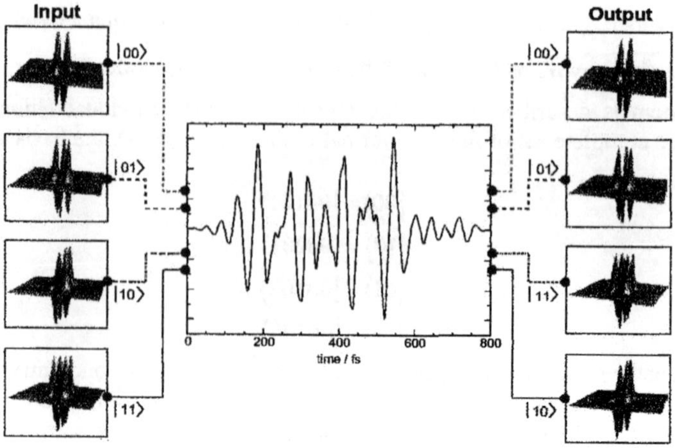

Figure 3. CNOT gate operating on the second qubit, only when the control quibt is in |1>.

This spatial variance of the basis states complicates the structure for the 1-qubit quantum gates, but it naturally provides the indispensable feature for the 2-qubit quantum gate, namely the controlled NOT (CNOT). Now the passive qubit takes over the function of a control qubit. The time evolution of a possible field for a CNOT gate operating on the second qubit is shown is in Fig. 3. Here, the first qubit is selected as control qubit and the CNOT gate switches the second qubit from $|0\rangle$ to $|1\rangle$ or vice versa only when the control qubit is in $|1\rangle$. The first two input states do not fulfill this condition and are left unchanged by the applied laserfield while the third and forth input state meet the condition and their second qubits are switched, respectively. The realization of the CNOT gate demonstrates the ability to implement 2-qubit gates in a molecular system using OCT algorithms and signifies a step ahead on the way to realize a molecular quantum computer.

6 Conclusion

The selected molecular processes are potential candidates to establish new reaction pathways, to develop concepts to drive molecular sized ultrafast switches which may become an issue for optical communication and to realize elements of molecular quantum computing by combining OCT concepts with molecular properties. With the help of new functionals we are able to provide realizable optimized laser fields, fulfilling multiple control aims at once (i.e. quantum computing) and possibly adapt the constraints from experiment without relevant loss of efficiency. The given examples underline the flexibility of the applied OCT algorithms and their applicability to a vast area of interest.

References

1. A.H. Zewail, Femtochemistry: Atomic-Scale Dynamics of the Chemical Bond, J. Phys. Chem. A, **104** (2000) 5660.
2. W. Zhu, J. Botina, and H. Rabitz, J. Chem. Phys., **108** (1998) 1953.
3. H. Rabitz, R. de Vivie-Riedle, M. Motzkus, K. Kompa, Science, **288** (2000) 824-828.
4. A. Assion, T. Baumert, T. Brixner, B. Kiefer, V. Seyfried, M. Strehle, and G. Gerber, Science, **282** (1998) 919.
5. R. de Vivie-Riedle and K. Sundermann, Appl. Phys. B, **71** (2000) 285-292 .
6. T. Hornung, M. Motzkus, and R. de Vivie-Riedle, J. Chem. Phys, **115** (2001) 3105-3110.
7. K. Sundermann and R. de Vivie-Riedle, J. Chem. Phys., **110** (1999) 1896-1904.
8. A. Hofmann and R. de Vivie-Riedle, J. Chem. Phys., **112** (2000) 5054-5059
9. Angelika Hofmann and Regina de Vivie-Riedle, Chem. Phys. Lett., **346** (2001) 299-304.
10. Dorothee Geppert, Diploma thesis, LMU-München, Dezember 2001.
11. R. de Vivie-Riedle, K. Sundermann, and Marcus Motzkus, Faraday Discuss., **113** (1999) 303.
12. A. Ekert, and R. Josza, Rev. Mod. Phys. **68**, 733 (1996)
13. M. Brune, F. Schmidt-Kaler, A. Maali, J. Dreyer, E. Hageley, J.M. Raimond, S, Haroche, Phys. Rev. Lettt. **76**, 1800 (1996).
14. J.I. Cirac, P. Zoller, Phys. Rev. Lett. **74**, 4091 (1995).
15. H.C. Nägerl, F. Schmidt-Kaler, J. Eschner, R. Blatt, W. Lange, H. Baldauf, H. Walther, in: D. Boumeester, A. Zeilinger (Eds.), The Physics of Quantum Information, Springer, Berlin, 2000, p. 163.
16. E. Biolatti, R. C. Iotti, P. Zanardi, and Fausto Rossi, Phys. Rev. Lett. **85**, (2000) 5647.
17. Carmen M. Tesch, Karl-Ludwig Kompa, and Regina de Vivie-Riedle, Chem. Phys., **267** (2001) 173-185.
18. Carmen M. Tesch, Lukas Kurtz, and Regina de Vivie-Riedle, Chem. Phys. Lett., **343** (2001) 633-641.

PREPARING ARBITRARY SUPERPOSITION STATES WITH MINIMUM NUMBER OF LASER PULSES: THE CARP METHOD

BO Y. CHANG, IGNACIO R. SOLÁ AND JESÚS SANTAMARÍA

Dpt. de Química Física I, Universidad Complutense, 28040 Madrid, Spain

VLADIMIR S. MALINOVSKY

Michigan Center for Theoretical Physics & FOCUS Center, Dpt. of Physics, University of Michigan, Ann Arbor, MI 48109

We study the dynamics of the Chirped Adiabatic Raman Passage scheme in a 3-state Λ system, where the final state is a manifold of closely spaced sublevels. Depending on the symmetry of the Hamiltonian, a single state of the manifold or a superposition state can be prepared. Despite working in a fully adiabatic regime, non-adiabatic population plays an important role between the sublevels of the manifold. We estimate the degree of control that is possible in preparing arbitrary superposition states in a simple and robust manner.

1 Introduction

The study of the quantum dynamics of simple systems is still a rich source of new discoveries. In the field of control of molecular motion it provides the essential models that allow us to understand how more complicated (for instance computationally optimized [1]) schemes work. Moreover, given a proper time-energy regime, these models can even be directly applied to the physical problem at hand. [2] In this contribution we study the dynamics of the simplest Raman scenario, modeled by a 3-level Λ system. We propose novel ways to control the population transfer to the final state when it consists of several closely spaced sublevels. The main novelty dwells in the target, which is not a single sublevel but the creation of an arbitrary superposition state. We show the degree of control that can be exerted with a single chirped pulse sweeping the frequencies of the different sublevels.

It is well known that using chirped pulses it is possible to completely invert the population between two coupled states in a very robust way. Unlike transformed limited pulses, that induce Rabi cycling between the ground and the excited state, the chirped pulses decouple the stimulated emission after transferring all the population to the excited state, provided the pulse area is much larger than π. In this case the population transfer at late times is approximately ruled by a Landau–Zener formula of the type $P_2 \approx 1 - \exp\left(-\pi\Omega^2/\chi\right)$, where $\Omega = E_0\mu_{12}/\hbar$ is the Rabi frequency of the transition (the coupling frequency given by the product of the amplitude of the field and the transition dipole moment) and χ is the rate of frequency change. Therefore if $\Omega^2 \gg \chi$ all the population will passage to the excited state at late times. This is the basis of any Rapid Adiabatic Passage (RAP) scheme. [3]

There are also other advantages in using chirped pulses. Since the frequency varies with time, there are no strict resonance requirements. The frequency swept can excite a broader sample of atoms or molecules from the Maxwell-Boltzmann distribution. Moreover, since the frequency components work at different times, the excitation can be selective even when the original transformed limited pulse bandwidth is much broader than the energy spacing between adjacent levels. [4] The population will be driven to the level whose energy first enters into resonance with the energy of the initial state. Therefore the method is both efficient, selective and robust.

When going to population transfer on 3-level systems by means of two laser pulses, which represents the simplest scenario of a stimulated Raman method, there are new factors that must be taken into account. Of course a "direct" sequence of transformed limited pulses with pulse areas that are odd multiples of π, can lead to population transfer from the ground state $|1\rangle$ to the electronically excited state $|2\rangle$, and then from $|2\rangle$ to the vibrationally excited state $|3\rangle$, by a resonant stimulated Stokes process. A more stable, non pulse area dependent transfer, would imply using two frequency chirped pulses. But in both cases the preparation of the final state is mediated by full population transfer to the excited intermediate state. Therefore fluorescence can drastically reduce the yield of the process for longer time excitations. Lighting the system with a counter-intuitive sequence, with the Stokes pulse preceding the pump pulse, instead of the direct sequence, induces population transfer through a dark resonance or trapped state, implying no excitation of the intermediate state. This is the basis of the reputed Stimulated Raman Adiabatic Passage (STIRAP) scheme. [5]

However, the STIRAP method cannot be implemented for Raman Λ systems using chirped pulses. [6] The conditions for trapping the population require two-photon resonance, which can only occur at one time for chirped pulses. Therefore, alternative schemes must be proposed to implement robust Raman chirped adiabatic passage that at the same time minimize the intermediate state population. Recently we have proposed two such schemes. [7] In the first one, named Chirped Raman Adiabatic Successive Passage (CRASP), we use two fully overlapping pulses that are chirped in parallel directions (either both positively or both negatively). Then the population is driven through two avoided crossings following the direct sequence $|1\rangle \rightarrow |2\rangle \rightarrow |3\rangle$. The population of the intermediate state is reduced by minimizing the time between the crossings and therefore the time of stay in $|2\rangle$. This is most efficiently done by small detunings at the time of the crossings.

In the second one, named Chirped Adiabatic Raman Passage (CARP), the pulses are chirped with opposite directions[a] and largely detuned with respect to any intermediate excited state. Then the population passage from $|1\rangle$ to $|3\rangle$ follows a single effective two-photon crossing, with Landau–Zener properties. The CARP scheme can also be implemented with strong pulses, inducing a continuous Raman Stokes vibrational excitation. [8] In CARP the population of the intermediate state is minimized by energy detuning. However, a careful examination shows that the coherent properties of the adiabatic transfer state subtly influence the population dynamics in $|2\rangle$. [7]

More surprises appear when the final state is a manifold of closely spaced (almost degenerate) sublevels. According to the selectivity rules of chirped adiabatic passage, the initially prepared adiabatic state determines the final sublevel selected by symmetry of the Hamiltonian. The question then appears: Can this selectivity rule be broken so that we can create controlled superposition states? In a recent paper we have partially anticipated the answer. [9] For some initially prepared adiabatic transfer states, non-adiabatic population passage can not be neglected, leading to lost of selectivity. Nevertheless, this non-adiabatic passage can be controlled allowing to prepare arbitrary superpositions of two states. In this contribution we further explore the possibilities of the method analyzing the flexibility to prepare superposition states with more than two final sublevels.

2 The CARP system

We study the CARP dynamics in a Λ system with a final manifold of sublevels $\{|3a\rangle, |3b\rangle, |3c\rangle, \ldots\}$ by numerically obtaining the dressed states and solving the time-dependent Schrödinger equation. For brevity we limit the equations for a system where the final state has three sublevels. In Fig. 1 we show a scheme of the energy levels of the system.

The following simplest Hamiltonian in reduced units [9] reproduces the majority of the facts of interest to us:

$$H = \frac{1}{2}\begin{pmatrix} 0 & -\Omega(t) & 0 & 0 & 0 \\ -\Omega(t) & 2\Delta & -\Omega(t) & -\Omega(t) & -\Omega(t) \\ 0 & -\Omega(t) & 2(\chi t - \Delta\omega_0 + \delta) & 0 & 0 \\ 0 & -\Omega(t) & 0 & 2\chi t + 2\delta & 0 \\ 0 & -\Omega(t) & 0 & 0 & 2(\chi t + \Delta\omega_0 + \delta) \end{pmatrix}. \quad (1)$$

[a] Alternatively one pulse could be frequency fixed and the other chirped.

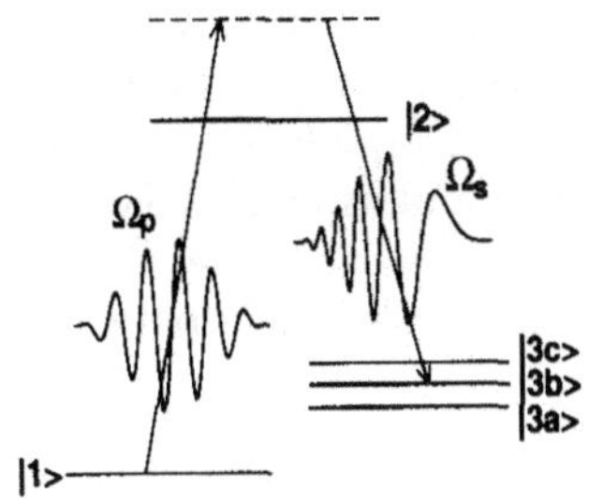

Figure 1. Scheme of a CARP method implemented on a Λ system with a final manifold of 3 sublevels.

In Eq. (1) all magnitudes, including time, are scaled with respect to the width of the pulse, σ. [9] We use two synchronous pulses with the same amplitude and shape and we also assume that the intermediate state has equal coupling strength with all the final sublevels, given by the Rabi frequency $\Omega(t) = \sigma A_0 S(t/\sigma)\mu_{2,3\alpha}/\hbar$ (where A_0 is the maximum amplitude of the field, $S(t)$ is the pulse shape, usually taken as $\cosh^{-2}(t)$, and $\mu_{2,3\alpha}$ is the corresponding transition dipole moment). In this system we choose the pump pulse as frequency fixed while the Stokes pulse is taken as linearly chirped, with chirp rate $\hbar\chi/\sigma^2$. Given the symmetry of the Hamiltonian, the same results are obtained for chirped pump pulses with opposite sign. Also we have taken $\Delta\omega_0 \ll \chi\sigma$ as a constant splitting between the sublevels, which is smaller than the bandwidth of the pulse. Finally, $\hbar\Delta/\sigma$ is the detuning with respect to the intermediate state, and $\hbar\delta/\sigma$ the two-photon detuning with respect to the final state, both at the time of the crossing. Therefore, if we choose $\delta = 0$ (as in all the simulations shown in this contribution) then the sublevel $|3b\rangle$ is being selected by resonance. However we will see that this condition does not determine the outcome of the dynamics.

3 Results and Conclusion

As depicted in Fig. 2, corresponding to the simple Hamiltonian of Eq. (1) there are four different scenarios of population transfer depending on the sign of chirp and detuning. In this figure we show the quasi-energies of the dressed states numerically calculated. For $\chi < 0$ and $\Delta > 0$ the system is initially prepared in the dressed state $|\Phi_-\rangle$, which adiabatically drives all the population to $|3a\rangle$. Reversing both signs we have the case $\chi > 0$ and $\Delta < 0$, which prepares the system initially in $|\Phi_+\rangle$, driving all the population to $|3c\rangle$. As in CRASP or other adiabatic passage schemes, in principle it should be expected that the sign of chirp selects the upper sublevel $|3c\rangle$ (with positive chirp) or the lower sublevel $|3a\rangle$ (with negative chirp) from all the members of the final manifold. However, rather than selecting a single level, the other two possible arrangements prepare a superposition state. For $\chi < 0$ and $\Delta < 0$ the initially prepared dressed state is $|\Phi_{0a}\rangle$. But near the encircled region (see Fig. 2) there are non-adiabatic couplings that lead to some population flow to $|\Phi_{0b}\rangle$ and to $|\Phi_{0c}\rangle$, building some final population in $|3b\rangle$ and $|3c\rangle$. On the same way, for $\chi > 0$ and $\Delta > 0$ we initially prepare the system in $|\Phi_{0c}\rangle$, but non adiabatic population transfer leads to some final population in $|3b\rangle$ and $|3a\rangle$. It must be stressed that this non-adiabatic population transfer is only between sublevels of the final manifold. The parameters of the CARP scheme are chosen to assure an adiabatic regime of population passage between the initial $|1\rangle$ and the final $\{|3a\rangle, |3b\rangle, |3c\rangle\}$ states.

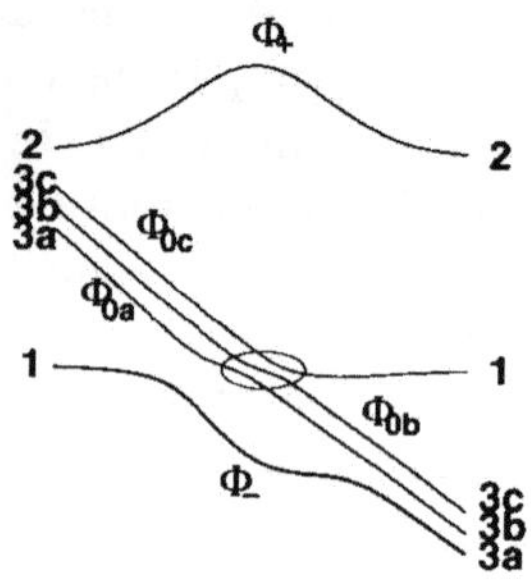

(a) $\chi < 0; \Delta > 0$

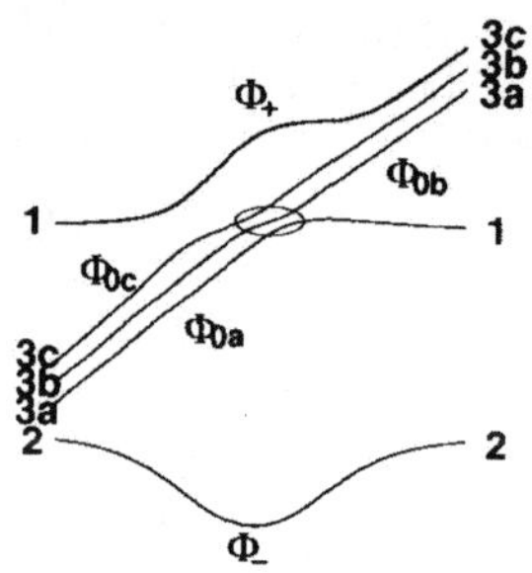

(b) $\chi > 0; \Delta < 0$

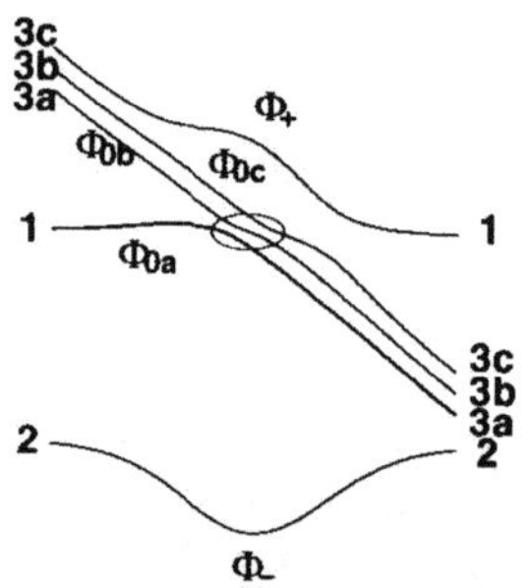

(c) $\chi < 0; \Delta < 0$

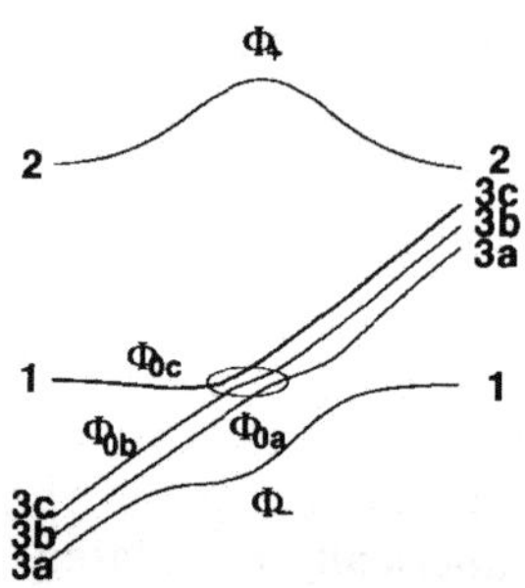

(d) $\chi > 0; \Delta > 0$

Figure 2. Dressed states evolution for the different scenarios of adiabatic passage in a Λ system with a final manifold of three sublevels. The thicker line represents the initially populated dressed state and the circle encloses the region with a higher probability of non adiabatic crossing. Depending on the sign of chirp and detuning the passage takes place on a different dressed state and a different final sublevel in the manifold ($|3a\rangle$ or $|3c\rangle$) is selected. The dressed states were obtained diagonalising the Hamiltonian of Eq. (1) with parameters $\Omega(0) = 120$, $\Delta = \pm 300$ and $\chi = \pm 50$.

Finally, the only way to drive all the population to the intermediate sublevel in the final state $|3b\rangle$ is by using pulses with small bandwidth so that $|\chi|\sigma < \Delta\omega_0$. In this case the effect of the remaining sublevels can be neglected and the structure of the Hamiltonian is simplified. Then every dressed state $|\Phi_{0a/b}\rangle$ or $|\Phi_\pm\rangle$ correlates at late times with $|3b\rangle$.

The properties of the selective adiabatic passage of CARP to a single level have been discussed in detail by Solá *et al*[7] and Chang *et al*[9], therefore we will concentrate more on the possibility of preparing any arbitrary superposition state, with any arbitrary number of sublevels coupled by just one pulse, the Stokes pulse. Although the non-adiabatic couplings are responsible for the relative populations on the superposition, the most interesting point of the method is that these populations are very robust with respect to changes in the Rabi frequencies. In Fig. 3 we show numerical results of the CARP dynamics via the $|\Phi_{0a}\rangle$ dressed state. For the parameters shown in the plot a superposition state is prepared with populations $p_{3a} \approx 0.71$, $p_{3b} \approx 0.21$ and $p_{3c} \approx 0.08$. However, although the remaining parameters are almost fixed, in the lower part of the figure the amplitude of the field is four times larger (therefore the pulse intensity is 16 times larger) yet the final result is almost identical. Although the Rabi oscillations of the different sublevels are increased, at final time the population of all the sublevels in the superposition is almost the same.

This result opens the door to the possibility of preparing superposition states in a controlled and simple way. The control parameter can be either the detuning or the chirp rate. [9] In Fig. 4 we show how the final populations of the sublevels change with variations in the (negative) chirp rate, that correspond to different initial transformed limited pulses. The width of the pulse and the Rabi frequency are fixed (and therefore also the pulse area, with $A \approx \Omega(0)\sigma \approx 40\pi$), and we show results for different blue detunings. The selected conditions ($\Delta < 0$, $\chi < 0$) correspond to the scenario where the initially prepared dressed state is $|\Phi_{0a}\rangle$ that creates a superposition state with larger component of the $|3a\rangle$ sublevel. By changing the signs of chirp and detuning the role of $|3a\rangle$ and $|3c\rangle$ would be symmetrically exchanged.

For small values of the chirp, the sublevel $|3b\rangle$ is selected for resonance considerations, as already explained. For $|\chi|\sigma$ sufficiently larger than $\Delta\omega_0$ but still small, the non-adiabatic couplings are not dominant and $|\Phi_{0a}\rangle$ almost correlates with $|3a\rangle$ at late times. However, increasing the chirp rate leads to a more balanced superposition. Finally, for larger values of the chirp the population in $|3b\rangle$ stays almost locked in one-third, while the population in the more separated $|3c\rangle$ sublevel exceeds that of $|3a\rangle$. However, for fixed Rabi frequency, the chirp rate cannot be increased more because then the

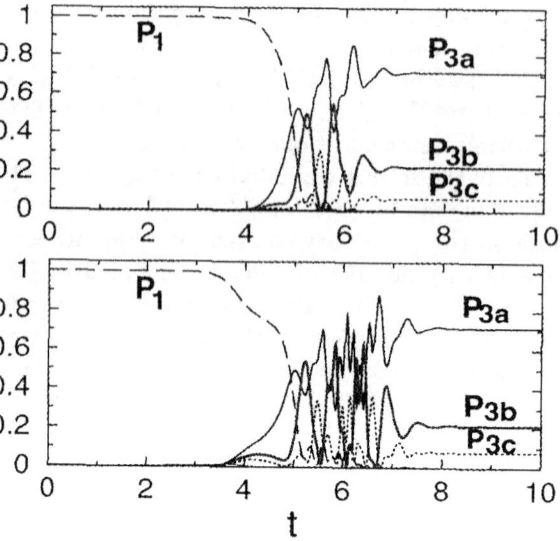

Figure 3. Population histories for the excitation of a superposition state using $\Omega(0) = 150$, $\Delta = -350$ and $\chi = -48$ for the upper plot, and $\Omega(0) = 600$, $\Delta = -350$ and $\chi = -65$ for the lower plot. Although the intensity is much higher and the dynamics are different, the final state is almost the same. The population of the intermediate state is always close to zero.

population passage from the initial to the final state is no longer adiabatic, and some population remains in $|1\rangle$.

From the results of Fig. 4 we can estimate the degree of control in the populations of the superposition. The population of $|3a\rangle$ or $|3c\rangle$ (depending on the initial symmetries) can vary almost in the whole range from 0% to 100%. The population of $|3b\rangle$ can only be safely controlled from 5% to 33% (in the regime not depending on resonance conditions). However, this does not allow for the preparation of arbitrary superposition states. Only the population of the dominant member of the superposition (either the upper or the lower sublevel) can be arbitrarily controlled from 40% to almost 100%. The populations of the remaining sublevels are not independent. Alternatively we could arbitrarily fix the relative population of the intermediate and the less dominant sublevels, but then the population of the predominant sublevel would be related to these populations.

This conclusion can be extended to the dynamics of systems with any

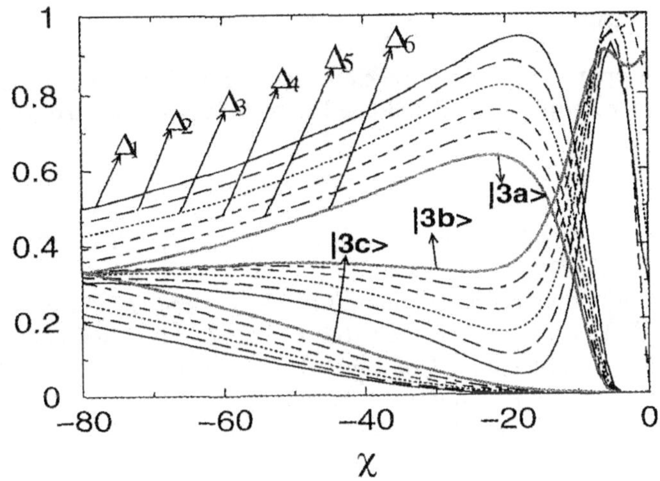

Figure 4. Preparation of arbitrary superposition states by changing the chirp rate for $\Omega(0) = 120$ and different values of detuning: $\Delta_1 = -300$, $\Delta_2 = -400$, $\Delta_3 = -500$, $\Delta_4 = -600$, $\Delta_5 = -700$ and $\Delta_6 = -800$. The first six lines correspond to the population in $|3a\rangle$, the following six to the population in $|3b\rangle$ and the final six to the population in $|3c\rangle$. For the populations in $|3b\rangle$ and $|3c\rangle$ the highest values correspond to the largest detunings, so that the order of the detunings is reversed with respect to that in the population of $|3a\rangle$.

number of final sublevels. We can control the population of the predominant member of the superposition, that can be either the upper or the lower sublevel. The remaining population is spread through the remaining sublevels. Only a small degree of control can be achieved between these relative populations. This is not surprising since in the CARP scheme only one single pulse is used to couple to all the final sublevels. Despite this limitation, the method is very simple to use and very robust. The populations of the superposition state barely change for different Rabi frequencies, pulse shapes or two-photon detunings. Of course, very different relative values of the dipole moments can reduce the degree of controllability. Finally, new scenarios of CARP appear when the initial state is also a manifold of closely spaced sublevels. But this will be the subject of a different contribution.

Acknowledgments

Financial support from the Dirección General de Investigación Científica y Técnica under Project No. PB98-0843 is gratefully acknowledged.

References

1. R.S.Judson and H.Rabitz, *Phys. Rev. Lett.* **68**, 1500 (1992); A.Assion *et al*, *Science* **282**, 919 (1998).
2. B.M. Garraway and K.-A. Suominen, *Rep. Prog. Phys.*, **58**, 365 (1995).
3. M. M. T. Loy, *Phys. Rev. Lett.* **32**, 814 (1974); L. Allen and J.H. Eberly, *Optical Resonance and Two Level Atoms* (Dover, New York, 1975).
4. J. S. Melinger, S. R. Gandhi, A. Hariharan, J. X. Tull, and W. S. Warren, *Phys. Rev. Lett.* **68**, 2000 (1992).
5. K. Bergmann, H. Theuer, and B. W. Shore, *Rev. Mod. Phys.* **70**, 1003 (1998).
6. B.W.Shore, K.Bergmann, A.Kuhn, S.Schiemann, J.Oreg and J.H.Eberly, *Phys. Rev.* A **45**, 5297 (1992).
7. I. R. Solá, V. S. Malinovsky, B. Y. Chang, J. Santamaría, and K. Bergmann, *Phys. Rev.* A **59**, 4494 (1999).
8. S. Chelkowsky and A. D. Bandrauk, *J. Raman Spectr.* **28**, 459 (1997); J. C. Davis and W. S. Warren, *J. Chem. Phys.* **110**, 4229 (1999); B. Y. Chang, I. R. Solá and J. Santamaría, *J. Phys. Chem.* A **105**, 8864 (2001); B. Y. Chang, I. R. Solá and J. Santamaría, *Chem. Phys. Lett.* **341**, 373 (2001).
9. B. Y. Chang, I. R. Solá, V. S. Malinovsky, and J. Santamaría, *Phys. Rev.* A **64**, 3420 (2001).

PERFORMANCE OF EVOLUTIONARY STRATEGIES IN COHERENT CONTROL STUDIES

D. ZEIDLER, S. FREY, K.-L. KOMPA, AND M. MOTZKUS

Max-Planck-Institut für Quantenoptik, D-85748 Garching, Germany
Tel: ++49 89 32905-216, Fax: ++49 89 32905-200
email: Marcus.Motzkus@mpq.mpg.de

Coherent control aims to steer a physical or chemical process via modulated femtosecond laser pulses. The characteristics of the pulses that drive the system to the desired target state are determined recursively in a self-learning loop, which connects a femtosecond pulse shaper, an optimization algorithm and an experimental feedback signal. The reliability of the feedback loop is essential to the significance of the optimization results, assigning the optimization algorithm an important role within these learning loops. The stability and effectiveness of the algorithm is investigated both in experiments and simulations with emphasis on the influence of number of configurations in search space and noise. The algorithm optimizes a set of variables parameterizing the electric field. This particular mapping greatly facilitates the dissection of the optimization goal.

1 Introduction

Laser-assisted control of chemical or physical processes aims to manipulate the dynamics of a system by means of a coherent laser field such that the system is steered towards a desired state or reaction channel[1]. Control schemes have been developed working either in the frequency domain[2] with cw lasers or in the time domain[3] with broadband ultrashort lasers. A generalization of the latter approach based on feedback-controlled shaping of ultrashort pulses[4] currently represents the most commonly pursued approach for coherent control[5].

The tools for the experimental realization of such a feedback loop comprise a source for ultrashort laser pulses, a computer-controlled pulse shaper and an optimization algorithm. The ultrashort laser pulse excites a sample, and some property of the system is monitored. This signal serves as feedback for the optimization algorithm which controls the shaper and proposes a new pulse and so forth until some convergence criterion is met. Since no a-priori knowledge of the physical system is necessary, this feedback scheme is applicable to all kinds of systems, even if intractable on a quantum mechanical level[6,7]. Of critical importance is a reliable optimization algorithm which searches for the global optimum in a multidimensional parameter space, since minimizing the data acquisition time in complicated control experiments, which may be running stably for only a short time, is essential.

The purpose of this article, which is a digest of a previously published paper[8], is to reveal the performance of evolutionary strategies (i.e. finding the global optimum, fast convergence) in coherent control studies under different optimization conditions. The convergence behavior with respect to the number of free variables and noise is exemplified and the influence of different parameterizations of phase and amplitude functions applied to the pulses is discussed.

2 Evolutionary algorithms

Evolutionary algorithms[9,10] attempt to imitate the process of evolution in a code which simulates selection, mutation and recombination within a feedback-controlled regulation loop. An initial set of vectors is generated at random. Each vector represents one search point in parameter space where the objective function is to be sampled. In analogy to evolution in nature, each search point is called an individual, the whole set of candidates is one population. At the onset of each iteration loop, the objective function for each search point in parameter space is determined. In biological terms, each individual is rated according to its fitness. Selection determines those vectors which will serve as a basis for the next loop iteration. The essential step within each iteration is replication, i.e. to generate new search points in parameter space by means of operators which mirror biological recombination and mutation processes. Numerical representations of recombination operators combine two parent vectors picked randomly from those selected in the preceding step to create a new child vector[8]. The mutation operator stands for a random change of the value of a vector element with a predefined mutation probability. The so-called step length represents the amount of change of the vector element during a mutation and has to be chosen suitably in order to achieve convergence. To this end, step-length control schemes are routinely used[9]. Finally, the new vectors are assembled to form a new population which is tested in the next iteration of the loop.

3 The role of steering parameters

The maximization of the time-integrated second harmonic (SH) generation of the output pulses delivered by a Ti:Sa pumped optical parametric amplifier (OPA) serves to demonstrate the performance of the algorithm under different conditions[8]. SH optimization has been extensively studied[11,12,13] and is attractive for several reasons: (a) the experimental realization is straightforward, (b) it is stable since there are no major noise sources other than the

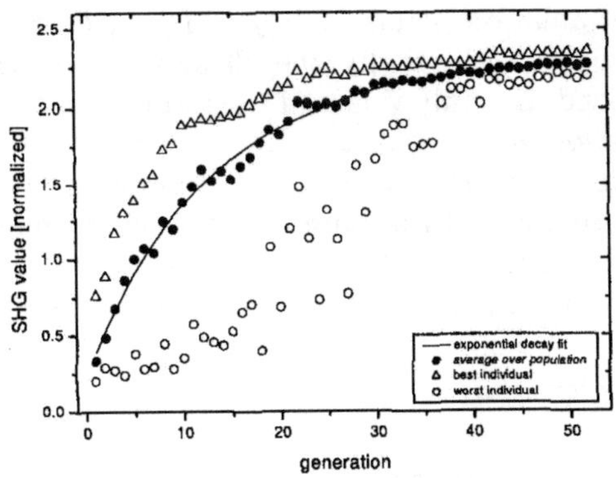

Figure 1. Evaluation of the optimization data. The average (•), the best (△) and the worst (○) fitness values are recorded for each generation n of an optimization run. An exponential rise $f(n) = a + b \cdot \exp(-c \cdot n)$ with the fit parameters a, b, and c is fitted to the average fitness values. The dimensionless parameter c provides a measure of the convergence speed.

laser, (c) the physical process can be simulated fairly easily, (d) it is sufficient to apply pure phase shaping.

In each optimization run, the best, the worst, and the average SH values were stored for each pass of the loop, i.e. for one set of 48 parameters to monitor the evolution of the SH signal versus the generation n. The convergence data of every scan were normalized with respect to the SH signal of the original OPA output. An exponential rise was fitted to the normalized mean SH values (Fig. 1). The terminal SH value was calculated by averaging the median SH value over the last 5 generations. In the experiments, one optimization run was performed for each data point (final value and convergence speed), and every run was terminated after 50 generations. For the simulations, the SH signal was derived analytically as described in a previous article[8]. In the simulations, a comparable increase of the SH signal as well as decrease of the pulse duration with respect to the Fourier limited pulse was observed as in the experiments. As in the experiments, each simulated optimization was equally aborted after 50 generations. The convergence data were averaged over 50 optimization runs for each single point.

3.1 *Performance of evolutionary strategies*

In order to be able to change the number of independent variables, the phase function to be impinged onto the spectrum was parameterized by defining a number of J interpolation points x_j, $j = 1, \ldots, J$ which were connected by linear interpolation. Additionally, the phase values at these interpolation points were discretized, i.e. only values of the form $i \cdot 2\pi/K$, $i = 0, \cdots, K-1$ with $K \in \mathbf{N}$ were allowed.

The optimization was run repeatedly for different sets of steering parameters. Out of the ensemble of these parameters (such as the number of phases, the number of independent variables, and the parameters which steer the step length control) only two were tuned whereas all the others were kept fixed.

3.2 Size of parameter space

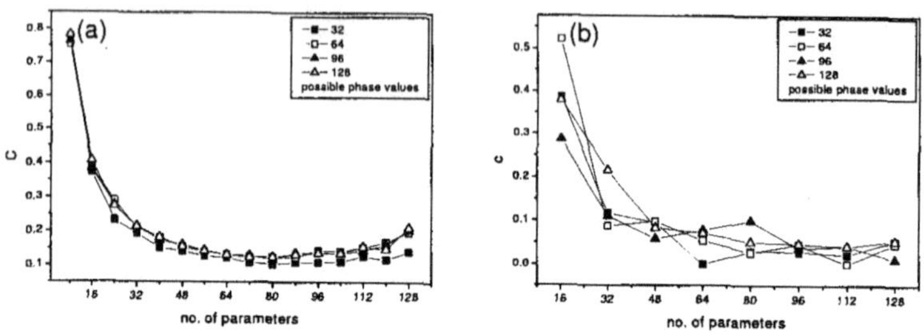

Figure 2. Influence of discretization of parameters and phases on convergence speed, (a) simulation, (b) experiment. The number of independent variables is scanned versus the number of possible phase values. The mutation rate is set to 0.1.

Our mapping of the phase uses interpolated phase functions and discrete phase values which easily allows for the alteration of the number of points in parameter space. To study the convergence behavior on the size of parameter space, the number of interpolation points was scanned from 16 to 128 in increments of 16. For each number of interpolation points, the scan range for the number of phases ran from 32 to 128 in steps of 32. Fig. 2 reveals that the convergence speed, as should be expected, decreases with increasing number of parameters, both in experiments and simulations.

3.3 Noise

The influence of noise was investigated only in simulations. Noise was implemented as a Gaussian-distributed fluctuation in the pulse energy. Its width was one of the scan parameters and is specified relative to the pulse energy. For each scan, the full width at half maximum (FWHM) of the pulse was calculated and normalized with respect to the FWHM of the Fourier limited pulse. As long as it is acceptable to approach the optimal relative FWHM

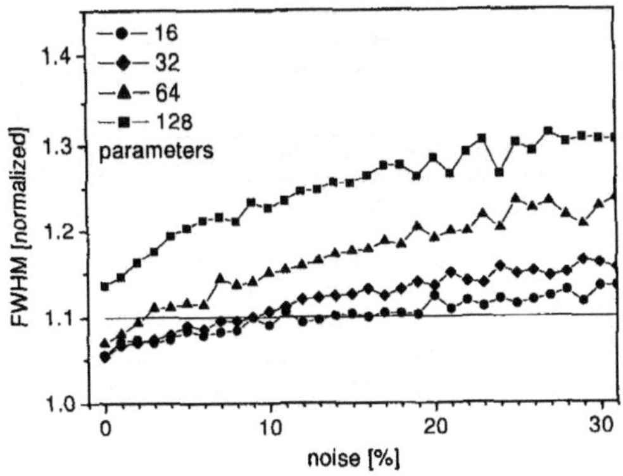

Figure 3. Noise values acceptable for an optimization run: The FWHM of the optimized pulse is normalized to the FWHM of the Fourier limited pulse. The noise which is implemented as energy fluctuation of the pulses is scanned from 0 to 30% for 16, 32, 64, and 128 parameters.

value with an accuracy of 10%, the threshold for noise is approximately 14%, 9%, and 3% for 16, 32, and 64 parameters, respectively (Fig. 3). For 128 parameters, convergence is clearly missed within 50 loop iterations even for the lowest noise rates.

4 The role of parameterization

The above described linear interpolation of the phase (and amplitude) is a universal approach for a parameterization of the electric field and includes any field which can be physically generated by a pulse shaper. Because this does not take into account any knowledge about the optimization process except that the phase (and amplitude) functions are smooth by construction, this mapping may as well be denoted as "free parameterization". A mapping between optimization parameters and experimental parameters (Fig. 4) which reflects the physical properties of the system is extremely useful.

4.1 Polynomial phase functions

The compensation of the residual phase of ultrashort pulses which consists of chirps of various orders suggests to restrict the algorithm to a search for

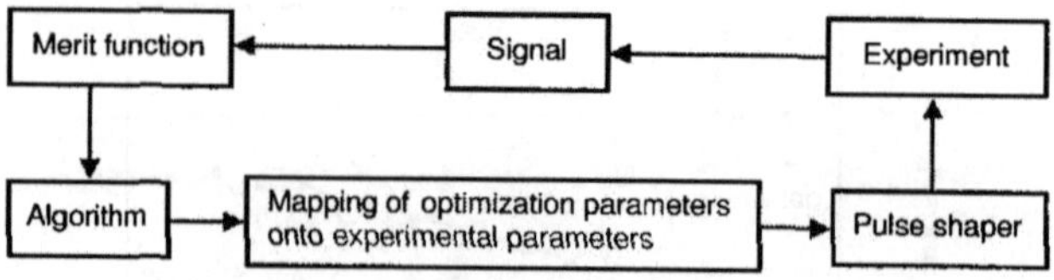

Figure 4. With an intelligent parameterization, the desired optimization effect can be dissected more efficiently from unwanted side effects. By investigating how a particular experiment responds to optimization runs with different parameterizations of the electric field, further insight into the physical process under study may be gained. A positive side effect is the reduction of the number of parameters which improves convergence.

polynomial phase functions. We thus chose a representation of the phase function,

$$\Phi_n = \sum_{k=2}^{K} c_k \left(\frac{n - N_0}{N}\right)^k \qquad n = 0, \cdots, N-1 = 127 \tag{1}$$

with quadratic terms as the lowest polynomial order k since constant ($k = 0$) or linear ($k = 1$) phase terms only produce a phase or time shift, respectively. The Taylor coefficients c_k and the offset N_0 are to be optimized by the algorithm. The reduction in the number of independent variables results in a rapid convergence. In our experiment, the adaptive compression of pulses from a non-collinear OPA from an initial 260 fs to below 16fs has been achieved[13] within approximately 10 generations which corresponds to below five minutes optimization time.

4.2 Periodic phase functions

For physical systems which are sensitive to symmetry properties in the phase functions, a periodic phase function

$$\Phi_n = a \cdot \cos(b \cdot n + c) \qquad n = 0, \ldots, 127 \tag{2}$$

n being the pixel number, is one suggestive parameterization. The parameters a, b, and c are optimized by the algorithm. It has been shown that the probability for a non-resonant two photon transition maximizes for any pulse with a spectral phase antisymmetric to the central wavelength (so-called bright pulses) and can be annulled by special pulses with a symmetric phase (dark pulses)[14]. The parameterization in Eq. 2 is sufficiently general to include both the generation of dark and bright pulses. In our experiment, the two photon transition in Na(3s $\rightarrow\rightarrow$ 5s) has been chosen to demonstrate the maximization and minimization of population transfer by tailoring the spectral phase of the pump[15].

4.3 Multiple pulse structures

Several experiments[16] and theoretical investigations followed the scheme by Tannor and Rice[3] where control has been achieved by the delay between a pump and a dump pulse. It is hence intuitive to use a parameterization in time domain where the interpulse delay is adjusted by the algorithm. With a mask capable of influencing both phase and amplitude it is possible to produce a train of pulses which can be expressed as

$$E_{out}(t) = \sum_{n=-\infty}^{\infty} c_n \cdot E_{in}(t + n \cdot \tau) \tag{3}$$

with the input pulse $E_{in}(t)$, complex coefficients c_n for amplitude and relative phase for each pulse and the minimum interpulse delay τ which depends on the physical properties of the mask and the input pulse. Only pulses at positions $n \cdot \tau$, $-N/2 \leq n < N/2$, can be independently influenced[17] where N is the number of pixels of the LC mask. This concept has been successfully applied in a study of the wavepacket dynamics in the prototype molecule K_2. The algorithm adjusted the index and value of a predefined small number of coefficients c_n. All other coefficients were set to zero. Thereby, the algorithm generated phase-related multiple pulses to manipulate the excitation of either the ground (X) or excited (A) state in potassium dimers[18].

5 Summary

Evolutionary strategies have proven to be practicably versatile and are ideally suited to spot global optima in coherent control studies. The optimization speed and final result strongly depends on the number of free parameters and/or a suitably chosen parameterization. The steering parameters which control the optimization behavior of the algorithm are non-critical if chosen within reasonable limits. It has been shown that in the presence of noise, the number of free parameters is inverse to the noise level which is acceptable for a satisfactory final convergence. The parameterizations can account for physical (or chemical) a-priori knowledge, and help to gain insight of the underlying process under study.

Acknowledgements

The authors thank D. Proch, T. Lang, R.Meier, and T. Hornung for many valuable discussions. Financial support by the DFG-Schwerpunkt "Femtosekunden - Spektroskopie" is gratefully acknowledged.

References

1. P. Gaspard, I. Burghardt, eds. *Chemical Reactions and Their Control on the Femtosecond Timescale*, vol. 101 of *Advances in Chemical Physics* (John Wiley & Sons, Bruxelles, 1997).
2. P. Brumer, M. Shapiro. *Chem. Phys. Lett.* **126**, 54 (1986).
3. D.J. Tannor, R. Kosloff, S.A. Rice. *J. Chem. Phys.* **85**, 5805 (1986).
4. R.S. Judson, H. Rabitz. *Phys. Rev. Lett.* **68**, 1500 (1992).
5. H. Rabitz, R. de Vivie-Riedle, M. Motzkus, K.-L. Kompa. *Science* **288**, 824 (2000).
6. A. Assion *et al. Science* **282**, 919 (1998).
7. C.J. Bardeen *et al. Chem. Phys. Lett.* **280**, 151 (1997).
8. D. Zeidler, S. Frey, K.-L. Kompa, M. Motzkus. *Phys. Rev. A* **64**, 023420 (2001).
9. H.-P. Schwefel. *Evolution and Optimum Seeking* (John Wiley & Sons, New York, 1995).
10. D.E. Goldberg. *Genetic Algorithms in Search, Optimization, & Machine Learning* (Addison-Wesley, Reading, 1997).
11. T. Baumert, T. Brixner, V. Seyfried, M. Strehle, G. Gerber. *Appl. Phys. B* **65**, 779 (1997).
12. D. Yelin, D. Meshulach, Y. Silberberg. *Opt. Lett.* **22**, 1793 (1997).
13. D. Zeidler, T. Hornung, D. Proch, M. Motzkus. *Appl. Phys. B* **70**, S125 (2000).
14. D. Meshulach, Y. Silberberg. *Phys. Rev. A* **60**, 1287 (1999).
15. T. Hornung, R. Meier, D. Zeidler, D. Proch, M. Motzkus. *Appl. Phys. B* **70**, 277 (2000).
16. A. Glaß, T. Rozgonyi, T. Feurer, R. Sauerbrey, G. Szabó. *Appl. Phys. B* **71**, 267 (2000).
17. M.M. Wefers, K.A. Nelson. *IEEE J. Quant. Electron.* **32**, 161 (1996).
18. T. Hornung, R. Meier, R. de Vivie-Riedle, M. Motzkus. *Chem. Phys.* **267**, 261 (2001).

MODE-SELECTIVE EXCITATION OF POLYDIACETYLENE BY FEEDBACK-CONTROLLED PULSE SHAPING

D. ZEIDLER, S. FREY, W. WOHLLEBEN, AND M. MOTZKUS
Max-Planck-Institut für Quantenoptik, D-85748 Garching, Germany
Fax: ++49 89 32905-200
email: Marcus.Motzkus@mpq.mpg.de

F. BUSCH, T. CHEN, W. KIEFER
Institut für Physikalische Chemie, Universität Würzburg,
Am Hubland, D-97074 Würzburg, Germany,
Fax: ++49 931 888 6341

A. MATERNY
School of Engineering and Science, International University Bremen,
P.O. Box 750 561, D-28725 Bremen, Germany

Optimal control of ground state dynamics in crystalline polydiacetylene is demonstrated in a feedback-controlled self-learning loop. The feedback signal is derived from the spectral distribution of a coherent anti-Stokes Raman scattering signal with a tailored Stokes pulse. The controlled excitation of one mode and simultaneous extinction of all other modes with high efficiency is demonstrated.

1 Introduction

Most coherent control studies so far have dealt with the control of electronically excited species[1]. Control of ground state dynamics, however, might be even more important since chemical reactions usually take place on the lowest potential energy surface. For mode-selective chemistry it is crucial to focus energy in a local vibrational motion of a certain bond (local mode) or, in a more general sense, to form a superposition of vibrations leading to dynamics along the reaction coordinate. Such dynamics can be described by a superposition of eigenfunctions (normal modes) resulting from the stationary Schrödinger equation. Due to the broad spectra of femtosecond laser pulses, these pulses are capable of coherently exciting more than one normal mode[2,3]. If the phases of these modes are chosen suitably, the enhancement of normal modes or the preparation of local modes should be possible.

In principle, the corresponding phase of the control field which drives the molecule into the desired direction can be computed by optimal control theory. However, this approach is restricted to simple prototype systems. In larger molecules, where the density of states is much higher and dynamics are determined by intramolecular vibrational relaxation (IVR) processes on a multi-dimensional potential energy surface, control fields are difficult to predict. In these cases, solutions can be achieved by the self-learning approach[4,5].

The realization of such a feedback controlled self-learning loop comprises a source for ultrashort laser pulses, a computer-controlled pulse shaper[6], and an optimization algorithm[7]. The ultrashort shaped laser pulse excites a sample and a suitable property of the system is monitored. This probe signal serves as feedback for the optimization algorithm that controls the shaper and generates new pulse shapes until a convergence criterion is met[8,9].

Here we demonstrate selective excitation of ground state vibrational modes in a polymer, polydiacetylene (PDA), by a stimulated Raman process. The process is optimized in a self-learning loop using the anti-Stokes Raman signal of a four-wave-mixing (FWM) process as feedback. Polydiacetelyne has been studied in detail with femtosecond time– and frequency–resolved coherent anti–Stokes Raman Scattering (fs–CARS)[10,11]. In those experiments, several normal modes were excited in a stimulated Raman process and the vibrational dynamics in the electronic ground and excited state were studied by analyzing the anti–Stokes signal. Additionally, two strategies to excite selectively vibrational motion of the $\tilde{\nu}_2$ mode in the electronic ground state of the PDAs were demonstrated. One technique made use of the so–called "Tannor–Rice" scheme[12], which is related to the vibrational dynamics in the excitonic states of the PDA. The other method included the change of the chirp of the ultrashort Stokes laser pulse[13], already pointing to the important role of the phase in the selective excitation of vibrational motions.

2 Experimental setup

The samples used in the experiments were FBS diacetylenes[14]. Less than 1% polymers (PDA) were embedded in the monomer single crystal. The backbones of the PDA molecules were aligned parallel to the b–axis of the crystal. The sample was cooled to approximately 15 K in a closed-cycle helium cryostat. Fig. 1 shows a part of the resonance Raman spectrum obtained from PDA at 10 K and the motions of the atoms resulting in the four vibrational normal modes $\tilde{\nu}_2/\tilde{\nu}_{2'}$ (Fermi–coupled modes), $\tilde{\nu}_3$, and $\tilde{\nu}_4$[15]. For simplicity, the $\tilde{\nu}_2/\tilde{\nu}_{2'}$ mode is denoted her as $\tilde{\nu}_2$. The modes are composed mainly of stretching and bending motions of the backbone bonds and were all excited by the pump and Stokes laser pulses of the fs–CARS experiment (Fig. 2a).

The experimental setup included a commercial Ti:Sapphire fs laser system which pumps two optical parametric amplifiers (OPA). The OPA outputs were tuned to 605 nm and 652 nm (FWHM=100 fs). The 605 nm pulses were

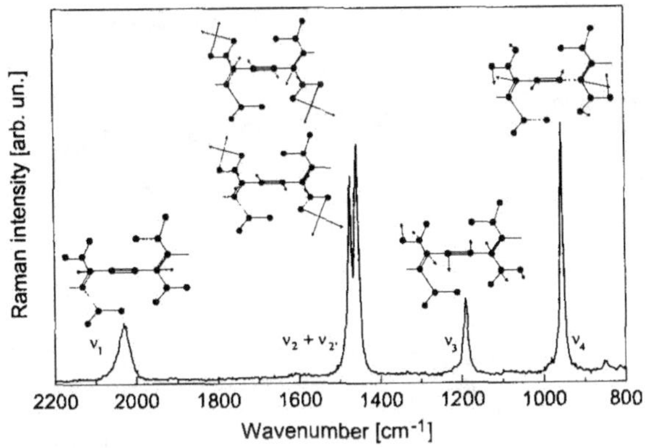

Figure 1. Resonance Raman spectrum displaying the vibrational modes excited by narrow-band pump and Stokes pulse. For each principal Raman vibrational mode (ν_1,ν_3, ν_4) as well as for the Fermi coupled vibrational modes (ν_2, $\nu_{2'}$) atomic displacements are given assuming a simplified side group structure.

compressed by a prism pair and split into pump and probe pulses. The 652 nm pulses served as Stokes beam and are fed through a shaping setup based on a spatial light modulator (SLM) in a $4f$–arrangement[6]. The design included gratings with $1/d = 1800/\text{mm}$ and cylindrical mirrors with f= 150 mm. Pump, Stokes and probe pulses with energies of $\approx$ 100 nJ were arranged in a folded BoxCARS configuration (Fig. 2c) and focused into the sample with a f = 60 cm concave mirror. The beams were oriented such that the sum of the wave vectors of pump and Stokes pulses points in the direction of the crystal b–axis, leading to effective excitation in the exciton band. Behind the sample, the CARS signal was collimated with an f = 60 cm lens, spectrally dispersed and detected with a cooled CCD line camera.

The spectral bands of the wavelength-resolved CARS signal which correspond to the different modes of PDA served as the feedback signal for the optimization algorithm. The optimization algorithm, based on an evolutionary strategy[7], searched for the optimal settings for the LC mask that enhance the desired mode.

In order to avoid features arising at zero time delay of pump, Stokes, and probe pulses, the delay of the unshaped Stokes pulse was chosen to be Δ= +100 fs which is well after the pump pulse (Fig. 2b). To overcome possible effects arising from the beating of the modes, the delay of the probe pulses was

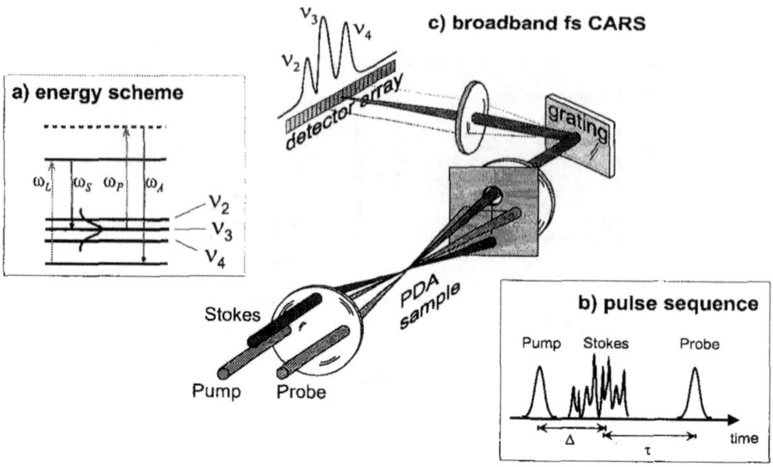

Figure 2. Setup of the fs-CARS experiment. a) energy diagram of the CARS pulses, b) pulse sequence of the three incoming pulses with Stokes pulse shaped, c) wavelength-resolved folded BoxCARS configuration. (A lens is shown instead of a focussing mirror for graphical reasons).

wobbled over several beatings with a frequency of about 5 Hz between +100 fs and +366 fs. The averaging was performed by simply recording spectra while the probe delay continuously alternated between the two boundary points.

3 Results and Discussion

Figure 3a shows the two-dimensional pattern obtained from the CCD detection of the transient anti–Stokes signal with nearly Fourier limited pulses. With a constant time delay of Δ between pump and Stokes pulse, the time delay τ between Stokes and probe pulses was varied. The relative intensities of the three Gaussian–like peaks, which are centered at about 1400 cm^{-1}, 1200 cm^{-1}, and 960 cm^{-1}, reflect the relative population of the three coherently excited modes. Cuts of the anti–Stokes signal along the time axis result in the transient behaviour of the three spectral features[13].

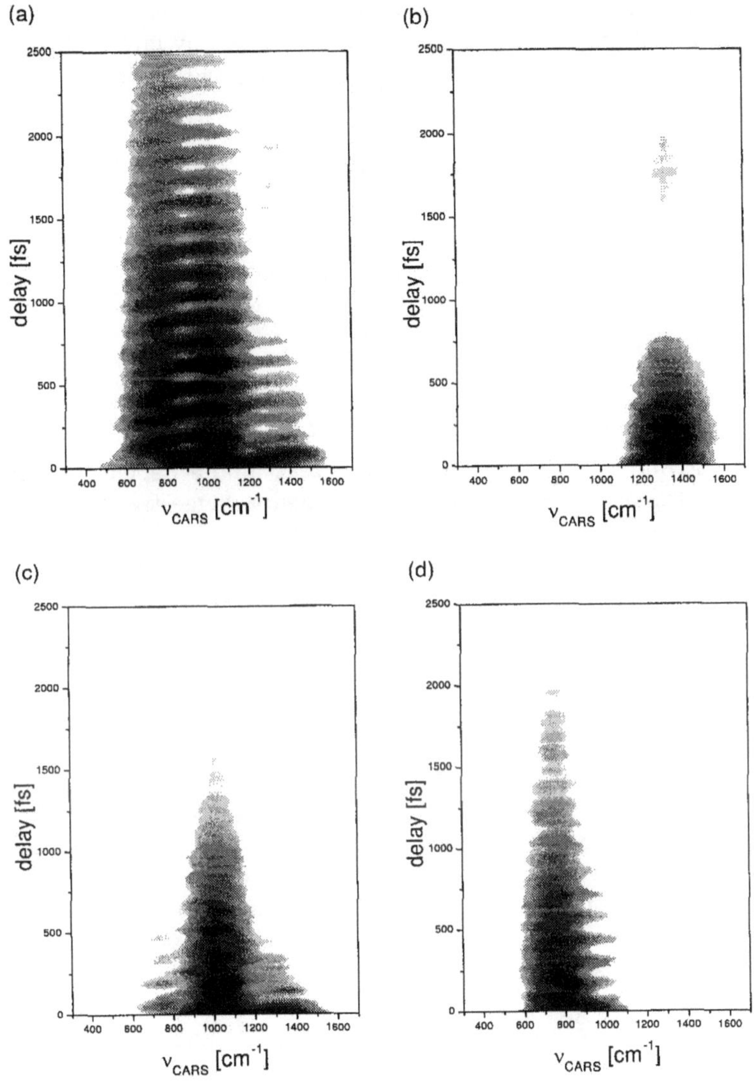

Figure 3. Spectrally resolved CARS transients on polydiacetelyne with shaped Stokes pulse for optimization of different modes: a) unshaped pulses (Fourier limited); b) Optimization of the $\tilde{\nu}_2$ mode; c) Optimization of the $\tilde{\nu}_3$ mode; d) Optimization of the $\tilde{\nu}_4$ mode.

Under the assumption of a Gaussian line shape for each mode, the corresponding population in each mode can be obtained by integrating over the whole spectral profile, which serves as an input for the merit function in the optimization algorithm:

$$P_{\tilde{\nu}_i} = \int_{\omega_i - \Delta\omega_0}^{\omega_i + \Delta\omega_0} \int_0^{t_0} I(\omega_i, t) d\omega \, . \tag{1}$$

The merit function $m_{\tilde{\nu}_i}$ for mode i was chosen as

$$m_{\tilde{\nu}_i} = \frac{P_{\tilde{\nu}_i}}{P_{\tilde{\nu}_j} + P_{\tilde{\nu}_k}} \, , \tag{2}$$

where $i, j, k = 2, 3, 4$ and $i \neq j \neq k$. In a former experiment[13], it was shown that the distribution of energy in the different modes depends strongly on chirp and delay of the Stokes pulse. It is hence suggestive to restrict the algorithm to a search for polynomial phase functions[7,16], without changing the spectrum of the pulses by an amplitude function. We thus chose a polynomial representation of the phase function in the first experiments,

$$\Phi_n = \sum_{k=2}^{K} c_k \left(\frac{n - N_0}{N} \right)^k \qquad n = 0, \cdots, N - 1 = 127 \tag{3}$$

with quadratic terms as the lowest polynomial order. The parameters c_k and N_0 are optimized by the algorithm such that the vibrational dynamics are only excited in the desired mode while all other modes are suppressed.

At first, $m_{\tilde{\nu}_2}$ served as the feedback signal in order to maximize the relative population of the $\tilde{\nu}_2$ mode. Figure 3b is the three dimensional time– and wavenumber–resolved CARS signal obtained after optimization of the $\tilde{\nu}_2$ mode. The two unwanted modes are almost completely suppressed.

The same method has been applied for the selective optimization of the $\tilde{\nu}_3$ and $\tilde{\nu}_4$ modes. The results are shown in Figs. 3c and 3d. In each case it was possible to suppress the two other modes. The strong beating between the modes, as observed in Fig. 3a, disappeared. In Fig. 4 the integrated mode intensities are shown, again indicating the high selectivity of excitation that has been achieved.

Applying defined phase functions in which only the coefficient c_2 (Eq. 3) has been varied, the beating structure could be changed in a defined manner giving rise to any superposition of normal modes.

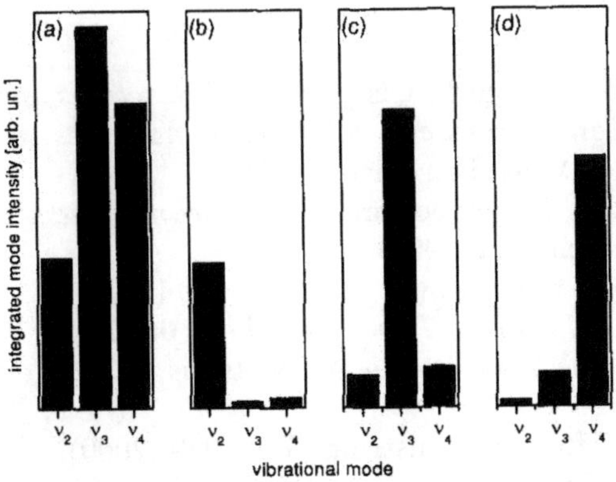

Figure 4. Integrated absolute mode intensities, as obtained from the transients in Fig. 3 (a) unshaped pulse (Fourier limited Stokes); (b) after optimization of the $\tilde{\nu}_2$ mode; (c) after optimization of the $\tilde{\nu}_3$ mode; (d) after optimization of the $\tilde{\nu}_4$ mode.

4 Conclusion

Optimal control of the ground state dynamics in PDA by a stimulated Raman process in a computer-controlled feedback loop has been demonstrated. We could show that selective excitation of one vibrational ground state mode and suppression of all other modes is possible in such a complex molecular system which is one of the prerequisites for mode-selective chemistry.

Acknowledgements

The authors thank K.-L. Kompa for continuous support of this study, D. Proch and T. Lang for many valuable discussions, and K. Bauer for his technical assistance. We thank Mrs. Irene Bauer and Prof. M. Schwoerer of the University of Bayreuth for supplying excellent samples of PDA. One of us (T.C.) thanks the Volkswagen Stiftung for a postgraduate scholarship. Financial support by the German Science Foundation (fs-Schwerpunkt: Mo-708/3-2, SFB 347, Project C-2 and SFB 410, Project C-3) as well as the Fonds der Chemischen Industrie arc acknowledged.

References

1. P. Gaspard, I. Burghardt, eds. *Chemical Reactions and Their Control on the Femtosecond Timescale*, vol. 101 of *Advances in Chemical Physics* (John Wiley & Sons, Bruxelles, 1997).
2. J. Manz and L. Wöste, editors. (*Femtosecond Chemistry*, VCH Verlagsgesellschaft, Weinheim 1995).
3. A.H. Zewail, J. Phys. Chem. A **104**, 5660 (2000).
4. R.S. Judson, H. Rabitz. *Phys. Rev. Lett.* **68**, 1500 (1992).
5. H. Rabitz, R. de Vivie-Riedle, M. Motzkus, K.-L. Kompa. *Science* **288**, 824 (2000).
6. A.M. Weiner. Rev. Sci. Instrum. **71**, 1929 (2000).
7. D. Zeidler, S. Frey, K.-L. Kompa, M. Motzkus. *Phys. Rev. A* **64**, 023420 (2001).
8. C.J. Bardeen *et al. Chem. Phys. Lett.* **280**, 151 (1997).
9. A. Assion *et al. Science* **282**, 919 (1998).
10. A. Vierheilig, T. Chen, P. Waltner, W. Kiefer, A. Materny, and A.H. Zewail. Chem. Phys. Lett. **312**, 349 (1999).
11. T. Chen, A. Vierheilig, P. Waltner, W. Kiefer and A. Materny. *Chem. Phys. Lett.* **325**, 176 (2000).
12. D.J. Tannor, R. Kosloff, S.A. Rice. *J. Chem. Phys.* **85**, 5805 (1986).
13. T. Chen, A. Vierheilig, P. Waltner, M. Heid, W. Kiefer, and A. Materny. *Chem. Phys. Lett.* **326**, 375 (2000).
14. A. Materny, and W. Kiefer. *Macromolecules* **25**, 5074 (1992).
15. D.N. Batchelder, and D. J. Bloor. *J. Polym. Sci. Polym. Phys. Ed.* **17**, 569 (1979)
16. T. Hornung, R. Meier, and M. Motzkus. *Chem. Phys. Lett.* **326**, 445 (2000).

A METHOD FOR ROBUST SOLUTIONS IN THE WEAK-INTENSITY LASER CONTROL OF MULTILEVEL SYSTEMS

ALESSANDRO LAMI

ICQEM-CNR, Area della Ricerca, Via Moruzzi 1, I-56100 Pisa, Italy; Email: lami@indigo.icqem.pi.cnr.it

FABRIZIO SANTORO

Dipartimento di Chimica dell'Università di Siena, Via Aldo Moro, I-53100 Siena, Italy; Email: fabrizio@indigo.icqem.pi.cnr.it

The laser control of complex multilevel systems is studied and a method is presented to ensure the robustness of the solution. We consider a weak multicolor gaussian pulse of given duration whose frequencies correspond to the main absorption lines of the system while the phases and amplitudes are taken as control parameters. The numerical example presented and discussed concerns the localization of a wave packet in a model system for a conical intersection.

1 Introduction

The researches in the field of the laser control of the yield of chemical processes have seen major advances in the last years. Reviews of the main achievements [1] and the perspectives [2] have recently appeared in literature. From a theoretical point of view the research has come to the settlement of what is referred as the optimal control theory [3,4] while from the experimental point of view impressive results [5,6] have been obtained by implementing the idea of Judson and Rabitz to use the output of the experiment as a feedback to improve the field through a learning algorithm [7].

Scarce attention have been devoted till very recently to the necessity that the result of the optimization procedure be robust with respect to small variations of the field. This feature is instead of primary importance when dealing with complex systems characterized by dense manifolds of states irregularly spaced, and with very different physical characteristics, like it is frequently the case for molecules showing conical intersections and strong nonadiabatic effects. In these cases, in fact, it is very common to get very good solutions of the control problem which then reveal definitely not robust [8]. One way to approach the problem is the analysis *a posteriori* of the field to understand the physical mechanism underlying the excitation and clean it up by unessential features which are often connected with the instability of the solution [9]. Hornung et al. have worked out a theoretical scheme to take into account during the optimization the experimental feasibility of the laser pulse to be prepared [10]. The scheme they depict, fulfilling their task, goes also in the right direction with respect to the robustness problem, since it ends up with cleaning the pulse form the unessential features which are often one of the causes of non robustness.

Very recently, Geremia et al. [11] showed how the robustness requirement can be taken into account during the optimization in learning loop procedures, simply adopting the suitable cost functionals. This paper has the main value of proposing a direct experimental implementation. From a theoretical point of view the method they adopted seems to be applicable only to simple systems, because of its computational cost. This limitation seems at the moment to be shared also by all the methods based on the optimal control theory with strong fields, because of its complex machinery.

We tackle the problem from a different perspective: we prefer to limit ourselves to weak-intensity fields to face the problem of the robust laser control of complex systems, as those with strong nonadiabatic effects. These systems often show a very rich ultrafast dynamics with many competitive processes, whose control is very challenging and can reveal very interesting for future applications.

The limitation to fields of weak intensity, gives us the possibility to use the first order perturbation theory, getting analytical solutions of the problem and recasting the robustness requirement in a very effective way, which allows for a direct introduction in the optimization procedure.

2 The method

2.1 *Preparation of a state by a weak-laser pulse*

Consider a multilevel system with eigenvectors $|\alpha\rangle$ and eigenvalues ω_α. Let us call $|g\rangle$ its ground state and $|d\rangle$ its doorway state, $|d\rangle \propto \hat{\mu}|g\rangle$ where $\hat{\mu}$ is the electric dipole moment. Here we want to address the problem of the preparation of a given target state $|v\rangle$,

$$|v\rangle = \sum_\alpha v_\alpha |\alpha\rangle \qquad (1)$$

by a weak laser pulse. The Hamiltonian of the system, where the light pulse is a classical electric field interacting with the molecular dipole, is:

$$H = H_0 + V(t); \quad V(t) = -\vec{\mu}\vec{E}(t). \qquad (2)$$

We use the following expression for the pulse [11]

$$E(t) = g(t-t_p)\sum(z_\beta \exp(-i\omega_\beta t) + c.c), \quad \omega_\beta = E_\beta - E_g \tag{3}$$

where the frequencies are equal to the excitation energies of the system, the complex amplitudes z_β are free parameters and

$$g(t-t_p) = \sqrt{\gamma/\pi}\exp[-\gamma(t-t_p)^2] \tag{4}$$

is a fixed gaussian envelope centered at $t=t_p$.

By first order perturbation theory it can be shown [8] that the pulse which creates a state whose excited part is coincident with the target state in Eq.(1), at a given time t_f (when the pulse is over), can be obtained by solving the linear system:

$$\mathbf{A}\,\mathbf{z} = \mathbf{w}, \tag{5}$$

where:

$$A_{\alpha\beta} = \exp(-i\Delta_{\alpha\beta}t_p - \Delta_{\alpha\beta}^2/4\gamma), \qquad \Delta_{\alpha\beta} = E_\alpha - E_\beta,$$
$$w_\alpha = (v_\alpha / d_\alpha)\exp(iE_\alpha t_f), \qquad d_\alpha = \langle\alpha \mid d\rangle, \tag{6}$$

$\mathbf{z}$ being the vector of the complex amplitudes in Eq. (3). The dimension of the linear system may be taken coincident, in principle, with the number of states of the selected basis set but we shall see that in order to get robust solutions it must be often drastically reduced.

2.2 *Estimate of the robustness of the solution*

Let us call $\mathbf{v}$ the state prepared by the field whose complex amplitudes are given by $\mathbf{z}$. From Eqs. (5) and (6) we have

$$\mathbf{v} = \mathbf{DAz} \qquad \text{with } D_{\alpha\beta} = d_a \exp\left(-i\omega_\alpha t_f\right)\delta_{\alpha\beta} \tag{7}$$

where $\delta_{\alpha\beta}$ is the Kronecker δ.

Both the frequencies and the amplitudes are subject to some experimental error. Form Eq.(6) it is easily seen that a small error on a frequency ω_α is equivalent to a small phase change of the corresponding amplitude z_α [8]. Thus we can consider only the real amplitudes and phases and study the effect of small static perturbations of $\mathbf{z}$ on $\mathbf{v}$. From Eq. (7) we have

$$\delta \mathbf{v} = \mathbf{DA}\delta \mathbf{z}\,. \tag{8}$$

A reasonable choice for $\delta \mathbf{z}$ is obtained imagining that the random variation of z_i is proportional to its modulus,

$$\delta \mathbf{z} = \lambda\, \mathbf{ZR} \tag{9}$$

where **R** is a vector of complex random numbers and **Z** is a diagonal matrix whose element are $Z_{\alpha\alpha} = | z_\alpha |$. Then $\delta \mathbf{v} = \lambda\, \mathbf{DAZR}$ and, after performing an average over the uncorrelated random variables, we get

$$| \delta \mathbf{v} |^2 = \lambda^2 R^2 \sum_\alpha | z_\alpha |^2 \sum_\beta | A_{\alpha\beta} |^2 d_\beta^2 = \lambda^2 R^2 \mathbf{z}_2 \cdot \mathbf{s} \tag{10}$$

where the two vectors $\mathbf{z}_2$ and **s** are defined as

$$z_{2\alpha} = | z_\alpha |^2 \text{, and } s_\alpha = \sum_\beta | A_{\alpha\beta} |^2 d_\beta^2\,. \tag{11}$$

The quantity in Eq.(10) is a measure of the robustness of the solution. It follows that each component s_α of the vector **s** can be associated with the robustness due to the component of the field with frequency ω_α. Some interesting *a priori* considerations can be driven by Eq. (10). In fact it shows that the greater is s_α the more unstable is the result and, since $| A_{\alpha\beta} |^2 = \exp(-\Delta_{\alpha\beta}^2 / 2\gamma)$ (Eq. (6)), it comes out that the frequencies ω_α which bring a greater instability are those corresponding to eigenstates $| \alpha \rangle$ close in energy to eigenstates $| \beta \rangle$ with a great transition dipole moment d_β. The effect is amplified if $| \alpha \rangle$ is weakly absorbing but well represented in the target state since this requires an high component z_α in the field. Eq.(10) is a double goal:

1. It leads to the a priori suggestion to use the frequencies corresponding to strongly absorbing states and disregard those corresponding to the weakly absorbing ones most of if close to strong absorbing ones.
2. Since it is very cheap to compute it allows for the introduction of a penalty function in the target function to ensure the robutness of the solution.

2.3 *Working algorithm*

1. We fix a threshold ε_d and we include in the pulse only the frequencies ω_α corresponding to eigenstates $|\alpha\rangle$ with $|d_\alpha| > \varepsilon_d$.
2. By random generation of fields (i.e. vectors **z**) we generate an orthonormal basis of states that can be prepared, solving Eq. (5).
3. The above working space $|\mathbf{v}\rangle$ is used to represent the observable $\hat{O}$ to be optimized. The eigenvector corresponding to the largest value of the target function (see point 4) gives a very good starting point for the optimization.
4. Fixed the time t_f, we search the state $|e\rangle = \sum_i x_i |v_i\rangle$ which maximizes the target function

$$f(\mathbf{x}) = w(F)\cdot g\left[\langle e(t_f)|\hat{O}|e(t_f)\rangle\right]\cdot S\left(|\delta\mathbf{v}|^2\right) \tag{12a}$$

$$w(F) = \exp[-(\log_{10} F/F_0)^3/100] \tag{12b}$$

$$S(|\delta\,\mathbf{v}|^2) = \exp[-(|\delta\,\mathbf{v}|^2 - \varepsilon_v)^2/\sigma_v^2] \tag{12c}$$

where g is a chosen function of the expectation value of the observable, $w(F)$ is a function of the fluence F of the pulse allowing to take into account both the selectivity and the absolute yield of the excitation (F_0 is the reference value for a vertical transition) and S is a penalty function to ensure the stability of the solution (ε_v and σ_v are adjusted during the optimization to drive gradually $|\delta\,\mathbf{v}|^2$ toward acceptable small values). The optimization is performed by a conjugate gradient method coupled with a scheme of random perturbations of the solution to escape from local minima.

3 A numerical example. Two coupled harmonic bidimensional diabatic surfaces

We consider a multilevel system mimicking a conical intersection in a polyatomic molecule. Conical intersections, which occur frequently in triatomic or larger molecules [12,13] play a key role in many photochemical reactions in radiationless decay process and also deeply influence intramolecular dynamics and spectroscopy [14,15]. The model Hamiltonian [12] involving two harmonic vibrational modes and two diabatic electronic states $|1\rangle$ and $|2\rangle$ is :

$$H = H_{11}\,|1\rangle\langle 1| + H_{22}\,|2\rangle\langle 2| + t_e y\left(|1\rangle\langle 2| + c.c\right) \quad (13a)$$

$$H_{11} = T + V(x) + V(y) = T + \frac{1}{2} m_x \omega_x x^2 + \frac{1}{2} m_y \omega_y y^2 \quad (13b)$$

$$H_{22} = T + V(x - \Delta x) + V(y) + \Delta E = H_{11} - gx + g^2 / 2m_x \omega_x^2 + \Delta E \quad (13c)$$

We also assume that the dipole operator is independent on nuclear coordinates in the above diabatic representation $\hat{\mu} = \mu\left(|1\rangle\langle 2| + c.c.\right)$.

The parameters are (in a.u.) ω_x= 2.1 10^{-3} and ω_y =4.0 10^{-3}, m_x =20000, m_y = 50000, g=-5 10^{-2}, t_e =3 10^{-2}, ΔE =0.0147 and $\mu = 1$. The profiles of the two potentials along x at y=0 are plotted in Fig. 1. With these parameters the energy of a vertical transition from the ground vibrational state is 5667 cm^{-1}.

The gaussian pulse (Eq(4)) which has a central time t_p =150 fs and a FWHM 100 fs is completely over at a final time t_f =300 fs.

Our goal here is to localize the wave packet in a given region of the space at a given time. The study of the dynamics of this model system after a δ excitation shows that, at t=300 fs, the 38% of the wave packet is on the ground diabatic state $|1\rangle$ and it is spread out along x as it can be seen in Fig.1 and Fig.2a. Thus we try to shape our pulse so to create at t=300 fs a wave packet mainly located on $|1\rangle$ and localized in some x-interval. The optimization of other observables have been reported elsewhere [8].

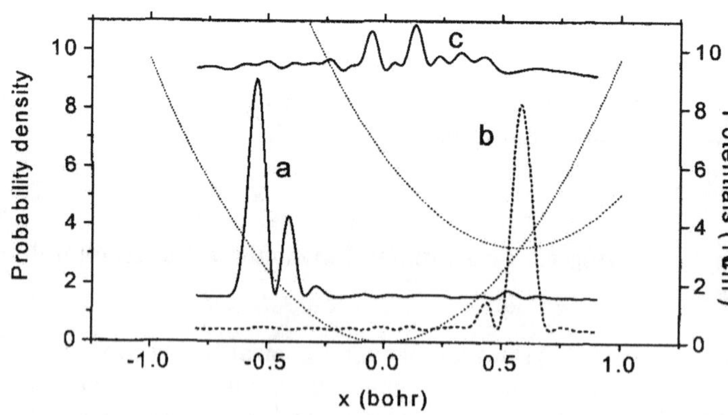

Fig.1 Energy profiles for the two diabatic surfaces at y=0; probability density on the electronic state $|1\rangle$ integrated on y at t=300 fs of the three wave packets prepared by: a) the pulse E^a, b) the pulse E^b. c) a δ excitation. The zero-ordinate of the wave packet are arbitrary. The probability density of a) and b) are scaled so to give the actual P_1 values, while the one of c) is multiplied by a factor 2.

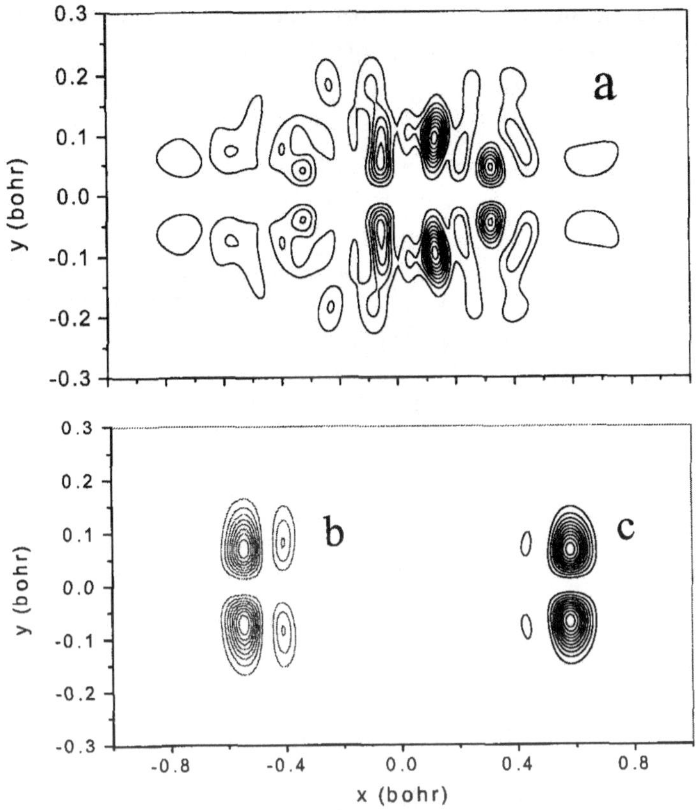

Fig.2 contour plots on $|1\rangle$ of: $|d\rangle$ (a), $|wp^a\rangle$ (b, gray lines) and $|wp^b\rangle$ (c, black lines).

In a first case we chose to localize it in a x region very far form the conical intersection I^a=(-0.6,-0.4) bohr, (Fig.2, (b) gray lines) and in a second case in the region of the conical intersection region I^b=(0.45,0.65) bohr, which is at about x=0.58 bohr, (Fig2, (c) black lines). In both cases the results are very satisfactory.

In the first case the optimal pulse E^a prepares a wave packet $|wp^a\rangle$ with a population P_1=0.93 on the $|1\rangle$ state whose 77% is in I^a. The average energy of $|wp^a\rangle$ is 5520 cm^{-1} above the ground state. The pulse populates very excited x modes since the average number of quanta on the x oscillator and the $|1\rangle$ state is $<n_{1x}>$=8.9. The stability of the solution has been tested by randomly perturbing the amplitudes and phases of E^a by a maximum amount of the 5% of their optimal values. After 1000 different random perturbations the observed maximum change in the observable was about 6.8%. The time dependence of the optimized pulse and the modulus of its Fourier transform are reported in Fig.3.

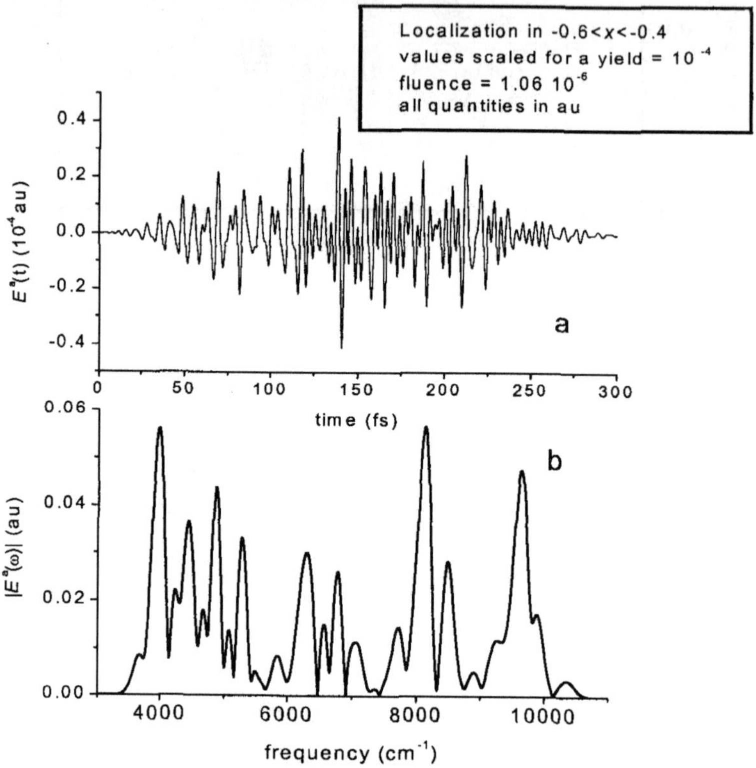

Fig. 3. Pulse E^a: (a) time dependence; (b) modulus versus the frequency

Comparable results have been obtained by localizing the wave in the ground state $|1\rangle$ and in the region I^b. In fact we get a pulse E^b that prepares a wave packet $|wp^b\rangle$ with P_1=0.89 whose 83% is in I^b. The average energy of $|wp^b\rangle$ is 5293 cm^{-1} above the ground state and $<n_{1x}>$=8.3. Thus the optimized pulse excites about the same eigenstates as in the previous case but, by a different phase structure, it leads to a localization of the wave packet in a very different region. The pulse E^b is reported in Fig.4, and clear differences can be seen with respect to that in Fig. 3, most of all as far as the modulus of its Fourier transform is considered. The stability of the solution has been tested has in the previous case getting a maximum change of about 9% after 1000 different random perturbations of the optimal field.

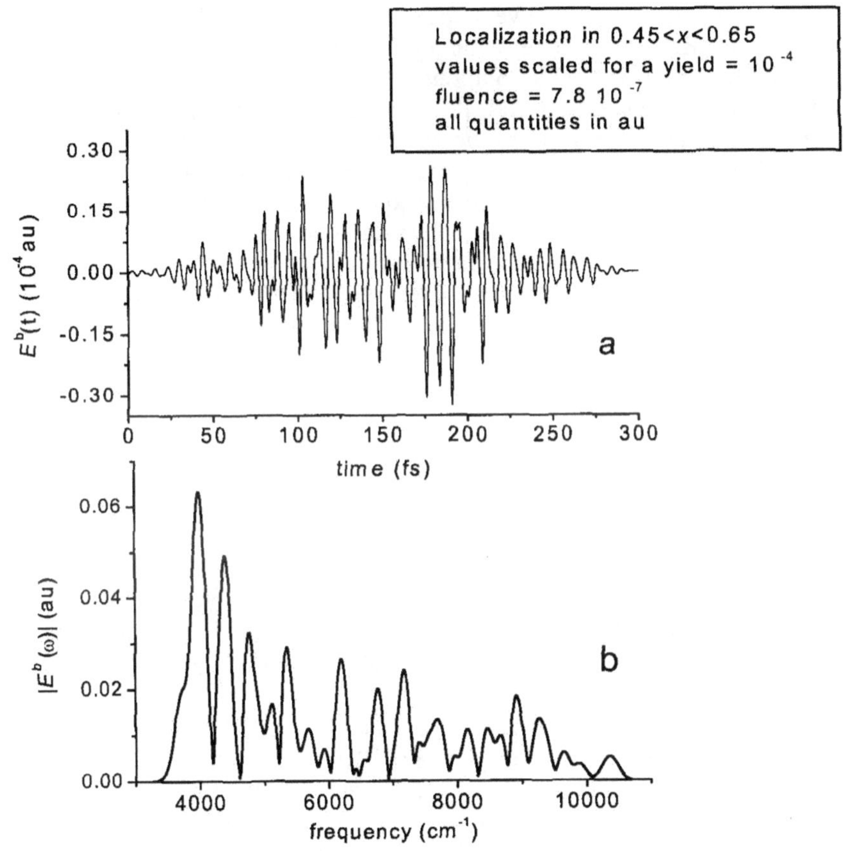

Fig. 4 Pulse E^b: (a) time dependence; (b) modulus versus the frequency

4 Conclusions

We have presented a theoretical scheme for the laser control in the weak field limit of multilevel systems that:
1) allows the treatment of complex multilevel systems
2) provides *a priori* suggestions for the reduction of the parameters of the field
3) provides a physically meaningful starting point for the optimization procedure
4) leads to the definition of a very cheap measure of the robustness of the solution which is used to build up a target function which automatically drives the optimization toward stable solutions

Further details are reported in Ref. [8]. The method have been applied to a model system for a conical intersection, showing that it is very effective in localizing the wave packet in different region of the space. The optimized pulses in Fig 3b and 4b are probably too wide in energy to be experimentally reproduced. However the analysis of the contributions of the eigenstates to the prepared states show that similar states could be produced by narrower pulses.

Some developments are now in progress with the aim of extending the treatment to systems for which we do not have an *a priori* knowledge of eigenstates/eigenvalues (only the results of wave packet numerical propagations are utilized).

References

1. Zare R. N., *Science* **279** (1998) p. 1875.
2. Rabitz H., de Vivie-Riedle R. and Motkus M., *Science* **288** (2000) p. 824.
3. Peirce P., Dahleh M. A. and Rabitz H., *Phys. Rev. A*. **37** (1998) p. 4950.
4. Kosloff R., Rice S. A., Gaspard P., Tersigni S. and Tannor D. J., *Chem. Phys* **139** (1989) p. 201.
5. Assion A., Baumert T., Bergt M., Brixner T., Keifer B., Seyfried W., Strehle M. and Gerber G., *Science* **282** (1998) p. 919.
6. Vajda S., Bartelt A., Kaposta E.-C., Leisner T., Lupulescu C., Minemoto S., Rosendo-Francisco P. and L. Wöste L., *Chem. Phys*. **267** (2001) p. 231.
7. R. S. Judson R. S. and Rabitz H., *Phys. Rev. Lett.* **68** (1992) p. 1500.
8. Lami A. and Santoro F., *Chem. Phys*., accepted.
9. Krause J. L., Whitnell R. M., Wilson K. R., Yan Y. and Mukamel S., *J. Chem. Phys*. **99** (1993) p. 6562.
10. Hornung T., Motzkus M. and de Vivie-Riedle R., *J. Chem. Phys*. **115** (2001) p. 3105.
11. Geremia J. M., W. Zhu W. and H. Rabitz H., *J. Chem. Phys*. **113** (2000) p.10841.

12. Köppel H., Domcke W. and Cederbaum L. S., *Adv. Chem. Phys.* **57** (1984) p.59.
13. Olivucci M., Bernardi F., Ottani S. and Robb M. A., *J. Am. Chem. Soc.* **116** (1994) 2034.
14. Domcke W. and Stock G., *Adv. Chem. Phys.* **100** (1997) p.1.
15. See eg. papers in: Conical Intersections in Photochemistry, Spectroscopy, and Chemical Dynamics, Y. Haas, M. Klessinger and S. Zilberg Guest Editors, *Chem. Phys.* **259** Nos 2-3, (2000) (special issue).

EFFECTS OF ELECTRONIC COHERENCE ON INTENSE ULTRASHORT CHIRPED PULSE EXCITATION OF MOLECULES COUPLED WITH A DISSIPATIVE ENVIRONMENT

B. D. FAINBERG

Holon Academic Institute of Technology, 52 Golomb Street,
Holon 58102, Israel
Raymond and Beverly Sackler Faculty of Exact Science, School of Chemistry,
Tel-Aviv University, Tel-Aviv 69978, Israel
E-mail: fainberg@hait.ac.il

V. A. GORBUNOV

Holon Academic Institute of Technology, 52 Golomb Street,
Holon 58102, Israel
E-mail: gorbunov@hait.ac.il

We have studied an intense chirped pulse excitation of a molecule coupled with a dissipative environment taking into account electronic coherence effects. We show that a strong chirp dependence of the excited state population can be explained not only by the pump-dump process for population wave packets based on incoherent consideration but the reduction of relaxation effects on electronic coherence for a positive chirp as well.

1 Introduction

The possibility of controlling molecular dynamics using properly tailored pulses has been the subject of intensive studies in the last few years [1,2,3,4,5,6,7,8]. Ultrashort pulses intrinsically consist of a broad range of frequency components. The relative phase of these frequency components can be systematically changed by introducing positive (PC) or negative chirp (NC). Chirped pulses can selectively excite coherent wave packet motion either on the ground electronic potential energy surface of a molecule or on the excited electronic potential energy surface [1,4,5,6]. In addition, they are very efficient for achieving population transfer in atomic and molecular systems. In this relation Cao, Bardeen and Wilson [9] have numerically shown that total electronic population inversion of molecules can be achieved with intense PC laser pulses similar to π pulse excitation of the atomic two-level systems [10].

At the present time the following processes are recognized to be responsible for population transfer in a two-level electronic molecular systems. 1) Intrapulse pump-dump process [1,4,5] which can be considered as a single pulse version of the "pump-dump" scheme, originally proposed by Tannor and Rice, using two distinct laser pulses [11]. The intrapulse pump-dump process in molecular systems coupled with a dissipative environment can be described in the framework of population wave packet picture [6]. Coupling with a dissipative environment gives rise to vibrational relaxation which explains the gain of the red edge of the NC pulse [6] observed in the experiment [5].

The following two processes are based on a coherent excitation of a molecule, and can, in principle, produce complete population transfer. 2) π pulses. This method makes use of the Rabi population oscillation [10]. The main disadvantage of the π-pulse excitation is the requirement for resonant laser light and the need of precise control of the pulse area [12].

3) Adiabatic rapid passage (ARP) [3,10,12]. The mechanism of ARP can be explained in the vector model [10] or by avoided crossing of dressed states as a function of the instantaneous laser pulse frequency $\omega(t)$ [12]. A scheme based on ARP is robust since it is insensitive to pulse area and to the precise location of the resonance.

The effect of relaxation processes on a coherent excitation of a molecule coupled with a dissipative environment is an intriguing question. It was studied numerically by Lin, Hayashi et al. with an one-dimensional displaced harmonic oscillator model with Markovian relaxation [7]. However, relaxation of large molecules in solutions is non-Markovian one. In Refs.[6,13] a model of Gaussian-Markovian modulation was used for description of relaxation processes in molecules excited with strong chirped pulses. However, these works were limited by first, incoherent light-matter interaction as to concern the electronic transition and second, the "contact" approximation (i.e. electronic transition occurs at instantaneous intersections of "photonic replication" and the corresponding term). In this work we discard these limitations. It allows us to consider coherent effects, including into consideration short and fast chirped pulses of large intensity.

2 Basic equations

Let us consider a molecule with two electronic states $n = 1$ and 2 in a solvent described by the Hamiltonian

$$H_0 = \sum_{n=1}^{2} |n\rangle \left[E_n + W_n(\mathbf{Q})\right] \langle n| \qquad (1)$$

where $E_2 > E_1, E_n$ is the energy of state $n, W_n(\mathbf{Q})$ is the adiabatic Hamiltonian of reservoir R (the vibrational subsystems of a molecule and a solvent interacting with the two-level electron system under consideration in state n).

The molecule is affected by phase modulated pulses of carrier frequency ω:

$$\mathbf{E}(t) = \frac{1}{2}\vec{\mathcal{E}}(t)\exp(-i\omega t + i\varphi(t)) + c.c. \tag{2}$$

where $\mathcal{E}(t)$ and $\varphi(t)$ are real functions of time, and $\varphi(t)$ describes the change of the pulse phase in a time t. The instantaneous pulse frequency is $\omega(t) = \omega - \frac{d\varphi}{dt}$.

The influence of the vibrational subsystems of a solute and a solvent on the electronic transition can be described as a modulation of this transition by low frequency (LF) vibrations $\{\omega_s\}$ [6]. In accordance with the Franck-Condon principle, an electronic transition takes place at a fixed nuclear configuration. Therefore, the quantity $u(\mathbf{Q}) = W_2(\mathbf{Q}) - W_1(\mathbf{Q}) - \langle W_2(\mathbf{Q}) - W_1(\mathbf{Q})\rangle_1$ is the disturbance of nuclear motion under electronic transition. Here $\langle\rangle_n \equiv Tr_R(\ldots\rho_{R_n})$ denotes the trace operation over the reservoir variables in the electronic state n,

$\rho_{R_n} = \exp(-\beta W_n)/Tr_R \exp(-\beta W_n)$, $\beta = 1/k_B T$.

Electronic transition relaxation stimulated by LF vibrations is described by the correlation function $K(t) = \langle u(0)u(t)\rangle$ of the corresponding vibrational disturbance with characteristic attenuation time τ_s [6]. We suppose that $\hbar\omega_s \ll k_B T$. Thus $\{\omega_s\}$ is an almost classical system and operators W_n are assumed to be stochastic functions of time in the Heisenberg representation. The quantity u can be considered as a stochastic Gaussian variable. We consider the Gaussian-Markovian process when $K(t)/K(0) \equiv S(t) = \exp(-|t|/\tau_s)$. In this case one can obtain for the elements of the density matrix by the generalization of the equations of Refs.[6,14,15,16]:

$$\begin{aligned}\frac{\partial}{\partial t}\rho_{12}(\alpha,t) &= i(\omega_{21}-\alpha)\rho_{12}(\alpha,t) - \frac{i}{2\hbar}D_{12}\mathcal{E}(t)\exp(i\omega t - i\varphi(t))\Delta'(\alpha,t)\\ &+\frac{L_{11}+L_{22}}{2}\rho_{12}(\alpha,t)\end{aligned} \tag{3}$$

$$\begin{aligned}\frac{\partial}{\partial t}\rho_{jj}(\alpha,t) &= \frac{(-1)^{j+1}}{\hbar}\mathrm{Im}[D_{21}\mathcal{E}(t)\exp(-i\omega t + i\varphi(t))\rho_{12}(\alpha,t)]\\ &+L_{jj}\rho_{jj}(\alpha,t)\end{aligned} \tag{4}$$

where $\alpha = -u/\hbar$, ω_{21} is the frequency of Franck-Condon transition $1 \to 2$; $i,j = 1,2$; D is the dipole moment operator, $\Delta'(\alpha,t) = \rho_{11}(\alpha,t) - \rho_{22}(\alpha,t)$. The terms

$$L_{jj} = \tau_s^{-1}\left[1 + (\alpha - \delta_{j2}\omega_{st})\frac{\partial}{\partial\alpha} + \sigma_{2s}\frac{\partial^2}{\partial\alpha^2}\right], \tag{5}$$

on the right-hand side of Eqs.(3,4) describe the diffusion with respect to the coordinate α in the corresponding effective parabolic potential [6]. Here δ_{ij} is the Kronecker delta, ω_{st} is the Stokes shift of the equilibrium absorption and luminescence spectra, $\hbar\beta\sigma_{2s} = \omega_{st}$.

The partial density matrix of the system $\rho_{ij}(\alpha, t)$ describes the system distribution with a given value of α at time t. The complete density matrix averaged over the stochastic process which modulates the system energy levels, is obtained by integration of $\rho_{ij}(\alpha, t)$ over α (the generalized solvent coordinate):

$$\langle \rho \rangle_{ij}(t) = \int \rho_{ij}(\alpha, t)\, d\alpha \tag{6}$$

where diagonal quantities $\langle \rho \rangle_{jj}(t)$ are nothing more nor less than the populations of the electronic states: $\langle \rho \rangle_{jj}(t) \equiv n_j$, $n_1 + n_2 = 1$.

3 Solution of equations

Solving Eq.(3) for $\rho_{12}(\alpha, t)$ and substituting the corresponding expression into Eq.(4) for $\rho_{jj}(\alpha, t)$, we obtain

$$\begin{aligned} \frac{\partial}{\partial t}\rho_{jj}(\alpha,t) = (-1)^j \frac{|D_{12}|^2}{2\hbar^2} \operatorname{Re} \int_{-\infty}^{\infty} d\alpha' \int_0^{\infty} dx \mathcal{E}(t)\mathcal{E}(t-x)\Delta'(\alpha', t-x) \\ \times G_{12}(\alpha, t; \alpha', t-x) \exp\{-i[\omega x - (\varphi(t) - \varphi(t-x))]\} \\ + L_{jj}\rho_{jj}(\alpha,t) \end{aligned} \tag{7}$$

if $\mathcal{E}(t) = 0$ for $t \leq 0$. Here $G_{12}(\alpha, t; \alpha', t')$ is the Green's function of Eq.(3). Below we describe a number of methods for solving Eqs.(3)-(4) (or Eqs.(7)). In general these equations can be solved numerically. Other methods of solution are approximate ones. They may be classified under two groups: 1)methods taking into account electronic coherence, and 2)equations for population wave packets.

3.1 Numerical solution of coupled differential equations

We solve coupled Eqs.(3)-(4), using a basis set expansion with eigenfunctions of diffusion operator $(L_{11}+L_{22})/2$ [15]. The corresponding solutions are termed "model 3" for short, bearing in mind that they take into account all the relaxations (diffusions) related to electronic coherence and populations in both electronic states.

3.2 Methods with consideration for electronic coherence

Semiclassical (Lax) approximation. For broad electronic spectra satisfying the "slow modulation" limit, we have $\sigma_{2s}\tau_s^2 \gg 1$, where $\sigma_{2s} = K(0)\hbar^{-2}$ is the LF vibration contribution to a second central moment of an absorption spectrum. In the last case electronic dephasing is fast. Therefore, the Green's function $G_{12}(\alpha, t; \alpha', t - x)$ in Eq.(7) can be approximated as $G_{12}(\alpha, t; \alpha', t - x) \approx \exp[i(\omega_{21} - \alpha)x]\delta(\alpha - \alpha')$. Substituting this equation into Eq.(7), we obtain Eq.(10) of Ref.[6]. The last equation is equivalent to Eqs.(3)-(4) if one ignores the last term $\rho_{12}(\alpha, t)(L_{11} + L_{22})/2$ on the right-hand side of Eq.(3). Therefore, the solutions which correspond to Eq.(10) of Ref.[6] are termed "model 2" for short.

Pulses shorter than τ_s. For pulses much shorter than τ_s one can ignore all the terms $\sim L_{jj}$ on the right-hand sides of Eqs.(3)-(4). It means that our system can be described as an ensemble of independent two-level systems with different transition frequencies corresponding to a pure inhomogeneously broadened electronic transition. There is an analytic solution of undamped Bloch equations for a chirped pulse of special form [17,18]. This solution is symmetric with respect to the sign of chirp. It is interesting from the point of view of evaluation of the greatest possible population of the excited state due to coherent effects, because this solution ignores all the irreversible relaxations destructing coherence. The solutions corresponding to ignoring all the terms $\sim L_{jj}$ on the right-hand sides of Eqs.(3)-(4) are termed "model 0" for short.

3.3 Equations for population wave packets

Let us consider again the "slow modulation" limit, and pulses longer than $T' = (\tau_s/\sigma_{2s})^{1/3}$ (i.e. $t_p >> T'$) with a moderate phase modulation rate when $|d\omega(t)/dt|\, T' < (T')^{-1}$ [6]. In addition, the pulse intensity is limited by the condition $\sigma_a(\omega_{21})\, J_{\max} << (T')^{-1}$ where $\sigma_a(\omega_{21})$ is the cross section at the maximum of the absorption band, $J(t)$ is the power density of the exciting radiation, $J_{\max}$ is its maximum value. Then the variable x in Eq.(7) is of the order of the relaxation time of the nondiagonal element of the density matrix which is about $x \sim T' \ll t$ for pulses under consideration. Therefore, one can disregard by changing the population difference $\Delta'(\alpha, t)$ during time x and take it outside the integral over x. Such a procedure corresponds to "model 4".

One can further simplify the "model 4" assuming that the $\alpha-$dependence of $\Delta'(\alpha, t)$ is much smoother than that of $G_{12}(\alpha, t; \alpha', t - x)$ [16,15]. Then we arrive to the following equation ("model 5")

$$\frac{\partial}{\partial t}\rho_{jj}(\alpha,t) = (-1)^j K(\alpha,t)\Delta'(\alpha,t) + L_{jj}\rho_{jj}(\alpha,t) \tag{8}$$

The first term on the right-hand side of Eq.(8) $\sim K(\alpha,t)$ describes radiation transitions between the electronic states. The magnitude $K(\alpha,t)$ has peaks at $\alpha = \omega_{21} - \omega(t)$ and is equal to

$$K(\alpha,t) = (2\hbar^2)^{-1}|D_{12}|^2 \int_0^\infty dx \mathcal{E}(t)\mathcal{E}(t-x)\,\mathrm{Re}\int_{-\infty}^\infty d\alpha' G_{12}(\alpha',t;\alpha,t-x) \times \exp\{-i[\omega x - (\varphi(t) - \varphi(t-x))]\} \tag{9}$$

4 Results

Let us study the influence of the chirp rate on the excited state population n_2 after the completion of pulse action. We consider linear chirped pulses of the form $E(t) = \mathcal{E}_0 exp[-\frac{1}{2}(\delta^2 - i\mu)(t-t_0)^2]$. If chirped pulses are obtained by changing the separation of pulse compression gratings, the parameters δ and μ are determined by the formulae [5,6]

$$\delta^2 = 2\{\tau_{p0}^2 + [2\Phi''(\omega)/\tau_{p0}]^2\}^{-1} \tag{10}$$

$$\mu = -4\Phi''(\omega)\left[\tau_{p0}^4 + 4\Phi''^2(\omega)\right]^{-1} \tag{11}$$

where $\tau_{p0} = t_{p0}/\sqrt{2\ln 2}$, t_{p0} is the pulse duration of the corresponding transform-limited pulse, and $\Phi''(\omega) = \Phi''(\nu)/(4\pi^2)$ is the phase term.

Fig.1 shows the calculation results of n_2 as a function of $\Phi''(\nu)$ for different detunings of the exciting pulse carrier frequency ω with respect to the frequency of Franck-Condon transition ω_{21}, corresponding to "coherent" models "0", "2" and "3". The values of parameters were the followings: $\tau_{p0} = 11$ fs, $\sqrt{\sigma_{2s}} = 546cm^{-1}$, $\tau_s = 70\ fs$, $Q' \equiv \sigma_a(\omega_{21}) J_{\max} t_p = 3$ (the saturation parameter).

One can see a strong dependence of the excited state population n_2 on chirp sign induced by relaxation (models "2" and "3" which give very similar results) for positive detunings $\omega - \omega_{21}$, whereas n_2 does not depend on chirp sign for model "0" (without relaxation). For negative detunings $\omega - \omega_{21}$ the asymmetry of $n_2(\Phi''(\nu))$ dependence with respect to the chirp sign diminishes. For large positive $\Phi''(\nu)$the excited state population n_2 tends to the constant value (that is about 1 for $Q' = 3$) which is very close to that given by model "0".

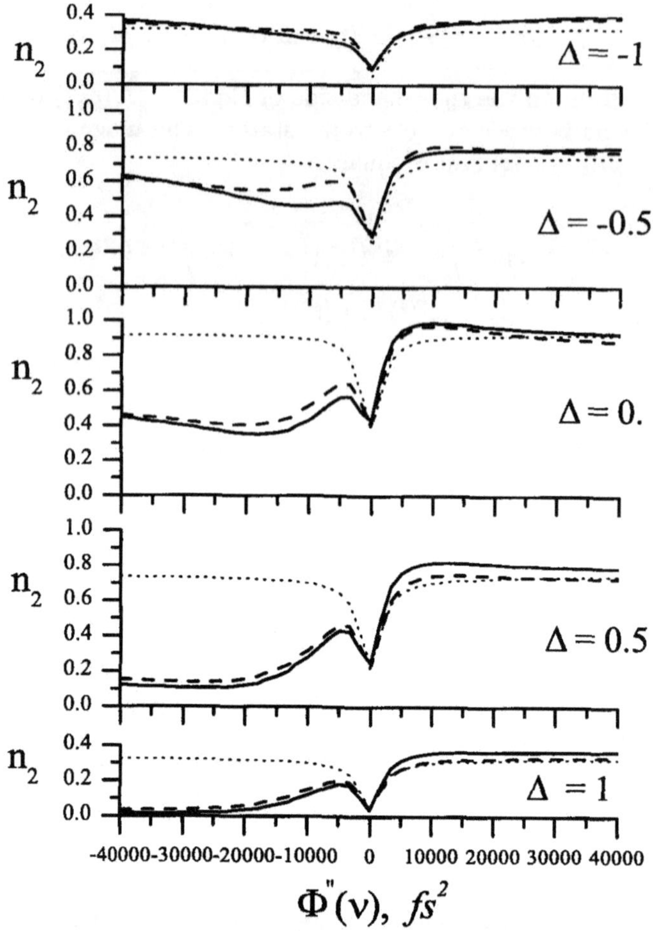

Figure 1. Excited state population versus $\Phi''(\nu)$ and frequency detuning $\Delta = (\omega - \omega_{21})/\omega_{st}$ for different relaxation models: model 3 (solid lines), model 2 (dashed lines), and model 0 (dotted lines).

5 Discussions

Let us consider first model "0" (without relaxation). It is possible to obtain the following criterion for ARP in this model

$$\sqrt{\frac{\sigma_{2s}}{\pi \ln 2}}\tau_{p0}Q' >> \frac{1}{2} \tag{12}$$

For $\tau_{p0} = 11\ fs$ and $\sqrt{\sigma_{2s}} = 546cm^{-1}$ it corresponds to the value of $Q' >> 1$ which conforms to the value of $Q' = 3$ used in our calculations. Thus, about a complete population transfer to the excited electronic state observed in our simulations for model "0", can be explained by ARP.

Models "2" and "3" give a qualitative dependence of n_2 on $\Phi''(\nu)$ similar to that given by "incoherent" models based on population wave packets picture (Ref.[6] and models "4" and "5"). However, the value of n_2 calculated by "incoherent" models for large positive $\Phi''(\nu)$, is essentially smaller than 1 when the saturation parameter $Q' = 3$. Thus, "incoherent" models fall down to explain about a complete population transfer to the excited state in conditions under consideration. In this relation one ought to take into consideration that models "2" and "3" give the limit value of n_2 for large $\Phi''(\nu) > 0$ that is very close to that given by model "0" (without relaxation). The last value is explained by ARP that is a coherent effect. Thus, a positive chirp is favorable to the reduction of relaxation effects on electronic coherence for detunings $\omega - \omega_{21} \approx 0$. This conclusion is supported by analytical considerations which will be published elsewhere.

In conclusion, we have considered strong chirped pulse excitation of a molecule in solution taking into account electronic coherence effects. The strong dependence of n_2 on chirp sign for positive detunings $\omega - \omega_{21}$ can be explained not only by pump-dump process for population wave packets based on incoherent consideration but the reduction of relaxation effects on electronic coherence for a positive chirp as well.

Acknowledgments

This work was supported by the Israel Science Foundation and the Ministry of absorption of Israel.

References

1. S. Ruhman and R. Kosloff, *J.Opt.Soc.Am.* B **83**, 5013 (1990).
2. J. L. Krause *et al.*, *J. Chem. Phys.* **99**, 6562 (1993).
3. J. S. Melinger *et al.*, *J. Chem. Phys.* **101**, 6439 (1994).
4. C. J. Bardeen, Q. Wang, and C. V. Shank, *Phys. Rev. Lett.* **75**, 3410 (1995).

5. G. Cerullo, C. J. Bardeen, Q. Wang, and C. V. Shank, *Chem. Phys. Lett.* **262**, 362 (1996).
6. B. D. Fainberg, *J. Chem. Phys.* **109**, 4523 (1998).
7. K. Mishima *et al.*, *Chem. Phys. Lett.* **309**, 279 (1999).
8. J. Manz, H. Naundorf, K. Yamashita, and Y. Zhao, *J. Chem. Phys.* **113**, 8969 (2000).
9. J. Cao, C. J. Bardeen, and K. R. Wilson, *Phys. Rev. Lett.* **80**, 1406 (1998).
10. L. Allen and J.-H. Eberly, *Optical Resonance and Two-Level Atoms* (John Wiley & Sons, New York, London, Sydney, Toronto, 1975).
11. D. J. Tannor and S. A. Rice, *J. Chem. Phys.* **83**, 5013 (1985).
12. B. W. Shore *et al.*, *Phys. Rev.* A **45**, 5297 (1992).
13. B. D. Fainberg and V. Narbaev, *J. Chem. Phys.* **113**, 8113 (2000).
14. L. D. Zusman, *Chem. Phys.* **49**, 295 (1980).
15. D. Y. Yang and R. I. Cukier, *J. Chem. Phys.* **91**, 281 (1989).
16. L. Hartmann, I. Goychuk, and P. Hanggi, *J. Chem. Phys.* **113**, 11159 (2000).
17. Y. N. Demkov and Kunike, *Vestn. Leningr. Univ. Fiz. Khim.* **16**, 39 (1969).
18. F. T. Hioe, *Phys. Rev.* A **30**, 2001 (1984).

EFFECT OF COHERENT TRANSIENTS ON A PUMP-PROBE TIME-RESOLVED SIGNAL

S. ZAMITH, J. DEGERT, S. STOCK, B. DE BEAUVOIR, V. BLANCHET, M. A. BOUCHENE AND B. GIRARD

Laboratoire de Collisions, Agrégats et Réactivité (CNRS UMR 5589), IRSAMC
Université Paul Sabatier,
31062 Toulouse, France
sebas@irsamc.ups-tlse.fr and val@irsamc.ups-tlse.fr

By simply varying the temporal shape of the pulses, different pump-probe signals can be obtained. We studied here theoretically and experimentally the effects of coherent excitation of a two and three level system with a linearly chirped pulse in the low field regime. The relevance of a pump detuning to the resonance is considered. The coherent transients are measured directly on the excited state population on an ultrashort time scale.

1 MOTIVATION

The UV-VUV wavelength range is a region of great interest to study the interstellar, planetary or atmospheric chemical processes such as radical formation, oxidation mechanisms and the reactivity of electronically excited species, or to control photochemical reactions on surface like surface bonds breaking[1]. Indeed, this energy domain allows excitations of large fraction of molecular species leading to informative dynamics as energy transfer of excited states of molecules in the upper atmosphere like bond breaking[2], or changes in the valence band of surface adsorbate[3]. The ultrafast laser sources are now stable enough in intensity [4,5] to get macroscopic conversion efficiencies into the UV-VUV region [6]. It is possible now to monitor the evolution of these reactions via pump-probe experiments. Furthermore the time scale of these dynamics is comparable with the laser duration [7,8,9,10]. However the characterization of these laser sources is far from complete especially in term of spectral phase. In this paper, we will consider the simplest case of linearly chirped pulses where the instantaneous frequency drifts in time. This excitation scheme is often encountered with UV-VUV wavelengths. Indeed in this wavelength region, the optic elements used introduce a non negligible dispersion. Depending on the pump frequency sweep, the population of the photoexcited state can depend crucially on the extend of Coherent Transients (CT). The CT result in oscillating population, signature of interferences between resonant and non-resonant excitation paths[11] that appear at different times with a chirped pump pulse. Most of the population transfer occurs at resonance without to lead to total population inversion[12]. The non-resonant excitation paths occurring after resonance can also lead to population transfer. However, these extra contributions have a phase which varies rapidly with respect to the phase of the dominant contribution. This results in the appearance of strong oscillations in the temporal profile of the population. This paper presents theoretical studies of the effects induced by CT on a pump-probe signal and illustrates experimentally the CT phenomenon on the simplest case of a two level system.

2 MODEL

In order to point out the effect of Coherent Transients on a pump-probe signal recorded to measure the lifetime τ_e of an intermediate state, we study three excitation schemes. We consider a positively chirped pump pulse and a Transform limited probe pulse. As shown on Figure 1, the three cases correspond to different detuning, which are defined as $\delta_p = (\omega_L - \omega_e)$ with ω_e the frequency of the transition and ω_L the carrier frequency of the pump pulse. In the on-resonance case (b) the resonance is reached at the temporal middle of the pump interaction, whilst in the two other cases, this phenomenon arises earlier for a positive detuning (a) and later for a negative one (c). The decay is assumed incoherent such that we do not treat any wave packet dynamic. In the low field regime, assuming a first order perturbation theory, the analytical expression of the final state population is thus :

$$|f(\tau)|^2 \propto e^{-\tau/\tau_e} \left| 1 - \phi\left(\frac{\gamma\sqrt{\beta_s + \beta_p}}{2\sqrt{\beta_s\beta_p}}\right)\right|^2 \tag{1}$$

where τ is the delay time between pump and probe and ϕ the error function. β_p and β_s characterize the pump and probe pulses respectively with

$$\beta_p = \frac{1}{\tau_{pc}^2} + i\alpha_p \quad \text{and} \quad \beta_s = 1/\tau_s^2 \tag{2}$$

where τ_{pc} and τ_s are the durations of pump and probe respectively and α_p corresponds to the linear chirp of the pump pulse defined from the quadratic dispersion ϕ" as follows :

$$\alpha_p = \frac{2\phi"}{\tau_p^4 + 4\phi"^2} \tag{3}$$

where τ_p is the transform limited pump pulse duration. The electric field of a pulse is then given by $E(t) = E_0 e^{(-\beta t^2 - i\omega_L t)}$. In equation 1 γ is a constant reflecting the excitation scheme with as the main parameter, the detuning δ_p and the lifetime τ_e

$$\gamma = \frac{-i\beta_s\delta_p - 2\beta_p\beta_s\tau}{\beta_s + \beta_p} + 1/(2\tau_e) \tag{4}$$

Figure 1 displays the pump-probe signal for the three detuning cases with a lifetime fixed at τ_e=500 fs, a probe duration (FWHM) of 100 fs and a pump pulse stretched linearly from 100 fs to 200 fs FWHM (ie ϕ"=6240 fs^2). In the panels (a) and (b) of the figure 1, the detuning is equal to a half of the pump pulse spectral width $\delta_p = \pm\frac{\Delta\omega_{pFWHM}}{2}$. This leads to $\delta_p \simeq 7/\tau_e$, assuming a homogeneous band. The pump-probe signals calculated via equation 1 are the straight line plots on fig.1. They can be compared to the cross-correlation trace between the two pulses displayed with dotted line on panel (b).

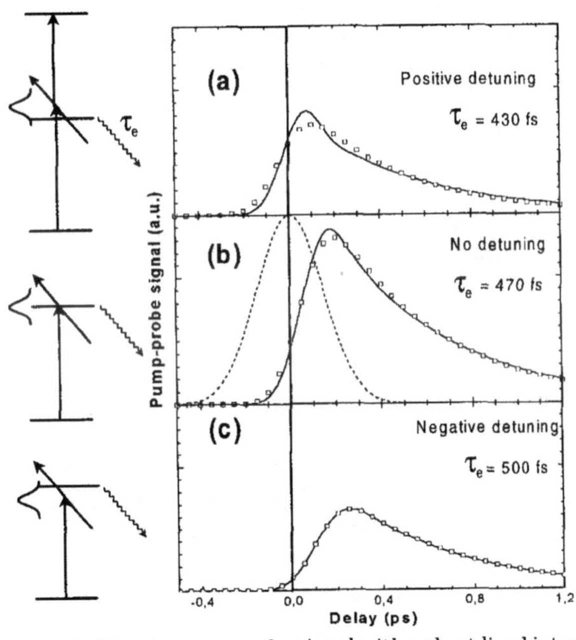

Figure 1: Simulation of a 3 levels pump-probe signal with a short lived intermediate state (500 fs) in three cases: (a) Positive detuning (b) No detuning (c) Negative detuning. The probe is a transform limited pulse of 100 fs. The dashed curve in panel (b) is the cross-correlation of the pump and probe pulses. The straight line plots correspond to the simulation of the pump probe signal whereas the square symbols plot are the fit of these curves realized with a rate equation model. The lifetime indicated are the results of this rate equation fit procedure.

For a resonant excitation, as long as the duration of the chirped pump pulse is shorter or comparable to the lifetime of the intermediate state, the coherent transients have no drastic effect on the pump-probe signal. But as soon as there is a detuning, the signal amplitude is strongly affected as shown on Fig.1(a,c). The amplitude damping decrease is mainly due to a weaker laser field at the resonance frequency. The maximum position shift corresponds simply to the temporal shift of the resonant path : the maximum arises earlier for a positive detuning and later for a negative detuning in comparison to the on-resonance case. Nevertheless, we need a more accurate understanding of CT to explain the decay changes as well as the difference on the amplitude signal when we compare each detuning case. In general, the lifetime is deduced from the curve fitting of the pump-probe signal by a rate equation model. This model is usefull in time-resolved studies and needs only the cross-correlation time and presupposes neither "chirp" or detuning. The fit represented by squared symbols in fig.1 gives different lifetimes τ_e for each detuning. Fit and pump-probe signal only agree in the case of a negative detuning (fig.1(c)). Indeed in this latter case, the result of CT are less important because the amount and weight of non-resonant paths are smaller than for the other detunings. The discrepancies observed between pump-probe signals and the equation rate fitting result from oscillations of the intermediate state population due to the existence of Coherent Transients. Theses CT are most often ignored in time-resolved experiments, although in some studies, they could mislead in the pump-probe signal interpretation.

3 EXPERIMENT

The main CT features can be impressively demonstrated in a pump-probe experiment involving a undamped intermediate state together with a significant chirp. To get a detailed understanding, we have illustrated experimentally the CT importance on the Rb ($5S \rightarrow 5P_{1/2}$) transition. The excitation scheme is depicted in Figure 2. Two detuning cases are illustrated experimentally : (a)no detuning with a pump carrier frequency at resonance ($\lambda = 795\,\text{nm}$) and (b) a negative detuning with a pump central wavelength ($\lambda \simeq 801\,\text{nm}$) detuned below resonance by half the laser bandwidth . To ensure that the final state populations ($5P_{1/2}$ - nS,n'D probe transition) reproduce the oscillating population of the intermediate state, the probe pulse duration must be shorter than the chirped pump pulse duration. In the experimental data displayed in figure 3, the pump pulse is negatively chirped from 130 fs to 11 ps with a pair of gratings and the transform limited probe pulse duration is 30 fs (NOPA set up). The pump-probe signal is detected by monitoring the fluorescence from the radiative cascade (nS,n'D) $\rightarrow$ 6p $\rightarrow$ 5s[11]. With a significant chirp and long lived intermediate state, we observe now the oscillations of the intermediate state population already mentioned. These large amplitude oscillations are the signature of the Coherent Transients.

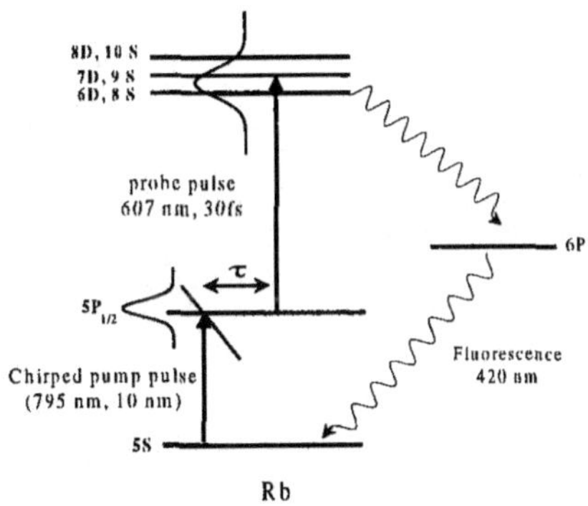

Figure 2: Principle of the Coherent Transient experiment realized in the Rubidium levels.

The experiment performed in the perturbative regime (fluence of $2\,\mu J/cm^2$) can be compared to the analytical solutions displayed in Fig.3 by the straight line curves. For a long lived intermediate state pumped by a long chirp pulse, the laser sweeping frequency goes through the atomic resonance at time $t_0 = -\delta_p\phi''$. The resonance takes places around $\tau = 0$ for no-detuning and $\tau < 0$ for negative detuning as observed in Fig.3(a) and (b) respectively. This is similare to the maxima shifts of pump-probe signals observed on Fig.1 except that the chirp is now negative. In the stationary phase approximation, the main contribution arises from delays within the interval $\left[t_0 - \sqrt{|\phi''|};\, t_0 + \sqrt{|\phi''|}\right]$ during which the laser frequency is resonant with the atomic transition. In the measurements presented here, this time window has a width of 1.3 ps. As time increases from t_0, contributions from non-resonant transitions are added to the probability amplitude of the intermediate state. These new contributions with smaller weights have a phase which increases more and more rapidly as the "instantaneous" frequency $\omega(t) = \omega_L + 2\alpha_p t$ varies. This leads alternatively to an increase and a decrease of the excited state amplitude with an increasing oscillation frequency. For a negative detuning and a negative chirp, the passage through resonance occurs at the beginning of the pulse with a maxima arising at negative delay time. Therefore the major part of the pump spectral components contributes after the resonance so that many oscillations result as can be seen on Fig.3(b). On the contrary, in the resonant case, only the second half of the pulse contributes to the interferences and less oscillations are observed on Fig.3(a). With a short intermediate state and a positive chirp as shown in figure (1), the positive detuning pump-probe signal (a) was more affected than the other detuning cases (b and c), because the oscillations were more important, similarly to the case shown on Fig.3(b).

An other way to explain the CT phenomenon is to examine the behaviour of

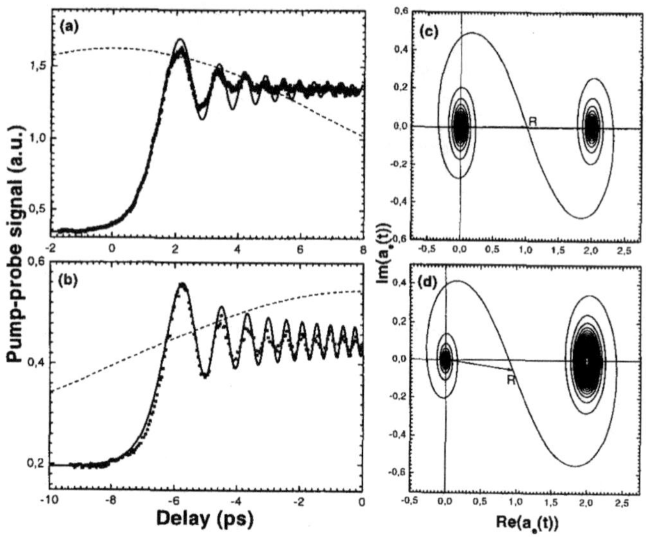

Figure 3: Experimental curves (squares) together with the theoretical ones (line) for (a) 795 nm (no detuning) and (b) 801 nm (negative detuning). The chirp parameter ϕ"=-4 10^5 fs^2, leading to a pump duration of 11 ps. The imaginary part of the probability amplitude of the excited state is plotted as a function of the real part in panels (c) and (d). The point R represents the passage through resonance. The dashed curves represent the envelope of the pump pulse.

the probability amplitude of the intermediate state $a_e(t)$ in the complex plane as displayed in Fig.3(c) and (d) for the experimental data presented on Fig.3(a) and (b) respectively. The probability amplitude follows a Cornu spiral starting from $(0,0)$ (no population in the intermediate state). The passage through resonance marked on fig.3 by the R vector, is characterized by a sharp increase of $|a_e(t)|$, in an almost "straight" direction as expected from the stationary phase approximation. This passage looks the same for the two detuning cases because it arises on the same time interval $\left[t_0 - \sqrt{|\phi''|}; t_0 + \sqrt{|\phi''|}\right]$ as already mentionned. The two spirals play different roles. The first one winds round the origin with an increasing radius. Therefore the resulting probability increases slowly and regularly. After resonance, the second spiral winds round its asymptotic value leading to strong oscillations. For a central laser frequency equal to the resonance frequency, the two spirals have the same weight (Fig.3c) and the resonance occurs in the "middle" of the Cornu spiral as indicated by the R vector. Detuning changes their relative weight (Fig.3b). For negative chirp and negative detuning, the second spiral (after resonance) has a larger weight since the resonance is reached during the first half of the pulse. This leads to a higher number of oscillations, as can be observed by comparing Fig.3a and 3c. Likewise, a positive chirp and a positive detuning lead to the same oscillations. Conversely, a negative (positive) chirp and positive (negative) detuning reduce the number of oscillations. These experiments provide a detailed understanding of the behaviour of Coherent Transients .

4 Quantum Beating

In the case of a chirped pump pulse centered in energy between two excited levels $|1\rangle$ and $|2\rangle$, the CT would not result in the same change on the two populations. Indeed the detuning of the two excited levels differs in sign. The positive detuning corresponds to the excited state $|1\rangle$ and the negative one to the excited state $|2\rangle$ as drawn in the inset of Figure 4 (c). We investigate this excitation scheme for a positively chirped pump pulse using by integrating numerically the Schrödinger equation. The two states have the same lifetime. The pump pulse is still chirped from 100 fs to 200 fs and the probe pulse stays Fourier limited to 100 fs. The pump and probe transition moments are set equal for the two states. The first resonant transition arises onto the excited state $|1\rangle$ and later into the excited state $|2\rangle$. In comparison with an excitation by a Fourier limited pump pulse, the maximum of the excited state population occurs thus earlier for the excited state $|1\rangle$ and later for the excited state $|2\rangle$, as shown on Figure 4(a) and (b) respectively. Since the lifetime τ_e is of the same magnitude than the pump duration, the CT have more impacts than in the case depicted in Figure 1. A first oscillation is even observed for the positive detuning case corresponding to the excited state $|1\rangle$. The pump-probe signal in this numerical calculation corresponds to the population of a higher excited state $|3\rangle$, coherent summation of $|1\rangle$ and $|2\rangle$ populations. The signal oscillation observed on Fig.4(c) are the well-known signature of the quantum beats of $|1\rangle$ and $|2\rangle$ states [13]. Let compare now the two pump-probe signals plotted on figure 4(c), one corresponding to a Fourier limited pump pulse and the other to the chirped one. The Coherent transients affects drastically the time-dependance of the signal both in intensity and contrast.

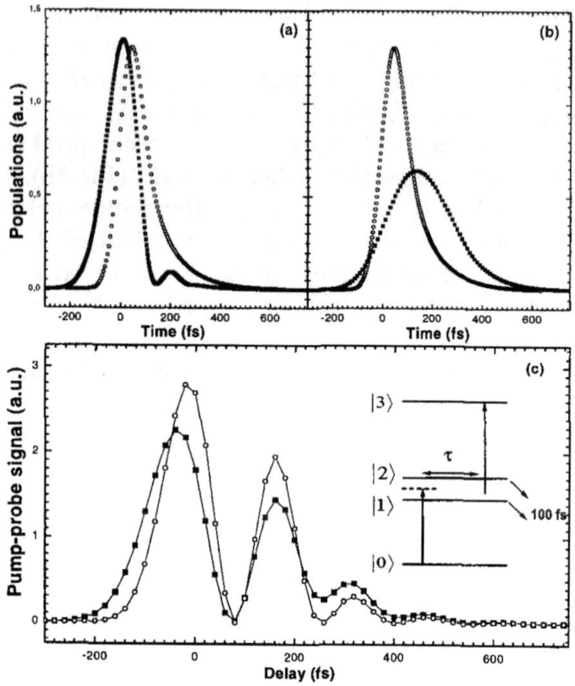

Figure 4: Simulation of a pump probe signal in a four levels system. The two intermediates states $|1\rangle$ and $|2\rangle$ have the same lifetime fixed at 100 fs. Population of (a) the $|1\rangle$ state and (b) the $|2\rangle$ state as a function of time.(c) Pump-probe signal as a function of the delay time τ between the pump and probe pulses. The open circles curves correspond to a Fourier transform limited pump pulse of 100 fs, the square curves correspond to a pump pulse stretched to 200 fs with a positive linear chirp. In both cases the probe is a 100 fs Fourier transform limited pulse.

5 CONCLUSION

It has been shown that Coherent transients might seriously affect the pump-probe experiments. Such effect can be deconvoluted to get the constants of dynamics using realistic laser fields. For this, the laser pulses have to be characterized accurately, especially in the wavelength range as UV-VUV where the refraction index can vary rapidly. We should also note that Coherent Transients resulting from a chirp of the probe pulse, can also be observed in the special case of a unique final state. This leads to the same behaviour as the ones depicted here for a chirped pump pulse. Finally a better understanding of such transient phenomena can improve the implementation of new laser control scheme.

References

1. A. Pusel, U. Wetterrauer and P. Hess, *Phys. Rev. Lett.* **81**, 645 (1998)
2. S.L. Sorensen *et al*, *J. Chem. Phys.* **112**, 8038 (2000)
3. M. Bauer *et al*, *Phys. Rev. Lett.* **87**, 025501 (2001)
4. C.G. Durfee *et al*, *Opt. Lett.* **22**, 1565 (1997)
5. E. Constant *et al*, *Phys. Rev. Lett.* **82**, 1668 (1999)
6. L. Misoguti *et al*, *Phys. Rev. Lett.* **87**, 013601 (2001)
7. H. Petek and S. Ogawa *et al*, Progress in Surface Science **56**, 239 (1998)
8. W. Fuβ, W.E. Schmid and S.A. Trushin,*J. Chem. Phys.* **112**, 8347 (2000)
9. M. Wittmann *et al*,*Opt. Commun.* **173**, 323 (2000)
10. V. Stert, P. Farmanara and W. Radloff,*J. Chem. Phys.* **112**, 4460 (2000)
11. S. Zamith *et al*, *Phys. Rev. Lett.* **87**, 033001 (2001)
12. N. V. Vitanov, *Phys. Rev. A* **59**, 988 (1999).
13. S. Zamith et al., *Eur. Phys. J. D.* **12**, 255 (2000).

FEMTOSECOND PUMP-PROBE PHOTOELECTRON-SPECTROSCOPY ON ELECTRONIC STATES OF NA_2: A TOOL TO STUDY ULTRAFAST CONTROL OF CHEMICAL REACTIONS

M. WOLLENHAUPT, A. ASSION, O. BAZHAN, D. LIESE, C. SARPE-TUDORAN, T. BAUMERT*

Experimentalphysik III, Fachbereich Physik, Universität Gesamthochschule Kassel,

Heinrich-Plett-Str. 40, D 34132 Kassel, Germany

(*) Fax: +49 (0)561 / 804 4453, E-mail baumert@physik.uni-kassel.de

Various one-parameter quantum control schemes applied to Na_2 serve as prototype for current multi-parameter control techniques in order to obtain physical insight into the underlying molecular dynamics.

1 Introduction

Microscopic control of the outcome of a chemical reaction is a long-standing dream in physical chemistry. Recently femtosecond lasers have emerged as a particularly suitable tool for quantum control of reaction dynamics. Assion et al. [1] were the first to demonstrate the use of tailored femtosecond laser pulses from a computer-controlled pulse shaper to control the branching ratios of various photodissociation channels in $CpFe(CO)_2Cl$. Moreover, quite recently Levis et al. [2] have shown that high intensity laser pulses permit to control a dissociative rearrangement reaction in which chemical bonds are not only selectively broken, but also newly formed. Quantum control over chemical reactions may be obtained employing various so-called one parameter schemes [3-6]. Currently much attention has been focused on the utilization of adaptive feedback controlled femtosecond pulse shaping making available a most versatile instrument for multi-parameter control schemes [1, 7-10]. These techniques have proven to be universal in the sense that they enable to optimize virtually any conceivable quantity and are capable to deliver the optimal electric field without the knowledge of the underlying potential energy surfaces (PES). However, the individual control mechanisms may be inferred, if at all, only for very simple systems. In order to get a better physical insight into the multi-parameter control driving such an experiment it is essential to investigate one parameter control schemes in detail on pertinent model systems.

2 The sodium dimer as model system for quantum control schemes

The choice of a suitable model system is governed by several requirements: in order to support the validity of the physical pictures of the control process full quantum mechanical simulations of the experiment should be feasible. Therefore one-dimensional systems such as diatomic molecules are most workable. Moreover, a detailed knowledge of all PES involved is prerequisite. Insight in the dynamics of the control can only be accomplished if the experiment allows structural and electronic changes in the configuration to be monitored with high spatial and temporal resolution. Measurements of kinetic photoelectrons using time-of-flight spectrometry (TOF) combined with a femtosecond multiphoton ionization pump-probe technique on Na_2 in a molecular beam allows to map molecular motion with sub-Angström spatial and femtosecond temporal resolution. Therefore they are ideally suited for detailed studies how quantum control can be achieved. Published and unpublished examples for such one parameter control schemes elucidating the mechanism of quantum control are presented.

3 The Tannor-Kosloff-Rice scheme

In the time domain the so called Tannor-Kosloff-Rice scheme [3] is a very illustrative way of how to obtain control over different products via photoexcitation of a molecule. In practice a pump laser prepares a vibrational wave packet on an electronic excited potential and a time-delayed probe laser populates the product state either by dumping the wave packet down again or by further excitation. In the case studied here we obtain a higher kinetic energy of photoelectrons at the inner turning point and a lower kinetic energy of the photoelectrons at the outer turning point from the ionization of a vibrational wave packet propagating on the electronic states (see Fig. 1a for relevant PES and Fig. 1b for measured data). By controlling the duration of the wave packet on the 2 $^1\Pi_g$ state we have access to different product states (not shown here). By this means we demonstrate the control of the photoproduct ratio Na^+/Na_2^+ in a pump-probe experiment [11].

4 Intensity of the laser pulses

In general, the wavelength range needed for the excitation of an intermediate state will differ from the wavelength range to reach the product state at a certain bond configuration. If the two wavelength regimes are not contained within the spectral width of the laser pulse a weak field approach may not be efficient. A solution can be found by utilizing intensity dependent effects in order to shift energy levels involved into the accessible spectral range. This was demonstrated on Na_2 [12] and also applied in the work of Levis [2]. In a set of measurements the pump laser

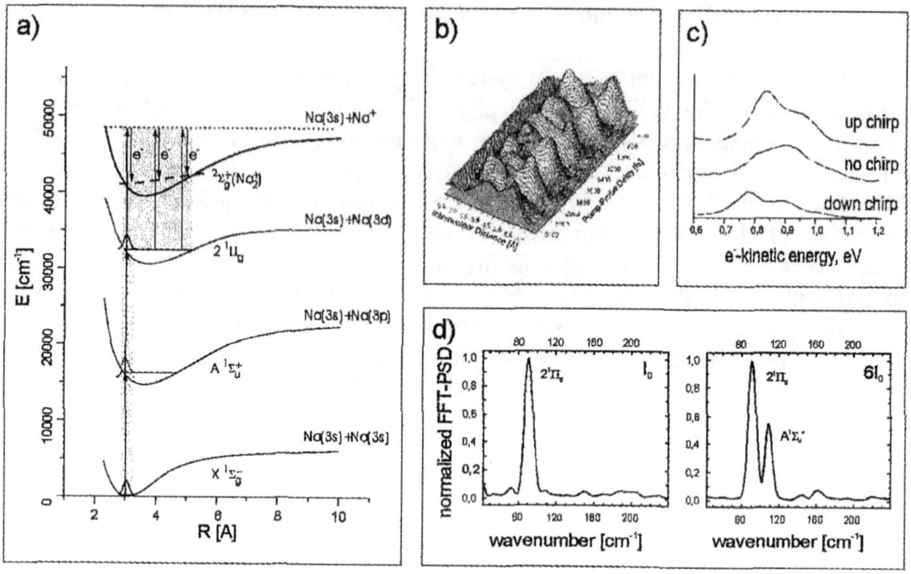

Fig. 1a: Excitation scheme of sodium dimer for 620 nm photons.

Fig. 1b: Mapping the wave packet with sub-Angström spatial resolution and femtosecond time resolution.

Fig. 1c: Electron spectra measured with single up-chirped (+3500 fs^2), unchirped and down-chirped (-3500 fs^2) laser pulse.

Fig. 1d: Cuts of FFT spectra at the outer turning point of the wave packet at I_0 (left) and 6 I_0 (right).

intensity is kept fixed whereas the probe laser intensity is increased from $I_0 \approx 10^{11}$ W/cm^2 to 10 I_0. The propagation of the vibrational wave packets over all allowed internuclear distances is monitored with a probe laser of variable intensity to investigate the influence of high laser intensities on the molecular potential sensitive to the internuclear distance. In Fig. 1d the FFT of the cuts at TOFs corresponding to the outer turning point of the wave packets are shown. With increasing probe laser intensity a contribution of the A $^1\Sigma_u^+$ is seen at the outer turning point. An explanation of this resonant behavior was given in terms of light induced potentials [13].

In order to steer a chemical reaction via wave packet motion on a particular electronic excited state it is advisable to achieve control over the population in the specific state of interest. This can be done by coherent coupling of the electronic states [18]. Fig. 2a displays results of a quantum calculation that are performed analogously to [19]. At the pump intensity I_3 the population of the 2 $^1\Pi_g$ state is involved in one Rabi cycle leading to the "2π" excitation of this state. In the upper part of fig. 2b the experimental results for $I_{pump} \approx 10^{11}$ W/cm^2 and $I_{probe} \approx 10^{12}$ W/cm^2 are displayed. Note the contributions of frequencies of the A $^1\Sigma_u^+$ and 2 $^1\Pi_g$ states to the Power Spectrum Density (PSD). The data in the lower part of fig. 2b are taken with twice the pump laser intensity. The observed reduction in amplitude of the 2 $^1\Pi_g$ frequencies to the PSD is attributed to the "2π" excitation of the 2 $^1\Pi_g$ state.

Concerning photodissociation reactions in intense fields, we have studied the transition from photodissociation on light-induced potentials to field ionization as a function of intensity on the Na_2^+ system [20].

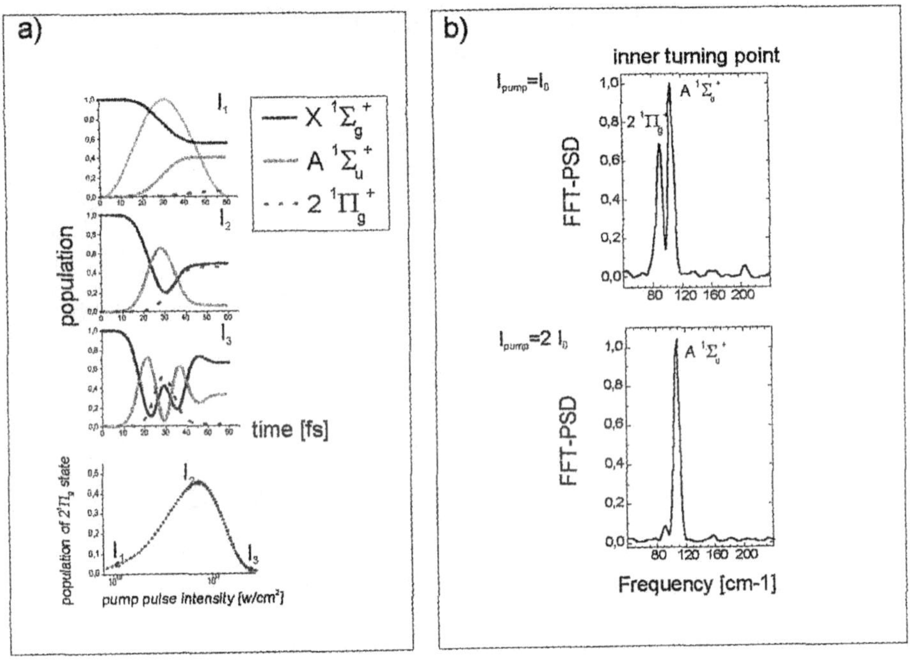

Fig.2a: Coherent coupling of electronic states (see fig.1a) at various laser intensities.

Fig.2b: Experimental verification of 2π excitation. FFT at inner turning point from mapping experiments under identical conditions: Upper: $I_{pump} \approx 10^{11}$ W/cm^2 ($=I_0$). Lower: $I_{pump}=2I_0$.

5 Chirp and duration of laser pulses

In the last couple of years an increasing number of experiments is devoted to the use of phase shaped laser pulses in order to control atomic and molecular properties [9]. Fig. 1c shows photoelectron spectra obtained from ionizing Na_2 with up-chirped, un-chirped and down-chirped laser pulses at 620 nm. The ionization yield is seen to double when the frequency order is switched from blue first to red first in the exciting laser pulse. Calculated electron spectra qualitatively reproduce the measured results. Form the analysis of this calculations the importance of synchronizing wave packet motion to chirp and intensity profile is derived for obtaining efficient population transfer and subsequent ionization.

6 Influence of electronic changes along the internuclear distance

Another aspect in quantum control is introduced by nonadiabatic coupling of the potential curves involved in the time evolution of a quantum system [14]. The

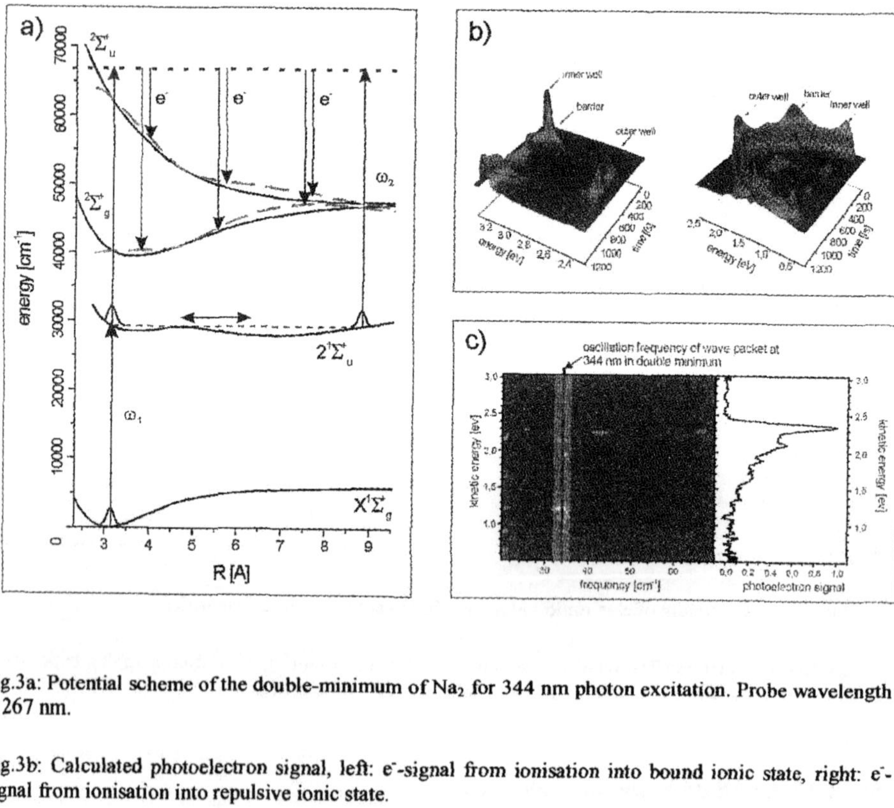

Fig.3a: Potential scheme of the double-minimum of Na_2 for 344 nm photon excitation. Probe wavelength is 267 nm.

Fig.3b: Calculated photoelectron signal, left: e^--signal from ionisation into bound ionic state, right: e^--signal from ionisation into repulsive ionic state.

Fig.3c: Preliminary experimental results: FFT from the measured photoelectron signal at a probe wavelength of 344 nm.

coupling of the covalent and ionic potentials in NaI is a prominent example [15 and references therein]. Electronic changes along the internuclear coordinate R are also predicted for the $^1\Sigma_u^+$ double minimum potential well in the Na_2 molecule (Fig. 3a). According to [16] the change of electronic structure near the avoided crossing at 4.7 Å leads to photoionization probabilities rapidly changing with geometry. We performed quantum mechanical calculations of the temporal evolution of a wave packet in the $^1\Sigma_u^+$ double minimum potential well and the resulting photoelectron signal assuming an R-independent ionization probability similar to [17] (Fig. 3b). More detailed calculations by Takatsuka et al. [16] taking into account the change in the electronic structure indicate no photoelectron yield from the inner well. Experiments with our newly commissioned femtosecond laser system at the University of Kassel are underway to focus on the mostly unstudied influence of R-dependent transition moment on coherent control. In this two-color experiment a wave packet is created at the inner turning point of the double minimum potential slightly above the barrier. During its temporal evolution the subsequent photoelectrons resulting from ionizing on the bound and repulsive ionic state are investigated. Preliminary results obtained at the University of Würzburg are shown in Fig. 3c.

7 Conclusion

Investigation of the wave packet motion of small molecules with high spatial and temporal resolution is the key feature to study a variety of one parameter control schemes in great detail. Based on this knowledge a comparison to results of feedback optimized laser pulses applied to such systems will provide valuable insights into automated multi parameter control schemes. Going to larger molecules, control schemes on isomerization, energy redistribution and molecular rearrangement can be studied using this method because of its sensitivity both to structural and electronic changes in a molecule upon excitation.

8 Acknowledgements

Stimulating discussions and collaboration with G. Gerber and T. Frohnmeyer at the University of Würzburg are gratefully acknowledged.

References

1. A. Assion, T. Baumert, M. Bergt, T. Brixner, B. Kiefer, V. Seyfried, M. Strehle, G. Gerber: Science 282, 919 (1998)
2. R.J. Levis, G. M. Menkir, H. Rabitz: Science 292, 709 (2001)
3. D.J. Tannor, R. Kosloff, S.A. Rice: J. Chem. Phys. 85, 5805 (1986)
4. P. Brumer, M. Shapiro: Chem. Phys. Lett. 126, 541 (1986)
5. U. Gaubatz, P. Rudecki, M. Becker, S. Schiemann, M. Külz, K. Bergmann: Chem. Phys. Lett. 149, 463 (1988)
6. C.J. Bardeen, Q. Wang, C.V. Shank: Phys. Rev. Lett. 75, 3410 (1995)
7. R.S. Judson, H. Rabitz: Phys. Rev. Lett. 68, 1500 (1992)
8. D. Meshulach, Y. Silberberg: Nature 396, 239 (1998)
9. C.J. Bardeen, V.V. Yakovlev, K.R. Wilson, S.D. Carpenter, P.M. Weber, W.S. Warren: Chem. Phys. Lett. 280, 151 (1997)
10. T. Hornung, R. Meier, D. Zeidler, K.-L. Kompa, D. Proch, M. Motzkus: Appl. Phys. B 71, 277 (2000)
11. T. Baumert, M. Grosser, R. Thalweiser, G. Gerber: Phys. Rev. Lett. 67, 3753 (1991)
12. T. Frohnmeyer, T. Baumert: Appl. Phys. B 71, 259 (2000)
13. M. Machholm, A. Suzor-Weiner: J. Chem. Phys. 105, 971 (1996)
14. V. Blanchet, M.Z. Zgierski, A. Stolow : J. Chem. Phys. 114, 1194 (2001)
15. P. Cong, G. Roberts, J.L. Herek, A. Mohktari, A.H. Zewail: J. Phys. Chem. 100, 7832 (1996)
16. K. Takatsuka, Y. Arasaki, K. Wang, V. McKoy: Faraday Discuss. 115, 1 (2000)
17. C. Meier, V. Engel: J. Chem. Phys. 101, 2673 (1994)
18. T. Baumert, V. Engel, C. Meier and G. Gerber: Chem. Phys. Lett. 200, 488 (1992)
19. Ch. Meier and V. Engel in Femtosecond Chemistry ed. by Manz and Wöste, VCH, 369 (1995)
20. A. Assion, T. Baumert, U. Weichmann and G. Gerber: Phys. Rev. Lett. 86, 5695 (2001)

AMPLIFICATION OF SHAPED ULTRASHORT PULSES IN THE VISIBLE SPECTRAL REGIME BY A TWO-STAGE NONCOLLINEAR OPTICAL PARAMETRIC PROCESS

ELMAR SCHREIBER, HOWE-SIANG TAN, AND WARREN S. WARREN

Center for Ultrafast Laser Applications and Department of Chemistry, Princeton University, Princeton, NJ 08544, USA

E-mail: eschreib@princeton.edu

A two-stage noncollinear optical parametric amplification (NOPA) process is used to generate and amplify ultrashort shaped pulses in the visible. The phase and amplitude profiles of the shaped pulses are conserved within our amplification scheme. This concept neatly eliminates the energy losses normally associated with the production of complex shaped pulses.

1 Introduction

The shaping of ultrashort laser pulses [1,2] is of increasing interest in a wide variety of optical applications, including quantum and optimal control [3,4], high speed communications [5,6], and material characterization [7]. Usually, pulse shaping is performed as a final step, although this mostly introduces significant energy losses. Recently, the amplification of shaped pulses has been demonstrated based on a Ti-sapphire chirped pulse regenerative amplification process [6,8], however, this limits the spectral range of the amplified shaped pulses to be in the vicinity of 800nm. To utilize shaped pulses in a wider range of applications and systems, methods and techniques have to be developed generating shaped optical pulses at widely tunable wavelengths.

The recent development of noncollinear optical parametric amplification (NOPA) has provided an ideal tool for generating ultrashort pulses throughout the visible regime [9-11]. These light sources have been employed in closed-loop pulse shaping experiments optimizing second harmonic generation and automating pulse compression [12], however, with the pulse shaping process after the parametric amplification step.

Preliminary studies of the amplification of shaped pulses employing the NOPA process (by shaping the white-light continuum prior to amplification) enabled the generation of intense visible optical pulses with simple phase details [13]. Here, we show that this method can be significantly enhanced by modifying a two-stage NOPA process [14], whereby the first stage NOPA generates a stable pulse, which is then shaped before being amplified again by a second stage NOPA with a stretched pump pulse. This scheme produces sufficient intensity in the pre-amplified shaped pulse to allow its phase and amplitude profile to be conveniently measured [15]. Direct comparison between the pulse shapes, prior and after the second stage amplification process can then be made. This approach, in particular, enables us to determine whether the pulse shape is adequately conserved in the process. The goal of such comparison facilitates the generation of stable amplified shaped pulses in the visible with much more complex phase and amplitude modulation.

2 Experimental

Figure 1 shows a schematic of our experimental setup. A two stage shaped pulse generator and amplifier is pumped by nearly transform-limited pulses of ~100fs (FWHM) duration. These pump pulses are generated in a regenerative amplifier system (<2.5μJ, 1kHz repetition rate with a peak-to-peak fluctuation of ~2%) with its wavelength centered at 800nm. The 800nm pulse trace is split into 4 portions:

- One portion (120μJ) is used in the pulse shape characterization as will be described later.
- Another portion (~10μJ) is used to generate a stable white light continuum by focusing into a 2mm thick sapphire plate.
- The major portion (680μJ) is used in a second harmonic generation process (type I, 0.5mm β-barium borate (BBO) crystal), generating pulses of 190μJ centered at 400nm.
- These pulses are split almost evenly to provide the pump pulses for the first and second stage NOPA processes.

The first stage amplification process closely follows the description offered by Riedle and co-workers [9]. The generated white light continuum is focused into a 2mm BBO crystal, while the pump pulses at 400nm (~90μJ) are focused to a point ~20mm in front of the BBO crystal with a spherical mirror. The energy of the resulting pulses ranges from 13μJ at 630nm to 9μJ at 515nm.

The pulses of the first NOPA stage are steered into an acousto-optic pulse shaper (AOPS). The AOPS setup [2,16] includes a pair of diffraction gratings (1200 grooves/mm) and a pair of spherical mirrors (f=200mm) set up in a 4-f configuration. The input pulses are dispersed onto the Fourier plane, where a TeO_2 crystal (42mm long, 10mm thick) is positioned. Arbitrarily shaped several-microsecond long acoustic waves act as a stationary transmission grating, as they propagate through the crystal with a velocity much slower than the speed of light. The time domain features of the acoustic wave transfer phase and amplitude modulation onto the first order diffracted optical beam in the spectral domain. This diffracted beam is then combined to deliver the shaped pulses.

A computer controlled arbitrary waveform generator produces the acoustic pulses at radio frequencies. A spectral window of 130nm is provided by the specific geometry of our pulse shaper setup. The wavelength tunability of the AOPS is achieved by turning the diffraction gratings to manipulate the diffraction angle. The typical power throughput of the AOPS assembly is 5-10%. This attenuation is due to the efficiency of the gratings as well as to our intention to work in the linear regime of the acousto-optic effect.

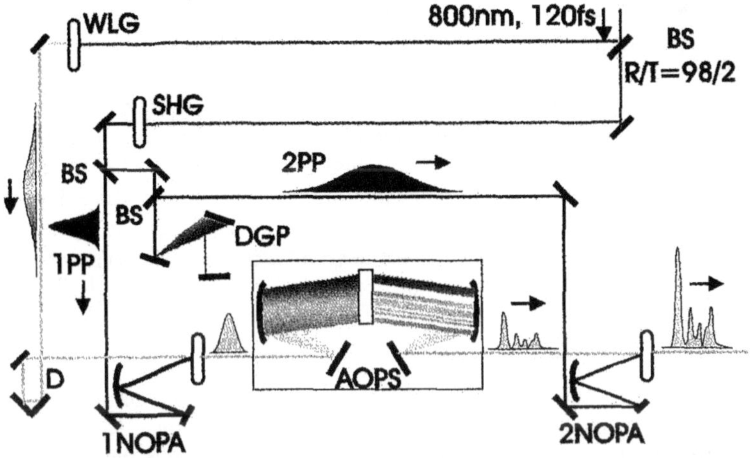

Figure 1. Experimental setup for the amplification of shaped pulses by a two-stage NOPA. BS: beam splitter. WLG: white light generation (Sapphire). SHG: second harmonic generation (type I, BBO). D: delay. DGP: diffraction grating pair. AOPS: acousto-optic pulse shaper. 1PP and 2PP: first and second stage pump pulse, respectively. 1NOPA and 2NOPA: first and second stage noncollinear parametric amplification process (type I, BBO), respectively.

We measure the spectra of the pulses from the first stage NOPA and produce a particular amplitude mask to match the experimental spectrum with the desired shaped pulse spectrum. Besides the residual chirp from the first stage NOPA, [9-11] the pulses passing through the pulse shaper acquire phase chirp due to material dispersion of the pulse shaper. Unmodulated pulses are characterized to determine the amount of chirp (typical parameters are group velocity dispersion ~$2000fs^2$ and cubic dispersion ~$22000fs^3$). Pre-compensation for this phase chirp is programmed into the radio-frequency arbitrary waveform generator, together with the desired phase profile and the above mentioned amplitude mask, to manufacture the radio frequency pulses needed to generate the desired pre-amplified shaped optical pulses.

The pre-amplified shaped pulses seed a second stage NOPA system. The setup of the second stage NOPA is, except for the duration of the pump pulses, similar to that of the first. In general, shaped pulses are not transform-limited and, hence, are longer than transform-limited pulses of similar spectral bandwidth. In order to amplify these shaped pulses entirely, it is necessary for the pump pulses to be temporally longer. Therefore, a diffraction grating pair (600 grooves/mm) is used to stretch the pump pulses temporally. We stretch these pump pulses from ~150fs to about 800fs by means of the two diffraction gratings with their perpendicular distance amounted 350mm. The losses in the diffraction and steering process attenuate the pump pulses, originally at 100μJ, to 30μJ per pulse.

We characterize both the pre-amplified and amplified shaped pulses with a variant of the spectrally and temporally resolved up-conversion technique (STRUT) [17,18]. A portion of the fundamental source pulse at 800nm as described earlier provides the transform-limited reference pulse (FWHM=8nm, centered at 800nm). We measured the spectra of the sum frequency generation (type I, 80μm BBO crystal) of the shaped pulse and the reference pulse as a function of delay between the two pulses. The obtained time-wavelength correlation or STRUT trace gives the plot of group delay versus frequency component of the shaped pulse. By plotting the maxima of the sum frequency spectra as a function of delay, the first derivative of the pulse phase with respect to angular frequency $\varphi'(\omega)$ is obtained.

3 Results and Discussion

With the setup just described, we are able to generate and amplify pulses with various complex shapes. Fig. 2 shows a hyperbolic secant pulse with a hyperbolic tangent frequency sweep as an example. This type of pulse is of particular interest, as it provides an analytical solution to the optical Bloch equation in inverting the population of two level systems that falls within a resonance window [19,20]. The field envelope of this pulse can be described by

$$E(t) = \mathrm{sech}(\alpha\, t)^{1+\mu i}$$

where $\alpha = 0.57 \times 300\text{fs}^{-1}$ determines the pulse duration and $\mu = 21$ parameterizes the frequency sweep. The theoretical pulse intensity, I(t) and the first derivative of pulse phase with time, $\dot{\phi}(t)$ is presented in Fig. 2a. The theoretical STRUT trace is shown in Fig. 2b. We generated the shaped pulse centered at 625nm. The spectrum of this pulse resembles a rectangular profile with a width of 50nm (Fig. 3b). The experimental STRUT trace of the pre-amplified shaped pulse and the amplified shaped pulse are depicted in Fig. 2c and 2d, respectively. The measured pulse energy of the pre-amplified shaped pulse was 0.15μJ. Upon amplification, the pulse energy became 4.8μJ. By measuring the standard deviation of the light induced voltage on a fast photodiode, we estimated the pulse-to-pulse energy fluctuation to be ~5%.

Figure 3a compares the first derivative of phase with respect to angular frequency $\varphi'(\omega)$, of pre-amplified and amplified shaped pulses (retrieved from the STRUT trace) with theory. From this comparison, we find that, besides a good agreement with theory, the complex phase profile of the shaped pulses is adequately preserved in the noncollinear parametric amplification process. Figure 3b compares the experimental electric field amplitude in the spectral domain of the pre-amplified and amplified pulses, obtained by a spectrometer, with the theoretical amplitude profile.

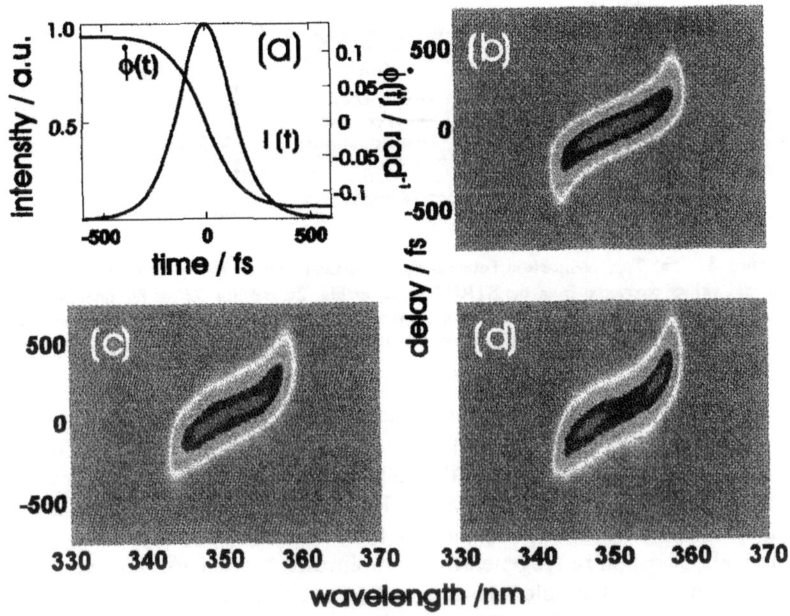

Figure 2. (a) Theoretical plot of the intensity I(t) and first time derivative of phase $\dot{\varphi}(t)$ of a $\mathrm{sech}(\alpha t)^{1+\mu i}$ pulse in time, with $\alpha = 0.57 \times 300\mathrm{fs}^{-1}$ and $\mu = 21$. (b),(c) and (d) are plots of the sum frequency generation spectra vs. delay (STRUT trace) of the shaped pulses (centered at 625nm) with a reference pulse (centered at 800nm). (b) Theoretical, (c) and (d) experimental pre-amplified and amplified, respectively.

Besides this we demonstrate (in Fig. 4) the amplification of a shaped pulse with strong amplitude modulation in the spectral domain. We started with a pulse with a rectangular spectrum of 70nm spectral width centered at 625nm with a cubic phase chirp of $30000\mathrm{fs}^3$. We subsequently imposed an amplitude modulation function of $\sin^2(\beta\lambda)$ in the wavelength domain with $\beta=0.51\mathrm{nm}^{-1}$. The pre-amplified pulse is amplified from 0.1μJ to 4μJ by the second stage NOPA process. From the comparison between the spectra of the pre-amplified and the amplified shaped pulses in Fig. 4a, we observe that the amplitude modulation of the shaped pulse is very well preserved in the NOPA process. Fig. 4b and 4c compares the STRUT traces of the pre-amplified and amplified shaped pulses and show that the phase profile (cubic phase chirp of $30000\mathrm{fs}^3$) is preserved adequately as well.

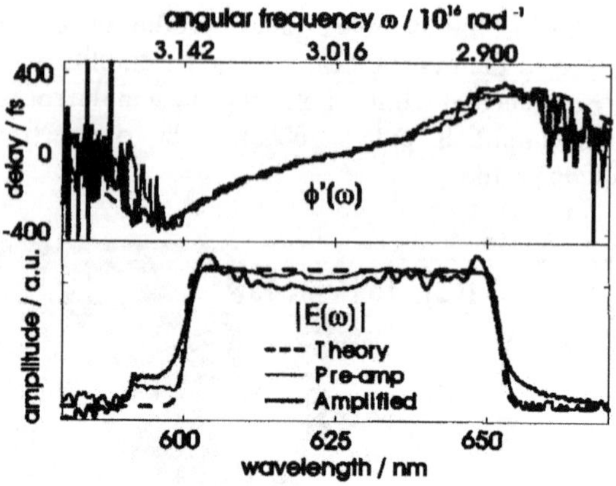

Figure 3. a) Top, comparison between the theoretical first derivative of phase φ'(ω) and the experimental values recovered from the STRUT traces in Fig. 2c and Fig. 2d for the pre-amplified and amplified sech(αt)$^{1+\mu i}$ pulses respectively. Bottom, comparison of the electric field amplitude |E(ω)| obtained theoretically and that of the experimental values, obtained from the spectrometer for the pre-amplified and amplified shaped pulses.

In the present study, for the second stage NOPA process, pump pulses with a duration of 800fs sufficed to amplify shaped pulses of duration ~400fs. The upper limit imposed upon the duration of the amplified shaped pulses increases with the temporal length of the pump pulses, as long as the pump pulse peak intensity exceeds the threshold needed to sustain the nonlinear optical process. Modern commercial Ti-sapphire regenerative amplification systems are capable of delivering up to 3mJ of ultrashort pulses at a 1kHz repetition rate. With the experimental minimum value of pump pulse peak intensity needed to sustain the NOPA process taken from recent literature [21], pump pulses as long as 10ps can still be expected to sustain the NOPA process for our present, and several-picoseconds long shaped pulses can be adequately amplified.

It should also be noted that the quadratic phase chirp of the pump pulses due to the stretching process does not distort the phase details of the amplified pulse, since only the absolute value of the temporal pump pulse profile matters. The ~5μJ energy per pulse output with a 30μJ pump is comparable in efficiency to NOPA processes reported in the visible [8-10]. The final amplified shape pulse energy can be increased by both increasing the pump pulse energy and the efficiency of the pump pulse stretching process.

Besides the signal (visible) beam of the NOPA the idler (infrared) can be used for pulse shaping experiments, e.g., for the real-time study of molecular vibrations in the electronic ground state. In an additional investigation we were able to generate ultrashort shaped idler pulses of high resolution [21]. Phase and amplitude profile of the shaped NOPA signal pulse have been transferred to the idler pulse. A similar scheme might be implemented for other optical parametric amplification systems pumped at near-infrared wavelengths to generate high resolution shaped pulses in the mid-infrared.

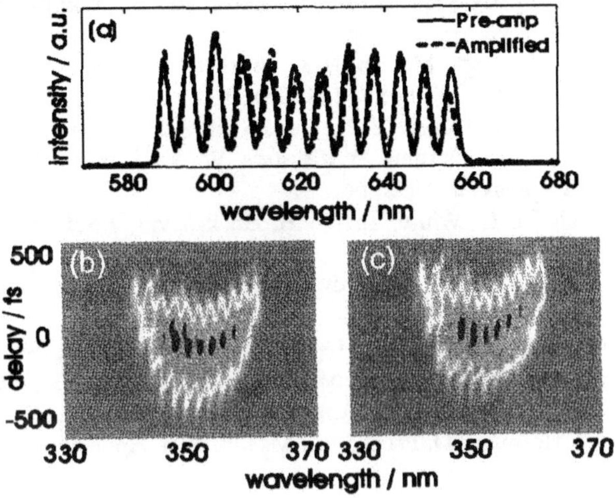

Figure 4. Comparison between the spectra of the pre-amplified and amplified shaped pulse with a sine-squared amplitude modulation across the 70nm spectrum. Experimental STRUT trace of (b) pre-amplified and (c) amplified shaped pulses with a cubic chirp of 30000fs^3, respectively.

4 Summary

We have demonstrated for the first time, the amplification of non-trivially shaped pulses by a noncollinear optical parametric process with the conservation in its phase and amplitude profile. Shaped pulses with complex phase structure (e.g. the hyperbolic tangent frequency sweep) and strong amplitude modulation (e.g. sine squared function) centered at 625nm with a bandwidth of up to 70nm are amplified from ~0.1μJ to ~5μJ. A generalization to shaped pulses centered at other wavelengths in the visible and infrared is straightforward, and is restricted only by the tunability of a NOPA system.

5 Acknowledgements

This work is supported by the New Jersey Commission on Science and Technology, the Air Force Office of Scientific Research, and the Department of Defense.

References

1. A. M. Weiner, Rev. Sci. Instr. **71** (2000) 1929.
2. J. X. Tull, M. A. Dugan, W. S.Warren, Adv. Magn. Opt. Res. **20** (1997) 1.
3. H. Rabitz, R. de Vivie-Riedle, M. Motzkus, and K. Kompa, Science **288** (2000) 824.
4. A. Assion, T. Baumert, M. Bergt, T. Brixner, B. Kiefer, V. Seyfried, M. Strehle, and G. Gerber, Science **282** (1998) 919.
5. F. Omenetto, B.P. Luce, and A.J. Taylor, J. Opt. Soc. Am. **B 16** (1999) 2005
6. M. R. Fetterman, D. Goswami, D. Keusters, W. Yang, J.-K. Rhee, and W. S. Warren, Opt. Express **33** (1998) 66.
7. T. C. Weinacht, J. L. White, and P. H. Bucksbaum, J. Chem. Phys. B **103** (1999) 10166.
8. A Haché, Y. Kostoulas, R. Atanasov, J.L.P. Hughes, J.E. Sipe, and H.M. van Driel, Phys. Rev. Lett. **78** (1997) 306
9. T. Wilhelm, J. Piel, and E. Riedle, Optics Letters **22** (1997) 1494.
10. G. Cerullo, M. Nisoli, and S. De Silvestri, Appl. Phys. Lett. **71** (1997) 3616.
11. A. Shirakawa, and T. Kobayashi, Appl. Phys. Lett. **72** (1998) 147.
12. D. Zeidler, T. Hornung, D. Proch, and M. Motzkus, Appl. Phys. **B70S** (2000) S125.
13. H.-S. Tan, W. S. Warren, and E. Schreiber, in Ultrafast Phenomena **XII**, T. Elsaesser, S. Mukamel, M. M. Murnane, and N. F. Scherer, eds. (Springer-Verlag, Berlin, 2001) 105.
14. J. Piel, M. Beutter, and E. Riedle, Opt. Lett. **25** (2000) 180.
15. H.S. Tan, W.S. Warren, and E. Schreiber, Optics Letters 2001 (in press)
16. C. Hillegas, J. X. Tull, D. Goswami, D. Strickland, and W. S. Warren, Opt. Lett. **19** (1994) 737.
17. J.-P. Foing, J.-P. Likforman, M. Joffre, and A. Migus, IEEE J. Quant. Electron. **28** (1992) 2285.
18. J.-K. Rhee, T. S. Sosnowski, T. B. Norris, J. A. Arns, and W. S. Colburn, Opt. Lett. **19** (1994) 1550.
19. L. Allen, and J. Eberly, in Optical Resonance and Two-Level Atoms (Dover, New York, 1987).
20. M. R. Fetterman, J. C. Davis, D. Goswami, W. Yang, and W. S. Warren, Phys. Rev. Lett. **82** (1999) 3984.
21. V. Krylov, O. Ollikainen, J. Gallus, U. Wild, A. Rebane, and A. Kalintsev, Opt. Lett. **23** (1998) 100.
22. H.S. Tan, E. Schreiber, and W.S. Warren, submitted to Opt. Lett.

SIMULATION OF BEST FIT PARAMETERS FOR FEMTOSECOND PULSE GENERATION IN FIGURE OF EIGHT LASER

R. JUSTIN RAJESH, R. RADHAN AND PREM B. BISHT*

Department of Physics, Indian Institute of Technology Madras

Chennai 600 036, India

**E-mail: bisht@acer.iitm.ernet.in*

Fibre lasers are small, compact and fall under the class of solid state lasers. The use of rare earth doped optical fibre lasers for application in optical communication systems has drastically increased the distance of long-haul high bit rate links, without the need of regenerators/repeaters. To increase further the distance, and to overcome the dispersion problem, the use of ultrashort, soliton pulses has been proposed. For such a scheme soliton laser is required. The erbium doped fibre laser in the figure 8 configuration is a convenient design that generates ultrashort soliton pulses. Best fit parameters for the design of the such a laser have been simulated here.

1 Introduction:

The figure of eight laser (F8L) is an all fiber mode locked laser configuration that consists of a nonlinear amplifying loop mirror (NALM) with its output connected to its input (Fig. 1). The F8L can be considered as an external loop that consists of a polarization-independent isolator, a 20% output coupler, a polarization controller and an internal loop that contains the erbium doped fiber (EDF), and an appropriate pump coupler, a length of fiber and a polarization controller [1-4]. The EDF that provides the gain is placed asymmetrically in the ring. A pulse entering NALM is divided equally by the coupler. The high intensity portions of the pulse are transmitted and amplified, while the low intensity portions are reflected and rejected by the optical isolator. The two components of the pulse counter propagate about the loop but experience different amounts of self phase modulation (SPM).

Depending upon the relative phase difference between them and therefore on the initial intensity, the two components recombine in the coupler to either reflect the pulse down the lead it initially came from, or switch the pulse to the other coupler lead. Pulse switching takes place when the relative phase difference between the counter propagating components are π.

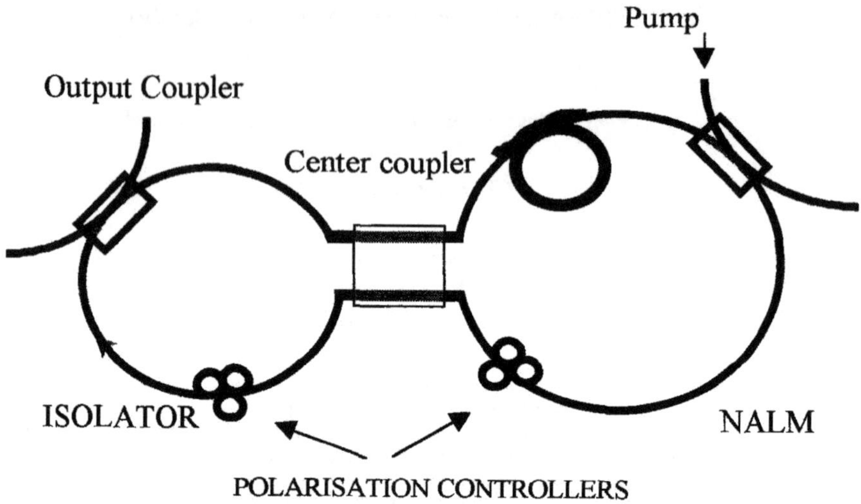

Fig. 1. Experimental arrangement of generating ultrafast pulse from a figure of eight laser.

Ultrashort pulses lead to pulse broadening due to group velocity dispersion (GVD) effects. Due to fibre dispersion different spectral components associated with the pulse travel at different speeds because of the frequency dependent refractive index [5,6]. The pulse envelope spreads under the influence of GVD whereas the spectrum remains invariant. Solitons are formed as an interplay between the negative GVD and SPM. The relative strength of nonlinearity is governed by the nonlinear index n_2. The effect of fibre nolinearity ($n_2 \neq 0$) on pulse propagation has been considered in recent years. When n_2 is combined with higher order dispersion [$\beta_3 \neq 0$) the effects are not familiar very much yet. Combination of n_2 and β_3 can lead to significant pulse distortions and compressions. In the present paper, aiming towards design optimization of the laser, theoretical simulations on fibre length and a range of higher order dispersion parameters have been carried out to achieve soliton-like ultrafast pulse generation in F8L. Best fit parameters for F8L to generate pulses of sub 500 fs are reported here.

2 THEORY

Mathematically the effects of fiber dispersion are accounted for by expanding the mode progression in a Taylor series about the central frequency ω_0 [6]

$$\beta(\omega) = \frac{n(\omega)\omega}{c} = \beta_0 + \beta_1(\omega - \omega_0) + \frac{1}{2}\beta_2(\omega - \omega_0)^2 + \ldots..$$

where

$$\beta m = \frac{\partial^m \beta}{\partial \varpi^m}, \qquad m = 0, 1, 2, \ldots\ldots\ldots$$

and the group velocity V_g is related to β_1 as

$$V_g = \frac{1}{\beta_1}$$

and β_2 is known as the group velocity dispersion (GVD) parameter. β_3 is the third order dispersion parameter. Intensity dependence of the refractive index leads to a large number of nonlinear effects; one of them being self phase modulation (SPM). It refers to the self induced phase shift experienced by an optical field during its propagation in optical fibers.

Non linear Schrödinger equation (NLSE) is derived from the basic Maxwell's wave equations. The dispersion and nonlinear terms are included in NLSE to study their effects on pulse evolution and method used to solve the NLSE for pulse propagation in nonlinear dispersive media is the split step Fourier transform (SSFT) method. The NLSE is given as

$$\frac{\partial A}{\partial z} + \alpha A/2 - \frac{i}{2}\beta_2 \frac{\partial^2 A}{\partial T^2} - \frac{1}{6}\beta_3 \frac{\partial^3 A}{\partial T^3} = \frac{i\gamma}{n_2}\left[(\Delta n)A + \frac{2i}{w_0}\frac{\partial}{\partial T}(\Delta n A)\right]$$

where $T = t - \beta_1 z$ and $\gamma = (n_2\omega_0)/(cA_{eff})$

and Δn is given by $T_R \frac{\partial \Delta n}{\partial T} + \Delta n = n_2 |A|^2$

where c is the velocity of light, A_{eff} is the effective mode area and T_R is Raman scattering coefficient.

The NLSE is also applicable in case of an active fiber (a pumped rare earth-doped fiber), where a gain term (g) is added to the loss (α). An approximate expression for the gain of the EDF amplifier is employed by taking into account gain dispersion and saturation [3].

$$g = G_0 \exp\left(\frac{-E_P}{E_{sat}}\right)\left[1+\left(\frac{2|\omega|}{\Delta\omega}\right)^{2m}\right]^{-1}$$

where m denotes the slope of the gain profile, Δω is the bandwidth of the amplifier, E_p is the pulse energy, E_{sat} is a saturation parameter, and G_0 stands for the peak unsaturated gain.

3 RESULTS and DISCUSSION

The pulses propagation and compression were simulated from the amplified spontaneous emission (ASE) noise in F8L with a mean power of 6-10 μW with the MATLAB software and by using the SSFT numerical method. The signal profile is monitored at the output coupler and when the pulse waveform converges to an almost constant profile, a steady state operation of the laser is assumed to have been reached [7].

The pulse evolution from ASE noise after a specific number of round trips around the loop shows that after 20 round trips a clean sech^2 shaped pulse is formed. This indicates that the NALM is operating in the nonlinear regime when it acts like a mirror, preferentially transmitting pulses of higher intensity. Further round trips of the pulses lead to compression and increase in peak power. At 25 round trips gain saturation occurs and there is no further increase in peak power. The peak power fluctuates with in the range of 135-150 W. The fluctuations decrease after each subsequent round trip. By 80th round trip the fluctuations disappear indicating that the laser had entered the stable soliton regime. Further, it is observed that even after 150 round trips, the profile remained similar to the one observed after 80 round trips confirming the fact that the laser had modelocked. For numerical simulations the following values for the fiber and EDFA pump parameters were taken initially, unless otherwise specified:

G_0 (EDFA peak gain) = 27 dB, $\Delta\omega$ (EDFA bandwidth) = 10 nm, m = 1,

E_{sat} (EDFA bandwidth) = 4.5pJ, C_r (Central coupler ratio) = 0.48,

C_k (Output coupling ratio) = 0.90, α (Loss) = 0.2 dB/Km, A_{eff} = 50 μm^2,

T_r (Raman scattering coefficient) = 6 fs, λ_0 (center wavelength) = 1.55 μm,

ln (NALM length) = 49m.

In previous studies [4,7], it was found that the gain of the EDFA should neither be too high nor too low in order to operate in the stable soliton regime. The value of E_{sat} should be in the range of 2-15 pJ, keeping the length of EDFA within 0.8-1.2 m. Sindhu et al [4] studied the effect of third order dispersion parameter (β_3) for a limited range. It was found that for β_2 = 17 ps^2/km and β_3 = 0.1 ps^3/km a sub 500 fs pulse was obtained. Although the pulse stability was not so good. In the following we present the numerical simulations for F8L for the varying values of β_3, C_r, and NALM length.

3.1 *Effect of nonlinear dispersion parameters (β_2 and β_3)*

Second order nonlinear dispersion parameter (β_2) was varied from –5 ps^2/km to –22 ps^2/km. However, the shortest pulse obtained was > 500 fs. The pulse was very unstable for β_2 values of -5 ps^2/km to -10 ps^2/km. The pulse becomes relatively stable as the β_2 value tends towards -15 ps^2/km. Stable and modelcoked pulses are obtained for the β_2 values around -19 ps^2/km.

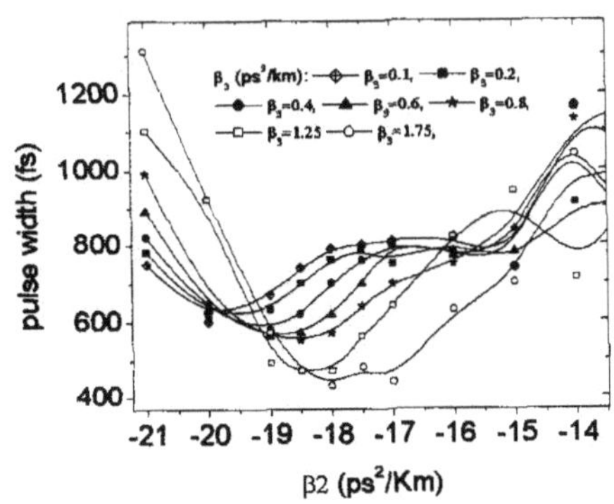

Fig 2. Plot of pulse width Vs β_2 values for different values of β_3 as indicated.

When β_3 was introduced and varied from 0.1 ps^3/ km to 2 ps^3/km the pulse duration is plotted in Fig. 2 for different β_2 values. The output pulse is highly unstable for β_2 values of >-16 ps^2/km and <-21 ps^2/km. A pattern with a crossing point near β_2 =-19 ps^2/ km is obtained, around which the pulses are stable. Fastest, stable pulse with a half width of 480 fs is obtained for β_2 =-18.5 ps^2/ km and β_3 =-1.75 ps^3/ km.

3.2 *Effect of NALM*

Fig. 3 (A) shows a plot of NALM length vs. the output pulse for the C_r = 0.48, β_2= -18.5 ps^2/km and β_3 = 0.8 ps^3/km. A stable modelocked fast pulse is obtained for the NALM value of 48.5 m. For other values of length of NALM the pulses are broad and unstable.

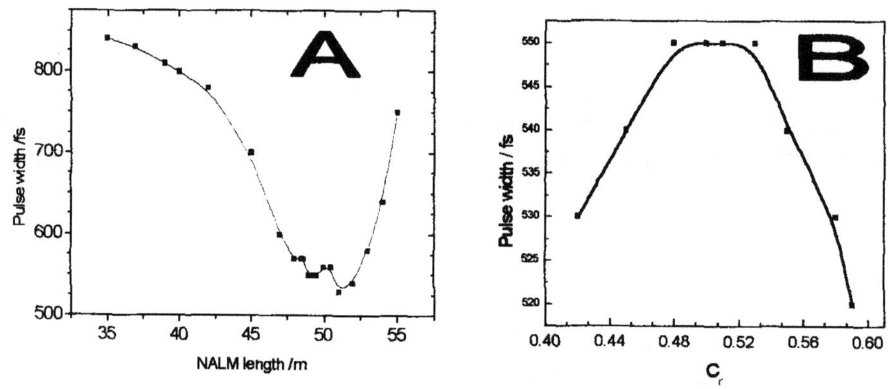

Fig. 3. Plot of pulse width vs. NALM length for β_2= -18.5 ps^2/km, β_3= 0.8 ps^3/km (Panel A)., and plot of pulse width vs. central coupler ratio (C_r) (Panel B).

3.3 Effect of central coupler ratio

The typical effect of central coupler ratio on the pulse formation is shown in Fig. 3(B). The stable and shortest pulse obtained for the variation of central coupler ratio is in the range of 0.48 to 0.52.

Although the pulse widths varies from 450 fs to 3.5 ps but pulses are stable only at C_T of 0.48 (within the flat region of the curve). Fig. 4 shows a best pulse obtained from the simulations. The half width of the pulse is ~480 fs for the values of β_2 =-19 ps^2/ km, β_3 =1.75 ps^3/ km and C_T = 0.48.

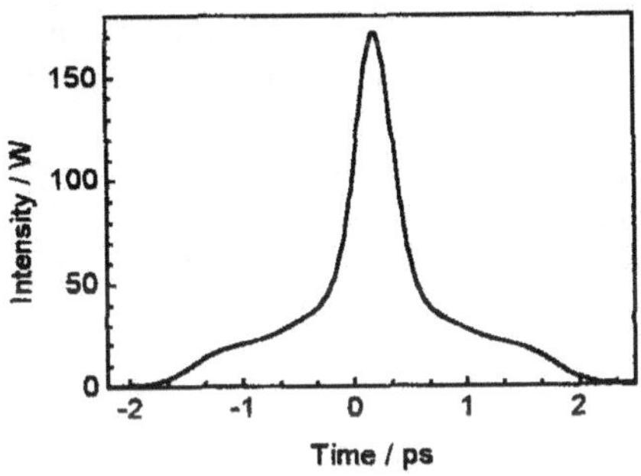

Fig. 4. A typical output pulse obtained in figure of eight laser.

In conclusion, we have theoretically simulated different parameters for an all fibre figure of eight laser by including the third order dispersion parameter. The best-fit parameters for generation of sub 500 fs pulse are given here.

4 Acknowledgements

We thank DST, New Delhi and DRDO, New Delhi for financial assistance.

References

1. I. N. Duling, Subpicosecond All fiber Erbium laser, Electronic lett 27 (1991), 544-545.
2. J. Theimer, J. W. Haus, Dispersion balanced figure-eight laser, Opt. Comm. 134 (1997), 161-164.
3. V. Tzelepis, S. Markatos, S. Kalpoginnais, Th. Sphicopoulos, and C. Caroubalos, Analysis of a passively modelocked self starting all fibrre soliton laser, J. Lightwave Technol 11 (1993) 1729-1735.
4. T. G. Sindhu, Prem B. Bisht, R. Justin Rajesh and M. V. Satyanaranyana, Microwave and Optical Technology Leters, 28 (2001) 196.
5. A. Hasegawa and F. Tappert, Transmission of stationary nonlinear optical pulse in dispersive optical fibre: I. Anomalous dispersion, Appl Phys Lett 23 (1973), 142-144.
6. G. P. Agrawal, Nonlinear fiber optics, 1989 (Acad. Press, San Diego).
7. P. Brijesh and M. Raghu, A study on the characteristic and optical pulse width dependence on group velocity dispersion in F8L, M. Sc. Thesis, Indian Institute of Technology Madras, 1999.

FEMTOSECOND PUMP-PROBE PHOTO-ION VELOCITY MAP IMAGING

W.G. ROETERDINK AND M.H.M. JANSSEN

Laser Centre and Department of Chemistry
Vrije Universiteit, DeBoelelaan 1083
1081 HV Amsterdam, The Netherlands
E-mail: mhmj@chem.vu.nl

The femtosecond pump-probe technique provides information on ultrafast dynamics in molecular systems. Combining this technique with photo-ion velocity map imaging provides direct information on the kinetic energy and angular distribution of the photofragments. We report experimental results on (multi) photon induced dynamics in Iodofluoralkanes.

1 Introduction

The fundamental goal of molecular reaction dynamics is understanding the elementary steps which make up a chemical reaction [1]. The possibility of real time studies opened up with the advent of lasers operating in the femtosecond regime. Especially femtosecond pump-probe experiments can reveal the ultrafast dynamics during the actual photodissociation, photodissociation being an uni-molecular reaction. Widely studied is the benchmark molecule NaI [2] in which an oscillatory behavior is observed. These oscillations arise from the resonances trapped in the potential well. Every time the wavepacket reaches the crossing point of the dissociative and the bound potential, a part of the molecules will dissociate, leading to a decrease in signal. The high field strength of amplified femtosecond pulses however easily induces multi-photon absorption processes in molecules. These multi-photon pathways can compete favorably and interfere with fast (pre)dissociation dynamics occurring at the one photon level. A competing multi-photon pathway was observed in a femtosecond pump-probe experiment on methyl iodide using time-of-flight (TOF) ion detection in a molecular beam [3]. These multi-photon pathways make it difficult to compare these ultrafast experiments with photodissociation experiments on the nanosecond timescale, which are usually on the one-photon level. In order to unravel in more detail the dynamics resulting from these competing pathways we have combined velocity map ion imaging detection [4] with the femtosecond pump-probe technique to study in real time photon induced dynamics in small molecular systems. In this contribution multi-photon pathways in CF_3I and CF_2I_2 shall be discussed.

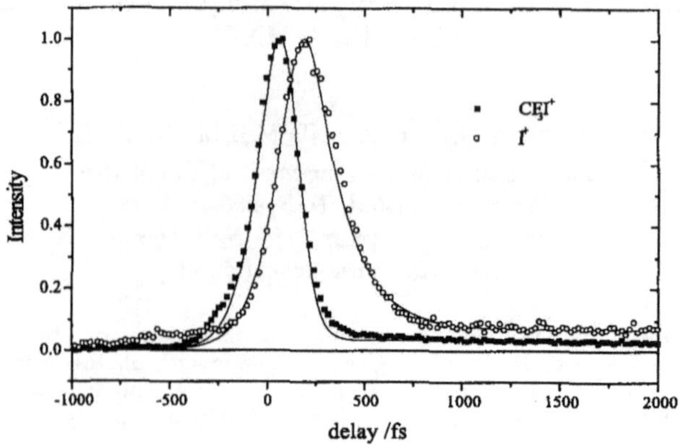

Figure 1. Pump-probe transient of the parent ion CF_3I^+ and the I^+ fragment.

2 Experimental

An actively mode-locked Titanium-Sapphire oscillator (Spectra Physics, Tsunami) is amplified in a chirped regen amplifier (Spectra Physics, Spitfire), which gives pulses centered around 800 nm at a repetition rate of 1 kHz. Typically, the pulses have a duration of 100-130 fs (FWHM) with an energy of 0.85 - 1 mJ. The amplified pulses are doubled an tripled in BBO crystals producing pulses around 396 nm (60 mJ/cm^2) and 264 nm (6 mJ/cm^2), which are used as probe and pump color, respectively. The molecular beam machine consist of three chambers, the source chamber containing the Gerlich piezo valve running at 1 kHz, a buffer chamber and the chamber containing the ion-optics. Halfway between the repeller and extractor plates, the molecular beam is crossed at right angles with the laser beams. The ions produced are accelerated into the time-of-flight (TOF) tube and at the end hit the micro-channel-plate (MCP) detector, which can be gated on a specific mass of one of the fragments. With the help of a fast phosphor screen and a CCD camera the velocity map images can be recorded. Typically 10 parent ions per laser shot are produced at $\tau = 0$ fs.

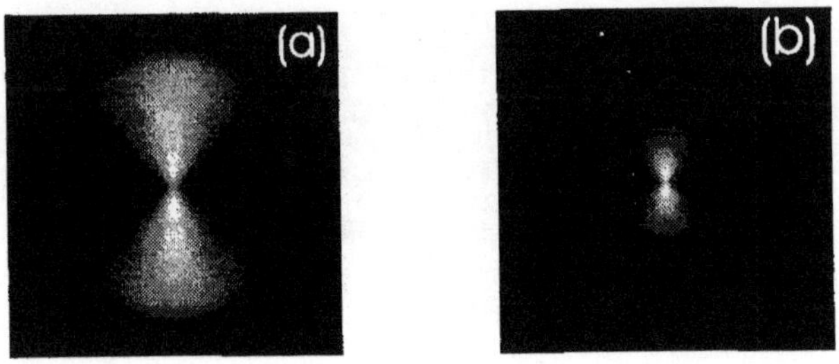

Figure 2. Velocity map ion images of the Iodine fragment (a) at 0 fs delay and (b) at 300 fs delay time, decrease in recoil velocity is clearly visible.

3 Velocity map imaging studies of ultrafast dissociative photodissociation dynamics of CF_3I

To characterize the newly commenced velocity map imaging apparatus a molecule with known photodynamics was chosen. The A band of CF_3I was examined many times by different researchers [5,6,7,8,9] with nanosecond lasers. After excitation at 266 nm the CF_3I molecule dissociates as a pseudo-diatomic molecule via the A band, yielding the fragments CF_3 and I. From the angular distributions of the Iodine fragment a lifetime of 250 fs can be estimated. In our experiments a pump laser at 266 nm and a probe laser at 400 nm was used. The integral ion transient for the fragments are shown in Figure 1. It shows an autocorrelation like shape around 0 fs delay and a long term enhancement is clearly visible. This long term enhancement must be caused by a long lived state ($\geq$ 50 ps) at the pump photon level and the A band clearly can be excluded. This would mean that at least two pump photons are absorbed to excite a high Rydberg state. The velocity map ion image radial distribution is directly related to the kinetic energy of the fragment. The appearance energies of the fragment ions of CF_3I are known and can be used in combination with the kinetic energy of the fragments to determine the number of photons absorbed. For the formation of the Iodine fragment first two pump photons are absorbed followed by two probe photons. At 300 fs delay time between pump and probe only very slow Iodine ions are detected, see Figure 2. We tried to attributed this to a different photon pathway and propose the following scheme. After the absorption of one pump photon, the A band is excited and the molecule starts dissociating. In this case three more probe photons need to be absorbed to produce slow Iodine fragments. At 300 fs pump-probe delay the transition strength can be largely enhanced by resonances due to ion pair states at large internuclear distances [10].

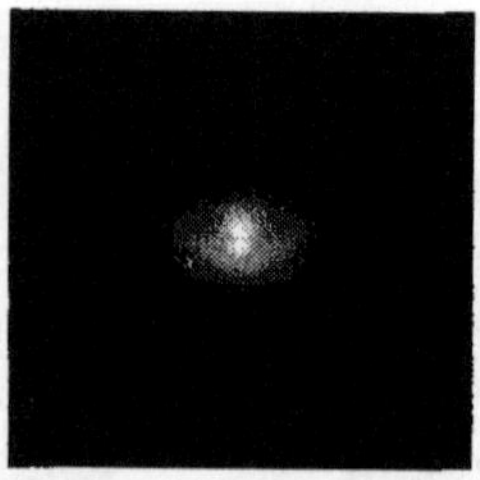

Figure 3. Velocity map image of the I_2 fragment at 0 fs delay time, in the middle the slow channel (parallel) can be seen, whereas the fast channel is more perpendicular.

4 Unraveling the photon induced molecular elimination dynamics of I_2 from CF_2I_2

In the case of CF_2I_2 specially interesting dynamics can be expected like three body decay and the elimination of molecular Iodine. In the case of elimination of molecular Iodine, two different ways of detachment can be distinguished, a concerted mechanism or a sequential mechanism. There is also special chemical interest in this molecular detachment reaction, namely it is the reverse of the insertion of a carbene into a system with a sigma bond. Studies employing nanosecond lasers around 266 nm show no production of molecular Iodine [11,12,13]. Huber *et al.* [14] performed measurements at 193 nm and observed the formation of molecular Iodine. Femtosecond pump-probe experiments using photo-ion photo-electron coincidence measurements were done by Radloff *et al.* [15]. Our results show that around 0 fs pump-probe delay two different channels for the production of molecular Iodine can be distinguished, see Figure 3. In the center of the image slow molecular Iodine can be seen produced in a multi-photon transition which is overall parallel like (β = 0.5) and the fast molecular Iodine is produced in a transition which is perpendicular like (β = -0.9). The channel producing the slow parallel type molecular Iodine has disappeared at 250 fs pump-probe delay. Assuming fast axial recoil, these angular distributions can be reproduced and the slow channel is attributed to a perpendicular one pump photon absorption to the dissociative A band of CF_2I_2 followed by the absorption of two probe photons in a parallel transition producing slow molecular Iodine. The angular distribution of the fast molecular Iodine is reproduced assuming the absorption of two pump photons via a perpendicular transition followed by the absorption of one probe photon in a parallel type transition. It can be concluded that the formation of molecular Iodine at long delay times is only possible via a two photon absorption, in agreement with nanosecond experiments. From the angular distribution at long delay times it can be concluded that molecular Iodine is formed in a asynchronous concerted mechanism.

5 Conclusion

The combination of velocity map imaging with the femtosecond pump-probe technique gives insight in the ultrafast dynamics during photodissociation. It provides details on the different multi-photon pathways leading to fragment ion formation. Unexpected dynamics was observed in the photodissociation process of CF_3I. At a delay time of 300 fs between pump and probe the kinetic energy of the I^+ fragment decreases to 0.05 eV, whereas at longer delay times it resembles the distribution at 0 fs. We attribute this decrease to a different photon pathway which becomes dominant around 300 fs pump-probe delay. Also the ultrafast dynamics of elimination of molecular iodine of CF_2I_2 is clarified. At short delay times upto 250 fs two channels are active at longer times only a (2+1') photon scheme gives rise to molecular Iodine.

Acknowledgments

The research has been financially supported by the councils for Chemical Sciences and Physical Sciences of the Netherlands Organization for Scientific Research (NWO-CW, NWO-FOM). The authors would like to thank prof. S. Stolte for his support. WGR gratefully acknowledges NWO-CW for a Ph.D fellowship within the Young Chemists Program.

References

1. R.D. Levine, R.B. Bernstein, Molecular reaction dynamics and chemical reactivity, Oxford University Press, New York (1987)
2. A.H. Zewail, Science 242, 1988, 1645
3. M.H.M. Janssen, M. Dantus, H. Guo, A.H. Zewail, Chem. Phys. Lett. 214, 281 (1993)
4. A.T.J.B. Eppink and D.H. Parker, Rev. Sci. Instrum. 68, 3477 (1997)
5. G.N.A. van Veen, T. Baller, A.E. de Vries, Chem. Phys., 93, 277 (1985)
6. P. Felder, Chem. Phys., 155, 435 (1991)
7. Y.S. Kim, W.K. Kang, K.-H. Jung, J. Chem. Phys., 105, 55 (1996)

8. A. Furlan, T. Gejo, J.R. Huber, J. Phys. Chem., 100, 7956 (1996)
9. K. Kavita, P.K. Das, J. Chem. Phys., 112, 8426 (2000)
10. W.G. Roeterdink, M.H.M. Janssen, Chem. Phys. Lett., 345, 72 (2001)
11. E.A.J. Wannenmacher, P. Felder, J.R. Huber, J. Chem. Phys., 95, 986 (1991)
12. G. Baum, P. Felder, J.R. Huber, J. Chem. Phys., 98, 1999 (1993)
13. K. Bergmann, R.T. Carter, G.E. Hall, J.R. Huber, J. Chem. Phys., 109, 474 (1998)
14. H.A. Scheld, A. Furlan, J.R. Huber, Chem. Phys. Lett. 326, 366 (2000)
15. P. Farmanara, V. Stert, H.-H. Ritze, W. Radloff, J. Chem. Phys., 113, 1705 (2000)

LOW-FREQUNENCY MOLECULAR VIBRATIONS IN ORGANIC MOLECULES STUDIED BY THZ-TDS

P. PLOCHOCKA

Warsaw University, Institute of Experimental Physics, Hoza 69, 00-681 Warsaw, Poland

E-mail: plochock@fuw.edu.pl

M. WALTHER, B. FISCHER AND P. JEPSEN

University of Freiburg, Faculty of Physics, Stefan-Meier-str. 19, D-79104 Freiburg, Germany

E-mail: jepsen@uni-freiburg.de

We report on a THz Time Domain Spectroscopy (THz-TDS) study of crystalline mono-substituted benzoic acids: benzoic acid, 2,3 or 4- hydroxybenzoic acid and acetylsalicylic acid (aspirin). The absorption spectra were recorded in the spectral region between 4 and 130 cm^{-1} (0.1 and 4 THz) at temperatures from 10K to 300K. The spectra are dominated by strong absorption lines originating from low-frequency vibrational-modes caused by inter-molecular collective motion. This kind of molecular movement is very sensitive to the positions of atoms forming molecules. Therefore THz-TDS can be used as a very sensitive tool for investigation of molecular conformation in crystal phase.

1 Introduction

Vibrational spectroscopy is known to be a powerful tool for investigation of the electronic structure and properties of molecules. Low-frequency dielectric properties of small and medium size molecules of importance in medicine and biochemistry are very interesting and relatively unexplored. This is specially true when one has to investigate the ground states of molecules with characteristic absorption features in the far-IR spectral region. Vibrations which involve many atoms are active in far-IR (the larger mass the lower the frequency). A good example is a torsion movement with typical resonant frequencies below 100 cm^{-1} or out-of-plane vibrations[1]. In molecules, which form dimers in crystal phase, not only intra, but also inter molecular vibrations are present in far-IR spectra. Also lattice vibrations in organic crystals might contribute to the far-IR spectra. Due to the collective nature of the lowest modes the far-IR spectra are very sensitive to molecular conformation and position of atoms forming molecules.

The THz-time domain spectroscopy (THz-TDS) [2] method allows one to measure the complex dielectric constant of different materials in the far infrared spectral range. Specially this experimental technique seems to be particulary useful for studies of molecules important in biology, chemistry or medicine.

The method relies on measuring the change in the shape of the electric field of a short (sub-picosecond) electromagnetic radiation pulse upon its propagation through a sample. Fourier analysis of the two traces recorded (one with and one without the sample) allows one to calculate the absorption coefficient and refractive index of the sample. A typical experimental setup is based on photoconductive emitters and detectors, exited and gated respectively by train of femtosecond pulses from a Ti:Sapphire laser. The setup covers the spectral range from 4 cm^{-1} up to approx. 167 cm^{-1} (0.1 THz – 5 THz). Subpicosecond duration time of the THz pulses makes it possible to obtain time resolution in femtosecond range in a time resolved experiment.

In this paper, we present and analyze absorption spectra of benzoic acid and its mono-substituted derivatives 2, 3, and 4- hydroxybenzoic acids. In spite of their similarity their far-IR spectra are very different. Therefore the THz-TDS is a very sensitive technique for investigation of molecular conformation[3]. We also demonstrate that the same is true for the molecule of acetylsalicylic acid (aspirin) and its breakdown product: salicylic acid (2-hydroxybenzoic acid).

2 Methods and Materials

Benzoic acid and 2 ,3 and 4 hydroxy-benzoic acid were purchased from Sigma-Aldrich Co. and used without further purification. Samples were prepared by mixing powdered sample substance with polyethylene (PE) powder at a mass ratio of 1:3 and pressing the mixture to discs of thickness between 450 and 600 μm by applying an approximate pressure of 400 bar. Since PE is nearly transparent in the far-infrared it is a suitable filling material for spectroscopic applications in this spectral region.

We use a standard THz spectrometer setup [2, 4] illustrated in Fig.1. Pulses of 40 fs duration from a Ti:sapphire oscillator are used to drive two photoconductive antennas, one which emits short pulses of THz radiation and the other, acting as a gated detector, which monitors the temporal shape of the THz field. The first two paraboloic mirrors in the THz beam path collimate and focus the THz beam to a frequency-independent spotsize of 2 mm at the position of the sample. The sample disc is mounted in a cryostat equipped with Teflon windows which are transparent to THz-radiation. The temperature is measured near the sample with a calibrated Si-diode. The cryostat can be moved so that the THz beam passes through either the sample or through an empty aperture of the identical size as the clear aperture of the sample. The two remaining parabolic mirrors recollimate and focus the THz beam onto the detector. The whole system is placed in a closed box purged with dry nitrogen in order to eliminate absorption by water vapor in the beam path. The setup has a useful bandwidth from 100 GHz to 4 THz (4 - 130 cm^{-1}). The inset in Fig.1 shows a typical THz pulse and its corresponding frequency spectrum before and after propagation through the salicylic acid sample.

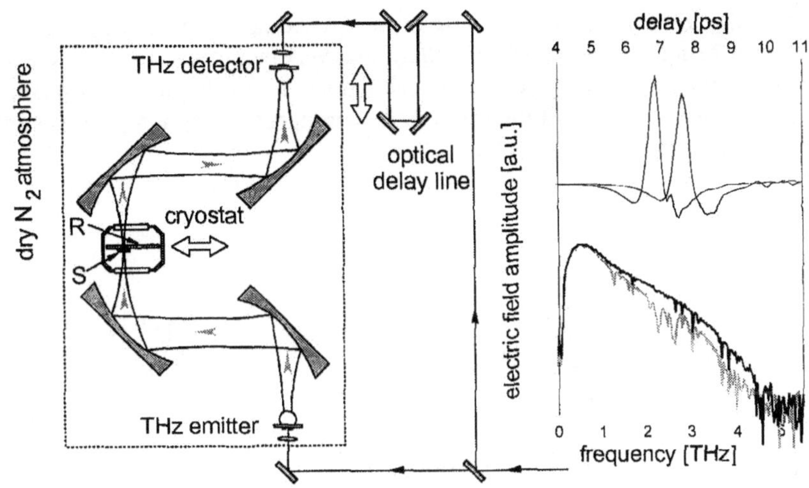

Figure 1. THz spectrometer setup with the cryostat, containing reference (R) and sample (S) aperture, mounted in the THz beam path. The inset shows a typical THz pulse and the corresponding frequency spectrum before and after passing the salicylic acid sample.

The field of the transmitted THz pulse is modified by the dispersion and absorption of the sample. The ratio of the electrical field strengths before (E_r) and after (E_s) the sample is given by

$$E_s(\omega)/\, E_r(\omega) = T(n) \exp\left(-\, \alpha d + i n \omega d/c\right)$$

where d is the thickness of the sample, ω is the frequency of the radiation, c is the speed of light in vacuum, and T(n) is a factor which accounts for reflection losses at the sample surfaces. Both the refractive index n and the absorption coefficient α can be determined from the ratio of the measured THz fields.

3 Results and discussion

We have investigated the absorption spectra of benzoic acid and its mono-substituted derivatives: 2, 3, and 4- hydroxybenzoic acids. The spectra obtained in our experiment are presented in Fig.2. All of the spectra were measured at the temperature of 10K. As one can see, all the spectra contain a bacground which increases with frequency and which we atribute to scattering in the samle. In addition each spectrum shows several well resolved and narrow resonances in the spectral range investigated in our experiment. Although the molecular structure of investigated samples is very similar there are very significant differences in the absorption spectra allowing to easily distinguish particular acid. However, one should also notice that for all the samples the strongest absorption peak is common and located at approximately 2 THz.

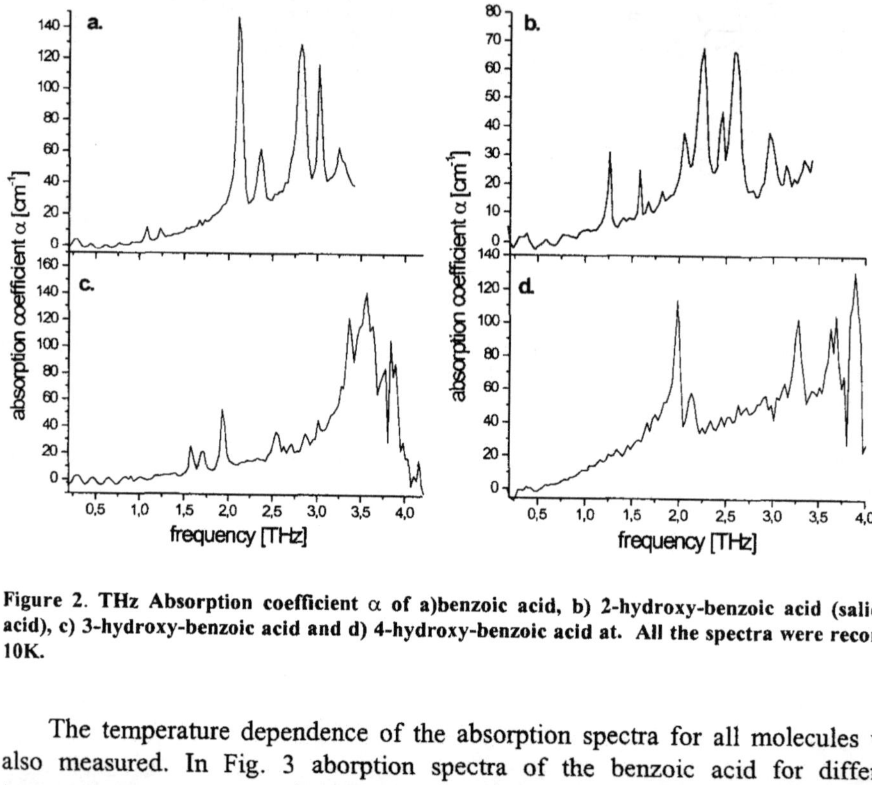

Figure 2. THz Absorption coefficient α of a)benzoic acid, b) 2-hydroxy-benzoic acid (salicylic acid), c) 3-hydroxy-benzoic acid and d) 4-hydroxy-benzoic acid at. All the spectra were recorded 10K.

The temperature dependence of the absorption spectra for all molecules was also measured. In Fig. 3 aborption spectra of the benzoic acid for different temperatures are presented. With the increase of the temperature all resonances become weaker, broader and move into lower frequencies. The temperature dependence of the individual lines can be explained by thermal excitation of the levels involved in low-frequency vibrational transitions in molecules. At the low temperatures only the lowest vibrational level is populated so only a single transition is excited thus leading to a very sharp and narrow line. At higher temperature populations of higher vibrational levels become significant and transitions starting from higher vibrational levels cannot be neglected. For example, at room temperature the kT value is the same order of magnitude as the vibrational frequencies (aprox. 180 cm^{-1}). In this case each absorption line results from a superposition of several transitions starting from different vibrational levels. The anharmonicity of the potential causes that the frequencies of individual transition decrease with increasing v (v vibrational quantum number). This, in turn, causes the absorption lines to broaden and shift towards lower frequencies with increasing temperature.

The identification of particular modes from the absorption spectra is difficult because of formation of dimer molecular complexes in the crystal phase[5]. Resonant frequencies for benzoic acid have been calculated by Zelsman et al. [6]. It was found that the modes below 72 cm^{-1} (2 THz) are due to translations of the dimers (lattice vibrations). Higher frequency modes result from inter- or intra-molecular vibrations. Intramolecular vibrations have higher frequencies than inter-molecular modes. Zelsman et al. have found that the lowest frequency for intramolecular vibrations is 188 cm^{-1} (6THz) which is beyond our experimental range. Therefore, we believe that all the resonanses observed in our experimental data correspond to inter-molecular vibrations. It is also known that very low modes

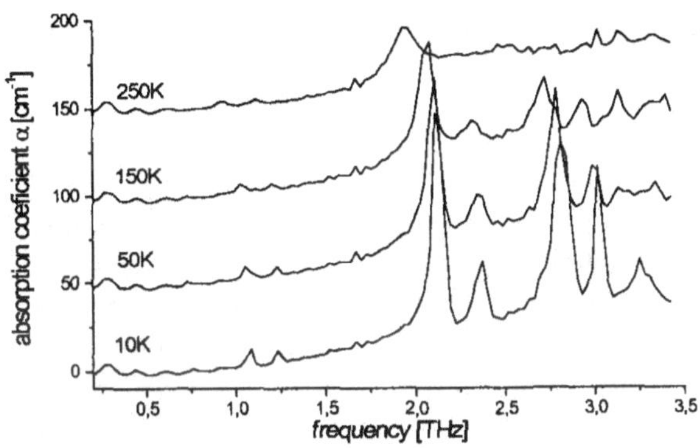

are out-of-plane torsion's.

Figure 3. Temperature dependent absorption spectra for the benzoic acid sample.

Comparison between the absorption spectra for aspirin and salicylic acid is presented in Fig. 4. Salicylic acid molecule is a breakdown product of acetylsalicylic acid (aspirin). The structure of the aborption spectrum of aspirin is richer than that for salicylic acid. The strongest aborption band at about 2 THz is similar for both investigated molecules. At higher frequencies there are several resonances for both crystals, but they are shifted. The main difference between the two spectra is the presence of a strong absorption band at aprox. 1.7 THz in the aspirin spectrum. This band probably comes from dynamics of acetyl group. Acetyl group might contribute to different and extra vibrations excited in aspirin.

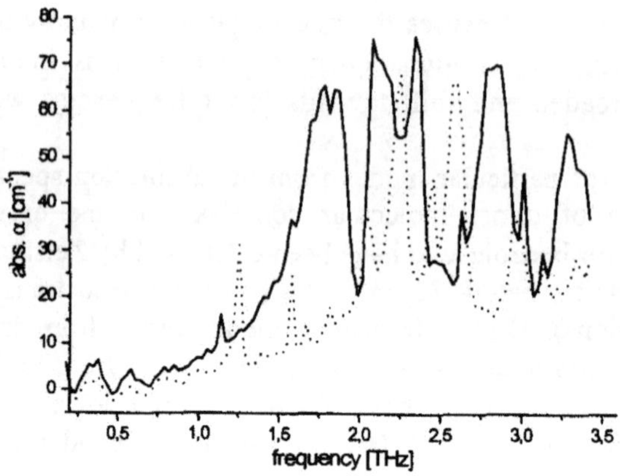

Figure 4. Absorption spectra for aspirin (solid line) and salicylic acid (dash line) at 10K.

4 Conclusion

In this paper we report on the measurements of FIR absorption spectra of benzoic acid and its derivatives: 2,3, or 4-hydroxybenzoic acids and acetylsalicylic acid by THz-time domain spectroscopy. All the spectra show are well-resolved resonances which we atribute to excitation of collective intermolecular motion. The experimental data indicate that far-IR spectroscopy is a sensitive tool which enables one to distinquish between different conformers of a given molecule in solid state samples. This property might be useful in drug tracing applications where the detection of small amounts of substance is needed.

5 Acknowledgements

This work was supported by the Deutsche Forschungsgemeinschaft, under SFB 276, TP C14.

References

1. P.I. Nagy, W.J. Dunn, G. Alagona, C. Ghio, *J. Phys. Chem.* **97** (1993) pp.4628-4628.
2. D. R. Grischkowsky, Ch. Fattinger. M. Van Exter, S.R. Keiding, *I. Opt. Soc. Am. B,* **7** (1990) pp. 2006-2015.
3. M. Walther, B. Fischer, M. Schall, H. Helm, P. Uhd Jepsen, *Chem. Phys. Lett.* **332** (2000) pp. 389-395.
4. M. Schall, H. Helm,, S.R. Keiding *Int. J. Infrared Milimeter Waves* **20** (1999) pp. 595-604.
5. S. Maheshvari, A. Chowdhury, N. Sathyamurthy, H. Mishra, H.B. Tripathi, M. Panda, J. Chandrasekhar *J. Phys. Chem.* **103** (1999) pp. 6257-6262.
6. H.R. Zelsmann, Z. Mielke *Chem. Phys. Lett.* **186** (1991) pp. 501-508.

OPTIMAL CONTROL THEORY OF MOLECULAR DYNAMICS: A COMPARATIVE APPROACH TO CHARACTERIZE THE COMPLEXITY OF THE CONTROL TASK

TOMÁŠ MANČAL AND VOLKHARD MAY

Institute of Physics, Humboldt-University at Berlin, Hausvogteiplatz 5-7, D-10117 Berlin, Germany, e-mail: mancal@physik.hu-berlin.de

The Optimal Control Theory is used to describe different types of laser pulse control scenarios for molecular systems. Once the pulse which solves the concrete control task has been computed, in a second step it is related to the experimental pulse shaping technique based on Liquid Crystal (LC) Spatial Light Modulators (SLM). Resulting from this procedure a quantity is introduce which will be interpreted as the complexity of the control task. Then this complexity is presented for different systems and control tasks including the control of dissipative dynamics.

1 Introduction

In recent years the Optimal Control Theory (OCT)[1] became a widely used theoretical tool to predict fs–laser pulse control of molecular dynamics. Simultaneously, highly flexible pulse shaping techniques have been developed[2] which enable one to realize the Optimal Control in the experiment[3]. In this contribution we report on an attempt to interconnect the experimental and theoretical approaches. This will be mainly achieved by introducing a quantity which characterizes the complexity of the control task[4].

2 Control tasks and the definition of complexity

To define such a complexity of the control task we compare systems which differ in their electron–vibrational spectrum. We take as examples systems characterized by a single electronic ground–state potential energy surface (PES) which is connected via the external laser field with a different number of PES in the excited state. The considerations include a small molecule with a single excited–state (model system I) and a larger systems showing photoinduced electron transfer among three different excited–state PES (model system II). All PES are defined versus a single harmonic vibrational coordinate. Alternatively, increasing complexity within the molecular systems is achieved via a dissipative coupling of the model system I to a thermal environment.

In a first step by utilizing OCT in a way valid for open quantum systems[5,6] the laser field $E_{opti}(t)$ is calculated which solves the control task (the optimal

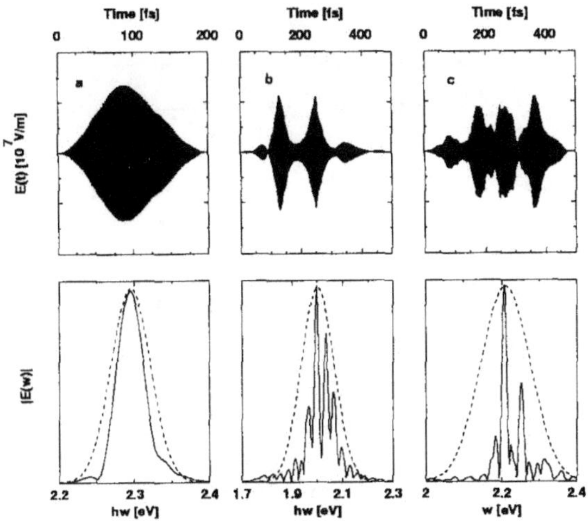

Figure 1. Shape of the optimal pulses which solve three different control tasks of increasing complexity. Upper line: pulse shape in the time domain, lower line: pulse shape in the frequency domain. Example (a): model system I (see text) and a single vibrational eigenstate in the excited–state PES as a target state, example (b): pump–dump scheme in the model system I, example (c): model system II and a single vibrational eigenstate of the acceptor PES as a target state. The Gaussian envelope of the frequency tranformed pulses constitutes the incoming pulse of the pulse–shaping scheme (see text).

field). We chose a version of OCT where one has to maximize the expectation value $O(T; E(t))$ of the projector on the target state $|\psi_{target}\rangle$. This state represents the aim of the control task and should be reached at a certain time T and under the constrain of a finite laser field–strength $E(t)$. The expectation value $O(T; E(t)) = tr\{\hat{\rho}(t = T; E)|\psi_{target}\rangle\langle\psi_{target}|\}$ is determined by the reduced density operator $\hat{\rho}(t)$ which has to be propagated under the action of the laser field.

Having at hand the optimal pulse $E_{opti}(t)$ for all mentioned systems, we prove the extend to which the experimental pulse shaping technique is able to reproduce the pulse and thus to solve the control task. Therefore, we compute the approximate pulses $\mathcal{E}_{opti}(t; N)$ which corresponds to the reproduction of the optimal pulse (following from OCT) by a pulse shaper with N LC cells. Afterwards $\mathcal{E}_{opti}(t; N)$ is used to drive the system and to calculate $O(T; \mathcal{E}_{opti}(t; N))$. This quantity describes the efficiency of the pulse shaping system with a number N of LC cells to solve the control task. To have a quantity at hand which is more appropriate for a comparison of different systems

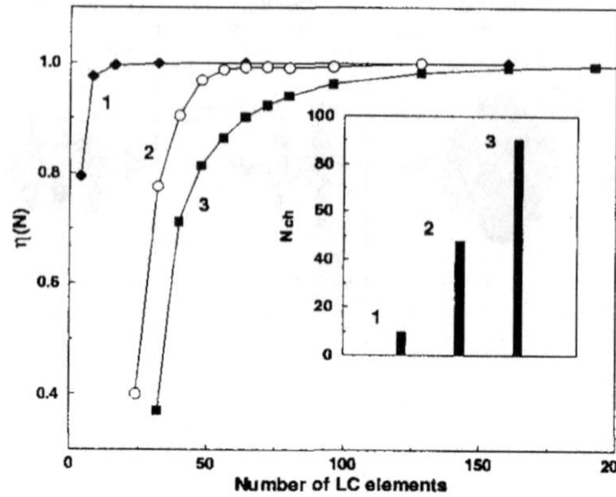

Figure 2. Dependency of the pulse reconstruction efficiency on the number of LC elements used by the shaper for the three model systems mentioned in the text. Insert: related values of N_{ch}.

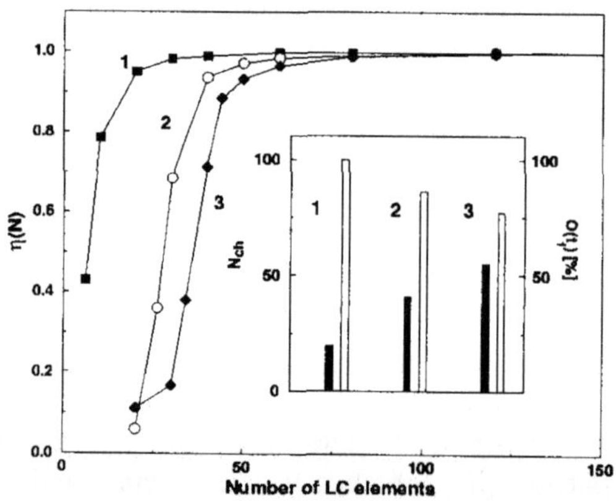

Figure 3. Dependency of the pulse reconstruction efficiency on the number of LC elements used by the shaper for model system I but including dissipation with increasing strength. Full columns of the insert: related values of N_{ch}, empty columns of the insert: target state population achieved with the unconstrained optimal pulse.

and different control tasks the ratio $\eta(N) = O(T;\mathcal{E}_{opti}(t;N))/O(T;\mathcal{E}_{opti}(t))$ is introduced. Then, the number $N_{\rm ch}$ for which $\eta(N)$ equals 0.95 can be taken as a measure for the complexity of a given control task.

3 Numerical results and conclusions

The optimal pulses $E_{opti}(t)$ which have been computed for the different types of control tasks mentioned above are shown on Fig. 1 (together with their Fourier transforms $E_{opti}(\omega)$). Figs. 2 displays the ration $\eta(N)$ for the examples of a control tasks which optimal pulse has been already drawn in Fig. 1. A smooth increase of $\eta(N)$ can be observed with the increase of the number N of LC elements. According to an increase of the characteristic number $N_{\rm ch}$ the different control tasks may be ordered. Indeed, this ordering reflects an increase of the complexity of the control task. And it is important to note that for all described control tasks the number of LC elements necessary for a nearly complete reconstruction of the optimal pulse lies among the parameters of commercially available shapers. In Fig. 3 similar results are shown but now including the dissipative influence of an environment. Here, the growing complexity of the system is a result of the increasing strength of dissipation. Again, we obtained a smooth dependency on the number of used LC elements.

In conclusion we would like to underline two points. First, if a control experiment has to be carried out on a system with a known energetic structure the approximated optimal field $\mathcal{E}_{opti}(t;N)$ as computed here, may serve as a good starting point for experimental feedback optimization procedures. And second, if less is known about the energy level structure of the considered molecule, Figs. 2 and 3 demonstrate that the slope of $\eta(N)$, or of $O(T;\mathcal{E}_{opti}(t;N))$ would be a helpful quantity. Once this quantity is measured in the experiment by varying the number of LC cells, one may get evidence for the degree the control task has already been solved.

References

1. H. Rabitz, R. de Vivie-Riedle, M. Motzkus and K. Kompa, Science **288**, 824 (2000).
2. M. M. Wegers and K. A. Nelson, J. Opt. Soc. Am. B **12**, 1343 (1995).
3. T. Baumert and G. Gerber, Isr. J. Chem. **34**, 103 (1995).
4. T. Mancal and V. May, *Phys. Rev. Letters* (submitted).
5. Y. Yan, R.E. Gillian, R.M. Whitnell, K.R. Wilson and S. Mukamel, J. Phys. Chem. **97**, 2320 (1993).
6. T. Mancal and V. May, Eur. Phys. J. D, Eur. Phys. J. D **14**, 173 (2001).

This page is intentionally left blank

4. Nanostructures, Polymers, Surfaces and Interfaces

This page is intentionally left blank

COHERENT EXCITATION OF ACOUSTIC PHONON MODES IN METAL NANOPARTICLES AND NANORODS

G. V. HARTLAND

Department of Chemistry and Biochemistry, University of Notre Dame
Notre Dame, IN 46556-5670, USA
E-mail: hartland.1@nd.edu

Ultrafast laser excitation of metal particles selectively deposits energy into the electronic degrees of freedom. Reasonable pulse energies can produce increases in the electronic temperature of several thousand K. The hot electrons subsequently equilibrate with the phonon modes on a picosecond timescale, producing an overall increase in temperature of hundreds of K. These rapid heating processes can impulsively excite the phonon modes that correlate with the expansion co-ordinate of the particles. In this paper the way the frequency of the impulsively excited phonon modes depends on the size and shape of the particles will be described. The relative contributions from the hot electrons and lattice heating to the impulsive excitation mechanism will also be discussed.

1 Introduction

Metals absorb light primarily through electronic transitions [1]. In metal particles three types of transitions are typically discussed: interband transitions (excitation of electrons from the filled valence band to empty levels in the conduction band); intraband transitions (transitions within the conduction band); and the plasmon band (a collective oscillation of the conduction electrons) [2]. Although the details of the excitation are different, in all three cases the absorbed energy is rapidly equilibrated amongst all the conduction electrons by electron-electron scattering processes. These events are extremely rapid, and the end result is the creation of a hot electron distribution within several hundred femtoseconds. Electrons have a very small heat capacity, thus, the electronic temperatures that can be created are very high. To put this into context, consider a 2 nm diameter Au particle — ca. 250 atoms. Absorption of a single 400 nm photon creates an electronic temperature of 1800 K (of course, in any real experiment the metal particles absorb many photons). The hot electrons subsequently equilibrate with the phonon modes on a picosecond timescale. This process has been extensively studied by ultrafast laser techniques, for recent reviews, see Refs. [3-8]. Because the lattice heat capacity is much greater than the electronic heat capacity, the overall temperature rise in the system, once the electrons and phonons have reached equilibrium, is much less than the initial electronic temperature.

The electron-phonon relaxation process can be theoretically described by the Two-Temperature Model (TTM) [9-12]. In this model the electrons and the lattice are assigned different temperatures, T_e and T_l, and the energy exchange between them is governed by the following coupled differential equations:

$$C_e \frac{dT_e}{dt} = -g(T_e - T_l) \qquad (1a)$$

$$C_l \frac{dT_l}{dt} = g(T_e - T_l) \qquad (1b)$$

where g is the electron-phonon coupling constant, C_l and $C_e = \kappa T_e$ are the lattice and electronic heat capacities, respectively, and κ is a constant that depends on the density of electronic states at the Fermi level [1]. The TTM model shows that the time scale for energy exchange between the electrons and phonons is roughly given by $\kappa T_e/g$. This means that higher laser powers (larger values of T_e) cause longer relaxation times. Note that energy exchange with the environment has been ignored in this analysis, in principle, this can be simply included by adding a decay term to Equation (1b). Figure 1 shows calculated electronic and lattice temperatures for an initial electronic temperature of 4700 K (absorption of approximately six 400 nm photons in a 2 nm diameter particle). The parameters used (g, C_e and C_l) are appropriate for Au.

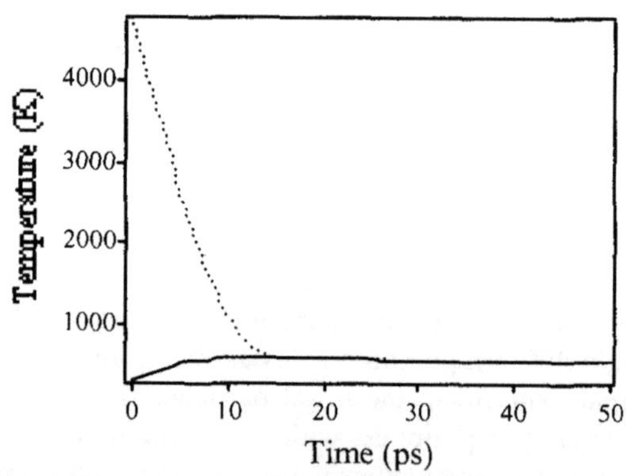

Figure 1. Time dependence of T_e and T_l calculated using the TTM for an initial electronic temperature of 4700 K.

The above discussion provides a background for the main topic of this paper, which is the change in particle volume induced by ultrafast laser excitation. Figure 1 shows that heat flows into the lattice in a few picoseconds after laser excitation (the exact value depends on the pump laser power). This will cause the lattice to expand, and the time scale for this response depends on the period of the phonon mode that correlates with the expansion co-ordinate. As will be shown below, the lattice response is slower than the heating time, which means that these phonon modes can be impulsively excited. The issues that will be addressed here are: (i) how the period of the coherently excited phonon modes depend on the size and shape of the particles, and (ii) the mechanism for impulsive excitation of the acoustic phonon modes. Specifically, we will examine the relative contributions of hot-electrons and lattice heating to the excitation process.

2 Experimental

The laser system used in these experiments is based on a regeneratively amplified Ti:Sapphire laser (Clark-MXR, CPA-1000). This laser produces 780 nm pulses with 0.5 mJ/pulse and a pulsewdith of 120-150 fs FWHM ($sech^2$ deconvolution). The output is split by a 90:10 beamsplitter to provide the pump and probe beams. The pump beam is frequency doubled in a 1 mm BBO crystal, and tunable probe pulses are produced by white-light generation in a 3 mm Sapphire window. The probe wavelength in a given experiment is selected by a Jobin-Yvon Spex H-10 monochromator placed after the sample. The spectral resolution of the monochromator is ca. 10 nm with the grating/slit combination used in our experiments. The pump laser power for the transient absorption experiments presented below was ~0.6 μJ per pulse, as determined by a Molectron J3-02 Energy Detector. Measurement of the amount of energy absorbed by the sample versus laser power show that the sample absorption is linear in this energy range. The pump laser spot size at the sample was measured by standard laser "burn paper," and was found to be $\sim 4.5\text{x}10^{-4}$ cm^2. A normalization scheme using gated integration, an analog division circuit, and lock-in detection of the signal was used to reduce noise from fluctuations in the probe laser. Normalization and a stable white light beam are crucial for the success of the experiments described below, as they allow good signal-to-noise data to be collected at relatively low pump laser power. The metal particle samples were kept in a sealed cuvette (2 mm path length) for the experiments. The samples were produced using recipes described in the literature [13-17].

3 Discussion and Results

3.1 Gold Nanoparticles

Transient absorption data for ca. 50 nm Au particles are presented in Figure 2. Modulations due to the coherently excited breathing modes can be clearly seen in both traces. The probe wavelengths used were 510 nm and 550 nm, which lie on the blue and red sides of the plasmon band of the sample, respectively. It is important to note that the period and damping time of the modulations are identical for the two traces, but they are 180° of phase with each other. This implies that the signal arises from a periodic shift in the position of the plasmon band [18-20]. Physically, this occurs because the wavelength of the plasmon band maximum is inversely related to the electron density of the particle: $\lambda_{max} \propto 1/\sqrt{n}$. Thus, changes in the particle size due to the coherently excited breathing mode change the electron density and, therefore, the position of the plasmon band. The relationship between λ_{max} and n can be used to quantitatively calculate the change in size versus time from the transient absorption data. An example calculation for the 50 nm diameter Au particles will be presented in Section 3.3 of this paper.

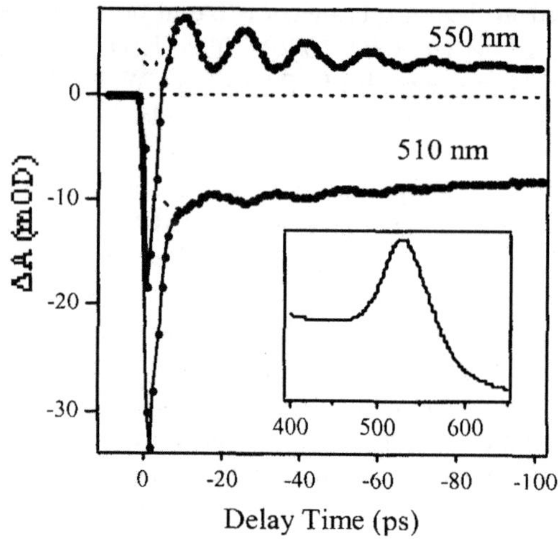

Figure 2. Transient absorption data for ca. 50 nm diameter Au particles collected with 510 and 550 nm probe laser pulses. The insert shows the absorption spectrum of the sample.

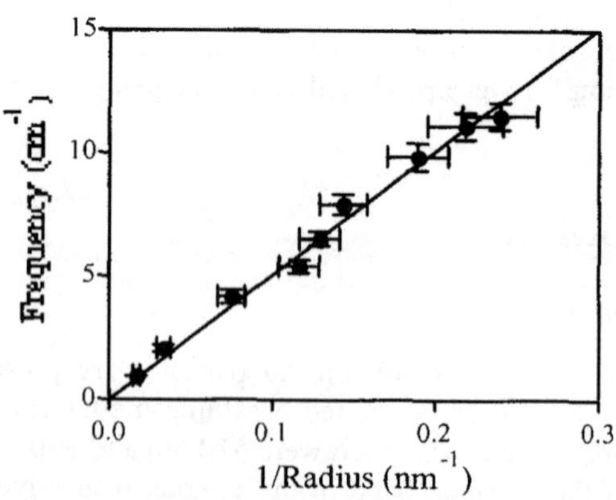

Figure 3. Frequency (cm^{-1}) versus 1/R (nm) for different Au particle samples.

Figure 3 shows the frequencies measured for different samples, plotted against the inverse of the average radius of the sample. The straight line in Figure 3 is the calculated frequency for the breathing mode of an elastic sphere:

$$\bar{\nu} = \frac{\eta c_l}{2\pi c R}, \tag{2}$$

derived through continuum mechanics [21,22]. In this equation c_l is the longitudinal speed of sound in Au, R is the radius, and η is an eigenvalue that depends on the longitudinal and transverse speeds of sound. The excellent agreement between the calculated and experimental results shows that continuum mechanics works extremely well for nanometer sized objects!

The dependence of the frequency on the radius means that the modulations will be inhomogeneously damped: different sized particles in the sample have different breathing mode frequencies, leading to a dephasing of the detected signal. Assuming that this inhomogeneous damping dominates the decay, and that the particle sizes are normally distributed, the signal can be written as [23]:

$$S(t) \approx Cos\left[\left(2\pi/\bar{T}\right)t + \phi\right] \times e^{-t^2/\tau^2} \tag{3}$$

where ϕ is the phase of the modulation, $\bar{T} = 2\pi\bar{R}/\eta c_l$ is the period corresponding to the average radius of the sample, and the damping time is given by $\tau = \bar{R}\,\bar{T}/\sqrt{2}\pi\sigma_R$. Fits to the transient absorption data using Equation (3) are included in Figure 2. This analysis yields an average period of $\bar{T}$ = 15.9 ± 0.1 ps, which corresponds to an average radius of $\bar{R}$ = 24.2 ± 0.1 nm The damping time is τ = 51 ± 1 ps, which implies a standard deviation of σ_R = 1.70 ± 0.05 nm. In comparing size distributions it is common to normalize the standard deviation by the average radius. For our sample we find $\sigma_R/\bar{R} = 7.0 \pm 0.2$ %, which corresponds to an extremely high quality, monodisperse sample. In principle, the decay of the modulations could contain contributions from energy relaxation to the environment. Thus, the value of σ_R derived from this analysis is strictly an upper limit to the true width of the distribution, i.e., $\sigma_R/\bar{R} \leq 7.0$ %.

3.2 Gold Nanorods

We have recently completed similar experiments with Au nanorods. These particles were made using an electrochemical technique, where a surfactant/co-surfactant mixture is used to template the shape of the growing particles [16,17]. Figure 4 shows results obtained from a rod sample with an average length of 35 nm and a width of 10 nm. The insert (again) shows the absorbance spectrum of the sample. Note that the rod shaped particles display two plasmon bands: a longitudinal band at

~740 nm corresponding to oscillation of the conduction electrons along the length of the rod, and a transverse band at ~ 520 nm corresponding to oscillations across the rod. The transient absorption experiments were performed with 400 nm pump pulses and probe laser wavelengths on the red and blue sides of the longitudinal plasmon band. There are several points to note. First, both the phase and the period of the modulations are different for the two probe laser wavelengths. This is in contrast to the experiments with spherical particles, where the period did not depend on the probe wavelength. Second, the measured periods (ca. 60 ps) are much longer than those observed for the spherical particles.

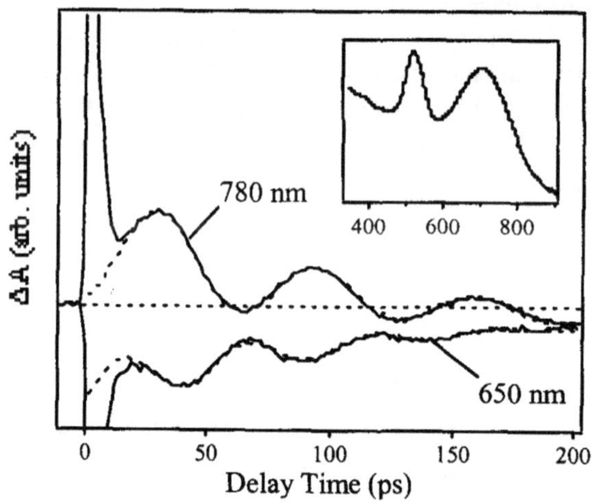

Figure 4. Transient absorption data recorded with 400 nm pump pulses and different probe wavelengths for Au nanorods with an aspect ratio of 3.5. The insert shows the absorbance spectra of the rods. Modulations are only observed when the probe laser is resonant to the longitudinal plasmon band.

Figure 5 shows the period versus the probe laser frequency for the sample in Figure 4. The period changes from 47 ps to 75 ps (i.e., by over 50%) as the probe laser is tuned across the longitudinal plasmon band. This effect occurs because the position of the longitudinal plasmon band depends on the length of the rod. Thus, different probe wavelengths interrogate different length rods in the sample. These results show that the period of the coherently excited vibrational mode is roughly proportional to the rod length. The straight line in Figure 5 is a fit to the data using $T = 2L/\alpha$, where α is an adjustable parameter. In this fitting process the well known relationship between the position of the longitudinal plasmon band and the length of the rod [16,17] was used to convert the probe laser wavelength to length. The value of α obtained from the fitting procedure was $\alpha = 1230$ ms^{-1}, which is very close to the transverse speed of sound in Au (1200 ms^{-1}) [24]. Note that this simple analysis does not take into account the finite size distribution of the sample, or the spectral widths of the probe laser and the transient absorption spectra of the rods.

These effects produce the deviation from the predicted straight line behavior at high and low wavelengths [25].

Similar quantum beat experiments have recently been performed for ellipsoidal Ag particles in a glass matrix [26]. These experiments showed modulations due to expansion and contraction along the long axis of the ellipse. The measured period was found to be approximately given by $2L/c_l$, where L is the length of the ellipse. For our sample, this simple formula predicts a period of 22 ps, which is much shorted than that observed in Figure 4. At this point it is not clear why the rod and ellipse systems give such different results. Two possibilities are: (i) The vibrational modes excited in the Au nanorods are different to those for the Ag ellipses. For example, the vibrational period for the nanorods is close to $2L/c_t$, which is approximately the period of a Rayleigh surface wave propagating along a cylinder [27]. (ii) The Au nanorods are single crystals [28], whereas, the Ag ellipses are polycrystalline. The elastic properties of the single crystal nanorods may be very different to those of polycrystalline rods. Further experiments and detailed continuum mechanics calculations are currently being performed to elucidate the vibrational dynamics of these systems.

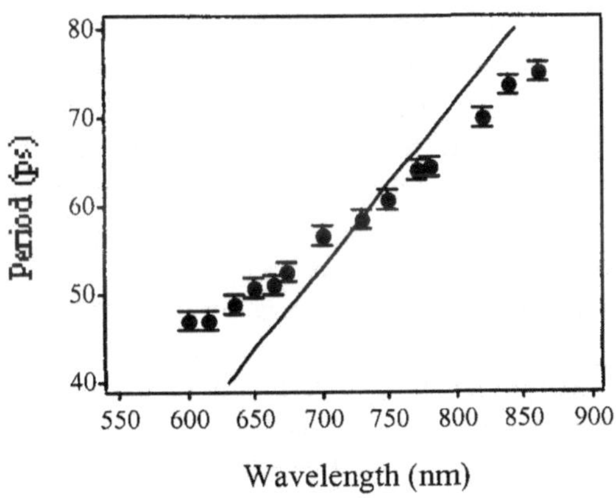

Figure 5. Measured period versus probe laser wavelength for a polydisperse gold nanorod sample with an average length of 35 nm and a width of 10 nm. The straight line is a fit to the data assuming that the period is proportional to the length of the rods.

3.3 *Impulsive Excitation*

The coherent vibrational motion of the particles is driven by the rapid electronic and lattice heating induced by the pump laser pulse. For spherical particles this process can be modeled by treating the breathing mode as a classical harmonic oscillator that oscillates around the equilibrium radius R_0 [19,20,29]. The value of the equilibrium radius depends on the temperature through the coefficient for thermal expansion for gold α, specifically,

$$\frac{d^2R}{dt^2}+\frac{2}{\tau}\frac{dR}{dt}+\left(\frac{2\pi}{T}\right)^2\left[R-\left\{R_0+\frac{R_0\alpha}{3}(\Delta T)\right\}\right]=0\ . \tag{4}$$

In this equation T is the period of the coherently excited breathing mode, τ is a phenomenological damping time and ΔT is the temperature increase of the particle, which depends on time through Equations (1a) and (1b). Note that the coefficient for thermal expansion is related to the heat capacity of the particle by [1]

$$\alpha=\frac{1}{B}\left(\gamma C_l+\frac{2}{3}C_e\right) \tag{5}$$

where B is the bulk modulus (which is nearly independent of temperature), C_l and C_e are the electronic and lattice heat capacities, and γ is the Grüneisen parameter for the lattice (the free electron value of 2/3 was used for the electronic Grüneisen parameter). For metals $C_l >> C_e$ when the electrons and lattice are in thermal equilibrium, which means that the change in the ion positions due to lattice anharmonicity dominates the expansion process. However, in ultrafast experiments $T_e >> T_l$ at early times (< 5 ps), as shown in Figure 1. This means that the electronic contribution to expansion cannot be ignored in our measurements.

Simulations of the change in radius versus time calculated using Equations (1), (4) and (5) are presented in Figure 6. The period and damping time for these calculations were chosen to match the data presented in Figure 2. The temperature rise in the system was calculated from the laser spot size, the power absorbed by the sample, and the concentration of Au in the solution. The other parameters (C_l, C_e, γ and g) where taken from the literature [1,24] or other measurements [30]. In Figure 6 calculations using both the electronic and lattice contributions to the expansion coefficient are compared to those where only the lattice contribution (i.e., the C_l term in Equation (5)) was included. The magnitude of the oscillations in ΔR are similar for the two calculations, but the phases are significantly different.

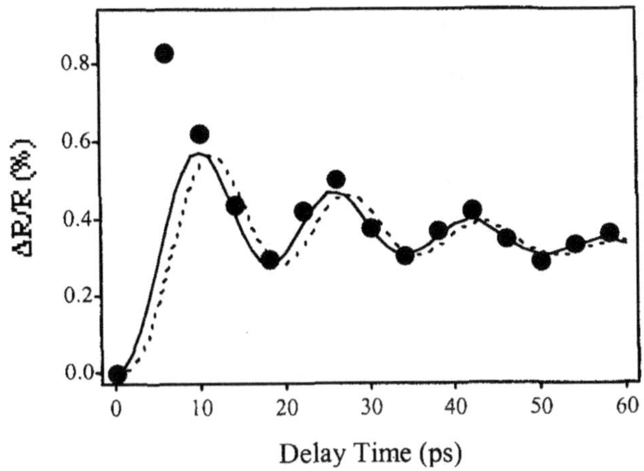

Figure 6. Change in radius ($\Delta R/R$) versus time for the Au particles in our experiments. (•) Determined from the transient absorption data; (—) calculated change in radius using Equations (1), (4) and (5). The dashed line (---) shows the results of the calculations without including hot-electron pressure.

Also shown in Figure 6 are values of $\Delta R/R$ derived from transient absorption measurements. In this analysis transient absorption traces recorded at different wavelengths (see Figure 2) were used to calculate the position of the plasmon band at different times. The position of the plasmon band depends on the free electron density and, therefore, the size of the particle. Thus, the experimental data can be directly scaled to calculate the radius versus time. In general, the experimental data and the calculations are in excellent agreement. Specifically, the magnitude of the oscillations is well reproduced for both calculations. However, only the simulation that includes contributions from the electrons to the driving force for expansion does a good job in reproducing the phase. These results show that the overall increase in size of the particles mainly depends on the thermal expansion of the lattice. However, at early times, when the electronic temperature is very high, the pressure from the hot electrons supplies a significant force to the nuclei, which affects the phase of the modulations. These "hot-electron pressure" effects have been previously observed for thin films [31], and Ag particles (both ellipses and spheres) in a glass matrix [26,32].

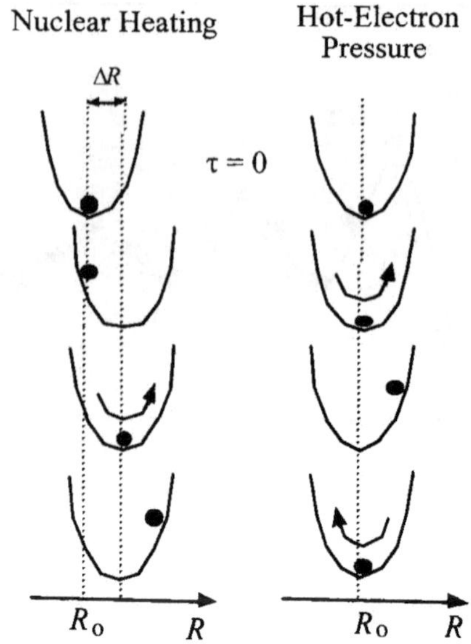

Figure 7. Cartoon of the coherent vibrational motion induced by nuclear heating (left) and hot-electron pressure (right).

A simple cartoon that shows the difference between hot-electron pressure and nuclear heating in our time-resolved experiments is shown in Figure 7. It is important to note that the force from the hot-electrons is only significant at high electronic temperatures. Thus, the hot-electrons provide a delta-function like kick to the nuclei. In contrast, the lattice heating contribution produces a permanent (on the time-scale of these experiments) change in the radius of the particle. Thus, this effect is a step-function like perturbation. The time-response of the nuclei to these two forces is very different. Specifically, the delta and step-function forces produce oscillations in the radius that have different phases. For true delta- and step-function perturbations the difference in phase is exactly 90°, on the other hand, for our experiments the finite electron-phonon coupling time smears out the response, so that the difference in phase is not quite 90°. However, as shown in Figure 6, the contribution to the phase of the modulations from hot-electron pressure can be clearly detected in transient absorption experiments.

4 Summary and Conclusions

Excitation of metal nanoparticles with an ultrafast laser pulse causes a rapid increase in the electronic temperature. The hot-electrons subsequently equilibrate with the phonon modes on a picosecond time scale. These rapid heating processes can impulsively excite the phonon mode that correlates with the expansion co-ordinate. For spherical particles the symmetric breathing mode is excited. For particles in the 8 nm to 120 nm diameter range, the frequency of this mode can be exactly calculated using continuum mechanics [5,19]. For the nanorod samples the period of the coherently excited phonon mode depends on the length of the rod, but the values obtained do not match those previously observed for Ag ellipses. At this stage it is not clear whether this is due to differences in the symmetry of the particles (cylindrical compared to ellipsoidal), or their material properties (the gold nanorods in these experiments are single crystals [28], whereas, the Ag particles in ref. [26] are polycrystalline).

By performing transient absorption experiments at a variety of wavelengths for a 50 nm diameter Au particles sample, we have been able to calculate how the radius changes with time. These results can be directly compared to model calculations where the expansion co-ordinate (the symmetric breathing mode) is treated as a harmonic oscillator. A careful analysis of the phase of the modulations shows that both hot-electron pressure and lattice heating contribute to the impulsive excitation process. The force due to the hot electrons is only significant at early times, when the electronic temperature is much greater than the lattice temperature. The overall expansion of the particles, and the magnitude of the oscillations in the particle radius, are controlled by the lattice heating contribution [1].

5 Acknowledgements

The work described in this paper was supported by the National Science Foundation by Grant No. CHE98-16164. The author is extremely grateful to his students for performing the work (primarily Jose Hodak), and to Profs. Arnim Henglein and Paul Mulvaney for providing the metal particle and nanorod samples.

References

1. Ashcroft N. W. and Mermin N. D., Solid State Physics (Harcourt Brace, Orlando, 1976).
2. Kreibig U. and Vollmer M., Optical Properties of Metal Clusters (Springer, Berlin, 1995).
3. Zhang J. Z., Ultrafast studies of electron dynamics in semiconductor and metal colloidal nanoparticles: Effects of size and surface. *Acc. Chem. Res.* **30** (1997) pp. 423-429.
4. Link S. and El-Sayed M. A., Spectral properties and relaxation dynamics of surface plasmon electronic oscillations in gold and silver nanodots and nanorods. *J. Phys. Chem. B* **103** (1999) pp. 8410-8426.

5. Hodak J. H., Henglein A. and Hartland G. V., Photophysics of nanometer sized metal particles: Electron-phonon coupling and coherent excitation of breathing vibrational modes. *J. Phys. Chem. B* **104** (2000) pp. 9954-9965.
6. Del Fatti N., Vallee F., Flytzanis C., Hamanaka Y. and Nakamura A., Electron dynamics and surface plasmon resonance nonlinearities in metal nanoparticles. *Chem. Phys.* **251** (2000) pp. 215-226.
7. Stagira S., Nisoli M., De Silvestri S., Stella A., Tognini P., Cheyssac P. and Kofman R., Ultrafast optical relaxation dynamics in metallic nanoparticles: from bulk-like toward spatial confinement regime. *Chem. Phys.* **251** (2000) pp. 259-267.
8. Voisin C., Del Fatti N., Christofilos D. and Vallee F., Ultrafast electron dynamics and optical nonlinearities in metal nanoparticles. *J. Phys. Chem. B* **105** (2001) pp. 2264-2280.
9. El-Sayed M. A., Some interesting properties of metals confined in time and nanometer space of different shapes, *Acc. Chem. Res.* **34** (2001) pp. 257-264 .
10. Elsayed-Ali H. E., Norris T. B., Pessot M. A. and Mourou G. A., Time-resolved observation of electron-phonon relaxation in copper, *Phys. Rev. Lett.* **58** (1987) pp. 1212-1215.
11. Schoenlein R. W., Lin W. Z., Fujimoto J. G. and Eesley G. L., Femtosecond studies of nonequilibrium electronic processes in metals, *Phys. Rev. Lett.* **58** (1987) pp. 1680-1683.
12. Sun C. K., Vallee F., Acioli L. H., Ippen E. P. and Fujimoto J. G., Femtosecond-tunable measurement of electron thermalization in gold, *Phys. Rev. B* **50** (1994) pp. 15337-15348.
13. Enüstün B. V. and Turkevich J., Coagulation of colloidal gold, *J. Am. Chem. Soc.* **85** (1963) pp. 3317-3328.
14. Henglein A. and Meisel D., Radiolytic control of the size of colloidal gold nanoparticles, *Langmuir* **14** (1998) pp. 7392-7396.
15. Henglein A., Radiolytic preparation of ultrafine colloidal gold particles in aqueous solution: Optical spectrum, controlled growth, and some chemical reactions, *Langmuir* **15** (1999) pp. 6738-6744.
16. Yu Y. Y., Chang S. S., Lee C. L. and Wang C. R. C., Gold nanorods: Electrochemical synthesis and optical properties, *J. Phys. Chem. B* **101** (1997) pp. 6661-6664.
17. Chang S. S., Shih C. W., Chen C. D., Lai W. C. and Wang C. R. C., The shape transition of gold nanorods, *Langmuir* **15** (1999) pp. 701-709.
18. Hodak J. H., Martini I. and Hartland G. V., Observation of acoustic quantum beats in nanometer sized Au particles, *J. Chem. Phys.* **108** (1998) pp. 9210-9213.
19. Hodak J. H., Henglein A. and Hartland G. V., Size dependent properties of Au particles: Coherent excitation and dephasing of acoustic vibrational modes, *J. Chem. Phys.* **111** (1999) pp. 8613-8621.

20. Del Fatti N., Voisin C., Christofilos D., Vallee F. and Flytzanis C., Acoustic vibration of metal films and nanoparticles, *J. Phys. Chem. A* **104** (2000) pp. 4321-4326.
21. Lamb H., On the vibrations of an elastic sphere, *Proc. Lond. Math. Soc.* **13** (1882) pp. 189-212.
22. Dubrovskiy V. A. and Morochnik V. S., Natural vibrations of a spherical inhomogeneity in an elastic medium, *Izv. Earth Phys.* **17** (1981) pp. 494-504.
23. Hartland G. V., Coherent vibrational motion in metal particles: determination of the vibrational amplitude and excitation mechanism, *J. Chem. Phys.*, submitted for publication.
24. CRC Handbook of Chemistry and Physics, 80th ed. (CRC Press, Boca Raton, FL, 1999).
25. Hartland, G. V., Hu M., Wilson O., Mulvaney P., and Sader J. E., Coherent excitation of vibrational modes in gold nanorods, *J. Phys. Chem. B*, submitted for publication.
26. Perner M., Gresillon S., Marz J., von Plessen G., Feldmann J., Porstendorfer J., Berg K. J. and Berg G., Observation of hot-electron pressure in the vibration dynamics of metal nanoparticles *Phys. Rev. Lett.* **85** (2000) pp. 792-795.
27. Pollard H. F., Sound Waves in Solids (Pion, London, 1977).
28. Wang Z. L., Mohamed M. B., Link S., and El-Sayed M. A., Crystallographic facets and shapes of gold nanorods of different aspect ratios *Surf. Sci.* **440** (1999) pp. L809-L814.
29. Del Fatti N., Voisin C., Chevy F., Vallee F. and Flytzanis C., Coherent acoustic mode oscillation and damping in silver nanoparticles, *J. Chem. Phys.* **110** (1999) pp. 11484-11487.
30. Hodak J. H., Henglein A. and Hartland G. V., Electron-phonon coupling dynamics in very small (between 2 and 8 nm diameter) Au nanoparticles, *J. Chem. Phys.* **112** (2000) pp. 5942-5947.
31. Tas G. and Maris H. J., Electron-diffusion in metals studied by picosecond ultrasonics, *Phys. Rev. B* **49** (1994) pp. 15046-15054.
32. Voisin C., Del Fatti N., Christofilos D. and Vallee F., Time-resolved investigation of the vibrational dynamics of metal nanoparticles, *App. Surf. Sci.* **164** (2000) pp. 131-139.

SUB-5-FS SPECTROSCOPY OF SEVERAL MOLECULAR SYSTEMS

TAKAYOSHI KOBAYASHI

Dept. of Phys., Univ. of Tokyo, Hongo 7-3-1, Bunkyo-ku, Tokyo, 113-0033 JAPAN
kobayashi@phys.s.u-tokyo.ac.jp

Sub-5-fs spectroscopy was applied to the study of dynamics in several molecular systems. New phenomena were found such as the dynamic intensity borrowing (DIB) in the J-aggregates, dynamic mode coupling in the polydiacetylene derivative, and dynamic Duschinsky rotation in cresyl violet molecules. In this paper we concentrate on the J-aggregate of porphyrin. Sub-5-fs spectroscopy reveals a coherent molecular vibration coupled to the Frenkel exciton. The bleaching and induced absorption signals show synchronous oscillations with the frequency of 244cm^{-1}. The coherent oscillation is explained in terms of a modulated transition dipole moment or in other words a dynamic intensity borrowing from the intense B-transition to the weak Q-transition through the ruffling mode with the 244cm^{-1}-frequency.

1 Dynamic intensity borrowing in porphyrin J-aggregates

1.1 Introduction

Molecular J-aggregates have attracted much attention because of their several characteristic features of their spectroscopic properties.[1–3] Intermolecular interaction between transition dipole moments results in the coherent delocalization of excitons over an aggregate.[2] Recently it was found that the aggregate has a hierarchical structure, which is described by a macroaggregate composed of many mesoaggregates.[4] The attractive forces in the latter and former aggregates are due to the van der Waals interaction and static dipole-dipole ($\mu_{st} - \mu_{st}$) interaction. The transition-dipole-transition dipole ($\mu_{tr} - \mu_{tr}$) interaction suffers much from thermal fluctuation because of weaker long-range dependence ($1/r^6$) than the static dipole-static dipole interaction ($1/r^3$). The thermal fluctuation of the order of phonons in the aggregate system with the wavelength λ_{ph} disturbs the $\mu_{tr} - \mu_{tr}$ interaction, but it does not deteriorate the $\mu_{tr} - \mu_{tr}$ interaction. The growth of the mesoaggregate stops at the aggregate size of $r \approx (kT/\mu_{tr}^2)^{1/3}$, where kT is the thermal energy. The mesoaggregates then have a large static dipole moment accumulated, and further aggregation of mesoaggregates forms a macroaggregate, since the static interaction is not much disturbed by the thermal fluctuation up to the size of $r \approx (kT/\mu_{tr}^2)^{1/3}$. It is very important to characterize the above hierarchic structure to discuss the experimental results obtained by coherent interaction with a coherent laser pulse. Hereafter the name ***J-aggregate*** is used as a mesoaggregate.

The excitonic transition of the J-aggregates is characterized by a sharp absorption band and is called *J-band* named after the scientist Jelly[5] who discovered the characteristic phenomena. Because of the concentration of the oscillator strength at the bottom of the band in j-aggregates, the transition energy of the J-band is lower than that of monomers by twice the dipole-dipole interaction energy, J [5,6] Among J-aggregates composed of various aromatic or macrocyclic compounds, tetraphenylporphine tetrasulfonic (TPPS) J-aggregates are of special interest since they are model aggregates of the light-harvesting antenna chlorophyll with a "storage-ring" configuration and of primary charge-separation systems in photosynthesis. Since the first studies of the formation of the J-aggregates independently reported by Fleischer *et al.*[7] and Pasternack *et al.*[8], a number of investigations have been performed using linear and nonlinear electronic spectroscopies which elucidated the optical properties characteristic of the excitonic transition in porphyrin J-aggregates.[9–17] Because of the three-level electronic structure of component monomer molecule, the absorption spectrum of TPPS aggregates shows a relatively weak band in the visible region and a strong peak in the near ultraviolet, denoted as Q- and B-bands, respectively. They form a two-band Frenkel exciton system composed of three states including the ground state. They are expected to have different properties from well-studied two-state systems. A stationary fluorescence of the J-aggregates is known to show no or very small Stokes shift. Although several studies have been made of exciton-phonon coupling,[18–20] few investigations related to the coherent molecular vibration coupled to the exciton in J-aggregates have been reported[21] in real-time domain because the excitons interact only weakly with vibrations and very high time-resolution is required for direct measurement of intramolecular vibrations. Recent progress of ultrashort laser pulses enables us to investigate directly the coherent molecular vibration induced by the electronic transition. Since wavepacket motions in dye molecules and bacteriorhodopsin were reported by Shank and coworkers,[22–24] various materials in condensed phase, including the chlorophyll J-aggregates,[21] have been studied by real-time spectroscopy.[25–31] Recently, we developed a novel optical parametric amplifier (OPA) which generates sub-5-fs visible pulses.[30,32] The parametric amplifier has a noncollinear geometry in a tilted pump pulse-front configuration. The noncollinear optical parametric amplifier (NOPA) has also two sets of chirped mirrors to compensate for the group velocity dispersion (GVD) between the signal and idler pulses before and after the OPA. In the present study, we apply the sub-5-fs real-time spectroscopy to study the exciton-

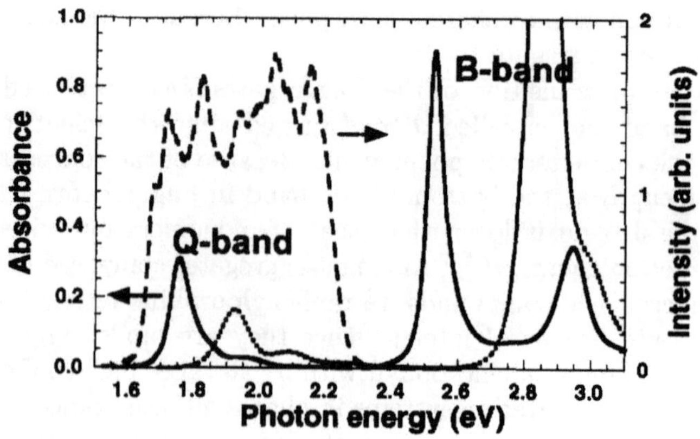

Figure 1. Linear absorption spectra of porphyrin J-aggregates (solid line) and monomer (dotted line). The dashed line shows the laser spectrum.

vibration coupling in the one-dimensional Frenkel exciton system. The coherent molecular vibration is clearly observed in the time dependence of the transmission change. The experimental results are well explained by a following new model; modulated Q-band-transition dipole moment which originates from the oscillator strength transfer from the intense B-band to the weak Q-band by dynamic intensity borrowing (DIB) mechanism through the vibronic interaction.

1.2 Experimental

Tetraphenylporphine tetrasulfonic acid (TPPS: Tokyo Kasei) and polyvinylalcohol (PVA: Kanto Chemical) purchased were used without further purification. After 200-mg PVA was dissolved into 4-ml distilled water at $\sim 100°C$, 20-mg TPPS was added into the hot solution. The concentration of TPPS was 5×10^{-3} mol/l, which is high enough for aggregate formation. A unidirectional oriented film of porphyrin J-aggregates was prepared using a vertical spin-coating technique developed previously by our group.[33] Figure 1 shows linear absorption spectra of porphyrin J-aggregates and monomer. Sharp absorption peaks at 2.53eV (S_2-state; B_x-band or Soret-band) and 1.75eV (S_1-state; Q_x-band) correspond to B- and Q-exciton bands, respectively, composing a quasi-two-band Frenkel-exciton system. Both B_x- and Q_x-band are red-shifted from those of the monomer (2.86 and 1.92eV) by 0.33. The former

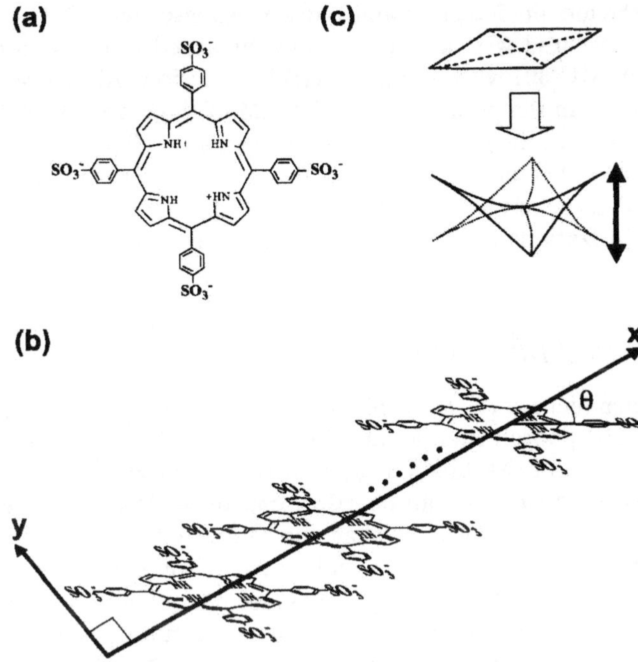

Figure 2. Structures of a porphyrin molecule (a) and J-aggregate (b). The x-axis corresponds to a stacking direction. The porphyrin molecules are making an angle θ with the x-axis. (c) Structural diagram of the ruffling motion of the porphyrin molecule.

and latter correspond to the lowest and second lowest singlet excited state (S_1- and S_2-state), respectively. Fine structures in the absorption bands are their polarization counterparts of the two corresponding bands, namely, the B_y- and Q_y -bands, which have peaks around 2.94 and 1.86eV, respectively. Because of the polarization dependence of the absorption, it can be stated that the x and y components originate from the transition dipole moments which are parallel and perpendicular, respectively, to the aggregate axis defined as a direction connecting the centers of neighboring molecules composing the aggregates. The x and y components are assigned to J-type and H-type atransitioons[11] The structure of the TPPS aggregate, which has a nearly coplanar structure between the phenyl rings and the porphyrin ring, is shown in Fig. 2.[13,14,16] The sulfonic group of one TPPS unit is directly above the diacid charge of the next one, which contributes to the stability of the aggregate.

The experimental setup of the sub-5-fs time-resolved pump-probe system is based on the lock-in detection and is described elsewhere.[28] The probe pulse intensity is measured with a Si photodiode attached to a 30-cm monochoromator (Ritsu, MC-30) with a 3nm spectral resolution. All measurements are performed at room temperature. The FWHM of the cross-correlation trace is about 8fs because of the geometrical smearing due to a finite angle between the pump and probe beams to avoid their interference. The pump pulse energy and photon density are about 14nJ and 5.9×10^{14}photons/cm^2, respectively. Both the polarizations of the pump and probe are parallel to the aggregate axis (x-axis).

1.3 RESULTS & DISCUSSION

The time dependence of the transmittance change in porphyrin J-aggregates at 7 different probe photon energies are shown in Fig. 3 (a). The laser spectrum covers the Q-band of the TPPS J-aggregates, while the B-band is completely outside of the spectrum. It can be safely concluded from the estimated two-photon absorption cross section that the Q-exciton is created by the one-photon resonant transition process and no B-exciton can be produced at the excitation photon flux of 1.2×10^{29}photons/sec$\cdot$cm^2. The predominant feature commonly found in the traces is high-frequency oscillations due to molecular vibration in addition to the underlying slow-dynamics of electronic transition. The oscillation period is 137 ± 4 fs and the oscillations persist for delay times longer than 2ps. As will be discussed in detail, the phase of the transient oscillation is reversed between the bleaching and IA, which is clearly shown in Fig. 4. Because of the sub-10-fs resolution, the phases of the oscillation are precisely determined to be -0.3 ± 0.1 (rad.) at both 1.72 and 1.80eV using a cosine function. The oscillation at $E_{\text{probe}} < 1.77$ eV is described by a plus-cosine function, whereas the oscillation at $E_{\text{probe}} > 1.77$ eV is expressed by a minus-cosine function. The oscillatory signal size is substantially reduced around 1.77eV.

The overall slow-dynamics of the transmittance change originates from the dynamics of the excited states in the J-aggregates by photoexcitation. The positive signal observed over the whole range of delay times for $E_{\text{probe}} <$ 1.77 eV is attributed to a bleaching between the Q-band and the ground state. At probe energies higher than 1.8eV, the signal becomes negative due to an induced absorption (IA). Additional complex oscillating features observed both at a negative delay and around the time origin are due to the pump-perturbed free induction decay,[34] cross-phase modulation,[35] and the interference left over between the pump and probe. Measured decay curves of the transmittance change are fitted by the sum of two exponential-decay functions, which

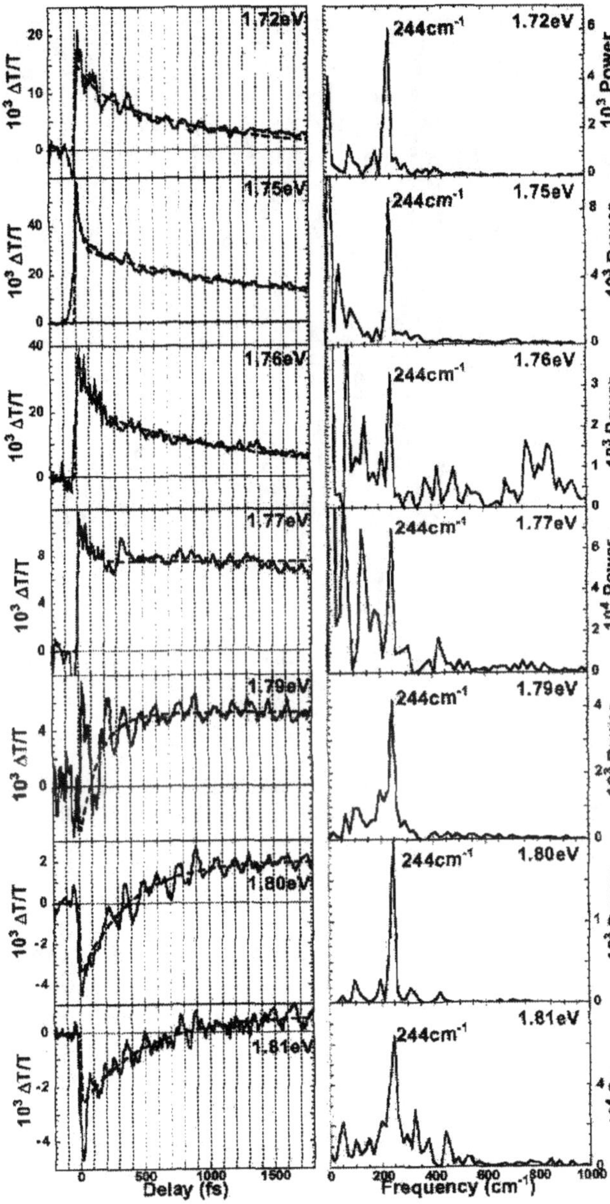

Figure 3. (a) Time-dependence of the transmittance change in porphyrin J-aggregates at 7 photon energies marked on the right. (b) Fourier-power spectra of real-time spectra integrated from 100fs to 1.8ps.

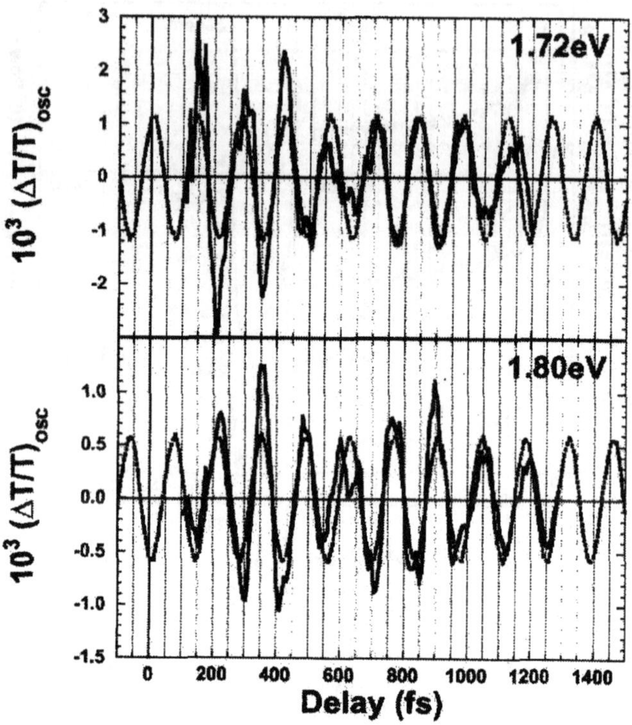

Figure 4. The oscillating component at two probe-photon energies (solid line) and results of the fitting using a cosine function (dashed line). Note that the phase is reversed with each other.

is convoluted with the instrument (cross-correlation) function. Since nearly chirp-free pulses are used for both pump and probe, the same function can be used for convolution with the response function over the whole range of the probe-photon energy. At the peak of the Q-band ($E_{\text{probe}} < 1.75\,\text{eV}$, the fast and slow decay components are determined to be $45 \pm 4\,\text{fs}$ and $1.5 \pm 0.1\,\text{ps}$, respectively.

The time dependent transmittance change is composed of the electronic and molecular vibrational contributions. The former and the latter correspond to the slow-dynamics component and the oscillating component, respectively. It is noteworthy that the oscillating component is observed only around the Q-band. In this subsection, we shall focus on the slow dynamics component.

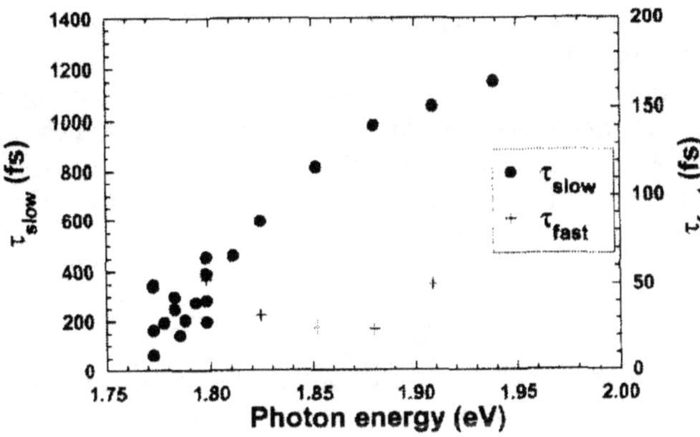

Figure 5. Probe-photon energy dependence of the slow (τ_{slow}) and fast (τ_{fast}) decay time constants in transient signals.

The signal is composed of three components, namely the bleaching due to $|1, S_1\rangle \rightarrow |G\rangle$, IA due to the transition to multi-exciton states (MES; $|n+1, S_1\rangle \leftarrow |n, S_1\rangle (n = 1, 2, \ldots)$), and IA due to the transition to higher excited exciton states ($|1, S_n\rangle \leftarrow |1, S_1\rangle (n = 2, 3, \ldots)$). Here n in $|n, S_1\rangle$ denotes the number of the S_1-excitons (Q-excitons) in a single mesoaggregate,[36–38] and $|1, S_1\rangle$ indicates one-exciton state of S_n-exciton, namely the excited state of an exciton originated from an S_n-excited state in a TPPS molecule. Although both of the induced emission ($|1, S_1\rangle \rightarrow |G\rangle$) and ground-state depopulation ($|1, S_1\rangle \leftarrow |G\rangle$) can contribute to the bleaching signal, the former is considered to be a dominant effect. This is because the oscillatory signal due to the latter, which corresponds to the impulsive stimulated Raman scattering, must be described by a sine function.

The positive difference transmittance $\Delta T/T$ around the Q-band is due to bleaching, while the negative one has two origins. First one is the induced absorption (IA) due to the transition to MES observed at the higher energy than the exciton peak because of the Pauli exclusion principle confirmed previously by using 100-fs-pump pulses.[39] Second one is the transition to the excited exciton, which is the deloclized excited states compose of the linear combination of the higher excited states in monomer. It is also supported by the fact that the absorption spectra of S_n-exciton states ($n \gg 1$) are expected to be broad and the oscillation is smeared.

Here at this point we shall discuss the origin of the slow and fast decay components. The lifetime of the Q-exciton is reported to be ~100 ps.[14,16,17]

Hence, both of the slow and fast decay components cannot be attributable to the Q-exciton lifetime. In our previous study, the decay time constant of the 1-ps component increases with the excitation power and the signal amplitude was super-linearly dependent on the pump power.[17] From these results, it was attributed to the S_2-exciton lifetime photogenerated by two-photon absorption. In our more recent report, however, it was found that the spectrum of the 1-ps component in IA shifted to the blue side as the delay time increased in the pump-probe experiment using a 100-fs-pump pulse.[39] Similar behavior is also observed in the IA ($|1, S_n\rangle \leftarrow |1, S_1\rangle (n = 2, 3, \ldots)$) signal in the present experiment. Figure 5 shows the probe-photon energy dependence of the decay time constants of the slow and fast decay components. The slow decay time constant of IA increases up to 1.2ps with probe-photon energy, which means blue-shifted IA. These results indicate that the slow decay component cannot be attributed to the two-photon generated S_2-exciton. This component is, therefore, ascribed to the non-thermal S_1-excitons (Q-excitons), which thermalize with time constant of about 1ps. In fact, the slow decay time constant increases slightly with the pump power[17], which also supports this assignment. On the other hand, the decay time of the fast decay component is insensitive to the probe-photon energy and the time constant is about 100fs. This component is attributed to the fast relaxation process from higher MESs ($|n + 1, S_1\rangle \leftarrow |n, S_1 (n = 1, 2, \ldots)$), which is expected to be faster than the recovery of the bleaching.

In this subsection, the bleaching and IA ($|n+1, S_1\rangle \leftarrow |n, S_1 (n = 1, 2, \ldots)$) signal with transient oscillation will be considered. For the discussion of the molecular vibration, the oscillating component is separated from the transient curves by using low-pass filter or smoothing. The fitted result of the slow component is subtracted from the transient signal and only the oscillating component can be extracted. Figure 3 (b) shows the Fourier-power spectra calculated from the oscillating components. The resolution is determined to be 16cm^{-1} (FWHM) by the width of the delay-time window (1.7ps) in the Fourier transformation analysis. The Fourier spectra clearly show an intense peak at 244 ± 8cm^{-1}, which corresponds to an 137 ± 4fs-oscillation period, in the whole spectral region of the Q-band. The intensity of the Fourier spectrum at 244 ± 8cm^{-1} is substantially reduced around 1.77eV, where the sign of the transmittance-difference amplitude is changed.

A conventional resonant Raman spectrum was measured for the sample used for sub-5-fs spectroscopy under the excitation of the B-band; the result is shown in Fig. 6. The peak at 244 cm^{-1} is clearly observed, which corresponds to the frequency obtained by the sub-5-fs real-time spectroscopy. The frequency of this peak is also detected at 245 cm^{-1} or at 241 cm^{-1} in

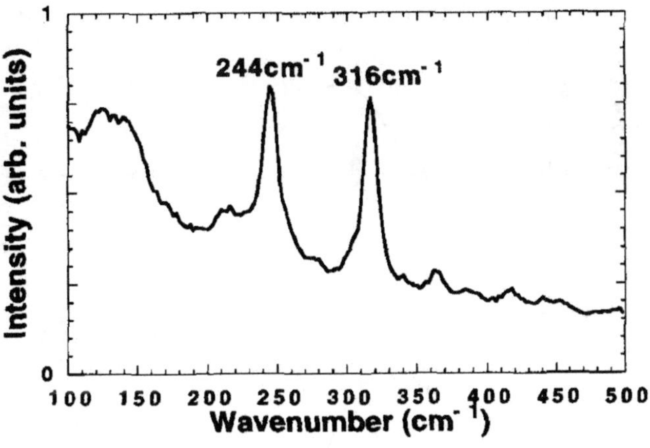

Figure 6. Resonance Raman spectrum of porphyrin J-aggregates. The excitation wavelength corresponds to 488nm.

the Raman spectrum in literature. This peak is assigned to a ruffling mode, which is more than 30 times enhanced by aggregation[14]. The reasoning of such n enhancement is puzzling. The changes in the molecular configuration associated with the ruffling mode are expected to strongly influence the dipole-dipole interaction. The ruffling motion is one of out-of-plane modes, which is defined by four alternate up- or down-positioned nitrogen atoms with respect to the average plane of the macrocycle (Fig. 2(c)).[14] The $316\,cm^{-1}$-doming mode, which is another aggregation-enhanced intense Raman mode in the stationary Raman spectrum, is not observed in the real-time spectra. Note that both modes are observed in the Raman scattering experiment under resonant excitation of the B-band. On the other hand, our experiment of the sub-5-fs real-time spectroscopy is performed under resonant excitation of the Q-band. The intensity ratio between the $244\,cm^{-1}$-ruffling mode and the $316\,cm^{-1}$-doming mode may depend on the excitation photon energy and the detection method. It is also of interest to study experimentally the stationary resonance Raman spectrum under the Q-band resonance condition.

The phase of the oscillation is evaluated by the complex Fourier transformation. Figure 7 shows the probe-photon energy dependence of the phase and amplitude of the 244 cm−−1-ruffling mode. As is mentioned before the oscillations in Fig. 4 are described by a plus-cosine or a minus-cosine function at $E_{probe} < 1.77\,eV$ or $E_{pump} > 1.77\,eV$, respectively, indicating that the sign of the amplitude is reversed around 1.77eV. Around 1.77eV, the transient

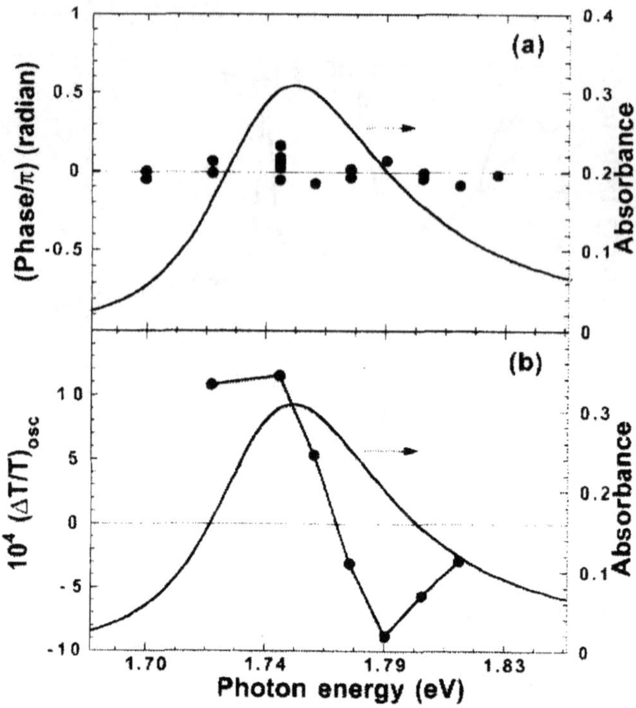

Figure 7. Probe-photon energy dependence of the phase (a) and amplitude (b) of the oscillation in the transient signals.

signal itself changes from the bleaching (positive signal) to the IA (negative signal) and the oscillation amplitude is extremely decreased. The plus-cosine oscillation in the bleaching and the minus-cosine oscillation in the IA indicate that the bleaching and IA increase (or decrease) synchronously (in-phase oscillation).

Several reports based on numerical calculation on the transmittance change in medium-size molecules photoexcited by an ultrashort pulse have attributed the transient oscillation to the dynamics of localized vibrational wavepackets which move on the electronic ground or excited potential energy surfaces.[22–24] In these cases, the phase of the transient oscillation is expected to be shifted continuously by changing the probe-photon energy, because the transition energy depends on the nuclear position, and different photon energies of the probe pulse are absorbed at different nuclear positions. Based on the conventional wavepacket motion on a harmonic potential of the vibronic excited state, we estimate the probe-photon energy dependence of the phase of the oscillation. The phase of the transient oscillation must show π-phase difference between $E_{\mathrm{probe}} = E_{\mathrm{abs}} = 1.75\,\mathrm{eV}$ and $E_{\mathrm{probe}} = 2E_{\mathrm{Fluo}} - E_{\mathrm{Abs}} = 1.71\,\mathrm{eV}$, where E_{Abs} and E_{Fluo} correspond to the peak photon energies of the absorption and fluorescence spectra, respectively. Although it is well resolved in the present experiment (~0.01eV), the π-phase difference is not observed. In addition, it is well known that the J-band associated with excitonic transition in J-aggregates interacts only weakly with molecular vibrations, which is the manifestation of the small Stokes shift in the stationary fluorescence spectrum. These results indicate that the vibronic coupling is much smaller than in the typical medium-size molecules such as Nile blue[22,23]. Because of these reasons, the oscillation does not originate from conventional vibronic coupling between the excitonic transition and relevant molecular vibrations represented by the wavepacket motion. Since the signal intensities of the bleaching and IA vary synchronously, and the oscillation is observed only around the Q-band, we propose a new mechanism which also induces the transient oscillation; the modulated transition dipole moment of the Q-transition in the TPPS molecules. The modulation frequency corresponds to the $244\,\mathrm{cm}^{-1}$-ruffling-mode vibration. Since the transition dipole moment depends on the vibrational coordinate this model is beyond the Condon approximation. The increase (or decrease) in the transition dipole moment is reflected by the increase (or decrease) in the signal intensity both of the bleaching and IA. The synchronous oscillations for the bleaching and IA are, therefore, interpreted by the modulation of the transition dipole moment.

This section focuses on the origin of the modulation in the Q-transition. Although the Q-transition is originally forbidden because of the symmetry of the electronic structure, the transition becomes allowed by mixing with the B-transition due to the configuration interaction and vibronic coupling.[9] Therefore, the modulation of the amount of intensity borrowing from the B-band may account for the modulation of the Q-transition. The additional mixing between the Q- and B-transition is induced by the pump pulse through the $244\,\mathrm{cm}^{-1}$-ruffling mode, and that the degree of the mixing is enhanced or reduced depending on the phase of the vibrational motion of the TPPS molecules which compose a J-aggregate. This is an outline of the dynamic intensity borrowing (DIB). DIB results in the transfer of the oscillator strength from the intense B-transition to the weak Q-transition.

The modulation depth of the transition dipole moment is expressed as $\delta\mu/\mu$, in which μ represents the Q-transition dipole moment of a TPPS molecule, and $\delta\mu$ denotes the change of μ induced by the molecular vibration. It must be noticed that the modulation of μ also changes the intermolecular dipole-dipole interaction, J, which is given by

$$J = \mu^2 \left(1 - 3\cos^2\theta\right) / \hbar a^3 \tag{1}$$

All transition dipoles in the TPPS molecules composing a J-aggregate are assumed to be parallel and making an angle θ with the aggregate axis. The molecules are equidistant with lattice constant a. Since the transition energy from the ground state to the first excited state of the aggregate is originally red-shifted by $2J$ in comparison with that of the monomer, the modulation of J gives rise to a peak shift of the J-band in the absorption spectrum. The amount of $\delta\mu$ and $\delta(2J)$ is estimated as follows. The induced absorption spectrum $\Delta A(\omega)$ is assumed to be modulated as $\Delta A'(\omega)$ by $\delta\mu$ and $\delta(2J)$, which is given by

$$\Delta A'(\omega) = \Delta A(\omega + \delta\omega(t))(1 + \delta u(t)), \tag{2}$$

where

$$\delta u(t) = \delta(\Delta A)/\Delta A \cos(\Omega t) = \delta(\mu^2)/\mu^2 \cos(\Omega t) \tag{3}$$

$$\delta\omega = \delta\omega \cos(\Omega t) = \delta(2J)\cos(\Omega t) \tag{4}$$

Here δu and $\delta\omega$ correspond to the amplitude modulation and the spectral shift of the signal, respectively. The modulation frequency, Ω, corresponds to the molecular vibration (244 cm^{-1}. Both δu and $\delta\omega$ are assumed to be independent of ω. Note that δu and $\delta\omega$ correspond $\delta\mu^2/\mu^2$ and $\delta(2J)$, respectively. Equation (2) is approximately expressed as

$$\Delta A'(\omega) \cong \Delta A(\omega) + \left(\delta A \cdot \Delta A(\omega) + \delta\omega \cdot \frac{d\Delta A(\omega)}{d\omega}\cos(\Omega t)\right) \tag{5}$$

$$\equiv \Delta A(\omega) + \Delta A_{\mathrm{osc}}(\omega)\cos(\Omega t). \tag{6}$$

Since the experimentally obtained transmittance change is composed of the slow-dynamics component ($\Delta A(\omega)$) and the oscillating component ($\Delta A_{\mathrm{osc}}(\omega)$), the former and the latter correspond to the first and second terms, respectively in Eq.(5). In order to evaluate the slow-dynamics component, we performed a multi-channel pump-probe measurement. The continuous spectrum of $\Delta A(\omega)$ at 20-fs-delay time is depicted in Fig. 8, which is measured by a polychromator and a CCD camera with sub-10-fs resolution. The signal amplitude of the continuous spectrum in Fig. 8 is scaled down by a factor of 0.59 because of the different concentration of the J-aggregates. The slow-dynamics component originated from the electronic contribution, $\Delta A(\omega)$, can be approximately evaluated using the continuous induced absorption spectra.

Assuming that $\Delta T/T(\omega)$is proportional to $\Delta A(\omega)$ in the weak signal approximation, the oscillating component is fitted by both of $\Delta A(\omega)$ and $d\Delta A(\omega)/d\omega$. As a result, we obtained $\delta u = (2.7\pm0.6)\times10^{-2}$ and $\delta\omega = (5.6\pm1.3)\times10^{-4}$(eV). Figure 8 shows the results of the fitting for $\Delta A_{\rm osc}(\omega)$ and $\Delta A(\omega)$. The value of $\delta\mu/\mu$ is derived from δu and has the value

$$\frac{\delta\mu}{\mu} = \frac{\delta(\mu^2)}{2\mu^2} = \delta u/2 = 1.4 \pm 0.3\% \tag{7}$$

On the other hand, the ratio of the frequency shift $\delta\omega$ (=$\delta(2J)$) to $2J$ is determined to be $\delta\omega/2J = 0.3 \pm 0.1\%$ assuming $2J \sim 170$meV from the stationary absorption spectrum. From eq. (1), the modulation of $2J$ can be derived as follows,

$$\frac{\delta\omega}{2J} \equiv \frac{\delta(2J)}{2J} = \frac{2}{\mu}\delta\mu - \frac{3\sin 2\theta}{3\cos^2\theta - 1}\delta\theta - \frac{3}{a}\delta a. \tag{8}$$

Since $\delta\mu/\mu$ and $\delta(2J)/2J$ are estimated to 1.4% and 0.3%, respectively, not only the modulation due to $\delta\mu$ but also the modulations due to $\delta\theta$ and δa should also be taken into account. In this model, it is assumed that all μ and J are effectively modulated. A simulation that takes into account a local disorder effect is now under progress.

1.4 Dynamic intensity borrowing

In this subsection, the amount of the DIB is estimated. As was discussed, the weak oscillator strength of the Q-band is caused by the mixing with the B-band due to electron-electron correlation (configuration interaction). We consider that the modulation of the transition dipole moment originates from DIB due to vibronic interaction between the Q- and B-transitions, which is induced by the pump pulse through the molecular vibration. Based on the vibronic coupling theory, following calculation is performed to estimate the amount of the oscillator strength transfer from the B-band to the Q-band. Although the symmetry is important for the vibronic coupling mechanism, the symmetry of the 244 cm^{-1} ruffling mode is not clearly defined. The perturbed transition dipole moment between the Q-band and the ground state, $\mu' \equiv \mu+\delta\mu$, is expressed as follows in the first order using vibronic coupling theory;

$$\mu' = \mu + \delta\mu = \mu + \frac{H_{\rm vib}}{E_B^0 - E_Q^0}\mu_B. \tag{9}$$

Here μ_B, $E_{\rm B}^0$, $E_{\rm Q}^0$, and $H_{\rm vib}$ denote the transition dipole moment due to the B-transition, the transition energies due to the B- and Q-transitions without the

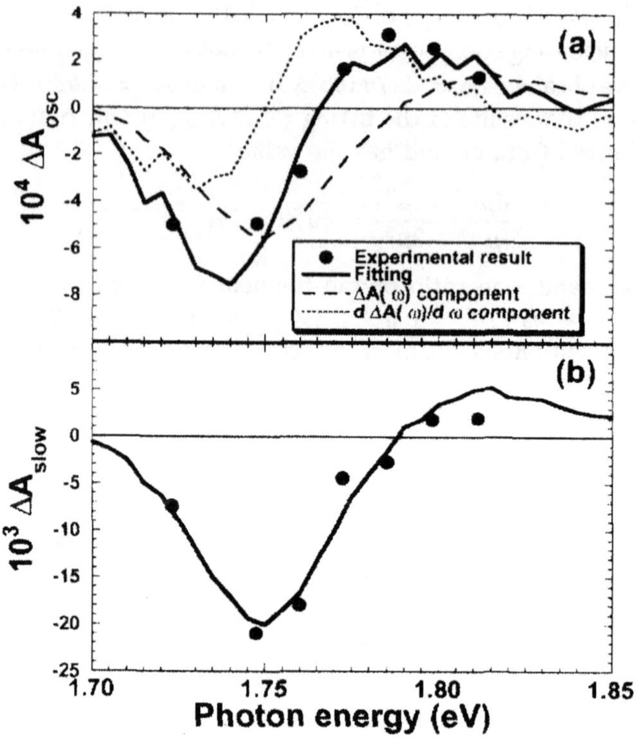

Figure 8. The results of the fitting for the oscillating component, $\Delta A_{\rm osc}(\omega)$ (a) and slow-dynamics component, $\Delta A(\omega)$ (b) using Eq.(5). The continuous spectrum, $\Delta A(\omega)$ indicates the induced absorption spectrum detected by the multi-channel pump-probe experiment. The oscillating component is fitted by $\Delta A(\omega)$ and $d\Delta A(\omega)/d\omega$.

perturbation by the ruffling mode, and the perturbation to the Hamiltonian, respectively. The normalized value of the modulation of μ is given by

$$\frac{\delta\mu}{\mu} = \frac{H_{\rm vib}}{E_B^0 - E_Q^0}\frac{\mu_B}{\mu} \tag{10}$$

$E_B^0 - E_Q^0 \sim 0.94\,\text{eV}$ and $\mu_B/\mu \sim 2.6$ are estimated from the stationary absorption spectrum of the monomer. The amount of $\delta\mu/\mu$ is calculated in the previous subsection as $\delta\mu/\mu = 1.4\%$. By substituting these coefficients, $H_{\rm vib} = 5.4\text{meV} = 41\text{cm}^{-1}$ is obtained, which is much less than the vibration frequency, $244\,\text{cm}^{-1}$. The modulated transition dipole moment is also

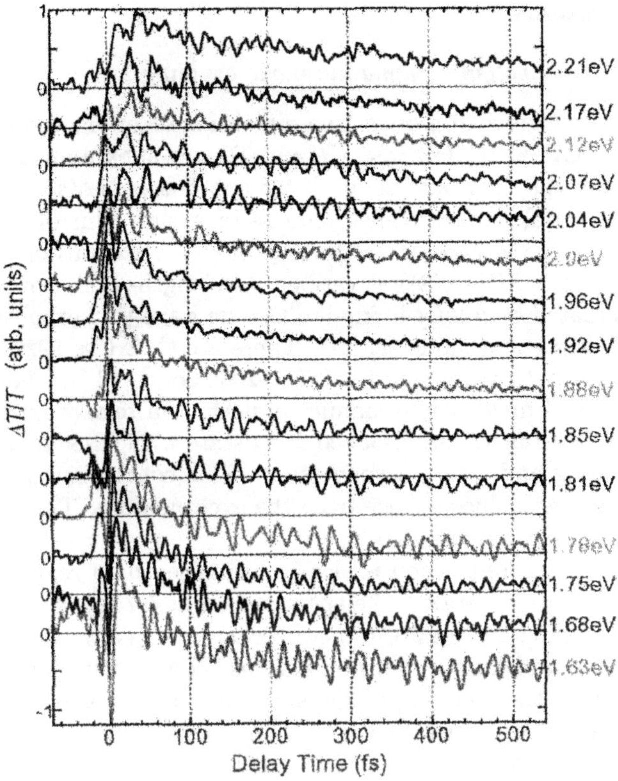

Figure 9. Time dependence of the normalized transmittance change at various photon energies. The intensties are normalized at their peaks.

estimated as $\mu' = \mu + 5.4 \times 10^{-3}\mu_B$. The similar intensity borrowing mechanism has also been discussed in the stationary resonance Raman spectrum of metalloporphyrins.[40–42] The resonance Raman spectra under the B- and Q-band excitations show dominant contributions of the Albrecht A-term and the B-term scattering, respectively. The dominant contribution of the B-term under the resonant Q-excitation is also explained by the intensity borrowing mechanism from the B-band to the Q-band.

2 Other systems

2.1 A conjugated polymer: dynamic mode coupling

The ultrafast dynamics of a quasi-one-dimensional conjugated polymer polydiacetylene $(=CR-C\equiv C-CR'=)_n$ (R and R' are substituents) was studied by pump-probe spectroscopy using visible sub-5-fs pulses(4). The spectrally resolved differential transmittance change in a thin film of a ladder polymer poly (5,7,17,19-tetracosatetraynylene bis(N-butoxycarbonylmethyl) carbamate) (PDA-4BCMU4A(8)) at various wavelengths is shown in Fig. 9. The signal is dominated by a long-living multimode wave-packet motion including three intense stretching mode signals. There are C–C ($\approx$ 1220 cm^{-1}, 27 fs), C=C ($\approx$ 1450 cm^{-1}, 23 fs), and C≡C ($\approx$ 2080 cm^{-1}, 16 fs). Both features of the electronic and molecular dynamics indicate the relaxation from a 1^1B_u-free exciton to a geometrically relaxed 2^1A_g state with 60-80 femtoseconds via vibronic coupling and internal conversion. This conclusion can be obtained from extremely weak fluorescence from the exciton state. The quantum yield is estimated to be lower than 10^{-5}.

The geometrical relaxation is clearly visualized by a real-time frequency analysis using a spectrogram. The spectrogram at 1.75 eV evidently exhibits the highly vibronic non-equilibrium state characterized by the instantaneous mode frequency and amplitude modulations. The feature of modulation is well explained by the coupling of the stretching modes via the bending motion in the geometrically relaxed butatriene-like backbone. This is the first observation of such diabatic molecular motions.

2.2 A dye molecule: dynamic Duschinsky rotation

In order to clarify the mechanism of molecular vibrational mode coupling discussed in the conjugated polymers in the previous section, we studied more simple system than conjugated polymers. A dye molecular cresyl violet (CV) was selected, since from the Raman spectrum in literature(10), it is shown that it has only one intense Raman signal.

Bleaching is observed throughout the probe wavelength region in the normalized transient differential transmittance as a function of delay time of the probe pulse after the pump pulse of CV doped in poly(vinyl alcohol) (PVA) against the pump-probe delay time.. The observed signal intensity does not decrease detectably within the probed delay time range of 1 ps because the lifetime of the excited singlet state, S_1, is reported to be as long as 3.2 ± 0.1 ns. The vibration of transmittance change of 57-fs period can be clearly seen for all the probe wavelengths between 600 and 640 nm. The oscillation period is corresponding to the ring-breathing mode and the period is corresponding to the frequency of 590 cm^{-1}observed in the Raman spectrum of CV. The Fourier transformation of the time-dependent transmittance change has a peak at this frequency. These vibrational signals are mostly due to the wave-packet motion of the excited state. The motion of the ground-state wave-packet is much more restricted than that of the excited state, because it is initially weakly displaced from the ground-state minimum and has a small initial momentum created by 6-fs-pulse excitation. In the present experiment the excitation pulse is shorter than 5 fs. Then the motion of the ground-state wave-packet is even more limited; and hence it can be concluded to be negligible and the vibrational structure can be attained to the wave-packet motion on the excited-state potential surface.

In order to investigate the evolution of the instantaneous frequency of the ring-breathing mode, we analyzed the transient differential transmittance using the spectrogram. It clearly shows that the amplitude and frequency of the ring-breathing mode are modulated. The modulation period and corresponding frequency are estimated to be about 370 fs and 91 cm^{-1}, respectively, by Fourier transformation of the time-dependent amplitude and instantaneous frequency. The modulations of the amplitude and phase are out of phase by about π. The modulation depths of the amplitude and frequency are about 13 and 4%, respectively. The same features are also observed at a probe wavelength of 620 nm. Such time dependent modulations of the ring-breathing mode parameters and their phase relationship can never be obtained by frequency-domain spectroscopy, and it will be discussed in detail in this paper. These modulations indicate that the ring-breathing mode is coupled to another vibrational mode. At other probe wavelengths, the transmittance change due to the ring-breathing mode is so weak that these modulations of such a small transmittance is difficult to be detected by the spectrograms. From the spectrograms we can calculate the potential curves of the excited and the ground states. Such a rotation of the potential curve associated with the electronic transition from the ground to the excited states is known as the Duschinsky effect. This effect was observed by the difference between the frequency of the ground state and that of the excited state. It is to be emphasize that this is the first dynamical observation of the Duschinsky effect by sub-5-fs real-time spectroscopy.

3 Conclusion

Several new phenomena were found by sub-5-fs pump-probe experiments. They are the dynamic intensity borrowing (DIB) in the J-aggregates, dynamic mode coupling in the polydiacetylene derivative, and dynamic Duschinsky rotation in cresyl violet molecules. The first one is mainly discussed in detail. The oscillation in the pump-probe time trace of porphyrin J-aggregates originates from the molecular vibration coupled to the Frenkel exciton and is assigned to the $244\,\mathrm{cm}^{-1}$-ruffling mode. It is interpreted by the modulated transition dipole moment of the Q-band, which is due to the DIB mechanism from the intense B-transition to the weak Q-transition through the molecular vibration. The amount of the modulation is estimated to be $\delta\mu/\mu = 1.4\%$. The DIB is a new mechanism of ultrafast nonlinearity and is observed for the first time in the present study.

Acknowledgements

The author would like to thank Drs. H. Kano, A. Shirakawa, and T. Fuji for their collaboration. He thanks H. F. Hofmann for enlightening discussions. Thanks are also due to Dr. Y. Takasu and Prof. I. Nishio of Aoyama Gauin University for Raman scattering measurement. The author appreciates Prof. Y. Toyozawa of Chuo University and Dr. S. Abe of Eletrotechnical Laboratory for stimulating discussion. The work is partly supported by Research for the Future of Japan Society for the Promotion of Science(JSPS-RFTF-97P-00101).

References

1. *J-aggregates* ; World Scientific, ed. by Kobayashi, T., 1996.
2. Mukamel, S.; *Principles of Nonlinear Optical Spectroscopy* ; Oxford, 1995.
3. Davydov, A.S.; *Theory of Molecular Excitons* ; Plenum, 1971.
4. Kobayashi,T. *Mol. Cryst. Liq. Cryst.* **1998**, 314, 1-11.
5. Jelly, E. E. *Nature* **1936**, 138, 1009-1010.
6. Sheibe, G. *Angew. Chem.* **1936**, 49, 563-564.
7. Fleisher, E. B.; Palmer, J. M.; Srivastava, T. S.; Chatterjee, A. *J. Am. Chem. Soc.* **1971**, 93, 3162-3167.
8. Pasternack, R. F.; Huber, P. R.; Boyd, P.; Engasser, G.; Francesconi, L.; Gibbs, E.; Fasella, P.; Venturo, G. C.; Hinds, L. de C. *J. Am. Chem. Soc.* **1972**, 94, 4511-4517.
9. Dolphin, D., Ed.; *The Porphyrins, volume III, Physical Chemistry, Part A* ; Academic Press, 1979.
10. Barber, D.; Fretag-Beeston, R. A.; Whitten, D.G. *J. Phys. Chem.* **1991**, 95, 4074-4086.

11. Ohno, O.; Kaizu, Y.; Kobayashi, H. *J. Chem. Phys.* **1993**, 99, 4128-4139.
12. Shick, G. A.; O'Grady, M. R.; Tiwari, R.K. *J. Chem. Phys.* **1993**, 97, 1339-1345.
13. Ribo, J. M.; Crusats, J.; Farrera, J. A.; Valero, M. L. *J. Chem. Soc. Chem. Commun.* 1994, 681-682.
14. Akins, D.L.; Ozcelik, S.; Zhu, H. R.; Guo, C. *J. Phys. Chem.* **1996**, 100, 14390-14396.
15. Chan, I. Y.; Hallock, A. J. *J. Chem. Phys.* **1997**, 107, 9297-9301.
16. Maiti, N. C.; Mazumdar, S.; Periasamy, N. J. *J. Phys. Chem.* **1998**, 102, 1528-1538.
17. Misawa, K.; Kobayashi, T. *J. Chem. Phys.* **1999**, 110, 5844-5850.
18. Scherer, P. O. J.; Knapp, E. W.; Fischer, S. F. *Chem. Phys. Lett.* **1984**, 106, 191-196.
19. Knapp, E. W.; Scherer, P. O. J.; Fischer, S. F. *Chem. Phys. Lett.* **1984**, 111, 481-486.
20. Spano, F. C.; Kuklinski, J. R.; Mukamel, S. *Phys. Rev. Lett.* **1990**, 65, 211-214.
21. Chachisvilis, M.; Pullerits, T.; Jones, M. R.; Hunter, C. N.; Sundstrom, V. *Chem. Phys. Lett.* **1994**, 224, 345-351.
22. Fragnito, H. L.; Bigot, J. Y.; Becker, P. C.; Shank, C. V. *Chem. Phys. Lett.* **1989**, 160, 101-104.
23. Pollard, W. T.; Fragnito, H. L.; Bigot, J. Y.; Shank, C. V.; Mathies, R. A. *Chem. Phys. Lett.* **1990**, 168, 239-245.
24. Pollard, W. T.; Dexheimer, S. L.; Wang, Q.; Peteanu, L. A.; Shank, C. V.; Mathies, R. A. *J. Phys. Chem.* **1992**, 96, 6147-6158.
25. Zhu, L. Y.; Sage, J. T.; Champion, P. M. *Science* **1994**, 266, 629-632.
26. Yeh, A. T.; Cerullo, G.; Banin, U.; Mews, A.; Alivisatos, A. P.; Shank, C. V. *Phys. Rev. B.* **1999**, 59, 4973-4979.
27. Cerullo, G.; Lanzani, G.; Muccini, M.; Taliani, C.; Silvestri, S. D. *Phys. Rev. Lett.* **1999**, 83, 231-234.
28. Kobayashi, T; Shirakawa, A. *Chem. Phys. Lett.* **2000**, 321, 385-393.
29. Takeuchi, S.; Tahara, T. *Chem. Phys. Lett.* **2000**, 326, 430-438.
30. Kobayashi, T; Shirakawa, A. *Appl. Phys. B* **2000**, 70, S239-S246.
31. Fuji, T.; Saito, T; Kobayashi, T *Chem. Phys. Lett.*, accepted.
32. Shirakawa, A.; Sakane, I.; Takasaka, M.; Kobayashi, T. *Appl. Phys. Lett.* **1999**, 74, 2268-2270.
33. Misawa, K.; Ono, H.; Minoshima, K.; Kobayashi, T. *Appl. Phys. Lett.* **1993**, 63, 577.
34. Brito-Cruz, H.; Gordon, J. P.; Becker, P. C.; Fork, R. L.; Shank, C. V.

IEEE J. Quantum Electron. **1988**, 24, 261-266.
35. Tokunaga, E.; Terasaki, A.; Kobayashi, T. *J. Opt. Soc. Am. B* **1995**, 12, 753-771.
36. Spano, F. C. *Phys. Rev. Lett.* **1991**, 67, 3424-3427.
37. Knoester, J. *Phys. Rev. A* **1993**, 47, 2083-2098.
38. Minoshima, K.; Taiji, M.; Misawa, K.; Kobayashi, T. *Chem. Phys. Lett.* **1994**, 218, 67-72.
39. Kano, H.; Kobayashi, T. *"Technical Digest of Quantum Electronics and Laser Science Conference 99"*, **1999**, p10.
40. Li, X. -Y.; Czernuszewicz, R. S.; Kincaid, J. R.; Spiro, T. G. *J. Am. Chem. Soc.* **1989**, 111, 7012-7023.
41. Czernuszewicz, R. S.; Li, X. -Y.; Spiro, T. G. *J. Am. Chem. Soc.* **1989**, 111, 7024-7031.
42. Schweitzer-Stenner, R.; Stichternath, A.; Dreybrodt, W; Jentzen, W.; Song, X. –Z.; Shelnutt, J. A.; Nielsen, O. F.; Medforth, C. J.; Smith, K. M. *J. Chem. Phys.* **1997**, 107, 1794-1815.

THE COMPLEX PUZZLE OF OPTICAL FEMTO-DYNAMICS IN POLY(P-PHENYLENEVINYLENE) - HIDDEN COHERENCES AND ENERGY-DISPERSIVE POPULATION TRANSFER

J. SPERLING, F. MILOTA, A. TORTSCHANOFF, CH. WARMUTH, AND H. F. KAUFFMANN

Institut für Physikalische Chemie, Universität Wien, A-1090 Wien, Austria

Fs-excitation tuning into the density-of-states (DOS) and thus typical, initial site-energy memories have been demonstrated for energy-dispersive, incoherent excitation transfer among the segmental sites of the conjugated polymer poly(p-phenylenevinylene). Low-energy excitation coherence in the ultimate regime of non-coupling DOS-bottom-states allows the preparation of discrete, few-component, electronic wavepackets. Their oscillatory relaxation has been probed, for the first time, by fs-luminescence interferometry. Intersite polarization interference may account for the fast beating (period ~100 fs) in the presence of strongly damping, inhomogeneous site dephasing

1 Introduction

The use of femtosecond lasers has strongly impacted numerous efforts toward revealing the pathways of generation and relaxation of optical excitations in conjugated main-chain polymers, which have been a topic of vivid debate over the last decade. The answers to this fundamental questions have enormous impact on the research and development of novel, molecular materials in optoelectronic and photonic applications. Conjugated polymers are polychromophoric systems, and the predominant photophysical events are controlled by intersite processes of electronic coupling and nuclear motion in typical many-body interactions, so that the macroscopic optical response is a complex convolution of microscopic coherent/incoherent evolutions.

A powerful model, more realistic than the idealized 1-D semiconductor approach[1], and established by numerous results based on site-selective fluorescence measurements[2], electroabsorption experiments[3] and stimulated emission pump-probe studies[4] rests upon a molecular approach. The molecular model, successfully applied, in particular, to poly(p-phenylenevinylene) (PPV) and its substitutional derivatives, treats the (rigid) PPV-conjugated chain as a collection of segmental sub-units brought about by a distribution of angular breaks, due to statistical, conformational disorder along the chain. The disrupture of the chain through breaks leads to a distribution of segments in length and self-energies. Thus, typically diagonal segmental site disorder causes an initial optical excitation to become site-oriented, with strong electron-hole correlation, presumably dressed by some structural deformation, but generally subject to weak electron-phonon coupling, only. Upon broad-band excitation through a fs-laser pulse, the dispersion of segments gives rise to a density-of-states (DOS) of energetically distributed, localized S_1-levels and, for time-scales longer than the inverse of global dephasing, to uncorrelated initial populations. In this time-regime ($t > 300$ fs), population relaxation is driven by a variation of space- and energy-dependent site-to-site, Förster-type couplings and the overall process is best described by incoherent, intersegmental population transfer steered by a pronounced energy-space-time relaxation toward the low-energy tail/bottom states of the DOS. The prominent features of this non-Markovian dynamics have been probed by previous[5] and more recent luminescence up-conversion measurements[6].

In this work, we report a quite special aspect of DOS-excitation by addressing the delicate balance between site population/energy transfer, on the one hand, and macroscopic site polarization/inter-site beating, on the other. Both the incoherent and (the quite novel) coherent process depend on the excitation energy and, therefore, show a pronounced site-energy memory effect in the subsequent fs-relaxation.

2 Experimental, Low-Temperature Spectra

To track the evolution of energy-selective optical excitations in the incoherent fs-regime, ultrafast luminescence experiments have been performed employing the luminescence-up-conversion technique. The experiments have been done with a commercial Ti:Sapphire oscillator and an amplifier system (Mira/RegA, Coherent). Using an OPA (Coherent), the central wavelengths of excitation pulses could be tuned across the absorption edge. The luminescence of the sample (thin solid film of PPV spin cast from solution) has been collected and focused, together with the gate pulse, on a BBO-crystal for sum frequency generation and detected with a monochromator and photomultipier; details of the setup are given elsewhere[7]. The low-phonon temperature experiments have been performed at T~1.4 K (in a liquid helium cryogenic workstation). For the study of the ultimate, low-energy region of bottom-sites and the measurement of their coherences the concept of *Coherence Observation by Interference Noise* (COIN) has been applied[8]. In COIN fs-luminescence interferometry, the interference noise of correlated populations is measured by using equivalent double pulses with random, relative phase[8, 9]. To begin, we show the absorption and luminescence of PPV at 1,4 K (Fig. 1). The onset of significant absorption is located at about 2.340eV which is slightly shifted to the blue at room temperature[6b]. Aside from the weakly resolved $S_1^{v=0} \leftarrow S_0^{v=0}$ transition peak (not detectable at room temperature), no tendency to substantial spectral narrowing can be observed in the (inhomogeneously broadened) absorption band. However, the low-temperature luminescence spectrum (excitation at ε = 2.583eV) consists of a predominant $S_1^{v=0} \rightarrow S_0^{v=0}$ band at ε = 2.313eV followed by the $S_1^{v=0} \rightarrow S_0^{v=n}$ transitions of the vibronic progression. These vibrations were already assigned to the C-C ring-stretching mode of the PPV backbone[10]. Generally, these luminescence spectra of PPV show the typical behaviour of fluorescent organic molecules and are similar to those observed in oligomeric model compounds of PPV[11] – the lack of mirror symmetry between luminescence and absorption indicating the ocurrence of spectral diffusion in the excited states.

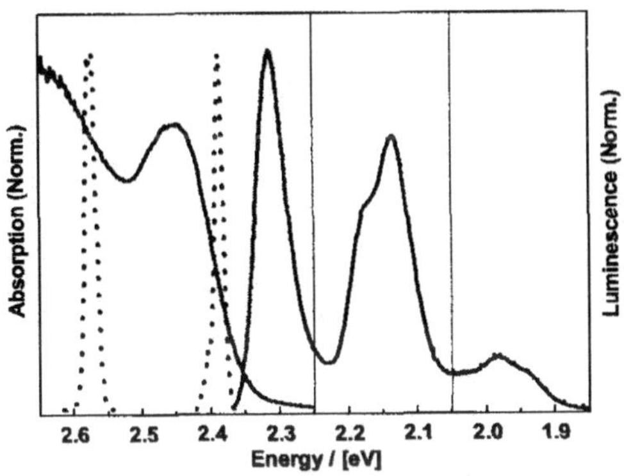

Figure 1. Steady-state absorption (left) and luminescence (right) spectrum of PPV at 1,4 K. The prominent vibronic progression $S_1^{v=0} \rightarrow S_0^{v=n}$ can be seen in the luminescence spectrum (excitation energy $\varepsilon = 2.583$ eV). Dotted lines represent the spectral shape of the laser excitation pulse at two different, central excitation energies, the two vertical lines indicating the upper and lower frequency bounds of the luminescence detection windows, i. e. the the $0 \rightarrow 1$ vibronic transition has been chosen for the accumulation of the transient spectra (Fig. 3).

3 DOS-Selective Excitation, Preliminary Results

Energy-dispersive excitation transport along a statistical energy-gradient in the course of intra-DOS population transfer leads one to expect a typical site-energy memory effect, when changing the initial fs-excitation boundaries. According to this picture, variation of pulse excitation energies, i.e. tuning of excitation wavelengths across the absorption edge, should allow selective excitation of site-subensembles in the DOS and thus to effect incoherent relaxational behaviour, particularly at low phonon temperatures, as excitations tuned into different energetic positions of the DOS (excitations *a, b, c* in Fig. 2) will map out quite different exploration volumes and excitation transfer routes toward the low-energy tail states. Thus, by measuring the central mean of the population distribution, i.e. the photons of any luminescence transition as a function of varying excitation wavelengths, it is expected that the evolution of the bathochromic frequency shift will proceed at different rates. The spectral excitation positions *a*, *b*, *c* inside the DOS (cf. sketch in Fig. 2) refer to transitions within the inhomogeneously broadened, longest-wavelength $S_1^{v=0} \leftarrow S_0^{v=0}$ absorption (Fig. 1) and correspond to excitation energies at $\varepsilon = 2.583$, 2.385 and 2.313eV, respectively.

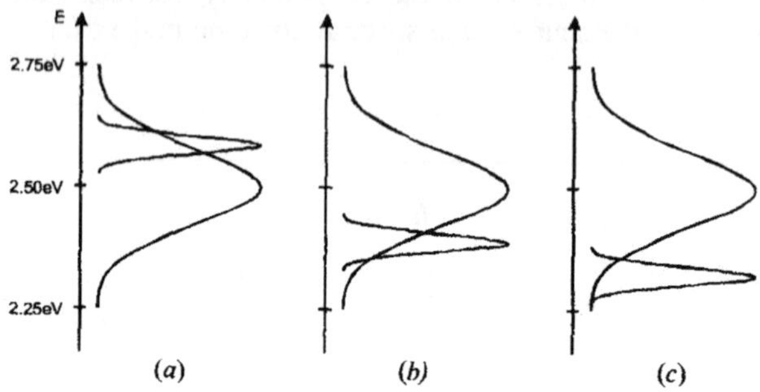

Figure 2. Density-of-states (DOS) and energetic positions (central mean) of initial fs-pulse excitation, $\varepsilon = 2.583$ (*a*), 2.385 (*b*), and 2.313 eV (*c*), schematic sketch

3.1 Population Transfer

In Fig. 3, the evolution of spectral luminescence transients for various time windows up to 20ps has been plotted for excitation at 2.583eV (a) and 2.385eV (b), respectively; the luminescence transitions of the $0 \to 1$ vibron ($2.05\text{eV} < \varepsilon < 2.25\text{eV}$) have been chosen for detection. Pulses centered at 2.583eV prepare excitations near the high-energy wing of the DOS (see Fig. 2, position a), corresponding to energetically high-lying transitions of the inhomogeneous $S_1^{v=0} \leftarrow S_0^{v=0}$ absorption band. On the contrary, a pulse with a mean energy of 2.385eV promotes an excitation that is located well below the center of the DOS (see Fig. 2, location b). Comparison between spectra 3(a) and 3(b) shows how the relaxational population dynamics is affected by varying excitation energies. The evolution of the spectral transients for excitation at $\varepsilon = 2.583$eV displays a substantial spectral narrowing on the high-energy side paralleled by the bathochromic shift dynamics. Both the pronounced spectral narrowing and the decrease of the central mean of the luminescence energy, clearly, indicate the generic features of energy-dispersive population transfer. The vanishing blue part of the transient spectra is the natural output of a migrative release of elementary S_1-excitations starting, first, with the very rapid jumps from a few wing states in the high-energy range of the DOS and reaching lower energy positions as time proceeds. For excitation at $\varepsilon = 2.385$eV (below the center of the DOS), the extent of bathochromic frequency shift is, in comparison to that at $\varepsilon = 2.583$eV, substantially decreased, but still significant. On the other hand, the spectra of both excitations show a pronounced bottleneck–effect on the red edge, i. e. they are spectrally fixed on the low-energy side and behave nearly invariant in time. This red-wavelength bottleneck, common to all spectral transients with excitation energies above a certain "demarcation line", can be reasonably interpreted as the kinetic signature of an ongoing migrational blockade which is, gradually, executed by the subensemble of low-energy wing-states.

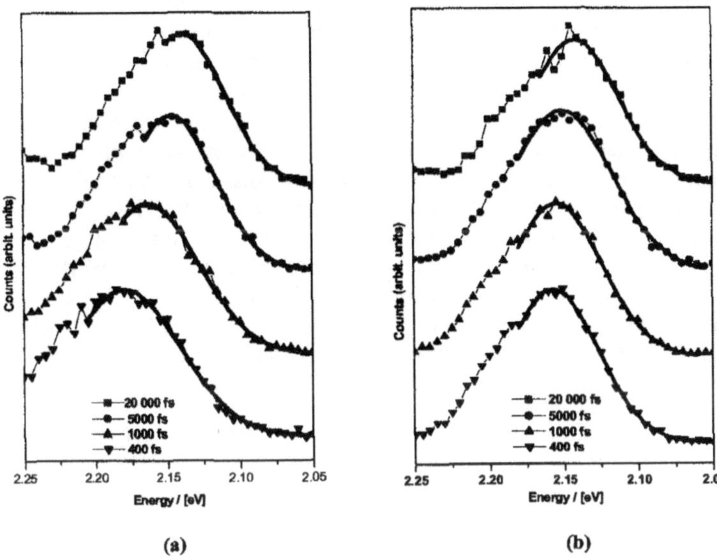

Figure 3. Time-resolved $S_1^{v=0} \rightarrow S_0^{v=1}$ luminescence of PPV measured by luminescence up-conversion after excitation with sub-200fs pulses at central energies of (a) 2.583eV and (b) 2.385eV respectively. Solid lines represent Gaussian fits to the lower-energy transition of the 1→0 vibronic dublet[11]. The pronounced bathochromic shift of luminescence is a consequence of intra-DOS downhill population transfer, the extent of the red shift, clearly, depending on the choice of the initial excitation energy (a > b)

For excitation energies below the demarcation line ($\varepsilon < 2.348$eV) the elementary excitons can no longer jump and reside on the sites, until the long-lived, intrinsic luminescence of the segments annihilates the populations. In this regime, where the site-density is extremely small, the probability of energy transfer drops to zero and the width of the dilute single-site Lorentzians becomes radiation-controlled without any substantial overlap at liquid helium temperatures[6b] (cf. Sec. 3.2).

3.2 Coherences

On early fs-time scales, the scenario of fs-free-evolution in PPV is even more complex and has extreme boundaries. This regime is initially controlled by the imprinting of phase coherence from the field onto the molecular, segmental states of the polymer. Since fs-pulse excitation, preferentially, near the center of the quasi-continuous DOS (including our experimental excitations *a* and *b* in Fig. 2), undergoes, quasi-instantaneously, inhomogeneous dephasing, in general, only nonlinear multi-wave mixing techniques may have the potential to look into the segmental Lorentzians (homogeneous broadening, HB) and hence into the electronic coherences of energy-transfer/delocalization inside the inhomogeneous band broadening, IHB *(vide infra)*. However, for excitations of segmental sites in the ultimate range of DOS-bottom states (Fig. 2c), experimental probing of electronic coherences must have a better prognosis. In this regime, the site density is low, the level-structure sparse, intersite-coupling marginal and, as a consequence, the (few) Lorentzians narrow and non-overlapping. For fs-pulse excitation, sufficiently below the demarcation line ($\varepsilon < 2.348$eV), it is, therefore, tempting to speculate that electronic coherent transients, in terms of typical polarization interferences from spatially and energetically distinct (decoupled) absorber segments, may survive on early femtosecond time scales. Provided the self-energy splittings of the low-dense segmental sites in the DOS are larger than their individual IHBs, one would expect – stimulated by previous work on binary crystals[9] and very recent studies on spatially interfering, resonant Rayleigh scattering waves from discrete PPV-segmental tail-states[12] - to detect their coherences by using (linear!) fs-luminescence interferometry.

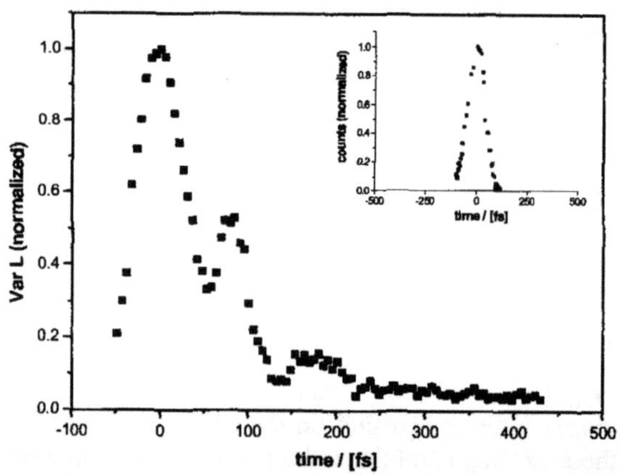

Figure 4. COIN-interferogram for excitation at 2.313eV (squares), luminescence detection at 2.033eV (inset: cross-correlation of the excitation pulses). Oscillatory pattern, due to inter-site beating in the sparse-level, bottom-state regime. See the text, for details.

Preliminary results from fs-luminescence correlation measurements with special emphasis placed on the selective excitation of the ultimate DOS tail states in PPV are very promising. Fig. 4 shows the luminescence interferogram measured in a COIN interferometric set-up, with excitation at $\varepsilon = 2.313$eV and detection at $\varepsilon = 2.033$eV. In COIN-interferometry a pair of 50/50 pulses (pulse 1, pulse 2) with random, relative phase ϕ is used[8]. Despite the randomness of the relative phases, the information about the coherences is preserved. COIN measurements collect the correlated noise of luminescence interferences F_{12} (τ) from the double-pulse excitations as a function of the delay-time τ between the two pulses and are analyzed through the variance of luminescence intensity fluctuations *var L* that arise from statistical population interferences. In Fig. 4, *var L* is plotted over the pulse-to-pulse delay time. The interferogram shows a rapid damping, but, quite obviously, a full cycle of a high-frequent oscillation survives. The raw data convolution can be fit, according to a damped oscillation, with a period of T~100fs and a damping constant $\Gamma[fs^{-1}]$ between $150<\Gamma<200$. From a first inspection of the data it becomes evident that the fs-damping must be related to inhomogeneous dephasing indicating strong inhomogeneous site broadening in the tail-states of the order of several hundred GHz (>400 GHz). In this strong damping regime, only one cycle of the high-frequent oscillation around 100fs can survive, whereas slower oscillations, quite obviously, are damped-out. The origin of the 100fs period is, in our opinion, some averaged polarization beating from two narrow two-level site (subensembles) *i, j* whose difference in site self-energy exceeds site–inhomogeneous broadening: $|\varepsilon_i - \varepsilon_j| > \sigma$. However, since our excitation pulses are of the same order of magnitude (~ 120fs) as those of the fit-parameters, care has to be taken in the definite assignment of this oscillatory relaxation. In forthcoming experiments, shorter pulses, typically down to 20fs will allow one to record the interferograms at shorter delay time, combined with a systematic excitation energy tuning at 1.4K. Such studies may help to deconvolve the COIN signal on time scales <100fs and to analyze the frequency-dynamics of the inter-site beats, more precisely. In addition, by using shorter excitation/probing pulses, higher frequency contributions and a broader range of beatings may be contained in the data, because the broader bandwidth will pick up more widely spaced features in the spectrum[13].

By closing, we note that for polymer segments in the high density regime cnstituting a vast number of broad and overlapping Lorentzians, non-linear photon-echoes are, nowadays, considered to be the coherent method of choice in revealing electronic couplings and delocalization. In particular, *two-dimensional (2-D)* optical photon-echoes[14] may have enormous impact on the search for hidden, electronic coherences, in a quasi model-free, unbiased fashion, for precisely the strongly disordered and spectrally congested polymer regime. Two-dimensional, heterodyne 3-pulse photon echoes[15] dispersed into two frequency axes (i) image complex absorption bands by projecting HB and IHB into two independent frequency channels[16] and (ii), in case of a distinct molecular interaction, give rise to typical cross-correlations and thus significant off-diagonal peaks. Simulated 2-D echo-spectra for a segmental pair in the PPV chain modelled by a 4-level two-site system and the calculation of its double Fourier transform response have demonstrated a great deal of essentials in the understanding of a low dimensional, excitonic-site system[17, 18]. Stimulated by these simulation experiments, we conjecture, therefore that, in particular, for smaller site ensembles at the oligomeric level, 2-D optical echo-spectra may provide new windows and snapshots into the nature of intermolecular coupling, in the near future. But also for polymeric multi-site organizations, such as PPV on well-defined nanometric templates, the emerging techniques of multi-dimensional echo-correlations may open, in the long run, strategies to reveal physical mechanisms of electronic coupling (Coulomb coupling versus short-range orbital overlap) as well as spatial information about excitonic segmental sites and through-space charge transfer in chain-to-chain site interactions.

Acknowledgement: This work has been funded within the framework of the Austrian Special Research Program F016 "ADLIS", Austrian Science Foundation Vienna/Austria.

---

[1] W. P. Su, R. J. Schrieffer, and A. J. Heeger, Phys. Rev. Lett. 42, 1698 (1979)
[2] H. Bässler, and B. Schweitzer, Acc. Chem. Res. 32 (2), 173 (1999)
[3] H. Bässler, M. Gailberger, R. F. Mahrt, J. M. Oberski, and G. Weiser, Synth. Met. 64, 295 (1994)
[4] M. Yan, L. J. Rothberg, F. Papadimitrakopoulos, M. E. Galvin, and T. M. Miller, Phys. Rev. Lett. 72, 1104 (1994)
[5] R. Kersting, U. Lemmer, R. F. Mahrt, K. Leo, H. Kurz, H. Bässler, and E. O. Göbel, Phys. Rev. Lett. 70, 3820 (1993); R. Kersting, B. Mollay, M. Rusch, J. Wenisch, G. Leising, and H. F. Kauffmann, J. Chem. Phys. 106, 2850 (1997)
[6] a) S. P. Kennedy, N. Garro, and R. T. Phillips, Phys. Rev. B 64, 115206 (2001); b) F. Milota, J. Sperling, A. Tortschanoff, H. Bässler, and H. F. Kauffmann, to be published
[7] B. Mollay, U. Lemmer, R. Kersting, R. F. Mahrt, H. Kurz, H. F. Kauffmann, and H. Bässler, Phys. Rev. B 50 (15), 10769 (1994)
[8] O. Kinrot, I. Sh. Averbukh, Y. Prior, Phys. Rev. Lett. 75, 3822 (1995)
[9] A. Tortschanoff, K. Brunner, Ch. Warmuth, and H. F. Kauffmann, J. Chem. Phys. 110, 4493 (1999)
[10] S. Heun, R. F. Mahrt, A. Greiner, U. Lemmer, H. Bässler, D. A. Halliday, D. D. C. Bradley, P. L. Burn, and A. B. Holmes, J. Phys. Condens. Mat. 5, 247 (1993)
[11] D. D. C. Bradley, R. H. Friend, H. Lindenberger, and S. Roth, Polymer, 27, 229 (1987)
[12] S. P. Kennedy, N. Garro, and R. T. Phillips, Phys. Rev. Lett. 86, 4148 (2001)
[13] F. Milota, J. Sperling, A. Tortschanoff, and H. F. Kauffmann, in preparation
[14] S. Mukamel, Annu. Rev. Phys. Chem. 51 (2000), 691
[15] W. P. de Boeij, M. S. Pshenichnikov, and D. W. Wiersma, Chem. Phys. Lett. 239, 1 (1995)
[16] J. D. Hybl, A. W. Albrecht, S. M. Galagher-Faeder, and D. M. Jonas, Chem. Phys. Lett 297 (1998), 307
[17] V. Szöcs, T. Palszegi, A. Tortschanoff, and H. F. Kauffmann, J. Chem. Phys., submitted
[18] V. Szöcs, T. Pálszegi, V. Lukes, A. Tortschanoff, and H. F. Kauffmann, this book

SURFACE FEMTOCHEMISTRY: ULTRAFAST REACTION DYNAMICS DRIVEN BY HOT ELECTRON MEDIATED REACTION PATHWAYS.

D.N. DENZLER, CH. HESS, S. FUNK, AND G. ERTL
Fritz-Haber-Institut der Max-Planck-Gesellschaft, Faradayweg 4-6, 14195 Berlin, Germany.

M. BONN
Leiden Institute of Chemistry, P.O. Box 9502, 2300 RA Leiden, The Netherlands.

CH. FRISCHKORN AND M. WOLF
Freie Universität Berlin, Fachbereich Physik, Arnimallee 14, 14195 Berlin, Germany.

Fundamental insights into the ultrafast dynamics of energy transfer processes at surfaces are of central importance for a microscopic understanding of chemical reactions at solid surfaces, e.g. in heterogeneous catalysis. The detailed investigation of the rates and pathways of energy flow in the adsorbate–substrate system as well as the chemical dynamics has become possible utilizing intense femtosecond laser pulses. Surprisingly, the strong non–equilibrium situation that arises upon irradiation with these pulses results in a new reaction channel caused by hot electron excitation for a number of systems investigated. This electron–mediated reaction mechanism is exemplified here for the model reaction $H_{ad}+H_{ad} \rightarrow H_{2,g}$ on Ru(001) and discussed in the broader context of other simple reactions such as the formation of H_2O, CO_2 and the CO desorption on the same surface.

1 Introduction

In comparison to the gas phase, where femtosecond (fs) laser pulses have been used to probe the ultrashort timescale of bond making and breaking for nearly two decades[1], insights into femtochemistry at solid surfaces have just started to emerge[2,3]. The extension of femtochemistry methods to the gas-solid interface is complicated considerably by additional experimental challenges: The well-defined preparation and characterization of the initial state of the sample, the low density of reactants of typically one monolayer, as well as the ultrashort lifetimes of excited states due to rapid quenching channels which have to be outpaced by sufficiently large reaction rates, especially on metal surfaces. Furthermore, the chemical mechanisms responsible for surface reactions to occur are entirely different to those in the gas-phase: The presence of a solid–state surface opens a variety of possible excitations and relaxation mechanisms for the adsorbed molecule, which lead to much more complex excitation mechanisms and pathways of energy flow compared to either gas- or solution–phase chemistry.

This article demonstrates that crucial information about these processes — accessible only on the femtosecond timescale but relevant for "slow", equilibrium chemistry — can be obtained by relatively simple means: The substrate–mediated breaking and formation of chemical bonds between adsorbates by intense fs-laser pulses and probing of the product yields by mass spectrometry.

2 Mechanisms for surface femtochemistry on metals

The key to a microscopic understanding of chemical reactions at surfaces lies in the dynamics of the energy transfer between the substrate and the (reacting) adsorbates. From a simplistic point of view, a metal substrate can be broken down into two subsystem[4]: The ion cores (lattice) and the surrounding electron gas. The energy content of the system can thus be described by two distributions, representing the degree of lattice excitation (collective ion core motions or phonons) and electron gas excitation, both characterized by the temperatures of their respective population distribution functions, T_{ph} for the phonon temperature and T_{el} for the electronic temperature. In thermal equilibrium (or when conventionally heated) $T_{el} = T_{ph}$ due to the electron–phonon coupling with a characteristic timescale in the picosecond range (see the illustration of the metal in Fig. 1).

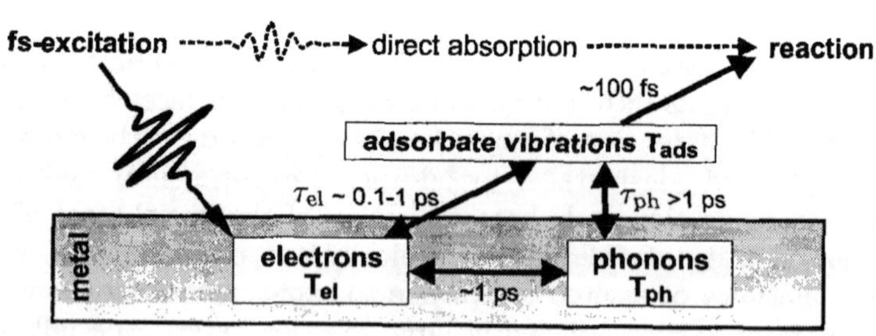

Figure 1. Schematic diagram of the fundamental energy transfer processes and their timescales in the surface chemistry on metals. In fs–photochemistry, the substrate is excited by a fs–laser pulse, which could in principle also lead to a reaction initiated by direct absorption in the adsorbate–substrate complex but is negligible in most cases (see text).

In principle, both of these energy reservoirs may couple energy independently into the adsorbate degrees of freedom relevant for a surface reaction. However, in the traditional picture of a thermal reaction on a metal surface, it is the phonons which drive the system across the activation barrier in the electronic ground state (Fig. 2, right panel). Nevertheless, electrons in the high–energy tail of a thermal Fermi–Dirac distribution can also induce chemical transformations by transient population of normally unoccupied states (Fig. 2, left panel). Such phonon- and electron–mediated processes are fundamentally different, but can not be distinguished by conventional means[5], where electrons and phonons are always in equilibrium.

The only possibility to gain distinctive information on these processes is to analyze the reactive behaviour on a timescale shorter than the electron–phonon coupling time, where a strong non–equilibrium between substrate electrons and phonons can be achieved by intense fs–laser pulses. These pulses provide a selective excitation by acting only on the electronic subsystem and therefore an ideal tool to gain insight into this so far largely unexplored dynamical behaviour[6], allowing to establish key reaction mechanisms and parameters such as the pathways and (subpicosecond) timescales of energy flow.

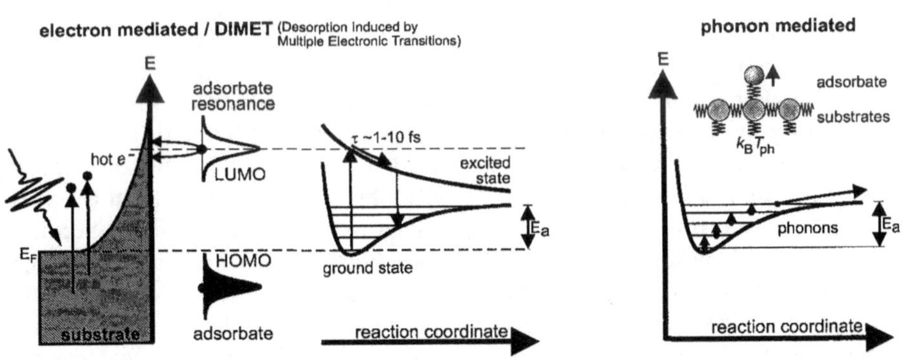

Figure 2. The two principally different substrate mediated excitation mechanisms in surface reactions illustrated for a simple desorption potential well. Left: Electron–mediated energy transfer into a vibrational coordinate of the adsorbate involves (repeated) transfer of electrons from the high–energy tail of a Fermi–Dirac distribution into an unoccupied molecular orbital of the adsorbate–substrate complex (HOMO/LUMO = Highest occupied/lowest unoccupied molecular orbital). This is a non–adiabatic process via an excited state potential energy surface. Right: In contrast, phonon–mediated energy transfer occurs adiabatically, i.e. on the ground state potential surface of the system.

2.2 Ultrafast photochemistry at metal surfaces

Fig. 1 schematically illustrates the fundamental energy transfer processes and their timescales when using an intense fs–laser pulse to photo–induce reactions at metal surfaces. The indirect, substrate-mediated mechanisms are dominant in most cases because of the small optical density of the adsorbate layer (typically a monolayer). Thus, direct absorption in the adlayer is negligible and the energy of the non–reflected part of the fs-laser pulse is absorbed entirely by the metal electrons, creating a hot electron distribution $n(E)$ within the optical penetration depth (~10–100 Å for $h\nu < 10\,\mathrm{eV}$). This hot electron distribution is initially non-thermal, i.e. the distribution can not be described by Fermi–Dirac statistics. However, very rapid electron–electron scattering leads to a thermalized distribution within several 10 fs for a d–band metal such as ruthenium in the strong excitation regime of laser fluences >10 J/m^2.[4] Therefore on a timescale >100 fs the energy deposited in the electronic subsystem can be described satisfyingly by an quasi–instantaneous electronic temperature T_{el} which characterizes the hot Fermi–Dirac distribution.

These very hot electrons (which may transiently reach thousands of Kelvin above the melting point of the metal[4]) may couple energy into the adsorbate degrees of freedom relevant for a surface reaction via the two completely different mechanisms discussed above in Fig. 2: (i) The hot substrate electrons can directly excite adsorbate vibrations by multiple resonant electron attachment (scattering) into a negative ion resonance. The overall process can be described by an energy transfer time, τ_{el}, typically several hundred fs, characterizing the energy transfer rate from the substrate electrons into the adsorbate nuclear coordinate relevant for reaction. (ii) The adsorbate vibrations can be excited on a somewhat longer timescale of τ_{ph} >1 picosecond (ps) via lattice phonons which equilibrate with the hot electrons within a few ps, due to the electron–phonon coupling. This second pathway corresponds to the traditional picture of thermal surface chemistry, in which lattice phonons couple vibrational energy into the reaction coordinate.

How can these fundamentally different mechanisms be distinguished? A direct way to discern the electron– and phonon–mediated adsorbate reaction pathways is provided by a two–pulse correlation (2PC) measurement, which allows for the analysis of the underlying mechanism: Two equally intense pulses are sent onto the adsorbate–covered surface and the time–integrated reaction yield is measured in a mass spectrometer as a function of the pulse–pulse delay. Due to the typical nonlinear dependence of the reaction rate from the incident laser fluence[7], a correlation function is obtained, whose width critically depends on the excitation pathway. As is illustrated in Fig. 3,

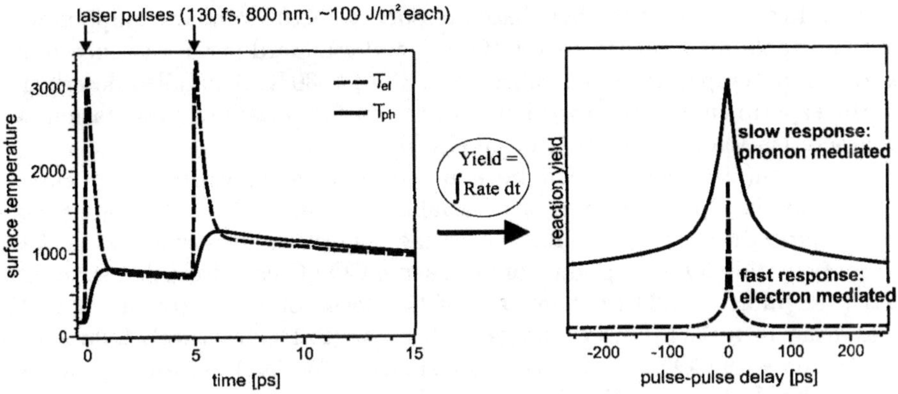

Figure 3. Principle of a two–pulse correlation measurement. Left: Surface temperature transients of Ru(001) calculated with the Two–Temperature–Model[4] for the excitation fluence used in the H_2 formation experiment (Fig. 4). Right: Schematic illustration of the expected correlation traces which allow to discern the two principal processes – the fast response for an electron–driven reaction and the slow response for a phonon–driven reaction.

a full width at half maximum (FWHM) of the correlation function on the timescale of the e–ph–equilibration time (for ruthenium ~1.5 ps) is observed for electron-mediated processes, whereas for phonon–driven processes a much longer timescale on the order of the slow cooling of the phonons into the ruthenium bulk of several 10's to 100's of ps is measured.

A quantitative analysis of such 2PC data can be performed using a friction model[8,9], which uses coupling ("friction") coefficients to describe the energy flow from either the electrons[9], the phonons, or both[8] into the relevant adsorbate nuclear mode. The energy in the reaction coordinate is then described with a transient adsorbate temperature T_{ads}, from which the reaction rate $R(t)$ can be calculated by use of an Arrhenius expression[10] according to $R(t) \propto exp(-E_{\mathrm{a}}/k_{\mathrm{B}}T_{\mathrm{ads}}(t))$. The time–integral over $R(t)$ then corresponds to the measured reaction yield in the experiment. Values for the coupling times τ_{el} and τ_{ph} (which are the inverse of the friction coefficients) and the activation energy E_{a} in the Arrhenius expression are obtained by fitting the results based on the friction model to the experimental 2PC–data.

3 Experimental

Briefly, our commercial laser system delivers 800 nm (1.5 eV), 110 fs pulses with an energy of up to 4.5 mJ per pulse. The Ru(001) sample is mounted in an ultrahigh vacuum chamber (base pressure $< 1\times10^{-10}$ mbar; equipped with a quadrupole mass spectrometer (QMS) and standard surface science tools) and can be temperature–controlled from 30 to 1530 K. A detailed description of the experimental setup together with the surface cleaning, preparation and characterization procedures is given elsewhere.[11]

The Ru(001) surface can be exposed to oxygen, hydrogen or carbon monoxide via a pinhole doser or a variable leak valve. The adsorbate preparation procedures are the following for the respective investigations: (i) CO_2 formation and CO desorption from the same CO+O co–adsorption structure: 5 L (Langmuir $= 1.33\cdot10^{-6}$ mbar s) of O_2 dosed onto the sample at 600 K (resulting in a 0.5 ML O–coverage with an O-(2x1)–structure), followed by saturation with CO at 95 K.[12] (ii) H_2 formation: H-(1x1)–saturation coverage (1 ML) by dosing 50 L H_2 at 170 K.[13] (iii) $H_2^{18}O$ formation: Preparation of an ^{18}O-(2x2)–precoverage by flashing the exposure of (i) shortly to 1450 K and subsequent saturation with H_2 at 90 K.[14] In all cases thermal desorption spectra of CO or H_2 from the structure were taken and carefully compared to the literature to ascertain the planned structure had been prepared.

In a typical experiment, the Ru(001) surface is covered with the adsorbate structure to be investigated and the reaction yield of the desorbing product species is measured with the mass spectrometer after fs-laser excitation. This is either done by by integrating a time–of–flight spectrum or by extrapolation of a depletion curve as shown in Fig. 5 to the first shot yield. Measurements for H_2 are "single-shot" generally, i.e. after each incident laser pulse the adsorbate is again prepared to its initial condition. The laser fluence is generally given as yield–weighted fluence $\langle F \rangle$, where the full spatial profile of the laser pulse as recorded by a CCD camera is taken into account.[11]

4 Results and Discussion

For the co–adsorption system of O and CO a new reaction channel — the formation of CO_2, which does not occur when the substrate is heated under equilibrium conditions — opens up under fs-laser excitation[7]. In this CO_2–formation channel, energy is transferred directly from the excited substrate electrons into the nuclear degrees of freedom of the oxygen adsorbate. This is in contrast to the "simple" CO–desorption from the same surface, where the conventional picture of heating via the substrate phonons is found to prevail. A friction model analysis gives a typical example for the different coupling times involved: τ_{el}=0.5 ps with an activation energy $E_a = 1.8$ eV for the electron–mediated oxidation, and τ_{ph}=1.0 ps with $E_a = 0.83$ eV for the phonon–mediated CO desorption describe the experimental 2PC data well[7].

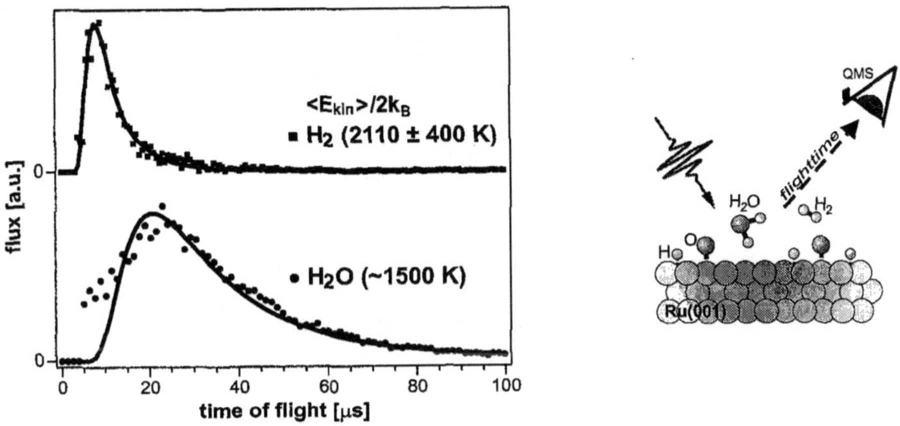

Figure 4. Time–of–flight distributions of the $H_2^{18}O$ and H_2 formation reactions from Ru(001) (incident fluences $\langle F \rangle \sim 110\,J/m^2$). The spectra are obtained by recording the number of molecules coming off the surface as a function of time after fs-laser excitation; the mean energy given in K (according to $\langle E_{trans} \rangle / 2\,k_B$) is calculated directly from the momenta of the measured velocity distributions. The solid line is a modified Maxwell–Boltzmann fit, which deviates considerably for H_2O at short times (future investigations should clarify this; therefore the value of the mean energy for this first dataset has to be treated with caution). The ^{18}O isotope is used to prevent any possible $H_2^{16}O$ background.

In the O+H co–adsorption system, water formation from the Ru(001) surface is observed after fs-laser excitation, a second example of a reaction which is not present with thermal heating. A first result is the time-of-flight (TOF) spectrum shown in Fig. 4. The measured high translational energy of the H_2O molecules (and also of the H_2 molecules) coming off the surface is in accordance with what is expected for an electron–mediated activation mechanism, where a potential barrier has to be overcome for product formation. In analogy to the CO_2 formation we propose for the mechanism of the H_2O–formation, that the O is so strongly bound to the ruthenium substrate that thermal heating is insufficient for a reactive weakening of the Ru-O–bond: Before oxygen activation can occur, the co–reactant (H_2 or CO, respectively) desorbs from the surface. Only by intense fs-laser pulses the Ru-O–bond can be sufficiently activated on an ultrashort timescale, such that a reaction, the formation of H_2O or CO_2, respectively, takes place. This oxygen activation is sufficiently fast to outpace the competing desorption of the co-reactant.

Fig. 5 shows the two pulse correlation data obtained for the formation of H_2 from a Ru(001) surface saturated with atomic hydrogen (a TOF-spectrum

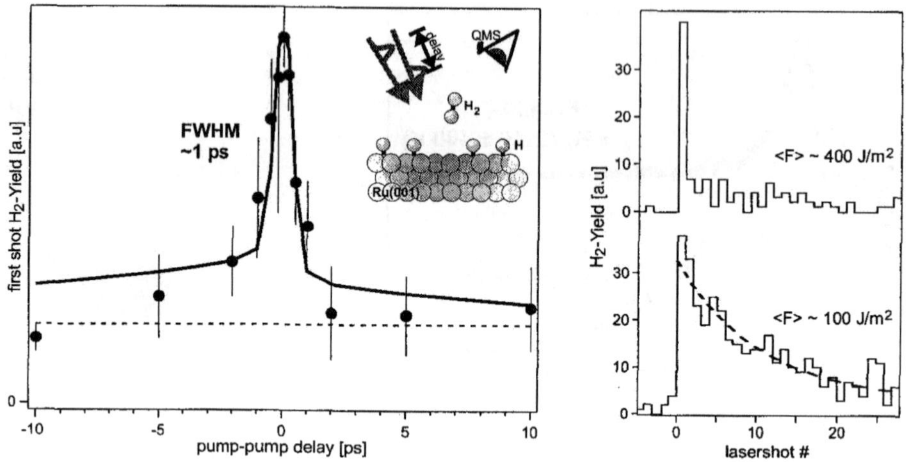

Figure 5. Left: Two–pulse correlation of the H_2 formation reaction yield. The FWHM which is shorter than the e–ph–equilibration time of 1.5 ps indicates that the process is driven by electron–mediated excitation. Right: Decay data of the yield as a function of shot number onto a freshly prepared surface for two different fluences. The 2PC on the left was measured in the low fluence regime, such that saturation of the H_2 formation reaction within one pulse is negligible (the dashed line in the lower panel is an exponential fit for comparison).

is also given in Fig. 4). The FWHM of the two–pulse correlation is on the e–ph–coupling timescale, clearly indicating that the H_2 formation reaction is driven by hot substrate electrons. The 2PC data are well described by the electronic friction model[9] (solid curve in Fig. 5), which yields a coupling time τ_{el}=160 fs with an activation energy of E_a=1.6 eV. However it is noteworthy, that this reaction can also be initiated via substrate phonons, as is the case when the surface is simply heated thermally. Hence, the H_2 formation can obviously proceed via two different pathways: One involving only the ground state potential energy surface (PES) — the phonon–mediated channel — and a second, completely different, reaction pathway proceeding very efficiently via non–adiabatic coupling to an excited state PES, whereby energy is transferred directly from substrate electrons into the nuclear degrees of freedom relevant for the reaction. It is only with intense fs-laser excitation that the latter channel completely overwhelms the phonon–mediated pathway and leads to very efficient H_2-formation, such that with high fluences nearly the whole H–layer can be desorbed within one laser shot (Fig. 5, right panel).

5 Summary and Outlook

Summarizing, several elementary reactions of adsorbates from Ru(001) have been investigated induced by fs–laser pulse excitation: Formation of H_2, H_2O and CO_2, as well as the desorption of CO. Some of the most relevant results for these systems are compared in Table 1.

It is found that after intense fs–laser excitation an electron–mediated reaction pathway is opened in all cases except for CO desorption, where the energy transfer into the reaction coordinate is of pure phononic origin. This demonstrates that hot electrons can chemically activate adsorbates — in contrast to the traditional picture of activation solely by phonons. One might argue that the extremely hot electron distributions which give rise to the electron–driven processes are unique to intense fs–laser excitation. Nevertheless, they persist only for the order of ~1 ps, whereas the interaction time for thermally heated (much colder) electrons is on the order of seconds. For example, according to Fermi–Dirac statistics, the electron population at 1.7 eV above the Fermi–level[15] for a typical electronic temperature of 6000 K after intense fs–laser excitation is ~12 orders of magnitude larger than that of a 600 K thermal distribution. If one defines some effective interaction strength of the hot electron distribution with the ion resonance relevant for the reaction as the typical timescale of the excitation (1 ps vs 1 s) times the typical available number of hot electrons (10^{12} vs 1), the result is quite similar for both cases, indicating that electron–mediated processes might also have to be considered in equilibrium chemistry. However, their relevance will further depend on (i) to which extend multiple electronic excitations of the adsorbate can compete with its vibrational relaxation and (ii) whether there is a competing phonon–driven reaction pathway; both will influence the importance of the respective excitation mechanisms on the "thermal" timescale of seconds and on the ps–timescale very differently. To draw more precise conclusions about the role of electronically non–adiabatic processes in equilibrium chem-

Table 1. Summary of experimental data of all adsorbate systems investigated on Ru(001).

adsorbate on Ru(001)	Product	TOF $\langle E\rangle/2k_B$	2PC FWHM	Ref.
O-(2x1) + CO	CO	640 K	20 ps	11
	CO_2	1590 K	3 ps	11
H	H_2	2110 K	1 ps	this work
O-(2x2) + H	H_2O	~1500 K		this work

istry, more definite knowledge of the electronic structure of the systems, i.e. exact positions of excited state PES, from theoretical modeling is necessary.

An especially interesting case is the H_2 formation reaction from atomic hydrogen, where a second electron–mediated reaction pathway exists in addition to the well–known thermal channel. Considering the principle of detailed balance, this suggests that in the reverse reaction — the dissociative adsorption of molecular hydrogen on transition metal surfaces — electronic excitations of the adsorbate–substrate system could play an important role in the relaxation processes leading to dissociative adsorption. The experiments presented here show the existence of electronic states which can couple energy in (and out of) the H–Ru(001)–system; their relevance in the dissociative adsorption process of hydrogen has yet to be investigated.

Furthermore, the existence and different timescale of the two energy transfer mechanisms from the excited substrate into the reaction coordinate allows for a rudimentary control of the chemical reactions involved. For example for a chemical reaction where there are two possible products, the outcome of the reaction could be influenced by changing the temporal width of the laser pulses if one of the two reaction pathways proceeds via an electron–mediated mechanism and the other by the slow phonon–mediated process (as e.g. in the CO–oxidation vs CO–desorption case). Depending on the reaction under investigation, this approach should allow to control the outcome of a chemical reaction at least to a certain degree. However, it should be noted that in the case of substrate–mediated excitation coherence is lost by the rapid e–e–scattering events and thus application of optimized laser pulses with a specific shaped phase does not provide a further degree of control. Nevertheless, if a direct excitation pathway is accessible in the system, as has been recently demonstrated for Cs/Cu(111)[16], this might open the perspective for optimum control scenarios of surface reactions.

Acknowledgments

This work was supported in part by the Deutsche Forschungsgemeinschaft through grant Wo 653/6 and Sonderforschungsbereich 450. M.B. acknowledges a fellowship of the Royal Netherlands Academy of Arts and Sciences.

References

1. A. Zewail, J. Phys. Chem. A **104**, 5660 (2000).
2. R.R. Cavanagh, D.S. King, J.C. Stephenson and T.F. Heinz, J. Phys. Chem. **97**, 786 (1993).

3. H.-L. Dai and W. Ho, Eds. *Laser Spectroscopy and Photochemistry on Metal Surfaces*, World Scientific, Singapore, 1995.
4. M. Bonn, D.N. Denzler, S. Funk, M. Wolf, S.-S. Wellershoff and J. Hohlfeld, Phys. Rev. B **61**, 1101 (2000) and references therein.
5. It is well known that promoters, such as alkali coadsorbates, can play an important role to enhance reaction rates (see e.g. G. Ertl, H. Knözinger, J. Weitkamp, Eds. *Handbook of heterogeneous catalysis*, VCH, Weinheim 1997). They modify the electronic structure of the system, i.e. change the height of the reaction barrier in the electronic ground state and may also lead to energetic shifts of excited state PES. Therefore, it is difficult to isolate contributions of only electronically non–adiabatic processes from such experiments.
6. Previous work focuses mainly on the desorption of diatomic molecules from surfaces: J.A. Prybyla, H.W.K. Tom and G.D. Aumiller, Phys. Rev. Lett. **68**, 503 (1992); J.A. Misewich, A. Kalamarides, T.F. Heinz, U. Höfer and M.M.T. Loy, J. Chem. Phys. **100**, 736 (1993); W. Ho, J. Phys. Chem. **100**, 13050 (1996); L.M. Struck, L.J. Richter, S.A. Buntin, R.R. Cavanagh, and J.C. Stephenson, Phys. Rev. Lett. **77**, 4576 (1996); R.J. Finlay, T.H. Her, C. Wu and E. Mazur, Chem. Phys. Lett. **274**, 499 (1997).
7. M. Bonn, S. Funk, Ch. Hess, D.N. Denzler, C. Stampfl, M. Scheffler, M. Wolf and G. Ertl, Science **285**, 1042 (1999).
8. F. Budde, T.F. Heinz, A. Kalamarides, M.M. Loy and J.A. Misewich, Surf. Sci. **283**, 143 (1993).
9. M. Brandbyge, P. Hedegard, T.F. Heinz, J.A. Misewich and and D.M. Newns, Phys. Rev. B **52**, 6042 (1995).
10. For the case of pure electronic friction the expression is slightly different, see Ref. 9.
11. S. Funk, M. Bonn, D.N. Denzler, Ch. Hess, M. Wolf and G. Ertl, J. Chem. Phys. **112**, 9888 (2000).
12. K.L. Kostov, H. Rauscher and D. Menzel, Surf. Sci. **278**, 62 (1992).
13. P. Feulner and D. Menzel, Surf. Sci. **154**, 465 (1985).
14. A. Schiffer, P. Jakob and D. Menzel, Surf. Sci. **465**, 198 (2000).
15. This is the value calculated for the relevant anti–bonding state of the O-(2x1)/Ru(001)–coverage in the CO_2–formation experiments, see Ref. 7.
16. H. Petek, H. Nagano, M.J. Weida and S. Ogawa, J. Phys. Chem. B **105**, 6767 (2001).

PUMP-PROBE SPECTROSCOPY OF A TRINUCLEAR TRANSITION METAL MIXED-VALENCE COMPLEX: SOLVENT DEPENDENCE OF BACK ELECTRON TRANSFER RATE

DAVID F. WATSON AND ANDREW B. BOCARSLY*

Department of Chemistry, Princeton University, Princeton, NJ 08544 USA

ELMAR SCHREIBER*

Center for Ultrafast Laser Applications (CULA), Princeton University, Princeton, NJ 08544 USA

One-color pump-probe experiments (400 nm) were performed on $[(NC)_5Ru^{II}\text{-}CN\text{-}Pt^{IV}(NH_3)_4\text{-}NC\text{-}Ru^{II}(CN)_5]^{4-}$. The timescale of back electron transfer (BET) following Ru(II)→Pt(IV) metal-to-metal charge transfer (MMCT) varied with solvent. The BET timescale of $[(NC)_5Ru^{II}\text{-}CN\text{-}Pt^{IV}(NH_3)_4\text{-}NC\text{-}Ru^{II}(CN)_5]^{4-}$ ranged from 476 ± 23 fs in water to 1110 ± 288 fs in a water/DMSO mixture (X_{DMSO} = 0.1). The timescale of ground state vibrational relaxation following BET was 1.4 - 3 ps. The variation of BET timescales is explained in terms of the solvent-dependence of the ground state free energy and the BET activation barrier.

1 Introduction

The trinuclear mixed-valence complex $[(NC)_5Ru^{II}\text{-}CN\text{-}Pt^{IV}(NH_3)_4\text{-}NC\text{-}Ru^{II}(CN)_5]^{4-}$ (herein referred to as Ru(II)-Pt(IV)-Ru(II)) undergoes photoinduced Ru(II)→Pt(IV) metal-to-metal charge transfer (MMCT) [13,14,22,24]. The MMCT transition generates the one-electron transfer intermediate, Ru(II)-Pt(III)-Ru(III). Decay of the intermediate occurs either through back electron transfer to yield the Ru(II)-Pt(IV)-Ru(II) starting material, or through thermal MMCT from the remaining Ru(II) center to the Pt(III) center to yield two equivalents of $Ru^{III}(CN)_6^{3-}$ and one equivalent of $Pt^{II}(NH_3)_4^{2+}$. The quantum yield of photodissociation is less than two percent.

Wu and coworkers studied the solvent-dependence of Fe(II)→Pt(IV) MMCT in the related complex, $[(NC)_5Fe^{II}\text{-}CN\text{-}Pt^{IV}(NH_3)_4\text{-}NC\text{-}Fe^{II}(CN)_5]^{4-}$ (herein referred to as Fe(II)-Pt(IV)-Fe(II)) [21,22]. In water/DMSO mixed solvent systems, the energy of photoinduced Fe(II)→Pt(IV) MMCT decreases linearly with increasing DMSO mole fraction (X_{DMSO}), while the quantum yield of photodissociation increases exponentially with increasing DMSO mole fraction. Increasing the DMSO mole fraction decreases the strength of Fe(II) t_{2g} → CN π^* backbonding, causing a negative shift in the $Fe^{III/II}$ reduction potential, or equivalently a shift of the Fe(II)-Pt(IV)-Fe(II) (ground electronic state) potential energy surface to higher energy.

The observed linear dependence of MMCT energy and exponential dependence of dissociative quantum yield on mole fraction DMSO suggest that the Fe(II)-Pt(IV)-Fe(II) potential energy shifts linearly with mole fraction DMSO.

This manuscript describes femtosecond pump-probe experiments on Ru(II)-Pt(IV)-Ru(II). Pump-probe data for Ru(II)-Pt(IV)-Ru(II) in water and water/DMSO mixtures are presented. The dependence of excited state lifetime on the activation barrier to back electron transfer is considered with regard to the well-understood dependence of the M(II)-Pt(IV)-M(II) potential energy on solvent.

2 Experimental

Synthesis. $Na_4[(NC)_5Ru^{II}\text{-}CN\text{-}Pt^{IV}(NH_3)_4\text{-}NC\text{-}Ru^{II}(CN)_5]$ was synthesized by adaptation of the method of Pfennig and Bocarsly [13]. An aqueous solution of $K_4Ru(CN)_6{\cdot}3H_2O$ (3.90 mmol, 46 mL) was oxidized to $Ru^{III}(CN)_6^{3-}$ by addition of aqueous $Ce(SO_4)_2$. The reaction mixture was filtered by gravity, then run through an Amberlite IR-120 sodium exchange column (Aldrich) charged with 1.0 M NaCl solution to remove Ce^{4+} and Ce^{3+}. To the $Ru^{III}(CN)_6^{3-}$ solution was added aqueous $Pt(NH_3)_4(NO_3)_2$ (3.53 mmol, 46 mL). The reaction mixture was stirred at room temperature under subdued lighting for 90 minutes, then concentrated and run through an Amberlite IR-120 sodium exchange column charged with 1.0 M NaCl solution to remove excess $Pt(NH_3)_4^{2+}$, then run through a Bio-Gel P-2 size exclusion column. The product was collected in fractions, which were analyzed by UV/visible absorption spectroscopy. Pure fractions were combined and air-dried, yielding the desired product.

Pump-probe spectroscopy. Pulses at 400 nm were generated by frequency doubling the 800 nm output of a regeneratively amplified Ti:sapphire oscillator (Spectra-Physics). The 400 nm beam was attenuated and split into pump (400 nm, 10 μJ, 100 fs, 1 kHz) and probe (400 nm, <10 nJ, 100 fs, 1kHz) beams. The probe beam was aligned through a high precision translation stage (Physik Instrumente C844-40) acting as a variable pathlength delay line with 0.2 fs resolution. The pump and probe beams were focused (f = 100 mm) into the flowing sample (0.25 mL/s in a 1 cm cuvet) at an angle of 6°. The pump beam was chopped (90 – 150 Hz), and the oscillatory component of the probe signal was measured with lock-in technique. Pump and probe polarizations were parallel.

3 Results and Discussion

3.1 Solvent dependence of MMCT energy in Ru(II)-Pt(IV)-Ru(II)

Steady-state UV/visible absorption spectra of Ru(II)-Pt(IV)-Ru(II) were obtained in water/DMSO mixtures of varying composition (Figure 1). The MMCT

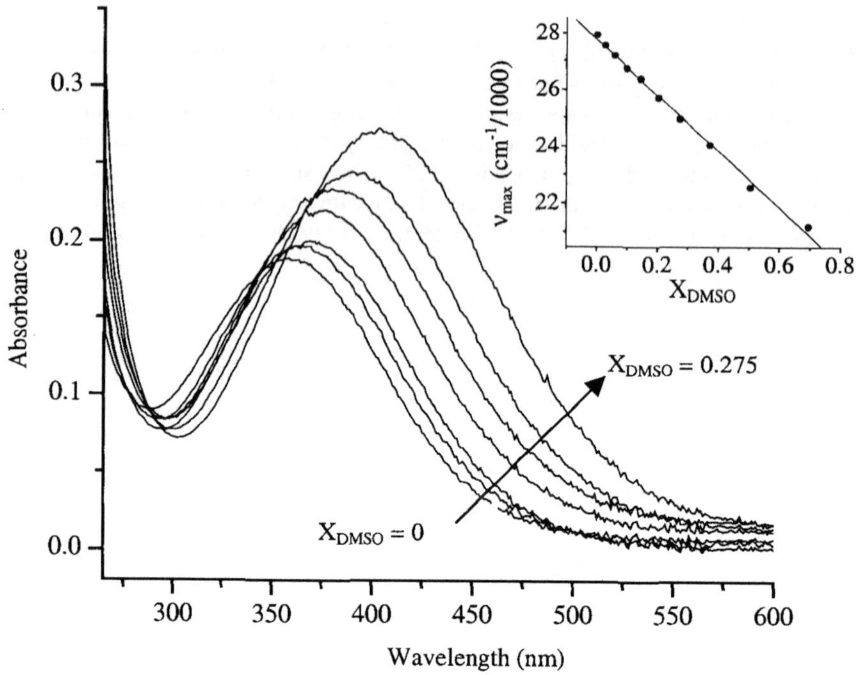

Figure 1. UV/visible absorption spectra of 0.1 mM Ru(II)-Pt(IV)-Ru(II) in water/DMSO mixed solvents with X_{DMSO} = 0, 0.0275, 0.06, 0.1, 0.15, 0.2025, 0.275. Cuvet pathlength = 1.0 cm. Inset: plot of MMCT absorption maximum (ν_{max}) vs. X_{DMSO} and fit to a straight line. (R^2 = 0.996).

band energy decreases linearly with increasing DMSO mole fraction (Figure 1 inset), while the absorption bandwidth remains the same. The linear red-shift of the MMCT band with mole fraction DMSO is consistent with a linear shift of the Ru(II)-Pt(IV)-Ru(II) potential energy surface to higher energy, as previously observed with Fe(II)-Pt(IV)-Fe(II) [22].

3.2 One-color pump-probe results

One-color pump-probe data (400 nm pump, 400 nm probe) were acquired for Ru(II)-Pt(IV)-Ru(II) in water and water/DMSO mixtures. Representative data are shown in Figure 2. The system response (Figure 2a) is well approximated by a Gaussian intensity profile, and is assigned as a cross phase modulation-induced artifact [4,5,12,19,20]. In the data for Ru(II)-Pt(IV)-Ru(II) in water (Figure 2b),

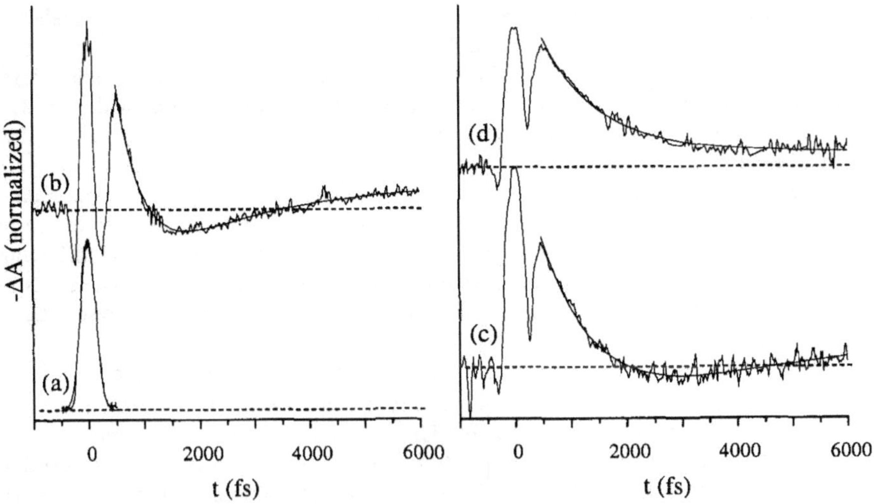

Figure 2. One-color pump-probe data. a) Cross phase modulation-induced artifactual signal with sample = water; b) Ru(II)-Pt(IV)-Ru(II), X_{DMSO} = 0; c) Ru(II)-Pt(IV)-Ru(II), X_{DMSO} = 0.1; d) Ru(II)-Pt(IV)-Ru(II), X_{DMSO} = 0.2. In each data set, the dashed line corresponds to $-\Delta A = 0$. Superimposed on the data are fits to equations 1 (b and c) and 2 (d).

two signals are present at very short delay times: the cross phase modulation-induced artifact [4,5,12,19,20] and an absorptive signal. The absorptive signal is assigned as an excited state absorption process originating from the vibrationally-excited one-electron charge transfer excited state, Ru(III)-Pt(III)-Ru(II)*. It is not unreasonable that the excited state undergoes absorption at 400 nm, because Ru(II)-Pt(IV)-Ru(II) absorbs strongly at 200 nm. Sequential 400 nm absorptions are energetically equivalent to a single 200 nm absorption.

Within 600 fs, the excited state absorption decays to a bleach, which is assigned as the vibrationally-relaxed one-electron charge transfer intermediate Ru(III)-Pt(III)-Ru(II). The decay of this bleach corresponds primarily to back electron transfer (BET) to yield Ru(II)-Pt(IV)-Ru(II), with a small percentage of population proceeding through forward electron transfer (FET) to the two-electron transfer product Ru(III)/Pt(II)/Ru(III). (It may be an oversimplification to assign the bleach at 600 fs as exclusively vibrationally-relaxed Ru(III)-Pt(III)-Ru(II), as BET and FET should occur from multiple vibrational states.)

At a delay time of approximately 1500 fs, the bleach decays to an absorptive signal, which is assigned as MMCT from the vibrationally-excited ground electronic state population. The decay of this absorptive signal corresponds to vibrational relaxation in the ground electronic state. The MMCT absorption from vibrationally-excited levels in the ground electronic state must be red-shifted and narrow relative to the MMCT absorption from primarily the ground vibrational state in the ground electronic state. The red shift is due to vibrational excitation in the ground state. The narrowing is due to the decreased number of vibrational states in the excited electronic state which are accessed by vertical transition, as a result of the displacement of the vibrationally-excited wavepacket along the nuclear coordinate axis. Since the 400 nm probe pulse is to the red of the static MMCT band maximum in aqueous solution, the red-shift of the band causes an increase of absorbance at 400 nm. Barbara and coworkers have observed a similar absorptive signal following decay of the excited-state bleach in time-resolved experiments on $[(H_3N)_5Ru^{III}$-NC-$Ru^{II}(CN)_5]^-$, with probe wavelengths to the red of the static absorption band maximum [9,10,16,17].

For Ru(II)-Pt(IV)-Ru(II) in water, the absorptive signal decays to a long-lived bleach ($\tau > 100$ ps), which is assigned to formation of the two-electron transfer product Ru(III)/Pt(II)/Ru(III) through FET. This assignment is justified by the absence of a long-lived bleach in $[(H_3N)_5Ru^{III}$-NC-$Ru^{II}(CN)_5]^-$. The absorptive signal in $[(H_3N)_5Ru^{III}$-NC-$Ru^{II}(CN)_5]^-$ decays to the baseline absorbance within 10 ps, completing the dynamics following photoinduced charge transfer [9,10,16,17]. The main difference between the two complexes is the presence of the third potential energy surface in Ru(II)-Pt(IV)-Ru(II), and therefore the second available decay pathway from the one-electron transfer intermediate, Ru(III)-Pt(III)-Ru(II). The interpretation of the one-color data for Ru(II)-Pt(IV)-Ru(II) in water is summarized in Figure 3.

The one-color pump-probe data for Ru(II)-Pt(IV)-Ru(II) in water/DMSO mixtures with $X_{DMSO} \leq 0.1$ exhibit similar features to the data for aqueous Ru(II)-Pt(IV)-Ru(II). The one-color data for Ru(II)-Pt(IV)-Ru(II) in water/DMSO mixtures with $X_{DMSO} > 0.1$ (Figure 2d) contain no absorptive signal following decay from the one-electron transfer intermediate. At sufficiently high X_{DMSO}, the static MMCT absorption maximum approaches 400 nm. Thus, the dynamic red shift of the MMCT band upon population of vibrationally-excited levels in the ground electronic state following BET causes a smaller (or negligible) increase in the absorbance at 400 nm. Therefore, the pump-probe signal (following the initial excited state absorption and cross phase modulation-induced artifact) consists of a single bleach, the decay of which represents a convolution of BET (and FET) dynamics with vibrational relaxation dynamics. The interpretation of the one-color pump-probe data for Ru(II)-Pt(IV)-Ru(II) in a water/DMSO mixture with $X_{DMSO} = 0.175$ is summarized in Figure 3.

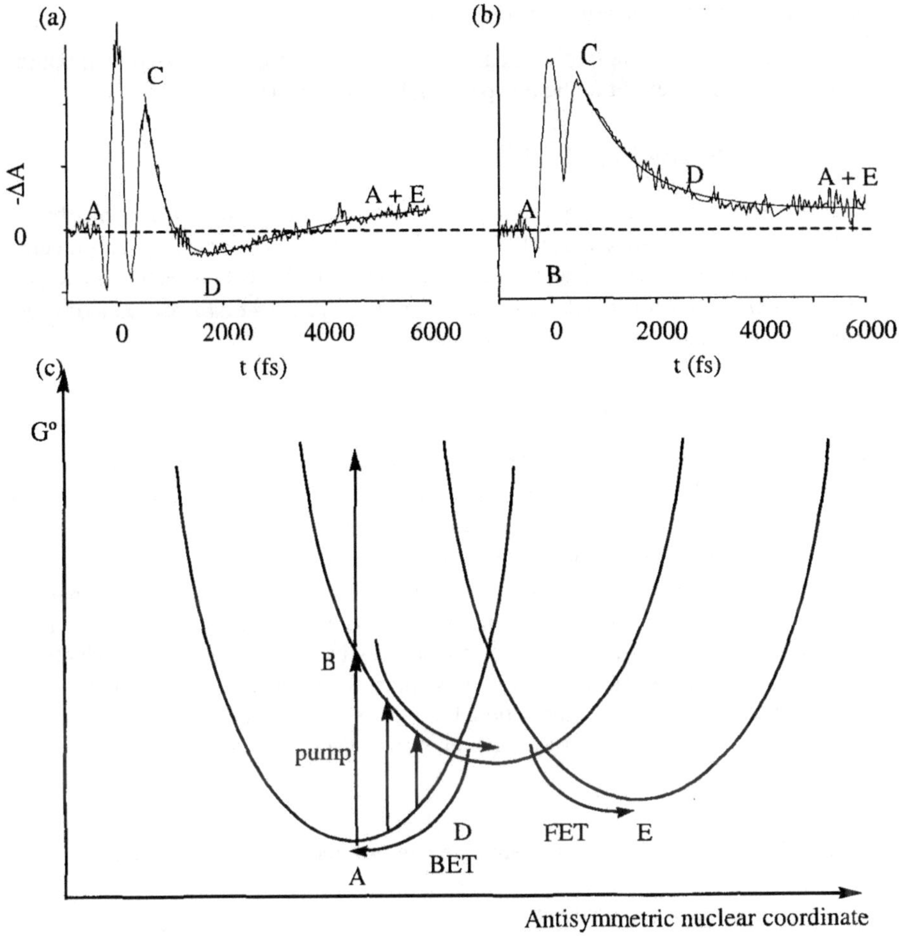

Figure 3. Interpretation of one-color pump-probe data for Ru(II)-Pt(IV)-Ru(II) in water (a) and in a water/DMSO mixture with X_{DMSO} = 0.2 (b). At t<0, the ground state Ru(II)-Pt(IV)-Ru(II) (A) is populated. Pump absorption creates population in the vibrationally-excited electronic excited state, Ru(II)-Pt(III)-Ru(III)* (B), from which an excited state absorption occurs (convolved with the artifactual signal). The excited state decays to the one-electron transfer state, Ru(II)-Pt(III)-Ru(III) (C), corresponding to a bleach. Decay through BET or FET yields Ru(II)-Pt(IV)-Ru(II) (A) or Ru(III) / Pt(II) / Ru(III) (E). Vibrational relaxation in the ground electronic state (D) corresponds to either an absorptive signal (as in a) or a bleach (as in b), depending on the relative energies of the probe beam and the MMCT transition from vibrationally excited levels in the ground electronic state. (See text.)

3.3 *Summary of pump-probe results*

The one-color pump-probe data for Ru(II)-Pt(IV)-Ru(II) in solvent mixtures with $X_{DMSO} \leq 0.1$ were fit to the biexponential decay function

$$\Delta A(t) = \Delta A_0 + C_1 e^{-t/\tau_1} + C_2 e^{-t/\tau_2} \tag{1}$$

where $\Delta A(t)$ is the time-dependent absorbance difference, ΔA_0 is the absorbance difference of the long-lived bleach, C_1 and C_2 are pre-exponential weighting factors, t is the delay time, and τ_1 and τ_2 are the lifetimes of the two decay processes. Mixed-solvent data for which $X_{DMSO} > 0.1$ obeyed psuedo-first order kinetics and were fit to the single exponential decay function

$$\Delta A(t) = \Delta A_0 + C_1 e^{-t/\tau_1} \tag{2}$$

Fit results are summarized in Table 1. Population dynamics are most easily extracted from the one-color data for Ru(II)-Pt(IV)-Ru(II) with $X_{DMSO} \leq 0.1$, for which deconvolution of the signals due to electron transfer dynamics (τ_1) and vibrational relaxation dynamics (τ_2) is straightforward.

The dependence of τ_1 on X_{DMSO} could arise from either changes in the dynamics of solvent reorganization or solvent-induced shifts of the Ru(II)-Pt(IV)-Ru(II) potential energy surface. A number of examples have been reported in which the rate of intramolecular electron transfer is limited by the rate of inertial solvation (uncoupled rotational and translational motion) [6,9,16-18,23]. The measured values of τ_1 for Ru(II)-Pt(IV)-Ru(II) in water/DMSO mixtures are significantly

Table 1. Results of fitting representative pump-probe data to eqs. 1 and 2.[a]

$\lambda_{pump} = 400$ nm, $\lambda_{probe} = 400$ nm

Sample	X_{DMSO}	τ_1 (fs)	τ_2 (fs)	C_1[b]	C_2[b]
Ru(II)-Pt(IV)-Ru(II)	0	476 ± 23	1890 ± 215	-0.80	0.20
	0.0125	796 ± 179	1395 ± 410	-0.61	0.39
	0.0375	861 ± 275	1465 ± 737	-0.62	0.38
	0.1	1110 ± 288	2720 ± 2130	-0.67	0.33
	0.2	1047 ± 23		-1	

[a] Data which contain the absorptive signal following bleach decay were fit to the biexponential function (equation 1). Data which exhibit pseudo-first order bleach decay kinetics were fit to the single exponential function (equation 2).
[b] Amplitudes are normalized such that $|C_1| + |C_2| = 1$.

longer than the inertial solvation timescales of water (50 fs) and DMSO (215 fs) [7-9]. Thus, it is unlikely that inertial solvent motion is the rate-determining process in the BET dynamics of Ru(II)-Pt(IV)-Ru(II).

As described above, the Ru(II)-Pt(IV)-Ru(II) potential energy surface shifts linearly to higher energy with X_{DMSO}, causing an increase in the activation barrier to BET ($\Delta G^{\ddagger}_{BET}$). Quantitative analysis of the dependence of τ_1 on $\Delta G^{\ddagger}_{BET}$ is complicated by several factors. The magnitude of the increase in $\Delta G^{\ddagger}_{BET}$ with X_{DMSO} depends on the slopes of the Ru(II)-Pt(IV)-Ru(II) and Ru(III)-Pt(III)-Ru(II) potential energy surfaces and on the degree of electronic coupling between them. (Thus, $\Delta G^{\ddagger}_{BET}$ is not necessarily linear with X_{DMSO}.) The increase of $\Delta G^{\ddagger}_{BET}$ causes a greater percentage of relaxation from Ru(III)-Pt(III)-Ru(II) to occur through FET, thereby increasing the quantum yield of the two-electron transfer product. In addition, a simple transition state kinetic model does not account for electron transfer by tunneling between potential energy surfaces. In spite of these complications, the BET rate constant depends exponentially on X_{DMSO}. The k_1 vs. X_{DMSO} data, where $k_1 = 1/\tau_1$, fit well ($R^2 = 0.988$) to the single exponential decay function

$$k_1 = Ae^{(-X_{DMSO}/t_1)} + C \qquad (3)$$

where $A = 1.11 \times 10^{12}\ s^{-1}$, $C = 9.46 \times 10^{12}\ s^{-1}$, $t_1 = 0.014$. The exponential dependence of k_1 on X_{DMSO} is consistent with Arrhenius-type behavior.

3.4 Conclusions

Systematic alteration of electron transfer dynamics through shifting of the relative energies of the Ru(II)-Pt(IV)-Ru(II) and Ru(III)-Pt(III)-Ru(II) potential energy surfaces has been demonstrated. Ground state vibrational relaxation timescales of 1.4 to 3 ps correspond well with previously-reported values in mixed-valence complexes [2,3,9,10,16]. BET timescales for Ru(II)-Pt(IV)-Ru(II) vary from 476 ± 23 fs in aqueous solution to 1110 ± 288 fs in a water/DMSO mixed solvent with $X_{DMSO} = 0.1$. Variation of the BET timescale with solvent is explained in terms of shifts of the Ru(II)-Pt(IV)-Ru(II) potential energy surface and the resulting change of $\Delta G^{\ddagger}_{BET}$. The observed increase of $\Delta G^{\ddagger}_{BET}$ with decreasing $|\Delta G^{\circ}_{BET}|$ clearly indicates that Ru(II)-Pt(III)-Ru(III) → Ru(II)-Pt(IV)-Ru(II) BET occurs with Marcus normal region behavior [1,11]. Normal region behavior is in agreement with the exceptionally large absorption bandwidths and total reorganization energies of Ru(II)-Pt(IV)-Ru(II) and Fe(II)-Pt(IV)-Fe(II) [13,15].

4 Acknowledgements

This work was funded by the National Science Foundation, the State of New Jersey, and the International Precious Metals Institute.

References

1. Barbara P. F., Meyer T. J. and Ratner M. A., Contemporary issues in electron transfer research, *J. Phys. Chem.* **100** (1996) pp. 13148-13168.
2. Doorn S. K., Dyer R. B., Stoutland P. O. and Woodruff W. H., Ultrafast electron-transfer and coupled vibrational dynamics in cyanide bridged mixed-valence transition-metal dimers, *J. Am. Chem. Soc.* **115** (1993) pp. 6398-6405.
3. Doorn S. K., Stoutland P. O., Dyer R. B. and Woodruff W. H., Picosecond infrared study of ultrafast electron transfer and vibrational energy relaxation in a mixed-valent ruthenium dimer, *J. Am. Chem. Soc.* **114** (1992) pp. 3133-3134.
4. Eichler H. J., Langhans D. and Massmann F., Coherence peaks in picosecond sampling experiments, *Opt. Commun.* **50** (1984) pp. 117-122.
5. Ekvall K., van der Meulen P., Dhollande C., Berg L.-E., Pommeret S., Naskrecki R. and Mialocq J.-C., Cross phase modulation artifact in liquid phase transient absorption spectroscopy, *J. Appl. Phys.* **87** (2000) pp. 2340-2352.
6. Heitele H., Dynamic solvent effects on electron-transfer reactions, *Angew. Chem., Int. Ed. Engl.* **32** (1993) pp. 359-377.
7. Horng M. L., Gardecki J. A., Papazyan A. and Maroncelli M., Subpicosecond measurements of polar solvation dynamics: Coumarin 153 revisited, *J. Phys. Chem.* **99** (1995) pp. 17311-17337.
8. Jimenez R., Fleming G. R., Kumar P. V. and Maroncelli M., Femtosecond solvation dynamics of water, *Nature (London)* **369** (1994) pp. 471-473.
9. Kambhampati P., Son D. H., Kee T. W. and Barbara P. F., Solvent effects on vibrational coherence and ultrafast reaction dynamics in the multicolor pump-probe spectroscopy of intervalence electron transfer, *J. Phys. Chem. A* **104** (2000) pp. 10637-10644.
10. Kliner D. A. V., Tominaga K., Walker G. C. and Barbara P. F., Comparison of experimental and theoretical absolute rates for intervalence electron transfer, *J. Am. Chem. Soc.* **114** (1992) pp. 8323-8325.
11. Marcus R. A., Electron-transfer reactions in chemistry - theory and experiment (nobel lecture), *Angew. Chem., Int. Ed. Engl.* **32** (1993) pp. 1111-1121.
12. Palfrey S. L. and Heinz T. F., Coherent interactions in pump-probe absorption measurements: The effect of phase gratings, *J. Opt. Soc. Am. B: Opt. Phys.* **2** (1985) pp. 674-679.
13. Pfennig B. W. and Bocarsly A. B., Optical and thermal charge-transfer processes occurring in a series of three-centered, cyanide-bridged intervalent charge-transfer complexes, *J. Phys. Chem.* **96** (1992) pp. 226-233.

14. Pfennig B. W. and Bocarsly A. B., Photoinduced multielectron charge transfer in the polymetallic intervalent complex $[(NC)_5Fe\text{-}CN\text{-}Pt(NH_3)_4\text{-}NC\text{-}Fe(CN)_5]^{4-}$ and related species, *Coord. Chem. Rev.* **111** (1991) pp. 91-96.
15. Pfennig B. W., Wu Y., Kumble R., Spiro T. G. and Bocarsly A. B., Time-dependent resonance raman analysis of a trinuclear mixed valence coordination complex, *J. Phys. Chem.* **100** (1996) pp. 5745-5750.
16. Reid P. J., Silva C., Barbara P. F., Karki L. and Hupp J. T., Electronic coherence, vibrational coherence, and solvent degrees of freedom in the femtosecond spectroscopy of mixed-valence metal dimers in h_2o and d_2o, *J. Phys. Chem.* **99** (1995) pp. 2609-2616.
17. Tominaga K., Kliner D. A. V., Johnson A. E., Levinger N. E. and Barbara P. F., Femtosecond experiments and absolute rate calculations on intervalence electron transfer of mixed-valence compounds, *J. Chem. Phys.* **98** (1993) pp. 1228-1243.
18. Tominaga K., Walker G. C., Kang T. J. and Barbara P. F., *J. Phys. Chem.* **95** (1991) pp. 10485-10492.
19. Vardeny Z. and Tauc J., Picosecond coherence coupling in the pump and probe technique, *Opt. Commun.* **39** (1981) pp. 396-400.
20. Weiner A. M. and Ippen E. P., Femtosecond excited state relaxation of dye molecules in solution, *Chem. Phys. Lett.* **114** (1985) pp. 456-460.
21. Wu Y., Cohran C. and Bocarsly A. B., The construction of molecular systems capable of photoinduced multielectron charge transfer: A photochemical analysis of the reactivity of $[(NC)_5Fe\text{-}CN\text{-}Pt(L)_4\text{-}NC\text{-}Fe(CN)_5]^{4-}$ (L=ethylenediamine), *Inorg. Chim. Acta* **226** (1994) pp. 251-258.
22. Wu Y., Pfennig B. W., Sharp S. L., Ludwig D. R., Warren C. J., Vicenzi E. P. and Bocarsly A. B., Light-induced multielectron charge transfer processes occurring in a series of group-8-platinum cyanobridged complexes, *Coord. Chem. Rev.* **159** (1997) pp. 245-255.
23. Yoshihara K., Tominaga K. and Nagasawa Y., Effects of the solvent dynamics and vibrational motions in electron-transfer, *Bull. Chem. Soc. Jpn.* **68** (1995) pp. 696-712.
24. Zhou M., Pfennig B. W., Steiger J., Van Engen D. and Bocarsly A. B., Multielectron transfer and single-crystal X-ray structure of a trinuclear cyanide-bridged platinum-iron species, *Inorg. Chem.* **29** (1990) pp. 2456-2460.

PROBING SOLVATION DYNAMICS AT AQUEOUS LIPID-MEMBRANE INTERFACES WITH 3-PULSE PHOTON-ECHO PEAK-SHIFTS

HELGE BÜRSING, SANTI KUNDU AND PETER VÖHRINGER

Max-Planck-Institute for biophysical Chemistry
Biomolecular and Chemical Dynamics Group, Am Faßberg 11, 37077 Göttingen, Germany
E-mail: pvoehri@gwdg.de

Three-pulse photon-echo peak-shift (3PEPS) experiments were performed to explore the time scales and the nature of structural relaxations relevant to solvation of a cyanine dye non-covalently anchored to aqueous interfaces of unilamellar phospholipid vesicles. The influence of the phospholipid polymorphism is investigated through the temperature-dependence of the 3PEPS decay. As opposed to the main order/disorder phase transition connected with packing phenomena in the hydrophobic membrane core, the rippled-to-gel pre-transition strongly affects the solvation dynamics on time scales below 2 ps. This finding indicates that the pre-transition is connected with structural rearrangements in the interfacial hydration shell surrounding the polar head groups and is consistent with a partial dehydration of the interface as found previously in multilamellar vesicles.

1 Introduction

Solvation dynamics at aqueous phospholipid membrane interfaces are of considerable importance for charge and mass transport across biological boundaries mediated either by diffusion or by integral membrane-bound transport proteins or ion channels. Solvent relaxation following electronic excitation of suitable membrane-bound optical chromophores has previously been studied on nanosecond time scales by reconstructing their time-dependent fluorescence Stokes shift from time-correlated single-photon counting data [6]. Similarly, solvation dynamics of probe molecules confined to the inner water pool of reversed micelles [8] or phospholipid vesicles [5] has also been studied. However, despite its particular relevance for biological transport phenomena, the ultrafast dynamics of interface-specific solvation in phospholipid membranes has neither been addressed experimentally nor by means of computer simulations.

We have recently been able to study the dynamics of solvation of a cyanine dye that was non-covalently anchored to the hydrophobic core of phospholipid SUVs by means of long-chain alkyl residues attached to its optical chromophore [2]. The latter itself has a very high specificity to the polar membrane interface as evidenced by high-resolution confocal microscopy [9]. Here, we present complementary 3-pulse photon-echo peak-shift (3PEPS) studies to elucidate in more detail the temperature dependence of biointerface-specific solvation dynamics. Phospholipid membranes exhibit a rich polymorphism that can be traced back primarily to ordering phenomena within the hydrophobic core [3]. The present study is our first attempt at exploring possible interrelations between the complex thermodynamic phase behavior of phospholipids and short-time structural relaxations occurring at their interface to the aqueous surroundings.

2 Experimental

Photon-echo experiments were carried out with a home-built non-collinearly phase-matched optical parametric amplifier (NOPA) driven by a Ti:Sapphire-regenerative amplifier at a repetition rate of 1 kHz. Following compensation for group velocity dispersion in a pair of fused silica prisms, the NOPA output was characterized by pulse durations of 20 fs and energies as high as 8 μJ when tuned to a center wavelength of 650 nm. A set of three Al-beamsplitters was used to prepare a sequence of three identical pulses whose timing was adjusted by computer controlled optical delay lines. Photon-echoes were generated in both triangular and Boxcar phase matching geometries as described previously [2]. SUVs of dimyristoyl (DMPC), dipalmitoyl (DPPC), and dioleyl (DOPC) phosphatidylcholine in aqueous solution were prepared and labeled with the chromophore, 1,1'-dioctadecyl-3,3,3',3'-tetramethylindodicarbocyanine chloride (C18-DiD) according to standard procedures. The molecular dye to lipid ratio was $\leq 1{:}10^3$ at a lipid concentration of 20 mM (DMPC) or 10 mM (DPPC). To avoid local heating and thermal lensing effects, the sample was contained in a home-built 1-mm pathlength optical cell capable of spinning at 1000 rpm.

3 Results and Discussion

To characterize the thermodynamic phase behavior of the labeled and unlabeled SUVs, differential scanning calorimetry was carried out. Representative differential specific heat data for DMPC and DPPC are displayed in Figure 1. For both lipids, the major feature reflects the gel-to-fluid phase transition at which the degree of ordering of the hydrocarbon tails in the membrane core dramatically changes (chain melting) [3]. For DMPC, this transition occurs at 23°C whereas for DPPC, the transition temperature is 41°C. Both systems exhibit a rather broad so-called pre-transition which is located roughly 12°C (for DMPC) or 9°C (for DPPC) below the main transition, i.e. fully in the gel phase [3]. According to freeze-fracture electron microscopy, this behavior is characteristic for formation of rippled vesicle structures on typical lengths scales of tens of nanometer. The phase intermediate to the two phase transitions is therefore denoted rippled gel [3].

3PEPS experiments were conducted in each of the phases of both lipids, i.e. in the gel, the rippled gel, and the fluid as indicated in Figure 1 by the vertical arrows. In a three-pulse echo experiment [1], the sample is subject to a sequence of three consecutive optical pulses which are on-resonance with an electronic transition of the optical chromophore. The first pulse prepares an electronic coherence in the chromophore which is converted into a population by the second pulse. The time delay between these two pulses is denoted the coherence time, t_{12}. The third pulse which is temporally separated from the second by a fixed waiting time, t_{23}, converts this population into a coherence upon which a photon echo will be emitted by the sample whose time-integrated intensity is monitored as a function of t_{12}. The primary observable of a 3PEPS experiment is the temporal position along t_{12} of maximal echo intensity as a function of increasing waiting time. In the limit of linear response, the 3PEPS experiment is capable of tracking the decay of the solvation correlation function relevant to the optical chromophore except on time scales comparable to the pulse durations where free induction decay (FID)-like effects from impulsive excitation into the quasi-continuum of intramolecular vibrational modes of the chromophore give rise to extraneous contributions [1].

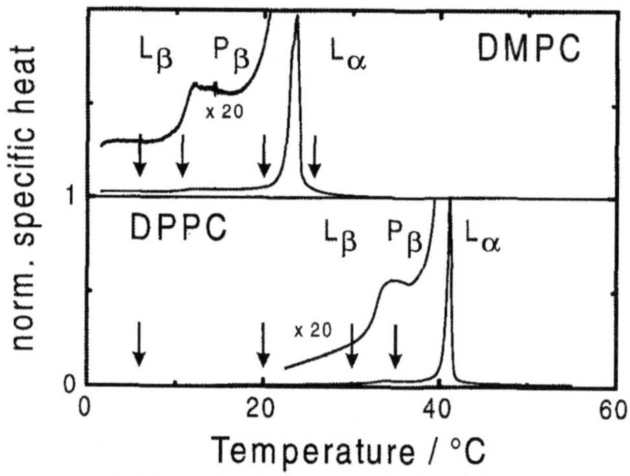

Figure 1. Differential scanning calorimetry of unilamellar vesicles of DMPC (top) and DPPC (bottom). The different phases are denoted L_β (gel), P_β (rippled gel), and L_α (fluid or liquid crystalline). Vertical arrows indicate the temperatures at which 3PEPS decays were recorded.

The resulting temperature dependent 3PEPS decays for C18-DiD in DMPC are shown in Figure 2. In the following, we will discuss the behavior of the 3PEPS data only for time scales longer than 80 fs (i.e. 4 times the pulse duration) to circumvent FID-like contributions. We also discard coherent intramolecular vibrational contributions giving rise to pronounced periodic modulations of the 3PEPS data on time scales below 1 ps. It can be seen that above the pre-transition, the temperature dependence of the overall 3PEPS decay is almost negligible. For these temperatures, a fit to a double-exponential decay can satisfactorily reproduce the data. Global time constants of 150 ps and 12 ps with amplitudes of 3.2 fs and 0.9 fs can be used independently of the temperature. We have previously assigned these two components to rotational self-diffusion of the chromophore and to rotational diffusion of water molecules in the immediate vicinity of the membrane interface at which the optical chromophore is embedded. Such a preliminary assignment was based on a qualitative agreement of the time constants with those obtained by other experimental techniques such as dielectric dispersion data for pure water and transient anisotropy data for cyanine dyes in liquid solution [2].

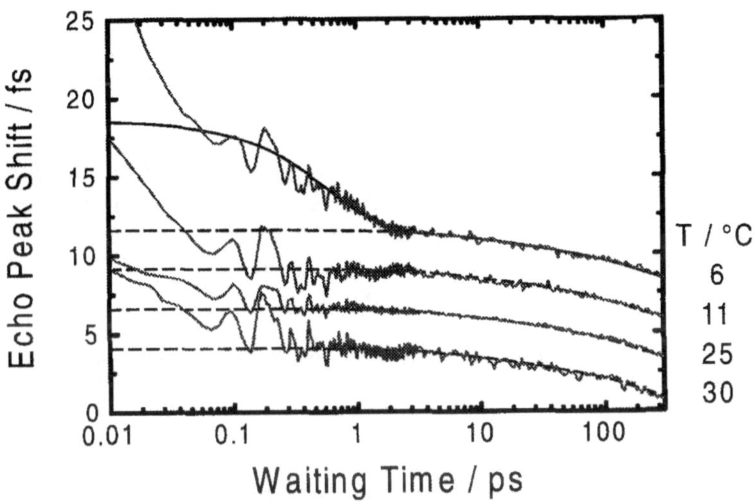

Figure 2. 3PEPS decays for C18-DiD in unilamellar vesicles of DMPC at different temperatures. The experimental data are displayed as gray curves. A fit to a temperature independent double-exponential decay is represented by the dashed curves. The solid curve represents a fit to the 6°C-data using a triple-exponential decay. For clarity, the data and the fits have been shifted vertically by successive increments of 2.5 fs.

Interestingly, the main chain-melting transition at 24°C does not affect the 3PEPS decay significantly. More importantly, however, for temperatures below the pre-transition, the 3PEPS decay contains an additional ultrafast component below 2 ps which is clearly absent in the rippled gel phase and the fluid phase. To adequately represent the 6°C-data, a triple-exponential decay is needed whose fastest component has a time constant of 600 fs with an amplitude of 7 fs while the remaining two components were taken from the previous double-exponential fit.

To explore in more detail whether this peculiar temperature dependence is indeed directly linked to phase transition-induced modifications of the interfacial solvation response or simply brought about by the presence of an interface regardless of the thermodynamic phase of the phospholipid itself, complementary experiments were performed in unilamellar vesicles of DPPC and DOPC. As shown in Figure 1, the phase diagram of DPPC is shifted to significantly higher temperatures. Accordingly, if the additional 600 fs-component observed in DMPC were truly specific to the gel phase, a similar but thermally shifted behavior can be expected for DPPC. More quantitatively, using the inflection points of the specific heat data, the gel-to-ripple phase transition of DMPC occurs at 11°C whereas in DPPC this transition is raised to a temperature of 32°C. Consequently upon cooling the rippled gel-phase, an additional gel phase-specific component would appear at much higher temperatures in comparison to DMPC provided the temperature dependence of such a component allows for its unambiguous identification from an experimental 3PEPS decay.

In contrast, the phase diagram of DOPC (not shown) is shifted to lower temperatures such that both the pre-transition as well as the main transition are located well below the freezing point of liquid water. Therefore, if the additional ultrafast component were indeed gel phase-specific, the 3PEPS decay should be more or less temperature independent as are the DMPC data above the pre-transition. If, on the other hand, the additional 600 fs-component were totally unrelated to the gel phase of phospholipids and were simply due to the presence of an aqueous lipid interface, the temperature dependence of the 3PEPS decay in DOPC should closely resemble that of DMPC.

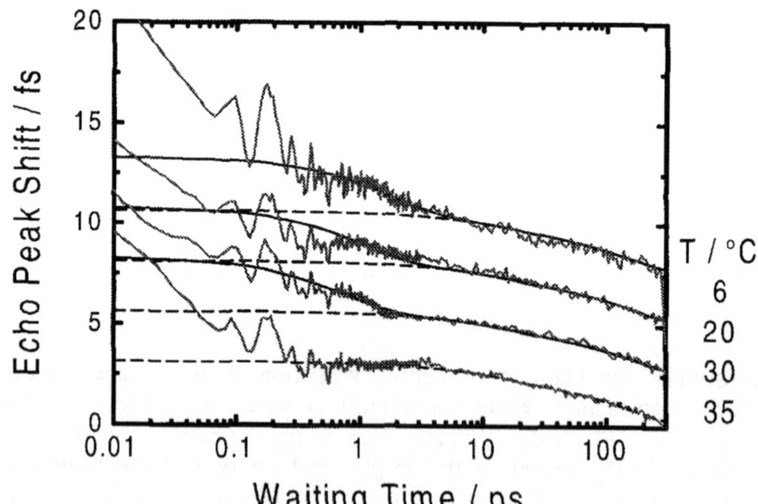

Figure 3. 3PEPS decays for C18-DiD in unilamellar vesicles of DPPC at different temperatures. The experimental data are displayed as gray curves. A fit to a temperature independent double-exponential decay is represented by the dashed curves. The solid curves represent fits using a triple-exponential decay whose fastest time constant is the only temperature dependent parameter. For clarity, the data and the fits have been shifted vertically by successive increments of 2.5 fs.

Figure 3 displays experimental 3PEPS data for C18-DiD anchored to DPPC vesicles. Again, for temperatures above the pre-transition, a double-exponential fit is sufficient to satisfactorily reproduce the data. The time constants are 160 and 12 ps with amplitudes of 2.3 and 0.85 fs. Again, a third ultrafast component with a time constant of 700 fs becomes evident but now at much higher temperatures as compared to DMPC. Upon further cooling of the sample, this component remains visible and its time constant increases with temperature (1.7°ps for 6°C, and 1.0 ps for 20°C). It should be noted, however, that its decay significantly deviates from an exponential and that a Gaussian can be used for fitting equally well. Most importantly, it should be emphasized that this third component is detected neither in the rippled phase nor in the fluid phase. This behavior strongly implies that this response is indeed entirely gel phase-specific. Such an interpretation is further strengthened by complementary 3PEPS experiments on C18-DID in DOPC (data not shown) which do not exhibit, within our signal-to-noise ratio, a significant temperature dependence between 5°C and 30°C. In addition, this finding also excludes a possible interpretation in which this component is assigned to the intramolecular response of the chromophore itself.

Having established its gel phase-specific nature, the microscopic origin of this ultrafast sub-2 ps response still needs to be clarified. Calorimetric studies on multilamellar vesicles of phosphatidylcholines as a function of the water content indicate that the pre-transition from the rippled-gel to the gel phase involves primarily a dehydration of the polar head groups [4]. Such an effect will also be experienced by the cyanine chromophore. It is quite possible that upon dehydration, the optical chromophore is slightly expulsed from the head group region while still being fully anchored with its alkyl tails to the hydrocarbon core. As a result, above the pre-transition, the chromophore is mostly exposed to the polar head groups and the strongly hydrogen-bonded interstitial water molecules. Below the pre-transition, dehydration of the head groups results in the formation of a water layer above the lipid membrane interface whose properties are reminiscent of bulk water. If the chromophore is able to penetrate into this bulk-like water layer, the solvation response is likely to become similar to the bulk.

In fact, we have previously shown that the room temperature solvation response of C18-DiD at DMPC/water interfaces differs from aqueous bulk solvation mostly for waiting times below 2 ps. It is on such short time scale where liquid water is able to respond with its restricted intermolecular translational modes typical for the random hydrogen-bonded bulk network. It is obvious that these bulk liquid modes are entirely missing in the rippled-gel or the fluid phase since interfacial water tends to be strongly bound to the polar head groups thereby completely disrupting the bulk network of hydrogen bonds. Finally, we note that this interface sensitivity of the cyanine membrane marker also explains why the solvation dynamics are unaffected by the main chain-melting transition where head-group dehydration is not important.

In conclusion, we have shown that the solvation response of a cyanine chromophore non-covalently anchored to aqueous interfaces of unilamellar phospholipid vesicles can be used to probe molecular level aspects of the lipid polymorphism. In comparison to the rippled-gel or the fluid phase, the gel phase is characterized by a loss of interfacial water molecules which is reflected in the short-time solvation of a probe molecule localized at the membrane surface. The interface-sensitivity of the membrane marker is responsible for significant modifications of the solvation response occurring exclusively at the phase transition from gel to rippled-gel.

4 Acknowledgements

We gratefully acknowledge Thomas Heimburg and his coworkers for performing calorimetric measurements as well as for technical assistance in preparing and labeling the phospholipid vesicles.

References

1. de Boeij W. P., Pshenichnikov M. S. and Wiersma D. A., Ultrafast solvation dynamics explored by femtosecond photon echo spectroscopies. *Ann. Rev. Phys. Chem.* **49** (1998) 99–123.
2. Bürsing H., Ouw D., Kundu S. and Vöhringer P., Probing solvation dynamics in liquid water and at phospholipid water interfaces with femtosecond photon-echo spectroscopies. *Phys. Chem. Chem. Phys.* 3 (2001) 2378–2386.
3. Cevc G. and Marsh D., Phospholipid bilayers: Physical principles and models. CRC press, Boca Raton (1990).
4. Cevc G., Polymorphism of the bilayer membranes in the ordered phase and the molecular origin of the lipid pretransition and rippled lamellae. *Biochem. Biophys. Acta* **1062** (1991) 59–69.
5. Datta A., Pal K., Mandal D. and Bhattacharyya K., Solvation dynamics of Coumarin 480 in vesicles. *J. Phys. Chem. B* **102** (1998) pp. 6114–6117.
6. Hutterer R., Schneider F. W., Sprinz H. and Hof M., Binding and relaxation behaviour of Prodan and Patman in phospholipid vesicles - A fluorescence and H-1 NMR study. *Biophys. J.* **61** (1996) pp. 151–160.
7. Koynova R. and Caffrey M., Phases and phase transitions of the phosphatidylcholines. *Biochem. Biophys. Acta* **1376** (1998) 91–145
8. Levinger N. E., Ultrafast dynamics in reversed micelles, microemulsions, and vesicles. *Curr. Opin. Coll. Interface Sci.* **5** (2000) pp.118–124.
9. Schwille P., Korlach J. and Webb W. W., Fluorescence correlation spectroscopy with single-molecule sensitivity on cell and model membranes. *Cytometry* **36** (1999) pp. 176–182.

ULTRAFAST TRANSIENT SECOND HARMONIC GENERATION FROM OXAZINE DYES AT THE AIR/WATER INTERFACE

DANIEL A. STEINHURST, ANDREW P. BARONAVSKI, AND JEFFREY C. OWRUTSKY

Chemistry Division, US Naval Research Laboratory, 4555 Overlook Avenue SW, Washington, DC 20375-5342, USA

We report single-color, ultrafast visible pump – second harmonic generation (SHG) probe measurements of oxazine dye aggregates (H-type) at the air/water interface. As we demonstrated previously [*J. Phys. Chem. B* **2001**, *105*, 3062] in our steady-state studies of aqueous oxazine dye solutions, the SHG signals are almost entirely due to dimers. The transient SHG recovery is biexponential (τ_1 = 4 – 8 ps, τ_2 = 22 – 44 ps) depending on the dye and the solution composition. Transient absorption measurements for the bulk dye solutions of water and methanol indicate a fast (2 ps) solvation dynamics component as well as a longer component due to the excited-state lifetime (>400 ps). The SHG recovery is single-photon resonant and is attributed to two parallel ground state recovery (GSR) mechanisms, restricting the possible explanations for the observed decay times. The shorter SHG decay time (~5 ps) could be due to ground state solvent relaxation, while the intermediate decay time (~25 ps) likely results from vibrational relaxation of "hot" ground state aggregates after internal conversion (IC) / intersystem crossing (ISC). The two decay times are different than what is seen for bulk solution and they resemble what has been previously reported for dye aggregates.

There has been interest in the dynamics of dye aggregates because of their importance to applications such as nonlinear optical materials and light-harvesting devices, including dye-capped semiconductors [1–3]. These systems often depend on molecular behavior at interfaces where aggregates often play a role. Also, the study of aggregate structure and dynamics offers a sensitive probe of the interface environment. The study of molecular dynamics at the air/water interface is greatly facilitated by the surface-specific technique of second harmonic generation (SHG) as demonstrated in the work of Eisenthal and coworkers [4]. As shown by Levinger et al. for IR-125 B-type aggregates [5] and in our previous steady-state SHG study of aqueous solutions of oxazine dyes (Nile blue, oxazine 720, and cresyl violet) [6], dimers (H-type for the oxazines) are the prevalent species at the air/water interface. The steady-state SHG signal intensity correlates with conditions that favor aggregation in bulk solutions. No SHG was detected from the interface of methanolic solutions or from other oxazines in aqueous solution for which negligible aggregation occurs in the bulk. It was also found that SHG is enhanced by adding salt, which is known to promote dimer formation in the bulk [7]. We report an ultrafast SHG investigation of oxazine dyes at the air/water interface, which is one of the first studies of dye aggregate dynamics at the air/water interface.

Single-color, femtosecond pump – SHG probe measurements were made at the air/water interface of aqueous dye solutions [8]. The laser pulses, centered in wavelength near 600 nm, are resonant with the first electronic transition of the aggregate and were derived from a regeneratively amplified Ti:Sapphire laser-pumped OPA [6] with a typical temporal resolution of 350 fs. Transient SHG (TSHG) recoveries for aqueous solutions of the three oxazine dyes with salt (LiCl) and for Nile blue without salt are shown in Figure 1 along with the calculated curves (solid lines) derived from a nonlinear fit to a biexponential decay convoluted with the instrument function.

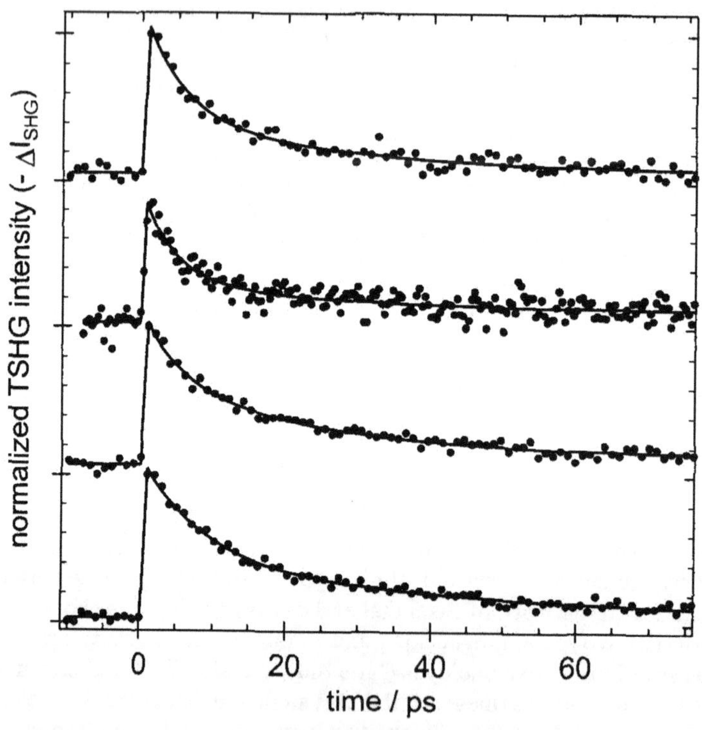

Figure 1. Normalized TSHG signal intensity scans (570 nm pump / 570 nm probe) for aqueous / LiCl solutions of (from top to bottom): oxazine 720, cresyl violet, and Nile blue. The bottom trace is for an aqueous Nile blue solution without LiCl. The solid lines represent biexponential fits to the data. The numerical results of the fits are found in Table 1.

In all cases, the pump pulse resulted in an instrument-limited decrease in the SHG signal followed by a partial recovery. The recovery is well described by a biexponential decay convoluted with the instrument function with decay times of 4 – 8 ps and 22 – 44 ps and a greater than 85% recovery in less than 100 ps. The derived decay constants are given in Table 1 along with the relative amplitudes of each component in parentheses. The SHG recovery is attributed to population changes since we found no evidence of orientational effects. The decay times are also probably too short to be due to rotational reorientation.

Table 1. Decay times and amplitudes for transient response of surface TSHG of aqueous oxazine dye solutions.[a]

dye	τ_1 / ps [b]	τ_2 / ps [b]	remaining [b]
oxazine 720	5 (57%)	23 (41%)	2%
cresyl violet	4 (64%)	23 (24%)	12%
Nile blue	5 (37%)	22 (55%)	8%
Nile blue w/o LiCl	8 (62%)	44 (38%)	<1%

[a] Derived from a nonlinear fit to a biexponential; amplitudes are given in parentheses.
[b] The uncertainty in the lifetimes and fractions remaining based on the nonlinear fitting procedure is ± 15%.

The results for the dyes at the interface are different that those observed for bulk dye solutions. Transient absorption (TA) / gain (TG) studies of the bulk dye solutions in both water and methanol were conducted and the results can be explained in terms of the well known mechanisms of solvation dynamics ($\tau \sim 2$ ps) [9] and excited state lifetimes ($\tau > 400$ ps) for the dye monomers [10]. No intermediate decay times (4 – 40 ps) were observed in the bulk measurements.

The fundamental wavelength from which the SHG probe is generated appears to be single-photon resonant with the dye aggregate $S_1 \rightarrow S_0$ transition and the recovery is attributed to GSR. The invocation of GSR implies that the TSHG recovery is due to aggregates in the ground state rather than in an excited state. This restricts the possible explanations for the observed biexponential form of the transients to two parallel mechanisms which repopulate the ground state. Probably both internal conversion (IC) and intersystem crossing (ISC) processes are involved as intermediate steps in the SHG recovery. The shorter SHG decay time could be due to ground state solvent relaxation after rapid IC from the aggregate excited state. The $S_1^+ \rightarrow S_1^-$ relaxation pathway is thought to be fast (sub-ps) [11], however rapid depopulation of the S_1^+ state via IC would not lead directly to ground state repopulation and therefore would not be observed in our experiment. Since the SHG probe depends on the ground state population, we do not attribute the observed decays to processes involving excited state population including excited state solvation dynamics. Probing via a two-photon resonance would be sensitive to population changes in both the ground and excited states, allowing the dynamics of the latter to be investigated if desired, as demonstrated by Zimdars et al. for coumarin 314 at the air/water interface [12].

Vibrational relaxation of "hot" aggregates after IC / ISC from the excited state is a probable explanation for the longer times. IC / ISC from the S_1^- / T_1 states to the ground state, S_o, has been proposed as the main relaxation route back to the ground state on silica and semiconductor particles for oxazine aggregates by Lui et al. [1] and Nasr and Hotchandani [2]. They report a single decay time for oxazine aggregates on silica in the range 2 – 250 ps. For aggregates on semiconductor particles, a second decay time in the same range due to back electron transfer is observed.

It is surprising that we find more complicated dynamics at the air/water interface than in the studies at water/solid interfaces. The bulk non-aqueous phase in our experiment (air) is less perturbative and we are only detecting ground state dynamics whereas the other studies are sensitive to both ground and excited state dynamics with TA/TG. These results are similar to those from previous reports of aggregate excited-state dynamics in solution and at solid interfaces, in which decays of 20 – 110 ps have been observed [13], but to our knowledge this is one of the first such reports for aggregates at the air/water interface.

Acknowledgements

This work was supported by the Office of Naval Research through the Naval Research Laboratory. This work was performed while DAS held a Naval Research Laboratory - National Research Council Research Associateship.

References

1. Liu, D; Kamat, P.V. *J. Chem. Phys.* **1996**, *103*, 209.
2. Nasr, C.; Hotchandani, S. *Chem. Mater.* **2000**, *12*, 1529.
3. Wachtveitl, J.; Huber, R.; Spörlein, S.; Moser, J.E.; Grätzel, M. *Int. Photoenergy* **1999**, *1*, 153.
4. Eisenthal, K.B. *Chem. Rev.* **1996**, *96*, 1343 – 1360.
5. Levinger, N.E.; Kung, K.Y.; Luther, B.M.; Willard, D.M. in *Laser Techniques for Surface Science II*, edited by Hicks, J.M.; Ho, W.; Dai, H.-L. *Proc. SPIE* **1995**, *2547*, 400.
6. Steinhurst, D.A.; Owrutsky, J.C. *J. Phys. Chem. B* **2001**, *105*, 3062.
7. Das, K.; Sarkar, N.; Das, S.; Datta, A; Nath, D.; Bhattacharyya, K. *J. Chem. Soc. Faraday Trans.* **1996**, *92*, 4993 – 4996.
8. Steinhurst, D.A.; Baronavski, A.P; Owrutsky, J.C. manuscript in preparation.
9. Jarzęba, W.; Walker, G.C.; Johnson, A.E., Kahlow, M.A.; Barbara, P.F. *J. Phys. Chem.* **1988**, *92*, 7039 – 7041.

10. Grofcsik, A.; Kubinyi, M.; Jones, W.J. *J. Molec. Struc.* **1995**, *348*, 197 – 200; Anfinrud, P.; Crackel, R.L.; Struve, W.S. *J. Phys. Chem.* **1984**, *88*, 5873 – 5882.
11. Martini, I.; Hartland, G.V.; Kamat, P.V. *J. Phys. Chem. B* **1997**, *101*, 4826 – 4830.
12. Zimdars, D.; Eisenthal, K.B. *J. Phys. Chem. A* **1999**, *103*, 10567.
13. Morgenthaler, M.J.E.; Meech, S.R. *J. Phys. Chem.* **1996**, *101*, 3323 – 3329.

LIGHT DRIVEN MULTISTEP ELECTRON TRANSFER IN A TYROSINE-RUTHENIUM-COMPLEX ANCHORED TO TiO_2 NANOPARTICLES

RAED GHANEM[1], JIE PAN[1], YUNHUA XU[2], JOHAN ANDERSSON[1], TOMÁŠ POLÍVKA[1], LICHENG SUN[2], VILLY SUNDSTRÖM[*,1]

[1]*Department of Chemical Physics, Lund University, P.O. Box 124, S-22100 Lund, Sweden*
E-mail: Villy.Sundstrom@chemphys.lu.se

[2]*Department of Organic Chemistry, Stockholm University, S-10691 Stockholm, Sweden*

We demonstrate a multistep electron transfer in a supramolecular complex adsorbed on the surface of nanocrystalline TiO_2. The complex mimics the function of the tyrosine$_Z$ and the chlorophyll unit P680 in natural Photosystem II (PSII). A ruthenium (II) tris-bipyridyl complex covalently linked to a L-tyrosine ethyl ester through an amide bond was anchored to the surface of nanocrystalline TiO_2 via four carboxylic acid groups linked to the bpy ligands. 450 nm excitation of the complex promotes an electron to a metal-to-ligand charge transfer MLCT excited state, from which the electron is injected into TiO_2. The photo-generation of Ru(III) is followed by an intramolecular electron transfer from tyrosine to Ru(III), regenerating the photosensitizer Ru(II) and forming the tyrosyl radical. The tyrosyl radical is formed in less than 5 μs with a yield of 15 %. This rather low yield is a result of a fast back electron transfer reaction from the nanocrystalline TiO_2 to the photooxidized Ru(III).

1 Introduction

The mechanism of photosynthetic water oxidation to molecular oxygen in Photosystem II (PSII) has been extensively studied [1,2]. Light is absorbed by the primary donor chlorophyll unit P680 and then, after several electron-transfer steps [3] the transfer of an electron to a quinone on the acceptor side of PSII leads to the oxidation of P680. Photooxidized $P680^+$ recaptures an electron by oxidizing a nearby tyrosine (Tyr_z) into a neutral tyrosyl radical ($Tyr_z^{\bullet}$) [1]. This radical in its turn oxidizes a tetranuclear Mn-cluster bound to PSII and after four consecutive turnovers, two water molecules are oxidized into molecular oxygen. Moreover, recent results proposed that the Tyr_z itself catalyze water oxidation by abstracting a hydrogen atom from a water molecule coordinated to the Mn-cluster [4]. This shows that the Tyr_z plays a crucial role in this process, although the exact mechanism of photosynthetic water oxidation is not known [3].

Ruthenium (II) tris-bipyridyl complexes were connected to L-tyrosine ethyl ester (to mimic the Tyr_z) and used to mimic the important parts of the light driven process in PSII in order to construct an artificial photosynthetic system for fuel production [5].

In previous studies it was shown that a model complex such as Ru(II)(bpy)(2)(4-Me-4'CONH-L-tyrosine ethyl ester-2,2'-bpy) $2PF_6$ in the presence of an external electron acceptor (methylviologen, MV^{2+}) mimics the Tyr_Z-P680 functional unit of PSII by generating a tyrosyl radical by means of intramolecular electron transfer from the tyrosine moiety to the photo-generated Ru(III) [4,6,7,8]. It was also demonstrated that the photo-generated tyrosyl radical can oxidize a binuclear manganese cluster from its Mn(III/III) state to the Mn(III/IV) state [7].

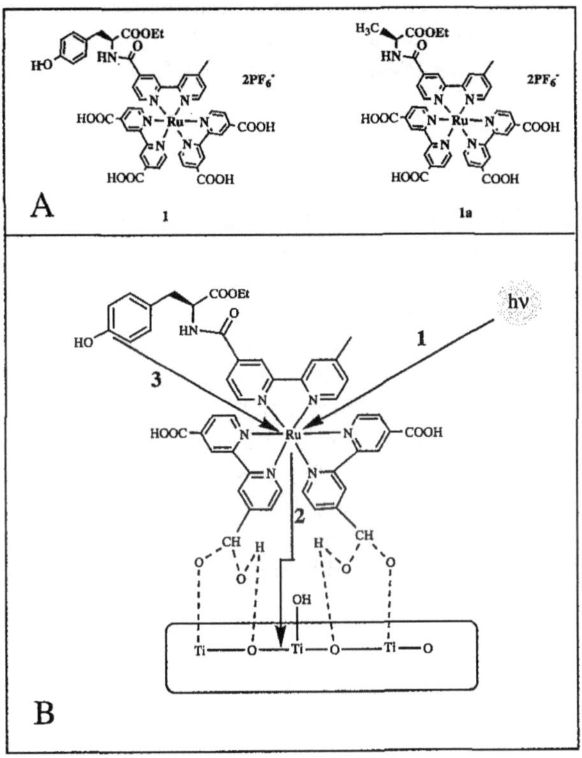

Figure 1. A. Complex **1** and its analogue complex **1a**. B. Reaction scheme proposed for the photo-induced electron transfer of the **1**-TiO_2 system. (1) Excitation of $Ru(II)(bpy)_3$. (2) Electron injection from the excited state of the Ru(II) complex to the TiO_2 producing Ru(III). (3) Intramolecular electron transfer from the tyrosine moiety to Ru(III) leading to recovery of the Ru(II).

Despite the proven functionality of the artificial system, use of an external electron acceptor such as MV^{2+} has the disadvantage of diffusion limited electron transfer from the excited state of the ruthenium complex to the external electron acceptor [4,6,7,8]. It is known that through dye molecule sensitization of a wide band gap semiconductor such as TiO_2, electrons can be injected from the excited state of the dye into the semiconductor on the femtosecond to picosecond time scale [9]. Thus, the use of a nanocrystalline TiO_2 semiconductor as an electron acceptor of the Ru-tyrosine complex will overcome the diffusion controlled nature of the electron transfer, and therefore enables to measure the true rate of the Tyrosine-to-Ru (III) electron transfer step.

New complexes **1** and **1a** (Figure1), which can be anchored to nanocrystalline TiO_2 via the presence of the four carboxylic acid groups attached to the bipyridine ligand have been synthesized. With this system (Figure 1) we will demonstrate the possibility of obtaining a multistep electron transfer in a supramolecular complex system incorporating tyrosine and a Ru-complex photosensitizer to mimic the function of the Tyr_Z and the P680 in the natural PSII.

2 Methods

Nanocrystalline colloidal TiO_2 particles were prepared by a controlled hydrolysis of $TiCl_4$ as proposed by Kormann et al [10]. Luminescence spectra were recorded using a SPEX Flurolog fluorimeter by exciting the sample at 450 nm. Absorption spectra were recorded on a Jasco-V-530 spectrophotometer in a 1 cm quartz cell.

Transient absorption experiments was conducted with a nanosecond laser flash photolysis setup. The excitation pulse at 450 nm was obtained from a Quanta-Ray Master Optical Parametric Oscillator (MOPO) pumped by a Quanta-Ray 230 Nd:YAG laser, having a pulse energy of 1-2 mJ and duration of ≈7 ns. The probe light from a Xenon arc lamp (75 W) was focused to a 1 mm diameter spot overlapping with the unfocused pump beam of 2.5 mm diameter. After the 1 cm quartz cell containing the sample, the probe light was passed through two single grating monochromators and detected by a photomultiplier Hamamatsu R928.

3 Result and Discussion

3.1 *Characterization of the Ru-TiO_2 system.*

Both complexes **1** and **1a** show characteristic photophysical properties found in other ruthenium complexes [5]. Quenching of the luminescence intensity of **1** with increasing of TiO_2 concentration was used to study the attachment of the complex on the surface of the TiO_2 particles.

The observed quenching of the MLCT excited state with increasing of TiO_2 concentration (Figure 2) is attributed to an efficient electron injection from the excited state of the complex to the conduction band of TiO_2 [11]. According to

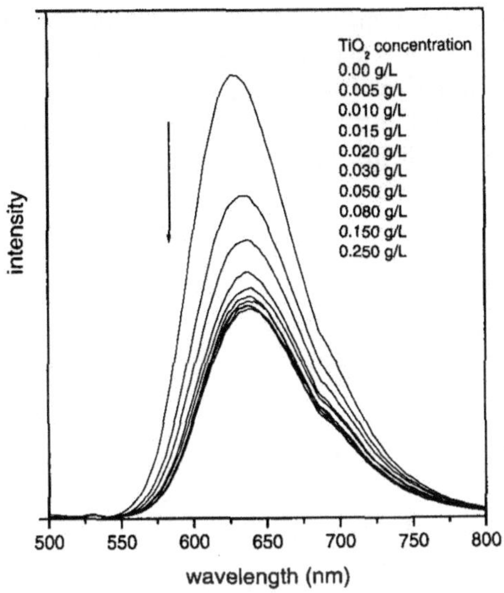

Figure 2. Luminescence emission spectra of **1** in the presence of various concentrations of TiO_2, [**1**] = 15 μM. Excitation at 450 nm

Kamat et al [12,13], the quenching of luminescence by a TiO_2 colloidal particle reflects the amount of chemisorbed dye on the surface of the TiO_2 particle. An apparent association constant K_a of 4.1 x 10^4 M^{-1} was calculated from analysis of the quenching of the MLCT excited state by TiO_2 particles. This analysis reveals that under our experimental conditions approximately 80 % of complex **1** or **1a** are adsorbed to the surface of TiO_2.

3.2 *Photoinduced electron transfer.*

Aqueous solutions containing 15 μM of **1** or **1a** and 120 μM of TiO_2 at pH = 2.4 were used for the flash photolysis measurements. The electron transfer processes were initiated when a pulse of visible light (λ = 450 nm) excites the MLCT band of Ru(II) and promotes an electron from a Ru d-orbital to a π^* orbital of the ligand, from which an electron can be injected into the conduction band of TiO_2. As a result, the dye cation Ru(III) is formed and the Ru(II) MLCT ground state absorption is bleached. Once the dye cation Ru(III) is formed, there are in principle two possible ways for the photogenerated Ru(III) to return to the Ru(II) ground state: either by back electron transfer from the nanorystalline TiO_2 (electron recombination), or by a second intramolecular electron transfer process from the linked tyrosine moiety [6,8]. Both reactions will result in repopulation of the Ru(II) ground state and thus in recovery of the Ru(II) ground state bleaching.

Analogously to other ruthenium dyes [14,15,16], electron injection from **1** and **1a** into TiO_2 is believed to occur on the subpicoseconds to picosecond time scale, which is significantly faster than the diffusion controlled electron transfer recorded in the presence of an external electron acceptor [8]. Therefore, for the complex **1** anchored to the nanocrystalline TiO_2, the second electron transfer from the tyrosine moiety to the Ru(III) is the rate-limiting step.

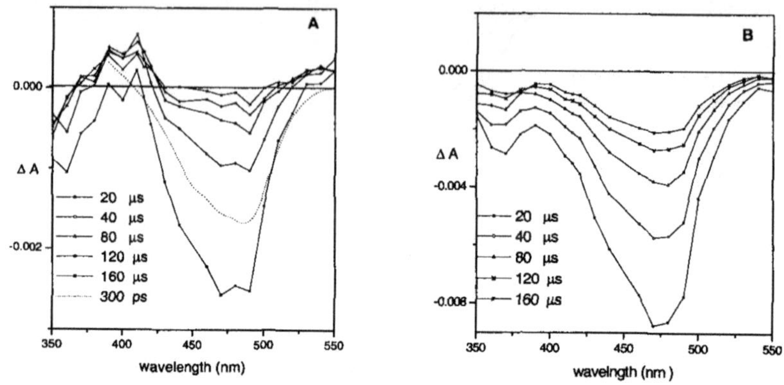

Figure 3. Transient absorption spectra of adsorbed complexes **1** (a) and **1a** (b) at pH 2.4 and 15 µM for both complexes **1** and **1a** with 120 µM of TiO_2 at different delay times after excitation at 450 nm. In panel (A), the transient absorption spectrum recorded at 300 ps was adjusted to allow a direct comparison with the spectra recorded at microsecond delays.

Figure 3 shows the temporal evolution of the transient absorption spectra of the **1**-TiO_2 and **1a**-TiO_2 systems. It is important to highlight two key points:

- While the bleaching recovery of Ru(II) for the **1**-TiO_2 system (Figure 3A) is accompanied by appearance of a new spectral feature located at 410 nm, which can be attributed to a formation of the tyrosyl radical (exhibits an absorption band at 410 nm [17]), the positive band at 410 nm is absent in the transient absorption spectra of **1a**-TiO_2.
- A direct comparison of the kinetics of the adsorbed complexes **1** and **1a** recorded at 470 nm (figure 4) shows that the bleaching recovery of the reference complex **1a** is slower than that observed for the complex **1**. Also, while the bleaching signal is almost gone after 100 µs in the case of complex **1**, the bleaching signal for complex **1a** did not recover to the base line within this time (Figure 4A).

This is a clear proof that the tyrosine moiety is responsible for the appearance of the positive band at 410 nm. In other words, the tyrosyl radical is formed as a result of the intramolecular electron transfer from the tyrosine part of **1** to the photogenerated Ru (III), while for the reference complex **1a,** which lacks the tyrosine part, this step no longer exists.

Taking into account the absorption coefficients of the tyrosyl radical (ε = 3000 $M^{-1}cm^{-1}$) and the complex **1** (ε = 15000 $M^{-1}cm^{-1}$), the yield of the Ru(III) to tyrosyl radical conversion can be estimated from the relative absorption magnitude of the Ru(II) ground state bleaching at early times (~0.5 μs) and the 410 nm band corresponding to the tyrosyl radical. Such estimation gives a tyrosine-ruthenium electron transfer yield of about 15%. This rather low yield of tyrosine-Ru(III) electron transfer is explained by a fast back electron recombination process from TiO_2 to the Ru(III), as can be deduced from the fitting of the kinetics in Figure 4.

Table 1. Time constants extracted from the multiexponential fitting of the 470 nm kinetics for both studied complexes.

System	τ_1(μs)	τ_1(μs)	τ_1(μs)	$k_{Effective}$
1a-TiO_2	0.5(73%)	4.4 ± 0.3 (13%)	20±3(14 %)	3.2x10^5 s^{-1}
1-TiO_2	0.5(73%)	4.2±0.3 (20 %)	20±7(7 %)	4.4x10^5 s^{-1}

The result of multiexponential fitting of the 470-nm kinetics of the complexes **1a** and **1** are summarized in Table 1. The fastest ≈ 0.5 μs component is assigned to the excited state lifetime of the unadsorbed dye. For the complex **1a,** lacking the tyrosine moiety, the microsecond time constants reflect the recombination between Ru(III) and electrons injected into TiO_2. Fitting of the 470 nm kinetic for the complex **1**, which necessarily contains also a component corresponding to the tyrosine-Ru(III) electron transfer, yields in fact almost the same time constants as for the complex **1a**. The amplitudes of the kinetic components are, however, different for the combined tyrosine-to-Ru(III) and TiO_2-to-Ru(III) electron transfer process. For the complex **1** the kinetic measured at 410 nm (Figure 4B) corresponds to the formation of the tyrosyl radical. For the complex **1a**, having no tyrosine moiety, the kinetics measured at 410 nm naturally reflects only the bleaching recovery due to the electron recombination between Ru(III) and TiO_2. Although the rates of the processes corresponding to the electron recombination and to the tyrosine-Ru(III) electron transfer are very similar, it is possible to calculate an effective rate constant from the amplitudes of the corresponding time constants. This effective rate is faster for the complex **1** due to an additional tyrosine-Ru electron transfer channel. The rate constant for the intramolecular electron transfer from tyrosine to Ru (III) observed here is faster than that obtained for a comparable complex in the presence of MV^{+2} as an external electron acceptor in solution [8].

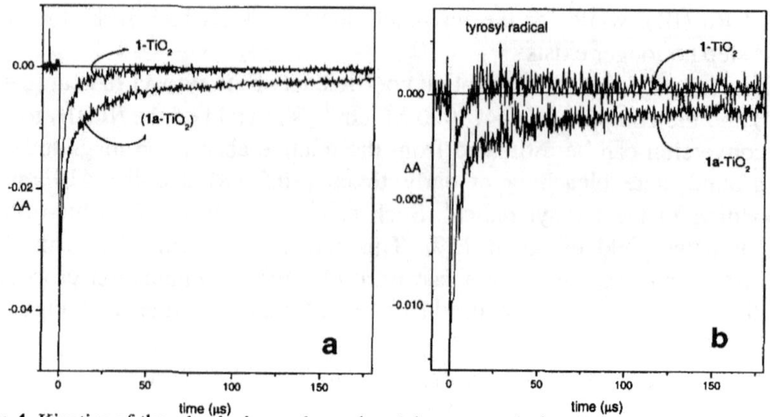

Figure 4. Kinetics of the adsorbed complexes **1** and **1a** measured after 450 nm excitation at pH 2.4 and concentration of 15 μM for both complexes with 120 μM of TiO_2. Kinetics were recorded at 470 nm (a) and at 410 nm (b).

4 Summary and conclusion

The presence of the four carboxylic acid groups in the complex **1** enables to anchor the complex to nanocrystalline TiO_2 with a good degree of association (up to 80%). This promotes an ultrafast injection of electrons from the excited MLCT state of the Ru-complex into the conduction band of TiO_2, which simplifies the study of the second intramolecular electron transfer step from the tyrosine moiety to the photogenerated Ru(III), since this process is now the rate limiting step. The tyrosine-to-Ru(III) electron transfer compete with back electron transfer from TiO_2 resulting to a rather low yield of the tyrosyl radical formation of ~ 15%.

5 Acknowledgements

We thank Dr Torbjörn Pascher for technical assistance. Financial support from the following agencies is gratefully acknowledged: Knut and Alice Wallenberg Foundation. Delegationen för Energiförsörjning i Sydsverige (DESS), the Swedish National Energy Administration, the Swedish Natural Science Research Council (NFR) and the Swedish Research Council for Engineering Sciences (TFR).

References

1. (a). Tommos, C., Tang, X. S., Warncke, K., Hoganson, C. W., Styring, S., McCracken, J., Diner, B. A and Babcock, G. T., Spin-density distribution, conformation, and hydrogen-bonding of the redox-active tyrosine y-z in photosystem-II from multiple electron magnetic-resonance spectroscopies - implications for photosynthetic oxygen evolution. J. Am. Chem. Soc. **117** (1995) pp, 10325-0335. (b). Hoganson, C. W., Lydakis-Simantiris, N., Tang, X. S., Tommos, C., Warncke, K., Babcock, G. T., Diner, B. A., McCracken, J and Styring, S., A hydrogen-atom abstraction model for function of Y-Z in photosynthetic oxygen evolution. Photosynth. Res. **46** (1995) pp, 177-184. (c). For recent reviews on photosynthetic water oxidation see the special issue of BBA-Bioenergetics BBA. **1503** (2001) Issue 1-2, 5
2. Yachandra, V. K., De Rose, V. J., Latimer, M. J., Mukerji, I., Sauer, K.an Klein, M. P.,Where plants make oxygen- A structural model for the photosynthetic oxygen-evolving manganese cluster. Science. **260** (1993), pp, 675-679.
3. (a). Hoganson, C. W and Babcock, G. T., A mettalloradical mechanism for the generation of oxygen from water in photosynthesis., Science. **277** (1997) pp, 1953-1956.
4. (a). Sun, L., Hammarström, L., Åkermark, B and Styring, S., Towards artificial photosynthesis: ruthenium-manganese chemistry for energy production. Chem. Soc. Rev. **30** (2001) pp, 36-49. (b) Hammarström, L., Sun, L., Åkermark, B and Styring, S., Artificial photosynthesis: towards functional mimics of photosystem II. BBA-Bioenergetics, **193** (1998) pp, 193-199.
5. (a). Kalyanasundaram, K. Photochemistry of Polypyridine and Porphyrin Complexes; Academic Press, London, (1992), Chapter 6 (b) Juris, A., Balzani, V., Barigelletti, F., Campagna, S., Belser, P and Vonzelewsky, A., Ru(II)

polypyridine complexses-photophysics, photochemistry, electrochemistry, and chemi-luminescence. Coord. Chem. Rev. **84** (1988) pp, 85-277.

6. Magnuson, A., Berglund, H., Korall, P., Hammarström, L., Åkermark, B., Styring, S and Sun, L., Mimicking electron transfer reactions in photosystem II: synthesis and photochemical characterization of a ruthenium(II) tris(bipyridyl) complex with a covalently linked tyrosine. J. Am. Chem. Soc. **119** (1997) pp, 10720-10725.
7. Magnuson, A., Frapart, Y., Abrahamsson, M., Horner, O., Berglund, H., Korall, P., Åkermark, B., Styring, S., Jean-Jacques, G., Hammarström, L and Sun, L., A biomimetic model system for the water oxidizing triad in photosystem II. J. Am. Chem. Soc. **121**(1999) pp , 89-96.
8. Sjödin, M., Styring, S., Åkermark, B., Sun, L. and Hammarström, L., Proton-coupled electron transfer from tyrosine in a tyrosine-ruthenium-tris-bipyridine complex: Comparison with tyrosineZ oxidation in photosystem II. J. Am. Chem. Soc. **122** (2000) pp, 3932-3936.
9. Hilgendorff, M and Sundström, V., Dynamics of electron injection and recombination of dye-sensitized TiO_2 particles. J. Phys. Chem B. **102** (1998), PP, 10505-10514.
10. Kormann, C., Bahnemann, W and Hoffmann, R., preparation and characterization of quantum-size Titanium dioxide. J. Phys. Chem. **92** (1988) pp, 5196-5201.
11. Thomspon, A., Smailes, M., Jeffery, J and Ward, M., Ruthenium tris-(bipyridyl) complexes with pendant protonable and deprotonatable moieties: pH sensitivity of electronic spectral and luminescence properities. J. Chem. Soc. Dalton Trans. (1997) pp, 737-743.
12. (a) Xiang, J., Chen, C., Zhou, B and Xu, G., Abnormal fluorescence behavior of some cyanine dys in a cyanine dye-TiO_2 colloids system. Chem. Phys. Lett. **315** (1999) pp, 371-378. (b) Kamat, P.,Photochemistry of nonreactive and reactive (semiconductor) surface. Chem. Rev. **93** (1993) pp, 267-300.
13. Kamat,V. , Photoelectrochemistry in particulate system. 9. Photosensitized reduction in a colloidal TiO_2 system using anthracen-9-carboxylic acid as sensitizer. J. Phys. Chem. **93** (1989) pp, 859-864.
14. Tachibana,Y., Haque, S., Mercer, I., Durrant, J and Klug, D., electroninjection and recombination in dye sensitized nanocrystalline titanium dioxide film : A comparison of ruthenium bipyridyl and prophyrin sensitizer dye.J. Phys. Chem. B. **104** (2000) pp, 1198-1205.

15. Asbury, J., Hao, E., Wang, Y., Ghosh, H. and Lian, T., Ultrafast electron transfer dynamics from molecular adsorbates to semiconductor nanocrystalline thin films. J. Phys Chem B. **105** (2001) pp, 4545-4557.
16. Argazzi, R., Bignozzi, C., Heimer, T., Castellano, F and Meyer, G., Light-induced charge separation across Ru(II)-modified nanocrystalline TiO_2 interfaces with phenolthiazine donors. J. Phys. Chem. B. **101** (1997) pp, 2591-2597.
17. Land, E. J. and Prutz, A., reaction of azide radicals with amino acidsand proteins. Int. J. Radiat. Bio. **36** (1979) pp, 75-83.

PHOTOINDUCED HETEROGENEOUS ELECTRON TRANSFER MODULATED BY A VIBRATIONAL WAVEPACKET

C. ZIMMERMANN, R. EICHBERGER, S. RAMAKRISHNA, W. STORCK, AND F. WILLIG

Hahn-Meitner-Institut, Glienicker Str. 100, D-14109 Berlin, Germany
E-mail: willig@hmi.de

Photoinduced heterogeneous electron transfer was measured with fs transient absorption for the perylene chromophore adsorbed on nanocrystals of TiO_2 that formed a sponge-type layer with a large surface area. Time constants between 35fs and around 200fs were found with spacer-cum-anchor groups of different length and chemical composition. Long flexible bridge groups allowed for several configurations and thus different distances of the perylene chromophore from the TiO_2 surface. They gave rise to different time constants for electron injection. Vibrational wavepackets of low energy modes (mainly $360cm^{-1}$ and $420cm^{-1}$) were generated in the perylene chromophore applying sub-30fs laser pulses. The wavepacket was generated in the donor state, survived the electron transfer reaction and continued its motion in the product state (ionized molecule) for several hundred femtoseconds. The rise of the product state showed a step-wise increase indicating the modulation of heterogeneous electron transfer with 75fs time constant due to the motion of a vibrational wavepacket ($480cm^{-1}$) in the donor state.

Ultrafast photoinduced electron transfer from large molecular adsorbates into a semiconductor is an interesting fundamental case study of interfacial reactions [1,2,3]. The application of fs transient absorption spectroscopy to such systems became possible through the development of high surface area semiconductor films, where the inner surface of a 1μm thick film accommodates about a hundred times more chromophore molecules in a monomolecular surface layer than a planar surface [4]. The perylene chromophore, adsorbed at a nanoporous sponge-like TiO_2 electrode, is well suited for studying laser pulse induced electron transfer with transient absorption signals since all the relevant absorption spectra are well separated in the visible range of the spectrum, i.e. for the ground state, the excited singlet state, and the ionized state (Fig. 1). The electronic coupling between the photoexcited donor orbital of the adsorbed chromophore and the unoccupied electronic levels in the semiconductor conduction band can be very strong in the case of a direct contact between atoms of the chromophore and atoms of the semiconductor. The time constant for electron transfer in such a system can be in the range of 10 femtoseconds and shorter. By inserting molecular spacer groups with saturated bonds (e.g. $-CH_2-$) between the chromophore and the semiconductor surface the electronic coupling strength can be decreased.

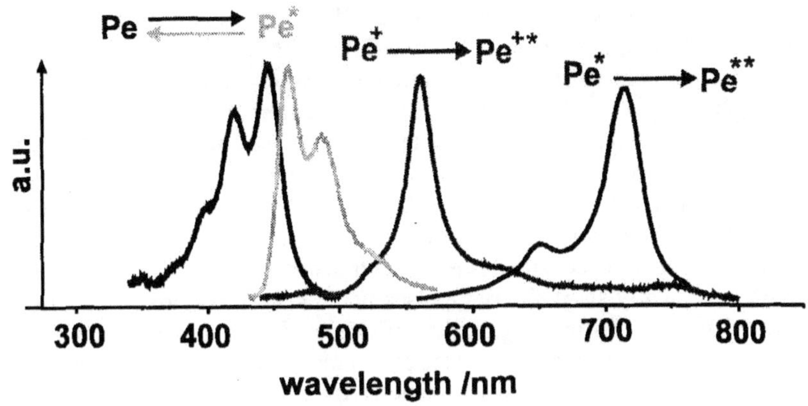

Figure 1. Emission spectrum and absorption spectra of perylene in its ground, first excited singlet, and ionized state.

There is a corresponding increase in the electron transfer time to several ten or even hundred femtoseconds, depending on length and the electron affinity levels of the spacer groups. In the case of a flexible molecular bridge various configurations of the spacer group can result in different distances and orientations of the chromophore at the semiconductor surface, and this gives rise to different electron transfer time constants.

Perylene compounds with four different spacer-cum-anchor-groups were used: 2,5-di-tertiary-butyl-9-perylenyl-*carboxylic acid* (a); 2,5-di-tertiary-butyl-9-perylenyl-*methyl-phosphonic acid* (b); 2,5-di-tertiary-butyl-9-perylenyl-*propionic acid* (c) and 2,5-di-tertiary-butyl-9-perylenyl-*methyl-(4-bicyclo[2.2.2]octane)-1-carboxylic acid* (d). Fig. 2 shows the formation of the perylene cation as the product state of the heterogeneous electron transfer reaction at the TiO_2 electrode, measured around 575nm. With a semi-flexible molecular bridge inserted between the perylene chromophore and the carboxylic anchor group (see dye (c) in Fig. 2) the electron transfer behavior on the surface of nanostructured anatase TiO_2 can be described with a bi-exponential fit. The short time constant is 45fs and the longer time constant is in the range of 200fs. Dye (d) showed nearly identical behavior to dye (c), therefore the transient absorption curve for dye (d) is not included in Fig. 2. The carboxylic group with the -CH_2-CH_2-bridge allows for a configuration where the chromophore is located close to the crystal surface. The configuration, where the perylene chromophore is separated by the tertiary butyl group from the TiO_2 surface, is ascribed to this short time constant. The longer time constant is interpreted as an average of different electron transfer times that represent different configurations of the molecular bridge with the chromophore further away from the surface. Monoexponential behavior with time constants of 35fs and 75fs, respectively, is found if the perylene chromophore forms a bond to TiO_2 with the carboxylic acid (a) or the -CH_2-phosphonic acid (b) as respective anchor groups [5,6,7]. In the case of the -CH_2-phosphonic acid anchor-cum-spacer group different configurations correspond to a rotation around the -C-C-bond. The latter does not affect the distance between the chromophore and a planar surface. The data show that completely rigid spacer-cum-anchor groups are required to achieve different well-defined distances at a planar surface. Semi-flexible molecular bridge groups will always lead to a distribution for the time constant of electron transfer.

With sufficiently short fs laser pulses and for an electron transfer time in the range of 100fs one can investigate the influence of vibrational wavepackets in the donor state on the heterogeneous electron transfer reaction. Coherent vibrational excitation can be achieved in the molecular donor state, and the wavepacket can modulate electron transfer. Excitation of the perylene chromophore in solution (substituted with tertiary butyl groups to prevent aggregation) with sub-30fs laser pulses results in strong oscillations of the stimulated emission signal at 475nm [6,8,9].

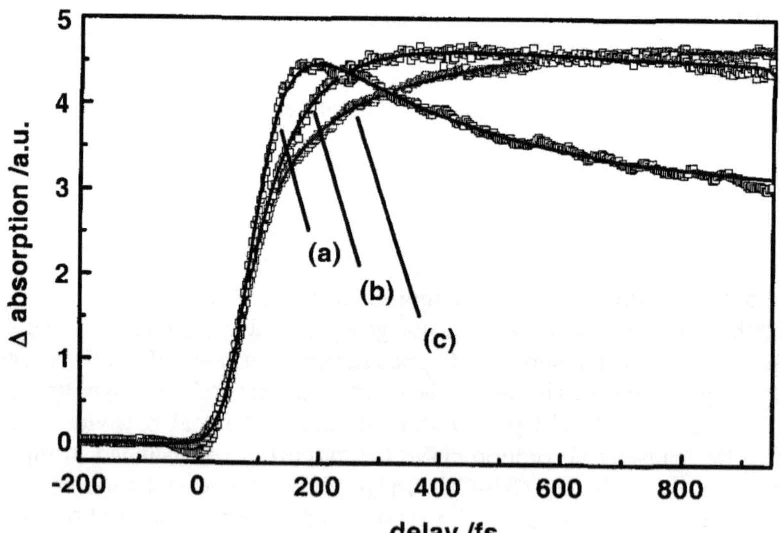

Figure 2. Formation of the perylene cation as the product state of the heterogeneous electron transfer reaction at the TiO_2 electrode, measured around 575nm / (a): 2,5-di-tertiary-butyl-9-perylenyl-*carboxylic acid*; (b): 2,5-di-tertiary-butyl-9-perylenyl-*methyl-phosphonic acid*; (c): 2,5-di-tertiary-butyl-9-perylenyl-*propionic acid.*

These oscillations last longer than two picoseconds. In the Fourier Transform of the oscillations three peaks can be identified, i.e. at 280cm^{-1}, 360cm^{-1} and 420cm^{-1}. While the vibrations at 360cm^{-1} and 420cm^{-1} are well known normal modes (i.e. -C-C-C-in-plane and out-of-plane bending modes) of the perylene chromophore [10] the mode at 280cm^{-1} is tentatively ascribed to an effect of the tertiary butyl substituents. In Raman measurements of this perylene compound the same frequency is observed [6,8,9]. Vibrational modes with comparable energies are found in the product state of dye (b), i.e. in the absorption signal of the oxidized dye that is formed in the heterogeneous electron transfer reaction [6,8,9]. It is well known that that there are only small shifts in the energies of vibrational modes for the oxidized state of an aromatic molecule compared to the neutral species. It should be noted that the oscillations still continue after the electron has completely left the molecule. That means the wavepacket moves in the potential of the product state long after electron transfer is completed. Oscillations indicating wavepacket motion in the product state cannot be identified, however, with periodic modulation of the electron transfer reaction. They only show that the vibrational wavepacket is surviving in the product state after electron transfer.

The donor state of compound (b) decays very fast (75fs) and therefore the modulation of the transient absorption signal can hardly be measured. Longer electron transfer times in the range of a few hundred femtoseconds appear to be required if clear oscillations due to the wavepacket motion are to be measured in the reactant state of the electron transfer reaction.

Closer inspection of the rising part of the transient cation absorption signal reveals a step-wise increase in the population of the product state of electron transfer (Fig. 3). Three distinct steps can be clearly recognized. These steps of decreasing height are attributed to the motion of a vibrational wavepacket in the molecular donor state, where curve crossing with the product state potential modulates electron transfer [8,9]. The temporal separation between the steps is about 70fs. This corresponds to a vibrational frequency of about 480cm^{-1}. This modulation effect can arise for a continuum of electronic acceptor states in the conduction band only if the density of states (DOS) is changing steeply over an energy range of the order of the reorganization energy below the molecular donor level [9,11]. The measured effect is well described by a recently developed theoretical model [11]. The periodic motion of the vibrational wavepacket in the potential curve of the donor state leads to modulations in the occupancy of electronic levels in the conduction band of the semiconductor and of vibrational levels in the ionized molecule [2,3]. The phase differences in population transfer to electronic states with different energies in the semiconductor, and to the corresponding vibrational states in the ionized molecule (energy conservation), depend on the specific molecular configurations where the wavepacket is reaching the respective crossing points with the potential curves for the different electronic product states [3].

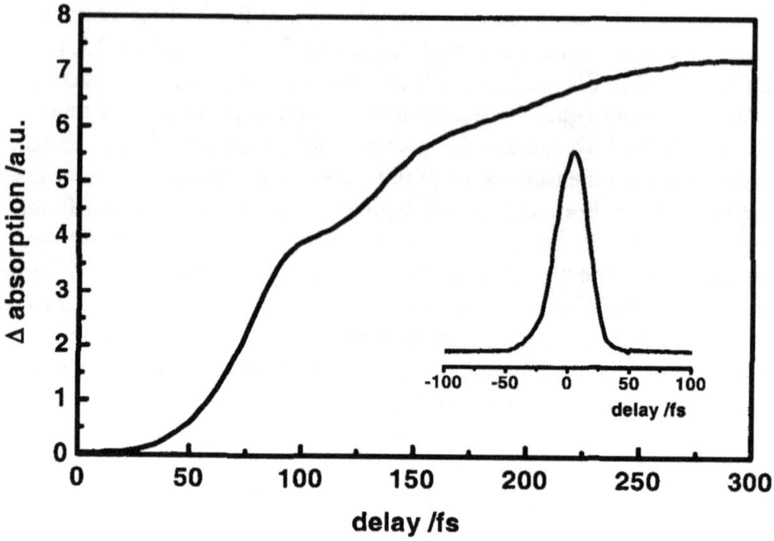

Figure 3. Transient absorption signal (585nm) probing the build-up of the population of the ionized perylene chromophore on the surface of TiO_2 after laser pulse excitation. Step-wise formation of this product state of electron transfer is ascribed to a curve crossing effect that is modulating electron transfer and is arising from the periodic motion of a vibrational wavepacket in the donor state. The small inset shows the instrumental response function (32fs fwhm).

Calculated curves for the build-up of the electronic and vibrational populations in the product states are given in the theoretical paper [3]. All the Franck-Condon factors and the corresponding different crossing points can be realized sequentially by the moving wavepacket provided the molecular donor state is located high enough above the bottom of the conduction band. This is the case in the present experimental system [12]. There is an ideal border case, the so-called wide band limit, where the density of states in the conduction band of the semiconductor is constant over an energy range spanning about twice the reorganization energy in the conduction band below the donor orbital. This constant DOS leads to complete cancellation of all the individual phase differences if one integrates over the occupancies in all the different electronic levels that are involved in the electron transfer. For constant DOS mono-exponential behavior with a smooth rise is predicted. All the cross terms cancel each other and the Franck-Condon factors contribute with their respective weight factors [2]. If the density of electronic band states (DOS) is not a constant the ideal behavior predicted for the wide band limit will not be observed. The step-wise increase in the population of the product states

(Fig. 3) reveals a similarity to the behavior of an electronic two level system. Contributions of certain Franck-Condon factors to electron transfer are enhanced where the DOS is high. Since the difference in DOS is important for the effect high energy vibrational modes will result in stronger modulation than low energy modes. Step-wise formation of product states has been discussed before by other groups in conjunction with experiments on homogeneous electron transfer [13,14] and also in related theoretical papers [15,16]. The generation of a vibrational wavepacket in the product state due to ultrafast electron transfer has to be distinguished from modulation of electron transfer by wavepacket motion in the molecular donor state in the case of a moderately fast electron transfer reaction. The latter scenario applies to the experimental signal shown in Fig. 3 with 75fs time constant.

As a corollary the signal shown in Fig. 3 provides direct evidence for electron transfer proceeding in this system from a wavepacket, i.e. from a hot vibrational state in the donor molecule.

References

1. R.D. Miller, G.L. McLendon, A.J. Nozik, W. Schmickler, and F. Willig, *Surface Electron Transfer Processes* chapter 5 (VCH Publishers, New York, 1995)
2. S. Ramakrishna and F. Willig, *J. Phys. Chem. B* **104** (2000) pp. 68
3. S. Ramakrishna, F. Willig, V. May, *Phys. Rev. B* **62** (2000) pp. R16330
4. B. O`Regan and M. Grätzel, *Nature* **353** (1991) pp. 737
5. B. Burfeindt, T. Hannappel, W. Storck, and F. Willig, *J. Phys. Chem.* **100** (1996) pp. 16463
6. F. Willig, C. Zimmermann, S. Ramakrishna, W. Storck, *Electrochimica Acta* **45** (2000) pp. 4565
7. C. Zimmermann, W. Storck, and F. Willig, *Chem. Phys. Lett.*, submitted
8. B. Burfeindt, C. Zimmermann, S. Ramakrishna, T. Hannappel, B. Meißner, W. Storck, and F. Willig, *Z. Phys. Chem.* **212** (1999) pp. 67
9. C. Zimmermann, F. Willig, S. Ramakrishna, B. Burfeindt, B. Pettinger,R. Eichberger, W. Storck, *J. Phys. Chem. B* **105** (2001) pp. 9245
10. K. K. Ong, J. O. Jensen, H. F. Hameka, *J. Mol. Struc.* **459** (1999) pp.131
11. R. Ramakrishna, F. Willig, V. May, *J. Chem. Phys.* **115** (2001) pp. 2743
12. T. Hannappel, PhD Thesis, TU Berlin, (1997)
13. K. Wynne, G. D. Reid, and R. M. Hochstrasser, *J. Chem. Phys.* **105** (1996) pp. 2287
14. S. Engleitner, M. Seel, and W. Zinth, *J. Phys. Chem. A* **103** (1999) pp. 3013
15. J. M. Jean and G. R. Fleming, *J. Chem. Phys.* **103** (1995) pp. 2092
16. J. Jortner and M. Bixon, *J. Chem. Phys.* **107** (1997) pp. 1470

PHONON CONFINEMENT IN ULTRATHIN METALLIC FILMS: INTERPRETATION OF RECENT FEMTOSECOND PUMP-PROBE EXPERIMENTAL DATA

A.MELIKYAN AND H.MINASSIAN

State Engineering University of Armenia, 105 Terian Street, Yerevan, 375009 Armenia
E-mail:physdep@seua.am

Recent femtosecond pump-probe experimental data related to lattice dynamics of ultrathin nickel films are interpreted exploring the idea of acoustic phonon spectrum size-quantization. Using the sound velocity in bulk as the only fitting parameter of the theory, excellent agreement with the experimental data is demonstrated.

New experimental data related to the dynamics of thermal expansion of ultrathin nickel films following femtosecond excitation are presented in [1]. Using femtosecond pump-probe technique (FPPT), it was demonstrated that the thermal expansion propagation speed decreases as film thickness decreases. In these experiments the 150 fs and 800 nm pump pulse was incident on the nickel film and excited the d-band electrons which was accompanied by the excitation of vibrational modes of the film resulting in thermal expansion. It was observed that the intensity of the second harmonic, generated by the probe in ultrathin film, oscillated depending on pump-probe delay. Another peculiarity, conditioned by extremely small thickness (less than 2 nm), was the trend towards smaller sound velocities in very thin films which is presented in Fig. 1. These peculiarities hint, that careful analysis is needed to reveal information about a thickness dependence of sound velocity and the speed of thermal expansion propagation in ultrathin films. The mechanism of this dependence is related to the size-quantization of the lattice vibrations frequency spectrum.

It is also worth to note that for manifestation of phonon confinement in electron-phonon interaction a greater time is necessary as compared with the time of flight t of phonon from one surface of the film to another – $t = L/v_s$. Just at this time a confined phonons are being formed, e.g.: at 10 nm thickness and sound velocity of $2\cdot10^3$ m/s t is ~ 5 ps, while at shorter times the effect can not be revealed because the phonons do not manage to reach surfaces. It was precisely the phenomenon observed in [2], when the thickness dependence of film reflectivity was found only for pump-probe delays larger than 5 ps. It is obviously clear that the independence of hot electron relaxation rates on film thickness for shorter delays is conditioned by the frequency-time uncertainty. Indeed, if the reciprocal of the delay is larger than phonon frequency quantization interval, the existence of the borders does not affect the electron-phonon interaction, and the phonon spectrum quantization is not revealed.

An obvious demonstration of phonon confinement can be found in recent pump-probe experiment [1] with nickel films of 40 nm to 1 nm thickness on copper substrate, where the dependence of sound velocity on thickness in the direction (001) was observed (Fig.27 in [1]).

The acoustic pulse propagates with a velocity equal to the phonon group velocity

$$V_g = \partial\omega / \partial q = \frac{1}{2}\omega_{max} \cdot a \cdot \cos\left(\frac{qa}{2}\right) \quad (1)$$

where ω_{max} is the maximum phonon frequency, q is the wave vector, a - the interplanar distance in the direction (001). Under the condition of size quantization the expression (1) for group velocity must be modified as the phonon frequency spectrum becomes discrete. Indeed, for the more simple case when in the confinement direction just two vibrational modes with frequencies ω_1 and ω_2, and wave vectors q_1 and q_2 exist, the group velocity, instead of (1), obviously becomes

$$V_g = \frac{\omega_2 - \omega_1}{q_2 - q_1} \quad (2)$$

In the case of bulk sample frequency spectrum is continuous and the ratio of finite differences in Eq. (2) reduces to the derivative as in (1). Let us find the phonon frequency spectrum in conditions of size quantization when the film is deposited on the substrate. Index 1 will be ascribed to the nearest to the substrate monolayer of the film, and the displacement of monolayer with number ν from the equilibrium state will be denoted as u_ν. The equations for the displacements are

$$\begin{aligned} m\ddot{u}_1 &= -ku_1 + k(u_2 - u_1) \\ m\ddot{u}_\nu &= -k(2u_\nu - u_{\nu+1} - u_{\nu-1}), \nu > 1 \end{aligned} \quad (3)$$

with boundary conditions $u_0 = 0$, $u_{N+1} = u_N$, where N is the number of monolayers, k – force constant. Here we neglect the small difference between force constants $k_{Ni:Ni} = 37.90 kg/s^2$ and $k_{Ni:Cu} = 33 kg/s^2$ [1]. The other important approximation is that we neglect the energy flow into the substrate. It is well known, that the leakage leads just to broadening, determined by the imaginary part of an eigenfrequency. The solution of the set (3) is sought in the form

$$u_{\nu} = u \cdot \sin(\nu\phi) \cdot e^{-i\omega t} \tag{4}$$

The frequency will be found by substituting the solution (4) into the initial set (3)

$$\omega_n = \omega_{\max} \left| \sin\left(\frac{2n-1}{2N+1} \cdot \frac{\pi}{2} \right) \right|, \; n = 1,2,3...N \tag{5}$$

One can see that group velocity depends on thickness and at $N >> n$, we obtain from (1)

$$V_g(L) \cong V_{bulk}\left(1 - \frac{\pi^2 n^2 a^2}{8L^2}\right) \tag{6}$$

where V_{bulk} is the group velocity in bulk. From this it follows that the group velocity decreases at the decrease of film thickness. It is interesting to note, that the lowest frequency in the case of 4 monolayers, i.e. n=1 and N=4, is very close to the value of 1.4 THz, observed in [1]. Indeed, adopting for V_{bulk} value of 4,200 m/s, and of 0.17 nm for interplanar distance a (both used in [1]), one can see from (11) and (16) that $\omega/2\pi \approx 1.39$ THz.

In [1], Fig. 27, the film thickness versus appearance time of the first peak of second harmonic generation in nickel film on Cu (100) was measured. For a thickness more than 4 nm, all points are in excellent agreement with an estimate of the longitudinal sound velocity, i.e. 4,200 m/s. For thickness less than 4 nm, the plotted curve shows a trend toward smaller values of velocity which is in qualitative agreement with (6). This clearly demonstrates that in experiment the group velocity of acoustic phonons is in fact measured. Thus, the experimental values on Fig. 27 of [1] should correspond to the curve described by equation

$$\frac{L}{V_g(L)} = \tau \tag{7}$$

where τ is time interval during which the acoustic pulse travels from the back wall to the front one. Indeed, the ratio L/τ, according to Fig. 27 of [1] decreases at the decrease of thickness. For the interpretation of experimental results we come from (5) and (7). The best agreement is achieved when the group velocity is calculated on the assumption that phonon modes with n = 4 and n = 5 are most essential ones, i.e.

$$V_g = \frac{\omega_5 - \omega_4}{q_5 - q_4} = \frac{2v_s}{\pi}(2N+1) \cdot \sin\left(\frac{\pi}{4N+2}\right) \cdot \cos\left(\frac{4\pi}{2N+1}\right) \tag{8}$$

The values calculated according to (7) and (8) are in excellent agreement with data obtained in [1], as it is shown in Fig.1.

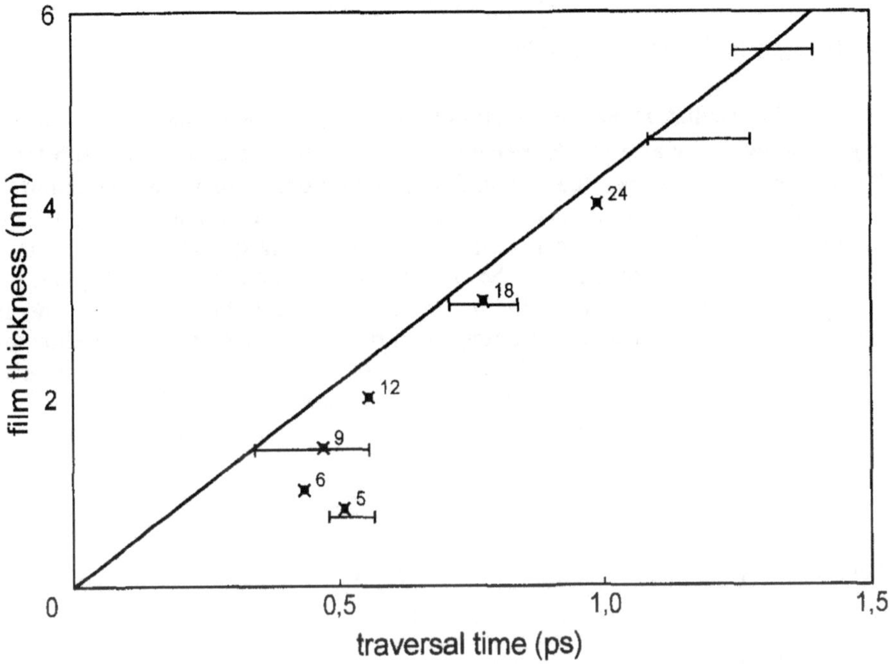

Figure 1. Film thickness versus the traversal time of the acoustic signal. Solid line corresponds to the velocity in bulk. Bars represent the experimental data. Crosses denote values, calculated in [24]. Integers neighboring crosses denote the number of monolayers in the sample.

One can see that with increasing the number of monolayers, crosses approach the straight line, i.e. the group velocity becomes insensitive to N for $n \leq 5$.

According to the measurements [1], the oscillations in a film with four ML are damped exponentially due to energy leakage into the substrate with a characteristic time of 1.6 ps. This means that the oscillation frequencies have a width of $\gamma \sim 0.6$ THz, which is very close to the value obtained by Fourier analysis of experimental data (inset of Fig. 24 in [1]).

In order to describe the thermal expansion resulting from the excitation of lattice vibrations, it is enough to take into account the lowest order anharmonism in ion displacements equations [3]:

$$\ddot{u}_\nu + \frac{k}{m}\left(2u_\nu - u_{\nu+1} - u_{\nu-1}\right) = \alpha\left[\left(u_{\nu+1} - u_\nu\right)^2 - \left(u_\nu - u_{\nu-1}\right)^2\right] + f(t) \tag{9}$$

For oscillations in the direction (100) u_ν is the displacement of crystalline plane with number ν as a whole, plane with $\nu = 0$ being adjacent to the substrate. The function $f(t)$ describes the excitation of vibrations during the interaction of pump with d-band electrons, and α is the anharmonism constant. Physically, the origin of function $f(t)$ is connected with ion recoil during the transition of d-band electron to the conduction band in the field of pump [5] and it is zero beyond the bounds of the pump pulse. For ultrathin films we neglect the vibrational energy flow from film to the substrate resulting in vibration damping. This approximation is justified by much more rapid thermal expansion of nickel in comparison with copper [1]. The duration of pump pulse in experiment [1] was 150 fs, while the duration of observation process was a few ps following the pump pulse. Hence, this results in the following form of the solution of Eq. (9):

$$u_\nu^{(0)} = \sum_n c_{\nu n} \sin \omega_n t \tag{10}$$

where $c_{\nu n}$ are determined from (21) at $t \to \infty$. This expression should be substituted to the first approximation equation:

$$\ddot{u}_\nu^{(1)} + \frac{k}{m}\left(2u_\nu^{(1)} - u_{\nu+1}^{(1)} - u_{\nu-1}^{(1)}\right) = \alpha\left[\left(u_{\nu+1}^{(0)} - u_\nu^{(0)}\right)^2 - \left(u_\nu^{(0)} - u_{\nu-1}^{(0)}\right)^2\right] \tag{11}$$

In the right hand side of this equation terms emerge which contain oscillations with all the possible combination frequencies $\omega_m \pm \omega_n$. If one or a few of them coincide with the eigenfrequencies of the left hand side of (11), which are simply the same frequencies ω_n, , then during the excitation of oscillations a resonant amplification of certain modes takes place. Let us find the resonance condition by representing the difference of two frequencies in the form:

$$\omega_m - \omega_n = 2\omega_{\max} \sin\left(\frac{m-n}{2N+1}\frac{\pi}{2}\right)\cos\left(\frac{m+n-1}{2N+1}\frac{\pi}{2}\right) \tag{12}$$

To have the right hand side equal to any eigenfrequency, m and n must have different parity, and the equality presented below must be satisfied:

$$2\cos\left(\frac{m+n-1}{2N+1}\frac{\pi}{2}\right)=1 \tag{13}$$

Since $m,n \leq N$, it follows from (13) that $3(m+n-1)=2(2N+1)$, which is possible only if $2N+1$ is divisible by 3. For example, in case of $N = 4$ and $N = 7$ (for which we have experimental data), the resonance conditions, according to (13), are as follows:

$$N=4,\ \omega_4=\omega_3+\omega_1;\ N=7,$$

$$\omega_4=\omega_7-\omega_3,\ \omega_5=\omega_6-\omega_1. \tag{14}$$

It should be noted that if $2N + 1$ is not divisible by 3, then resonance condition can be satisfied only approximately. Thus, for $N = 5$, N=6 and N=9 we have respectively

$$2\omega_2-\omega_4=0{,}01\omega_{max},\ \omega_5-(\omega_4+\omega_1)=0.02\omega_{max} \tag{15}$$

$$\omega_5-(\omega_4+\omega_1)=0{,}015\omega_{max} \tag{16}$$

$$\omega_5-(\omega_4+\omega_1)=0.008\omega_{max},$$

$$\omega_8-(\omega_4+\omega_3)=-0.003\omega_{max} \tag{17}$$

It is easy to see from (13), that further increase of N leads to increase of number of resonances. For example, at $N = 10$ we have as much as three exact resonances, though they are realized for higher modes. It is natural that with the increase of thickness the film's properties approach those of bulk, and the resonances are smeared.

From (13) - (17) one can see that the modes with n=4 and n=5 are resonant ones at small thicknesses, which can provide a basis for the approach described in [4]. At larger thicknesses, the sound velocity becomes insensitive to the thickness. As it was shown in [1], the minimum vibrational frequency ω_1 in case of four ML is very close to the value of 1.4 THz observed in [1]. However, the frequencies ω_3 and ω_4 , which satisfy the resonance condition (14), have not been observed experimentally. We attribute this to the fact that at duration of probe pulses $\tau = 150$ fs, the vibrations with frequencies higher than $1/\tau = 6.7$ THz are averaged and cannot be detected. At thickness 7 ML no selected frequency was observed, which was conditioned by superposition of vibrations with as much as six incommensurable frequencies.

Thus, considering the anharmonicity of vibrations as a source of thermal expansion and solving the corresponding equations for vibrations with lowest order in anharmonicity we obtain quantitative agreement with the experimental data [1] where for 4 monolayer thick nickel film the value ~ 1200 m/s for the speed of thermal expansion is obtained. Considering the sound velocity in bulk (~4200 m/s) as the only fitting parameter of our theory, we have extracted with a great degree of accuracy the observed vibrational frequencies of nickel film on copper substrate as well .

References

1. J.Hohlfeld, S.-S. Wellershoff, J. Guedde, U. Conrad, V. Jaehnke, E. Matthias, Electron and lattice dynamics dfollowing optical excitation of metals. *Chem. Phys.* **251**(2000) pp. 237 -258
2. J.Hohfeld, J.G.Müller, S.S.Wellershoff, E.Matthias. Time-resolved thermoreflectivity of thin gold films and its dependence on film thickness *Appl.Phys.***B64,** pp.387-390.
3. Charles Kittel, Introduction to Solid State Physics, Fourth Edition, (John Wiley and Sons, Inc.)
4. A.Melikyan, H.Minassian, Phonon confinement in ultrathin nickel films. *Chem. Phys. Lett.* **331** (2000) pp.115-118.
5. A.Melikyan, H.Minassian, Interaction of vibrational modes in ultrathin metallic films following femtosecond excitation. *Solid State Commun.* **119** (2001) pp. 497-499

5. Biological Systems

This page is intentionally left blank

THE CAROTENOID S_1 STATE IN LH2 COMPLEXES OF PURPLE BACTERIA

T. POLÍVKA, D. ZIGMANTAS, J.L. HEREK

Chemical Physics, Lund University, Box 124, S-221 00 Lund, Sweden
E-mail: tomas.polivka@chemphys.lu.se

R. J. COGDELL

Division of Biochemistry and Molecular Biology, University of Glasgow, G12 8QQ Glasgow, United Kingdom

V. SUNDSTRÖM

Chemical Physics, Lund University, Box 124, S-221 00 Lund, Sweden

Using near-infrared femtosecond absorption spectroscopy, we have determined the S_1 energies of the carotenoids spheroidene and rhodopin glucoside in LH2 complexes of purple bacteria. The S_1 energies are 13400±100 cm^{-1} for spheroidene and 12550±200 cm^{-1} for rhodopin glucoside. The 850 cm^{-1} shift between the S_1 energy of these two carotenoids significantly affects the energy transfer pathways within the LH2 complex. The S_1 energy of spheroidene is high enough to allow efficient energy transfer from the S_1 state to bacteriochlorophylls, resulting in a substantial shortening of the spheroidene S_1 lifetime in LH2 complex (1.7 ps) as compared with the lifetime in solution (8.5 ps). Rhodopin glucoside, however, has its S_1 energy in LH2 complex too low for efficient S_1 energy transfer and rather the carotenoid S_2 state becomes the main energy donor in LH2 complexes containing this carotenoid.

1 Introduction

Carotenoids are a class of natural pigments that play important roles in many biological systems. In photosynthetic organisms, they assist (bacterio)chlorophylls in the light-harvesting function, efficiently covering the spectral region 450-550 nm [1]. The light-harvesting function of carotenoids relies on their ability to transfer energy from low-lying excited states, including S_1 ($2A_g^-$ in the idealized C_{2h} group) and S_2 ($1B_u^+$). Due to symmetry reasons, the transition from the ground state S_0 to the S_1 state is forbidden, preventing direct detection of the S_1 state [2]. The energy of this dark state, however, is one of the most important parameters underlying energy transfer processes between carotenoids and (bacterio)chlorophylls.

A substantial effort has been made to determine the energy of this dark state in solution [3,4,5], but for understanding its relevance to energy transfer processes, it is necessary to determine the S_1 energy of carotenoids in their natural protein environment. One of the promising approaches to achieve this goal is based on a two-photon absorption technique, which was already used to obtain the S_1 energy of the carotenoid spheroidene in a peripheral light-harvesting protein LH2 of purple bacteria [6].

Recently, we introduced a different experimental approach to determine the S_1 energies of carotenoids based on scanning the S_1-S_2 excited state absorption shortly after excitation [4]. In contrast to the S_0-S_1 transition, the S_1-S_2 transition is allowed, providing a possibility to overcome problems with the direct detection of the S_1 state. Knowing the energies of both S_1-S_2 and S_0-S_2 transitions, the S_1 energy can be then determined by simple subtraction.. This method was successfully used to obtain S_1 energies of a few carotenoids in solution [4,7]. In this paper, we present application of this technique to locate the S_1 energies of two carotenoids in the LH2 complexes of *Rhodobacter* (*Rb.*) *sphaeroides* (containing the carotenoid spheroidene) and *Rhodopseudomonas* (*Rps.*) *acidophila* (rhodopin glucoside).

2 Methods

The LH2 complexes of both *Rb. sphaeroides* and *Rps. acidophila* were prepared according to Ref. 8 and stored in the dark at –55°C. Prior to the experiments, the isolated LH2 complexes were dissolved in a buffer solution (50 mM Tris, pH 8, 0.1 % LDAO). All measurements were carried out at room temperature with the sample placed into a 2 mm pathlength quartz rotational cuvette to avoid sample degradation during the course of the measurement.

The femtosecond spectrometer used in these studies is based on an amplified Ti:Sapphire laser system, with tunable pulses obtained from two optical parametric amplifiers and/or white-light continuum generation. The femtosecond pulses obtained from the Ti:Sapphire oscillator operating at a repetition rate of 82 MHz were amplified by a regenerative Ti:Sapphire amplifier pumped by a Nd:YLF laser with a repetition rate of 5 kHz and producing ~120 fs pulses with an average output power of ~1 W at a central wavelength of 800 nm. The amplifier output was divided by a 70/30 beam-splitter to pump two independent parametric amplifiers for generation of the pump and probe pulses. The parametric amplifier used for generation of probe pulses was controlled by a computer, enabling direct scanning of the wavelength of the probe pulses over the spectral region 850-1800 nm. The excitation pulses were attenuated using neutral density filters to typical energies of 40 nJ/pulse. Furthermore, absorption spectra were measured before and after measurements to ensure that no permanent photochemical changes occurred over the duration of the experiment.

3 Results and Discussion

Absorption spectra of the LH2 complexes of *Rb. sphaeroides* and *Rps. acidophila* shown in Fig. 1 exhibit characteristic spectral features in the near-infrared region due to B800 and B850 bands of BChl-a [9]. A well-known three-peak feature typical for the S_0-S_2 transition of carotenoids is observed at wavelengths below 550 nm. The 0-0 band of this transition is located at 19530 cm^{-1} for *Rb. sphaeroides* containing the carotenoid spheroidene, which has 10 conjugated C=C bonds, while the 0-0 band in *Rps. acidophila* is shifted towards lower energies peaking at 19050 cm^{-1}, since rhodopin glucoside has a conjugation length of 11. A slightly larger difference of ~700 cm^{-1} is also observed for these carotenoids in solution (Fig. 1). The S_0-S_2 transitions of both carotenoids in LH2 are

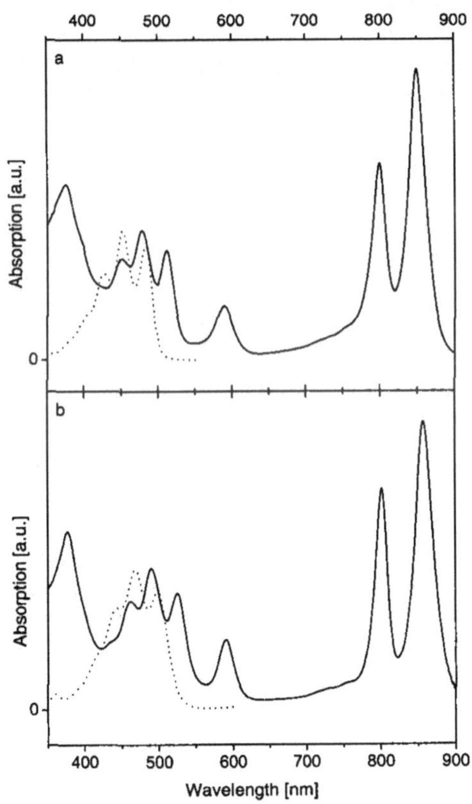

Figure. 1. Steady-state absorption spectra of LH2 complexes (solid lines) from *Rb. sphaeroides* (a) and *Rps. acidophila* (b), together with absorption spectra of both carotenoids in solution (dashed lines): spheroidene in n-hexane (a) and rhodopin glucoside in methanol (b).

shifted to lower energies by about 1000 cm^{-1} due to the interaction with the protein environment.

To obtain the spectral profile of the carotenoid S_1-S_2 transition, which is crucial to calculate the S_1 energy, a transient absorption spectrum in the near-infrared spectral region recorded within a time window given by the S_1 lifetime is necessary [4,7]. The S_1 lifetimes of both carotenoids in LH2 complexes were measured using the well-known S_1-S_N transition occurring in the visible region [10], yielding 1.5 ps for spheroidene and 4 ps for rhodopin glucoside. While the 4 ps lifetime of rhodopin glucoside is actually almost the same as the S_1 lifetime of this carotenoid in solution, the S_1 lifetime of spheroidene in LH2 is markedly shorter than that observed in solution [11]. This reflects the known fact that *Rps. acidophila* utilizes predominantly the S_2-Q_x energy transfer pathway, while in case of *Rb. sphaeroides* the S_1 pathway plays a significant role.

Thus, knowing the S_1 lifetime of both carotenoids in LH2, the near-infrared transient absorption spectra were measured at 2 ps after excitation. At this delay, the S_1 states of both carotenoids are still at least partially populated allowing detection of the carotenoid S_1-S_2 transition. The transient absorption spectra after excitation into the 0-0 vibrational band of the S_0-S_2 transition, at 515 nm for *Rb. sphaeroides* and 525 nm for *Rps. acidophila*, are shown in Fig. 2a,b. Since the S_1-S_2 transition is much weaker than the S_1-S_N transition and some contribution from the BChl-a excited state absorption may be expected, the transient absorption spectra after excitation of the B850 BChl-a band are shown for comparison. The transient absorption spectra of both carotenoids exhibit a pronounced feature located at around 6200 cm^{-1} (*Rb. sphaeroides*) and at ~6500 cm^{-1} (*Rps. acidophila*). Since this band is not observed either when the B850 band is excited directly or at 20 ps after carotenoid excitation, we assign this feature to the 0-0 band of the S_1-S_2 transition of carotenoids. The higher vibrational bands of this transition, which are clearly visible for carotenoids in solution [4,7], are hampered by the strong BChl-a excited state absorption below 1300 nm.

To support our assignment of this band, the differential transient absorption spectra after subtraction of the BChl-a excited absorption of both LH2 complexes are shown in Fig. 2 c,d. Subtracting of BChl-a contribution clearly reveals the 0-1 vibrational band in the transient absorption spectrum in the LH2 of *Rb. sphaeroides*. The energy distance between these two bands observed in the transient spectrum corresponds well to the known energy difference between vibrational bands of the carotenoid S_2 state (~1300 cm^{-1}), supporting our assignment of the spectral feature in the transient absorption spectrum to the spectral profile of the lowest vibrational band of the S_1-S_2 transition. In the case of *Rps. acidophila* the transient spectrum is shifted to higher energies and suffers significantly from overlap with BChl-a contribution, which precludes observation of higher vibrational bands. Also, a direct comparison with the S_1-S_2 spectra of both carotenoids in solution (Fig. 2c,d) shows strong similarity, especially for spheroidene, although the S_1-S_2 profile measured in LH2 complexes is shifted to lower energies indicating a smaller gap between the S_1 and S_2 states in the protein. The S_1-S_2 transient spectrum of rhodopin glucoside in solution is markedly broader than that of spheroidene, most likely due to larger spectral inhomogeneity induced by the presence of a glucoside ring at the end of conjugated backbone of the carotenoid.

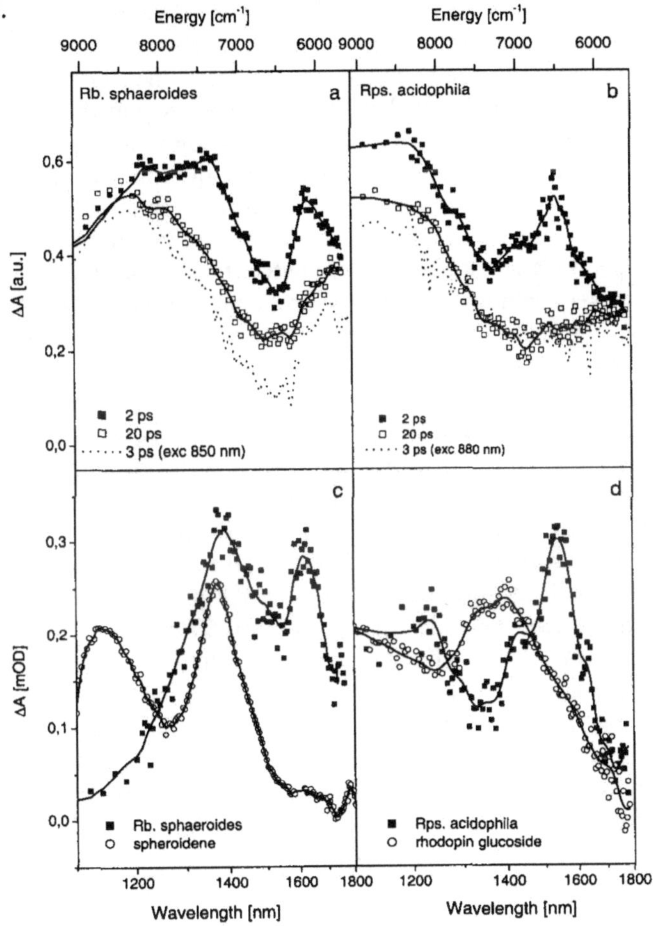

Figure. 2. (a, b) Near-infrared transient absorption spectra of LH2 complexes from *Rb. sphaeroides* and *Rps. acidophila* recorded at 2 ps (full squares) and 20 ps (open squares). The transient absorption spectrum measured after excitation of B850 band is shown for comparison (dotted line). (c, d) Differential transient absorption spectra after subtraction of BChl-a signal of LH2 complexes from both species (full squares) together with the S_1-S_2 profiles of the corresponding carotenoids in solution (open circles).

To be certain that the observed transient absorption spectra indeed correspond to the S_1-S_2 transition, we measured kinetics at the maxima of the lowest energy band of both carotenoids in the LH2 complex. The measured kinetics are shown in Fig. 3 and the time components extracted from multiexponential fitting of the kinetics are summarized in Table 1. In case of *Rb. sphaeroides*, the kinetic signal measured at 550 nm (corresponding to the S_1-S_N transition) is in perfect match with that recorded at 1600 nm, supporting our assignment of the bands. The changes in amplitudes of the decay components are caused by much stronger contribution from the BChl-a excited state absorption in the infrared. Also, the slight difference in the rise component is due to the fact that in the infrared region the rise component interferes with a S_2-S_N excited state absorption occurring in this spectral region [7]. The low-amplitude component of 8 ps observed in both visible and infrared spectral regions is very close to the S_1 lifetime of spheroidene in solution [7,11] and can be therefore ascribed to the S_1 lifetime of non-transferring carotenoids in the LH2 complex. The S_1-S_N kinetic trace for rhodopin glucoside in the LH2 complex of *Rps. acidophila*, however, exhibits a deviation from that measured at 1500 nm: the kinetic recorded in the infrared region contains an additional subpicosecond component, which is clearly not present when measured at 550 nm. This apparent discrepancy can be explained by the effective S_2-Q_x pathway operating in the LH2 complexes from *Rps. acidophila*. It was shown that the energy transferred from the S_2 state of rhodopin glucoside is accepted by the Q_x states of both B800 and B850 bacteriochlorophylls with a similar efficiency [12]. Then, the subpicosecond component observed in the 1500 nm kinetic is most likely due to a decay of the B800 excited state absorption reflecting B800-B850 energy transfer. The dynamics of this process is well documented [13] and the 0.8 ps component observed here matches well the known dynamics of the B800-B850 energy transfer in LH2 of *Rps. acidophila*. Thus, although the kinetic measured at maxima of the S_1-S_2 transition has an additional component due to B800-B850 energy transfer in the case of LH2 from *Rps. acidophila*, we can conclude that the infrared transient spectrum measured at 2 ps delay shown in Fig. 2 indeed reflects the spectral profile of the S_1-S_2 transition of rhodopin glucoside, since most of the signal at 2 ps originates from the carotenoid.

Table 1. The time constants and their amplitudes extracted from multiexponential fits of kinetics measured in visible and infrared parts of the transient spectrum for LH2 complexes of *Rps. acidophila* and *Rb. sphaeroides*.

	λ_{probe} [nm]	A_1[%] τ_1<200fs	τ_2 [ps]	A_2 [%]	τ_3 [ps]	A_3 [%]	A_4 [%] τ_4 ~ 1 ns
Rps.	580	-100	-	-	4	92	8
acidophila	1500	-	0.8	32	4	37	31
Rb.	560	-100	1.5	84	7.7	10	6
sphaeroides	1600	14	1.5	38	8	2	46

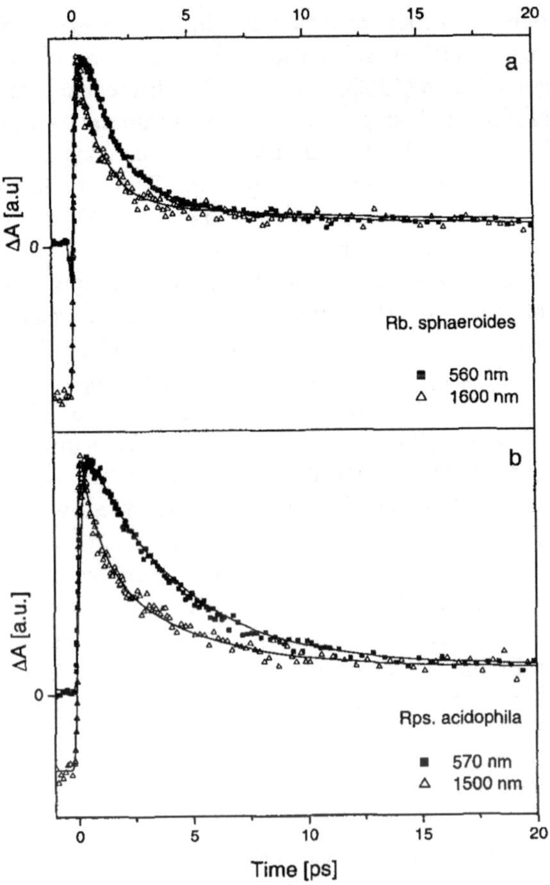

Figure. 3. Kinetics recorded at maximum of the S_1-S_N transition (full squares) and the 0-0 band of the S_1-S_2 transition (open triangles) for the LH2 complexes from *Rb. sphaeroides* (a) and *Rps. acidophila* (b). The corresponding fits of the kinetics are represented by solid lines.

Having assigned the bands in the infrared transient absorption spectra, we can calculate the S_1 energy of both carotenoids in LH2 complexes by subtracting the 0-0 energies of the S_0-S_2 and S_1-S_2 transitions. This procedure yields the S_1 energies of 13400 ± 100 cm^{-1} (spheroidene) and 12550 ± 200 cm^{-1} (rhodopin glucoside). It is interesting to compare these values with those obtained for both carotenoids in solution. The S_1 energy of spheroidene in n-hexane was determined to be 13400 cm^{-1} [7], thus exactly the same as in the LH2 complex. This result is not surprising, since the negligible dipole moment of the S_0 state makes the S_1 state almost insensitive to the changes in local environment. In case of *Rps. acidophila* the S_1 state in LH2 complex is slightly shifted to lower energy as compared with that measured in methanol solution (12800 cm^{-1}), but this discrepancy can be due to the rather broad S_1-S_2 profile of rhodopin glucoside in solution. As noted above, this broadness is most likely a result of a mixture of various conformers occurring in solution, thus it is not straightforward to compare the S_1 energy in solvent with the energy measured in the LH2 complex, since in the LH2 complex only one particular conformation of rhodopin glucoside is present.

Using the S_1 energies of both carotenoids in the LH2 complex, the different energy transfer pathways in *Rb. sphaeroides* and *Rps. acidophila* can be rationalized. As shown very recently using time-dependent density functional theory (TDDFT), the S_1 energy transfer rate in the LH2 complex is fast and rather insensitive to the position of the S_1 state in the energy range of 13500-15000 cm^{-1}, while for energies below 13000 cm^{-1} the S_1 energy transfer pathway becomes quite inefficient yielding carotenoid S_1 to BChl energy transfer rates lower than $(30\ ps)^{-1}$ [14]. The result of the TDDFT theory is in perfect agreement with the S_1 energies obtained here: the S_1 energy 13400 cm^{-1} of spheroidene in LH2 is at the edge of the high efficiency interval, which results in quite effective energy transfer from the S_1 state of spheroidene to bacteriochlorophylls. In LH2 of *Rps. acidophila*, however, the S_1 state of rhodopin glucoside is shifted out of this interval and the S_1-Q_y pathway becomes negligible, as demonstrated by the same S_1 lifetimes of rhodopin glucoside in both solution and the LH2 complex.

4 Acknowledgements

Authors thank to Harry Frank and James Bautista from University of Connecticut for kindly providing us purified spheroidene. The work at Lund University was performed with funding from the Swedish National Research Council. The work of R.J.C was funded by the Biotechnology and Biological Sciences Research Council.

References

1 Frank H. A. and Cogdell R. J., Carotenoids in photosynthesis. *Photochem. Photobiol.* **63** (1996) pp. 257–264.

2 Christensen R. L., The electronic states of carotenoids. In *Photochemistry of Carotenoids*, ed. by H. A. Frank, A. J. Young, G. Britton and R. J. Cogdell (Kluwer Academic Publisher, Dodrecht, 1999) pp. 137–159.

3 Fujii R., Onaka K., Kuki M., Koyama Y. and Watanabe Y., The $2A_g^-$ energies of all-trans-neurosporene and spheroidene as determined by fluorescence spectroscopy. *Chem. Phys. Lett.* **288** (1998) pp. 847-853.

4 Polívka T., Herek J. L., Zigmantas D., Åkerlund H.-E. and Sundström V., Direct observation of the (forbidden) S_1 state in carotenoids. *Proc. Natl. Acad. Sci. U.S.A.* **96** (1999) pp. 4914-4917.

5 Frank H. A., Bautista J. A., Josue J. S. and Young A. J., Mechanism of nonphotochemical quenching in green plants: Energies of the lowest excited singlet states of violaxanthin and zeaxanthin. *Biochemistry* **39** (2000) pp. 2831-2837.

6 Krueger B. P., Yom, J., Walla P. J. and Fleming G. R., Observation of the S_1 state of spheroidene in LH2 by two-photon fluorescence excitation. *Chem. Phys. Lett* **310** (1999) pp. 57-64.

7 Polívka T., Zigmantas D., Frank H. A., Bautista J. A., Herek J. L., Koyama Y., Fujii R. and Sundström V., Near-infrared time-resolved study of the S_1 state dynamics of the carotenoid spheroidene. *J. Phys. Chem. B* **105** (2001) pp. 1072-1080.

8 Fraser N. J., Dominy, P. J., Ücker, B., Siminin, I., Scheer, H. and Cogdell, R. J., Selective release, removal, and reconstitution of bacteriochlorophyll a molecules into the B800 sites of LH2 complexes from *Rhodopseudomonas acidophila* 10050 *Biochemistry* **38** (1999), pp. 9684-9692.

9 Kennis J. T. M., Streltsov A. M., Vulto S. I. E., Aartsma T. J., Nozawa T. and Amesz J., Femtosecond dynamics in isolated LH2 complexes of various species of purple bacteria. *J. Phys. Chem B* **101** (1997) pp. 7827-7834.

10 Desamero R. Z. B., Chynwat V., van der Hoef I., Jansen F. J., Lugtenburg J., Gosztola D., Wasielewski M. R., Cua A., Bocian D. F. and Frank H. A., Mechanism of energy transfer from carotenoids to bacteriochlorophyll: Light-harvesting by carotenoids having different extents of p-electron conjugation incorporated into the B850 antenna complex from the carotenoidless bacterium *Rhodobacter sphaeroides* R-26.1. *J. Phys. Chem B* **102** (1998) pp. 8151-8162.

11 Zhang J.-P., Fujii R., Qian P., Inaba T., Mizoguchi T., Koyama Y., Onaka K. and Watanabe Y., Mechanism of the carotenoid-to-bacteriochlorophyll energy transfer via the S_1 state in the LH2 complexes from purple bacteria. *J. Phys. Chem B* **104** (2000) pp. 3683-3691.

12 Macpherson A. N., Arellano J. B., Fraser N. J., Cogdell R. J. and Gillbro T., Efficient energy transfer from the carotenoid S_2 state in a photosynthetic light-harvesting complex. *Biophys. J.* **80** (2001) pp. 923-930.

13 Sundström V., Pullerits T. and van Grondelle R., Photosynthetic light harvesting: Reconciling dynamics and structure of purple bacterial LH2 reveals function of photosynthetic unit. *J. Phys. Chem B* **103** (1999) pp. 2327-2346.

14 Hsu C-P., Walla P. J., Head-Gordon M. and Fleming G. R., The role of the S_1 state of carotenoids in photosynthetic energy transfer: The light-harvesting complex II of purple bacteria. *J. Phys Chem B* **105** (2001) pp. 11016-11025.

INVESTIGATION OF THE CARBONMOXIDE AND OXY HEMOGLOBIN I COMPLEXES FROM *LUCINA PECTINATA* BY FEMTOSECOND TRANSIENT ABSORPTION

CACIMAR RAMOS, EUNICE RAMÍREZ CARLOS RUIZ, RUTH PIETRI AND JUAN LÓPEZ GARRIGA

Chemistry Department. University of Puerto Rico, PO. Box 9019, Mayagüez, P.R. 00680.
E-mail: jlopez@sonw.uprm.edu

JAN HELBING, AND MAJED CHERGUI

Institut de Physique Expérimentale, Université de Lausanne, CH-1015 Lausanne-Dorigny, Switzerland.
E-mail: Majed.Chergui@ipmc.unil.ch

Hemoglobin I (HbI) is an invertebrate monomeric hemeprotein from *Lucina pectinata* that contains 142 amino acid residues. The HbI heme pocket structural composition shows an arrangement of phenylalanyl residues (Phe (29), Phe (43) and Phe (68)) that has not been observed in the globin families. The photochemistry of HbI ligand complexes can yield relevant information about invertebrate globins in terms of ligands rebinding and photophysical processes. We have performed ultrafast measurements to determine the extent of O_2 and CO recombination subsequent to their photodissociation from the HbI moiety. To separate the contributions of electronic, thermal and conformational relaxation, from ligand rebinding in the spectroscopic signals, measurements of the deoxyHbI were also performed. The transient absorption signals in the Soret region of HbICO and HIO_2 are composed of similar time constants as observed for other heme proteins. The fast component (300 fs) is due to the formation of the unligated deoxy species from the excited heme state Hb_I*. The second time constant (2.2-2.5 ps) corresponds to the decay of the excited heme state Hb_{II}*. The contribution of this component to the transient signal near the Soret band maximum appears to be greater in HbI complexes than in other myoglobins. Its presence in the decay of the Soret bleaching signals has been interpreted as a signature of fast geminate recombination. Thus, these results could suggest that the differences in the transient signal between HbI and Mb are influenced by the different heme-pocket structures. Photoexcitation of the HbI complexes led to the formation of a short-lived species in the 630 nm region. Formation of this species appears to be instantaneously, while it decays with a 300 fs time constant. This new transient is related to an iron-to-porphyrin charge transfer state.

1 Introduction

Hemoglobin and myoglobin are metalloproteins that are involved in the transport of O_2 in vertebrates systems. While great progress has been made recently in the analysis of ligand rebinding in these globins, the elementary aspects of this process is still not fully understood. Invertebrate globins constitute an excellent system to study because of their ability to transport ligands other that O_2 [1,2]. For instance, *Lucina pectinata* is a mollusk that contains different hemoglobins, which are believed to be involved in oxygen and/or hydrogen sulfide transportation [3]. One of these hemoglobins, designated as hemoglobin I (HbI), is the monomeric protein responsible for the transportation of H_2S to an intracellular symbiotic bacterium [4]. The HbI heme-pocket structural organization is different from that observed for vertebrate myoglobins. Hemoglobin I pocket reveals a peculiar arrangement of phenylalanine residues (Phe (29), Phe (43) and Phe (68)) and Gln (64) instead of the highly conserved vertebrate His (64) [5]. The ligand-binding properties of HbI are related to its unusual heme pocket amino acids. Indeed, resonance Raman, proton nuclear magnetic resonance and infrared spectroscopic studies of different HbI derivatives have suggested that synergetic effects of the aromatic residues in the heme pocket modulate the ligand-binding properties in this system [6-8]. Dynamics in the HbI heme pocket after ligand photodissociation, in an ultrashort time scale, can yield a better understanding of the role of the distal residues in modulating ligand-binding properties. Ligand dissociation and recombination from vertebrate hemoglobins and myoglobins have been extensively studied using femtosecond pump-probe spectroscopy [9-14]. These studies have shown that upon ligand photodissociation with femtosecond laser pulses two short-lived species are generated. Both species, which are formed in less than 50 fs, have been assigned to electronic excited states of the heme. These species, designated Hb_I^* and Hb_{II}^*, relax to a ground state either by a parallel [11] or a sequential path [14]. In the parallel process the unligated deoxy species in the ground state is formed on a timescale of 300 fs from Hb_I^*. The ligand subsequently rebinds to the deoxy species and this process is usually referred to as geminate recombination. Alternatively, a different intermediate (Hb_{II}^*) may be formed upon photoexcitation, which directly recombines to the ligated complex on a 2-3 ps time scale without formation of the deoxy ground state [11]. Recently, new relaxation sequential pathways have been suggested [14]. The model indicates that Hb_I^* is an iron to porphyrin charge transfer state with the iron atom in its ferric state, that relaxes in 300fs to form Hb_{II}^*. This relaxation process was characterized by a porphyrin to iron back charge transfer reaction generating the unligated ferrous ground state species (Hb_{II}^*) in ~3ps. Ligand recombination then occurs with this species. Despite the differences in the electronic relaxation pathway suggested by these models, ligands such as CO, O_2 and NO recombine with the relaxed heme on a picosecond to nanosecond time scale [15-17]. On this basis, it has been proposed that the extent of ligand recombination with the ground states unligated species is sensitive to the interaction of a ligand with the distal amino acid residues and to differences in ligand reactivity toward the heme iron [16,17]. Therefore, a study of *Lucina pectinata* HbI ligand complexes can contribute to a better understanding of the synergetic effects of the heme pocket amino acids aromatic residues.

In the present contribution, ultrafast spectral evidence, which indicates that the unique phe-cage arrangement is responsible for the significant O_2 and CO recombination in *L. pectinata* HbI moiety, is offered. Evidence for the formation of a new short-lived species, which absorb near 630nm, is also presented.

2 Methods

The methodology for isolation, separation and purification of Hemoglobin I from *Lucina pectinata* followed a guideline that has been previously reported [3]. In the laser experimental arrangement, a diode laser at 532 nm was used to pump a mode-locked Ti:Sapphire laser to generated 80 fs pulses at 800 nm. These pulses were amplified using the Ti:Sapphire amplifier region from the Multikilohertz Amplifier pumped by cw doubled Q-switched Nd:YLF laser. The system yielded pulses of 100 fs with energies of 700 μJ at 800 nm and at 1kHz of repetition rate. A portion of this output was focused into a 500 μm BBO crystal for second harmonic generation of the fundamental to produce pump pulses at 400 nm. This beam was routed through an optical delay line and a chopper before reaching the sample. The other portion of the fundamental at 800 nm was focused into a 1-cm water cuvette to generate a white light continuum that was used as the probe beam. Pump- and probe-beam were focused and overlapped into a 1 mm thick motorized spinning cell containing the sample. To detected changes in sample after pump beam excitation at various delay times, the transmission of probe beam was focused onto a fiber bundle and passed into a monochromator where it was dispersed. The intensity of the white light probe was detected by a photomultiplier in combination with a lock-in amplifier, for the time-resolved pump-probe scans at fixed wavelength. The transient signals were fitted to a mono or biexponential function.

3 Results

The UV-vis absorbance spectra of deoxyHbI, $HbIO_2$ and HbICO at room temperature are shown in Figure 1. A strong Soret band dominates these spectra at 433, 416 and 421 nm respectively. The other bands present in the spectra are the weaker Q bands peaking at 557 nm for deoxyHbI, 541 and 576 nm for $HbIO_2$, and 539 and 570 nm for HbICO. To follow the spectral evolution of the HbI species, ultrafast transient spectra were taken in the 400–630 nm region after Soret excitation.

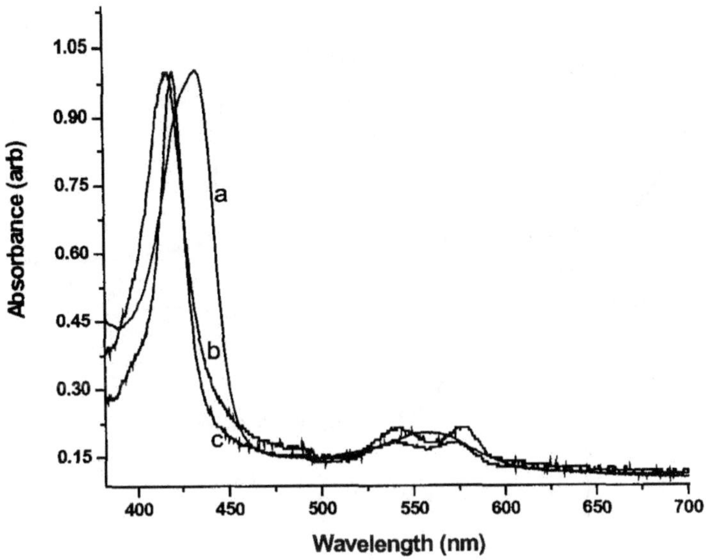

Figure 1. Equilibrium absorbance spectra of a) deoxyHbI, b) $HbIO_2$ and c) HbICO

Transient Bleaching in the Soret Region. Figure 2 shows the transient bleaching of $HbIO_2$, deoxyHbI and HbICO. The signals (Figure 2a-2c) near the maximum of the Soret band of the respective HbI complexes consist of a bleaching component due to the loss of the ground state species. The sharp rise in probe transmission indicates that ligand photodissociation occurs within the time resolution of our experiment. The $HbIO_2$ and HbICO transients (Figure 2a, 2c) present a significant signal recovery during the first six picoseconds. However, in the HbICO case, as opposed to $HbIO_2$, a complete signal recovery was not observed in 200 ps. The solid line in Figure 2 represents the fit to the data with a biexponential function. These fits can be very well described by two relaxation time constants of ~300 fs and 2.2 ps with constant terms of 0.03 and 0.66 for $HbIO_2$ and HbICO, respectively.

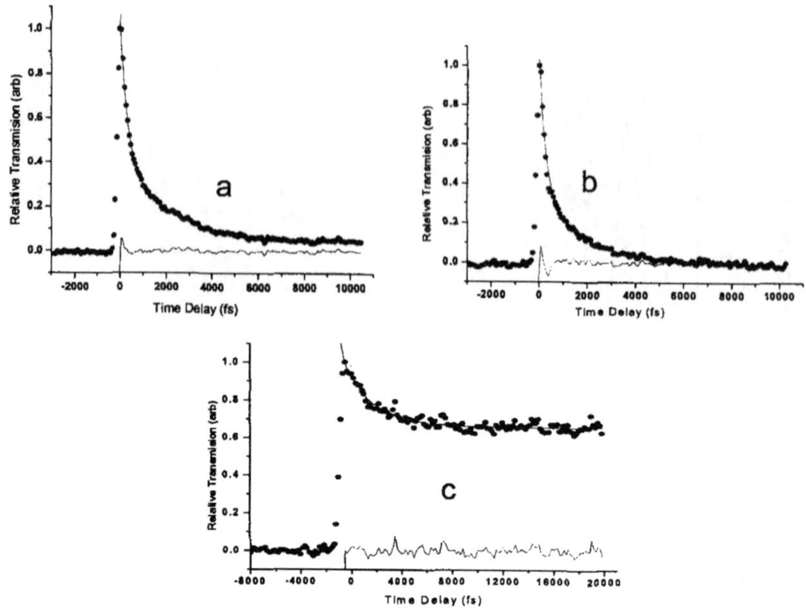

Figure 2. Transient bleaching signals at probe wavelengths a) 408 nm, b) 430 nm, c) 410 nm after excitation of $HbIO_2$, HbIdeoxy and HbICO respectively. The solid lines represent fits to the data with a biexponential function. For $HbIO_2$, $I(t)/Io$ = 0.64 exp(-t/300) + 0.33 exp(-t/2.2) + 0.03 while for HbICO $I(t)/Io$ = 0.09 exp(-t/300) + 0.25 exp(-t/2.2) + 0.66. HbIdeoxy is very similar to Mb, $I(t)/Io$ = 0.70 exp(-t/300) + 30 exp(-t/2.2).

Transient Absorption in the Soret and Q-Regions. Figure 3 shows that beyond 435 nm the transient signals change sings due to product absorption states arising from photoexcited species. At 435 nm, the transient signals for photoexcited $HbIO_2$ (Figure 3a) and HbICO (Figure not shown) present a transient absorbing species that appears at ~300 fs. For $HbIO_2$ this species decay in about 2.6 ps while no substantial decaying was observed for HbICO. Figure 3b and 3c shows that near 455 nm and 490 nm two-product absorption transients are also observed in the HbI complexes together with the deoxy derivative. The absorbing species near 490 nm can be fitted by a monoexpotential decay with a 300 fs constant, whereas for the 450 nm species the signal was best fitted with a 2.2 ps constant. To further investigate the ligand recombination and photophysical processes, the Q-band region of the HbI complexes were examined. In the 500-600 nm Q-region a superposition of HbI-ligated ground state bleaching and Q-band absorption of the unligated photoproducts was observed. At 630 nm, however, a very short –lived excited state absorption signal was observed. Figure 4b presents this transient absorption signal. While formation of this species appears to be instantaneously in the transient signal, its decay can be very well fitted by a monoexponetial function with a 300 fs constant.

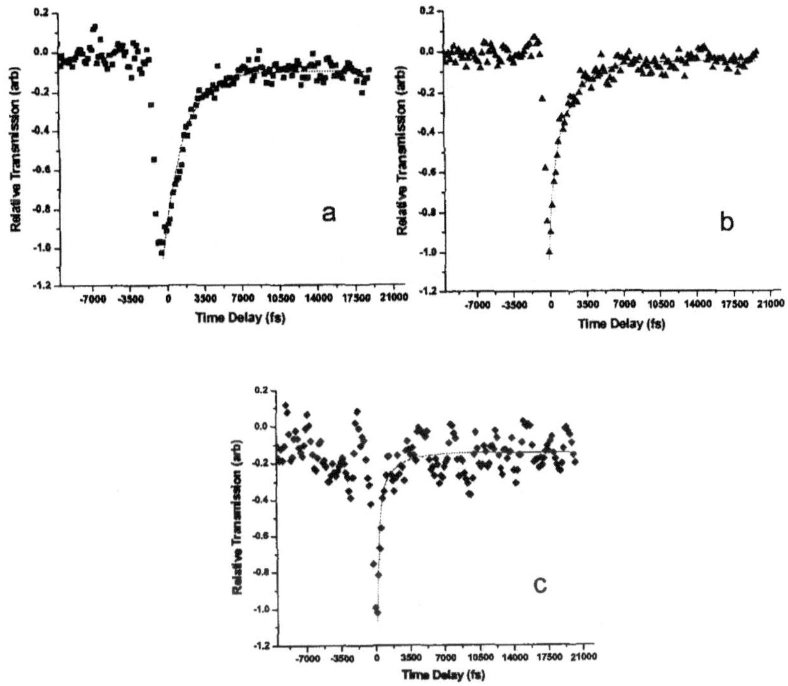

Figure 3. Transient absorption signals of $HbIO_2$ at probe wavelengths a) 435 nm, (biexponential function with a growth of 300 fs and a decay of 2.6 ps). b) 455 nm and c) 490 nm, (monoexponential decay with a 300 fs and 2.2ps constant repectively). The solid lines represent fits to the data as described in the text.

4 Discussion

Photoexcitation of the HbI complexes evidently leads to bleaching signals due to the loss of their respective ground states and the formation of four absorption signals arising from photoexcited species.

Transient Bleaching. The two relaxation time constants of ~300 fs and 2.2 ps observed in Figure 2 are in good agreement with previous femtosecond studies of other hemeprotein complexes [11,12]. The femtosecond component has been implicated in the formation of the five-coordinated, high-spin deoxy species from the excited state Hb_I^* while the picosecond component represents the decay of the excited Hb_{II}^* [11-13]. This species along with the constant term are believe to be involved in the recovery of the Soret signal and hence ligand recombination. Considering then, the recovery of the Soret bands only in terms of the population in the picosecond decay and in the long relaxation constant, we can estimate that in photodissociated $HbIO_2$ and HbICO the Soret signals recover 90% and 30%, respectively. The significant recombination of the dissociated HbI complexes, as oppose to vertebrate heme-O_2 and CO complexes, can be interpreted in terms of its heme pocket amino acids composition. The bulky benzyl side chain of the unusual aromatic residues surrounding the HbI binding site (Phe (29), Phe (43) and Phe (68)) may obstruct diffusion of the photodissociated ligands away from the heme. Consequently, this aromatic cage could exert a restraining force on the O_2 and CO, which in turn minimizes the ligand-metal distance. The attractive Fe-ligand coordinate is thereby maximized leading to the ligated forms of the HbI complexes. This interpretation is supported by the observation that mutating the position 68 for a phenylalanine residue in cobalt Mb and iron Mb yield a much larger recombination of NO than the wild type proteins containing a valine in that position [19]. Moreover, the π faces of the aromatic residues in *L. pectinata* HbI pocket could exert a strong synergetic effects that modulate its ligand-binding properties and hence its biological function. The differences in the protein environment surrounding the metal center in vertebrate myoglobins and HbI indicate the important role of the heme pocket structure in ligand dynamics.

Transient Absorption. Figure 3a shows that probing at 435 nm, a wavelength characteristic of the deoxyHbI species, yield a transient absorbing signal that appears at ~300 fs in both $HbIO_2$ and HbICO. In previous studies a deoxy like Hb and Mb also appears in this region with a 300 fs time constant [9-11]. It has been suggested that this species is formed from Hb_I^* and that the 300 fs decay in the bleaching transients (Figure 2) is due to formation of this deoxyHb that has a no negligible absorption in Soret region. The results of the present study support this suggestion since the 300 fs component is also observed in the HbI-ligated and unligated bleaching transients. Furthermore, the presence of an absorbing species at 490 nm demonstrates that this is the excited state Hb_I^* that appears instantaneously and decays in 300 fs to yield the deoxy like HbI. The study also confirms the presence of the other excited state, Hb_{II}^*, which absorbs near 450 nm and also forms upon photodissociation but relaxes in 2.2 ps.

In addition, photoexcitation of the HbI complexes leads to the formation of a short-lived species at 630 nm that to our knowledge has not been observed before. The analysis of this new absorbance may appear to be difficult since spectroscopic changes occurring on ultrafast time scales can arise from electronic relaxation, thermal relaxation, ligand rebinding and conformational relaxation [18]. However, the 630 nm feature is present in the high spin, unligated deoxy form of HbI. This suggests that ligand rebinding and conformational relaxation, arising from the iron movement out of the heme plane, do not contribute to the spectral evolution of this transient absorbing species. Moreover, the relaxation constant of the 630 nm transient is too rapid to invoke contribution from thermal relaxation, which occurs from 3 ps to tens of ps [14,18]. Therefore, the feature at 630 nm can be attributed to electronic relaxation dynamics of the heme. Regarding this, polarized single-crystal absorption spectroscopic studies of several hemeproteins have identified a spectral band in both the ligated and unligated forms of the heme near 630 nm [19]. The intensity of this band was ascribed to a z-polarized iron-to-porphyrin charge transfer transition. For the ligated heme complexes the transition was assigned to iron $e_g(d_{xz},d_{yz}) \rightarrow$ porphyrin $e_g(\pi^*)$promotion while for the unligated high spin form an iron $a_{1g}(d_z^2) \rightarrow$ porphyrin $e_g(\pi^*)$ transition was invoked. Femtosecond near-IR spectroscopy [18], time-resolved resonance Raman and transient absorption spectroscopy [14] have identified spectral features in the near-infrared region consistent with charge transfer transitions. Therefore, it can be suggested that the new 630 nm transient is an iron-to-porphyrin promotion. An iron-to-porphyrin charge transfer transition at 500 nm has recently been invoked for the excited heme state Hb_I^*, which interestingly relaxes with a 300 fs constant similar to the 630 nm feature (Figure 4). The resemblance of lifetimes in both transients may suggest different relaxation pathway other than Hb_I^* and Hb_{II}^*. However, because the experimental error of the 630 nm spectra reported here is ±50 fs, it is possible that this excited state is one of the low-lying bottleneck states that populates faster than photodissociation can occur and hence faster than Hb_I^* [11]. Although the possibility of an additional relaxation pathway cannot be rule-out the assignment of a low-lying state is favored here because of the many distinct futures that has been observed in the near infrared region for virtually all hemeproteins [18].

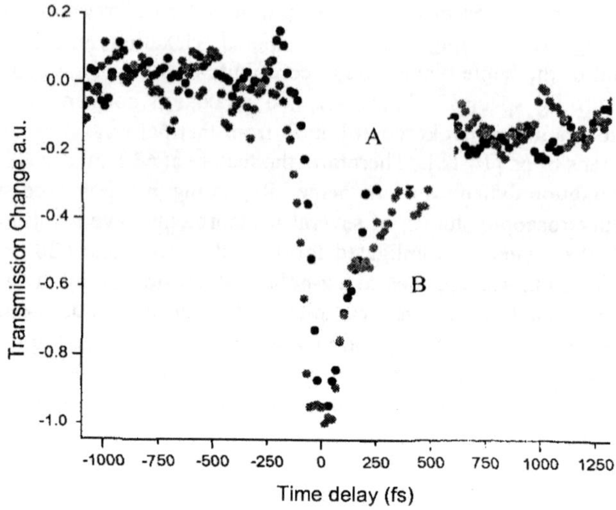

Figure 4. Transient absorption signal of the $HbIO_2$ complex at a) 490 nm and b) 630 nm.

5 Acknowledgements

This research was supported in part by the NSF-RIMI (HRD 9550705 and NSF (MCB-9974961) programs.

References

1. Terwilliger N. M., Functional adaptations of oxygen-transport proteins. *The Journal of Experimental Biology.* 201 (1998) pp. 1085-1098
2. Weber R. E. and Vinogradov S.N., Nonvertebrate hemoglobins: Functions and molecular adaptations. *Physiol Rev.* **81** (2001) pp. 569-628
3. Kraus D. W. and Wittenberg J. B., Hemoglobins of *Lucina pectinata* bacterial symbiosis. I. Molecular properties, kinetics and equilibria of reactions with ligands. *J. of Biol. Chem.* **265** (1990) pp. 16043-16053.
4. Antommattei F. M., Rosado-Ruiz T., Cadilla C. L. and López-Garriga J., The cDNA-derived amino acid sequence of hemoglobin I from *Lucina pectinata*. *J. Protein Chem.* **18** (1999) pp. 831-836.

5. Rizzi M., Wittenberg J. B., Coda A., Fosano M., Ascenzi P. and Bolognesi M., Structure of the sulfide-reactive hemoglobin from the clam *Lucina pectinata. J. Mol. Biol.* **244** (1994) pp. 86-99.
6. Cerda J., Echevaría Y., Morales E. and Garriga J. L., Resonance Raman studies of the heme-ligand active site of hemoglobin I from *Lucina pectinata. Biospectroscopy.* **5** (1999) pp. 289-301.
7. Silfa E., Almeida M., Cerda J., Wu S. and Garriga J. L., Orientation of the heme vinyl groups in the hydrogen sulfide-binding hemoglobin I from *Lucina pectinata. Biospectroscopy.* **4** (1998) 311-326.
8. Navarro A. M., Maldonado M., Gonzáles-Lagoa J., López-Mejía R., Garriga J. L. and Colón J. L., Control of carbon monoxide binding states and dynamics in hemoglobin I of *Lucina pectinata* by nearby aromatic residues. *Inor. Chim. Acta.* **243** (1996) pp. 161-166.
9. Martin J. L., Mingus A., Poyart C., Lecarpentier Y., Astier R. and Antonetti A., Spectral evidence for sub-picoseconds iron displacement after ligand detachment from hemoproteins by femtosecond light pulses. *EMBO J.* **2** (1983) pp. 1815-1819.
10. Martin J. L., Mingus A., Poyart C., Lecarpentier Y., Astier R. and Antonetti A., Femtosencond photolysis of CO-ligated protoheme and hemoproteins: Appearance of deoxy species with a 350-fsec time constant. *Biophysics.* **80** (1983) pp. 173-177.
11. Petich J. W., Poyart C. and Martin J. L., Photophysics and reactivity of heme proteins: A femtosecond absorption study of hemoglobin, myoglobin, and protoheme. *Biochemisty.* **27** (1988) pp. 4049-4060.
12. Martin J. L., and Vos M. H., Femtosecond measurements of geminate recombination in heme proteins. *Methods in Enzyology*, **232** (1994) pp. 418-431. Academic Press, Inc., Oxford, England.
13. Lim M., Anfinrud T. A. and Anfinrud P. A., Mid-infrared vibrational spectrum of CO after photodissociation from haem: Evidence for a ligand docking site in the haem pocket of haemoglobin and myoglobin. *J. Chem. Phys.* **102** (1995) pp. 4355-4366.
14. Franzen S., Kiger L., Poyart C. and Martin J. L., Heme photolysis occurs by ultrafast excited state metal-to-ring charge-transfer. *Biophys. J.* **80** (2001) pp. 2372-2385.
15. Petich, J. W., Lambry J. C., Kuczera K., Larplus M., Poyart C. and Martin J. L., Ligand binding and protein relaxation in heme proteins: A room temperature analysis of NO geminate recombination. *Biochemisty.* **30** (1991) pp. 3975-3987.
16. Petrich, J. W., Lambry J. C., Balasubramanian S., Lambright. D. G., Boxer S. G. and Martin J. L., Ultrafast Measurements of Geminate Recombination of NO with Site-specific Mutants of Human Myoglobin. *J. Mol. Biol.* **238** (1994) pp. 437-444.

17. Kholodenko Y., Gooding E. A., Dov Y., Ikeda M. and Hochstradder R. M., Heme proteins dynamics revealed by geminate nitric oxide recombination in mutants of iron and cobalt myoglobin. *Biochemistry.* **38** (1999) pp. 5918-5924.
18. Lim M., Jackson T. A. and Anfinrud P. A., Femtosecond Near-IR Absorbance Study of Photoexcited Myoglobin: Dynamics of Electronic and Thermal Relaxation. *J. Phys. Chem.* **100** (1996) pp. 12043-12051.
19. Makinen M. W. and Churg A. K. Structural and Analytical Aspects of Electronic Spectra of Hemeproteins. Edited by Lever A.B.P., and Gray H.B. pp141-235. Addison Wesley, Massachusetts.

FEMTOSECOND STUDIES OF PROTEIN-LIGAND HYDROPHOBIC BINDING AND DYNAMICS: HUMAN SERUM ALBUMIN*

A. Douhal[1], D. Zhong[2], and A.H. Zewail[2]

[1]*Departamento de Química Física, Sección de Químicas, Facultad de Ciencias del Medio Ambiente, Campus Tecnológico de Toledo, Avenida Carlos III, S.N., Universidad de Castilla-La Mancha, 45071 Toledo, Spain.*
[2]*Laboratory for Molecular Sciences, Arthur Amos Noyes Laboratory of Chemical Physics, California Institute of Technology, Pasadena, CA 91125.*

In this contribution, we are concerned with the dynamics of chemical bond breaking, making and twisting, and the direct relationship of the foregoing to the involved confined nanostructures. With femtosecond resolution, we examined the proton transfer and orientational motion of a ligand upon ultrafast electronic excitation, and the lifetime change in hydrophobic phases ranging from flexible spheres (micelle), cones (cyclodextrins) to rigid three-dimensional networks provided by a protein (Human Serum Albumin). Free dynamics observed in neat solvent was examined in order to compare with the chemical and biological caged dynamics.

* This work is based mainly on reference 1.

Introduction

In recent studies, we have reported on femtochemistry in nanocavities providing new insight into chemical bond dynamics.[2-6] Here, we continue our efforts in this direction by exploring the interplay between the ultrafast dynamics of chemical bond and the nanostructure of confined chemical and biological media. It is well known that weak, noncovalent interactions govern the ligand-binding process during protein-ligand complexation[7]. Therefore, elucidating the role of these interactions and the time scales involved provides insights into the mechanism of molecular recognition and the role of binding cooperativity in protein dynamics. 2-(2'-Hydroxyphenyl)-4-methyloxazole (HPMO), a proton-transfer fluorescent dye, was chosen as a probe ligand, and Human Serum Albumin (HSA) was chosen as a prototype protein (Fig. 1). HSA recognizes a wide variety of agents and transports them in the blood stream.[8] The principal regions of ligand bindings to HSA are located in hydrophobic cavities in subdomains IIA (binding site I) and IIIA (binding site II).[9]

HPMO is a hydrophobic aromatic molecule which barely dissolves in water.[10] Its binding to HSA upon complexation by the strong hydrophobic interactions at site I enhances its fluorescence intensity.[10] Fig. 1 illustrates this recognition process. Upon UV electronic excitation, HPMO undergoes an intramolecular proton transfer (IPT),[11] which takes place in less than 300 fs.[2] The process of IPT results from an ultrafast electronic redistribution[12] initiated by the excitation and a vibrational coherence of the elementary modes modulating the H-bond coordinate(s).[13,14] The phototransformation of the initial enol structure to the keto one results in a large Stokes shift (~ 10,000 cm^{-1}) observed in the steady-state emission spectrum.[10,11]

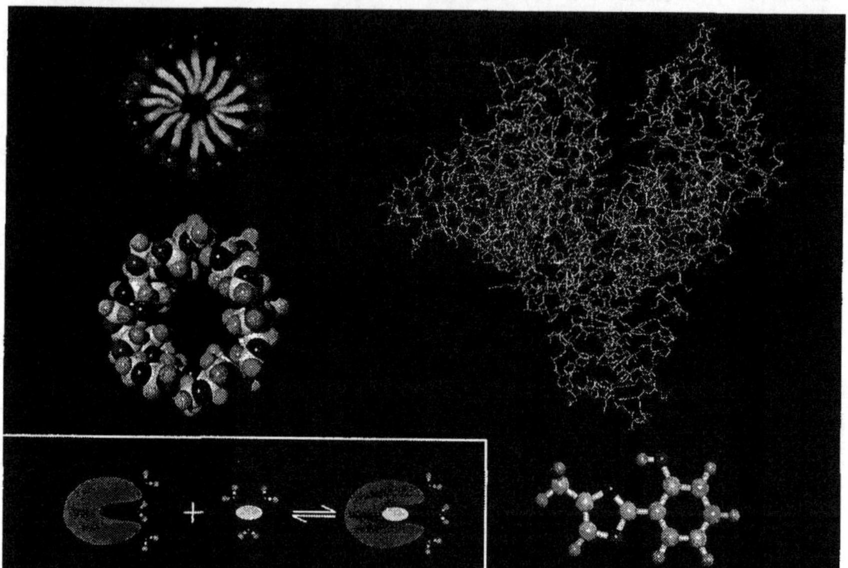

Figure 1: Right: X-ray structure of the HSA protein and molecular structure of the ligand HPMO. Left: Schematic representation of a micelle structure (left top); X-ray structure of β-cyclodextrin (left middle), and illustration of a protein-ligand recognition process (left bottom).

By binding HPMO to the HSA protein in water solution, we studied proton transfer dynamics and the orientational motions of HPMO in the biological environment. For comparison, we carried out similar studies using chemical caging nanostructures of micelles (1-O-octylβ-D-glycopyranoside) and cyclodextrins (heptakis-(2,6-di-O-methyl)-β-cyclodextrin, CD) (Fig. 1). These confined structures, ranging from "spheres" (micelles) and "cones" (cyclodextrins) to the more complex, three-dimensional networks of structures (proteins), provide a unique opportunity for studies of dynamics and binding, and the interplay with the confinement offered by the nanostructure. The dynamics of the probe in *p*-dioxane were examined for comparison with a cavity-free environment.

Results and Discussion

All experimental measurements used the femtosecond-resolved fluorescence up-conversion nonlinear technique. The apparatus has been already described.[1,2] The binding constant of HSA-HPMO complexes is 47000 ± 11000 M^{-1} at 300 K (I. Gracia-Ochoa, and A.D., unpublished results). Compared with the same concentration of HPMO dissolved in *p*-dioxane, the steady-state fluorescence intensity in the protein was enhanced by several hundred times, and the emission peak (455 nm) was blue-shifted by ~10 nm. This indicates the change in the local environment of HPMO upon recognition by HSA and confirms the binding of HPMO to HSA. The steady-state fluorescence emission of HPMO/micelle and HPMO/CD complexes was more than 100 times stronger than in the free solvent (water), and the spectra also were blue-shifted by 5-10 nm. These findings are consistent with the inclusion of HPMO into the hydrophobic cavities of CD and the micelle.

Time-resolved anisotropy: Binding and Structures. The femtosecond-resolved fluorescence anisotropy of HPMO at 470 nm, $r(t)$, is given in Fig. 2 for the four studied environments. Fig. 3 shows the corresponding femtosecond transients obtained at the parallel and perpendicular polarization using HSA and *p*-dioxane. In the nonpolar solvent, *p*-dioxane, the major part (74%) of the anisotropy decays to 0 with a time constant of 45 ps; the minor component (26%) has a time constant of 5.3 ps. In the micelle, these decay constants become 97 ps (83%) and 3.7 ps (17%). For CD, they are 154 ps (84%) and 2.6 ps (16%). The anisotropy of HPMO in HSA protein is dramatically different from all others. It lacks a short-time behaviour and has a slight drop (8%) with a lifetime of 93 ps, but the major component (92%) stays with no decay at a large constant value of 0.304. The minor short-decay component is due to structural relaxation.

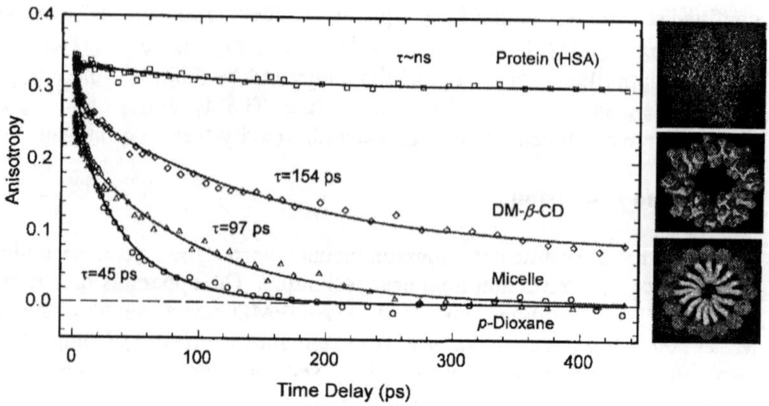

Figure 2. Femtosecond-resolved fluorescence anisotropy decay in four typical environments at 470-nm emission. Left: illustration of the used nanocavities.

Let us analyse these observations starting from the free behaviour, and ending by the biologically caged dynamics. In *p*-dioxane, the observed 45-ps decay of the anisotropy represents the complete orientational relaxation of the free probe. The calculated orientation relaxation times are 64 and 23 ps for the stick- and slip-boundary condition limit, respectively. Our experimental value of 45 ps lies in the middle.

Into the micelle, the diffusive orientation relaxation of the guest (97 ps) becomes two times longer than in *p*-dioxane (1.18 cP), so the corresponding effective viscosity reaches 2.55 cP. The guest is buried into the nonpolar hydrophobic core of the micelle which hinders its rotation in comparison with a *p*-dioxane solution. HPMO trapped in the micellar organized core (octyl chains) needs more time to change its orientation. The observed anisotropy in the micelle indeed decays to zero, reflecting a complete orientation relaxation and indicating an "isotropic" environment in the core. The rotational relaxation of the whole guest-host complex occurs on a much longer time scale, and such slow motion does not influence the anisotropy of the guest in the picosecond range.

Into the CD's molecular box, the situation is different from that observed using micelles. The anisotropy decay (154 ps) is more than three times slower than in *p*-dioxane (45 ps) and up to 500 ps the anisotropy does not decay to 0; it stays at

0.074. The persistence of the anisotropy at the value of 0.074 (not zero) indicates that the orientation relaxation is hindered. This is consistent with the more rigid, confined geometry of CD. Proton NMR studies of HPMO in CD's aqueous solutions suggested that the oxazole ring of the guest molecule is inserted into the cavity, while the other aromatic moiety is exposed to water solution.[10] Because the cavity inner diameter (6.2 Å) is smaller than the probe molecular length (9 Å long), the complete reorientation of the trapped guest is hampered. The corresponding average change in direction of the transition moments of the guest from the initial anisotropy of 0.34 to the final value of 0.074 is about 30°. Compared with the core (>13 Å in diameter) of the micelle, the hydrophobic cavity of CD (~8 Å in inner large diameter; 6 Å in length) is smaller and more rigid. Thus, for HPMO, it takes more time to reorient in the CD cavity.

Into the biological cavity, HSA, the persistence of the anisotropy over 500 ps indicates that the observed diffusive orientational motion in solution essentially is

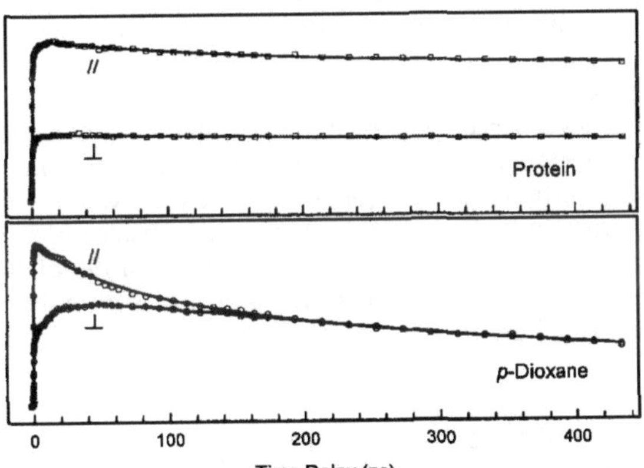

Figure 3: Femtosecond-resolved fluorescence transients analysed at parallel and perpendicular detection of emission using the HSA protein and *p*-dioxane.

absent in the protein, and hence, the angular locking of the probe in it. The rotational relaxation of HSA in water is on the time scale of tens of nanoseconds[10] and makes no contribution to the anisotropy of the ligand in the picosecond region. The high value of the initial anisotropy and the lack of any initial ultrafast decay indicate the strong hydrophobic attraction by molecular recognition, the hindrance of molecular structure relaxation, and the rigidity of the local structure of the binding complex. These results are consistent with recent x-ray structural studies,[9] suggesting a strong hydrophobic interaction between the ligand's aromatic ring and the residues located at sites I of the protein.

The true time scale of molecular recognition is masked by diffusion, which takes, at the concentration of 10^{-4} M, about 1-100 μs. The two rates critical for the final recognition are those describing the reorientation of the ligand (τ_r) in its solvent shell and the direct "reaction" in a single encounter (τ_e). In our case, $\tau_r \sim 50$ ps (taken to be similar to that of ground-state ligand in water), suggesting the regime *in which* $\tau_D > \tau_r$ (τ_D *is the average diffusion time*) and, therefore, the recognition by the ligand is limited by (τ_r/τ_e), which may reach a factor of 100. This same quantity is relevant to the degree of order in the solvent environment that controls the extent of hydrophobicity.

Lifetime at Different Wavelengths. Table 1 lists the lifetimes and relative contributions in the observed transients for different wavelengths at the magic angle (54.7°) and in the four environments studied. The contrast and trends for the lifetimes (for more detail about the decays see reference 1) are remarkable and suggest the following. 1) The lifetimes in *p*-dioxane solution are similar, and independent on the observation wavelength. 2) The lifetimes change upon inclusion in CD and micelle, from a single-exponential decay in the free solvent to multiple-exponentials. The change in nonradiative decay reveals the heterogeneous local structure which results from the anisotropic hydrophobic interactions between the probe and the surroundings.[10,15] 3) In the HSA protein the lifetimes change differently from the blue side to the red one when compared with the other environments studied. This shows the influence of the rigid binding on the population distribution and nonradiative decays in the biological nanocavity. The lengthening of lifetime upon inclusion probably is from the suppression of the nonradiative channel induced by torsional motions.[16]

Table 1. Lifetimes (ns) and relative contributions[a]

λ(nm)	dioxane	CD	micelle[b]	HSA protein
420		0.61(0.88);3.92	0.31 (rise);2.47	0.10(0.33);6.20
430	0.52	0.58(0.80);2.73	0.25 (rise);1.81	0.13(0.25);2.24
470	0.48	0.36(0.56);1.38	0.27(0.08);1.72	0.17(0.12);2.67
510	0.47	0.19(0.35);0.93	0.21(0.24);1.39	0.09(0.12);~10
580	0.45	0.17(0.35);0.83	0.11(0.28);1.21	0.11(0.28);1.89

a. *The relative contributions are listed in parentheses.*

b. The rise is probably from a relaxation process due to a unique weak coupling between HPMO and the alkyl chain of the hydrophobic core.

Directed Dynamics in Confined Structure: Intramolecular Proton Transfer and Twisting Motion. The transients in the studied media (not shown, see reference for more detail) lead to two groups of time-resolved fluorescence emission, which overlap around 430 nm. Relevant data are listed in table 1. For the observation at wavelengths less than 430 nm, the transients show two initial decay components. In dioxane and at 420 nm for example, these times are 300 fs and 3.1 ps. In the biological nanocavity, these times become much different and increase to 1.1 ps and 20 ps, respectively. However, nearly no changes were observed for the guest-host complexes using CD and micelle. The fluorescence intensity at 430 nm is very weak, and the emission at 420 nm is about 15% of the one at 470 nm. For detections at wavelengths longer than 430 nm, we also observed two short-time components: one that appears within 100-250 fs for all wavelengths and the other component (20-30%) that rises at the blue side and decays at the red side of the spectrum. The time constants change systematically. In p-dioxane for example, from a decay of 6.3 ps at 580 nm to a rise of 6.5 ps at 470 nm. No major change of time constants was observed for all complexes, including that with the protein.

To explain the above behaviour, we provide the following picture. The two groups of emission are from two types of trajectories. A schematic potential energy contour is shown in Fig. 4*a*. In the ground state, the stable structure is planar in the enol structure.[11] The other isomers, such as the keto tautomer and the enol rotamers, lie at higher energy. After the initial excitation, the excited-state enol form suffers some loss of aromaticity in the six-member ring and the charge rapidly redistributes, which results in a steep potential surface. This constitutes the driving force for the femtosecond proton transfer from N··· H--O to N---H·· O configuration.[12] The charge redistribution barely changes the direction of transition moments as suggested by the observed initial anisotropy (0.34) which is close to an ideal parallel transition (0.40). Because the minimum structure of the keto tautomer is also planar, the proton-transfer process mostly involves the stretch motion of the heavy atoms and is expected to be ultrafast. Experimentally, we observed a rising component (100-250 fs) for all transients recorded at wavelengths longer than 430 nm. This indicates that vibrational energy redistribution also occurs simultaneously on a similar time scale. This prompt dynamic process evolves along the proton coordinate and is described as trajectory I in Fig. 4*a*. Clearly, the initial wave packet moves quickly along the repulsive potential toward the keto structure (Fig. 4*b*), and the emission during the transformation is centred around 360 nm (Fig. 5). In this region, we observed a weak decay signal giving a time constant of 100 fs or less (Fig. 5), similar to those found in other systems.[12,17-19]

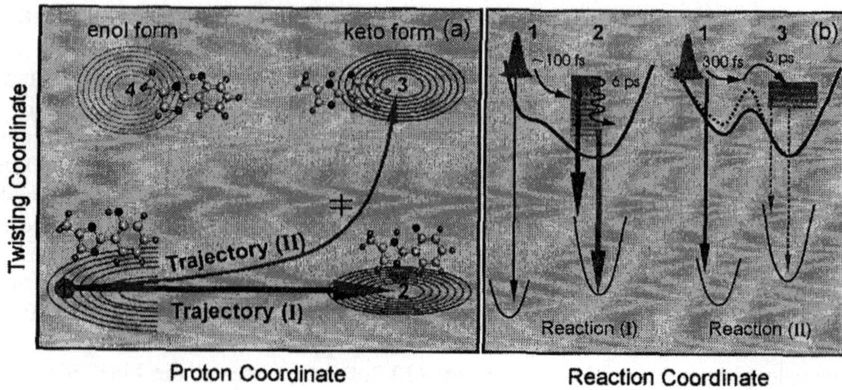

Figure 4: (a) Schematic illustration for two trajectories of the direct proton-transfer (I) and the one involving twisting motion (II). (b) Potential energy curves along the reaction coordinate for trajectories I and II with the observed emission indicated schematically. The dashed line represents the potential energy in the protein environment. Note that rotamers of non-planar enol ground state can undergo proton transfer on the upper surface following an initial twist towards planarity. See text for detail.

Besides the dominant, direct proton-transfer process (trajectory I), the launched wave packet can bifurcate[19] and some family of trajectories evolve along a different route to a keto rotamer minimum, which lies in the same energy range, trajectory II in Fig. 4a. Such a trajectory involves two types of motions: one that is the proton motion at earlier times and the other that is the twisting motion of the two heterocyclic moieties at a later time. Because the potential energy in the rotamer ground state is higher than in the corresponding enol form, the rotamer emission during the twisting shifts to the long wavelength at around 410 nm (Fig. 5). This process involves a twisting motion of the two moieties of excited HPMO, and along this twisting reaction coordinate, the tautomerization usually involves an energy barrier, as predicted by theoretical calculations.[11] Therefore, in the rigid protein-ligand complex, this barrier increases and the reaction time lengthens significantly. Experimentally, we observed a fast decay component of 200 fs in *p*-dioxane but 640 fs in HSA protein at 415 nm. This component represents the initial motion of the wave packet out of the repulsive potential surface (Fig. 4b), and this time is longer than the 100 fs obtained for trajectory I because of the twisting motion. The second decay component of 3 ps in *p*-dioxane represents the barrier-crossing dynamics. Remarkably, this time constant increases to 8 ps at 415 nm, 20 ps at 420 nm, and 37 ps at 430 nm in the rigid protein-ligand complex. In contrast, the guest-host complexes using CD and micelle do not possess such protein rigidity, and in their cavities the twist angle may appear similar to that in *p*-dioxane. Indeed, in these media we observed only a 10-20% change of time constants upon going from *p*-dioxane to the guest-host complexes. For the dominant trajectory I, we observed that the resulting keto structure decays (2-6 ps, depending on the gated region) at wavelengths longer than 520 nm and rises (2-6 ps) at shorter than 520 nm; at 520 nm, we observed flattening of the transient. Therefore, the red-side emission is from the vibrationally hot structures after proton transfer, with transitions to the high vibrational levels of the ground state (Fig. 4b). The time constant of 2-6 ps gives the rate at which these structures are relaxing to achieve the equilibrium configuration of the keto form. Because the process of proton transfer is barrierless (<100 fs) for this planar-type structure, the picosecond structural relaxation is expected to be the same as for solution or the protein environment, as observed experimentally.

Finally, for all systems studied here, we did not observe significant solvation kinetics. Although the dipole moment changes on the time scale of proton transfer, electronic redistribution occurs in less than 60 fs,[12] shorter than the time scale of complete solvation. Even using methanol or acetonitrile (not shown), no apparent solvation dynamics were observed, consistent with this picture.

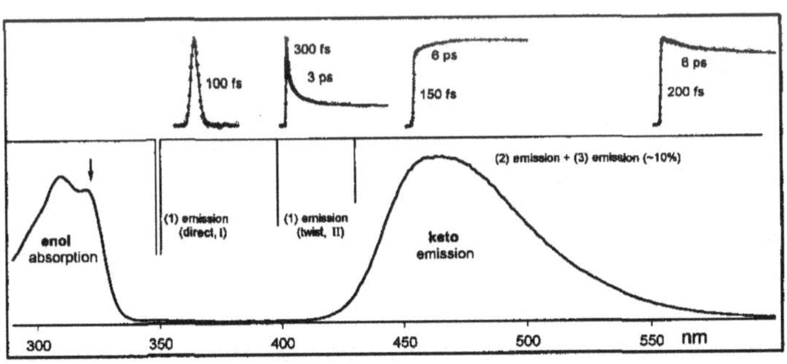

Figure 5: A dynamical dissection of the fluorescence corresponding to the different dynamical regions described in Fig.4. The corresponding transients are shown on the top. The excitation marked by the arrow was at 325 nm.

Conclusion

The results reported here, with femtosecond time resolution, elucidate the *important role of dynamics in the molecular recognition process* and the interplay to structure. The protein host exhibits the strongest binding in the hydrophobic phase of the structure with optimum rigidity having its roots in the chelating forces of the cavity. The persistence of the measured anisotropy of the ligand over hundreds of picoseconds indicates that the maximum motion is ~3^0, as measured by rotation of the transition dipoles. The nuclear motions of the ligand and its structural relaxation in the protein elucidate another point: the directed dynamics of motion imposed by the structural rigidity in the recognition process. The protein suppresses the "inefficient" picosecond proton transfer whereas the direct femtosecond channel observed in isolated systems remains robust.[12,20-22] Such rigidity of conformations and directed dynamics are less exhibited in the cavities of the guest-host complexes of micelles or CD and may be a unique feature of the biological protein environment for the function and in this case for drug delivery.[19]

Acknowledgements. This work was supported by the National Science Foundation (USA). A.D. was supported by the Ministry of Science and Technology (Spain) through projects PB98-0310, and 1FD1997-1658, and by the U.S.A.-Spain cooperative program.

References

1. Zhong, D.; Douhal, A.; Zewail, A.H. *Proc. Natl. Acad. Sci. (USA)* 97(2000)14052.
2. Douhal, A.; Fiebig, T.; Chachisvilis, M.; Zewail, A.H. *J. Phys. Chem. A* 102(1998)1657.
3. Chachisvilis, M.; Garcia-Ochoa, I.; Douhal, A.; Zewail, A.H. *Chem. Phys. Lett.* 293(1999)153.
4. Douhal, A. in *Femtochemistry*, De Schryver, F.C.; De Feyter, S.; Schweitzer, G. (Edits.), Wiley-VCH, Chapt. 15(2001)267.
5. Zhong, D.; Pal, S.K.; Zewail, A.H. *ChemPhysChem.* 2(2001)219.
6. Zhong, D.; Zewail, A.H. *Proc. Natl. Acad. Sci. (USA)* 98(2001)11867.
7. Gellman, S.H. (Ed.) *Chem. Rev.* 97(1997)1231.
8. Berde, C.B.; Hudson, B.S.; Simon, R.D.; Sklar, L.A. *J. Biol. Chem.* 254(1979)391.
9. He, X.M.; Carter D.C. *Nature (London)* 358(1992)209.
10. García-Ochoa, I.; López, M.D.; Viñas, M.H.; Santos, L.; Martínez, E.; Amat-Guerri, F.; Douhal, A. *Chem. Eur. J.* 5(1999)897.

11. Guallar, V.; Moreno, M.; Lluch, J.; Amat-Guerri, F.; Douhal, A. *J. Phys. Chem.* 100(1996)19789.

12. Herek, J.L.; Pedersen, S.; Bañares, L.; Zewail, A.H. *J. Chem. Phys.* 97(1992)9046.

13. Douhal, A. *Science* 276(1997)221.

14. Chudoba, C.; Riedle, E.; Pfeiffer, M.; Elsaesser, T. *Chem. Phys. Lett.* 263(1996)622.

15. Bright, F.V. *Appl. Spectrosc.* 49(1995)14A.

16. Organero, J.A.; Santos, L.; Douhal, A. See the article in the Section Liquids of this book.

17. Schwartz, B.J.; Peteanu, L.A.; Harris, C.B. *J. Phys. Chem.* 96(1992)3591.

18. Chudoba, C.; Lutgen, S.; Jentzsch, T.; Riedle, E.; Woerner, M.; Elsaesser, T. *Chem. Phys. Lett.* 240(1995)35.

19. Zewail, A.H. *J. Phys. Chem. A* 104(2000)5660, and references therein.

20. Douhal, A.; Lahmani, F.; Zewail, A.H. *Chem. Phys.* 207(1996)477.

21. Organero, J.A.; Moreno, M.; Santos, L.; Lluch, J.M.; Douhal, A. *J. Phys. Chem. A* 104(2000)8424.

22. Baskin, S.J.; Zewail, A.H. see Section Gas Phase and Clusters of this book.

TWO DIMENSIONAL VIBRATIONAL SPECTROSCOPY OF ANTAMANIDE; A SIMULATION STUDY

S. M. PARK, C. SCHEURER, AND S. MUKAMEL

Department of Chemistry, University of Rochester, Rochester, NY 14627, USA
E-mail: mukamel@chem.rochester.edu

Abstract

Infrared pulse sequences may be used to provide ultrafast snapshots of molecular structural fluctuations. The two-dimensional (2D) vibrational echo and reverse transient grating signals from the amide I band of the four conformational states of the cyclic decapeptide antamanide are computed. We demonstrate that by combining a method of calculating the vibrational Hamiltonian from MD snapshots, a clustering algorithm for structures, and the nonlinear exciton equations (NEE) it is possible to simulate realistic multidimensional IR spectra of chemically and biologically interesting systems.

1 Introduction

The cyclic decapeptide antamanide [-V1–P2–P3–A4–F5–F6–P7–P8–F9–F10-] (Fig. 1) is a small oligopeptide which serves as a model for proteins and was studied intensively by NMR as well as theoretically by MD simulations (see[1] and references therein). Antamanide has biological activity as an antidote against the poisonous phalloidins that occur in the mushroom *Amanita phalloides.*[2] Remarkably, antamanide is contained in small quantities in the same mushroom. In mammals, phalloidins block the depolymerization of F–actin into G–actin in the liver cell membrane. This process can be prevented by the presence of antamanide.

Antamanide contains a network of intramolecular hydrogen bonds[3,4] (Fig. 1 b, c). It has been suggested, initially from 1 D NMR measurements[5] and subsequently from ultrasonic absorption,[6] that it exists in solution in an equilibrium of rapidly interchanging conformers ($k \approx 10^6\, s^{-1}$). In more recent studies, two–dimensional NMR, relaxation measurements in the rotating frame, and homo– and heteronuclear coupling constant determinations have been used to characterize the dynamic properties of the molecule.[4] Direct evidence for the presence of a conformational exchange process with an activation energy of $\approx 20\,\mathrm{kJ\,mol^{-1}}$ and an exchange lifetime of $\approx 25\,\mu s$ at 320 K has been obtained from rotating frame relaxation $T_{1\rho}$ measurements which suggest rapid hydrogen bond exchange dynamics at ^{1}Val HN and ^{6}Phe HN.

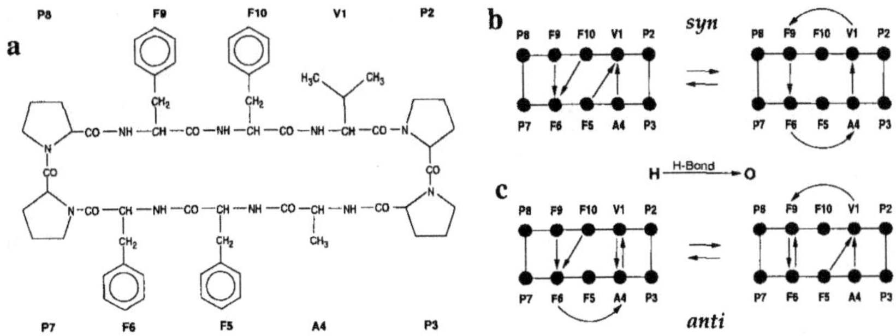

Figure 1. **a** The cyclic decapeptide antamanide. **b** Schematic representation of the hydrogen bonding network in the structural *syn*–pair (E 116, G 129) of the antamanide exchange system. **c** Schematic representation of the hydrogen bonding network in the structural *anti*–pair (A 128, G 193) of the antamanide exchange system.

Local backbone dihedral angle changes accompany these processes. Two of the most likely pairs of dynamic exchange systems have been identified, designated as the *syn*–pair (Fig. 1 b) and the *anti*–pair (Fig. 1 c) on the basis of the *syn*–correlated and *anti*–correlated structural changes at the two mentioned residues.[4] It was pointed out that even though the two preferred exchange systems which emerged from the analysis led to the best fit of the available experimental data, further pair structures could not be excluded for antamanide in solution. Further measurements would be necessary to characterize the ensemble of possible structure in more detail.

Multidimensional nonlinear optical experiments which are able to probe the respective dynamics in real-time and not indirectly via relaxation measurements as in NMR could complement these NMR studies. Two–dimensional IR spectroscopy of the amide I band proved to be sensitive to hydrogen bonding involving the carbonyl oxygen.[7–9] This band originates from the stretching motion of the C=O bond coupled to the bending of the N-H bond and the stretching of the C-N bond (see Fig. 2).[10] Conformers of the decapeptide with differing hydrogen bond networks should give rise to different 2D–spectra that allow the experimental identification of the species in solution.

In addition to hydrogen bond dynamics, sidechain dynamics,[11] backbone dynamics,[12] and proline ring puckering[1,13,14] in antamanide were studied by NMR relaxation measurements as well as by MD simulations. The study of sidechain dynamics was performed on the four phenylalanine sidechains[11] (see Fig. 3 a). Based on a MD trajectory a motional model for the angular degrees of freedom (χ_1, χ_2 in Fig. 3 b) was proposed. The analytical model describes

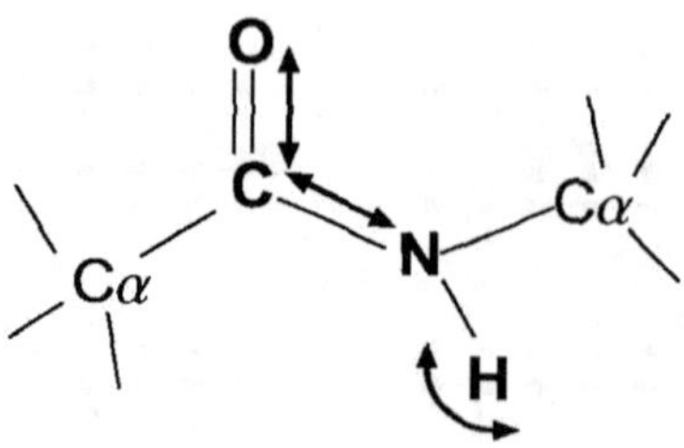

Figure 2. Schematic representation of a peptide bond in proteins. The arrows denote the stretching motion of the C=O bond coupled to the stretching of the C-N bond and the bending of the N-H bond, thereby contributing to the amide I band .

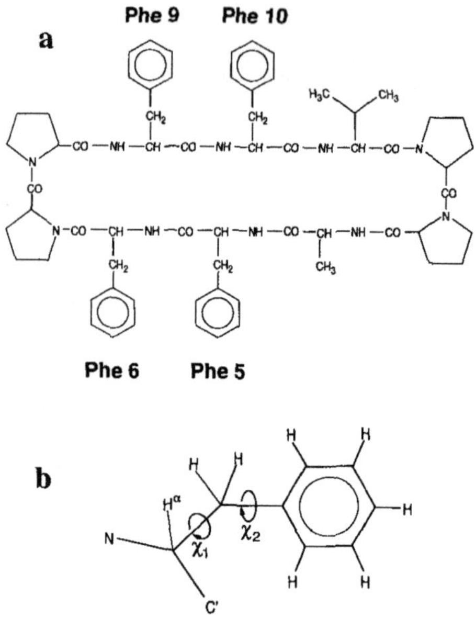

Figure 3. **a** The four phenylalanine sidechains in the cyclic decapeptide antamanide. **b** Rotational degrees of freedom of the phenylalanine sidechains.

Gaussian axial fluctuation (GAF) within localized potential wells, interrupted by occasional jumps between different rotameric states over relatively high barriers. The relevant motional parameters like jump rates and fluctuation amplitudes were extracted by fitting a large set of NMR relaxation data for the different sidechains to the analytical model.

Pump–probe and optical fluorescence spectroscopy can directly probe the dynamics of the phenylalanine chromophores in real–time.[15] The validity of the dynamical model obtained from NMR relaxation studies could be tested and the respective motional parameters determined. NMR and optical spectroscopy can give complementary information on the dynamics and allow a

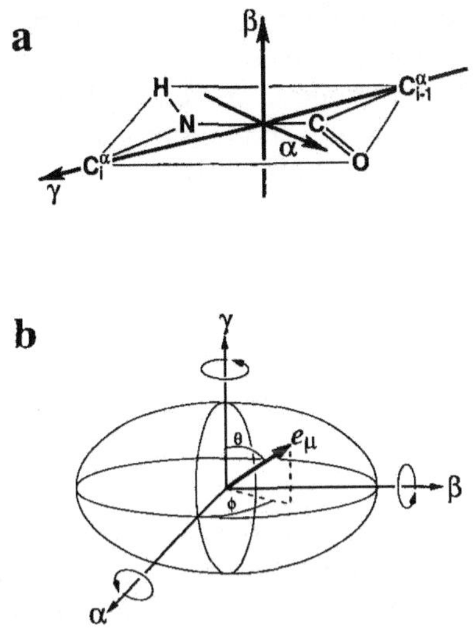

Figure 4. **a** Definition of the local reference coordinate system for the description of anisotropic motion of the peptide planes in proteins. **b** Fluctuation ellipsoid describing anisotropic intramolecular motion.

cross–validation of the models derived from experimental data.

An extension of the GAF model to three dimensions has been used to describe the anisotropic peptide–plane motion in the backbone.[12] The model describes the reorientation in terms of rotations of the mostly rigid peptide plane about the three axes α, β, and γ (see Fig. 4 a) with Gaussian fluctuations about an average angle. The fluctuation amplitudes can be visualized in form of reorientational fluctuation tensors (Fig. 4 b). Fitting of the motional model to a set of ^{13}C' T_1, ^{15}N T_1, and $\{^1H\}^{15}$N NOE values resulted in motional parameters for the six non–proline peptide planes.

Multidimensional IR spectroscopy of the amide I band provides an independent observation of the same fast dynamics which is free of the assumptions underlying the interpretation of NMR relaxation data. The determination of the set of peptide planes in a protein that can be described by fast small–amplitude motion is important for the identification of the peptide groups involved in large–amplitude motions that dominate the conformational changes of the protein underlying its biological function. Only after a sound understanding of the small–amplitude motion and its separation from dynamics leading to conformational changes is achieved, it shall be possible to develop motional models for the more complex large-amplitude motions.

Different visible or IR nonlinear spectroscopic techniques can cover the whole range of relevant timescales (from ps to μs) occuring in the dynamics of Antamanide.[15,16] Comparison with the more indirect NMR relaxation experiments should lead to a better understanding of the relevant dynamical processes in this model compound. Detailed motional models can thereby be tested with completely independent experimental methods. The improved understanding of the dynamics in a small model compound should facilitate investigations on larger proteins and help to identify limitations and strengths of either spectroscopic technique.

2 Molecular Dynamics Simulation

The advent of femtosecond laser pulses has made it possible to apply various multidimensional spectroscopic techniques to probe electronic and vibrational dynamics on the femtosecond timescale. Spectroscopic techniques based on the application of sequences of carefully shaped and timed femtosecond pulses provide a multidimensional view of molecular structure as well as vibrational and electronic motions, interactions, and relaxation processes.[17–29] For example, the intensities and profiles of cross-peaks give a direct signature of molecular structure (distances between chromophores) and dynamics (the spectral density of the chromophores' environments). By combining MD simulation, a fluctuating exciton Hamiltonian for the amide-I band [reference], and the NEE approach[29,30] for the calculation of 2D vibrational spectra 2D vibrational echo signals of the cyclic decapeptide antamanide was calculated. For the MD simulation, the dihedral angles of four most common conformers (E116, G129, A128, and G193) were used to construct antamanide using the CHARMM 27 force field, topologies, and parameters. For each structure constructed, SHAKE algorithm[31] was applied to constrain bond lengths, and Adopted Basis Newton-Raphson (ABNR) algorithm was used to minimize the energy. After equilibration in CHARMM a 50 nanosecond Langevin dynamics MD trajectory with a timestep of 1 femtosecond was generated for each structure, and snapshots of the structure were taken every 2 picosecond. 25,000 snapshots of antamanide have been obtained for each conformer. Bath temperature for the MD simulation was set at 300 K and closely monitored for any abrupt fluctuation; each simulation was completed successfully with a temperature variation of less than, or equal to ± 10 K from the set value. Solvent was not included explicitely in the simulation, instead the stochastic models for force - Langevin dynamics- was applied to antamanide to simulate the environment of chloroform.

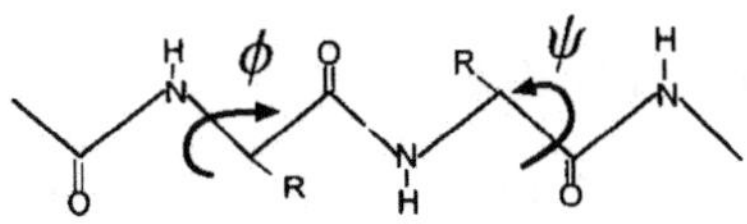

Figure 5. Schematic illustration of the protein backbone dihedral angles ϕ and ψ

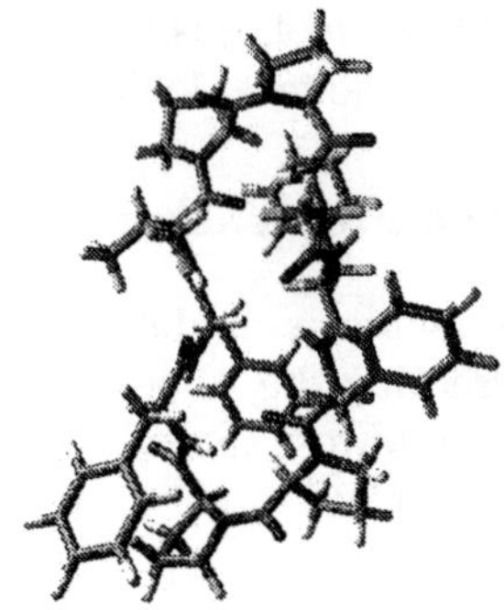

Figure 6. 3D representation of Antamanide

3 The Clustering Algorithm

We have used a clustering algorithm, written in C, to classify each structure generated by the simulation into a cluster of either E116, G129, A128, or G193. The input for the algorithm is the coordinate information from the MD simulation snapshot. The output shows which cluster the snapshot belongs to and lists the fluctuations of the transition frequencies and dipole-dipole couplings. It consists of following steps:

1. The backbone dihedral angles ϕ and ψ are measured for each residue (see Fig. 5).

2. Hydrogen bond energy is calculated using Kabsch-Sander electrostatic energy criterion[32] in which $f = 332\, e^{-2}\, \mathring{A}$ kcal/mol, $q_1 = 0.42\, e$, and $q_2 = 0.2\, e$.

$$E_{ks} = f q_1 q_2 \left(\frac{1}{d_{ON}} + \frac{1}{d_{CH}} - \frac{1}{d_{CN}} - \frac{1}{d_{OH}}\right) \qquad (1)$$

   - A hydrogen bond is deemed to exist if $E_k s$ gives a value less than the cutoff - 0.5 kcal/mol.
   - The two strongest H-bonds per acceptor Oxygen are determined.

3. The mean square deviation of ϕ, ψ, and hydrogen bond energy is calculated with respect to E116, G129, A128 and G193 (see Fig. 1 b,c).

4. The snapshot is put into one of the four cluster which showed the smallest mean square deviation.

4 The Fluctuating Vibrational Exciton Model

Antamanide is modelled as a system of $N = 10$ interacting local vibrations representing the 10 peptide bonds (see Fig. 7). For third-order spectroscopies, only the lowest three levels of each peptide group need to be taken into consideration.[9,33]

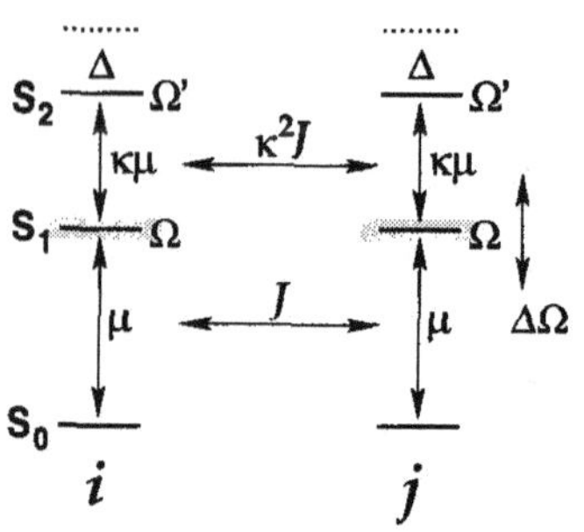

Figure 7. The three-level systems representing the peptide plane groups in the exciton Hamiltonian. The levels in each system, i and j, are denoted S_0, S_1, S_2 with energies $0, \Omega, \Omega'$ for each chromophore. ($\Delta = \Omega' - 2\Omega$) The levels are shifted by hydrogen bonds to the carbonyl oxygen giving rise to a fluctuating term$\Delta\Omega$. The only nonvanishing elements of the transition dipole operator are μ and $\mu' = \kappa$. The carbonyl vibrations are dipole-coupled where the coupling matrix J_{ij} depends on the local geometries.

In the present MD simulation, only the fluctuations of transition frequencies (Ω_i) and dipole-dipole couplings (J_{ij}),[34] which are expected to dominate the spectra, are taken into account in computing the Hamiltonian. Fluctuations of other parameters, such as the dipole coupling to the laser field are neglected. The quadratic part of the Hamiltonian can be represented in the form of coupled harmonic oscillator:[9]

$$\hat{H}_0 = \sum_i \epsilon_i B_i^\dagger B_i + \sum_{i \neq j} J_{ij} B_i^\dagger B_j \tag{2}$$

The transition dipole-dipole coupling constant between each residues (J_{ij}) is given by:[9]

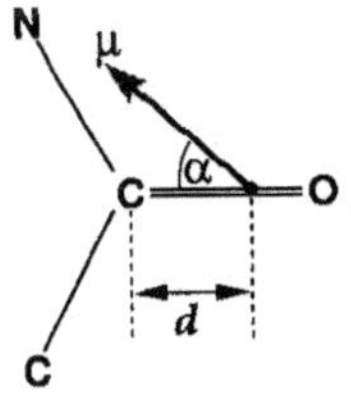

$$J_{ij} = \frac{1}{r_{ij}^3}(\vec{\mu}_i \cdot \vec{\mu}_j - 3(\vec{n}_{ij} \cdot \vec{\mu}_i)(\vec{n}_{ij} \cdot \vec{\mu}_j) \quad (3)$$

in which $\alpha = 25°$, $|\mu| = 0.37\,\mathrm{D}$, $d = 0.868\,\mathrm{Å}$. The fluctuation in the vibrational frequencies induced by the hydrogen bond of carbonyl group (C_i) to NH groups of different residues (NH_j) is given by:

$$\delta\epsilon_i = \alpha_H(2.6 - d_{OH}) \quad (4)$$

Here d_{OH} is the O$\cdots$H distance in Å and the vlaue of $\alpha_H = -30\,\mathrm{cm}^{-1}/\mathrm{Å}$ results in a frequency downshift of $\sim 20\,\mathrm{cm}^{-1}$ for a typical hydrogen bond length of $1.9 - 2.0\,\mathrm{Å}$.[7,35]

5 2D Vibrational Echo and Transient Grating Spectra

The lowest-order nonlinear IR spectroscopy in non-centrosymmetric media requires three pulses (see Fig. 8).[36] In femtosecond 2D three-pulse techniques, two of the three pulses are time-coincident and only differ by their wave vector k.[33]

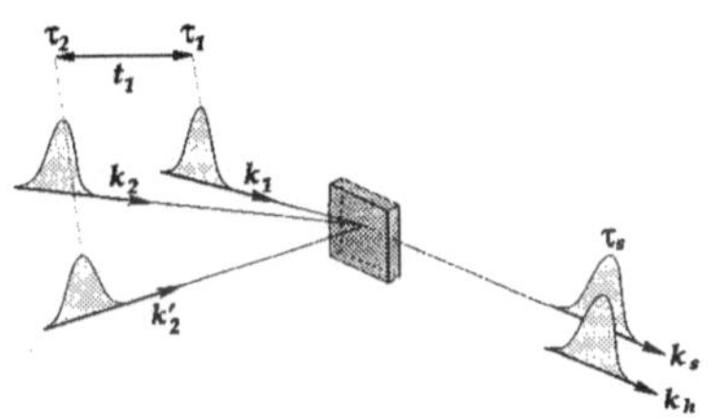

Figure 8. Nonlinear IR spectroscopy (photon echo technique)

The system interacts with a single pulse with wave vector k_1 at time τ_1 and twice with the pulses of a pulse pair $\mathbf{k_s} = \mathbf{k'_2} + \mathbf{k_2} - \mathbf{k_1}$ (photon echo), $\mathbf{k_s} = -\mathbf{k'_2} + \mathbf{k_2} - \mathbf{k_1}$ (reverse transient gating) at $t = \tau_2$. The delay between the single pulse and the pulse pair will be denoted $t_1 = \tau_2 - \tau_1$. The centered signal is detected by combining it with an additional heterodyne pulse k_h at time τ_s. This gives the second time variable $t_2 = \tau_s - \tau_2$ of the 2D correlation plot.

Closed expressions for the third–order nonlinear response function $R(t, \tau_3, \tau_2, \tau_1)$ were obtained in Ref.[30] by solving the Nonlinear Exciton Equations (NEE) for the exciton Hamiltonian. These can be used to obtain expres-

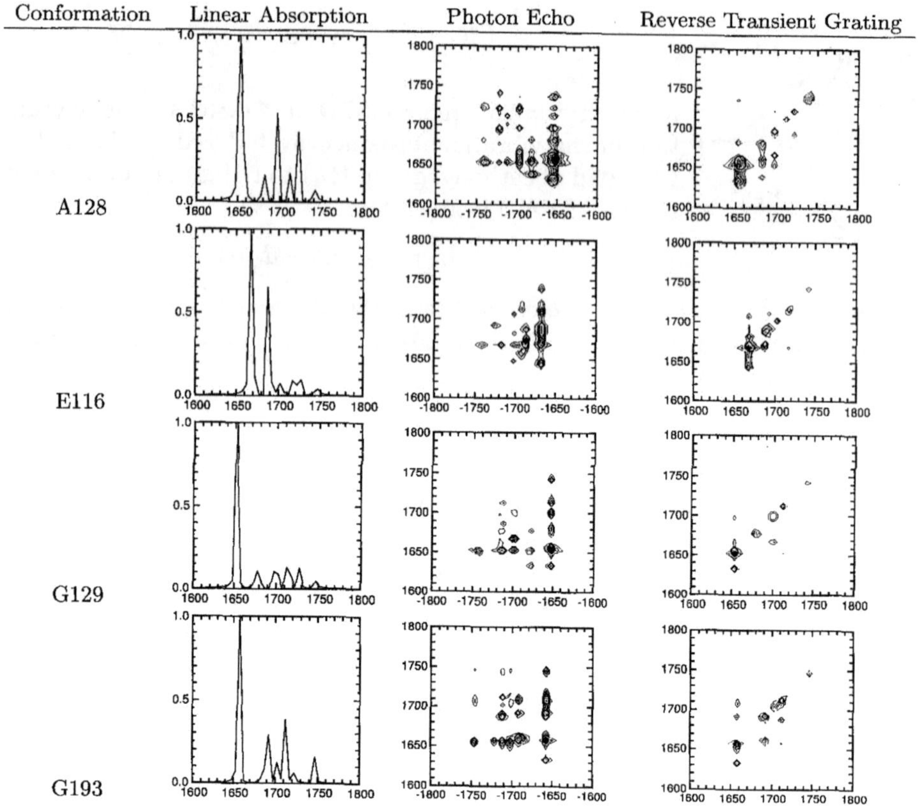

Figure 9. Signal intensities are given in arbitrary units. Frequencies are in cm^{-1}. Transition frequencies are between 1682 – 1736 cm^{-1} for A128, 1700 – 1735 cm^{-1} for E116, 1686 – 1735 cm^{-1} for G129, and 1685 – 1738 cm^{-1} for G193. Fluctuations in the transition frequencies of hydrogen bonded CO groups are between 5 – 24 cm^{-1}

sions for 2D signals $S(\Omega_2, \Omega_1)$ for different spectroscopies.[37] Double Fourier transformation of the time domain signal yields the signal in the frequency domain

$$S(\Omega_2, \Omega_1) = \int_0^\infty dt_2 \int_0^\infty dt_1 \exp\left(i\left(\Omega_1 t_1 + \Omega_2 t_2\right)\right) S(t_2, t_1)$$

Figure 9 shows the linear absorption, photon echo and reverse transient grating spectra of a single snapshot from A128, E116, G129, and G193. All four linear absorption spectra show a peak with maximum intensity at 1650 cm^{-1} and other smaller peaks that is due to the amide I band absorption. Unlike the linear absorption, the photon echo and reverse transient grating spectra not only show diagonal peaks at $1650 - 1650\,\mathrm{cm}^{-1}$ but several off-diagonal peaks that provide characteristic signatures for each conformation. The average over an ensemble of spectra needs to be carried out in order to compare with experiment

Acknowledgements

The support of the National Institutes of Health (GM59230–01A2). C.S. thanks the Alexander von Humboldt Foundation for a Feodor Lynen fellowship.

References

1. J. W. Peng, C. A. Schiffer, P. Xu, W. F. van Gunsteren, and R. R. Ernst. Investigations of peptide hydration using NMR and molecular dynamics simulations: A study of effects of water on the conformation and dynamics of antamanide. *J. Biol. NMR*, 8:453–476, 1996.
2. T. Wieland and H. Faulstich. Amatoxins, phallotoxins, phallolysin, and antamanide — biologically–active components of poisonous amanita mushrooms. *Crit. Rev. Biochem.*, pages 185–260, 1978.
3. Rafael Brüschweiler, Martin Blackledge, and Richard R. Ernst. Multi–conformational peptide dynamics derived from NMR data: A new search algorithm and its application to antamanide. *J. Biol. NMR*, 1:3–11, 1991.
4. Martin Blackledge, Rafael Brüschweiler, Christian Griesinger, Jürgen M. Schmidt, Ping Xu, and Richard R. Ernst. Conformational backbone dynamics of the cyclic decapeptide antamanide. application of a new multiconformational search algorithm based on NMR data. *Biochemistry*, 32:10960–10974, 1993.
5. D. J. Patel. Antamanide conformations in non–aqueous media dependence on hydrogen–bond acceptor properties of solvent. *Biochemistry*,

12:667–676, 1973.

6. W. Burgermeister, T. Wieland, and R. Winkler. Antamanide — relaxation study of conformational equilibria. *Eur. J. Biochem.*, 44:311–316, 1974.
7. H. Torii, T. Tatsumi, and M. Tasumi. Effects of hydrogen bonding and solvation in dielectric media on the amide I frequencies: Ab initio molecular orbital study. *Mikrochim. Acta Suppl.*, 14:531–533, 1997.
8. P. Hamm, M. H. Lim, and R. M. Hochstrasser. Structure of the amide I band of peptides measured by femtosecond nonlinear–infrared spectroscopy. *J. Chem. Phys. B*, 102:6123–6138, 1998.
9. C. Scheurer, A. Piryatinski, and S. Mukamel. Signatures of β–peptide unfolding in two–dimensional vibrational echo spectroscopy: A simulation study. *J. Am. Chem. Soc.*, 123(13):3114–3124, 2001.
10. Samuel Krimm and Jagdeesh Bandekar. Vibrational spectroscopy and conformation of peptides, polypeptides, and proteins. *J. Adv. Protein Chem.*, 38:181–364, 1986.
11. Tobias Bremi, Rafael Brüschweiler, and Richard R. Ernst. A protocol for the interpretation of side–chain dynamics based on NMR relaxation: Application to phenylalanines in antamanide. *J. Am. Chem. Soc.*, 119:4272–4284, 1997.
12. Tobias Bremi and Rafael Brüschweiler. Locally anisotropic internal polypeptide backbone dynamics by NMR relaxation. *J. Am. Chem. Soc.*, 119:6672–6673, 1997.
13. Z. L. Madi, C. Griesinger, and R. R. Ernst. Conformational dynamics of proline residues in antamanide. j coupling analysis of strongly coupled spin systems based on E.COSY spectra. *J. Am. Chem. Soc.*, 112:2908–2914, 1990.
14. Jürgen M. Schmidt, Rafael Brüschweiler, Richard R. Ernst, Roland L. Dunbrack, Diane Joseph, and Martin Karplus. Molecular dynamics simulation of the proline conformational equilibrium and dynamics in antamanide using the charmm force field. *J. Am. Chem. Soc.*, 115:8747–8756, 1993.
15. T. Elsaesser, S. Mukamel, M. M. Murnane, and N. F. Scherer, editors. *Ultrafast Phenomena XII*, Berlin, 2001. Springer.
16. S. Mukamel and R. Hochstrasser. *Chemical Physics (Special Issue)*, 266(2–3):135, 2001.
17. M. C. Asplund, M. T. Zanni, and R. M. Hochstrasser. Two-dimensional infrared spectroscopy of peptides by phasecontrolled femtosecond vibrational photon echoes. *Proc. Nat. Acad. Sci.*, 97:8219–8224, 2000.
18. R. Zadoyan and V. A. Apkarian. Imaging the molecular rovibrational co-

herence through time-gated, frequency-resolved coherent anti-stokes raman scattering. *Chem. Phys. Lett.*, 326:1–10, 2000.
19. K. A. Merchant, David E. Thompson, and M. D. Fayer. Two–dimensional time–frequency ultrafast infrared vibrational echo spectroscopy. *submitted*, 2000.
20. P. Hamm, M. Lim, W. DeGrado, and R. Hochstrasser. Pump/probe self heterodyned 2D spectroscopy of vibrational transitions of a small globular peptide. *J. Chem. Phys.*, 112:1907–1916, 2000.
21. K. Okumura, A. Tokmakoff, and Y. Tanimura. Structural information from two-dimensional fifth–order raman spectroscopy. *J. Chem. Phys.*, 111(2):492–503, 1999.
22. K. D. Rector, A. S. Kwok, C. Ferrante, A. Tokmakoff, C. W. Rella, and M. D. Fayer. Vibrational anharmonicity and multilevel vibrational dephasing from vibrational echo beats. *J. Chem. Phys.*, 106:10027–10036, 1997.
23. T. H. Joo, Y. W. Jia, J. Y. Yu, D. M. Jonas, and G. R. Fleming. Dynamics in isolated bacterial light harvesting antenna (LH2) of rhodobacter sphaeroides at room temperature. *J. Phys. Chem.*, 100:2399–2409, 1996.
24. E. J. A. Brown, I. Pastirk, B. I. Grimberg, V. V. Lozovoy, and M. Dantus. Population and coherence control by three-pulse four-wave mixing. *J. Chem. Phys.*, 111:3779–82, 1999.
25. John D. Hybl, Yannick Christophe, and David M. Jonas. Peakshapes in femtosecond 2D correlation spectroscopy. *Chem. Phys.*, 266:295–309, 2001.
26. O. Golonzka, M. Khalil, N. Demirdöven, and A. Tokmakoff. Vibrational anharmonicities revealed by coherent two-dimensional infrared spectroscopy. *Phys. Rev. Lett.*, 86(10):2154–2157, 2001.
27. M. Khalil and A. Tokmakoff. Signatures of vibrational interactions in coherent two–dimensional infrared spectroscopy. *Chem. Phys.*, 266:213–230, 2001.
28. K. A. Merchant, David E. Thompson, and M. D. Fayer. Two-dimensional time-frequency ultrafast infrared vibrational echo spectroscopy. *Phys. Rev. Lett.*, 86(17):3899–3902, 2001.
29. S. Mukamel. Multidimensional femtosecond correlation spectroscopies of electronic and vibrational excitations. *Annu. Rev. Phys. Chem.*, 51:691–729, 2000.
30. V. Chernyak, W. M. Zhang, and S. Mukamel. Multidimensional femtosecond spectroscopies of molecular aggregates and semiconductor nanostructures; the nonlinear exciton equations. *J. Chem. Phys.*, 109:9587, 1998.
31. J. P. Ryckaert, G. Ciccotti, and H. J. C. Berendsen. Numerical integra-

tion of the cartesian equations of motion of a system with constraints: molecular dynamics of n-alkanes. *J. Comp. Phys.*, 23:327–341, 1977.
32. W. Kabsch and C. Sander. Dictionary of protein secondary structure: pattern recognition of hydrogen-bonded and geometrical features. *Biopolymers*, 22:2577–2637, 1983.
33. Wei Min Zhang, T. Meier, V. Chernyak, and S. Mukamel. Exciton-migration and three-pulse femtosecond optical spectroscopies of photosynthetic antenna complexes. *J. Chem. Phys.*, 108:7763–74, 1998.
34. S. Krimm and Yasuaki Abe. Intermolecular interaction effects in the amide I vibrations of β polypeptides. *Proc. Nat. Acad. Sci.*, 69:2788–2792, 1972.
35. S. Krimm and W. C. Reisdorf. Understanding normal-modes of proteins. *Faraday Discuss.*, 99:181–197, 1994.
36. S. Mukamel. *Principles of Nonlinear Optical Spectroscopy.* Oxford University Press, New York, 1995.
37. W. M. Zhang, V. Chernyak, and S. Mukamel. Multidimensional femtosecond correlation spectroscopies of electronic and vibrational excitons. *J. Chem. Phys.*, 110:5011, 1999.

ULTRA FAST LASERS TO IDENTIFY MOLECULAR BIOLOGICAL STRUCTURES

LOTFIA EL-NADI

Faculty of Science, Laser Phys. Lab.& NILES,Cairo University, EGYPT

E-mail: lotfianadi@hotmail.com

LATIFA EL-HOUTY

Faculty of Science, Physics Dept., Qatar University, Doha, QATAR

YOSR E.GAMAL

National Institute of Laser Enhanced Sciences, Cairo university, EGYPT

We report here an experimental technique in which we developed a commercial 100 fs Ti-sapphire Laser in conjunction with a photomicrographic microscope to study multiphoton absorption microscopy. The microscope is provided by a piezo-stage scanner sample holder and a high precision sensitive detection system. With the optimized system we determined the cross section of two-photon absorption for new dyes; pyridine L3, Comarine L12 and L13, that could be excited by two-photon absorption from fs laser of wavelength 760+- 10 nm nearly without tuning. To our knowledge this is the simplest multiphoton absorption microscopic technique that combine the advantages of multiphoton excitation as well as localizing the respective position of the fluorescing molecules in the sample. The intensity of fluorescence is recorded for each piezo-stage step with high resolution better than 12 nm. Some biological structures are imaged spatially for span dimensions of 280 X 200 um. where only protein molecules could be recognized.

1 Introduction

Imaging biological molecules could well be realized by tagging the biological samples with suitable dye molecules that can absorb UV light photons and fluoresce in the visible region of the spectrum. Applying sensitive microscopic techniques one can then detect the fluorescent light and subsequently image the biological molecules. The facts that UV photons have low penetration and highly damage the living cells, limited the studies of the biological molecules behavior in real time and at depth. Two photon absorption (TPA) techniques of near infra red photon (NIR) to excite the dye molecular levels could well solve the problem, since one can achieve the UV excitation as well as decrease the damage problem and increase the penetration depth of the NIR photons. Focused ultra fast NIR lasers could over come this last problem by providing the high power photon flux required to overcome the cross-section limitation of TPA processes. Physicists are highly occupied to utilize femtosecond (NIR) lasers combined with optical techniques and photo detectors in order to isolate and image single molecules in both space and time. Watt Webb et al [1] were the first to report the advantages of applying TPA to excite fluophores over single photon excitation. In conjunction with a scanning laser microscope, Moeiner and Orrit [2] used TPA to study single molecules in condensed matter. Since the dispersion of the microscope and the laser beam broadening limit the resolution of such studies, Wolleschauky et al [3] were urged to characterize and optimize fs laser-scanning microscope.

We introduce for the first time here a simple multiphoton absorption system that combines the advantages of (TPA) process, microscopy and special sample holder stage that could be precisely driven piezoelectrically in the X,Y,Z directions. The physical properties of the system using new fluorescent compounds and some standard compounds are reported in this study. We determined the TPA cross-section of these compounds in order to use them as tagging chromophores to biological samples. We report images of protein molecules in XYZ plane of biological samples from botanic origins, tagged with the new fluorescent dye molecules.

2 Experiment

The experimental set up is shown in (Fig.1) where imaging microscopy of the emitted fluorescence is attained without confocal pinholes

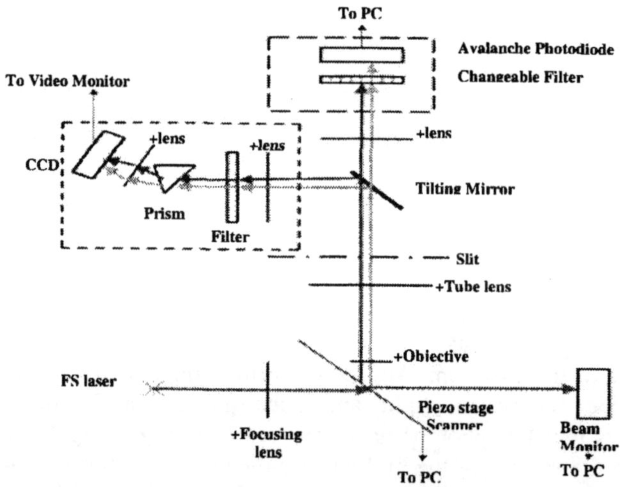

Figure 1 High resolution setup for multi-photon Excitation with nanometer resolution piezo stage scanner.

The main advantage of TPA process over one photon excitation lies in the fact that TPA is restricted to the focal volume of the laser pulse at the sample, since out of focus fluorescence is eliminated and since the fluorescent molecules do not absorb NIR laser photons and have low scattering cross-section to such photons. The setup consists of :

1. The Ti-sapphire fs laser (spectra physics) being the main most important component of the set up is tunable in wavelength range of 750 – 950 nm., 100 MHz rep. rate, 1 nJ energy per pulse and spectral width of ~ 6nm. The initial pulse width was monitored by a commercial autocorrelator. The direct focusing of the laser beam to a spot of diameter 10+- 5 um on the sample is accomplished by only one lens of high N.A. value 1.65 type UPLAPO60*W/IR lens. The geometrical path of the laser beam allowed power and temporal monitoring of the emerging beam from the sample. The fs. Laser wavelength was adjusted at 760 +- 10 nm during all measurements.
2. The originality of the set up lies in utilizing a Thorlab piezo controled XYZ fluxre improved system complete with driver and computer control on which we fixed a 45° mounted sample stage.It has external and internal voltage control for the three axes through three piezo drive actuators controlled independently from 0 –150 volts in steps of Vmin = 0.1 V. The piezo crystals which are modeled as capacitors provide the capacitance C of this piezo system = 1.4 uF, the displacement factor df. is 0.116 um/V and the output current I has a value up to 60mA. The motion span of the axes lies between 100 nm.- 2mm.
3. A commercial olympus photomirographic microscope was developed according to the suggested setup to detect the fluorescence emission of the samples in synchronization with the position of the piezo controlled XYZ flexure system, it has both photo detector and CCD intensified camera. The microscope is in conjunction with an image processing computer system. The use of some software packages allowed simple adjustment of contrast, color balance, color transformation and sharpness to improve the image and define the fluorescence intensity and the positions of the fluorescent molecules.

The fluorescent dyes were synthesized by colleagues at the chemistry department from new pyridene and comarine dye compounds named as L3, L12 and L13. Their synthesis and chemical structure will be published elsewhere shortly. They have max. absorption at 368+-20, 364+-20 and 350+-25 nm respectively, that could be excited by TPA from fs laser at wavelength 760+-10 nm nearly without tuning. They have max. fluorescence at wavelength 415-450+-10, 525+-15 and 470+-20 nm respectively, so they could be distinguished if used simultaneously as probes in a sample.

Using a sensitive CCD video camera combined to digital still image capture adaptor, the TPA fluorsence intensity as a function of the dye concentration were recorded. The spatial distribution of the blinking molecules were then estimated by the computer processing of the above data at concentrations of 10^{-2}, 10^{-4} and 10^{-6} mole of the dye solutions in DMF.

The TPA fluorescence intensity and the signal to noise ratio were measured and estimated for the dye samples at concentrations of 10^{-2} mole and as a function of the excitation incident power of the laser beam. The measured signals are average values of 8 – 10 detection events each integrated over 1 ms with accuracy of 5 – 6%. The noise values were measured when the laser beam was operated at the cw regime.

3 Results and Discussion

Applying the above experimental techniques we were able to determine three main main features: the physical properties of the experimental setup, the properties of the new fluorescent dyes and the preliminary results of imaging protein rich samples.

3.1 The physical properties of the setup

The most important property parameter is to deal with the stability of the insident fs laser beam both from the point of view of the power and the pulse duration. Since the geometrical path of the laser beam allowed power monitoring of the emerging beam from the sample, normalization of the data could be easily carried out. While pulse duration should be optimized. We were carful to keep stable temporal width during all measurements.

3.1.1. Stability of the fs pulse duration

The spectrum of the fs pulse was measured before and after the focusing lens. No change of the spectrum at the sample position was observed, but measuring the pulse duration a value of 82+- 5 fs. was recorded after the lens, compared to 80 +-5 fs. before the lens. This indicated very small non linear optical effect due to the focusing lens material which could easily be neglected. Also the direct beam irradiation of the sample limited the temporal broadening since laser beam scanning was avoided and the beam position was fixed.

3.1.2. The dynamic performance of the piezo-stage scanner.

One can calculate the minimum scanning step distance dmin and the fastest time tmin of scanning a step or a distance span by substituting the above values of Vmin, df, C and I the properties of the piezo-crystals driving the XYZ flexure in the following equations .

$$d_{min} = V_{min} * df = 11.6 \text{ nm} \qquad t_{min} = C * V_{min} / I = 2.3 \text{ ns.}$$

These are the limits of using the system when controlled by a PC running an ASC 11 terminal emulator program which remotely access the piezo flexure features. The resolution of the system ~ 12 nm. Since the focal volume diameter is of the order of 15um it is not necessary to scan the distance to better them 10 – 20 um. steps *i.e.* Vmin = 8.1 – 16.2 V and could be scanned at the fastest rate of t = 0.2 – 0.4 us and consequently a span of 1 mm would be accomplished app. within 2 – 4 ms.

3.2. Properties of the new fluorescent dyes

order to study the TPA processes in the new fluorescent dyes one should adopt an timum dye concentration and study the florescence intensity as a function of :ident fs laser intensity.

3.2.1. Optimum dye concentration

Experimental spatial fluorescence images integrated over 1 ms. For each frame arising due to TPA in Comarine L13 at different concentrations in DMF are shown in (Fig.2).This was also measured for Pyridine L3 and Comarine L12. We accordingly adopted the concentration 10^{-2} mole for all next measurements due to the attained better statistics.

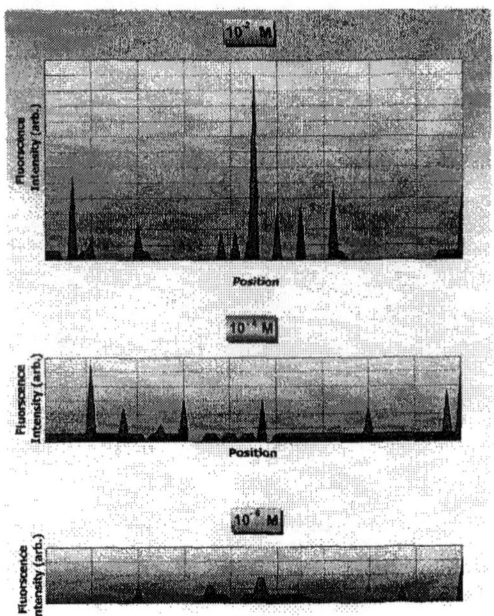

Figure.2 Integrated TPA fluorescence intensity over 1ms as a function of position along interaction vol. of 82 fs, 0.15W, 760 nm laser beam with Comarine L13 at different conc. DMF.

3.2.2. The cross-section of TPA for the new fluorescent dyes

(Fig3a) represent the experimental fluorescence signals relative to that of Comarine L12 in DMF at incident laser power 240 mW. The figure shows clearly that the TPA process is only possible above certain threshold power estimated as 70, 85 and 95 mW for the L12, L13 and L3 dye molecules respectively. The behavior of the curves indicate that the TPA process for all dyes is characterized by 3 distinct regions, namely linear, transition and saturation regions.

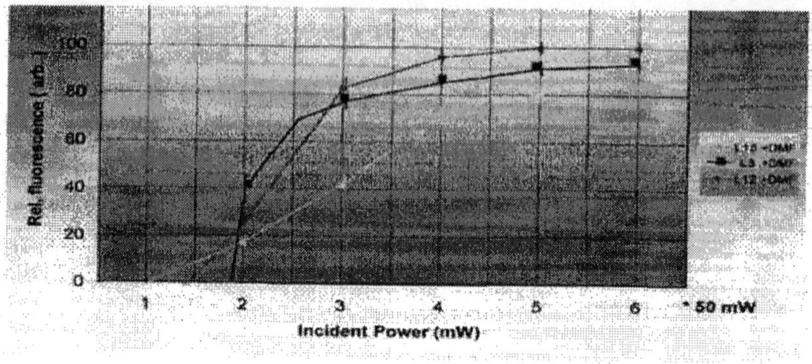

a.

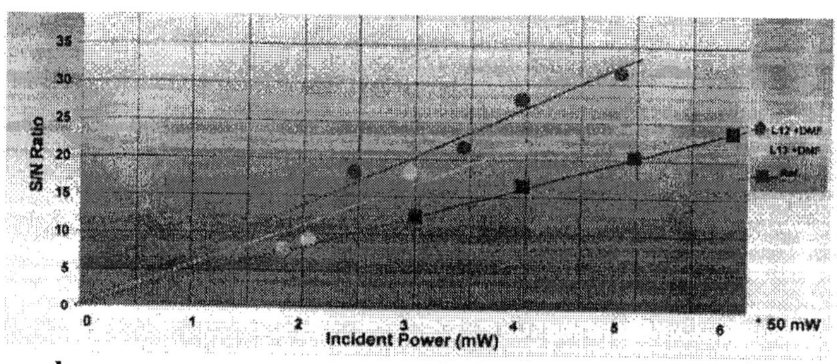

b.

Figure 3 Fluorescence signals (S) and florescence signal to noise ratio (S/N) as afunction of incident fs laser power **a.** S relative to that of L12 at 240 mW for L3, L12 and L13 as a function of incident fs laser power, shows thresheld process and three distinct regions: linear transition and saturation regions respectively. **b)**. S/N for L12,L13 and Rodamine B as afunction of incident fs laser power shows linear relation for the three dyes.

The linear region of L13 and L12 lies between 100 and 200 mW laser power while for L3 it is relatively absent. The transition region for all dyes has a span of 50 mW and saturation region starts at 175, 200 and 225 mW incident laser power for L3, L12 and L13 respectively. The linear region indicate domination of TPA absorption, the transition region indicate competition of absorption and emission processes while the saturation region indicate domination of emission process. The differences may be due to different absorption cross-sections for the TPA processes in the three dyes at wavelength corresponding to 760+- 10nm wavelength at which the fs laser was fixed without tuning.

The measured fluorescent signal to noise ratio as a function of laser beam incident power is shown in figure 3b for L12, L13 and Rhodamine B (as a reference dye) are used to estimate the TPA cross-section. The transition probability for TPA is known to be proportional to the square of the incident photon flux according to the equation

$$S = C_o [q\,\sigma\, P2\,(t)] / 2$$

Where S is the fluorescence signal C_o is the concentration of the dye
q is the quantum efficiency σ is the cross-section of TPA
P(t) is the photon flux = I(t) / h ν

And assuming poisson distribution for the emitted photons then the noise due to TPF

$$N = \sqrt{S} = \sqrt{C_o [q\,\sigma P2\,(t)] / 2}$$

therefore

$$S / N = \sqrt{C_o\, q\, \sigma/2}\;\; P(t)$$

I.e. Liniear relation between S / N and P(t)

(Fig.3b) proves the linear relation of S / N against incident laser beam power and all the assumptions given above. The slope of the lines will be equal to $(C_o\, q\, \sigma/2)^{1/2}$

Using the standard Rhodamine B having a known σ = 27X10-40 cm2/photon [4],the cross-section of TPA for L12 and L13 are estimated to be 43 X 10-40 cm2/photon and 34.7 X 10-40 cm2/photon respectively.

3.3 Imaging of living protein rich samples

The new fluorescent dyes were used to tag some biological protein rich botanic samples namely castor oil seeds and living bean. The samples were ~ 50 micron thick sections on microscope slides prepared without tagging and with L13 or L12 or L3 tagging. (Fig.4) show images where only protein molecules in 3D could be clearly recognized when tagged with comarine L13 and using blue filtering to detect only the fluorescence of the tagging L13 molecules. Further studies for spatial and temporal behavior of protein molecules are in progress.

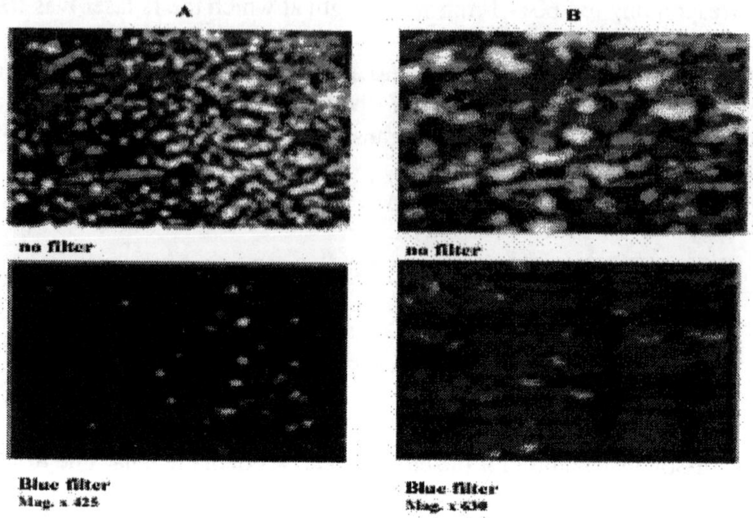

Figure 4 3D two photon fluorescence images of protein rich samples tagged with comarine L13 dye with and without blue filter (A). samples of castor oil seed (B). sample of living bean. It is clear to assign the positions and dimensions of the protein molecules from the lower slides using the magnification factors shown above.

4 Conclusion

The fs laser TPA microscope reported here for the first time to our knowledge proves to be a promising sensitive imaging technique for biological molecules. Fixed focal position of the fs laser beam and sample scanning allowed stable beam properties and accurate high resolution optical geometry. The TPA cross-sections for new chromophares (whose chemical structures are to be published elsewhere) were determined using single exciting laser beam wavelength. Further studies are in progress for temporal and in depth behaviour of biological protein molecules.

5 Acknowledgement

We would like to thank our colleagues at NILES particularly Prof. Yehia Badr the dean for providing instrumental facilities. One of us Lotfia El-Nadi would like to thank Prof. Dr. Ahmed Zewail for encouraging the attendance of the conference,Prof. Abd EL-RAZZAK Douhal chairman of the conference for partial support and Cairo university for providing travel expenses.

References

1. C.Xu, W.W.Webb, Two photon absorption in Rohdamin B. *J.Opt.Soc.Am.***B13** (1996) 481.
2. W.E. Moeiner and Micheal Orrit, Illuminating single molecules in condensed matter. *Science* **283**, (1999) 1670.
3. R.W.Heschensky, J.Feurer, R.Sauerbrey, U.Simon, Characterization and optimization of laser sanning microscope in femto-second regime *Appl.phys.***B 67** (1998) 87.
   Wolfgang Becker, Photon counting : higher dimensaions in microscopy, *Biophotonics*, Dec.(2000) 54.
4. J.D.Bhawalker, A.Shih, S.I.Pan, W.S.Liou, J.Swiatkiewicz, B.A.Reinhardt, P.N.Prasad, P.C.Cheng, Two photon excitation for standard dye solutions at 800 nm. excitation wavelength, *Bioimaging* **4** (1996) 3.

PRIMARY PROCESSES IN THE PHOTOISOMERIZATION OF BACTERIORHODOPSIN STUDIED BY FEMTOSECOND INFRARED SPECTROSCOPY

JOHANNES HERBST, KARSTEN HEYNE* AND ROLF DILLER

Institut für Experimentalphysik, Freie Universität Berlin, Arnimallee 14, D-14195 Germany
** present address: Max-Born-Institut, Max-Born-Straße 2A, 12489 Berlin, Germany*
E-mail: diller@physik.fu-berlin

An overview is given about the progress during the past years in time-resolving the primary, photoinduced all-trans to 13-cis isomerization-process in Bacteriorhodopsin. New results are presented on the basis of femtosecond time-resolved infrared vibrational spectroscopy that allows to monitor directly the structural dynamics. For the first time it is shown that vibrational marker-bands in the fingerprint region around 1200 cm^{-1} rise with a time constant of 500 fs, indicating the formation of a 13-cis product on the same time-scale as the formation of the electronic ground state.

1 Introduction

Ultrafast nuclear rearrangement accompanies numerous photoinduced reactions in condensed phase and lies at the heart of femtochemistry and femtobiology. For the dynamics and outcome of the reaction the environment plays an essential role.

In solution based reactions the solvent molecules fluctuate freely around the reacting solute, whereas the "solvent" in protein based reactions shows a different behaviour. Here, the global three-dimensional structure allows spatial fluctuations only around the static equilibrium geometry. Thus, the protein environment provides for unique static as well as dynamic constraints for the chromophore (cofactor) and leads to an altered, new photochemistry not observed in solution. As a consequence, the protein environment can catalyse specific reaction channels that are essential for the respective biological function.

A model system in this respect is the light-driven proton pump Bacteriorhodopsin (bR). It represents a class of transmembrane proteins with different biological functions but of similar three-dimensional structure, employing the same chromophore, although in different configurations. The initial photoreaction in each of these systems is optimized to the respective function by the specifically taylored protein environment in the chromophore binding pocket. In bR the initial, photoinduced process is an all-trans to 13-cis isomerization of the retinal chromophore, which is covalently bound to a Lysine residue of the protein by a protonated Schiff Base (SB). The governing control mechanisms that shape the electronic potential energy surfaces of the chromophore are by far not yet understood, but the hope is, that with accumulating experimental data, e. g. structural information from X-ray and NMR techniques as well as dynamic information from laser-spectroscopic methods, this goal can be accomplished.

2 Primary events in the Bacteriorhodopsin photocycle

Experimental information concerning the ultrafast primary events are obtained primarily from femtosecond (fs) transient absorption spectroscopy, probing the electronic state dynamics. In the late eighties a model was established [1,2] for the photoinduced isomerization process. According to this model the Franck-Condon region is left within 200 fs along a barrierless S_1-surface and the first product electronic ground state is reached with a time constant of 500 fs (J-state). Relaxation processes lead within 3-4 ps to the K-state which can be trapped at low temperature. At ambient temperature K decays on the μs time scale. More recently, studies at time resolution of up to 30 fs or better [3-8] provided a more detailed understanding of the processes preceeding the S_1 - S_0 - transition. It is believed that the Franck-Condon region is left along high frequency stretching modes of the retinal chromophore. Then a region on the S_1 surface is reached where motion along torsional modes is favoured and preludes trans - cis isomerization. Model calculations [9,10] have given support for this scheme.
Although optical fs-spectroscopy probes electronic state dynamics very well it does not directly reveal information about (incoherent) structural dynamics. Therefore, the exact timing of the isomerization process remained unclear. However, it was believed to occur along the S_1 - S_0 - transition on the ground of observed spectral shifts in the electronic absorption spectra and the knowledge that 13-cis isomerization does occur at some early time after photoexcitation and is established in the K-state, i. e. after 3-4 ps. The latter was concluded by Resonance-Raman (RR) and Infrared (IR) spectroscopic methods, employing isotopically substituted retinals in solution and bound to the protein as well as model calculations [11-14]. It turned out that the 13-cis state (K) is characterized by a vibrational band with dominant C-C - single-bond stretching amplitude in the fingerprint region at 1190 cm^{-1} with a shoulder at 1180 cm^{-1} whereas the all-trans ground state (BR_{570}, the subscript denotes the absorption maximum in nm) shows bands at 1171, 1202 and 1217 cm^{-1}. To clarify the question of the exact timing of the isomerization transient vibrational spectra at higher time resolution are required.

Using RR-techniques, here a problem arises. Application of shorter laser-pulses leads to fewer scattered photons (smaller signal-to-noise-ratio) and an increase of the detected vibrational linewidth, thereby blurring the spectra and complicating the interpretation. Nevertheless, ps laser-pulses were used to obtain RR-spectra of the early K-state [15,16]. Subtraction methods were applied to obtain RR-spectra of J from spectra composed of a mixture of BR_{570}, J and K [15,16]. With pulses of 1 ps and less [17,18] it was possible to obtain spectra of precursor states of K. In [18], spectral features in the fingerprint region, indicating a 13-cis configuration of the chromophore, were reported. However, recent ps time-resolved CARS experiments [19] claimed J accomodating a chromophore very similar to the all-trans chromophore of BR_{570} and 13-cis isomerization taking place possibly as late as K-formation.

3 Ultrafast infrared spectroscopy on Bacteriorhodopsin

In contrary to RR methods which address predominantly vibrational bands of the resonantly excited chromophore, infrared (IR) difference methods are susceptible to all vibrational transitions - of chromophore and protein - that experience a change in absorption cross section or frequency upon photoexcitation of the reacting chromophor. Fourier-Tranform-Infrared (FTIR) methods enable an extremely high signal/noise but the time resolution is now limited to the nanosecond time scale [20], not high enough to study the isomerization process itself.

A breakthrough came with the first highly time resolved IR experiments on bR in 1991 [21,22], opening up an entirely new approach to investigations of the early states of the bR photocycle at room temperature, including dynamic contribution of the protein. The time resolution of 10 ps was achieved by optically gating (upconversion) a continuous-wave (cw) IR-source, a Carbon-Monoxide (CO) laser. For the first time a ps time-resolved IR difference spectrum between BR_{570} and K was obtained in the region around 1640 cm^{-1}, where the C=NH - stretch vibration of the chromophore absorbs. This band is sensitive to the (hydrogen) bonding environment of the Schiff base linkage between chromophore and protein. It was shown that the difference spectrum was not identical but showed great similarities to the static, low temperatur FTIR bR-K-spectrum [23]. Disadvantages of this experiment was the limited time resolution and limited spectral range of the CO-laser used. Thus, important spectral regions could not be accessed.

The time resolution in upconversion schemes is predominatly dependent on the pulse duration of the gating laser pulse. The strength of the detected pulses depends on the number of upconverted photons, here directly related to the cw-power of the CO-laser. If the gate pulse is too short this number becomes very small and leads to shot-noise limited signals. An improvement was made by using a liquid alcohol cooled CO-laser [24] with higher cw-power. This laser was combined with a femtosecond laser light source, thereby making VIS-pump/IR-probe experiments feasible at a time resolution of 400 fs and in the spectral region down to 1580 cm^{-1}. It was shown that the spectral pattern in the C=NH stretch region, already found in [21,22] at 10 ps, rises within less than 1 ps. This indicated that a photoproduct is formed long before K-formation with a chromophore in a Schiff base environment similar to that in K. Interestingly, a band appearing at 1660 cm^{-1}, which could not be assigned to a chromophore vibrational mode, pointed to a fast protein response to the photoreaction.

The same experimental method was used again [25,26], however a liquid-nitrogen cooled CO-laser with broadened emission spectrum was employed. Thereby the ethylenic stretch (C=C - double bond) vibrational modes around 1530 cm^{-1} became accessible. The frequency of these modes depends on the extent of double-bond character between the ethylenic carbon-atoms, i .e π-electron delocalization. The rise of product bands around 1510 cm^{-1} could be shown to occur within less than 1 ps, indicating a red-shifted optical absorption maximum for this intermediate state (J). However, the fingerprint region, i. e. the 13-cis marker bands, could still not be accessed.

As it is described in the remainder of this article this was achieved now with an apparatus employing nonlinear optical methods to generate short and spectrally broad laser pulses in the mid-IR (will be described somewhere else). In short, the IR pulses are generated at a repetition rate of 700 Hz by difference frequency generation, mixing signal- and idler- pulses derived from a Titanium:Sapphire amplifier-system running at 780 nm. The IR pulses have a duration of typically 170 fs (FWHM) and are tunable between 4 and > 12 μm. Optical excitation pulses are generated via a nonlinear optical parametric amplifier (NOPA) with typically 140 fs (FWHM) pulses at the absorption maximum of bR, λ_{max} = 570 nm, leading to the experimental system response (cross correlation) of typically 220 fs.

Since water is essential for protein function but exhibits strong and broad absorption bands in the region of interest (mid-IR), the sample is a hydrated film of bR on a CaF_2-substrate. Still, the typical photoinduced absorbance difference signals are in the range of 10^{-3} OD-units and have to be detected against a strong absorption background due to protein vibrational bands, requiring a high signal-to-noise-ratio.

Fig. 1 shows IR-transients (absorbance difference between non-excited and excited sample) in the fingerprint region around 1200 cm^{-1} and in the ethylenic stretch region around 1530 cm^{-1}. In general, negative signals (bleach) indicate population of the bR ground state with the chromophore in all-trans configuration (BR_{570}). Positive signals are due to population created as a consequence of electronic excitation, i. e. vibrational bands of the excited electronic state or the electronic ground state of photoproducts. At positive delay times the kinetic traces are fitted by a model of a sum of exponentials.

Note that at negative delay times between excitation and probe event (probe comes first, electronic excitation second), the signal evolves much slower than the time evolution of the experimental system response of 220 fs. This is due to a coherent effect. The IR probe pulse induces a polarization when it spectrally overlaps with a vibrational transition. The polarization emits in the direction of the probe pulse and decays with the dephasing-time T_2 of the vibrational transition (corresponding to the inverse of its spectral width), typically 1 ps or less. If the sequential electronic excitation does not affect the respective vibrational transition, no net effect is observed along the axis of time-delay. However, if the excitation pulse does alter the frequency or absorption cross section of the vibrational transition, oscillations in the time- and frequency-domain are observed. In the degenerate case when the vibrational oscillator is spectrally completely removed from the probe pulse frequency, the expected negative (bleach) signal does not rise as determined by the system response but instead with the time constant T_2. This effect (perturbed free induction decay) is kown also from optical transient absorption spectroscopy [27] and is modelled here according to [27-29].

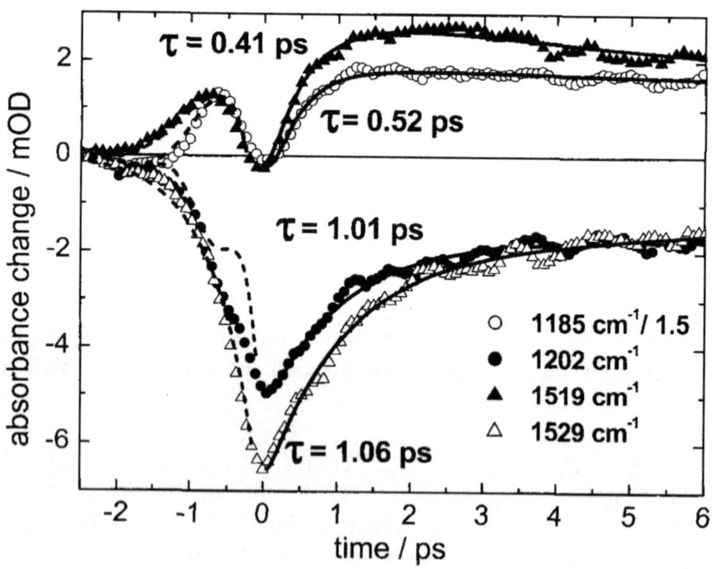

Figure 1. Selected infrared transients (absorbance difference between unexcited and excited bR-sample, taken at room temperature; excitation at 570 nm) in the fingerprint region around 1200 cm^{-1} (C-C-stretch) and the ethylenic (C=C-stretch) region around 1520 cm^{-1}. Given are the time constants of bleach recovery and the rise of the positive signals as fitted by a sum of exponentials (——). At negative delay times, the signals are modelled (------) according to [27-29].

The positive signals in Fig. 1 rise with a time constant of ca. 0.5 ps. They indicate the formation of photoproduct bands in the respective spectral region. The negative bands display the photoinduced bleach that recovers partially with a time constant of about 1 ps. Of special interest is the fingerprint region, i. e. the time course of the C-C-stretch vibrations that indicate the all-trans to 13-cis chromophore isomerization. Transient difference spectra of this region at three delay times (0, 1 and 6 ps) are shown in Fig. 2. Evidently, the positive band with peak at 1190 cm^{-1} is fully developed already at 1 ps. Its rise with 0.5 ps is represented by the trace taken at 1185 cm^{-1} as displayed in Fig. 1. This band appears between the bleach signals at 1200 and 1170 cm^{-1}. Together they form the pattern typical for the formation of a 13-cis product. Note that this pattern is established clearly before K-formation (3-4 ps).

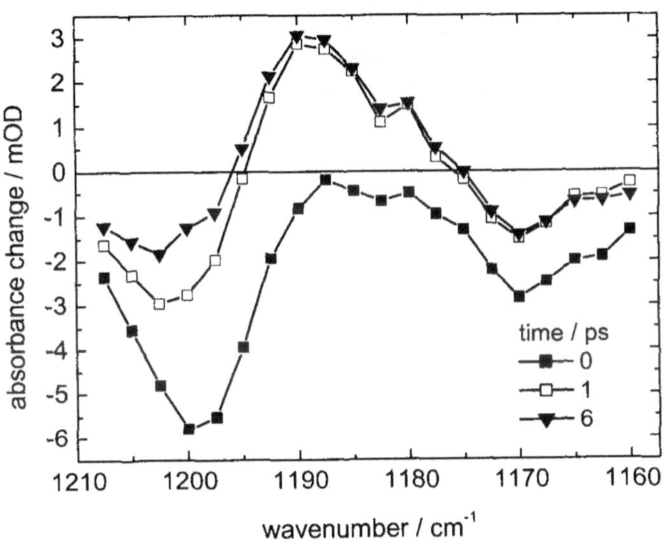

Figure 2. Selected transient IR difference spectra (absorbance difference between unexcited and excited bR-sample, taken at room temperature) in the fingerprint region around 1200 cm^{-1} (C-C-stretch) at 0, 1 and 6 ps after photoexciatation at 570 nm.

The ethylenic stretch region exhibits a similar time behaviour. This is shown by the kinetic curves taken at 1519 cm^{-1} (product band) and at 1529 cm^{-1} (bleach of BR_{570}). Data taken in the C=NH-stretch region (not shown) as well show similar time constants, thereby being consistent with results reported in [24]. The new results presented here clearly show that with a time constant of ca. 500 fs the retinal chromophore in bR

1.is in 13-cis configuration;

2.exhibits an ethylenic stretch vibration at about 1512 cm^{-1} corresponding to a red shifted electronic absorption and
3.experiences a drastically changed Schiff base environment as indicated by the positive bands between 1610 and 1620 cm^{-1}.

Most of the spectral patterns observed at 0.5 ps are much alike those found for the K-state, but small changes in spectral position or band amplitude are discernible on the ps time scale.

The observed kinetics yield more information about the fast dynamics along the isomerization process. Note that the partial recovery of the negative signals with a time constant of about 1 ps is slower than the product formation (0.5 ps) but faster than K-formation (3-4 ps). The recovery is caused by those electronically excited bR-molecules that are fed back to the all-trans BR_{570} corresponding to the bR reaction quantum yield of 64 % [30-32]. A simple branching reaction would lead to only one apparent time constant observed for product formation and educt re-formation. Two kinetic schemes should be mentioned here that take into account the observation of two time constants. First, a branching reaction occurs very early after excitation, opening a reactive (fast) and a non reactive (slower) path to respective S_0 - surfaces. Second, the educt re-formation happens in more than one single kinetic step, possibly involving vibrational relaxation, e. g. vibrational cooling of a vibrationally hot, electronic ground state chromophore [33].

In conclusion, fs time-resolved IR spectroscopy has been used to unravel the ultrafast isomerization dynamics in Bacteriorhodopsin by directly monitoring vibrational states during the formation of the electronic ground state. More molecular processes will be addressed by this method. This concerns e. g. the dynamics of vibrational states on the S_1 - surface and photoinduced protein dynamics.

4 Acknowledgements

We thank D. Stehlik for continuous support and M. Heyn, U. Alexiev and I. Wallat for supply of Purple Membrane and characterization of its ms kinetics. This work was partially supported by Deutsche Forschungsgemeinschaft (Sfb 450).

References

1. Dobler J., Zinth W., Kaiser W. and Oesterhelt D., Excited-state dynamics of bacteriorhodopsin studied by femtosecond spectroscopy. *Chem. Phys. Lett.* **144** (1988) pp. 215-220.
2. Mathies R., Cruz B., Pollard W. and Shank C., Direct observation of the femtosecond excited-state *cis-trans* isomerization in bacteriorhodopsin. *Science* **240** (1988) pp. 777-779.
3. Hasson K. C., Gai F. and Anfinrud P. A., The photoisomerization of retinal in bacteriorhodopsin: Experimental evidence for a three-state model. *Proc. Natl. Acad. Sci. USA* **93** (1996) pp. 15124-15129.
4. Gai F., Hasson K. C., Cooper McDonald J. and Anfinrud P. A., Chemical dynamics in proteins: The photoisomerization of retinal in bacteriorhodopsin. *Science* **279** (1998) pp. 1886-1891.
5. Zinth W., Sieg A. O., Huppmann P., Blankenhorn T., Oesterhelt D. and Nonella M., Femtosecond spectroscopy and model calculations for an understanding of the primary reaction in bacteriorhodopsin. Ultrafast Phenomena XII, Springer Series in *Chem. Phys.* **66**, eds. Elsaesser T., Mukamel S., Murnane M. M., Scherer N. F. (2000) pp. 680-682.
6. Ye T., Gershgoren E., Friedman N., Ottolenghi M., Sheves M. and Ruhman S., Resolving the primary dynamics of bacteriorhodopsin, and of a '$C_{13} = C_{14}$ locked' analog, in the reactive excited state. *Chem. Phys. Lett.* **314** (1999) pp. 429-434.
7. Ye T., Friedman N., Gat Y., Atkinson G. H., Sheves M., Ottolenghi M. and Ruhman S., On the nature of the primary light-induced events in bacteriorhodopsin: Ultrafast spectroscopy of native and C_{13}=C_{14} locked Pigments. *J. Phys. Chem. B* **103** (1999) pp. 5122-5130.
8. Haran G., Wynne K., Xie A., He Q., Chance M. and Hochstrasser R. M., Excited state dynamics of bacteriorhodopsin revealed by transient stimulated emission spectra. *Chem. Phys. Lett.* **261** (1996) pp. 389-395.
9. Garavelli M., Celani P., Bernardi F., Robb M. A. and Olivucci M., The C5H6NH2+ protonated Schiff base - an ab initio minimal model for retinal photoisomerization. *J. Am. Chem. Soc.* **119** (1997) pp. 6891-6901.
10. Garavelli M., Negri F. and Olivucci M., Initial excited-state relaxation of the isolated 11-cis protonated Schiff base of retinal: Evidence for in-plane motion from ab initio quantum chemical simulation of the resonance Raman spectrum. *J. Am. Chem. Soc.* **121** (1999) pp. 1023-1029.
11. Braiman M. and Mathies R. A., Resonance Raman spectra of bacteriorhodopsin's primary photoproduct: Evidence for a distorted 13-*cis* retinal chromophore. *Proc. Natl. Acad. Sci. USA* **79** (1982) pp. 403-407.

12. Curry B., Palings I., Broek A. D., Pardoen J. A., Lugtenburg J. and Mathies R. A., Vibrational analysis of the retinal isomers. *Adv. Infrared Raman Spectrosc.* **12** (1985) pp. 115-178.
13. Smith S. O., Pardoen J. A., Lugtenburg J. and Mathies R. A., Vibrational analysis of the 13-*cis*-retinal chromophore in dark-adapted bacteriorhodopsin. J. *Phys. Chem.* **91** (1987) pp. 804-819.
14. Smith S. O., Braiman M. S., Myers A. B., Pardoen J. A., Courtin J. M. L., Winkel C., Lugtenburg J. and Mathies R. A., Vibrational analysis of the *all-trans*-retinal chromophore in light-adapted bacteriorhodopsin. J. *Am. Chem. Soc.* **109** (1987) pp. 3108-3125.
15. Atkinson G. H., Brack T. L., Blanchard D., and Rumbles G., Picosecond time-resolved resonance Raman spectroscopy of the initial trans to cis isomerization in the bacteriorhodopsin photocycle. *Chem. Phys.* **131**, (1989) pp. 1-15.
16. Doig S. J., Reid P. J. and Mathies R. A., Picosecond time-resolved resonance Raman spectroscopy of bacteriorhodopsin's J, K, and KL intermediates. *J. Phys. Chem.* **95** (1991) pp. 6372-6379.
17. Song L. and El-Sayed M. A., Primary step in bacteriorhodopsin photosynthesis: Bond stretch rather than angle twist of its retinal excited-state structure. *J. Am. Chem. Soc.* **120** (1998) pp. 8889-8890.
18. van den Berg R., Du-Jeon-Jang, Bitting H. C., and El-Sayed M. A., Subpicosecond resonance Raman spectra of the early intermediates in the photocycle of bacteriorhodopsin. *Biophys. J.* **58** (1990) pp. 135-141.
19. Atkinson G. H., Ujj L. and Zhou Y., Vibrational spectrum of the J-625 intermediate in the room temperature bacteriorhodopsin photocycle. *J. Phys. Chem. A* **104** (2000) pp. 4130-4139.
20. Rödig C., Chizhov I., Weidlich O. and Siebert F., Time-resolved step-scan Fourier transform infrared spectroscopy reveals differences between early and late M intermediates of bacteriorhodopsin. *Biophys. J.* **76** (1999) pp. 2687-2701.
21. Diller R., Iannone M., Bogomolni R. and Hochstrasser R. M., Ultrafast infrared spectroscopy of bacteriorhodopsin. *Biophys. J.* **60** (1991) pp. 286-289.
22. Diller R., Iannone M., Cowen B. R., Maiti S., Bogomolni R. A. and Hochstrasser R. M., Picosecond dynamics of bacteriorhodopsin, probed by time- resolved infrared spectroscopy. Biochem. **31** (1992) pp. 5567-5572.
23. Siebert F. and Mäntele W., Investigation of the primary photochemistry of bacteriorhodopsin by low-temperature Fourier-transform infrared spectroscopy. *Eur. J. Biochem.* **130** (1983) pp. 565-573.
24. Diller R., Maiti S., Walker G. C., Cowen B. R., Pippenger R., Bogomolni R. A. and Hochstrasser R. M., Femtosecond time-resolved infrared laser study of the J-K transition of bacteriorhodopsin. *Chem. Phys. Lett.* **241** (1995) pp. 109-115.

25. Dziewior R. and Diller R., Femtosecond infrared spectroscopy on bacteriorhodopsin using a broad band carbon monoxide laser, *Ber. Bunsenges. Phys. Chem.* **100** (1996) pp. 2103-2106.
26. Diller R., Vibrational relaxation during the retinal isomerization in bacteriorhodopsin. *Chem. Phys. Lett.* **295** (1998) pp. 47-55.
27. Joffre M., Hulin D., Migus A., Antonetti A., Benoit à la Guillaume C., Peyghambarian N., Lindberg M. and Koch S. W., Coherent effects in pump-probe spectroscopy of excitons. *Optics Lett.* **13** (1988) pp. 276-278.
28. Wynne K. and Hochstrasser R. M., The theory of ultrafast vibrational spectroscopy. *Chem. Phys.* **193** (1995) pp. 211-236.
29. Hamm P., Coherent effects in femtosecond infrared spectroscopy. *Chem. Phys.* **200** (1995) pp. 415-429.
30. Schneider G., Diller R. and Stockburger M., Photochemical quantum yield of bacteriorhodopsin from resonance Raman scattering as a probe for photolysis. *Chem. Phys.* **131** (1989) pp. 17-29.
31. Logunov S. L. and El-Sayed M. A., Redetermination of the quantum yield of photoisomerization and energy content in the K-intermediate of bacteriorhodopsin photocycle and its mutants by the photoacoustic technique. *J. Phys. Chem B.* **101** (1997) pp. 6629-6633.
32. Tittor J. and Oesterhelt D., The quantum yield of bacteriorhodopsin. *FEBS Lett.* **263** (1990) 269-273.
33. Hamm P., Zurek M., Röschinger T., Patzelt H., Oesterhelt D. and Zinth W., Subpicosecond infrared spectroscopy on the photoisomerization of the protonated Schiff base of all-trans retinal. *Chem. Phys. Lett.* **268** (1997) pp. 180-186.

MUTATION-PHOTOPHYSICS RELATIONSHIP IN INTRINSICALLY FLUORESCENT PROTEINS

AHMED A. HEIKAL[§], SAMUEL T. HESS[§], ERIN D. SHEETS[†‡], AND WATT W. WEBB[§‡]

[§]Applied and Engineering Physics, [†]Department of Chemistry & Chemical Biology, and [‡]Nanobiotechnology Center, Cornell University, Ithaca NY 14853, USA

E-mail: aah14@cornell.edu and www2@cornell.edu

Green fluorescent proteins (GFP), isolated from *Aequorea victoria* jellyfish, have been utilized as noninvasive, site-specific fluorescent markers and intracellular *pH* indicators in biochemistry, cell biology, and molecular genetics. Numerous mutations, aimed at optimizing various spectroscopic and thermodynamic properties of GFP, have been created. A fundamental understanding of the effects of pH, mutations, and excitation wavelength on the photophysics of these biopolymers is essential to advance their applications in biological research. We have utilized different spectroscopic techniques to elucidate the molecular dynamics and spectroscopy of both the neutral and anionic forms of selected GFP mutants. The results are compared with the newly discovered red fluorescent protein (DsRed) from a corallimorpharian of the *Discosoma* genus.

1 Introduction

The chromophore in wild-type (wt) green fluorescent proteins (GFP) is a product of autocatalytic cyclization of the amino acid residues S65, T66, and G67 and a subsequent dehydrogenation under the action of molecular oxygen [1,2]. The β-barrel surrounding the chromophore provides the environment necessary for the chromophore formation and enhancement of its fluorescence quantum yield. The wt-GFP absorption maxima at 396 and 475 nm are attributed to two different ionic states of the chromophore, one with a neutral phenol and the other a phenolate group. The first excited anionic state lifetime is 3.3 ns with maximum emission at 504 nm. Denatured GFP [2] and a model system of the fluorophore [3,4] are non-fluorescent at room temperature due to an efficient *cis/trans* isomerization, which is suppressed in the native folded protein environment. Mutations of wt-GFP aim at enhancing and red-shifting the anionic (deprotonated) state transition (strongly fluorescent) while minimizing the neutral (protonated) state population (weakly fluorescent). The spectral red shift of GFP is critical to avoid cellular autofluorescence in biological applications. We describe our work on the spectroscopy and excited state dynamics of selected mutants as a function of pH and excitation wavelength. Our results suggest possible strategies for creating GFP mutations with desired fluorescence and thermodynamic properties. As described by Tsien [1], the chromophores of GFP mutants studied in this report can be classified according to their chemical structures. A mixture of neutral phenol and anionic phenolate is present (class 1) as for wt-GFP. Class 2 includes mutants that exist primarily as phenolate anions (e.g., S65T and EGFP), whereas T203Y, T203F, and Citrine belong to class 3, with a stacked π-electron system of the phenolate anion with Y66. In such complex molecular systems, only S65 and T203 substitutions play prominent roles in their photophysics. The results are compared with the newly discovered red fluorescent protein DsRed (cloned from a corallimorpharian of the *Discosoma* genus) that exhibits the most red-shifted fluorescence [5,6,7] of any GFPs, thereby avoiding autofluorescence in cellular applications. Surprisingly, the chromophore of DsRed and GFP is fairly similar [7]. We have also investigated intracellular photophysics of the enhanced green fluorescent protein (EGFP) and DsRed as noninvasive fluorescent markers for quantitative *in vivo* studies.

2 Methods

Wild-type GFP, S65T, and EGFP were purchased from Clontech (see Table 1 for specific mutations). Citrine, DsRed, T203Y, and T203F were generous gifts from Professor R. Tsien, G. Baird, and A. Miyawaki (University of California, San Diego, CA). These mutants were maintained in 100 mM potassium phosphate at the indicated pH. The absorption spectra were measured with a single beam, diode array spectrophotometer, while the fluorescence spectra were recorded using a PTI-QM-1 fluorometer. A standard single photon counting technique was used for time- and polarization-resolved fluorescence measurements, as described elsewhere [8].

3 Results

3.1 Steady-state spectroscopy

At pH 9, the absorption spectrum of wt-GFP shows a dominant peak at 396 nm (neutral state transition) with a weak absorption around 475 nm (anionic state transition) (Fig. 1a). Except for wt-GFP, the intrinsically fluorescent proteins (IFPs) described here exist primarily in the anionic form with negligible neutral state absorption. The maximum absorption (λ_a) of these anionic IFPs ranges from 488 nm (e.g., S65T) to 553 nm (DsRed) (Table 1). The peak emission wavelength (λ_{fl}) of wt-GFP at 503 nm (Fig. 1b) is independent of the excitation wavelength (λ_x=396 to 475 nm). The maximum emission of T203Y and Citrine occurs around 514 nm, while DsRed emit maximally at 558 nm. According to tandem mass spectrometry, the extended (by $-C=N-C=O$) DsRed chromophore, at the 2-position of the imidazolidinone, quantitatively accounts for the observed red shift.

Fluorescence correlation spectroscopy (FCS) of the anionic EGFP shows that the chromophore undergoes an external proton exchange with buffer on a ~45 to 340 μs time scale [9], causing pH-dependent fluorescence flicker (Fig.2a). The estimated pK_a for this protonation process is 5.8±0.1, with a standard free energy $\Delta_r G^0$=33±1 kJ/mol. Similar observations were also reported for S65T [9], T203Y and T203F [10], and Citrine (Fig. 2a) [5]. An additional light-driven fluorescence flicker, at a rate that is linearly proportional to the excitation rate below saturation, was also observed in these mutants independent of pH (Fig. 2a) [5,10]. As a result of the external protonation, a new absorption band appears at 396 nm (412 nm for Citrine) at pH 5 (Fig. 2b), whose intensity increases with the concomitant decrease of the anionic state absorption. The new absorption band resembles that of the neutral state transition in wt-GFP (except for Citrine). In contrast, DsRed does not undergo external protonation and the absorption around 396 nm is negligible (Fig. 2b) [5,6]. On the other hand, a weak emission shoulder around 457 nm was observed in the dispersed fluorescence of S65T and EGFP and was assigned to the neutral state transition [5]. This shoulder intensity is sensitive both to mutations (e.g., ~25 times larger for S65T and EGFP than wt-GFP) and pH.

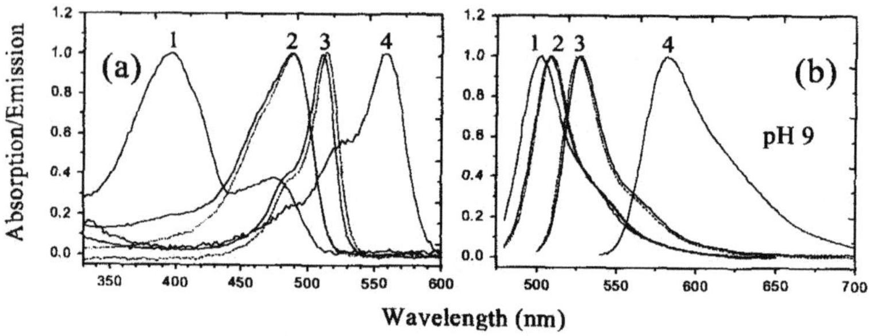

Figure 1. (a) Absorption and (b) anionic state fluorescence spectra of (1) wt-GFP, (2) S65T (dotted)& EGFP (solid), (3) T203Y (solid) and Citrine (dashed), and (4) DsRed at pH 9.

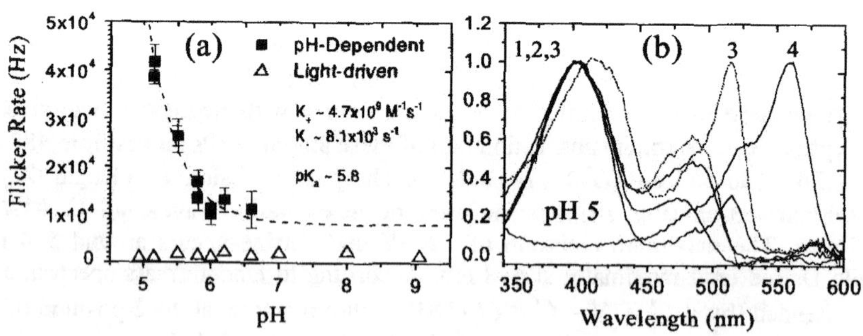

Figure 2. (a) The kinetics of external proton exchange in Citrine probed via pH-dependent (squares) fluorescence flicker using FCS [5]. Light-driven fluorescence flicker was also observed (triangles) that is pH-independent and excitation rate (or intensity) dependent. Concomitantly, the neutral state population is enhanced as shown in (b) the absorption spectra of IFP (normalized for the 396 nm band) at pH 5. Numbers in (b) are as described in figure caption of Fig. 1.

3.2 *First excited anionic state fluorescence decay*

The first excited anionic state fluorescence (510 nm) of wt-GFP (at high pH) is fairly independent of λ_x (405 and 475 nm), with a fluorescence lifetime of 3.3 ns, in agreement with previous solution [11] and crystal [2] studies. The fluorescence of S65T, EGFP and T203Y (Fig. 3a) decays primarily as a single exponential with 2.9, 2.8, and 3.7 ns time constants (Table 1), respectively [8,10]. A minor second decay component (with an amplitude c_2=0.16 and time constant τ_2~1.6 ns) was observed only for EGFP. The excited state relaxation is independent of λ_{fl} over the 510 to 560 nm range with negligible sensitivity to pH (pH 5 to 9). One would have expected that the excited state fluorescence lifetime of DsRed to be affected by the extended electron conjugation of its chromophore, contrary to the experimental observation (Table 1). As a result, the effects of the oligomer configuration (see below) on the observed photophysics of DsRed must not be ruled out.

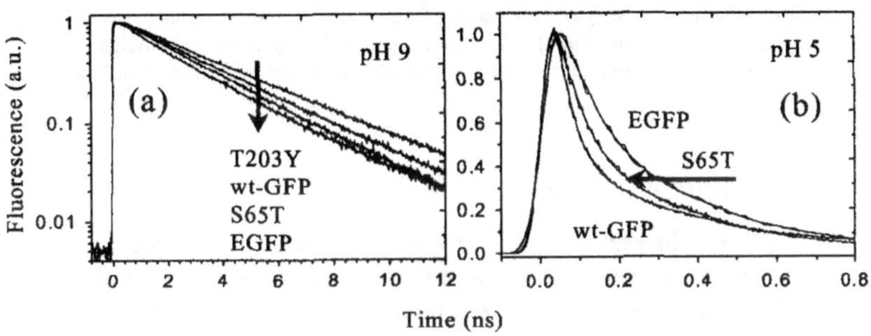

Figure 3. Fluorescence decays of the first excited (a) anionic (λ_x=488 nm) and (b) neutral state (λ_x=405 nm) of selected IFPs. The excited anionic and neutral state fluorescence was detected at 510 and 460 nm, respectively. Note the difference in the time scale for the neutral and anionic state relaxation [8].

3.3 *First excited neutral state fluorescence decay*

The excited neutral state fluorescence of wt-GFP (pH 5), excited at 405-nm and probed via 460-nm emission, decays as a multiexponential with τ_1=24 ps, c_1=0.78, τ_2=184 ps, c_2=0.18, τ_3=832 ps, and c_3=0.04. The average fluorescence decay time constant of EGFP (148±12 ps) and S65T (~126 ps) are slightly larger than that of wt-GFP (~84 ps) at pH 5. Interestingly, the opposite trend was observed for the anionic state decay (Fig. 3a) in these mutants. Furthermore, two distinct fluorescence regimes of the neutral and anionic state transition (420 to 470 nm and $\lambda_{fl}\geq$500 nm) are identified with a transition region (480 to 490 nm) [9].

Under 405 nm illumination, the anionic state fluorescence (510 nm) of wt-GFP (pH 5) builds up on a time scale of 15±5 ps (Fig. 4). This rise time of the excited anionic state fluorescence is comparable to the major decay component of the excited neutral state (~24 ps) at 460 nm, and is attributed to an internal excited state proton transfer, ESPT, [8,11,12]. In contrast, the 510-nm fluorescence of EGFP builds up instantaneously with no rise time observed [8,11].

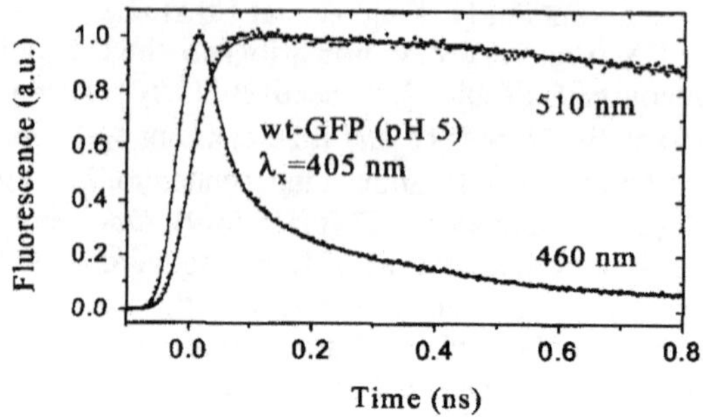

Figure 4. Internal proton transfer on the excited state potential energy surface of wt-GFP (pH 5). Following 405 nm excitation, the neutral state fluorescence (460 nm) decays as a multiexponential (see text) while the anionic state fluorescence (510 nm) builds up on a ~15 ps time scale and decays on ~3.2 ns time scale.

3.4 Rotational diffusion of IFP in solution

The time-resolved fluorescence anisotropy of anionic EGFP decays as a single exponential with an estimated rotational diffusion time of ϕ~15 ns, similar to Citrine [5], wt-GFP and S65T [13]. These rotational times are consistent with theoretical predictions based on the molecular weight of GFP (~27 kDa) [8]. In contrast, the anisotropy of DsRed decays as a biexponential with ϕ_1~210 ps (c_1~0.05) and ϕ_2~60 ns (c_2~0.27), following 490 nm excitation, with an estimated initial anisotropy r_0~0.3 [5]. While the DsRed monomer is similar in molecular weight to GFP, the slower overall rotational time (~60 ns) suggests an oligomer configuration (most likely a tetramer) [5] in agreement with analytical ultra-centrifugation [6,7] and x-ray crystallography [14]. Our diffusion coefficient measurements using FCS provide further evidence for oligomerization with a dissociation constant $\leq 10^{-9}$ M. The fast (~210 ps) internal depolarization and low initial anisotropy of DsRed are consistent with intraoligomer energy transfer between non-parallel transition dipoles [5,14]. The depolarization angle between the absorbing and emitting dipole is 24±3°.

4 Discussion

Systematic studies on the spectroscopy and excited state dynamics of selected GFP mutants enable us to relate the effects of specific individual mutations on the photophysical properties. These results also provide a guideline for future mutation strategies aimed at creating specific fluorescence and thermodynamic characteristics. The amino acid substitution S65→T in S65T and EGFP mutants suppresses the neutral state transition (Fig. 1a). While there are multiple sites for possible protonation within the GFP chromophore, the heterocyclic nitrogen likely plays a key role by accepting a proton from S65 [1,2]. The S65→G mutation is even more efficient than S65→T at suppressing the neutral state transition in the T203Y, T203F, and Citrine mutants as revealed by the negligible absorption around 396 nm (Fig. 1a). Furthermore, the S65→T mutation slightly stabilizes (i.e., red shift) the anionic state transition energy in S65T and EGFP. Our comparative excited state dynamics of wt-GFP, S65T, and EGFP also indicate that mutations affect the excited state barriers separating the neutral and anionic state potential energy surfaces (PES) and explain the absence of ESPT in EGFP. Further support comes from our observation of the slight increase in the neutral state time constants of EGFP in contrast to wt-GFP. This is also consistent with the enhanced neutral state emission around 475 nm in S65T and EGFP (~25 times larger than wt-GFP) [8].

The bathochromic shift (~38 nm) of the anionic state transition in class 3 (*e.g.*, T203Y, T203F, and Citrine) with respect to wt-GFP, is attributed to π-π electron stacking with Y66 due to the T203 amino acid mutations [1]. Single T203 site mutation studies [15] attribute the red shift of the anionic state transition to an increase in the ground state energy as compared to wt-GFP. However, the excited state fluorescence lifetime of T203 mutants [5,10] is distinct from that in wt-GFP and class 2 mutants (Table 1) [8]. This observation supports the red shift argument based on π-π electron stacking with Y66 due to the T203 amino acid mutations. It is reasonable to conclude that the observed red shift in class 3, in which both S65 and T203 substitutions are present, is likely due to cooperative effects. The Q69M substitution in Citrine confers several useful properties, such as its resistance to acid quenching, reduced sensitivity to chloride, and enhanced expression in cells at 37 °C [5] compared to other T203 mutants. The local environment of the neutral chromophore in Citrine is affected by Q69M mutation (~17 nm red shift, Fig. 2b) of the neutral absorption (at low pH), compared with wt-GFP. Therefore, Q69M substitution may affect both the spectroscopy and pK_a of the neutral state transition, based on the decreased polarity of Met compared with Gln [5].

The negligible pH effects on the excited anionic state relaxation imply that the external protonation of GFP, observed with FCS [5,9,10], takes place on the ground state PES rather than during the excited state lifetime. Consequently, the environmental pH affects mainly the anionic and neutral ground state distribution but not the excited anionic state dynamics (*i.e.*, fluorescence quantum yield). The neutral state transition in most GFP mutants can be probed via the blue emission tail (~457 nm), which reveals a complex relaxation pathway and effective nonradiative channels [8] consistent with the negligible contribution to the overall fluorescence and the observed ultrafast decay rates. In addition to excited state proton transfer, nonradiative mechanisms might also include photoisomerization due to conformational rearrangements [4,8] and excited state intramolecular energy transfer.

Understanding the cellular environmental effects on the photophysics of the IFPs is critical for advancing their biological applications. We have studied multi-photon excited state fluorescence spectroscopy and dynamics of EGFP and DsRed constructs in both the cytoplasm and plasma membrane of RBL mast cells. Our results indicate that the chromophore in these IFPs is well protected by the β-barrel, which minimizes its sensitivity to the cellular environment. According to our time-resolved fluorescence anisotropy results, the overall rotational diffusion of these proteins indicates a higher viscosity inside the cells, as compared with solution. Furthermore, the oligomer configuration of DsRed is also preserved in cellular environment similar to solution, as will be described elsewhere.

5 Acknowledgements

The financial support of this work was provided by NIH (P41-RR04224) and NSF (BIR 8800278 and DBI-0080792). EDS was supported in part by the STC Program of NSF (ECS-9876771) and by NIH (AI22449 and AI18306). STH also benefited from NSF Graduate Research and NIH-MBTG (GM08627) Fellowships.

Table 1. Mutations of intrinsically fluorescent proteins and their anionic state photophysical properties.

Protein	Mutation	λ_a/nm	λ_{fl}/nm	τ_{fl}/ns	pK_a (ref)
wt-GFP	Parent	475	503	3.3	4.5 (1)
S65T	S65T	488	509	2.9	5.8 (9)
EGFP	S65T/F64L	490	509	2.8	5.8 (3,9)
T203Y	S65G/S72A/T203Y	512	524	3.7	5.3 (10)
T203F	S65G/S72A/T203F	514	527	3.7	5.2 (10)
Citrine	S65G/V68L/Q69M/S72A/T203Y	514	527	3.6	5.7 (5)
DsRed	N/A	558	583	3.6	--- (5,6,7)

References

1. Tsien R. Y., The green fluorescent protein, *Annu. Rev. Biochem.* **67** (1998) pp. 509-544; and references therein.
2. Chalfie M. and Kain S., Green fluorescent proteins: properties, applications, and protocols (Wiley-Liss, New York, 1998).
3. Voityuk A. A., Michel-Beyerle M. E., and Rosch N., Structure and rotation barriers for ground and excited states of the isolated chromophore of the green fluorescent protein, *Chem. Phys. Lett.* **296** (1998) pp. 269-276.

4. Weber W., Helms V., McCammon J. A., and Langhoff P. W., Shedding light on the dark and weakly fluorescent states of green fluorescent proteins, *Proc. Natl. Acad. Sci. USA* **96** (1999) pp. 6177-6182.
5. Heikal A. A., Hess S. T., Baird G. S., Tsien R. Y., and Watt W. W., Molecular spectroscopy and dynamics of intrinsically fluorescent proteins: coral red (dsRed) and yellow (Citrine), *Proc. Natl. Acad. Sci. USA* **97** (2000) pp. 11996-12001.
6. Baird G. S., Zacharias D. A., Gross L. A., Hoffman R. C., Baldridge K. K., and Tsien R. Y., Biochemistry, mutagenesis, and oligomerization of DsRed, a red fluorescent protein from coral, *Proc. Natl. Acad. Sci. USA* **97** (2000) pp. 11984-11989.
7. Gross L. A., Baird G. S., Hoffman R. C., Baldridge K. K., and Tsien R. Y., The structure of the chromophore within DsRed, a red fluorescent protein from coral, *Proc. Natl. Acad. Sci. USA* **97** (2000) pp. 11990-11995.
8. Heikal A. A., Hess S. T., and Webb W. W., Multiphoton molecular spectroscopy and excited state dynamics of enhanced green fluorescent protein (EGFP): acid-base specificity, *Chem. Phys.* **274** (2001) pp. 37-55.
9. Haupts U., Maiti S., Schwille P., and Webb W. W., Dynamics of fluorescence fluctuations in green fluorescent protein observed by fluorescence correlation spectroscopy, *Proc. Natl. Acad. Sci. USA* **95** (1998) pp. 13573-13578.
10. Schwille P., Kummer S., Heikal A. A., Moerner W. E. and Webb W. W., Fluorescence correlation spectroscopy reveals fast optical excitation-driven intramolecular dynamics of yellow fluorescent proteins, *Proc. Natl. Acad. Sci. USA* **97** (2000) pp. 151-156.
11. Lossau H., Kummer A., Heinecke R., Pöllinger-Dammer F., Kompa C., Bieser G., Jonsson T., Silva C. M., Yang M. M., Youvan D. C., and Michel-Beyerle M. E., Time-resolved spectroscopy of wild-type and mutant green fluorescent proteins reveals excited state deprotonation consistent with fluorophore-protein interactions, *Chem. Phys.* **213** (1996) pp. 1-16.
12. Chattoraj M., King B. A., Bublitz G. U., and Boxer S. G., Ultra-fast excited state dynamics in green fluorescent protein: multiple states and proton transfer, *Proc. Natl. Acad. Sci. USA* **93** (1996) pp. 8362-8367.
13. Volkmer A., Subramaniam V., Birch D. J. S., and Jovin T. M., One- and two-photon excited fluorescence lifetimes and anisotropy decays of green fluorescent proteins, *Biophys. J.* **78** (2000) pp. 1589-1598.
14. Yarbrough D., Wachter R. M., Kallio K. Matz M. V., and Remington, S. J., Refined crystal structure of DsRed, a red fluorescent protein from coral, at 2.0-angstrom resolution, *Proc. Natl. Acad. Sci. USA* **98** (2001) pp. 462-467.
15. Kummer A. D., Wiehler J., Rehaber H., Kompa C., Steipe B., and Michel-Beyerle M. E., Effects of threonine 203 replacements on excited-state dynamics and fluorescence properties of the green fluorescent protein (GFP), *J. Phys. Chem. B* **104** (2000) pp. 4791-4798.

ENERGY TRANSFER FROM SUPRAMOLECULAR ASSEMBLIES OF SYNTHETIC ZINC CHLORINS TO ATTACHED ENERGY TRAPS

VALENTIN I. PROKHORENKO[1)], MARTIN KATTERLE[1)], ALFRED R. HOLZWARTH[1)], KURT SCHAFFNER[1)], TOMOHIRO MIYATAKE[2)], AND HITOSHI TAMIAKI[3)]

[1)] *Max-Planck-Institut für Strahlenchemie, Postfach 101365, D-45413, Mülheim a.d. Ruhr, GERMANY*

E-mail: prokh@mpi-muelheim.mpg.de

[2)] *Department of Material Chemistry, Faculty of Science and Technology, Ryukoku University, Otsu, Shiga, 520-2194, JAPAN*

[3)] *Department of Bioscience and Biotechnology, Ritsumeikan University, Kusatsu, Shiga, 525-8577, JAPAN*

Singlet energy transfer from supramolecular zinc chlorin aggregates to co-aggregated energy traps is explored using time-resolved fluorescence spectroscopy. The excitation energy is transferred within 7-9 ps with efficiency reaching up to 70%. The excitation in these aggregates is delocalized over at least 10-15 molecules even at room temperature, which indicates promising perspectives for their use as antennae in artificial photosynthetic devices for solar energy conversion.

1 Introduction

Special natural chlorins (bacteriochlorophylls, BChl *c*, *d*, and *e*) as well as chemically modified zinc chlorins, which are more stable, by self-aggregation form large supramolecular assemblies containing several thousands of chlorins [1-3], arranged in rod-like micelles. The mechanism of self-organization of these special BChls contained in natural chlorosomes [4] and their zinc analogs is now quite well established [1,5]. Strong excitonic interaction between neighboring molecules (ca. 400 cm^{-1}) and their regular organization lead to concentration of large dipole strength in only a few allowed transitions [2] which is very favorable for highly efficient light collection. Such supramolecular assemblies could provide an elegant way to attach an antenna function to an artificial reaction center unit [6,7]. However, for further utilization of the absorbed light energy in a potential antenna device, the energy should be concentrated rapidly in an energy trap, to which an artificial reaction center might be attached. In native chlorosomes the excitation is transferred within a few picoseconds to the so-called baseplate containing BChl *a* molecules which function as the first energy trap [8]. In artificial systems the metal-free bacteriochlorins, which can be effectively linked by self-aggregation, have been shown to be effective as traps, forming an artificial light-harvesting unit (LHU) [3]. The excited state of metal-free bacteriochlorin is in resonance with the Q_y absorption band of the aggregates (~750 nm). After co-aggregation of a small amount of trap molecules this additional absorption appears in the absorption spectrum as a red shoulder, located at ~800 nm (Fig. 1a). The red shift is most likely due to excitonic interaction with the aggregate molecules. Strong excitonic interaction is favorable for fast and efficient energy transfer (ET) to the trap molecule. The energy trapped there can be monitored easily using steady-state fluorescence (Fig. 1,b).

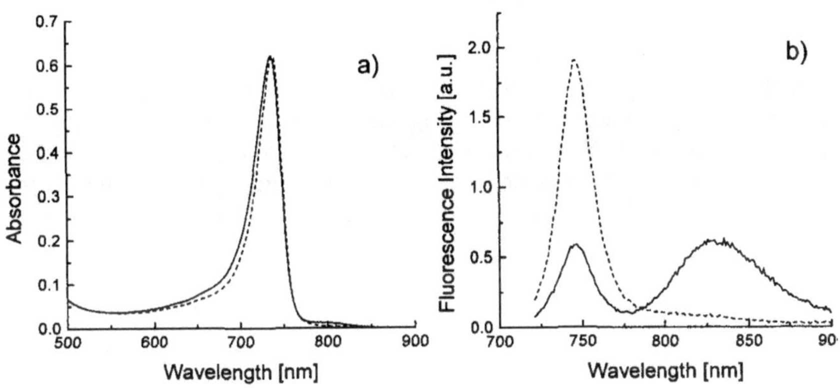

Figure 1. Absorption (a) and emission spectra of zinc chlorin aggregate (dashed) and LHU (solid). The ratio of zinc chlorin / metal-free bacteriochlorin is 25:1.

In this work we present a time-resolved fluorescence study of the ET-processes occurring in the supramolecular co-aggregates.

2 Experimental

The energy transfer kinetics was measured using fluorescence single photon timing technique. Samples were excited at 710 nm with laser pulses of ~10 ps fwhm at a 4 MHz repetition rate provided by a laser system, consisting of a Spectra Physics 2080-15S mode-locked argon ion pumping laser, a Spectra Physics model 375 dye laser (DCM was used), and a Spectra Physics model 344 cavity dumper. At the excitation density of $\sim 2 \cdot 10^{11}$ photons/cm^2 no annihilation effects were observed in the fluorescence kinetics. The fluorescence was passed through a monochromator and detected under magic angle polarization with a fast microchannel plate photomultiplier R3809U-51 (Hamamatsu). The system response fwhm was ~30 ps, allowing time resolution for lifetimes up to 2 ps using deconvolution procedures. The decay traces $fl(t,\lambda)$ were treated in two different ways. By a global lifetime analysis method (assuming equal lifetimes τ_i over the whole spectrum), resulting in decay-associated spectra $A_i(\lambda)$ (DAS), the traces were fitted as:

$$fl(t,\lambda) = \sum_i A_i(\lambda) \int_0^\infty Pr(t-x)\exp(-x/\tau_i)dx, \qquad (1)$$

where $Pr(t)$ is the response function of the system. Alternatively single decay traces were treated using a numerical inverse Laplace transform (NILT) technique [2] that uses a continuous lifetime scale ranging from $min(\tau)$=2 ps up to $max(\tau)$=2 ns:

$$fl(t,\lambda) = \int_{min(\tau)}^{max(\tau)} d\tau \int_0^\infty dx A(\lambda,\tau)\, Pr(t-x)\exp(-x/\tau), \qquad (2)$$

Synthetic chlorophyll derivatives of antenna pigments (zinc-methyl-3-devinyl-3-hydroxymethylpyropheophorbide *a*, Fig. 2a) and the energy trap (methyl bacterio-pyropheophorbide *a*, Fig. 2b) were prepared according to reported procedures [9,10]. The LHU with concentration ratio of the aggregated zinc chlorins to energy trap molecules of 25:1 was prepared as described in [11] in aqueous solution using

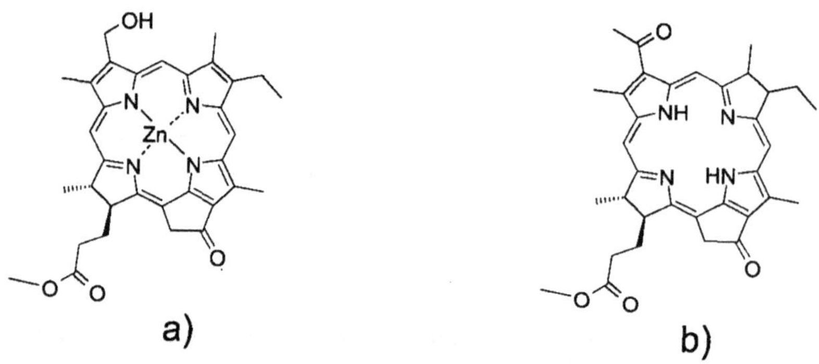

Figure 2. Structural formulae of used synthetic compounds for aggregate (a) and energy trap (b).

α-lecithin as a lipid to form vesicles. The aqueous suspension was pumped through a flow cell with 1.5x1.5 mm^2 active area under a nitrogen atmosphere. In order to avoid oxidation conditions, a small amount of sodium dithionite was added to the solution before the measurements. All experiments were carried out at room temperature.

3 Results

The excitation kinetics of the antenna aggregate without energy trap can be described by 4 exponentials, ranging from 16 ps up to ~1 ns (Fig. 3a). The DAS of these components have very similar spectral shapes and maxima. The shortest lifetimes which carry the largest amplitudes reflects most likely internal ET between different excitonic levels of the aggregate, characterized by different radiative lifetimes. Due to strongly overlapping the DAS no ET components with negative amplitudes can be observed, however.

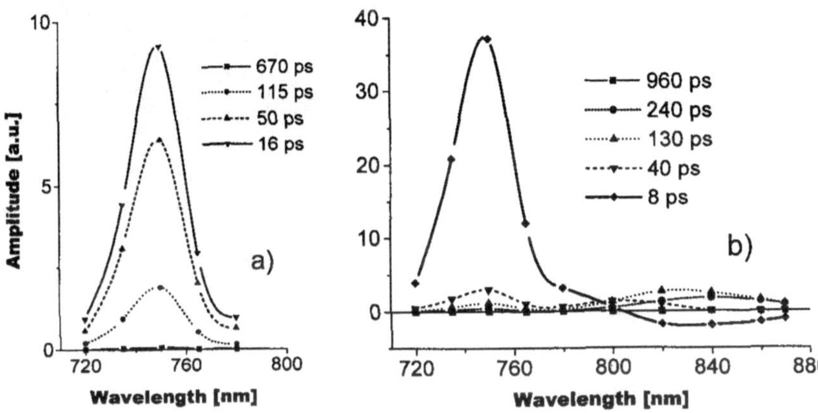

Figure 3. Decay-associated fluorescence spectra of zinc-chlorin aggregate (a) and LHU (b).

The average lifetime of the aggregate is 41 ps independently on the wavelength.

The fluorescence of the LHU shows two emission bands (Fig. 1b). The 750-nm band reflects the fluorescence of the zinc-chlorin aggregate, but with significantly lower intensity than for the pure aggregate (the average lifetime decreases to 17 ps). A broader second band located at 825 nm corresponds to fluorescence from the attached energy trap and has an average lifetime of 180 ps. In the DAS of the LHU (Fig. 3b) the ET-component is clearly resolved. Energy transfer occurs with a ca. 8 ps lifetime. The maximum of the negative ET amplitude is located at ~835 nm. The ratio between negative and positive amplitudes of the ET-component (8 ps) is in the range of 1:20 - 1:30, whereas the ratio between the amplitudes of decay components in the aggregate emission region and in the trap region is ca. 1:3 to 1:5. Similar results are obtained using NILT-analysis for single decay traces. Figure 4 (left) shows the lifetime spectra corresponding to fluorescence decay traces at 735 nm (top) and at 840 nm (bottom). The shortest lifetime component changes the sign from positive to negative. However, their gravity centers are not equal - decay at 735 nm occurs within 8 ps (shortest component), whereas at 840 nm the fluorescence growths within ~10 ps. This dispersive character of fluorescence lifetimes can more clear by seen in the so-called lifetime density map (Fig. 4, right), where the lifetime spectra are plotted versus wavelength (note that the lifetime coordinate is given on a logarithmic scale), The physical mechanism of this lifetime dispersion is unknown at present.

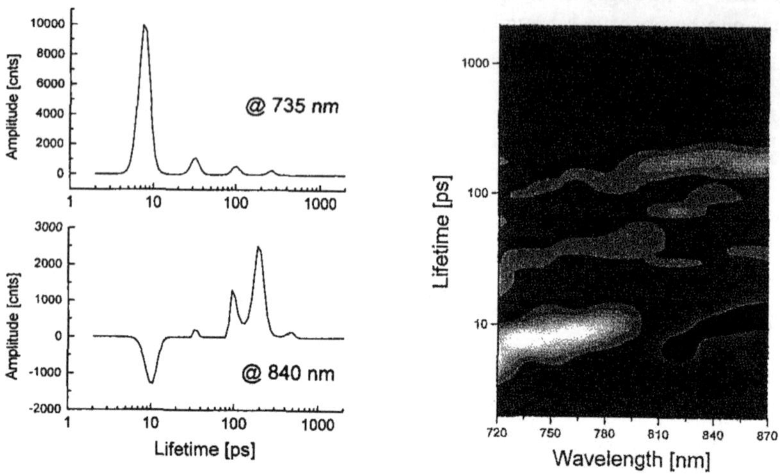

Figure 4. Lifetime spectra for LHU of single fluorescence decay traces (left), and lifetime density map of whole fluorescence spectrum (right). The density corresponds to the amplitudes (from dark color - negative amplitude - through gray - zero amplitude - to bright - positive amplitude). Note that the amplitudes are not corrected to the steady-state fluorescence spectrum.

The discrete nature of the lifetime spectra indicates that the spectrum of excitonic states has also a discrete nature, i.e., only a few excitonic transitions are allowed as proposed earlier [2].

The fluorescence from the energy trap region is characterized by several lifetime components as well (Fig. 3b, and Fig. 4). This multi-exponential character of the fluorescence decay is possibly caused by heterogeneity of the co-aggregates, e.g. due to location of energy traps at different positions in the aggregate body. Furthermore the concentration ratio between trap molecules and aggregated zinc chlorins of 1:25 implies several trap molecules per aggregate body which is expected also to cause a multi-exponential decay of fluorescence. In that case the small deviations between ET lifetimes for different energy trap molecules cannot be resolved in our analysis.

4 Modeling and discussion

The large ratio between positive and negative amplitudes of the ET-component is at first glance unexpected and surprising, given the small energy difference of the bands and the aggregate/trap ratio. However, by applying a simple kinetic model for a coupled donor (aggregate) - acceptor (trap) pair, we can describe these features quite well. In the simplest case, assuming that only one acceptor is excited, the change of population for weak excitation can be given as:

$$\begin{aligned} \dot{n}_d &= \sigma_d n_0 I_{exc} - k_{fl}^d n_d - k_{et} + \beta k_{et} n_a \\ \dot{n}_a &= k_{et} n_d - \beta k_{et} n_a - k_{fl}^a n_a \end{aligned} \quad , \tag{3}$$

where n_d, n_a are the excited state concentrations of donor and acceptor, respectively, n_0 is the ground state concentration, σ_d the absorption cross-section for the donor, $I_{exc}(t)$ is excitation pulse, $\beta = exp(-\Delta E / kT)$ is the Boltzmann factor that describes the uphill energy transfer, k_{et} is the forward ET-rate, and k^d_{fl}, k^a_{fl} are fluorescence rate constants (which includes radiative and non-radiative rates $k_{fl} = k_{rad} + k_{nr}$) for donor and acceptor, respectively. The intensity of fluorescence is proportional to the radiative rate constant and to the concentration of excited donors/acceptors: $I_{fl}^{d(a)} \propto k_{rad}^{d(a)} \cdot n_{d(a)}$. Solving the system Eq. 3 for the stationary case (I_{exc}=*const*) gives the intensity ratio of aggregate/trap fluorescence:

$$I_{fl}^d / I_{fl}^a = \left(k_{rad}^d / k_{rad}^a\right) \cdot \left(k_{fl}^a / k_{et}\right) \tag{4}$$

In our case the fluorescence rate constant of the acceptor is much smaller than the ET rate constant, thus fluorescence from these bands with equal intensities can only be observed if $k_{rad}^d >> k_{rad}^a$. Solving Eq. 3 for impulsive excitation gives the ratio between amplitudes A of the ET-component in the aggregate and trap regions:

$$A^d / A^a = -\left(k_{rad}^d / k_{rad}^a\right) \cdot \left[1 + \left(k_{fl}^d - k_{fl}^a\right) / k_{et}\right] \approx -k_{rad}^d / k_{rad}^a \tag{5}$$

Taking into account the difference in spectral widths for the aggregate and trap fluorescence bands (1:2 on a wavenumber scale), both Eqs. 4 and 5 can be reconciled with the experimental data only if $k_{rad}^d \geq 15 k_{rad}^a$. That is, the excitonic transition in the aggregate has a ~15 times higher dipole strength than the energy trap. If we assume that the dipole strength of the energy trap in the LHU is roughly that of the isolated molecule, our result implies that at least 15 molecules in the aggregate are coherently coupled even at room temperature. Note that this number represents a lower limit due to our simplified analysis. The actual number of coupled molecules might be substantially larger. Nevertheless, such a large degree of delocalization leads to a high absorption cross-section and makes the zinc-chlorin aggregates particularly suitable for the design of artificial LHUs. The estimated efficiency of energy transfer/energy storage is very high, reaching 70%.

Acknowledgements

This work was supported by the EU-TMR network on "Green bacterial photosynthesis", grant # FM RX-CT960081 by the European Comission, and the Human Frontier Science Organization, HFSPO Award # RG 0184/1997-M awarded to ARH.

References

1. A.R. Holzwarth and K. Schaffner, Photosynth. Res. *41* (1994) 225.
2. V.I. Prokhorenko, D.B. Steensgaard, and A.R. Holzwarth, Biophys. J. *79* (2000) 2105.
3. H. Tamiaki, T. Miyatake, R. Tanikaga, A.R. Holzwarth, and K. Schaffner, Angew. Chem. Int. Ed. Engl. *35* (1996) 772.
4. J.M. Olson, Photochem. Photobiol. *67* (1998) 61.
5. S. Yagai, T. Miyatake, Y. Shimono, and H. Tamiaki, Photochem. Photobiol. *73* (2001) 153.
6. D.I. Shuster, P. Cheng, S.R. Wilson, V. Prokhorenko, M. Katterle, A.R. Holzwarth, S.E. Braslavsky, G.Klihm, R. Williams, and C. Luo, J. Am. Chem. Soc. *121* (1999) 11599.
7. A.R. Holzwarth, M. Katterle, M.G. Müller, Y.-Z. Ma, and V.I. Prokhorenko, Pure Appl. Chem. *73* (2001) 469.
8. M.G. Müller, K. Griebenow, and A.R. Holzwarth, Biochim. Biophys. Acta *1144* (1993) 161.
9. H. Tamiaki, M. Amakawa, Y. Shimono, R. Tanikaga, A. R. Holzwarth, and K. Schaffner, Photochem. Photobiol. *63* (1996) 92.
10. H. Tamiaki, M. Kouraba, K. Takeda, S. Kondo, and R.Tanikaga, Tetrahedron: Asymmetry *9* (1998) 2101.
11. T. Miyatake, H. Tamiaki, A.R. Holzwarth, and K. Schaffner, Helv. Chim. Acta *82* (1999) 797.

ULTRAFAST EXCITED STATE DYNAMICS OF MULTIPORPHYRIN ARRAYS

ANA MORANDEIRA[1], ERIC VAUTHEY*[1], OLIVIER MONGIN[2] AND ALBERT GOSSAUER[2]

1) Department of Physical Chemistry of the University of Geneva, 30 quai Ernest Ansermet, CH-1211 Genève 4, Switzerland

2) Department of Chemistry of the University of Fribourg, Pérolles, CH-1700 Fribourg, Switzerland

The excited state dynamics upon Soret band excitation of arrays containing three and six Zn tetraphenylporphyrin moieties has been investigated using fluorescence up-conversion and transient grating. These arrays exhibit several dynamic features that differ substantially from those of the monomer: strong shortening of the S_2 state lifetime, ultrafast depolarisation, additional decay components of the S_1 fluorescence. Moreover at high excitation intensities, S_1-S_1 annihilation becomes the dominant decay mechanism. A model involving S_2 exciton coupling is proposed to account for these differences.

1 Introduction

Over the past few years, intensive experimental and theoretical efforts have been invested to understand the energy transport mechanism in vegetal and bacterial light harvesting complexes. Although all the fine details of the excitation energy transfer (EET) are still not completely understood, it is well known that the overall EET efficiency of these antenna complexes is very high [1].

This high efficiency has stimulated many chemists to design and prepare synthetic antenna systems, which are often based on porphyrin pigments as well. Moreover the proteins which, ensure a very precise spatial arrangement of the pigments in the natural systems, are replaced by rigid spacers.

The Coulomb interaction has been found to be the dominant EET mechanism in natural antennae [2]. In the weak interaction limit, the exciton is localised on a single pigment and the EET can be discussed in terms of the Förster theory. In some antenna complexes, a small interpigment distance results in a strong interaction. This leads to excitonic states with different energies and where the excitation is delocalised over several chromophores (coherent exciton). Broadening and/or splitting of the absorption band of the aggregate compared to that of monomer is an indication of the strength of this coupling.

In porphyrin arrays, both the Coulomb and the exchange interactions seem to be operative, depending on the interchromophoric distance, on the nature of the spacer [3] and on the position of the linker on the porphyrin rings.

We have recently reported on an investigation of the EET upon Q-band excitation in arrays composed of three porphyrins linked by phenylethynyl groups of different length, allowing to vary the interchromophoric centre to centre distance from 23 to 46 Å [4]. It was found that in these arrays, through-bond exchange interaction was prevailing. The data were also predicting a predominance of the Coulomb interaction upon decreasing the distance below 20 Å.

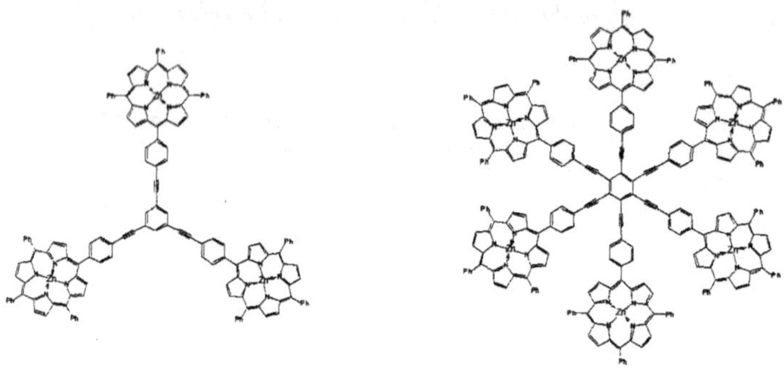

Figure 1: non-optimised structures of TrZn (left) and StZn (right)

We report here on an investigation of the excited state dynamics of a tri- and a hexaporphyrin arrays upon Soret band excitation (Figure 1). The centre to centre interchromophoric distance is 23 Å and 13 Å in the trimer and the hexamer, respectively. We will show that Soret band excitation leads to the occurrence of new dynamical features.

2 Experimental

Fluorescence dynamics measurements were performed around 400 nm with the frequency doubled output of a Ti:Sapphire laser at 82 MHz. For the slow dynamics, time-correlated single photon counting (TCSPC) was used [5]. The full width at half maximum (FWHM) of the instrument response function (IRF) was less than 200 ps. Ultrafast fluorescence dynamics was measured using up-conversion. The FWHM of the IRF was around 200 fs.

Population dynamics was also investigated using the multiplex transient grating (TG) technique [6,7]. Excitation was performed around 400 nm using the frequency-doubled output of a 1 kHz Ti:Sapphire amplifier. The FWHM of the IRF was about 250 fs.

The synthesis of the porphyrin arrays has been described in [8]. All measurements were performed in toluene (Tol, Fluka UV grade). The concentration of the arrays was inferior to 10^{-4} M.

3 Results

3.1 *Steady state measurements.*

In the Q band region (500-650 nm), the spectrum of TrZn is identical to that of the monomer with a three times larger absorption coefficient. On the other hand, the Soret band (400-450 nm) is slightly broader and less intense. This effect is more pronounced with StZn. The difference between monomer and hexamer is much weaker in the Q band region. The Q(1,0) and Q(0,0) bands are slightly shifted to longer wavelengths and their relative intensities are somewhat different.
The S_1 fluorescence spectrum of ZnTPP consists in two bands at 606 and 648 nm with relative magnitudes of 1:0.85. In the trimer, these bands are shifted to 613 and 655 nm, respectively, and their relative magnitudes are 1:0.7. Those of StZn are at the same wavelength, but their relative magnitudes are 1:0.5. The fluorescence quantum efficiencies of TrZn and StZn are essentially the same as that of ZnTPP.

Both ZnTPP and TrZn show additionally a band at 430 and 450 nm, respectively, due to S_2 fluorescence. The S_2 band TrZn is about four times smaller than that of the monomer. No S_2 fluorescence was detected with StZn.

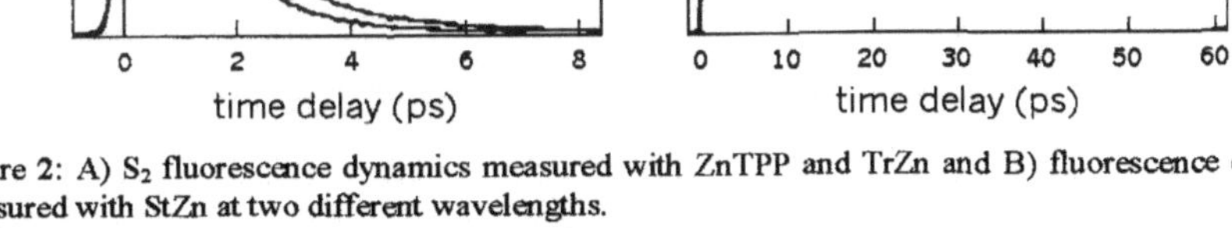

Figure 2: A) S_2 fluorescence dynamics measured with ZnTPP and TrZn and B) fluorescence dynamics measured with StZn at two different wavelengths.

3.2 *Isotropic time-resolved fluorescence.*

ZnTPP. The S_1 fluorescence lifetime of ZnTPP in Tol amounts to 1.8 ns [4]. Up-conversion measurement show that, upon Soret band excitation, the S_1 fluorescence exhibits first a rise with a time constant of 1.5 ps, which corresponds to S_2-S_1 internal conversion. The same time constant is obtained from the decay of the S_2 fluorescence at 430 nm (Figure 2A).
TrZn. The S_1 fluorescence lifetime of TrZn measured by TCSPC is the same as that of ZnTPP [4]. However, the fluorescence rise time is sensibly shorter (1.0±0.1 ps). The same time constant is obtained from the decay of the S_2 fluorescence at 450 nm

(Figure 2A). The decay of the S_1 emission exhibits, additionally to the 1.8 ns component, a 40 ps component with 35% of the total amplitude at 613 nm and 28% at 655 nm.

StZn. The fluorescence lifetime of StZn in the 600-670 nm region measured by TCSPC is the same as that of ZnTPP and TrZn. The ultrafast dynamics is however more complex. Between 450 and 560 nm, the fluorescence time profiles are best described using a triexponential decay with time constants of 150-300 fs, 2-3 ps and 15-20 ps. At 450 nm, their amplitudes are 90%, 5% and 5% of the total amplitude, respectively. They vary continuously with the wavelength to about 20%, 40% and 40% at 560 nm.

Above 580 nm, the fast decay component becomes a rising component with a similar time constant. On the other hand, the fit of the fluorescence decay requires at least three exponential functions with time constant around 20 ps, 90 ps and 1.8 ns (Figure 2B).

The faster component is dominant with 66% and 51% of the total amplitude at 610 and 650 nm, respectively. The amplitude of the 90 ps component changes from 16% to 21%, while that of the slower component increases from 18% to 28%. This dynamics is independent on the excitation intensity used in the up-conversion experiments ($<2\times10^{13}$ photons cm^{-2} $pulse^{-1}$).

3.3 *Anisotropic time-resolved fluorescence*

ZnTPP. The polarisation anisotropy of the S_2 fluorescence decays from an initial value of $r\approx0.4$ to 0.1 with a time constant of about 300 fs. The value of 0.4 indicates that the transition dipole involved in the emission is parallel to that involved in the excitation. The S_0-S_2 transition dipole is doubly degenerate and in this case the initial polarisation anisotropy is predicted to be 0.1. Therefore, the 0.4 to 0.1 decay corresponds to the randomisation of the population between the degenerate states. The initial decay of $r=0.7$ to 0.4 predicted by theory [9] must be too fast to be observed here.

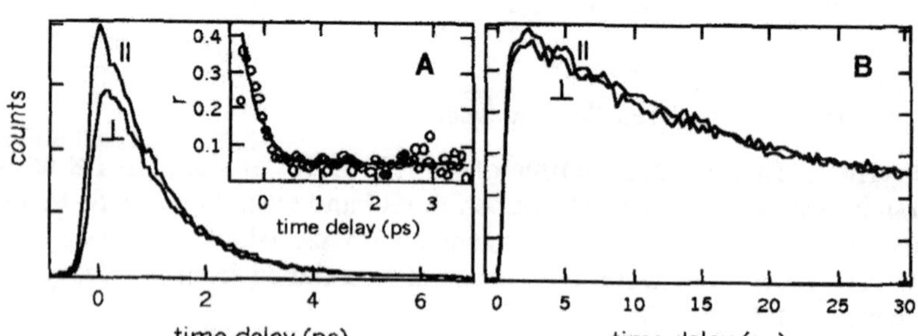

Figure 3: A) S_2 fluorescence dynamics polarised parallel and perpendicular to the excitation and anisotropy decay (inset) measured with TrZn; B) same as A but measured with StZn at 610 nm.

The polarisation anisotropy of the S_1 fluorescence upon Soret band excitation decays exponentially from 0.1 to zero with a time constant of 130±20 ps. This process occurs through rotational diffusion of ZnTPP.

TrZn. As for ZnTPP, the initial polarisation anisotropy of S_2 fluorescence of TrZn amount to ≈0.4. However, its decays to a value of less than 0.1 (≈0.05) with a ≈300 fs time constant (Figure 3A). The S_1 fluorescence anisotropy starts at a 0.04±.01 and decays to zero with a time constant of 100±30 ps. This value compares rather well with that of 75 ps obtained by polarisation grating measurements at 532 nm [4]. In that case, the initial anisotropy was 0.1 as expected for a doubly degenerate transition dipole. The anisotropy decay is not due to rotational diffusion, the molecular volume of TrZn being much larger than that of ZnTPP, but to energy hopping as discussed in [4].

StZn. The S_1 fluorescence of StZn is essentially unpolarised (r<0.03). No significant difference between the fluorescence kinetics polarised parallel and perpendicular to the polarisation of the excitation was detected (Figure 3B).

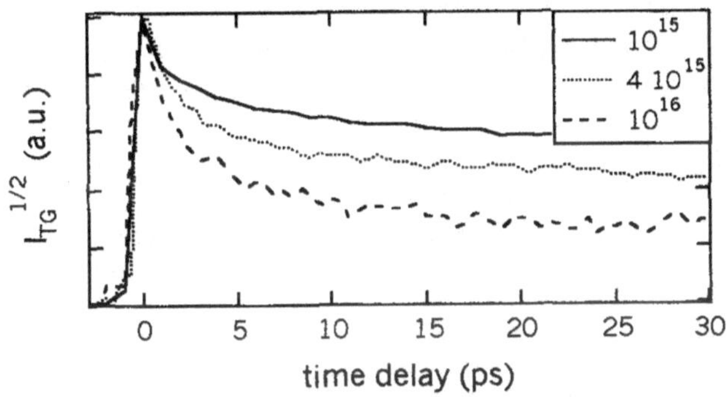

Figure 4: time profiles of the square root of the TG intensity at 460 nm measured with StZn at different pump intensities (in photons $cm^{-2}pulse^{-1}$).

3.4 Multiplex TG measurements

The TG spectra of ZnTPP and TrZn upon 355 nm excitation with 30 ps pulses have been discussed in detail in [4]. The population dynamics deduced from the TG intensity is similar to that measured by TCSPC.

The TG spectra of StZn measured with 100 fs pulses show essentially the same features, which can be assigned to S_1-S_n transitions. No significant spectral dynamics was observed in the 420-700 nm range. The time dependence of the TG intensity differs strongly from that measured with ZnTPP, and exhibits an additional fast non-exponential decay, which depends strongly on the pump intensity (Figure 4). With a pump intensity of 10^{16} photons·cm^{-2}·$pulse^{-1}$, the initial signal intensity decreases by 66% in 10 ps, while with a lower intensity (10^{15} photons·cm^{-2}·$pulse^{-1}$) this decrease is smaller (40%) and slower.

4 Discussion

4.1 Exciton annihilation

The strong variation of the TG decay dynamics with pump intensity can be ascribed to exciton annihilation in the StZn array. This is confirmed by the substantial decrease of the fluorescence yield as the pump intensity increases. This dependence was analysed with a model developed by Paillotin et al. [10] for exciton annihilation in photosynthetic membranes and which indicates that the average number of excitation per array is one at about 10^{15} photons·cm^{-2}·pulse^{-1}. Therefore, a much lower intensity has to be used to avoid exciton annihilation. This is the case in the up-conversion measurements where, the intensity was below $2 \cdot 10^{13}$ photons·cm^{-2}·pulse^{-1}, corresponding to less than 0.02 excitation per array.

4.2 Exciton coupling

The broadening of StZn Soret band is most probably due to exciton coupling. The S_0-S_2 transition dipole of ZnTPP is very large and the coupling between such dipoles located at 1.3 nm is of the order of 150 cm^{-1}, assuming a value for the orientational factor of unity. This is too weak to generate a splitting of the Soret band but can explain the substantial broadening.

The coupling between the S_0-S_1 transition dipole is much weaker, around 15 cm^{-1}, as expected from the small change of the Q band by going from ZnTPP to StZn.

4.3 Ultrafast fluorescence depolarisation

Exciton coupling might also explain the absence of polarisation anisotropy of the S_1 fluorescence with StZn and the weak one with TrZn. Indeed, a delocalisation of the S_2 exciton over several chromophores can lead to a substantial loss of anisotropy. Moreover, fluctuations may result to a subsequent relocalisation of the exciton on a chromophore that is different from the originally excited one.

On the other hand, if the S_2 exciton is localised, the anisotropy can also decay by energy hopping. The time constant for this process, estimated using the Förster equation, is about 200 fs.

However, the absence of a S_2 fluorescence with StZn and the substantial coupling favour the picture of a delocalised exciton. Indeed, this coupling leads to excitonic states with different energies and oscillator strengths, which can act as efficient ladder for the non-radiative transition to lower excited states.

4.4 Early fluorescence dynamics

The fluorescence observed with StZn below 580 nm most certainly originates from vibrational excited states and its ultrafast decay dynamics corresponds to vibrational relaxation. A similar wavelength dependence of the fluorescence dynamics has been observed with free-base tetraphenylporphine [11]. This is in agreement with the similar risetime observed above 580 nm. Emission from an upper electronic state, a localised or delocalised excitonic S_2 state, could also contribute to the signal.

The origin of the 20 and 90 ps components, observed above 580 nm, is not clear. These decays are not due to S_1 annihilation and do not lead to a significant decrease of the integrated fluorescence yield compared to that of ZnTPP. Moreover, S_1 energy hopping seems not to be responsible for these components because they are present on both sides of the emission band, albeit with different amplitude.

The data could be explained by invoking an excited state different from the localised S_1 state, which would be populated by internal conversion from a S_2 excitonic state. Its emission spectrum is blue shifted relatively to the localised S_1 fluorescence and its dipole strength is larger. Apart from fluorescence, the main deactivation channel of this state is the conversion in a few tens of ps to the localised S_1 states, which itself decays in 1.8 ns. The exact nature of this state is not known and additional measurements with arrays with larger interchromophoric distances, as well as calculations of the molecular structures are in progress.

1. V. Sundström, T. Pullerits, and R. van Grondelle, J. Phys. Chem. B *103* (1999) 2327.
2. H. van Amerongen, L. Valkunas, and R. van Grondelle, *Photosynthetic Excitons*, World Scientific, Singapore, 2000.
3. H. S. Cho, D. H. Jeong, M.-C. Yoon, Y. H. Kim, Y.-R. Kim, D. Kim, S. C. Jeoung, S. K. Kim, N. Aratani, H. Shinmori, and A. Osuka, J. Phys. Chem. A *105* (2001) 4200.
4. P. Brodard, S. Matzinger, E. Vauthey, O. Mongin, C. Papamicaël, and A. Gossauer, J. Phys. Chem. A *103* (1999) 5858.
5. P.-A. Muller, C. Högemann, X. Allonas, P. Jacques, and E. Vauthey, Chem. Phys. Lett. *236* (2000) 321.
6. C. Högemann, M. Pauchard, and E. Vauthey, Rev. Sci. Instrum. *67* (1996) 3449.
7. E. Vauthey, J. Phys. Chem. A *105* (2001) 340.
8. O. Mongin, N. Hoyler, and A. Gossauer, Eur. J. Org. Chem. (2000) 1193.
9. K. Wynne and R. M. Hochstrasser, Chem. Phys. *171* (1993) 179.
10. G. Paillotin, C. E. Swenberg, J. Breton, and N. E. Geacintov, Biophys. J. *25* (1979) 513.
11. Y. Shibata, H. Chosrowjan, N. Mataga, N. Yoshida, and A. Osuka, in: T. Elsaesser et al.(Eds.) *Ultrafast Phenomena XII*, Springer, Berlin, 2000, p. 689.

THE 7-AZAINDOLE BASE PAIR PHOTOTAUTOMERIZATION REVISITED: AN AB INITIO STUDY OF THE POTENTIAL ENERGY SURFACE[1]

MIQUEL MORENO AND JOSÉ MARÍA LLUCH

Unitat de Química Física, Departament de Química. Universitat Autònoma de Barcelona, 08193 Bellaterra (Catalonia), Spain
E-mail: mmf@klingon.uab.es

ABDERRAZZAK DOUHAL

Departamento de Química Física, Sección de Químicas, Facultad de Ciencias del Medio Ambiente, Avenida Carlos III, SN, Campus Tecnológico de Toledo, Universidad de Castilla-La Mancha, 45071 Toledo, Spain
E-mail: adouhal@amb-to.uclm.es

Ab initio full geometry optimizations were carried out for the ground and first excited singlet electronic states of the 7-azaindole base pair. Results indicate that the excited state has to be described as a dimer of an excited base unit and another one in the ground state. This asymmetry leads to an asynchronic double proton transfer process that is stepwise and passes through a very shallow zwitterionic intermediate that may not possess any bound state. Our theoretical calculations have shown the existence of another intermediate where the transfer of a single proton is compensated by a charge-transfer electronic excitation. This neutral intermediate is very stable and could be responsible for the stepwise process observed in several recent femtochemistry experiments. Inclusion of a polar solvent in the calculations leads to a stabilization of the product and a lowering of the energy barrier so that an increasing of the rate constant in condensed phase is predicted.

1 Introduction

Multiproton transfer reactions in electronic excited states have attracted considerable interest in the last years.[1-3] Among them, the double proton transfer in the 7-Azaindole dimer $(7\text{-AI})_2$ ranks one of the most intensely studied systems as it is easily formed in isolated and condensed conditions and it is suggested as a prototype to mimic the elementary reactions involved in the photomutagenesis phenomenon in the DNA base pairs.[4-23]

Studies in the gas phase have revealed that the double proton transfer in the first excited singlet electronic state (S_1) of $(7\text{-AI})_2$ involves two steps that take place in different time scales.[11,14,16] According to these works, the first process near the zero-

[1]This work is based on references 12, 22, 23 and 38.

point energy takes place in ca. 650 fs and involves the jump of one proton producing a picosecond living intermediate which has the two H-atoms in one side of the dimer. The second proton-transfer reaction takes place more slowly (ca. 3.3 ps at vibrational energy near zero in the dimer) and leads to the full tautomer as the final product. Both steps have energy barrier and can occur through tunneling. The whole process is depicted in the Scheme below. In solution, the mechanism might be affected by solvation dynamics and thermalization as it is well known when comparing gas phase results with solution ones.[17,20]

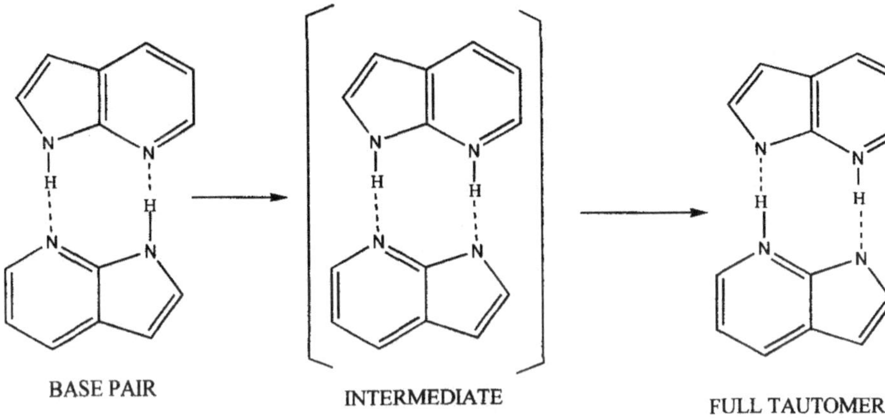

SCHEME

From the point of view of theory, this system has also been extensively studied.[12,19-23] Our previous results based on ab initio methods by using the single configuration interaction method (CIS) and optimizing the geometry of the S_1 state confirmed the stepwise mechanism.[12] Using the small split-valence 4-31G basis set a zwitterionic intermediate with the two hydrogen atoms in the same 7-AI unit was found though the minimum located was very shallow with a very low energy barrier to yield the second H-atom transfer. Further theoretical works by the groups of Miller[19] and Zewail,[20] using a similar methodology with a slightly larger basis set confirmed our previous results. Also in those works, the presence of a different intermediate was disclosed. This intermediate is not zwitterionic as the charge transfer produced by the jump of the proton is compensated by the electronic excitation that also involves a negative charge transfer in the same direction.

By using a so-called hybrid density functional methodology (HDF) Catalan and co-workers claimed that the double proton transfer in S_1 has to be synchronic with the two hydrogen atoms transferring in a symmetric manner.[21] In recent contributions we have pointed the inadequacy of the results obtained using the HDF methodology for the excited state.[22,23] In short, the most serious drawback arises from the fact that within the HDF method geometry optimizations are not feasible so that all the calculations are based on *ground state* geometries and so, the results are not relevant to the phototautomerization of 7-AI dimer where geometry changes occur upon electronic excitation.

Here, we focus on the mechanism for the double proton transfer in the S_1 state by doing an extensive analysis of the potential-energy surface (PES). The synchronic mechanism is also considered and compared with the stepwise one. The role played by a polar solvent in this mechanism is also considered by using a continuum method that considers the bulk effect of a solvent.

2 Methods

Quantum-chemical calculations of excited electronic states of large molecules are still in their infancy. This situation contrasts with what is found in the ground state where sophisticated and quite reliable methods can be used to deal with medium-sized molecules. Up to now there are only two alternatives to analyze the PES of electronic excited states, the CIS which is equivalent to the Hartree-Fock SCF methodology by virtue of the Brillouin's theorem,[24] and the complete active space (CAS) SCF calculations.[25] The latter is supposed to be somewhat better though it has the problem of being computationally more demanding. When used for a reactivity problem it can be difficult to define the active space as it may change quite suddenly along the reaction coordinate.

Both CIS and CAS methods suffer from the lack of dynamic correlation. There is no reliable way to introduce the dynamic correlation in a CIS calculation (there is a CIS-MP2 procedure proposed in the first presentation of the CIS method[24] but it has been seriously questioned since).[26] As for the CAS, there are perturbative schemes developed to introduce the correlation. Among them the CAS-PT2 procedure seems quite promising.[27] Anyway up to now it is not possible to optimize at the CAS-PT2 level so that this procedure cannot be used to explore the PES of the excited state.

Other methods based on the Density Functional Theory (DFT)[28] are unable to deal with electronic excited states unless the symmetry of the excited state is different from that of the ground state. The recently developed time-dependent methods[29] are a promising way to overcome this difficulty but, again, they are still not able to optimize the geometry.
In this limited context of available methods we have chosen the CIS one for our study as the CAS procedure is too computationally demanding in a system as big as the base dimer. In any case for other related systems it has been seen that CAS and CIS results lead to qualitatively comparable geometries and energies of the stationary points.[30]

All calculations were performed with the double-> quality 6-31G(d) basis set which includes a set of d-polarization functions on heavy atoms.[31] The ground electronic state geometries and energies were obtained with the B3LYP hybrid density functional[32] in conjunction with the 6-31G(d) basis set. The CI all-single excitations with a spin-restricted Hartree-Fock reference ground state (CIS) was used to optimize the geometries and calculate the energies of the excited S_1 state.[24] Stationary points were located using the minimization procedure of Schlegel by using redundant internal coordinates.[33] The nature of the stationary points was disclosed through diagonalization of the energy second-derivatives matrix. Intrinsic reaction coordinate calculations (IRC)[34] were also performed for each located transition state in order to ascertain which minima are connected through each transition state.

Diagonalization of the second-derivatives matrix also provides the vibrational harmonic frequencies which were used to obtain the thermodynamic functions at 1 atm and 298.15 K assuming that the system behaves as an ideal gas and that the rigid rotor and harmonic models hold for the rotational and vibrational degrees of freedom, respectively.[35]

The bulk effect of the solvent was introduced through the isodensity surface-polarized continuum model (IPCM).[36] We used an electronic density of 0.0001 au to define the cavity. The dielectric constant of acetonitrile (ε= 36.2) was used. IPCM calculations were carried out both in the ground and the excited electronic states without reoptimization of the geometries.

All the calculations presented here were done with the Gaussian 98 series of programs.[37]

3 Results and Discussion

The main results of our calculations are displayed in Figures 1 and 2. They show schematically the energies and geometries of all the located stationary points in the ground (S_0) and first excited singlet (S_1) states, respectively.

It is to be noted the great qualitative differences in the energy profiles of both states. In S_0 the process is concerted as only one transition state is located. The IRC confirms that this transition state connects reactants (the base pair) with products (the full tautomer). However it is to be remarked that both H-atom transfers are taking place in quite different stages of the reaction as indicated by the fact that in the transition state one of the protons has been almost transferred whereas the other is in the middle of its "flight". Therefore, the two H-atoms are far from transferring in a synchronic way. In any case, the endothermicity as well as the high energy barrier make the whole process quite improbable in the ground state.

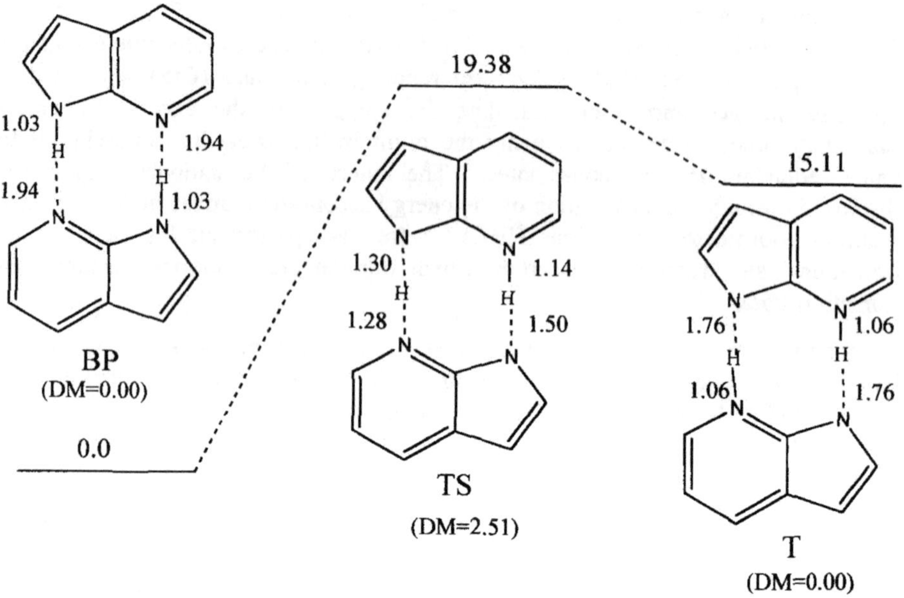

Figure 1. Geometries and relative potential energies (in kcal/mol) for the stationary points corresponding to the double proton-transfer process in the ground electronic state. Distances between the transferring hydrogens and the acceptor/donor nitrogens are given in Å. Dipole moments (DM) are also posted (in debye).

The proton motions and related dynamics and thermodynamics are different for the excited state (Figure 2). To begin with, the whole process is now exothermic and the energy barrier to overcome is also considerably diminished. Comparison of geometries reveals that in the excited state the base pair minimum is not of two identical monomeric units (as in the ground state). This comes from the fact that the electronic excitation is totally localized in one of the 7-AI moieties. In a sense the excited state is a dimer formed by one electronically excited base unit and another molecule still having the ground state character. In a previous paper[38] we showed that this asymmetry enables the formation of stronger hydrogen bonds between both monomers as an important charge redistribution takes place in the excited unit. As a consequence of this asymmetry, the whole double-proton transfer process becomes stepwise in nature as a zwitterionic intermediate INT(Z)' appears. IRC calculations confirm that the reaction takes place following the sequence of stationary points: BP'-TS1'-INT(Z)'-TS2'-T', a process which has to be contrasted with the concerted one obtained for the S_0 state.

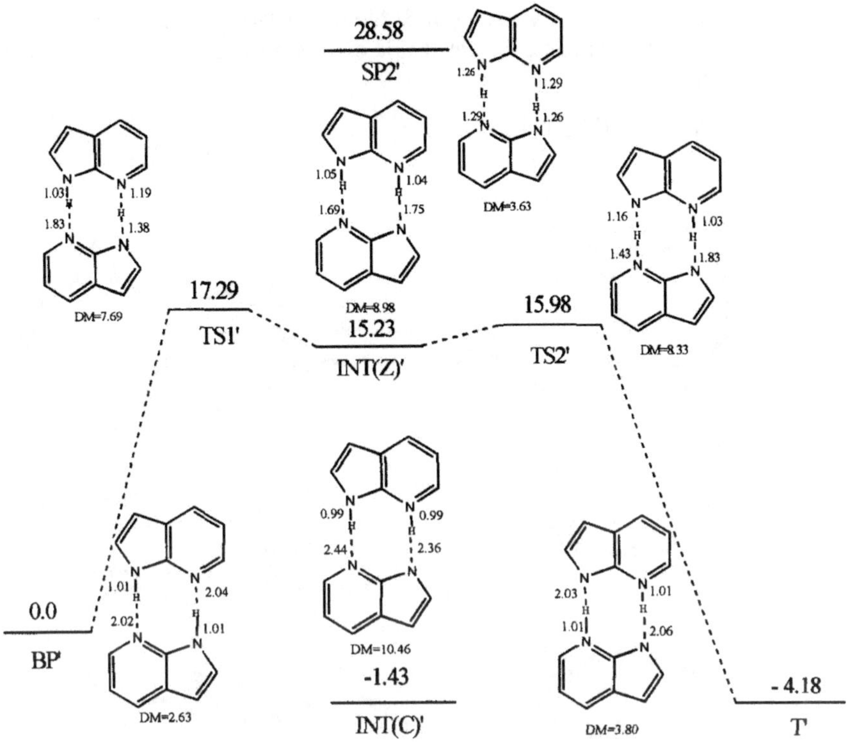

Figure 2. Geometries and relative potential energies (in kcal/mol) for the stationary points corresponding to the double proton-transfer process in the excited electronic state. Distances between the transferring hydrogens and the acceptor/donor nitrogens are given in Å. Dipole moments (DM) are also posted (in debye).

In Figure 2, two additional structures are shown that are not part of this mechanism. One is labeled SP2' and corresponds to the synchronic double-proton transfer. It is clear that this point is higher in energy than the stepwise mechanism. In addition to that, this stationary point is not a transition state as it has two imaginary frequencies so that it is a saddle point of second order. The highest imaginary frequency indicates the double proton transfer whereas the smaller one leads to a breaking of the synchronicity of the two proton motions. Given that the Murrell-Laidler theorem[39] states that when a saddle point of second order is present there is a true transition state with lower energy to be found by following the direction of the eigenvector that has the smaller (in absolute value) imaginary frequency, it is clear that our results indicate that the stepwise mechanism is the only allowed mechanism for the double proton transfer within the pair at S_1. This asymmetry induced by electronic excitation has also been seen in the base pairs of DNA[40]. In that case the localization of the excitation is even more favored as the two monomers are different. In fact, only when the hydrogen bonds (or any other intermolecular force linking the two units) are very strong in S_0 the dimer will probably prefer to act as a unique entity and, in that case, the electronic excitation will not affect the symmetry of the whole dimer.[38]

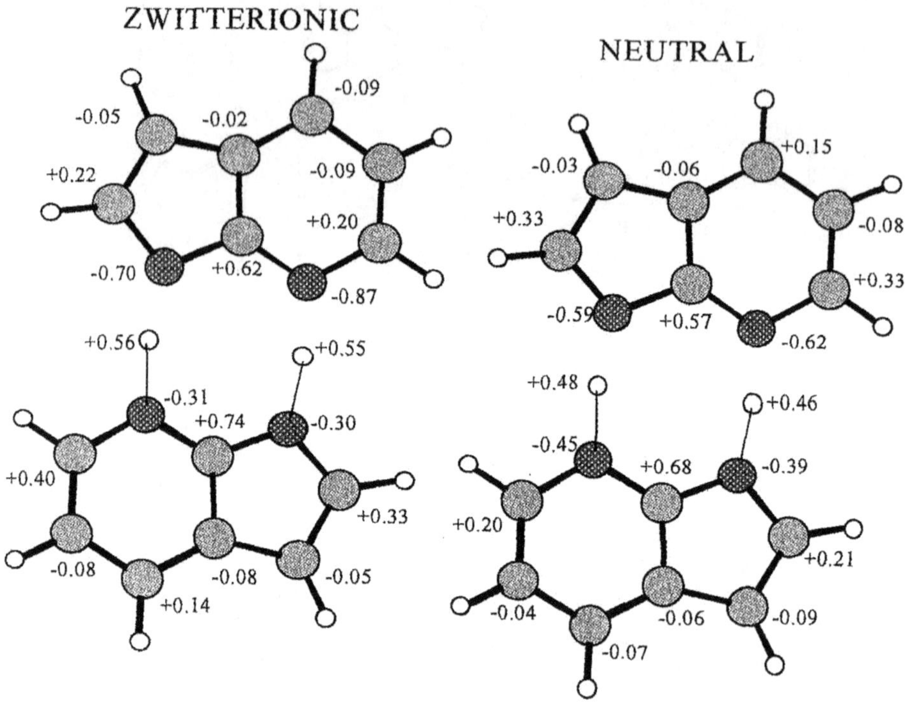

Figure 3. Mulliken atomic charges for the zwitterionic and netutral intermediates. Charges are given in atomic units. The charge of the hydrogen atoms, except the ones involved in the double proton-transfer, has been added to the carbon atom to which it is bonded.

The other structure that is not part of the "regular" mechanism at S_1 is labeled INT(C)' in Figure 2, and has a quite low energy being in fact below both reactant and product structures. This point corresponds also to an intermediate where only one of the H atoms has been transferred as seen in Figure 2. However its geometry is quite different from that of the other intermediate INT(Z)'. Its charge distribution is also quite different as can be seen in Figure 3 where the Mulliken charges for both intermediates are compared. For clarity the charges of hydrogen atoms not implied in the double proton transfer are included in the total charge of the carbon atom to which they are bonded. By adding up the charges corresponding to the atoms of each base unit it can be disclosed that there is no charges separation between both monomers. Thus, this new intermediate is not a zwitterion. This is so because in that structure the positive charge transfer produced by the jump of one proton from one base unit to the other is compensated by an electronic excitation that produces a negative electronic charge in the same direction. In another way, INT(C)' belongs to another excited state PES where the electronic excitation is not localized in one monomer but it involves a charge transfer from one base unit to the other. This electronic state has a quite high energy in the zones corresponding to both reactants and products as the electronic excitation is producing there a quite unstable charge separation. In fact, our calculations of the base pair minimum do not reveal the appearance of any charge-transfer excited state among the first 20 excited singlet states calculated with the Gaussian 98 program. This does not mean that such a state is not present but it puts a lower limit to the energy where it is to be found, 98.63 kcal/mol above the lowest excited state BP'. Therefore, at some point along the proton-transfer coordinates the PES of this charge-transfer state must cross with the S_1 state PES. As the energy of this charge-transfer state raises quickly when going away from the intermediate, the crossing is not probably far away from that zone. Because the different nature of the two excited states, this crossing involves a nonadiabatic transition so that the probability to jump from one state to the other should be relatively small.

Let us finally consider the free energies both in gas and condensed phases. Figure 4 depicts schematically the corresponding energy profiles in S_1. Figure 4(a) shows the results in gas phase. It is noted that by adding up the thermodynamic (thermic and entropic) contributions, as indicated in the previous section, leads to the disappearance of the zwitterionic intermediate that now is the higher point in energy along the regular mechanism in S_1. Thermodynamic corrections do not modify the higher energy of the synchronic path, nor the fact that the neutral intermediate is the most stable structure shown in the Figure. On the other hand, Figure 4(b) depicts the free energy profile when the bulk effect of a polar solvent (acetonitrile) is considered. By comparing this profile with the gas-phase one it is clear that solvent greatly modifies the stability pattern of the stationary points so that the mechanism in condensed phase may be quite different from what it is found in isolated conditions. In general, the solvent stabilizes the full tautomer and reduces the energy barriers. It is also noteworthy that the zwitterionic intermediate is again present as a such though the energy barrier for the second step is quite low. The fact that such a zwitterionic structure is stabilized in a polar solvent was to be expected. Surprising at first sight is the dramatic lowering in energy of the neutral intermediate. The charge distribution of the neutral intermediate depicted in Figure 3 is a first clue to realize that even if there is not a net charge-transfer between both base units, there are important charge separations between different parts of the whole dimer. This can also be seen by examining the dipole moments which are the magnitudes more directly related to the stabilization of the related structures by a polar solvent. The dipole moment (in Debyes) for the zwitterionic intermediate is ~9 and for the neutral one 10.5. This higher value accounts for the higher stabilization of the INT(C)' structure.

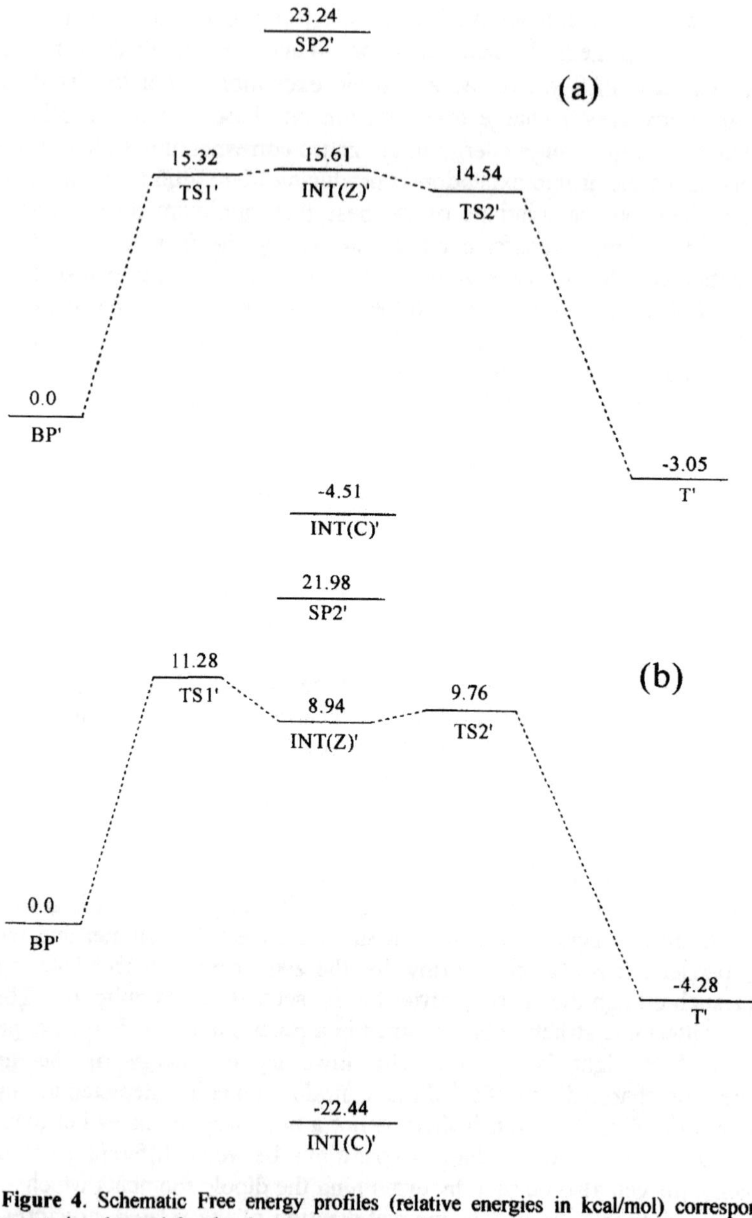

Figure 4. Schematic Free energy profiles (relative energies in kcal/mol) corresponding to the stationary points located for the double proton-transfer process in the excited electronic state. (a) Gas phase results. (b) Condensed phase (acetonitrile) results.

4 Conclusions

The first excited singlet electronic state of 7-azaindole dimer has the excitation located in one base unit. Our analysis has revealed that the mechanism for the double proton transfer in this state is stepwise and involves a zwitterionic intermediate whit only one H-atom transferred. When zero point energy and the rest of thermodynamic corrections are accounted for, the intermediate has a higher energy than the first transition state (TS1').

Our results also point to the relevance of a different excited state that results from an excitation between one base unit and the other one of the dimer. Such a charge-transfer state is very high in energy near the reactants and products zone but it is very stable in the region of the single proton transfer as there the negative charge transfer is compensated by the transfer of the proton. This intermediate can only be accessed through a nonadiabatic transition from the initially obtained excited state.

Finally the bulk effect of the solvent has been studied concluding that a polar solvent stabilizes the product and reduces the energy barrier so that solvation will affect the rate constants of the elementary processes, a result that agrees with recent experimental observations.[17,20]

5 Acknowledgements

Financial support from DGES through projects PB98-0915, PB98-0310, PM97-0074 and 1FD1997-1658 is acknowledged. The use of computational facilities of the CESCA and the CEPBA coordinated by the C^4 is also gratefully acknowledged.

References

1. Limbach, H.; Manz, J. (Eds.) Hydrogen Transfer: Experiment and Theory. *Ber. Bunsengens. Phys. Chem.* **1998,** *102*.
2. Agmon, N.; Gutman, M. (Eds.) Proton Solvation and Proton Mobility. *Isr. J. Chem.* **1999**, *39*(3/4).
3. Schowen, R.L. *Angew. Chem. Int. Ed. Engl.* **1997**, *13/14*, 1434, and references therein
4. (a) Taylor, C. A.; El-Bayoumi, M. A.; Kasha, M. *Proc. Natl. Acad. Sci. U.S.A.* **1969**, *63*, 253. (b) Ingham, K. C.; Abu-Elgheit, M.; El-Bayoumi, M. A. *J. Am. Chem. Soc.* **1971**, *93*, 5023. (c) Ingham, K. C.; El-Bayoumi, M. A. *J. Am. Chem. Soc.* **1974**, *96*, 1674.
5. Hetherington III, W. M.; Micheels, R. H.; Eisenthal, K. B. *Chem. Phys. Lett.* **1979**, *66*, 230.
6. Bulska, G.; Grabowska, A.; Pakula, B.; Sepiol, J.; Waluk, J.; Wild, U. P. *J. Lumin.* **1984**, *29*, 65.

7. (a) Tokomura, K.; Watanabe, Y.; Itoh, M. *Chem. Phys. Lett.* **1984**, *111*, 379. (b) Tokomura, K.; Watanabe, Y.; Udagawa, M.; Itoh, M. *J. Am. Chem. Soc.* **1987**, *109*, 1346.
8. (a) Fuke, K.; Yoshiuchi, H.; Kaya, K. *J. Phys. Chem.* **1984**, *88*, 5840. (b) Fuke, K.; Kaya, K. *J. Phys. Chem.* **1989**, *93*, 614.
9. Share, P.; Pereira, M.; Sarisky, M.; Repinec, S.; Hochstrasser, R. M. *J. Luminesc.* **1991**, *48/49*, 204.
10. Chen, Y.; Rich, R. L.; Gai, F.; Petrich, J. W. *J. Phys. Chem.* **1993**, *97*, 1770.
11. Douhal, A.; Kim, S. K.; Zewail, A. H. *Nature* **1995**, *378*, 260.
12. Douhal, A.; Guallar, V.; Moreno, M.; Lluch, J. M. *Chem. Phys. Lett.* **1996**, *256*, 370.
13. Nakajima, A.; Hirano, M.; Hasumi, R.; Kaya, K.; Watanabe, H.; Carter, C. C.; Williamson, J. M.; Miller, T. A. *J. Phys. Chem. A* **1997**, *101*, 392.
14. Lopez-Martens, R.; Long, P.; Sogaldi, D.; Soep, B.; Syage, J.; Millie, Ph. *Chem. Phys. Lett.* **1997**, *273*, 219.
15. (a) Takeuchi, S.; Tahara, T. *Chem. Phys. Lett.* **1997**, *227*, 340. (b) Takeuchi, S. Tahara, T. *J. Phys. Chem. A* **1998**, *102*, 7740.
16. (a) Folmer, D. E.; Poth, L.; Wisniewski, E. S.; Castleman Jr., A. W. *Chem. Phys. Lett.* **1998**, *287*, 1. (b) Folmer, D. E.; Wisniewski, E. S.; Castleman Jr., A. W. *Chem. Phys. Lett.* **2000**, *318*, 637. (c) Wisniewski, E. S.; Stairs, J. R.; Folmer, D. E.; and Castleman Jr., A. W. In Femtochemistry, De Schryver, F. C.; DE Feyter, S. and Schweitzer, G; (Edits.), Wiley-VCH, **2001**, 113, and references therein.
17. Chachisvilis, M.; Fiebig, T.; Douhal, A.; Zewail, A. H. *J. Phys. Chem. A* **1998**, *102*, 669.
18. Mente, S.; Maroncelli, M. *J. Phys. Chem. A* **1998**, *102*, 3860.
19. Guallar, V.; Batista, V.; Miller, W. H. *J. Chem. Phys.* **1999**, *110*, 9922.
20. Fiebig, T.; Chachisvilis, M.; Manger, M.; Zewail, A. H.; Douhal, A.; Garcia-Ochoa, I.; De la Hoz Ayuso, A. *J. Phys. Chem. A* **1999**, *103*, 7419.
21. (a) Catalán, J.; del Valle, J. C.; Kasha, M. *Proc. Natl. Acad. Sci. (USA)* **1999**, *96*, 8338. (b) Catalán, J.; del Valle, J. C.; Kasha, M. *Chem. Phys. Lett.* **2000**, *318*, 629. (c) Del Valle, J. C.; Kasha, M.; Catalán, J. *Int. J. Quantum Chem.* **2000**, *77*, 118.
22. (a) Douhal, A.; Moreno, M.; Lluch, J. M. *Chem. Phys. Lett.* **2000**, *324*, 75. (b) Douhal, A.; Moreno, M.; Lluch, J. M. *Chem. Phys. Lett.* **2000**, *324*, 81.
23. Moreno, M.; Douhal, A.; Lluch, J.M.; Castaño, O.; Frutos, L.M. *J. Phys. Chem. A* **2001**, *105*, 3887
24. Foresman, J. B. Head-Gordon, M.; Pople, J. A.; Frisch, M. J. *J. Phys. Chem.* **1992**, *96*, 135.
25. (a) Hegarty, D.; Robb, M.A. *Mol. Phys.* **1979**, 38, 1795. (b) Eade, R.H.E.; Robb, M.A.; *Chem. Phys. Lett.* **1981**, *83, 362.* (c) Schlegel, H.B.; Robb, M.A. *Chem. Phys. Lett.* **1982**, *93,* 43.

26. Head-Gordon, M.; Rico, R. J.; Oumi, M.; Lee, T. J. *Chem. Phys. Lett.* **1994,** *219*, 21.
27. Andersson, K.; Malmqvist, P.-Å.; Roos, B.O.; Sadlej, A.J.; Wolinski, K. *J. Phys. Chem.* **1990,** *94*, 5483. (b) Andersson, K.; Malmqvist, P.-Å.; Roos, B.O *J. Chem. P.***1992**, *96*, 1218.
28. Parr, R.G., Yang, W. *Density Functional Theory of Atoms and Molecules*, Oxford Universuty Press, Oxford, 1989.
29. Stratmann, R.E.; Scuseria, G.E.; Frisch, M.J. *J. Chem. Phys.* **1998**, *109*, 8218.
30. (a) Paz, J. J.; Moreno, M.; Lluch, J. M. *J. Chem. Phys.* **1995**, *103*, 353. (b) Scheiner, S. *J. Phys. Chem. A* **2000**, *104*, 5898.
31. Francl, M. M.; Pietro, W. J.; Hehre, W. J.; Binkley, J. S.; Gordon, M. S.; DeFrees, D. J.; Pople, J. A. *J. Chem. Phys.* **1982**, *77*, 3654.
32. (a) Lee, C.; Yang, W.; Parr, R. G. *Phys. Rev. B* **1988**, *37*, 785. (b) Becke, A. D. *J. Chem. Phys.* **1993**, *98*, 5648.
33. Peng, C.; Ayala, P. Y.; Schlegel, H. B.; Frisch, M. J. *J. Comput. Chem.* **1996**, *17*, 49.
34. (a) Gonzalez, C.; Schlegel, H. B. *J. Chem. Phys.* **1989**, *90*, 2154. (b) Gonzalez, C.; Schlegel, H. B. *J. Phys. Chem* **1990**, *94*, 5523.
35. McQuarrie, D. A. *Statsitical Thermodynamics*; University Science Books: Mill Valley, CA, 1973
36. Foresman, J. B.; Keit, T. A.; Wiberg, K. B.; Snoonian, J.; Frisch, M. J. *J. Phys. Chem.* **1996**, *100*, 16098.
37. Frisch, M. J.; Trucks, G. W.; Schlegel, H. B.; Scuseria, G. E.; Robb, M. A.; Cheeseman, J. R.; Zakrzewski, V. G.; Montgomery, J. A.; Stratmann, R. E., Burant, J. C.; Dapprich, S., Millam, J. M.; Daniels, A. D.; Kudin, K. N.; Strain, M. C.; Farkas, O.; Tomasi, J.; Barone, V.; Cossi, M.; Cammi, R.; Mennuci, B.; Pomelli, C.; Adamo, C.; Clifford, S.; Ochterski, J.; Petersson, G. A.; Ayala, P. Y.; Cui, Q.; Morokuma, K.; Malick, D. K.; Rabuck, A. D.; Raghavachari, K.; Foresman, J. B.; Cioslowski, J.; Ortiz, J. V.; Stefanov, B. B.; Liu, G.; Liashenko, A.; Piskorz, P.; Komaromi, I.; Gomperts, R.; Martin, R. L.; Fox, D. J ; Keith, T.; Al-Laham, M. A.; Peng, C. Y.;. Nanayakkara, A.; Gonzalez, C.; Challacombe, M.; Gill, P. M .W.; Johnson, B. G.; Chen, W.; Wong, M. W.; Andrés, J. L.; Head-Gordon, M.; Replogle, E. S.; Pople, J. A. *Gaussian 98*; Gaussian Inc.: Pittsburgh, PA, 1998.
38. Guallar, V.; Moreno, M.; Lluch, J. M. *Chem. Phys.* **1998**, *228*, 1.
39. Murrell, J. N.; Laidler, K. J. *Trans. Faraday Soc.* **1968**, *64*, 371.
40. Guallar, V.; Douhal, A.; Moreno, M.; Lluch, J.M. *J. Phys Chem. A* **1999**, *103*, 6251.

COHERENT DEACTIVATION PROCESS OF PSEUDOAZURIN

YUTAKA NAGASAWA, SATORU NAKASHIMA, CHIHIRO EGAMI, TADASHI OKADA

Department of Chemistry, Graduate School of Engineering Science and Research Center for Materials Science at Extreme Conditions, Osaka University, 1-3 Toyonaka, Osaka 560-8531, Japan
E-mail: nagasawa@chem.es.osaka-u.ac.jp

TAKAMITU KOHZUMA, RAN MATUKURA, AND SACHIKO YANAGISAWA

Faculty of Science, Ibaragi University, Mito, Ibaragi, 310-8512, Japan
E-mail: kohzuma@mito.ipc.ibaraki.ac.jp

Ultrafast deactivation process of a blue copper protein, pseudoazurin (Paz), was measured by means of pump-probe (PP) spectroscopy. The deactivation process was as fast as 250 fs and a low frequency mode at 30 cm^{-1} was observed. The deactivation process did not slow down even at 30 K indicating that the process is ballistic rather than diffusive, which also indicates that such an ultrafast process can cause coherent nuclear motion. Cooling process on the ground state was also observed with a time constant of 1.6 ps.

1 Introduction

Electron transfer (ET) in proteins is one of the vital molecular processes of organic life [1]. Blue copper proteins (BCP) function as mobile electron carriers in a wide range of biological systems. BCPs are sometimes a very complex enzyme with more than two copper ions involved. Here, we consider the ones with simple structure with only one copper ion, such as plastocyanin, azurin, and pseudoazurin. The protein forms a β-barrel-type structure with a activation site, which consists of a distorted tetrahedral structure with a copper (II) ion in the center, i.e. the type I Cu(II) site. The protein backbone is sufficiently rigid to hold the instable structure of the activation site, which results in an unusual strong absorption around 600 nm. The extinction coefficient of the type I Cu(II) site at ~600 nm is about 5000 $M^{-1}cm^{-1}$ which is much stronger than that of the usual simple copper compound, i.e. 5-10 $M^{-1}cm^{-1}$. This strong absorption, which is the origin of the blue color of BCPs, is considered to be the result of ligand-to-metal charge transfer (LMCT), which occurs from the sulfur atom of the ligating cysteine to the central Cu(II) ion. Vast numbers of resonant Raman scattering experiments have been carried out to understand the structural reorganization involved in this unusual electronic transition [2], on the other hand, only a few ultrafast experiments have been carried out to understand the dynamical aspects of the LMCT [3-6]. Only three sub-picosecond studies have been reported for the deactivation process of BCPs after direct excitation in the LMCT band [3-5]. Two of these studies concluded that the deactivation process to the ground state is indirect involving an intermediate d-d ligand-field excited state [3, 4]. The third report on plastocyanin by our own group indicates that the ultrafast deactivation process occurs through a coherent nuclear motion of a low frequency mode at 33 cm^{-1}, which can be the skeletal motion of the protein [5]. However, this conclusion was met by strong objections. Book et al. have measured ultrafast deactivation processes of some BCPs with excitation in the d-d transition region around 800 nm and observed a similar low frequency modes in the range of 35-55 cm^{-1} [6]. It was also tempting for them to assign these low frequency modes to the protein phonon modes, however, they concluded that such assignment is premature and considered other possibilities such as difference frequencies and artifacts.

Here we report a study on ultrafast deactivation process of another simple BCP, pseudoazurin (Paz), by means of pump-probe (PP) spectroscopy. The deactivation process was as fast as 250 fs and we have also observed a low frequency mode at 30 cm^{-1}. The major result of the experiments is that the deactivation process did not slow down even at 30 K indicating that the process is ballistic rather than diffusive, which also indicates that such process can cause coherent nuclear motion.

2 Experimental Section

The details of the homemade cavity-dumped Kerr lens mode-locked chromium-doped forsterite laser and pump-probe (PP) measurement system were reported elsewhere [7]. The difference from the previous report is that the laser was pumped by Spectra Physics Millenia IR and the pulse duration became slightly shorter. PP measurements were carried out using the second harmonic pulse of the laser, which was centered at 635 nm with a pulse duration of 26 fs and a repetition rate of 100 kHz. To our knowledge, this is the shortest pulse ever generated from a cavity-dumped Cr: forsterite laser. The PP signal was detected by a photodiode and a lock-in amplifier with and without a monochrometer in front of the detector. *Achromobacter cycloclastes* pseudoazurin (Paz) was expressed in *E. coli*. The optical density of the sample was about 0.6 in a 1 mm thick cell for room temperature experiments and was 0.3 for low temperature experiments. For room temperature measurements, the sample cell was rotated at 2000 rpm. Closed-cycle cryostat was use for the low temperature experiment. The phosphate buffer sample solution was mixed with glycerol (50 % in volume) in order to form glass at low temperature. Convolution fitting was carried out for the room temperature PP signal with several different fitting functions. The best fit of the decay was achieved with a combination of an exponential decay and a critically damped low frequency oscillation (Fig. 1 (b)).

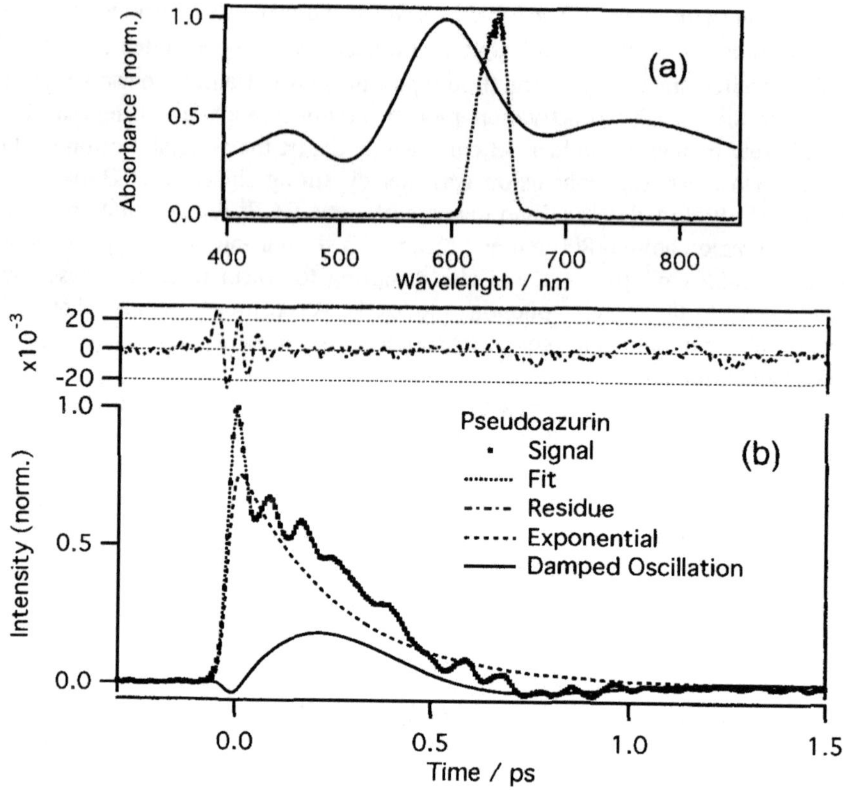

Figure 1. (a) Absorption spectrum of Paz and the spectrum of the second harmonic pulse of Cr; forsterite laser. (b) The PP signal of Paz and the results of the convolution fitting.

3 Results and Discussion

The absorption spectrum of Paz is shown in Fig. 1 (a) with the laser spectrum. It can be seen that the laser pulse excites the red side of the LMCT band around 600 nm. The measured PP signal is shown in Fig. 1 (b) with the results of the convolution fitting analysis. The impulsive stimulated Raman scattering causes the small amplitude high frequency modulation of the signal. The PP signal of Paz is very similar to the previously reported signal of plastocyanin [5]. The residue shows that the fitting was quite successful with the chosen fitting function. The exponential decay component had a time constant of 250 fs and the frequency, dephasing time, and phase of the low frequency oscillation was 30 cm^{-1}, 340 fs, and

100 degrees, respectively. The intensity of the oscillation was about half of that of the decay component. Despite the good result of the fitting, we would like to mention that there could be a much appropriate nonexponential fitting function, such as delayed oscillation. However, we have used the phase-shifted critically damped oscillation for the time being, because we have not developed any theoretical model, which supports a particular decay function.

The non-exponential feature of the signal becomes more apparent when the signal intensity is logarithmically plotted as in Fig. 2 (a). After the initial spike, the signal forms a plateau-like feature, which was previously understood as the result of the crossing to the intermediate d-d ligand field state [4]. However, existence of an intermediate state cannot explain the sudden drop of the signal intensity around 0.4-0.7 fs. This type of feature is sometimes an artifact arising from excess subtraction of the background. However, it is clear from Fig. 2 (a) that this is not the case, because for this figure, background is not completely subtracted. When the exponential component is subtracted from the signal, oscillatory feature clearly appears as seen in Fig. 2 (b). From these results, we concluded that the low frequency oscillation could hardly be an artifact.

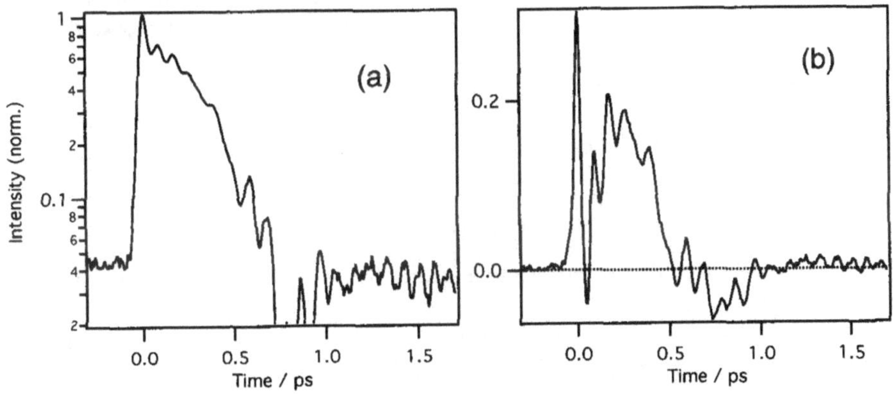

Figure 2. (a) Logarithmic plot of the PP signal of Paz at room temperature. The background is not completely subtracted. (b) The PP signal of Paz with the low frequency mode subtracted.

We believe that the intensity of the low frequency mode is too strong for a difference frequency of the higher frequency modes. The intensity of the difference frequency should not exceed the original fundamental modes. In our previous work, we concluded that the low frequency mode is not just an oscillation caused by impulsive stimulated resonance Raman scattering but a coherent nuclear motion caused by the back-LMCT process [5]. This conclusion was reached by the fact that the mode had a large phase shift, i.e. ~120 degrees, compared to those of the other Raman modes. Our new result on temperature dependent experiment suggests that such a process is possible, because the deactivation is ballistic rather than diffusive.

Fig. 3 (a) shows that the decay of PP signal did not slow down at lower temperatures. This is a strong indication that the process is not diffusive which should slow down with temperature.

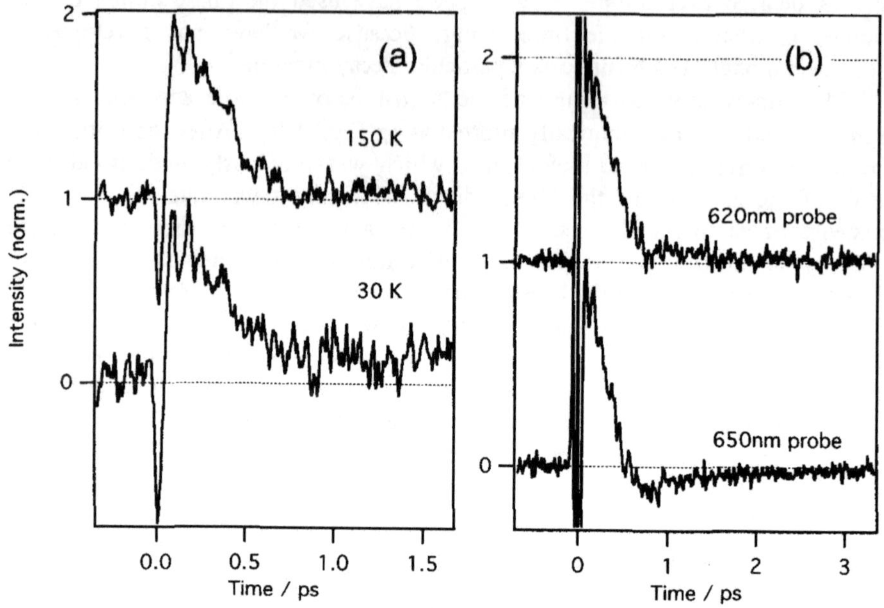

Figure 3. (a) PP signals of Paz at 150 K and 30 K in 50 % glycerol solution. (b) The wavelength resolved PP signal of Paz measured with a monochrometer at room temperature.

If the recovery of the ground state is ultrafast and ballistic, what happens to the excess energy? There must be a very effective way of energy dissipation into the heat bath. When a monochrometer was set in front of the detector and the probe beam was wavelength resolved, negative signal appeared on the red edge as seen in Fig. 3 (b). Since transmittance is monitored in our experiment, negative signal means increase of absorption. We attributed this transient absorption to the vibrationally hot ground state, because hot band only appears on the red edge of the ground state absorption. The transient absorption decayed with a lifetime of 1.6 ps, which can be attributed to the cooling process in the ground state. We would like to note that this negative signal is not the origin of the nonexponetial decay. The oscillatory feature still remained even after the negative component was subtracted from the signal.

The final question to be answered is the origin of the low frequency mode. In our previous report, we assigned it to a delocalized protein motion. This assignment seemed reasonable because the active site of Paz is directly connected to the residues of the protein backbone. However, assignment to some specific motion is still premature. Simply thinking, it might be a wagging motion of some portion of the protein or a breathing mode of the entire protein. Similar low frequency modes were observed in the nonresonant Raman spectra of lysozyme [8] and α-chymotrypsin [9] and in neutron scattering of a low temperature myoglobin [10]. There are also theoretical studies that α-helix, β-sheet, or β-barrel can exhibit such an oscillation [11, 12]. In liquid studies, such a low frequency mode is considered to be the librational motion of the liquid molecules [13]. Similar libration can be expected for proteins and because of the rigid nature of BCPs, such motion may exhibit stronger oscillatory feature compared to that of the fluid liquid molecules. Liberations or low frequency modes, such as Boson peak, were also commonly observed among artificial polymers, especially at low temperatures [14]. It is temping to consider the possibility of the low frequency mode to perform an important role in the physiological electron transfer. However, such a consideration is beyond the scope of our present experiment, because photoexcitation does not play any role in the actual electron transfer, and secondly, the LMCT excitation induces only 20 % of the charge to redistribute. There is no doubt that the static function of the protein is to retain the unstable distorted tetrahedral structure of the active site, however, the dynamical function is still unclear.

As a conclusion, we have observed a nonexponential ultrafast deactivation process of a blue copper protein, pseudoazurin, which can be reasonably reproduced by a fitting function constructed by a combination of a exponential decay with a time constant of 250 fs and a oscillation with a frequency of 30 cm^{-1}, a dephasing time of 340 fs, and a phase shift of 100 degrees. The deactivation process did not slow down even at 30 K indicating that the process is ballistic rather than diffusive, which also indicates that such an ultrafast process can cause coherent nuclear motion. Dissipation process of the excess energy in the ground state was also observed with a time constant of 1.6 ps. Specific assignment of the low frequency mode and whether such a mode can play an important role in the physiological electron transfer is the subject of our future work.

4 Acknowledgements

Y. N., S. N., and T. O. were supported by Grants-in-Aid for Scientific Research on Specially Promoted Research from the Ministry of Education, Science, Sports, and Culture, Japan. S. N. was also supported by Novartis Foundation (Japan) for the promotion of science. Y. N. was also supported by Sumitomo Foundation. T. K. was supported by Grants-in-Aid for Scientific Research on Priority Areas from the Ministry of Education, Science, Sports, and Culture, Japan, and by the Grants-in-Aid for the Ground Experiments for the Space Utilization from the Japan Space Forum and National Space Development Agency of Japan.

References

1. Lippard S. J. and Berg J. M., *Principles of Bioinorganic Chemistry* (University Science Books, Herndon, 1994).
2. Woodruff W. H., Dyer R. B. and Schoonover J. R. *Biological Applications of Raman Spectroscopy* **3** (Wiley, New York, 1988) pp. 413-438.
3. Wiesenfeld J. M., Ippen E. P., Corin A. and Bersohn R., Electronic relaxation in azurin: picosecond reverse charge transfer. *J. Am. Chem. Soc.* **102** (1980) pp. 7256-7258.
4. Edington M. C., Diffey W. M., Doria W. J., Riter R. E. and Beck W. F., Radiationless decay from the ligand-to-metal charge-transfer state in the blue copper protein plastocyanin, *Chem. Phys. Lett.* **275** (1997) pp. 119-126.
5. Nakashima S, Nagasawa Y, Seike K., Okada T., Sato M. and Kohzuma T., Coherent dynamics in ultrafast charge-transfer reaction of plastocyanin. *Chem. Phys. Lett.***331** (2000) pp. 396-402.
6. Book L. D, Arnett D. C., Hu H. and Scherer N. F., Ultrafast pump-probe studies of excited-state charge-transfer dynamics in blue copper proteins. *J. Phys. Chem.* **102** (1998) pp. 4350-4359.
7. Nagasawa Y., Ando Y., Watanabe A. and Okada T., Generation of 30 fs pulses at 635 nm by frequency doubling of cavity-dumped chromium-doped forsterite laser, and its application to spectroscopy. *Appl. Phys. B* **70[Suppl.]** (2000) pp. S33–S36.
8. Genzel L., Keilmann F., Martin T. P., Winterling G., Yacoby Y., Fröhlich H. and Makinen M. W., Low-frequency Raman spectra of lysozyme. *Biopolymers* **15** (1976) pp. 219-225.
9. Brown K. G., Erfurth S. C., Small E. W. and Peticolas W. L., Conformationally dependent low-frequency motions of proteins by laser Raman spectroscopy. *Proc. Nat. Acad. Sci. USA* **69** (1972) pp. 1467-1469.
10. Cusack S. and Doster W., Temperature dependence of the low frequency dynamics of myoglobin. *Biophys. J.* **58** (1990) pp. 243-251.
11. Chou K.-C., Biological functions of low-frequency vibrations (phonons). *Biophys. J.***45** (1984) pp. 881-890.
12. Chou K.-C., Low-frequency motions in protein molecules. *Biophys. J.***48** (1985) pp. 289-297.
13. Kinoshita S., Kai Y., Ariyoshi T. and Shimada Y., Low frequency modes probed by time-domain optical Kerr effect spectroscopy. *Int. J. Mod. Phys. B* **10** (1996) pp. 1229-1272.
14. Frick B. and Richter D., The microscopic basis of the glass transition in polymers from neutron scattering studies. *Science* **267** (1995) pp. 1939-1945.

THREE-PULSE TWO-COLOUR PHOTON ECHO AND TRANSIENT GRATING STUDIES OF MYOGLOBIN

C. N. LINCOLN, L. V. DAO, R. M. LOWE, W. J. ROWLANDS AND P. HANNAFORD

Centre for Atom Optics and Ultrafast Spectroscopy, BSEE, Swinburne University of Technology, PO Box 218, Hawthorn, Australia 3122

E-mail: clincoln@swin.edu.au

Coherent transient spectroscopic techniques have been applied to the model systems, carbonmonoxy myoglobin (MbCO) and myoglobin (Mb), to study the population and vibrational dynamics during photodissociation. Both transient grating (TG) and three-pulse photon echo peak shift (3PEPS) measurements using a two-colour configuration, are reported. TG results agree well with the newly proposed mechanism for the photodissociation of MbCO [6]: $^1Q \rightarrow Mb_I^* \rightarrow Mb_{II}^* \rightarrow Mb$ with measured lifetimes ≤50 fs, ~250 fs and ~2 ps for ^1Q, Mb_I^* and Mb_{II}^*, respectively. 3PEPS results show peak shifts indicative of inhomogeneous broadening.

1 Introduction

The use of nonlinear coherence spectroscopic techniques such as photon echoes and transient gratings induced by femtosecond laser pulses provides a powerful method for investigating ultrafast processes, such as electronic and energy transfer, in biological molecules on timescales down to of order 10^{-13} second [1-5]. We report here the use of three-pulse two-colour photon echoes and transient grating techniques to investigate ultrafast transient processes that occur during the photodissociation of carbonmonoxy myoglobin (MbCO). Myoglobin (Mb) is the single-globin analogue of the more complex hemoglobin and is responsible for the storage of oxygen in animals and plants.

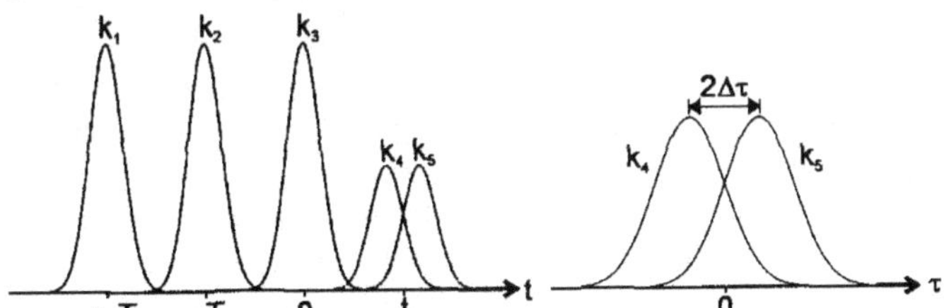

Figure 1. Pulse sequence for TG and 3PEPS experiments showing population time T and coherence time τ separations of applied pulses $\mathbf{k}_{1,2,3}$ and evolution of the echo signal pulses $\mathbf{k}_{4,5}$. Note for transient grating experiments coherence time τ is set to zero.

In a three-pulse photon echo experiment, two pump pulses and a third (probe) pulse with wavevectors $\mathbf{k}_1$, $\mathbf{k}_2$ and $\mathbf{k}_3$ impinge on the sample to generate a nonlinear macroscopic polarisation, and phase-matching conditions require the generated electric field to be created in two unique directions $\mathbf{k}_{4,5} = \mathbf{k}_3 \pm (\mathbf{k}_1 - \mathbf{k}_2)$. The photon echo signal may be recorded by scanning against coherence time τ for a range of fixed population times T (Fig. 1). The additional degree of freedom allowed by the use of three pulses of variable time separations enables one to extract both the optical dipole dephasing time T_2 and the population relaxation time T_1. In the special case of coincident pump pulses $\mathbf{k}_1$ and $\mathbf{k}_2$ ($\tau = 0$), the measurement becomes a transient grating experiment, which provides information on the population dynamics and laser-induced coherences between, for example, vibrational levels of the electronically excited molecule. Measurement of the shift of the peak position ($\Delta\tau$) of the two echo signals $\mathbf{k}_4$ and $\mathbf{k}_5$ for fixed values of T (which we refer to as the three-pulse photon echo peak shift - 3PEPS) provides information on the inhomogeneous broadening due to coupling with the solvent bath. The use of a probe pulse k_3 of different wavelength to that of $\mathbf{k}_1$ and $\mathbf{k}_2$, to create a two-colour three-pulse photon echo experiment, allows one to systematically probe the population and coherences between electronic and/or vibrational levels which the originally excited molecule relaxes to during photodissociation.

In our three-pulse two-colour photon echo and transient grating experiments, we use ultrashort (~70 fs) pulses from two independently tunable optical parametric amplifiers pumped by an ultrafast titanium sapphire regenerative amplifier. The pump pulses k_1 and k_2 are centred at 550 nm to examine the Q-band region of myoglobin and the probe pulse k_3 is tuned in the region of 625 nm.

2 Results and Discussion

2.1 *Transient Grating (TG)*

The TG signal for MbCO and Mb are shown in Fig. 2. A two-exponential fit to the Mb curve gives decay times of ~250 fs and ~2 ps, whereas a three-exponential fit to the MbCO signal results in $\leq$50 fs, ~500 fs and ~12 ps. Also a difference curve was generated for MbCO-Mb and compares well with transient absorption curves from a pump-probe experiment on MbCO. The difference curve then has time constants of $\leq$50 fs, ~250 fs and >>2 ps.

Recently a new mechanism for the photodissociation of MbCO was proposed [6]. The scheme is a sequential decay from $^1Q \rightarrow Hb_I^* \rightarrow Hb_{II}^* \rightarrow$Hb ground state. The lifetimes for the three states ^1Q, Hb_I^* and Hb_{II}^* were determined to be $\leq$50 fs, ~300 fs and ~2 ps for both Hb and HbCO. If it is assumed Mb dynamics are analogues to hemoglobin (Hb) dynamics then the two decays in the TG signal for Mb can be assigned as $Mb_I^* \rightarrow Mb_{II}^*$ (~250 fs) and $Mb_{II}^* \rightarrow Mb$ (~2 ps) lifetimes. Also once the absorption changes for the dissociation MbCO to Mb are removed the TG signal for MbCO also contains the same lifetimes ~250 fs and ~2 ps. The resultant lifetimes in the difference curve for MbCO can be assigned to $^1Q \rightarrow Mb_I^*$ ($\leq$50 fs), $Mb_I^* \rightarrow Mb$ or $Mb_I^* \rightarrow MbCO$ (~250 fs) and finally association $Mb \rightarrow MbCO$ (>>2 ps).

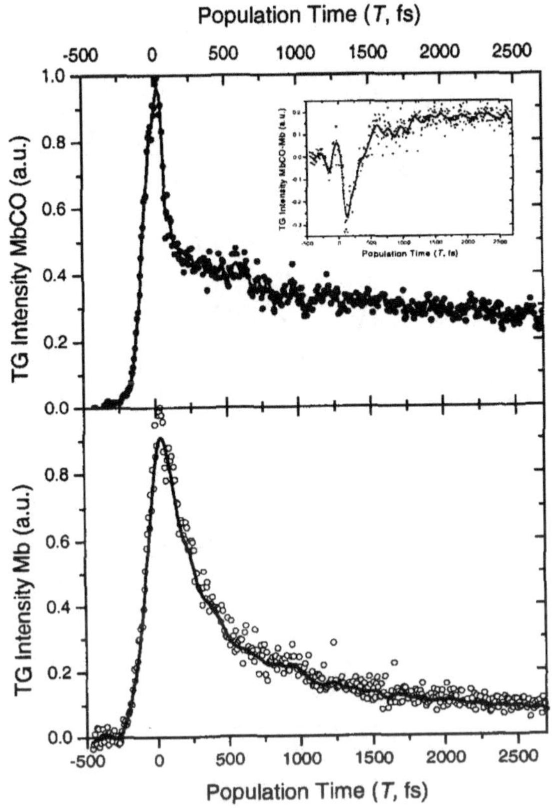

Figure 2. a) Transient grating data MbCO (closed circles) and difference curve for MbCO-Mb TG signals (inset). b) Transient grating data Mb (open circles). Smoothed experimental data (solid curves).

2.2 *Three Pulse Photon Echo Peak Shift (3PEPS)*

Preliminary 3PEPS results for MbCO and Mb (Fig. 3) show strong peak shifts, characteristic of strong inhomogeneous broadening. The peak shift at $T = 0$ fs is near zero and the period for the oscillation of $\Delta\tau$ is ~350 fs for MbCO and Mb.

Typically a one-colour 3PEPS has two distinct components, a large fast decaying peak shift due mainly to intermolecular inhomogeneous broadening and a weaker, slow decaying oscillation of the peak shift due mainly to intramolecular inhomogeneous broadening [7,8]. The absence of the fast decaying peak shift in the two-colour 3PEPS data (Fig. 3) suggests weak coupling between the two vibrational levels probed by the two wavelengths [9].

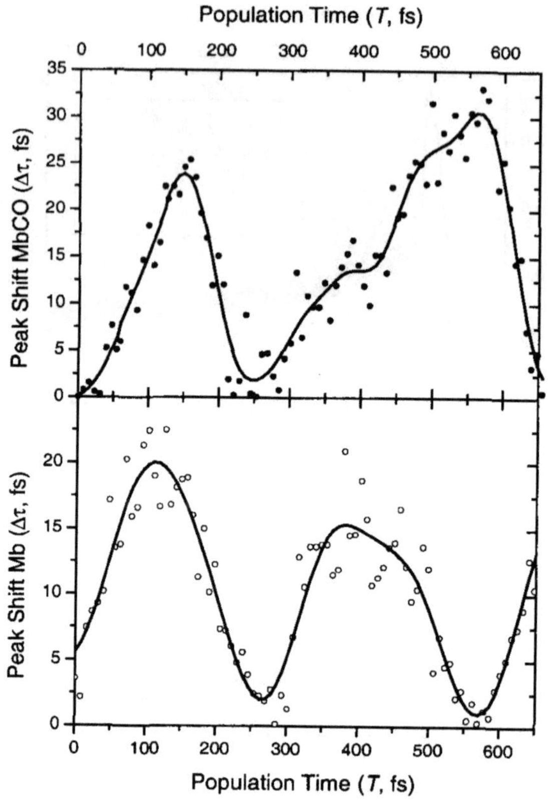

Figure 3. a) 3PEPS data MbCO (closed circles). b) 3PEPS data Mb (open circles). Smoothed experimental data (solid curves)

3 Summary

The population decay times for the TG experiment were found to agree well with the new model proposed for photodissociation of myoglobin. Unfortunately the incomplete 3PEPS data makes it difficult to describe any inhomogeneous broadening characteristics of the MbCO and Mb systems. However it was apparent that there was weak or no coupling between the two vibrational levels probed at 550 nm and 625 nm for MbCO and Mb.

References

1. Xu Q-H., Scholes G. D., Yang M. and Fleming G. R., Probing Solvation Dynamics and Reaction Coordinates of Ultrafast Photoinduced Electron-Transfer Reactions Using Nonlinear Spectroscopies: Rhodamine 6G in Electron-Donating Solvents. *J. Phys Chem. A* **103** (1999) pp. 10348-10358.
2. Jordanides X. J., Lang M. J., Song X. and Fleming G. R, Solvation Dynamics in Protein Environments Studied by Photon Echo Spectroscopy. *J. Phys. Chem. B* **103** (1999) pp. 7995-8005.
3. Sakakura M., Yamaguchi S., Hirota N. and Terazima M., Dynamics of Structure and Energy of Horse Carboxymyoglobin after Photodissociation of Carbon Monoxide. *J. Am. Chem. Soc.* **123** (2001) pp. 4286-4294.
4. Berg M. A., Rector K. D. and Fayer M. D., Two-pulse echo experiments in the spectral diffusion regime. *J. Chem. Phys.* **113** (2000) pp. 3233-3242.
5. Leeson D. T., Wiersma D. A., Fritsch K. and Friedrich J., The Energy Landscape of Myoglobin: An Optical Study. *J. Phys. Chem. B.* **101** (1997) pp. 6331-6340.
6. Franzen S., Kiger L., Poyart C. and Martin J-L., Heme Photolysis Occurs by Ultrafast Excited State Metal-to-Ring Charge Transfer. *Biophys. J.* **80** (2001) pp. 2372-2385.
7. Larsen D. S., Ohta K., Xu Q-H., Cyrier M. and Fleming G. R., Influence of intramolecular vibrations in third-order, time-domain resonant spectroscopies. I. Experiments. *J. Chem. Phys.* **114** (2001) pp. 8008-8019.
8. Ohta K., Larsen D. S., Yang M. and Fleming G. R., Influence of intramolecular vibrations in third-order, time-domain resonant spectroscopies. II. Numerical calculations. *J. Chem. Phys.* **114** (2001) pp. 8020-8039.
9. Yang M. and Fleming G. R., Two-colour three-pulse photon echoes as a probe of electronic coupling in molecular complexes. *J. Chem. Phys.* **110** (1999) pp. 2983-2990.

MICROSCOPIC THEORY OF EXCITON ANNIHILATION: APPLICATION TO THE LH2 ANTENNA SYSTEM

VOLKHARD MAY AND BEN BRÜGGEMANN

Institute of Physics, Humboldt–University at Berlin, Hausvogteiplatz 5-7, D-10117 Berlin, Germany,
e-mail: may@physik.hu-berlin.de

To get insight into the excited–state structure of chromophore complexes forming biological antenna systems such as the LH2, pump–probe experiments including a variation of the pump–pulse intensity are analyzed. This is achieved by a generalization of the multi–exciton density matrix theory to include exciton–exciton annihilation (EEA) which proceeds at higher pump–pulse intensities. The simulated kinetics fit the measured transient absorption data very well. Details of the EEA are visualized in drawing the spatial redistribution of the excitation energy among the chromophores of the LH2 complex.

1 Introduction

Antenna systems are a part of photosynthetically active membranes of bacteria and higher plants which absorb sun light and transfer the respective excitation energy via a Frenkel exciton mechanism to the photosynthetic reaction center[1]. Because of their outstanding importance antenna systems are a subject of an ongoing and intensive research, not only concentrating on those processes taking place under physiological conditions. For example, it is a common practice to vary the intensity of the exciting laser pulse to study higher excited states of antenna systems and to investigate relaxation channels such as EEA[2]. In the EEA process the energy of two excitons is used to form a higher excited electronic state at a single Bacteriochlorophyll (BChl). Afterwards an internal conversion process de–excites the Bchl to its first excited state, just ending up with the presence of a single exciton only. To achieve a microscopic description of such an EEA process the present note introduces a generalization of the multi-exciton theory[4]. The calculations will concentrate on the B850 ring of the LH2 complex with 18 BChls.

2 Multi-Exciton Density Matrix Theory

The multi-exciton theory[4] is based on a certain ordering scheme of the electronic eigenstates $| \alpha_N >$ which belong to the ensemble of BChl molecules forming, e.g. the LH2. Beside the ground–state ($N = 0$) the scheme covers the various exciton manifolds ordered with respect to an increase of the related

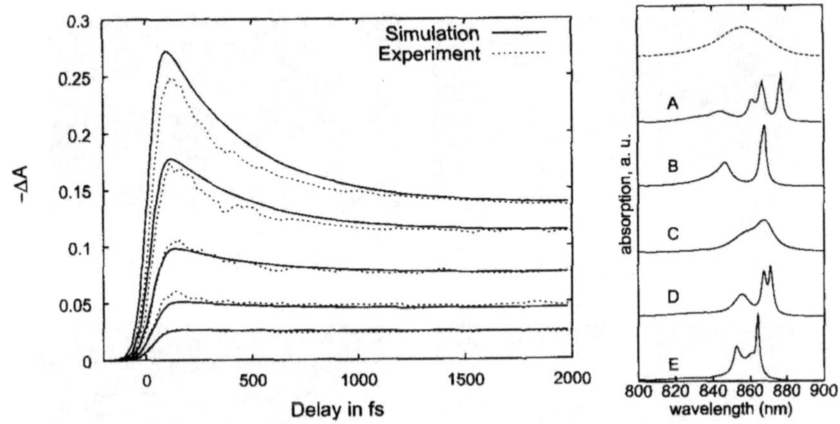

Figure 1. Ensemble and single–molecule absorption spectra of the B850 ring. Left figure: Transient (ensemble) absorption kinetics versus delay time for pump pulses with 100 fs duration and energies of 128, 64, 32, 16 and 8 $\mu J/cm^2$ (from top to bottom). Pump pulse wavelength 850 nm, probe pulse wavelength 860 nm. Experimental data[2] (dotted curves) and results of simulations[3] (solid curves) are shown. Right figure: Single–molecule linear absorption spectra (curves A – E) calculated for such disorder configurations which roughly reproduce the single–molecule fluorescence excitation spectra of Ref.[5]. For comparison, the ensemble average is shown in the upper part.

excitation energy, starting with the single–exciton states ($N = 1$) followed by two–exciton states ($N = 2$), and so on. Optical excitation and detection of these states is considered by introducing a coupling to the various laser pulses used in the experiment, whereas excitation energy dissipation is accounted for by a proper introduction of an exciton–vibrational coupling. Slow conformational transitions within the protein matrix keeping the BChls together are described via static disorder. Finally, the EEA is included by the incorporation of internal conversion within every BChl from a higher excited electronic state to the first excited one[3]. All mechanisms are put together to compute the time–dependence of the multi–exciton density matrix $\rho(\alpha_M, \beta_N; t)$.

3 Results

Neglecting static disorder in a first step, a nice agreement with the measured data[2] can be achieved for all pump–intensities resulting in an inverse of the BChl internal conversion rate of about 70 fs (cf. Fig. 1)[3]. This value decreases somewhat if static disorder is considered. All parameters of the simulation

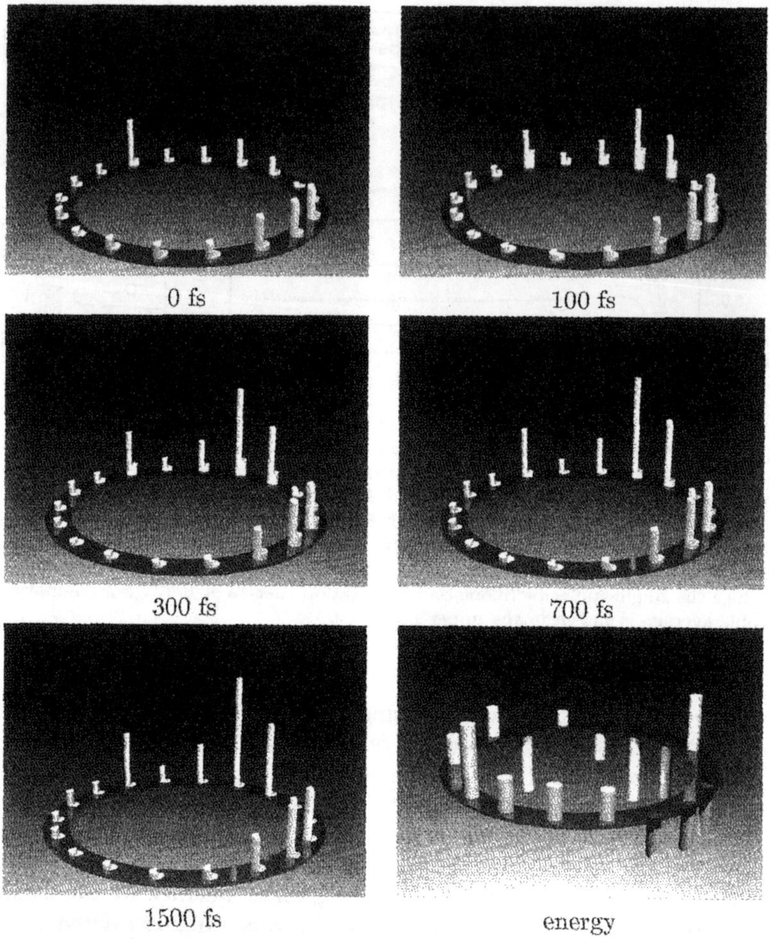

Figure 2. Spatial excitation energy redistribution in the B850 part of the LH2 (a side view is given on the B850 part of the antenna where the 18 BChls are arranged perpendicular to the shown ring). Site–populations of the one–exciton states (left part of each column) and the reduced two–exciton states (right part) are drawn for times up to 1500 fs after optical excitation (corresponding to the pump–pulse of Fig. 1). The highest column of the picture at 0 fs equals a population of 0.043. Right down corner: diagonal disorder shown by deviations of the site–energies in the BChls of LH2 from their mean value (the largest positive deviation corresponds to 436 cm^{-1}). The disorder configuration results in the single–molecule absorption A of Fig. 1.

of the transient absorption spectra except the internal conversion rate have been determined via a fit of the low–intensity absorption. The successful reproduction of the measured spectra based on the multi–exciton theory offers an access to a more detailed visualization of the internal ultrafast and dissipative excitation energy dynamics connected with the EEA. The redistribution of the overall populations $P_N(t)= \sum_{\alpha_N} \rho(\alpha_N, \alpha_N; t)$ referring to different multi–exciton states as well as all elements of the density matrix $\rho(\alpha_M, \beta_N; t)$ belonging to the first and second manifold have been given in Ref.[3]. Here, the ***spatial*** redistribution of excitation energy is considered. Fig. 2 shows two different types of site–populations (with site–index m), the reduced two–exciton population $P_m^{(ee)} = \sum_n \sum_{\alpha_2,\beta_2} C_{\alpha_2}(me, ne) C^*_{\beta_2}(me, nf)\ \rho(\alpha_2, \beta_2; t)$ to characterize the reactant state of the EEA and the one–exciton distribution $P_m^{(e)} = \sum_{\alpha_1,\beta_1} C_{\alpha_1}(me) C^*_{\beta_1}(me)\ \rho(\alpha_1, \beta_1; t)$, which refers to the EEA product. (The $C_{\alpha_2}(me, ne)$, and $C_{\alpha_1}(me)$ denote the two–exciton and single–exciton expansion coefficients, respectively.) To have spatially inhomogeneous distributions the calculations have been carried out for a certain diagonal disorder configuration corresponding to that what has been measured in Ref.[5].

After a population of the one– and two–exciton states in the course of the optical excitation (0 fs and 100 fs–picture in Fig. 2) there is a disproportionate increase of the one–exciton population at those sites with negative deviations of their energies form the mean value (300 fs and 700 fs–picture in Fig. 2). This behavior gives a clear indication for the ongoing EEA. In conclusion, our computations offer an impressive picture of sub–ps excitation energy dynamics and demonstrate how the chosen disorder configurations favor or disfavor the EEA process.

Acknowledgments

References

1. H. van Amerongen, R. van Grondelle and L.Valkunas, Photosynthetic Excitons, (World Scientific, Singapore, 2000).
2. Trinkunas, G.; Herek, J. L.; Polívka, T.; Sundström, V.; Pullerits, T., Phys. Rev. Lett. 86 (2001) 4167.
3. B. Brüggemann, J. L. Herek, V. Sundström, T. Pullerits, and V. May, J. Phys. Chem. B. (in press)
4. Th. Renger, V. May, and O. Kühn, Phys. Rep. 343 (2001) 137.
5. A. M. van Oijen, M. Ketelaars, J. Köhler, T. J. Aartsma, and J. Schmidt, Science 285 (1999) 400.

SOLVENT EFFECT ON THE DYNAMICS OF AN ANESTHETIC ANALOGOUS

L. TORMO, J.A. ORGANERO AND A. DOUHAL*

[1]*Departamento de Química Física, Sección de Químicas, Facultad de Ciencias del Medio Ambiente, Campus Tecnológico de Toledo, Avenida Carlos III, S.N., Universidad de Castilla-La Mancha, 45071 Toledo, Spain.*
**Corresponding author: adouhal@amb-to.uclm.es, fax: + 34-925-268840.*

In this contribution, we report studies of solvent effect on emission spectra and dynamics of an anesthetic analogous. We show that H-bond interaction at both ground and excited state largely affects the electronic properties. We also show that encapsulation by the hydrophobic pocket of Human Serum Albumin protein affects such characteristics. We anticipate that the result might be relevant to the behaviour of medicines and drugs used in phototherapy. To elucidate the experimental observation, we performed ab-initio calculations on free and water bonded dye at the ground state.

Introduction

The study of molecular fluorescent probes has provided a wealth of knowledge about protein site physical chemistry and on the local structure of artificial and natural mebranes.[1-4] Therefore, molecular probes having amino groups and which might be considered as model of anesthetics have been widely studied under different experimental conditions. These amine local anesthetics include the para aminobenzoate derivatives, procaine and tetracaine. We have developed a new ortho aminobenzoate derivative, an anesthetic analogous to procaine and tetracaine. The analogous, 2-amino-4,5-dimethoxy methylbenzoate) (ADMB), is an aromatic molecule with two groups able to form an intramolecular H-bond (Figure 1). This bond is comparable to that found in methyl salicylate, where the hydroxyl group has been substituted by the amino one.

Theoretical calculations

The calculations were performed with the Density Functional Theory using the B3LYP functional and the 6-31+G* basis set which includes a set of d polarisation functions and a set of diffuse functions on heavy atoms as implemented in Gaussian 98 program.[5-7] The full geometry optimisation has been carried out by means of the Schlegel gradient optimisation algorithm by using redundant internal coordinates (Gaussian 98 package).[7]

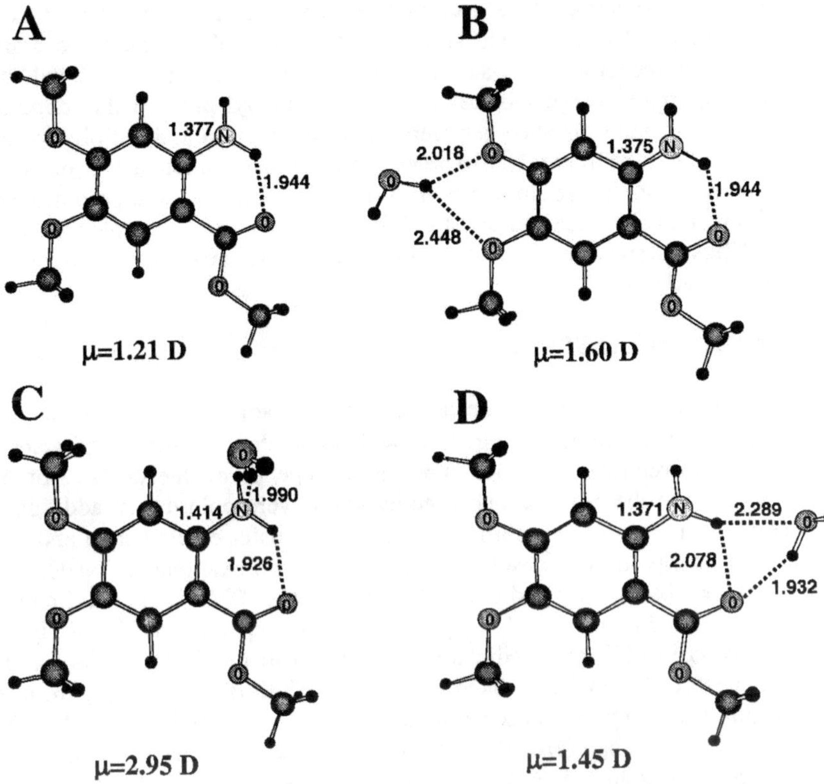

Figure 1: Molecular structures of 2-amino-4,5-dimethoxymethylbenzoate (ADMB). A: free and B-D: water bounded ADMB molecules at the S_0 state. The indicated distances are in Å. The values of the dipole moments of the structures are also given.

The results show that at the S_0 state, ADMB has a short intramolecular H-bond (1.944 Å) between the amino group and the carboxylate one (structure A in figure 1). This H-bond enhances the charge transfer character in the ground state, induced by the amino group. The double bond character between the nitrogen atom and the phenyl part (distance of 1.337 Å) reflects the charge transfer process at the ground state. Other local minima with the methoxy group(s) rotated have also been located (not shown). Their energy is about 0.5 kcal/mol higher than that of structure A (figure 1). In these structures, the rotated methoxy groups are almost perpendicular to the plane defined by the phenyl ring.

We have also calculated the energy of different H-bonded complexes (1:1) of ADMB with water (figure 1). The results suggest several types of 1:1

complex where the specific interaction might be through the amino, the methoxy or the ester groups. The electronic structure of the complex, where the amino group acts as an acceptor and water as a donor of the involved H-bond, has suffered an important charge redistribution in the phenyl part. In this complex (Fig. 1C), the N-C bond distance between the nitrogen atom and the phenyl part increases by 0.077 A° (figure 1). However, in the other complexes, it does not suffer any significant change. In addition to that, the value of the dipole moment changes from 1.21 D (free probe) to 2.95 D upon water complexation (C complex). Thus interaction of water with the amino group will affect the photophysics of ADMB.

Polarity and H-bonding Effects

The maximum of the intensity of the uv-absorption band to the first electronically excited state (π,π^* transition with some charge transfer character at S_1) is located between 335 and 345 nm, and it depends on the polarity of the medium. It shifts to the red side upon increasing solvent polarity. In addition to that, H-bonding interaction destabilizes the charge-transfer excited state and thus increases the energy of the absorption transition. For example, in water, the maximum of the absorption band is at 329 nm, while in DMSO the maximum is located at 346 nm. Figures 2 and 3 show the emission spectra and the decays of ADMB emission in different media upon excitation to the S_1 state. As for the absorption spectra, the emission band exhibits a red shift when going from an inert medium (cyclohexane, maximum at 398 nm) to a polar one (DMSO, maximum 424 nm). Therefore, the difference in spectral shift of emission is interpreted in terms of an intramolecular charge transfer character of ADMB at S_1. However, specific interactions such as hydrogen bonding abilities of the solvent (water, maximum of emission at 437 nm) also play an important role in the spectroscopy of ADMB. This observation agrees with the theoretical calculations. So, both polarity and H-bond ability of the medium are important for the relaxation of the excited anesthetic analogous. Note that the calculations indicate that the dipole moment of water-bounded ADMB (2.95 D) is larger than that of the free one (1.21 D) (figure 1).

The emission decays (Fig. 3) in these solvents clearly show that the H-bond with the medium plays a crucial role in the dynamics of ADMB, leading to an ultrafast quenching of the emission. With a femtosecond time resolution (300 fs), we could not observe any rising component. Water relaxation around ADMB is very fast. When using non-hydrogen bonding donor solvents (cyclohexane, dioxane and DMSO) the emission lifetime (between 6.5 and 8.9 ns) can be fitted by a single exponential function and ranges in the 10-ns time scale. However, using water, a bi-exponential function is need to fit the decay giving components of 60 and 190 ps. The contribution of the first one is about 93 % and 90 % at 425 and 480 nm, respectively. In methanol, these lifetimes are 930 ps (70%) and 2.20 ns (30%) at 425 nm. The ultrafast quenching is due to the formation of a strong H-bond between the NH_2 group (acceptor) and the OH group of water (or methanol) molecule. This interaction will affect the charge transfer process from the amino group to the ester one. Ab initio calculations found three stable H-bonded complexes (B,C and D in Figure 1) with small energy differences (about 0.5 kcal/mol) between them. The short H-bond (about 2 Å) between the amino group (acceptor) and water (donor) in complex C will destabilise the charge transfer character and thus will enhance the non radiative process.

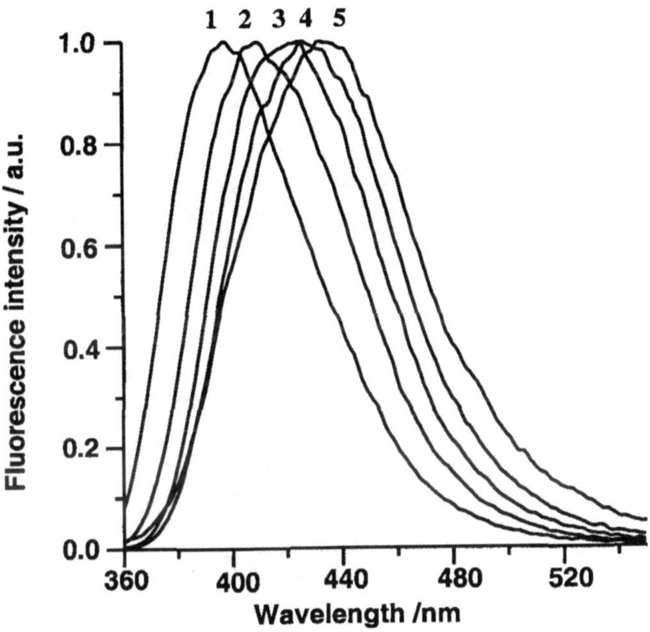

Figure 2: Normalised steady-state emission spectra of ADMB in cyclohexane (1), 1,4-dioxane (2), DMSO (3), methanol (4) and neutral water (5) at 298 K, and excited at 340 nm.

Hydrophobic Effect of HSA Protein Pocket

To obtain some information about the dynamics of electronically excited ADMB in a hydrophobic pocket of a protein, we gated its emission in neutral water and in Human Serum Albumin (HSA) protein. Upon addition of micromoles of HSA to an aqueous solution of ADMB, the emission intensity is largely enhanced (by a factor of 50) as it was also observed in another system.[2,4] So the binding of this anesthetic analogous to the HSA protein is very strong, and it has an association constant of 1.8×10^6 M^{-1} at 300 K. Furthermore, the emission

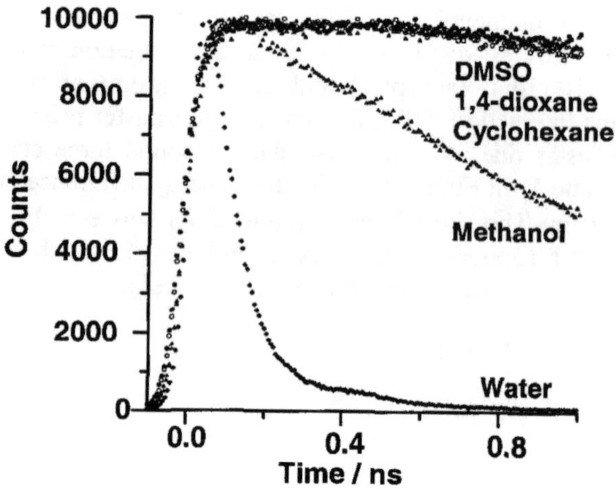

Figure 3: Room temperature normalised emission decays of ADBM in the indicated media, when excited at 393 nm and gated at 425 nm (magic angle).

decay in presence of the protein is greatly affected showing the influence of the hydrophobic and biological molecular pocket on the dynamics of ADMB (Figure 4). The fluorescence lifetime has now a nanosecond component (about 10 ns). The strong confinement of the new anesthetic analogous by the HSA protein is also confirmed by the large value of the stationary anisotropy (0.25).

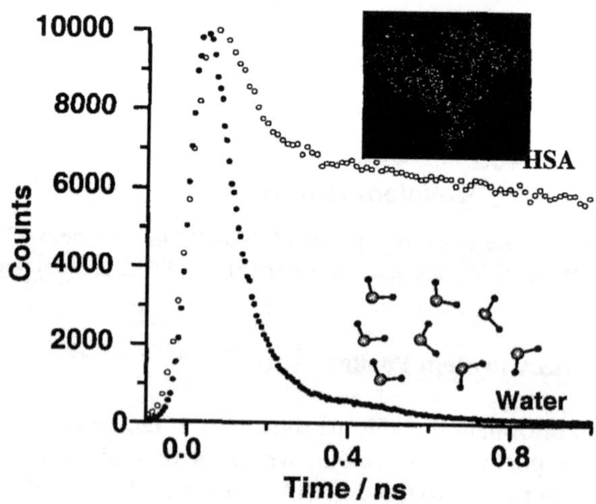

Figure 4: Emission decays of ADBM in neutral water and upon addition of 20 micromoles of HSA protein. Excitation at 393 nm and gating at 425 nm (magic angle). Insets show water and HSA protein molecular structures.

Conclusion

The experimental observation shows the importance of polarity and H-bonding interaction of the solvent; and the hydrophobic effect of HSA protein pocket on the electronic and structure properties of ADMB: an anesthetic analogous to procaine. These effects are enhanced upon electronic excitation, and are reflected on the photodynamics. Both ground state and photoinduced behaviours of this anesthetic analogous may be relevant to those of drugs and medicines used in phototherapy.

This work was supported by the Ministry of Science and Technology (Spain) through projects PB98-0310 and 1FD1997-1658.

References

1. Mateo, C.R.; Douhal, A. Proc. Natl. Acad. Sci. (USA), 95 (1998) 7245, and references therein.
2. Zhong, D.P.; Douhal, A.; Zewail, A.H. Proc. Natl. Acad. Sci.(USA) 97 (2000) 14052.
3. Zhong, D.; Pal, S.K.; Zewail, A.H. ChemPhysChem. 2 (2001) 219, and references therein.
4. Zhong, D.; Zewail, A.H. Proc. Natl. Acad. Sci. (USA) 98 (2001) 11867.
5. Becke, A.D.; *J. Chem. Phys.* 98 (1993) 5648.
6. Lee, C.; Yang, W.; Parr, R. G.; *Phys. Rev. B* 37 (1988) 785.
7. Frisch, M. J.; et al. Gaussian 98; Gaussian Inc.: Pittsburgh, PA, 1998.

Author Index

CPSIA information can be obtained
at www.ICGtesting.com
Printed in the USA
BVHW040942131219
566540BV00007B/43/P

9 789810 248666